MW00760896

Ultrafast Dynamics of Quantum Systems

Physical Processes and Spectroscopic Techniques

NATO ASI Series

Advanced Science Institutes Series

A series presenting the results of activities sponsored by the NATO Science Committee, which aims at the dissemination of advanced scientific and technological knowledge, with a view to strengthening links between scientific communities.

The series is published by an international board of publishers in conjunction with the NATO Scientific Affairs Division

A	**Life Sciences**	Plenum Publishing Corporation
B	**Physics**	New York and London
C	**Mathematical and Physical Sciences**	Kluwer Academic Publishers
D	**Behavioral and Social Sciences**	Dordrecht, Boston, and London
E	**Applied Sciences**	
F	**Computer and Systems Sciences**	Springer-Verlag
G	**Ecological Sciences**	Berlin, Heidelberg, New York, London,
H	**Cell Biology**	Paris, Tokyo, Hong Kong, and Barcelona
I	**Global Environmental Change**	

PARTNERSHIP SUB-SERIES

1. Disarmament Technologies	Kluwer Academic Publishers
2. Environment	Springer-Verlag
3. High Technology	Kluwer Academic Publishers
4. Science and Technology Policy	Kluwer Academic Publishers
5. Computer Networking	Kluwer Academic Publishers

The Partnership Sub-Series incorporates activities undertaken in collaboration with NATO's Cooperation Partners, the countries of the CIS and Central and Eastern Europe, in Priority Areas of concern to those countries.

Recent Volumes in this Series:

Volume 369 — Beam Shaping and Control with Nonlinear Optics
edited by F. Kajzar and R. Reinisch

Volume 370 — Supersymmetry and Trace Formulae: Chaos and Disorder
edited by Igor V. Lerner, Jonathan P. Keating, and David E. Khmelnitskii

Volume 371 — The Gap Symmetry and Fluctuations in High-T_C Superconductors
edited by Julien Bok, Guy Deutscher, Davor Pavuna, and Stuart A. Wolf

Volume 372 — Ultrafast Dynamics of Quantum Systems: Physical Processes and Spectroscopic Techniques
edited by Baldassare Di Bartolo

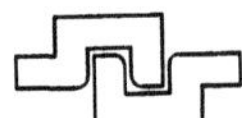

Series B: Physics

Ultrafast Dynamics of Quantum Systems

Physical Processes and Spectroscopic Techniques

Edited by

Baldassare Di Bartolo

Boston College
Chestnut Hill, Massachusetts

Assistant Editor

Giulio Gambarota

Boston College
Chestnut Hill, Massachusetts

Plenum Press
New York and London
Published in cooperation with NATO Scientific Affairs Division

Proceedings of a NATO Advanced Study Institute on
Ultrafast Dynamics of Quantum Systems: Physical Processor and
Spectroscopic Techniques,
held June 15 – 30, 1997,
in Erice, Italy

NATO-PCO-DATA BASE

The electronic index to the NATO ASI Series provides full bibliographical references (with keywords and/or abstracts) to about 50,000 contributions from international scientists published in all sections of the NATO ASI Series. Access to the NATO-PCO-DATA BASE is possible via a CD-ROM "NATO Science and Technology Disk" with user-friendly retrieval software in English, French, and German (©WTV GmbH and DATAWARE Technologies, Inc. 1989). The CD-ROM contains the AGARD Aerospace Database.

The CD-ROM can be ordered through any member of the Board of Publishers or through NATO-PCO, Overijse, Belgium.

Library of Congress Cataloging-in-Publication Data

Ultrafast dynamics of quantum systems : physical processes and spectroscopic techniques / edited by Baldassare Di Bartolo ; assistant editor, Giulio Gambarota.
p. cm. -- (NATO ASI series. Series B, Physics ; v. 372)
"Proceedings of a NATO Advanced Study Institute on Ultrafast Dynamics of Quantum Systems: Physical Processes and Spectroscopic Techniques, held June 15-30, 1997, in Erice, Italy"--T.p. verso.
"Published in cooperation with NATO Scientific Affairs Division."
Includes bibliographical references and index.
ISBN 0-306-45929-9
1. Laser pulses, Ultrashort--Congresses. 2. Picosecond pulses--Congresses. 3. Laser spectroscopy--Congresses. 4. Electrooptics--Congresses. 5. Quantum electronics--Congresses. I. Di Bartolo, Baldassare. II. Gambarota, Giulio. III. NATO Advanced Study Institute on Ultrafast Dynamics of Quantum Systems: Physical Processes and Spectroscopic Techniques (1997 : Erice, Italy) IV. Series.
QC689.5.L37U464 1998
535--dc21 98-42551
CIP

ISBN 0-306-45929-9

A Division of Plenum Publishing Corporation
233 Spring Street, New York, N.Y. 10013

http://www.plenum.com

10 9 8 7 6 5 4 3 2 1

Printed in the United States of America

Quel sol che pria d'amor mi scaldò 'l petto
Di bella verità m'avea scoverto,
Provando e riprovando, il doce aspetto.

—Dante, *Paradiso, III Canto*

PREFACE

This book presents the Proceedings of the course "Ultrafast Dynamics of Quantum Systems: Physical Processes and Spectroscopic Techniques" held in Erice, Italy from June 15 to June 30, 1997. This meeting was organized by the International School of Atomic and Molecular Spectroscopy of the "Ettore Majorana" Centre for Scientific Culture.

This Institute was devoted to the study of ultrafast physical phenomena, such as nonequilibrium phonon effects, evolution of excitations at surfaces, dynamics of electrons in semiconductors and metals, rapid conformational changes in molecules and optical control of the quantum states of molecules and solids. A total of 74 participants came from 54 laboratories and 14 different countries (the Czech Republic, Denmark, Finland, France, Germany, Italy, The Netherlands, Portugal, Russia, Spain, Switzerland, Turkey, The United Kingdom, and The United States).

The secretaries of the course were Giulio Gambarota and Daniel Di Bartolo.

50 lectures divided in 14 series were given. In addition 9 (one or two-hour) "long seminars," 1 interdisciplinary lecture, 14 "short seminars" and 17 posters were presented. The sequence of the lectures was in accordance with the logical development of the subject of the meeting. Each lecturer started at a rather fundamental level and ultimately reached the frontier of knowledge in the field.

Two round-table discussions were held. The first round-table discussion took place after 4 days of lectures in order to evaluate the work done in the first days of the course and consider suggestions and proposals regarding the organization, format and presentation of the lectures. The second one was held at the conclusion of the course, so that the participants could comment on the work done during the entire meeting and discuss various proposals for the next course of the International School of Atomic and Molecular Spectroscopy.

I wish to express my sincere gratitude to Dr. Alberto Gabriele, and Mr. Pino Aceto and to all the personnel of the "Ettore Majorana" Centre, who contributed to create a congenial atmosphere for our meeting. I wish to acknowledge the sponsorship of the meeting by the NATO Scientific Affairs Division, Boston College, the ENEA Organization, the European Physical Society, the Italian Ministry of Education, the Italian Ministry of University and Scientific Research, and the Sicilian Regional Government.

I wish also to acknowledge the National Science Foundation for providing travel grants for some of our participants.

I would like to thank Ms. Maria Zaini of the Ettore Majorana Centre, the members of the Organizing Committee (Prof. Claus Klingshirn, Dr. Cees Ronda, and Dr. Giuseppe Baldacchini), the secretaries of the course (Giulio Gambarota and Daniel Di Bartolo), Prof. John Collins, Dr. Aliki Collins, Prof. Xuesheng Chen, Ms. Daniela Alba, Mr. Giuseppe Corrao and Dr. Brian Walsh for their valuable help, and Dr. John Di Bartolo for putting the program in the Internet.

As in the past I felt privileged to direct a NATO Institute, to encounter in my native land of Trapani so many fine people, and to share with them the Erice experience. The meeting of this year provided me with a particularly joyous experience, for I had the singular pleasure and honor to receive all the participants in the house of my family in Trapani, which my brother Francesco has recently restored.

I am already looking forward to the next 1999 meeting of the International School of Atomic and Molecular Spectroscopy, where I am sure I will see again many of you, my friends. Arrivederci a presto!

Baldassare (Rino) Di Bartolo
Director of the International School of Atomic and Molecular Spectroscopy of the "Ettore Majorana" Centre

CONTENTS

LONG SEMINARS

INTERDISCIPLINARY LECTURE

SHORT SEMINARS

POSTERS

Ultrafast Dynamics of Quantum Systems

Physical Processes and Spectroscopic Techniques

LINEAR AND NONLINEAR PROPAGATION OF SHORT LIGHT PULSES

B. Di Bartolo

Department of Physics
Boston College
Chestnut Hill, MA, USA

. . . Nil sine magno
Vita labore dedit mortalibus.
Horace

ABSTRACT

Over the pat decades techniques for the production and manipulation of short optical pulses have progressed at an accelerated rate, reducing the time interval accessible to measurements from the millisecond in ~1949 to the femtosecond regime in ~1985. In particular, in the last ten years, major developments have included versatile and stable laser sources with wide spectral coverage, and new methods for shaping and controlling light pulses. This contribution presents some of the background necessary for the treatment of the subject of the book, by considering the basic physical principles and phenomena.

We shall first deal with pulses that do not change the characteristics of the media in which they propagate. This case will provide the occasion for the introduction of many fundamental concepts and for the examination of the interplay of dispersion and absorption, which is embodied in the Kramers-Kronig relations.

We shall then consider pulses so intense that the characteristics of the media they go through are affected by them. This case will give the opportunity to tie nonlinear optics to pulse shaping and propagation and to deal with such phenomena as pulse break-up and soliton waves in nonlinear dispersive media.

I. LINEARITY AND LINEAR MEDIA

Nulla dies sine linea.
Apelle

1.A. Definition of Linearity

The linearity of a system is a property related to the interaction of some wave perturbation to it. We shall concentrate on the response of materials to short pulses of electromagnetic radiation, but many of our considerations will have wider implications.

We shall define *linearity* in two equivalent ways:

1. If a system is linear the parameters that define its response to signals passing through it are independent of the intensity of the signals.

Ultrafast Dynamics of Quantum Systems: Physical Processes and Spectroscopic Techniques, Edited by Di Bartolo and Gambarota, Plenum Press, New York, 1998

2. If a system is linear, each Fourier component of a signal passes through it *independently*, without influencing the passage of any other component. In particular, each component can be attenuated or amplified, but no frequency component can be present at the output, if no such component exists at the input.

We can make the following points:

1. A linear circuit consists of elements whose values do not vary with the current that goes through them. The equations describing the response of the circuit are linear differential equations with constant coefficients. In such a case, the steady-state conditions correspond to certain particular integrals of these equations.

2. We shall call *linear* a device when, if we introduce in it a signal $f_i(t)$, the waveform that we obtain at the output differs from $f_i(t)$ only for a variation of amplitudes and phases of the harmonics of $f_i(t)$.

3. The transmission of a signal $f(t)$ through a linear device takes place *without distortion* when the output waveform preserves the same shape of $f(t)$ (only the amplitude has changed) and is delayed in time.

4. Call

$f_i(t)$ = input waveform

$F_i(\omega) = |F_i(\omega)|\, e^{i\varphi_i(\omega)}$ = Fourier transform of $f_i(t)$

$f_u(t)$ = output waveform

$F_u(t) = |F_u(\omega)|\, e^{i\varphi_u(\omega)}$ = Fourier tranform of $f_u(t)$

In order to have *distortionless* passage of the signal through the device it is necessary and sufficient that:

$$\frac{|F_u(\omega)|}{|F_i(\omega)|} = K = \text{constant} \tag{1.1}$$

$$\varphi_u(\omega) = \varphi_i(\omega) - \omega\tau \tag{1.2}$$

If the conditions (1.1) is not respected, we have *amplitude distortion*; if condition (1.2) is not respected, we have *phase distortion*. If both conditions are respected, then:

$$F_u(\omega) = |F_u(\omega)|\, e^{i\varphi_u(\omega)} = K\,|F_i(\omega)|\, e^{i[\varphi_u(\omega)-\omega\tau]} = KF_i(\omega)e^{-i\omega\tau} \tag{1.3}$$

$$\frac{F_u(\omega)}{F_i(\omega)} = Ke^{-i\omega\tau} \tag{1.4}$$

Therefore

$$f_u(t) = \frac{1}{2\pi}\int_{-\infty}^{+\infty} F_u(\omega)e^{i\omega t}d\omega = \frac{1}{2\pi}\int_{-\infty}^{+\infty} KF_i(\omega)e^{-i\omega t}e^{i\omega t}d\omega = \frac{K}{2\pi}\int_{-\infty}^{+\infty} F_i(\omega)e^{i\omega(t-\tau)}d\omega = Kf_i(t-\tau) \tag{1.5}$$

5. A *nonlinear* device does not follow the principle of superposition, but it introduces in the waveform that passes through it new harmonics and produces interaction among the harmonics that compose the waveform.
In this case we speak of *nonlinear distortion.*

Summary

In this section we have given two equivalent definitions of linearity of general systems. We present then a definition of **linearity** for circuits and consider the concepts of linear and nonlinear distortion.

I.B. Characterization of Linear Systems

Consider a system such that, $f_i(t)$ being an input signal, the output signal is $f_u(t)$. The system is characterizaed by a rule that relates $f_u(t)$ to $f_i(t)$. For example:

$$f_u(t) = \sqrt{f_i(t)} \tag{1.6}$$

The response of a *linear* system to the sum of two inputs is the sum of the responses to the individual inputs; therefore,

$$f_u(t) = \int_{-\infty}^{+\infty} h(t,\tau) f_i(\tau) d\tau \tag{1.7}$$

where $h(t,\tau)$ = weighting function which represents the contribution of the input at time τ to the output at time t.

Let

$$f_i(t) = \delta(t-T) \tag{1.8}$$

Then

$$f_u(t) = \int_{-\infty}^{+\infty} h(t,\tau) f_i(\tau) d\tau = \int_{-\infty}^{+\infty} h(t,\tau)\delta(t-T) d\tau = h(t,T) \tag{1.9}$$

Thus, the function $h(t,\tau)$ is the *impulse-response function,* also known as the *Green's function* of the system.

A linear system is said to be *time-invariant* if, when the input is shifted in time, the output is shifted by an equal time, but otherwise does not change. For such a system

$$h(t,\tau) = h(t-\tau) \tag{1.10}$$

Then

$$f_u(t) = \int_{-\infty}^{+\infty} h(t-\tau) f_i(\tau) d\tau \tag{1.11}$$

namely the *output is the convolution of the input with the impulse response function.*

If

$$f_i(t) = \delta(t), \tag{1.12}$$

then

$$f_u(t) = \int_{-\infty}^{+\infty} h(t-\tau) f_i(\tau) d\tau = \int_{-\infty}^{+\infty} h(t-\tau)\delta(\tau) d\tau = h(t) \tag{1.13}$$

If

$$f_i(t) = \delta(t-T), \tag{1.14}$$

then

$$f_u(t) = \int_{-\infty}^{+\infty} h(t-\tau)\delta(\tau - T) d\tau = h(t-T) \tag{1.15}$$

Let

$F_i(\omega)$ = Fourier transform of $f_i(t)$

$F_u(\omega)$ = Fourier transform of $f_u(t)$

$H(\omega)$ = Fourier transform of $h(t)$

Figure 1.1 represents the situation. The convolution property of Fourier transforms allows us to write

$$f_u(t) = \int_{-\infty}^{+\infty} h(t-\tau) f_i(\tau) d\tau \tag{1.16}$$

$$F_u(\omega) = H(\omega) F_i(\omega) \tag{1.17}$$

If the input signal is a harmonic function,

$$f_i(t) = F_i(\omega) e^{i\omega t} \tag{1.18}$$

then the output signal is a harmonic function:

$$f_u(t) = F_u(\omega) e^{i\omega t} = H(\omega) F_i(\omega) e^{i\omega t} \tag{1.19}$$

$H(\omega)$ is called the *transfer function* of the linear time-invariant system.

Note that the transfer function is the Fourier-transform of the impulse-response function.

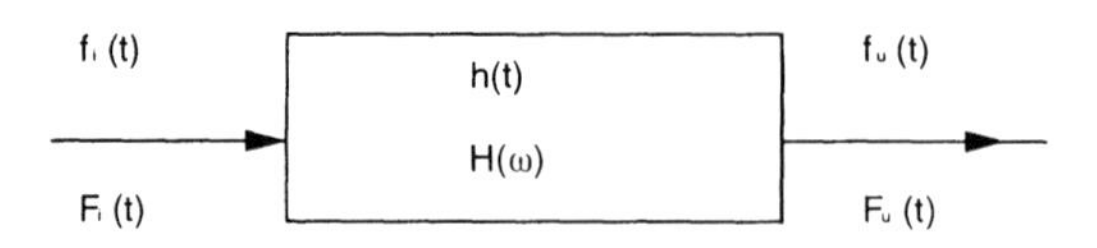

Fig. 1.1 Schematic view of the input and output functions and of the system.

Examples

Ideal System

$H(\omega) = 1; \; h(t) = \delta(t)$

$f_u(t) = f_i(t)$

Ideal system with delay

$H(\omega) = e^{-i\omega T}; \; h(t) = \delta(t - T)$

$$f_u(t) = \int_{-\infty}^{+\infty} h(t-\tau) f_i(\tau) d\tau = \int_{-\infty}^{+\infty} \delta(t - T - \tau) f_i(\tau) d\tau = f_i(t - T)$$

System with exponential response

$h(t) = e^{\frac{-t}{T}}, \; t \geq 0$ and $h(t) = 0, \; t \langle 0$

$$H(\omega) = \frac{T}{1 + i\omega T}$$

If $f_i(t) = \delta(t)$

$$f_u(t) = \int_{-\infty}^{+\infty} h(t-\tau)\delta(\tau) d\tau = h(t) = e^{\frac{-t}{\tau}} \qquad (t \geq 0)$$

Summary

In this section, we have introduced a parameter which may characterize linear systems: the **impulse response function**, the response of the systems to a delta-function input. The **time-invariance** of a linear system is reflected in the fact that, given a certain input, the output is the convolution of the input with the impulse response function.

The so-called **transfer function** of the linear system is the Fourier transform of the impulse response function: because of the convolution property of Fourier transforms, the Fourier transform of the output is equal to the product of the transfer-function and of the Fourier transform of the input.

II. CAUSALITY AND THE KRAMERS-KRONIG RELATIONS

It may indeed be true that the first words spoken by the best theoretical physicists while in their cots are not "Ma", "Da" or "more", but rather the magic words "Kramers-Kronig"

J. Hamilton

Christ's College, Cambridge (UK)

II.A. Response of an Ideal Low-Pass Filter to a Rectangular Pulse Waveform

The response of a linear circuit to a waveform is given by the sum of the responses of the circuit to all the harmonic components of the waveform.

Let

$f_i(t)$ = input waveform

$F_i(\omega)$ = spectral function of $f_i(t)$

$f_u(t)$ = output waveform

$F_u(\omega)$ = spectral function of $f_u(t)$

Then

$$\frac{F_u(\omega)}{F_i(\omega)} = Y(\omega) = |Y(\omega)|\, e^{i\varphi(\omega)} = \text{transfer function of the circuit} \tag{2.1}$$

Let us assume that $f_i(t)$ is the rectangular waveform represented in Fig. 2.1, and that the system is a low-pass filter, also represented in Fig. 2.1. The circuit lets through the harmonics between o and ω_t (eliminating all others) and introduces in them phase shifts proportional to ω:

$$\begin{cases} Y(\omega) = 0 & \omega \rangle -\omega_t \\ Y(\omega) = Ke^{-i\omega t} & -\omega_t \langle \omega \langle \omega_t \\ Y(\omega) = 0 & \omega \rangle \omega_t \end{cases} \tag{2.2}$$

We now proceed with the detailed calculations:

1) Calculation of $F_i(\omega)$

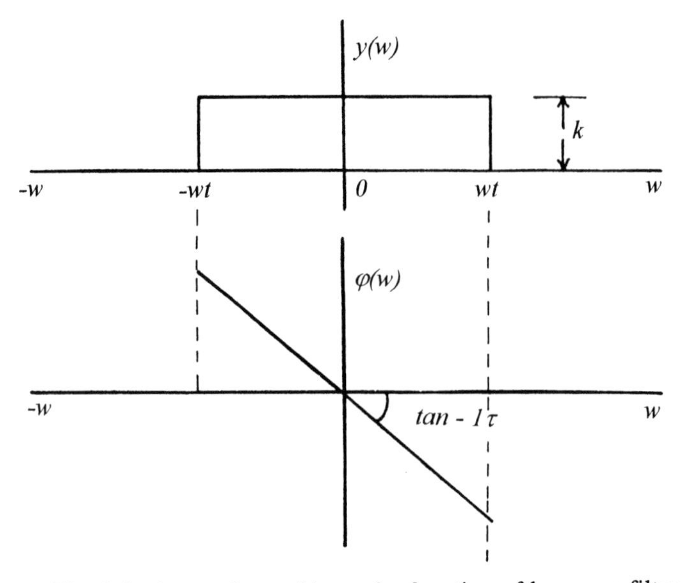

Fig. 2.1 a) waveform; b)transfer function of low-pass filter.

$$F_i(\omega) = \int_{-\infty}^{+\infty} f_i(t) e^{-i\omega t} dt = \frac{e^{-i\omega t_2} - e^{-i\omega t_1}}{-i\omega} \tag{2.3}$$

2) Calculation of $F_u(\omega)$

$$\begin{cases} F_u(\omega) = Y(\omega) F_i(\omega) = K e^{-i\omega\tau} \dfrac{e^{-i\omega t_2} - e^{-i\omega t_1}}{-i\omega} & \text{for} \quad -\omega_t \langle \omega \langle \omega_t \\ F_u(\omega) = 0 & \text{for} \quad \omega \rangle \omega_t \quad \text{and} \quad \omega \langle \omega_t \end{cases} \tag{2.4}$$

3) Calculation of $f_u(t)$

$$\begin{aligned} f_u(t) &= \frac{1}{2\pi} \int_{-\infty}^{+\infty} F_u(\omega) e^{i\omega t} d\omega \\ &= \frac{1}{2\pi} \int_{-\infty}^{+\infty} [K e^{-i\omega\tau} \frac{e^{-i\omega t_2} - e^{-i\omega t_1}}{-i\omega}] e^{i\omega t} d\omega \\ &= \frac{iK}{2\pi} \int_{-\omega_t}^{\omega_t} [\frac{\cos\omega(t-\tau-t_2)}{\omega} - \frac{\cos\omega(t-\tau-t_1)}{\omega} \\ &+ \frac{i\sin\omega(t-\tau-t_2)}{\omega} - \frac{i\sin\omega(t-\tau-t_1)}{\omega}] d\omega \end{aligned} \tag{2.5}$$

But

$$\int_{-\omega_t}^{\omega_t} \frac{\cos\omega(t-\tau-t_2)}{\omega} d\omega = \int_{-\omega_t}^{\omega_t} \frac{\cos\omega(t-\tau-t_1)}{\omega} d\omega = 0$$

and

$$\int_{-\omega_t}^{\omega_t} \frac{\sin\omega(t-\tau-t_2)}{\omega} d\omega = 2\int_{0}^{\omega_t} \frac{\sin\omega(t-\tau-t_2)}{\omega} d\omega$$

$$\int_{-\omega_t}^{\omega_t} \frac{\sin\omega(t-\tau-t_1)}{\omega} d\omega = 2\int_{0}^{\omega_t} \frac{\sin\omega(t-\tau-t_1)}{\omega} d\omega$$

Then

$$f_u(t) = -\frac{K}{\pi} \int_{0}^{\omega_t} [\frac{\sin\omega(t-\tau-t_2)}{\omega} - \frac{\sin\omega(t-\tau-t_1)}{\omega}] d\omega \tag{2.6}$$

Let

$$\begin{cases} u_1 = \omega(t-\tau-t_1) & u_{1t} = \omega_t(t-\tau-t_1) \\ u_2 = \omega(t-\tau-t_2) & u_{2t} = \omega_t(t-\tau-t_2) \end{cases} \tag{2.7}$$

Then

$$d_{u1} = (t-\tau-t_1)d\omega \qquad d\omega = \frac{du_1}{t-\tau-t_1}$$
$$d_{u2} = (t-\tau-t_2) \qquad d\omega = \frac{du_2}{t-\tau-t_2}$$

and

$$f_u(t) = \frac{K}{\pi}\int_0^{u_{1t}}\left[\frac{\sin u_1}{u_1}du_1 - \int_0^{u_{2t}}\frac{\sin u_2}{u_2}du_2\right] \tag{2.8}$$

Let

$$S_i(x) = \int_0^x \frac{\sin y}{y}dy = \textit{function sine-integral of } x \tag{2.9}$$

Then

$$f_u(t) = \frac{K}{\pi}S_i(u_{1t}) - \frac{K}{\pi}S_i(u_{2t}) \tag{2.10}$$

where

$$\frac{K}{\pi}S_i(u_{1t}) = \frac{K}{\pi}\int_0^{\cot(t-\tau-t_1)}\frac{\sin x}{x}dx \tag{2.11}$$

$$\frac{K}{\pi}S_i(u_{2t}) = \frac{K}{\pi}\int_0^{\cot(t-\tau-t_2)}\frac{\sin x}{x}dx \tag{2.12}$$

The function $\frac{K}{\pi}S_i(u_{1t})$, represented in Figure 2.3,

a) is zero for $t = t_1 + \tau$,

b) for $t \to \pm\infty$ tends asympotically to $\pm\frac{K}{2}$

c) has at $t = t_1 + \tau$ a slope proportional to ω_t:

$$\frac{K}{\pi}\left[\frac{d}{dt}\int_0^{\omega_t(t-\tau-t_1)}\frac{\sin x}{x}dx\right]_{t=t_1+\tau} = \omega_t\frac{K}{\pi} \tag{2.13}$$

d) presents oscillations around the values $+\frac{K}{2}$ and $-\frac{K}{2}$ that decrease rapidly as $t \to \pm\infty$.

The function $\frac{K}{\pi}S_i(u_{2t})$, represented in Figure 2.4,

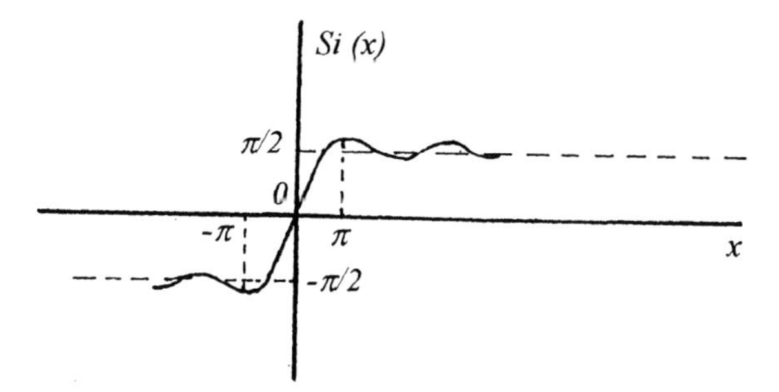

Fig. 2.2 The function Si(x)

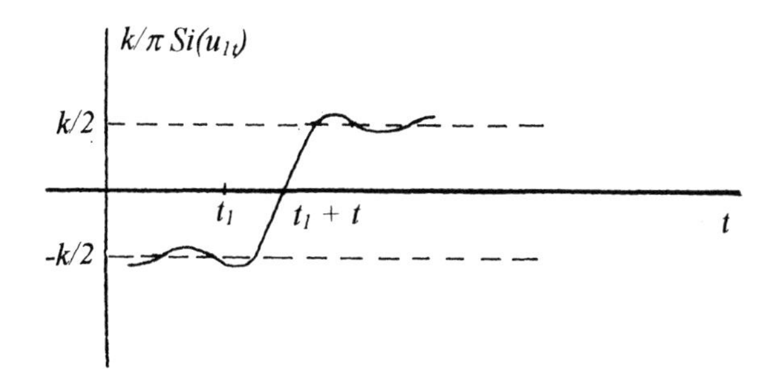

Fig. 2.3 The function $\frac{K}{\pi}Si(u_{1t})$

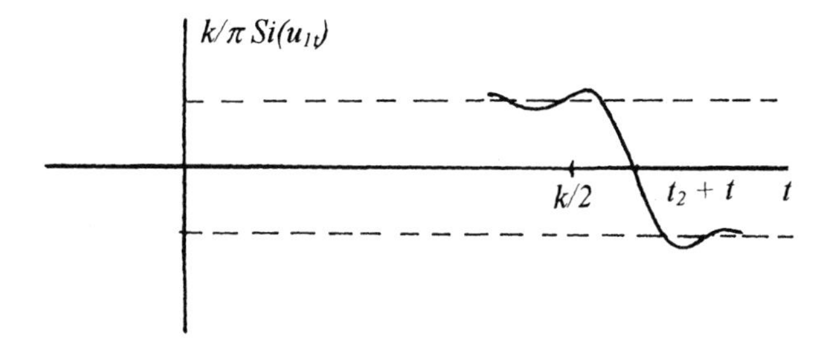

Fig. 2.4 The function $\frac{K}{\pi}S_i(u_{2t})$

a) is zero for $t = t_2 + \tau$,

b) for $t \to \pm\infty$ tends asympotically to $\pm\frac{K}{2}$

c) has at $t = t_2 + \tau$ a slope proportional to $-\omega_t$, and

d) presents oscillations around the values $+\frac{K}{2}$ and $-\frac{K}{2}$ that decrease rapidly as $t \to \pm\infty$.

The function (2.10), sum of (2.11) and (2.12) is represented in Fig. 2.5, together with the original waveform.

The following observations can be made:

1) The filter responds with a certain delay to the waveform applied to it. This delay, if we consider it given by the time at which $f_u(t)$ reaches one half the value of K, is given by τ.

2) The waveform $f_u(t)$ has at times $t = t_1 + \tau$ and $t = t_2 + \tau$ slopes proportional to ω_t, and $-\omega_t$, respectively. The larger is ω_t, the steeper are the sides of $f_u(t)$.

3) If the filter lets through all the frequencies, i.e., if

$$Y(\omega) = Ke^{-i\omega\tau}$$

for any value of ω, from $-\infty$ to $+\infty$, then $f_u(t)$ is equal, apart amplitude, to $f_i(t)$, but is retarded by τ with respect to the latter (no distortion).

4) Transients may occur at times earlier than t_1; this seems to contradict the principle of causality.

5) An interesting explanation of this contradiction is given in the following terms:

A filter like the one considered, eliminating sharply all the frequencies $\rangle\omega_t$ is realizable *in practice* only with an infinite number of sections. A filter of this type would respond to a waveform with an infinite delay, and no $f_u(t)$ would have transients preceding in time t_1.

This explanation seems to explain the contradiction by using technical arguments, where a more fundamental one is needed. Indeed the needed explanation is that the real and imaginary parts of a transfer function cannot be chosen arbitrarily, as it has been done in the present case.

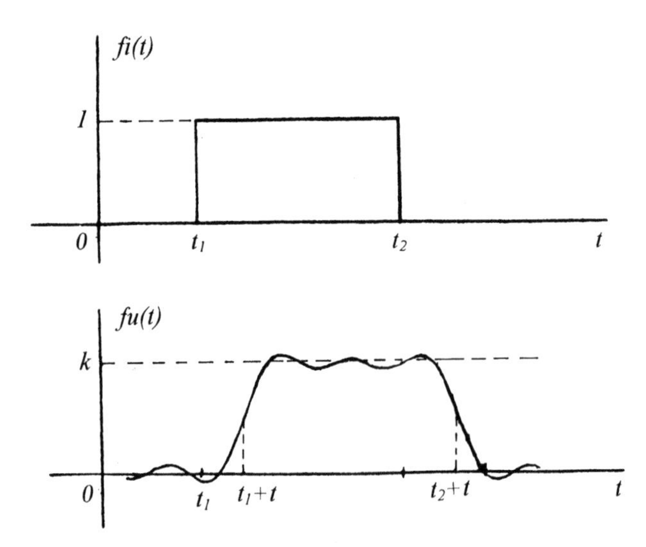

Fig. 2.5 Rectangular waveform and the response of a low-pass filter to it.

Summary

An ideal **low-pass filter** has a frequency transfer function that is flat in amplitude and linear in phase. The transfer function of the filter is K $\exp(-i\omega\tau)$. A rectangular pulse is the input signal.

The filter responds with a delay τ to the signal, but transients occur at time earlier than the time at which the pulse enters the filter.

This example demonstrates that for an **arbitrary** linear system (a specific filter in this case), the response to a square pulse can have transients that arrive at negative times (i.e. they precede the input pulse). This violates **causality**, and clearly illustrates the fact that the real and imaginary parts of a transfer function cannot be chosen arbitrarily: a filter for which such a choice is made may not be physically realizable.

II.B. Response of a Molecular Gas to an Applied Field

1. Polarization. Let us consider an ensemble of molecules under the action of a "real" electric field, and let $E(t)$ be the value of the field at a certain point. We can write:

$$E(t) = \int_{-\infty}^{+\infty} E(\omega) e^{-i\omega t} d\omega \tag{2.14}$$

where

$$E(\omega) = \frac{1}{2\pi} \int_{-\infty}^{+\infty} E(t) e^{i\omega t} dt \tag{2.15}$$

Since $E(t)$ is real,

$$E(-\omega) = E^*(\omega) \tag{2.16}$$

The field $E(t)$ can be considered a "stimulus" applied to the molecules which in turn "respond" by become polarized. If $P(t)$ is the polarization at time t,

$$P(t) = \int_{-\infty}^{+\infty} P(\omega) e^{-i\omega t} d\omega \tag{2.17}$$

where

$$P(\omega) = \chi(\omega) E(\omega) \tag{2.18}$$

Then

$$P(t) = \int_{-\infty}^{+\infty} \chi(\omega) E(\omega) e^{-i\omega t} d\omega \tag{2.19}$$

A real field must produce a real polarization. For $P(t)$ to be real, we must have

$$\chi(-\omega) = \chi^*(\omega) \tag{2.20}$$

If we set

$$\chi(\omega) = \chi_r(\omega) + i\chi_i(\omega) \tag{2.21}$$

relation (2.20) is written

$$\chi_r(-\omega) + i\chi_i(-\omega) = \chi_r(\omega) - i\chi_i(\omega) \tag{2.22}$$

or

$$\begin{cases} \chi_r(\omega) = \chi_r(-\omega) \\ \chi_i(\omega) = -\chi_i(-\omega) \end{cases} \tag{2.23}$$

These relations, called *crossing relations for the susceptibility*, are derived from the condition that the response to the real field must be a real polarization.

To proceed further with this treatment, it is necessary to specify the model we are using. We shall assume that each molecule can be represented by a particle of mass m and charge e undergoing possible natural oscillations of angular frequency

$$\omega_o = \sqrt{\frac{K}{m}} \tag{2.24}$$

about an equilibrium position where a charge of equal value and opposite sign resides. The equation of motion of the individual molecule under the action of a field component $E_o(\mathbf{x})e^{-i\omega t}$, which we shall assume polarized in the z-direction, is

$$\begin{aligned} \ddot{z} &= -\frac{K}{m}z + \frac{e}{m}E_o(\mathbf{x})e^{-i\omega t} - \gamma\dot{z} \\ &= -\omega_o{}^2 z + \frac{e}{m}E_o(\mathbf{x})e^{-i\omega t} - \gamma\dot{z} \end{aligned} \tag{2.25}$$

where $-\gamma\dot{z}$ is the *damping* term. We shall make the assumption that the z displacements are always much smaller than the wavelength of the radiation and that γ is independent of ω. The steady-state solution is

$$z(\mathbf{x},t) = z_o(\mathbf{x})e^{-i\omega t} \tag{2.26}$$

where

$$z_o(\mathbf{x}) = \frac{e/m}{\omega_o - \omega^2 - i\gamma\omega}E_o(\mathbf{x}) \tag{2.27}$$

The induced dipole moment is

$$ez(\mathbf{x},t) = \frac{e^2/m}{\omega_o^2 - \omega^2 - i\gamma\omega}E_o(\mathbf{x})e^{-i\omega t} = \alpha(\omega)E_o(\mathbf{x})e^{-i\omega t} \tag{2.28}$$

where the polarizability $\alpha(\omega)$ is given by

$$\alpha(\omega) = \frac{e^2/m}{\omega_o^2 - \omega^2 - i\gamma\omega} \tag{2.29}$$

The susceptibility is given by

$$\chi(\omega) = n_o\alpha(\omega) = \frac{(n_o e^2)/m}{\omega_o^2 - \omega^2 - i\gamma\omega} \tag{2.30}$$

where n_o = density of molecules. We can also write

$$\chi(\omega) = \frac{(n_o e^2)/m}{\omega_o^2 - \omega^2 - i\gamma\omega} = -\frac{(n_o e^2)/m}{\omega_1 - \omega_2}\left[\frac{1}{\omega - \omega_1} - \frac{1}{\omega - \omega_2}\right] \tag{2.31}$$

where

$$\begin{cases} \omega_1 = -\frac{1}{2}i\left[\gamma - \sqrt{\gamma^2 - 4\omega_o^2}\right] \\ \omega_2 = -\frac{1}{2}i\left[\gamma + \sqrt{\gamma^2 - 4\omega_o^2}\right] \end{cases} \tag{2.32}$$

The positions of ω_1 and ω_2 in the complex plane are depicted in Figure 2.6.

2. Derivation of the Kramers-Kronig Relations. Consider the integral [1]

$$I = P\int_{-\infty}^{+\infty} \frac{\chi(\omega')}{\omega' - \omega} d\omega' = \lim_{\delta \to 0} \int_{-\infty}^{\omega-\delta} \frac{\chi(\omega')}{\omega' - \omega} d\omega' + \lim_{\delta \to 0} \int_{\omega+\delta}^{+\infty} \frac{\chi(\omega')}{\omega' - \omega} d\omega' \tag{2.33}$$

where P indicates the *principal value.* The path of this integral is represented in Figure 2.7. The integral over this path is shown in Figure 2.8 to be equal to the integral over the path A minus the two integrals over paths B and C.

The integrand in I has three poles:

$$\begin{aligned} \omega' &= \omega \\ \omega' &= \omega_1 \\ \omega' &= \omega_2 \end{aligned} \tag{2.34}$$

We now calculate the various parts of I.

Part A. This path for this integral is a closed contour. Given a certain function of a complex variable $z, f(z)$, if L is the boundary of a region in which $f(z)$ is analytic except at a finite number n of poles, a_k, then

$$\oint_L f(z)dz = 2\pi i \sum_{k=1}^{n} Res(a_k) \tag{2.35}$$

where

$$Res(a_k) = \frac{1}{(m-1)!}\left[\frac{d^{m-1}}{dz^{m-1}}(z - a_k)^m f(z)\right]_{z=a_k} \tag{2.36}$$

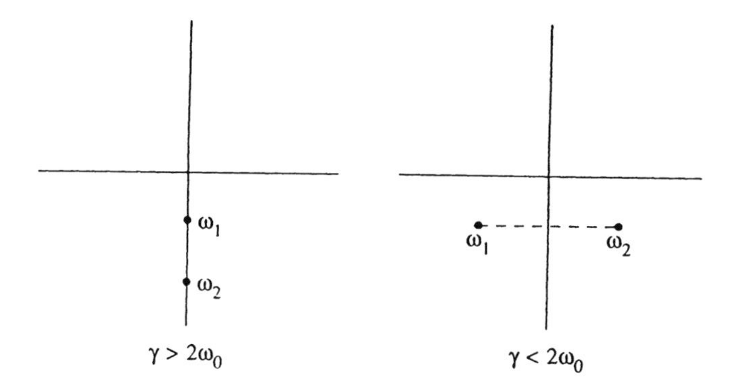

Fig. 2.6 Positions of ω_1 and ω_2 in the complex plane

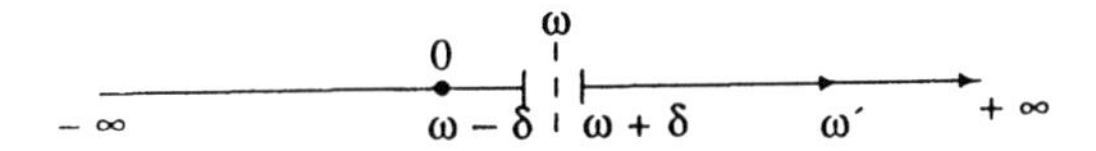

Fig. 2.7 Path of integral (2.33)

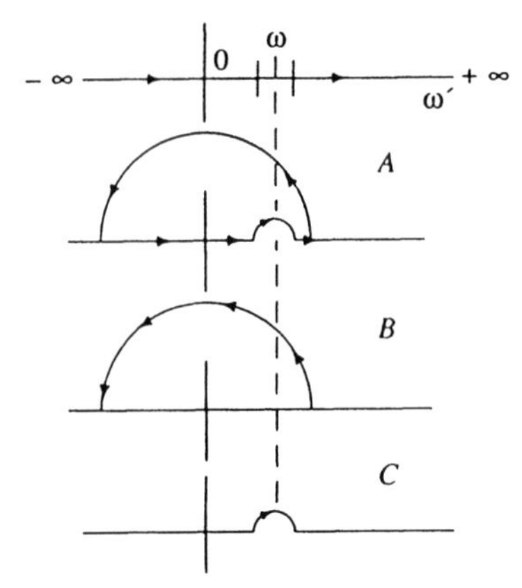

Fig. 2.8 Path (2.33) as a result of the integral over path A minus integrals over paths B and C.

This is known as *Cauchy's Residue Theorem* [2]. In the present case the part A is zero, because there are no poles in the upper plane.

Part B. The contour for this integral involves very large values of ω'.

$$\chi(\omega') \sim \left(\frac{1}{\omega' - \omega_1} - \frac{1}{\omega' - \omega_2} \right) \tag{2.37}$$

The integrand has an $(\omega')^{-2}$ dependence for large ω'. The length of the contour is $\pi\omega'$. The integral has then an

$$(\omega')^{-2} \times \pi\omega' \approx (\omega')^{-1} \tag{2.38}$$

dependence for large ω'. For an infinite contour, the part B of the integral is zero.

Part C. If $f(z)$ has a simple pole at $z = a$ with the residue $Res(a)$, and if c_p is a circular arc of radius ρ and center at $z = a$ intercepting an angle α at $z = a$, then

$$\lim_{\rho \to 0} \int_{C_p} f(z)dz = \alpha i Res(a) \tag{2.39}$$

α is positive if the integration is carried out in the counterclockwise way. In our case,

$$\lim_{\rho\to 0}\int_{C_p}\frac{\chi(\omega')}{\omega'-\omega}d\omega = -i\pi Res(\omega) \tag{2.40}$$

$$Res(\omega) = \left[\frac{\chi(\omega')}{\omega'-\omega}(\omega'-\omega)\right]_{\omega'-\omega} = \chi(\omega) \tag{2.41}$$

and the part C is

$$-i\pi\chi(\omega)$$

On collecting terms,

$$-i\pi\chi(\omega) = P\int_{-\infty}^{+\infty}\frac{\chi(\omega')}{\omega'-\omega}d\omega' \tag{2.42}$$

$$i\pi(\chi_r + i\chi_i) = P\int_{-\infty}^{+\infty}\frac{\chi_r(\omega')}{\omega'-\omega}d\omega' + iP\int_{-\infty}^{+\infty}\frac{\chi_i(\omega')}{\omega'-\omega}d\omega' \tag{2.43}$$

$$\begin{cases} \chi_r(\omega) = \dfrac{1}{\pi}P\displaystyle\int_{-\infty}^{+\infty}\frac{\chi_i(\omega')}{\omega'-\omega}d\omega' \\ \chi_i(\omega) = -\dfrac{1}{\pi}P\displaystyle\int_{-\infty}^{+\infty}\frac{\chi_r(\omega')}{\omega'-\omega}d\omega' \end{cases} \tag{2.44}$$

These relations must be in accord with the crossing relations:

$$\begin{aligned}
\chi_r(\omega) &= \frac{1}{\pi}P\int_{-\infty}^{+\infty}\frac{\chi_i(\omega')}{\omega'-\omega}d\omega' \\
&= \frac{1}{\pi}P\left[\int_{-\infty}^{0}\frac{\chi_i(\omega')}{\omega'-\omega}d\omega' + \int_{0}^{+\infty}\frac{\chi_i(\omega')}{\omega'-\omega}d\omega'\right] \\
&= \frac{1}{\pi}P\left[\int_{0}^{\infty}\frac{\chi_i(-\omega')}{-\omega'-\omega}d\omega' + \int_{0}^{\infty}\frac{\chi_i(\omega')}{\omega'-\omega}d\omega'\right] \\
&= \frac{1}{\pi}P\left[\int_{0}^{\infty}\frac{\chi_i(\omega')}{\omega'+\omega}d\omega' + \int_{0}^{\infty}\frac{\chi_i(\omega')}{\omega'-\omega}d\omega'\right] \\
&= \frac{1}{\pi}P\int_{0}^{\infty}\chi_i(\omega')\left[\frac{1}{\omega'+\omega} + \frac{1}{\omega'-\omega}\right] = \frac{2}{\pi}P\int_{0}^{\infty}\frac{\chi_i(\omega')\omega'}{\omega'^2+\omega^2}
\end{aligned}$$

Also,

$$\begin{aligned}
\chi_i(\omega) &= -\frac{1}{\pi}P\int_{-\infty}^{+\infty}\frac{\chi_r(\omega')}{\omega'-\omega}d\omega' \\
&= -\frac{1}{\pi}P\left[\int_{-\infty}^{0}\frac{\chi_r(\omega')}{\omega'-\omega}d\omega' + \int_{0}^{+\infty}\frac{\chi_r(\omega')}{\omega'-\omega}d\omega'\right]
\end{aligned}$$

$$= -\frac{1}{\pi} P \left[\int_0^\infty \frac{\chi_r(\omega')}{-\omega' - \omega} d\omega' + \int_0^\infty \frac{\chi_r(\omega')}{\omega' - \omega} d\omega' \right]$$

$$= -\frac{1}{\pi} P \int_0^\infty \chi_r(\omega') \left[\frac{1}{\omega' \quad \omega} + \frac{1}{\omega' \quad \omega} \right] d\omega'$$

$$= -\frac{2\omega}{\pi} P \int_0^\infty \frac{\chi_r(\omega')\omega'}{\omega'^2 - \omega^2}$$

We can now write

$$\begin{cases} \chi_r(\omega) = \dfrac{2}{\pi} P \displaystyle\int_0^\infty \frac{\omega' \chi_i(\omega')}{\omega'^2 - \omega^2} d\omega' \\ \chi_i(\omega) = -\dfrac{2\omega}{\pi} P \displaystyle\int_0^\infty \frac{\chi_r(\omega')\omega'}{\omega'^2 - \omega^2} d\omega' \end{cases} \tag{2.45}$$

These relations are known as *Kramers-Kronig relations* (and also as *dispersion relations*).

On the other hand, the complex index of refraction is given by

$$n(\omega) = \sqrt{K(\omega)} = \sqrt{1 + 4\pi\chi(\omega)} \approx 1 + 2\pi\chi(\omega) = n_r(\omega) + i n_i(\omega) \tag{2.46}$$

$$\begin{aligned} n_r(\omega) &= 1 + 2\pi\chi_r(\omega) \\ n_i(\omega) &= 2\pi\chi_i(\omega) \end{aligned} \tag{2.47}$$

Then the Kramers-Kronig relations give us

$$\begin{cases} n_r(\omega) - 1 = \dfrac{2}{\pi} P \displaystyle\int_0^\infty \frac{\omega' n_i(\omega')}{\omega'^2 - \omega^2} d\omega' \\ n_i(\omega) = -\dfrac{2\omega}{\pi} P \displaystyle\int_0^\infty \frac{n_r(\omega')}{\omega'^2 - \omega^2} d\omega' \end{cases} \tag{2.48}$$

since

$$P \int_0^\infty \frac{1}{\omega'^2 - \omega^2} d\omega' = 0 \tag{2.49}$$

The proof of (2.49) is left as a problem for the reader.

Summary

The **polarization** of a system is its response to an applied electric field. We consider a gas of molecules with a single resonant frequency and some damping in order to derive an expression for the complex susceptibility, which fully characterizes the response of the system in the frequency domain. Such a susceptibility has two poles in the lower half of the complex plane.

The complex susceptibility is then used to derive **Kramers-Kronig (KK) relations** by using contour integration in the complex plane and the Cauchy Residue theorem.

II.C Beyond the Molecular Model. General Observations on the Kramers-Kronig Relations

The Kramers-Kronig relations have been derived by using a specific expression for the susceptibility:

$$\chi(\omega) = \frac{(n_o e^2)/m}{\omega_o^2 - \omega^2 - i\gamma\omega} \tag{2.50}$$

This expression is based on a particular model of the "molecule".
However, only two properties of $\chi(\omega)$ were used:

1. The positions of the poles ω_1 and ω_2 *below* the real axis, and
2. the decrease of $\chi(\omega')(\omega' - \omega)$ faster than $(\omega')^{-1}$ for large ω'.

In general, the poles of a response function must lie in the lower half of the complex plane. The principle of causality introduces an asymmetry between $t \rangle t_o$ and $t \langle t_o$, which mathematically leads to the asymmetry in the frequency domain, that is, *poles* below the real axis [3]. For the classical oscillator model, the causality principle is related to the condition that γ in (2.30) must be positive. The real part of the induced dipole is given by

$$Re[ez(\mathbf{x},t)] = \frac{e^2}{m} \frac{E_o(\mathbf{x})}{\sqrt{(\omega_o^2 - \omega^2)^2 + \gamma^2\omega^2}} \cos(\omega t - \phi) \tag{2.51}$$

where

$$\tan\phi = \frac{\gamma\omega}{\omega_o^2 - \omega^2} \tag{2.52}$$

and

$$\begin{array}{ll} \omega \approx 0, \quad \tan\phi = 0, \quad \phi = 0; & \text{the dipole follows the field} \\ \omega = \omega_o, \quad \tan\phi = \infty, \quad \phi = \pi/2; & \text{the dipole moment lags by } 90^\circ \\ \omega = \infty, \quad \tan\phi = 0_-, \quad \phi = \pi; & \text{the dipole moment lags by } 180^\circ \end{array} \tag{2.53}$$

The positive sign of γ places ω_1 and ω_2 below the real axis.

The Kramers-Kronig relations show that the real and imaginary parts of the susceptibility are very intimately connected. Indeed, if we know one part at all the frequencies, we know by means of the integrals the other part at *all* the frequencies. This may be very useful when it is easier to measure one part of $\chi(\omega)$ rather than the other.

Note that a frequency-dependent real part $\chi_r(\omega)$ implies a nonzero imaginary part. Also, a sharp maximum in $\chi_i(\omega)$ produces a sharp change in the slope of $\chi_r(\omega)$.

Summary

The derivation of Kramers-Kronig (KK) relations in the preceding section was made by using a specific expression of the susceptibility based on a particular model of a molecular gas. However, only two properties were used: 1) The complex function representing the susceptibility has no poles in the upper half-plane, and 2) it goes to zero as (frequency)$^{-2}$ in the large frequency limit. The position of the poles in the lower half-plane is due to the sign of the damping term (a passive rather than an active system). The two properties above have a certain generality, and therefore, the KK relations are found to have wide applications.

II.D Relaxation

The mechanisms that affect the index of refraction may be of various kinds. We can have, for example, radiation damping. If we have a liquid with molecules that have dipole moments (for instance, water or organic substances of the benzene group) at low frequencies, the greatest contribution derives from the orientation of the elementary dipole moments; the main damping in this case is due to the *viscosity* of the medium.

Assume that we have a number of n_o molecules per unit volume, each molecule having an electric dipole moment. Let P_s be the value that the polarization ultimately takes after the application of a constant E field. If the E field is applied at time $t=0$, the polarization will follow the law

$$P(t) = P_s(1 - e^{-t/\tau}) \tag{2.54}$$

where τ = *relaxation time*. We can also write

$$\frac{dP(t)}{dt} = \frac{1}{\tau} P_s(1 - e^{-t/\tau}) = \frac{1}{\tau}\left[P_s - P(t)\right] \tag{2.55}$$

The evolution of *P(t)* is represented in Figure 2.9.

For a time-dependent field, $E_o e^{-i\omega t}$,

$$\frac{dP(t)}{dt} = \frac{1}{\tau}\left[n_o \alpha_o E_o e^{-i\omega t} - P(t)\right] \tag{2.56}$$

where α_o = orientational polarizability and $n_o \alpha_o E_o e^{-i\omega t}$ is the saturation value of the polarization for a stationary electric field equal in value to $E_o e^{-i\omega t}$. The solution of equation (2.56), apart from a transient $Ce^{-t/\tau}$ in which we are not interested here, is given by

$$P(t) = \frac{n_o \alpha_o}{1 - i\omega\tau} E_o e^{-i\omega t} = \chi(\omega) E_o e^{-i\omega t} \tag{2.57}$$

Therefore,

$$\chi(\omega) = \frac{n_o \alpha_o}{1 - i\omega\tau} = \frac{n_o \alpha_o}{1 + \omega^2\tau^2}(1 + i\omega\tau) \tag{2.58}$$

and

$$\begin{cases} \chi_r(\omega) = \dfrac{n_o \alpha_o}{1 + \omega^2\tau^2} \\ \chi_i(\omega) = \dfrac{n_o \alpha_o \omega\tau}{1 + \omega^2\tau^2} \end{cases} \tag{2.59}$$

These two functions are depicted in Figure 2.10. An angle ϕ, a measure of the phase lag of *P(t)* with respect to *E(t)*, is defined by the relation

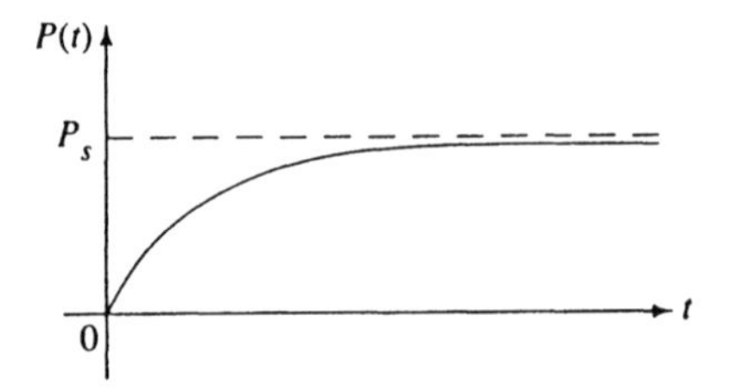

Fig. 2.9. Response of a gas of dipolar molecules to an applied field

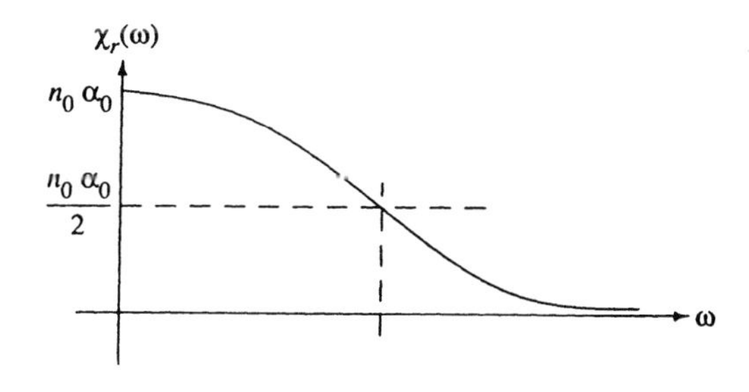

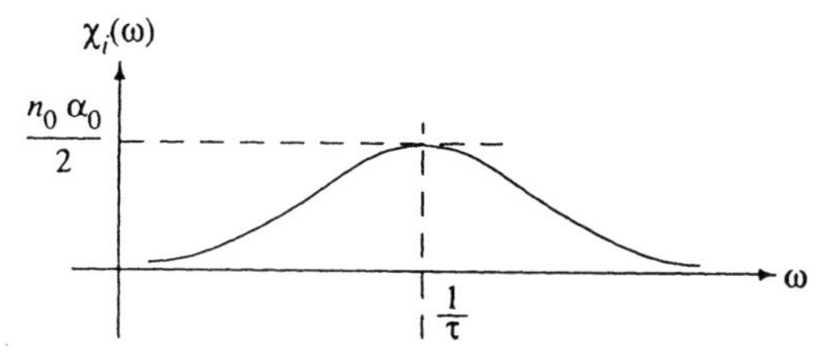

Fig. 2.10. Real and imaginary parts of the orientational susceptibility of a gas of dipolar molecules

$$\tan\phi = \omega\tau \tag{2.60}$$

and

$$\begin{cases} \phi = 0, & \text{for} \quad \omega = 0 \\ \phi = \dfrac{\pi}{4}, & \text{for} \quad \omega = \dfrac{1}{\tau} \\ \phi = \dfrac{\pi}{2}, & \text{for} \quad \omega = \infty \end{cases} \tag{2.61}$$

In liquids [4], the relaxation time is related to the viscosity η by the approximate relation

$$\tau = \frac{4\pi\eta a^3}{k_B T} \tag{2.62}$$

where

k_B = Boltzmann's constant
a = "radius" of the molecule
η = inner friction constant of the liquid

For water at T = 300K

$$\eta = 0.01 \; gm \; cm^{-1} s^{-1}$$
$$a = 2 \times 10^{-8} \; cm$$
$$\tau = \frac{4\pi \times 0.01 \times (2 \times 10^{-8})^3}{1.38 \times 10^{-16} \times 300} \approx 1.9 \times 10^{-11} s$$

Relaxation times in solids are much longer than in liquids, because a solid presents a more rigid barrier to the internal motion of molecules, leading to their orientation in the direction of an applied field.

Summary

We consider the case of **relaxation**, exemplified by, say, water. At low frequency the largest contribution to the susceptibility derives from the orientation of the dipole moments represented by the molecules of water: viscosity provides the damping mechanism. We verify that in this case the susceptibility presents the properties noted in the section II.C; therefore, the real part and the imaginary part of the susceptibility are tied by the KK relations.

II.E. Relations Between $n_r(\omega)$ and $n_i(\omega)$

The relations between $n_r(\omega)$ and $n_i(\omega)$ also contain interesting information. We have

$$n_r(\omega) = 1 + \frac{2}{\pi} P \int_0^{\infty} d\omega' \frac{\omega' n_i(\omega')}{\omega'^2 - \omega^2} \tag{2.63}$$

We define the quantity

$$\mu(\omega) = \frac{2\omega n_i(\omega)}{c} \tag{2.64}$$

as the *absorption coefficient*, and write

$$\omega' n_i(\omega') = \frac{c\mu(\omega')}{2} \tag{2.65}$$

Then

$$n_r(\omega) = 1 + \frac{c}{\pi} P \int_0^{\infty} d\omega' \frac{\mu(\omega')}{\omega'^2 - \omega^2} \tag{2.66}$$

and

$$n_r(0) = 1 + \frac{c}{\pi} P \int_0^{\infty} d\omega \frac{\mu(\omega)}{\omega^2} = \sqrt{K(0)} \tag{2.67}$$

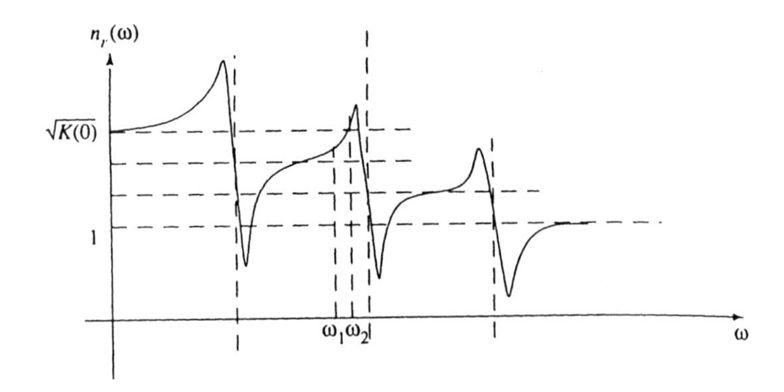

Fig. 2.11 General behavior of the real part of the index of refraction

where *K(0)* is the dielectric constant for stationary fields. Then

$$K(0) = 1 + \frac{2c}{\pi} P \int_0^\infty d\omega \frac{\mu(\omega)}{\omega^2} \tag{2.68}$$

But

$$\mu(\omega) = \sigma(\omega) n_0 \tag{2.69}$$

where $\sigma(\omega)$ = absorption cross section and n_o = number of molecules per unit volume.

Therefore,

$$K(0) - 1 = \frac{2cn_0}{\pi} P \int_0^\infty d\omega \frac{\sigma(\omega)}{\omega^2} \tag{2.70}$$

and

$$\begin{aligned} n_r(0) = \sqrt{K(0)} &= 1 + \frac{c}{\pi} P \int_0^\infty d\omega \frac{\mu(\omega)}{\omega^2} \\ &= 1 + \frac{cn_o}{\pi} P \int_0^\infty d\omega \frac{\sigma(\omega)}{\omega^2} \end{aligned} \tag{2.71}$$

These two relations indicate that, if $n_r(0)$ and consequently *K(0)* are different from 1, the system absorbs in some frequency regions. We can also write

$$\begin{aligned} n_r(\omega) - n_r(0) &= \frac{c}{\pi} P \int_0^\infty \mu(\omega') \left[\frac{1}{\omega'^2 - \omega^2} - \frac{1}{\omega'^2} \right] d\omega' \\ &= \frac{c\omega^2}{\pi} P \int_0^\infty \frac{\mu(\omega')}{\omega'^2 (\omega'^2 - \omega^2)} d\omega' \end{aligned} \tag{2.72}$$

The index of refraction $n_r(\omega)$ increases with the frequency ω as long as ω does not fall in the region where a resonance occurs.

The general behavior of the index of refraction n_r of a system with several resonances is depicted in Figure 2.11. Going up in frequency, if, *after* a resonance, the index of refraction increases again and exceeds the value of 1, at least another resonance is present at a higher frequency. If, after a resonance, the index of refraction goes to 1, no resonance occurs at greater frequencies.

We note that in the frequency regions where n_r changes, n_i also changes, so we can say with Debye, *Every dispersion has to be accompanied by absorption* [5].

We note also that all the properties considered above are derived simply by the application of the causality argument.

Summary

When the susceptibility is complex, the **dielectric constant** and the **index of refraction** are also complex.

The real part of the index of refraction $n_r(\omega)$ is related to the dispersion properties of the system; the imaginary part $n_i(\omega)$ to the absorption properties.

$n_r(\omega)$ generally increase with frequency (**normal dispersion**), except in the frequency regions where resonances occur (**anomalous dispersion**); $n_i(\omega)$ is appreciable only in regions of anomalous dispersion.

Every dispersion has to be accompanied by absorption.

II.F. Revisiting the Low-Pass Filter

We would like now to go back to the low-pass filter of section II.A. We shall write the transfer function as follows

$$Y(\omega) = Ke^{i\omega\tau} \tag{2.73}$$

where we have changed the sign of ω to make the notation consistent with that of section II.B.

We shall assume that $\omega_t \to \infty$. When this is the case, as we already noted in II.A., the output signal is retarded with respect to the input signal by τ, but, apart from amplitude, is equal to it; of course in such a case the transients have disappeared together with the contradiction with the principle of causality.

Consider now the integral

$$I = P\int_{-\infty}^{+\infty} \frac{e^{i\tau\omega'}}{\omega' - \omega} \tag{2.74}$$

Let

$$\omega' - \omega = z, \qquad \omega' = \omega + z$$

Then

$$I = P\int_{-\infty}^{+\infty} \frac{e^{i\tau(\omega+z)}}{z} dz \tag{2.75}$$

Let the integrand of *I* be called *F(z):*

$$F(z) = \frac{e^{i\tau(\omega+z)}}{z} \tag{2.76}$$

and let us integrate *F(z)* over the contour in Figure 2.12. Since *F(z)* is analytic inside the contour, the Cauchy's residue theorem applies:

$$\int_{-R}^{\rho} \frac{e^{i\tau(\omega+z)}}{z} dz + \int_{C_\rho} \frac{e^{i\tau(\omega+z)}}{z} dz + \int_{\rho}^{R} \frac{e^{i\tau(\omega+z)}}{z} dz + \int_{C_R} \frac{e^{i\tau(\omega+z)}}{z} dz = 0 \tag{2.77}$$

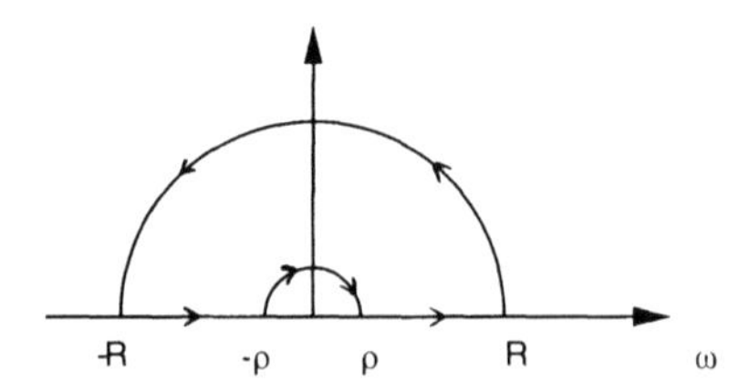

Fig. 2.12. Contour for the integration of *F(z)*

As $R \to \infty$, the third term above is equal to $-i\pi\tau\omega$; the fourth term is zero because of the *Jordan lemma* [6]. We can then write:

$$\lim_{\substack{\rho\to 0\\ R\to\infty}} \left\{ \int_{-R}^{\rho} \frac{e^{i\tau(\omega+z)}}{z} dz + \int_{\rho}^{R} \frac{e^{i\tau(\omega+z)}}{z} dz \right\} - i\pi e^{i\tau\omega} = 0 \tag{2.78}$$

or

$$P\int_{-\infty}^{+\infty} \frac{e^{i\tau(\omega+z)}}{z} dz = P\int_{-\infty}^{+\infty} \frac{e^{i\tau\omega'}}{\omega' - \omega} d\omega' = i\pi e^{i\tau\omega} \tag{2.79}$$

and

$$\begin{aligned} P\int_{-\infty}^{+\infty} \frac{\cos\tau\omega' + i\sin\tau\omega'}{\omega' - \omega} d\omega' &= i\pi(\cos\tau\omega + i\sin\tau\omega) \\ &= i\pi\cos\tau\omega - \pi\sin\tau\omega \end{aligned} \tag{2.80}$$

Let

$$\begin{cases} Y_r(\omega) = \cos\tau\omega \\ Y_i(\omega) = \sin\tau\omega \end{cases} \tag{2.81}$$

Then

$$\begin{cases} P\displaystyle\int_{-\infty}^{+\infty} \frac{Y_r(\omega')}{\omega' - \omega} d\omega' = -\pi Y_i \\ P\displaystyle\int_{-\infty}^{+\infty} \frac{Y_i(\omega')}{\omega' - \omega} d\omega' = \pi Y_r \end{cases} \tag{2.82}$$

in agreement with the Kramers-Kronig relations.

Summary

In this section we revisit the low-pass ideal filter, with the proviso that the cut-off frequency ω_t is made to go to infinity. In such a case, the pulse waveform goes through the filter scaled in amplitude by a factor K and delayed by a time interval τ, but preserves its rectangular shape: no transients appear at times earlier than the time at which the pulse enters the filter. No violation of causality occurs.

Detailed calculations confirm that in this case, the real and imaginary parts of the transfer function are tied together by the KK relations.

II.G. Dispersion Relations of a Simple Circuit

We shall now consider the simple circuit of Figure 2.13 in light of the Kramers-Kronig relations. Let $\zeta(t)$ be the applied emf:

$$\zeta(t) = \int_{-\infty}^{+\infty} \zeta(\omega) e^{-i\omega t} d\omega \tag{2.83}$$

Because $\zeta(t)$ is real we require

$$\zeta^*(-\omega) = \zeta(\omega).$$

The current in the circuit $I(t)$ can be expressed as follows:

$$I(t) = \int_{-\infty}^{+\infty} I(\omega) e^{-i\omega t} d\omega \tag{2.85}$$

where

$$I^*(-\omega) = I(\omega) \tag{2.86}$$

From circuit theory we know that

$$\left(R - i\omega L + \frac{i}{\omega c}\right) I(\omega) = \zeta(\omega) \tag{2.87}$$

or

$$I(\omega) = \frac{\zeta(\omega)}{R - i\omega L + \frac{i}{\omega C}} = \frac{\zeta(\omega)}{\Omega(\omega)} \tag{2.88}$$

where

$$\Omega(\omega) = \frac{1}{R - i\omega L + \frac{i}{\omega C}} \tag{2.89}$$

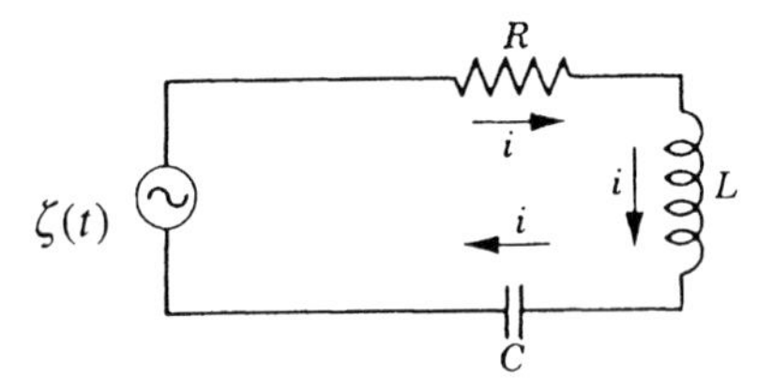

Fig. 2.13 A series RLC circuit under the action of an applied emf

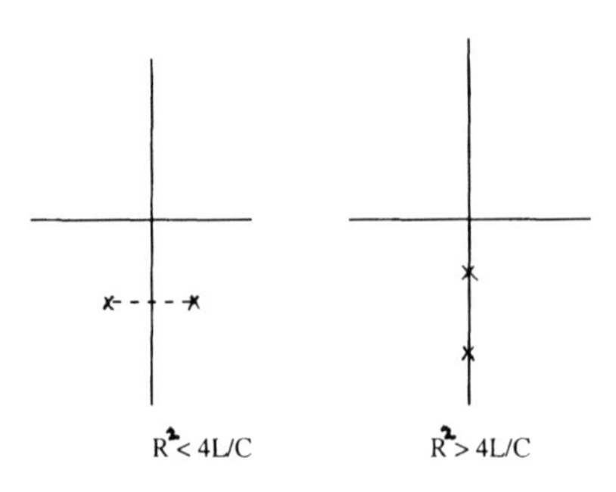

Fig. 2.14 Positions of the poles of $\Omega(\omega)$ in the complex plane

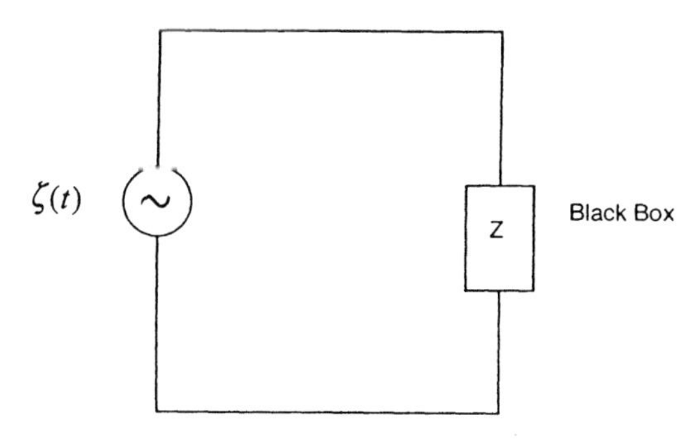

Fig. 2.15 General circuit with a black box

is the *admittance* of the circuit. Dispersion relations can be derived by using the explicit form of $\Omega(\omega)$. When we regard ω as a complex variable, then $\Omega(\omega)$ is an analytic function of ω with the *simple* poles at

$$\omega = \frac{-iR + \sqrt{4\frac{L}{C} - R^2}}{2L} \quad \text{and} \quad \omega = \frac{-iR - \sqrt{4\frac{L}{C} - R^2}}{2L} \tag{2.90}$$

The poles lie in the lower half plane because $R\rangle 0$ (see Figure 2.14). The resistance has an *aborptive* effect on the current flow.
By performing a contour integration as we did for the molecular susceptibility, we find

$$P\int_{-\infty}^{+\infty} \frac{\Omega(\omega')}{\omega' - \omega} d\omega' = i\pi\Omega(\omega) \tag{2.91}$$

where the integration is performed over real values only. We then derive the dispersion relations

$$\begin{cases} \mathrm{Re}\,\Omega(\omega) = \dfrac{1}{\pi} P\displaystyle\int_{-\infty}^{+\infty} \frac{\mathrm{Im}\,\Omega(\omega)}{\omega' - \omega} d\omega' \\ \mathrm{Im}\,\Omega(\omega) = -\dfrac{1}{\pi} P\displaystyle\int_{-\infty}^{+\infty} \frac{\mathrm{Re}\,\Omega(\omega)}{\omega' - \omega} d\omega' \end{cases} \tag{2.92}$$

which are similar to the ones (2.44) obtained for the susceptibility. As for the latter, they express the fact that the system obeys causality: no current flows before some emf is applied.

The dispersion relations apply also to the more general circuit reported in Figure 2.15, in which the *black box* contains any combinations of resistances, inductances, and capacitances, but no source of emf. Let

$$\Omega(\omega) = \frac{1}{Z(\omega)} \tag{2.93}$$

Causality requires that all the poles of $\Omega(\omega)$ should lie in the lower half plane. We write

$$\Omega(\omega) = \frac{R(\omega) + iX(\omega)}{|Z(\omega)|^2} \tag{2.94}$$

where $R(\omega)$ and $X(\omega)$ are real. Let

$$\zeta(t) = emf = \zeta_o e^{-i\omega_o t} \tag{2.95}$$

Then

$$I(\omega_o)e^{-i\omega_o t} = \zeta_o \Omega(\omega_o) e^{-i\omega_o t} \tag{2.96}$$

The in-phase part of the current is given by

$$\zeta_o \,\mathrm{Re}\left[\Omega(\omega_o) e^{-i\omega_o t}\right] \tag{2.97}$$

The average rate of absorption of energy is then

$$\frac{1}{2}\zeta_o^2 \,\mathrm{Re}\,\Omega(\omega_o) \tag{2.98}$$

We can then know $\mathrm{Re}\,\Omega(\omega)$ at all frequencies by measuring the absorption of energy at all frequencies. Once we know $\mathrm{Re}\,\Omega(\omega)$ we can use this knowledge to find $\mathrm{Im}\,\Omega(\omega)$ with the help of the dispersion relations. These relations allow us to find $Z(\omega)$ if we know how the power input into the circuit varies with frequency.

Summary

In this section we examine the dispersion relations of a simple series RLC circuit. The applied emf is the input and the current flowing the circuit is the output. The admittance of the circuit plays a role similar to that of the susceptibility in previous examples.

The real and the imaginary part of the admittance are tied by relations of the KK type.

III. PULSE CHARACTERIZATION

Be what you seem to be, and
seem to be what you really are.
Mother Ann of the Shakers

III.A. Time Duration and Spectral Widths of Waveforms

Regardless of the definition of time duration of a waveform, the spectral width is inversely proportional to it, in accordance with the scaling property of Fourier transforms.

The *power rms width* of a waveform f(t) is given by

$$\sigma_t^2 = \frac{\int_{-\infty}^{+\infty} (t-\bar{t})^2 |f(t)|^2 dt}{\int_{-\infty}^{+\infty} |f(t)|^2 dt} \tag{3.1}$$

where

$$\bar{t} = \frac{\int_{-\infty}^{+\infty} t|f(t)|^2 dt}{\int_{-\infty}^{+\infty} |f(t)|^2 dt} \tag{3.2}$$

The *spectral width* is given by

$$\sigma_\omega{}^2 = \frac{\int\limits_{-\infty}^{+\infty} (\omega - \overline{\omega})^2 |F(\omega)|^2 d\omega}{\int\limits_{-\infty}^{+\infty} |F(\omega)|^2 d\omega} \tag{3.3}$$

where

$$\overline{\omega} = \frac{\int\limits_{-\infty}^{+\infty} \omega |F(\omega)|^2 d\omega}{\int\limits_{-\infty}^{+\infty} |F(\omega)|^2 d\omega} \tag{3.4}$$

If we use the Schwarz inequality:

$$\left(\int |f|^2 d\tau\right)\left(\int |g|^2 d\tau\right) \geq \left|\int f * g d\tau\right|^2 \tag{3.5}$$

we find

$$\sigma_t \sigma_\omega \geq \frac{1}{2} \tag{3.6}$$

For a Gaussian function

$$f(t) = e^{-\frac{t^2}{4\sigma_t{}^2}} \tag{3.7}$$

the power rms width is σ_t. The Fourier transform is given by

$$F(\omega) = \frac{\sqrt{\pi}}{\sigma_\omega} e^{\frac{-\omega^2}{4\sigma_\omega{}^2}} \tag{3.8}$$

where

$$\sigma_\omega = \frac{1}{2\sigma_t} \tag{3.9}$$

Then for Gaussian functions

$$\sigma_t \sigma_\omega = \frac{1}{2} \tag{3.10}$$

We shall now give some examples.

Exponential function

$$\begin{cases} f(t) = e^{-\frac{t}{\tau}}, & t \geq 0 \\ f(t) = 0, & t \langle 0 \end{cases} \tag{3.11}$$

The width at $1/e$ of the maximum $\Delta t_{1/e}$ is τ . The Fourier transform of the function is

$$F(\omega) = \frac{\tau}{1 + i\omega\tau} \tag{3.12}$$

with

$$\Delta\omega_{3dB} = \frac{1}{\tau} \tag{3.13}$$

<u>Double-sided exponential function</u>

$$f(t) = e^{-\frac{|t|}{\tau}} \tag{3.14}$$

The width at $1/e$ of the maximum is 2τ. The Fourier transform of the function is

$$F(\omega) = \frac{2\tau}{1+(\omega\tau)^2} = \frac{\Delta\omega}{\omega^2 + \left(\frac{\Delta\omega}{2}\right)^2} \tag{3.15}$$

where

$$\Delta\omega = \Delta\omega_{1/2} = \frac{2}{\tau} \tag{3.16}$$

<u>Gaussian function</u>

$$f(t) = e^{-\frac{t^2}{2\tau^2}} \tag{3.17}$$

$\Delta t_{1/2} = 2\sqrt{2}\tau$. The Fourier transform is

$$F(\omega) = \sqrt{2\pi}\tau e^{\frac{\omega^2\tau^2}{2}} \tag{3.18}$$

with

$$\Delta\omega_{1/e} = \frac{2\sqrt{2}}{\tau} \tag{3.19}$$

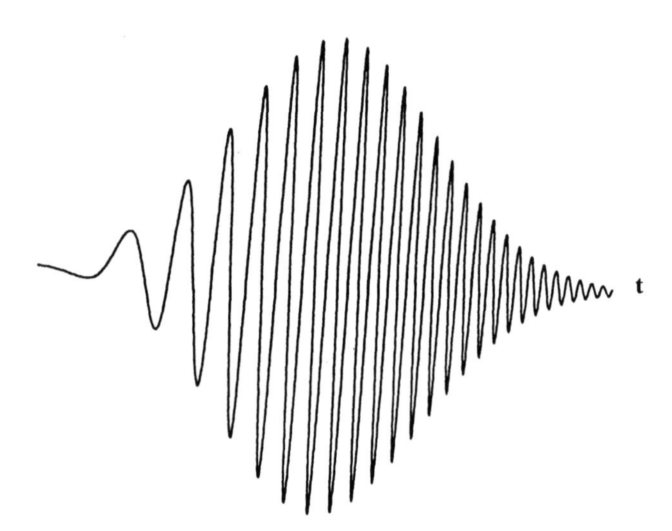

Fig. 3.1 A chirped Gaussian pulse with $b\rangle 0$

and

$$\Delta\omega_{1/2} = \frac{2\sqrt{2\ln 2}}{\tau} \tag{3.20}$$

Note that

$$\Delta\omega_{1/2} = \sqrt{\ln 2}\Delta\omega_{1/e} = 0.833\Delta\omega_{1/e} \tag{3.21}$$

Summary

This section is used to define the time duration Δt of a waveform and its relation to the frequency spread $\Delta\omega$. Regardless of the manner in which these two parameters are defined, they are inversely proportional.

Definitions of importance are the following:

σ_t for the **signal power rms width**
σ_ω for the **spectral power rms width**

The proportionality cited above can then be expressed by the relation $\sigma_t\sigma_\omega \geq \frac{1}{2}$.
The equality applies to **Gaussian pulses.**

III.B. Gaussian Pulses

A Gaussian pulse has a carrier frequency ω_o and a complex Gaussian envelope

$$E(t) = e^{-at^2 + i(\omega_o t + bt^2)} = e^{-\Gamma t^2} e^{i\omega_o t} \tag{3.22}$$

where

$$\Gamma = a - ib$$

is a complex parameter describing the pulse. The instantaneous intensity is proportional to

$$|E(t)|^2 = e^{-2at^2} = e^{-4\ln 2(t/\tau_p)^2} \tag{3.23}$$

where

$$\tau_p = \sqrt{\frac{2\ln 2}{a}} \tag{3.24}$$

is the time at which the intensity of the pulse is reduced to 1/2. The instantaneous phase of the signal is

$$\phi(t) = \omega_o t + bt^2 \tag{3.25}$$

Then the instantaneous frequency is

$$\omega_i t = \frac{d\phi(t)}{dt} = \omega_o + 2bt \tag{3.26}$$

where b is the measure of the *chirp* of the pulse.

The Fourier transform of the pulse (3.22) is given by

$$E(\omega) = \int_{-\infty}^{+\infty} \zeta(t) e^{-i\omega t} dt = \int_{-\infty}^{+\infty} e^{-\Gamma t^2 + i(\omega_o - \omega)t} dt \propto e^{\frac{-(\omega-\omega_o)^2}{4\Gamma}} \tag{3.27}$$

We can write

$$\frac{(\omega-\omega_o)^2}{4\Gamma}=\frac{(\omega-\omega_o)^2}{4(a-ib)}=\frac{(\omega-\omega_o)^2}{4}\left[\frac{a}{a^2+b^2}+i\frac{b}{a^2+b^2}\right] \tag{3.28}$$

The power spectrum of the pulse is then given by

$$|E(\omega)|^2 \propto e^{-\left[\frac{a(\omega-\omega_o)^2}{2(a^2+b^2)}\right]}=e^{-4\ln 2\left(\frac{\omega-\omega_o}{\Delta\omega_p}\right)^2} \tag{3.29}$$

where

$$\Delta\omega_p=2\sqrt{2\ln 2}\sqrt{a\left(1+\frac{b^2}{a^2}\right)}=\Delta\omega_{1/2} \tag{3.30}$$

Then

$$\Delta f_p=\frac{\Delta\omega_p}{2\pi}=\frac{\sqrt{2\ln 2}}{\pi}\sqrt{a\left(1+\frac{b^2}{a^2}\right)} \tag{3.31}$$

For a given pulsewidth in time as determined by the real parameter a, the presence of a frequency chirp as determined by ib increases the spectral bandwidth by the factor

$$\left(1+\frac{b^2}{a^2}\right)^{1/2}.$$

Now we have

$$\tau_p=\sqrt{\frac{2\ln 2}{a}} \tag{3.32}$$

$$\Delta\omega_p=2\sqrt{2\ln 2}\sqrt{a\left(1+\frac{b^2}{a^2}\right)} \tag{3.33}$$

and

$$\tau_p\Delta\omega_p=4\ln 2\sqrt{1+\frac{b^2}{a^2}} \tag{3.34}$$

For a Gaussian pulse the relation between σ and $FWHM$ is

$$FWHM=2\sqrt{2\ln 2}\,\sigma \tag{3.35}$$

We can then write the power rms width of the pulse *in time*:

$$\sigma_t=\frac{\tau_p}{2\sqrt{2\ln 2}}=\frac{\sqrt{2\ln 2}}{a}\frac{1}{2\sqrt{2\ln 2}}=\frac{1}{2\sqrt{a}} \tag{3.36}$$

Power rms width *in angular frequency*:

$$\sigma_\omega=\frac{\Delta\omega_p}{2\sqrt{2\ln 2}}=\sqrt{a\left(1+\frac{b^2}{a^2}\right)} \tag{3.37}$$

We then find for Gaussian pulse with chirp:

$$\sigma_t \sigma_\omega = \frac{1}{2}\sqrt{1+\frac{b^2}{a^2}} \tag{3.38}$$

For *b=0*, we reobtain the relation (3.10).

For an arbitrary pulse, the above product depends on the exact shape of the pulse or on the amount of *chirp* or substructure present within the pulse.

As a general rule

$$\sigma_t \sigma_\omega \geq \frac{1}{2} \tag{3.39}$$

Pulses which have little chirp or other internal substructure are called *transform-limited.* For them $\sigma_t \sigma_\omega \approx 0.5$.

Summary

In this section we introduce pulses that we are going to use often: pulses with a carrier frequency ω_o and a Gaussian envelope. The **pulsewidth** in time is determined by a parameter **a**: the full width at half intensity is $(2\ln 2/a)^{1/2}$. The **instantaneous frequency** within the pulse varies linearly with time: $\omega = \omega_o + 2bt$, where **b** is a measure of the **chirp** present in the pulse, the substructure under the Gaussian envelope. The presences of the frequency chirp increases the spectral bandwidth by the factor $(1+b^2/a^2)^{1/2}$. Pulses with very little chirp or other internal substructure are called **transform-limited**. For such pulses, $\sigma_t \sigma_\omega \approx 1/2$.

The product $[\Delta t \Delta\omega]$ for a pulse depends in general on the following:

(i) the exact shape of the pulse
(ii) how we define Δt and $\Delta\omega$, and
(iii) the amount of chirp or substructure present in the pulse.

IV. PROPAGATION OF PULSES IN LINEAR MEDIA

Defiez-vous de vos premiers mouvements:
ils sont presque toujours bons.
Talleyrand

IV.A. Chromatic Dispersion

The response of a medium to the passage of electromagnetic waves manifests itself through the frequency dependence of the *index of refraction*, i.e. the *chromatic dispersion.*

The propagation in vacuum is characterized by a parameter β called the *propagation constant*, which represents the rate at which the phase changes with distance. The propagation constant in a medium is given by

$$\beta(\omega) = n(\omega)\frac{\omega}{c} \tag{4.1}$$

where $n(\omega)$ is the index of refraction. It can be expanded in a Taylor series about the *center* frequency ω_o of a train of waves:

$$\beta(\omega) = n(\omega)\frac{\omega}{c} = \beta_o + \beta_1(\omega - \omega_o) + \frac{1}{2}\beta_2(\omega - \omega_o)^2 + \ldots \tag{4.2}$$

where

$$\beta_m = \left(\frac{d^m \beta}{d\omega^m}\right)_{\omega=\omega_o} \tag{4.3}$$

Then

$$\beta_1 = \left(\frac{d\beta}{d\omega}\right)_{\omega=\omega_o} = \left(\frac{n}{c} + \frac{\omega}{c}\frac{dn}{d\omega}\right)_{\omega=\omega_o} = \frac{1}{c}\left(n + \omega\frac{dn}{d\omega}\right)_{\omega=\omega_o} = \frac{1}{v_g} \tag{4.4}$$

where v_g=group velocity. Also

$$\beta_2 = \left(\frac{d^2\beta}{d\omega^2}\right)_{\omega=\omega_o} = \frac{1}{c}\left(2\frac{dn}{d\omega} + \omega\frac{d^2n}{d\omega^2}\right)_{\omega=\omega_o} \approx \left(\frac{\omega}{c}\frac{d^2n}{d\omega^2}\right)_{\omega=\omega_o} \tag{4.5}$$

We know that

$$\frac{d\omega}{d\lambda} = -\frac{2\pi c}{\lambda^2} \tag{4.6}$$

Then

$$d\omega^2 = \frac{4\pi^2c^2}{\lambda^4}d\lambda^2 \tag{4.7}$$

and

$$\begin{aligned}\beta_2 &\approx \left(\frac{\omega}{c}\frac{d^2n}{d\omega^2}\right)_{\omega=\omega_o} = \left(\frac{2\pi}{\lambda}\frac{\lambda^4}{4\pi^2c^2}\frac{d^2n}{d\lambda^2}\right)_{\lambda=\lambda_o} \\ &= \left(\frac{\lambda^3}{2\pi c^2}\frac{d^2n}{d\lambda^2}\right)_{\lambda=\lambda_o} = -\left(\frac{1}{v_g{}^2}\frac{dv_g}{d\omega}\right)_{\omega=\omega_o}\end{aligned} \tag{4.8}$$

Most materials have positive β_2 in the visible region, and negative β_2 in the infrared. When $\beta_2 \rangle 0$, $\frac{dv_g}{d\omega} \langle 0$ and the high frequency components have lower v_g.

<u>Group Velocity</u>

We know already that

$$\frac{1}{v_g} = \frac{n_g}{c} = \frac{1}{c}\left(n + \omega\frac{dn}{d\omega}\right) \tag{4.4'}$$

Then

$$v_g = \frac{c}{n + \omega\frac{dn}{d\omega}} = \frac{c}{n + v\frac{dn}{dv}} \tag{4.9}$$

But

$$\frac{dn}{d\omega} = -\frac{dn}{d\lambda}\frac{2\pi c}{\omega^2} \tag{4.10}$$

Then

$$v_g = \frac{c}{n + \omega\left(-\frac{2\pi c}{\omega^2}\frac{dn}{d\lambda}\right)} = \frac{c}{n - \frac{2\pi c}{\omega}\frac{dn}{d\lambda}} = \frac{c}{n - \lambda\frac{dn}{d\lambda}} \tag{4.11}$$

or

$$v_g = \frac{c}{n_g}$$

where n_g is called *group index* and is given by

$$n_g = n - \lambda\frac{dn}{d\lambda} = n + \omega\frac{dn}{d\omega} = n + v\frac{dn}{dv} \tag{4.12}$$

Dispersion Coefficient

We define the *dispersion coefficient* as follows

$$D_v = \frac{d}{dv}\left(\frac{1}{v_g}\right) = 2\pi\frac{d}{d\omega}\left(\frac{1}{v_g}\right) = 2\pi\frac{d}{d\omega}\left(\frac{d\beta}{d\omega}\right) = 2\pi\frac{d}{d\omega}\left(\frac{d^2\beta}{d\omega^2}\right) \tag{4.13}$$

But

$$\beta_2 = \frac{d^2\beta}{d\omega^2} \approx \frac{\omega}{c}\frac{d^2n}{d\omega^2} = \frac{\lambda^3}{2\pi c^2}\frac{d^2n}{d\lambda^2} \tag{4.8'}$$

Then

$$D_v = \frac{\lambda^3}{c^2}\frac{d^2n}{d\lambda^2} \tag{4.14}$$

We define a dispersion coefficient in terms of λ as follows:

$$D_\lambda d\lambda = D_v dv \tag{4.15}$$

Then

$$\begin{aligned} D_\lambda &= D_v\frac{dv}{d\lambda} = D_\lambda\left(-\frac{c}{\lambda^2}\right) \\ &= \frac{\lambda^3}{c^2}\frac{d^2n}{d\lambda^2}\left(-\frac{c}{\lambda^2}\right) = -\frac{\lambda}{c}\frac{d^2n}{d\lambda^2} \end{aligned} \tag{4.16}$$

But

$$\beta_2 = \frac{d^2\beta}{d\omega^2} = \frac{\lambda^3}{2\pi c^2}\frac{d^2n}{d\lambda^2} \tag{4.8''}$$

Then

$$-\frac{2\pi c}{\lambda^2}\beta_2 = -\frac{2\pi c}{\lambda^2}\frac{\lambda^3}{2\pi c^2}\frac{d^2n}{d\lambda^2} = -\frac{\lambda}{c}\frac{d^2n}{d\lambda^2} \tag{4.17}$$

and

$$D_\lambda = -\frac{2\pi c}{\lambda^2}\beta_2 \tag{4.18}$$

Also we can write, from (4.8)

$$\frac{d\beta_1}{d\lambda} = \frac{d\beta_1}{d\omega}\frac{d\omega}{d\lambda} = -\frac{2\pi c}{\lambda^2}\beta_2 = -\frac{2\pi c}{\lambda^2}\frac{\lambda^3}{2\pi c^2}\frac{d^2 n}{d\lambda^2} = -\frac{\lambda}{c}\frac{d^2 n}{d\lambda^2} \tag{4.19}$$

Therefore

$$D_\lambda = \frac{d\beta_1}{d\lambda} \tag{4.20}$$

Summary

This section is used to define some important parameters that characterize the propagation of trains of electromagnetic waves in linear media, such as **index of refraction, propagation constant, phase velocity,** and **group velocity**. In addition **dispersion coefficients** in terms of frequency and wavelength are defined.

The most important formulae are now reported:

$$\beta(\omega) = n(\omega)\frac{\omega}{c}$$

$$\beta_1 = \frac{n_g}{c} = \frac{1}{v_g}, \qquad n_g = n + \omega\frac{dn}{d\omega} = n + \nu\frac{dn}{d\nu} = n - \lambda\frac{dn}{d\lambda}$$

$$\beta_2 = \frac{d\beta_1}{d\omega} = \frac{\omega}{c}\frac{d^2 n}{d\lambda^2} = \frac{\lambda^3}{2\pi c^2}\frac{d^2 n}{d\lambda^2} = -\frac{1}{{v_g}^2}\frac{dv_g}{d\omega}$$

$$D_\nu = \frac{d}{d\nu}\left(\frac{1}{v_g}\right) = 2\pi\frac{d^2\beta}{d\omega^2} = \frac{\lambda^3}{c^2}\frac{d^2 n}{d\lambda^2}$$

$$D_\lambda = -\frac{\lambda}{c}\frac{d^2 n}{d\lambda^2} = -\frac{2\pi c}{\lambda^2}\beta_2 = \frac{d\beta_1}{d\lambda}$$

IV.B. Propagation of Gaussian Pulses

1. Effects of Phase Velocity and Group Velocity. Let the input signal at z=0 be a Gaussian pulse represented by

$$E(0,t) = e^{-\Gamma_o t^2 + i\omega_o t} \tag{4.21}$$

where

$$\Gamma_o = a_o - ib_o \tag{4.22}$$

If $f(t)$ has the Fourier transform $F(\omega)$, then $f(t)e^{i\omega_o t}$ has the Fourier transform $F(\omega - \omega_o)$. Then

$$E(0,\omega) \propto e^{\frac{-(\omega-\omega_o)^2}{4\Gamma_o}} \tag{4.23}$$

After traveling a distance z through the medium, the pulse will have the spectrum

$$E(z,\omega) = E(0,\omega)e^{-i\beta(\omega)z} \propto$$

$$\propto \exp\left[\frac{-(\omega-\omega_o)^2}{4\Gamma_o}\right]\exp\left[-i\beta(\omega)z - i\beta_1 z(\omega-\omega_o) - \frac{i\beta_2 z}{2}(\omega-\omega_o)^2 + \ldots\right] \qquad (4.24)$$

$$= \exp\left[-i\beta(\omega_o)z - i\beta_1 z(\omega-\omega_o) - \left(\frac{1}{4\Gamma_o} + \frac{i\beta_2 z}{2}\right)(\omega-\omega_o)^2 + \ldots\right]$$

The signal at distance z is given by

$$E(z,t) = \frac{1}{2\pi}\int E(z,\omega)e^{i\omega t}d\omega$$

$$= \frac{e^{i[\omega_o t - \beta(\omega_o)z]}}{2\pi}\int_{-\infty}^{+\infty}\exp\left[-\frac{(\omega-\omega_o)^2}{4\Gamma_o} + i(\omega-\omega_o)t - i(\omega-\omega_o)\beta_1 z\right]d(\omega-\omega_o) \qquad (4.25)$$

$$= \frac{1}{2\pi}\{\exp[i\omega_o t - i\beta(\omega_o)z]\}\int_{-\infty}^{+\infty}\exp\left[-\frac{(\omega-\omega_o)^2}{4\Gamma(z)} + i(\omega-\omega_o)(t-\beta_1 z)\right]d(\omega-\omega_o)$$

where

$$\frac{1}{\Gamma(z)} = \frac{1}{\Gamma_o} + i2\beta_2 z \qquad (4.26)$$

and where the {} brackets contain the time and space dependences of the carrier frequency. The integral, on the other hand, gives the time and space dependences of the pulse envelope.

The integral represents the inverse Fourier transform of a Gaussian pulse of the form $\exp(-\Gamma t^2)$ but with a shift in time $t-\beta_1 z$. We can then write

$$E(z,t) = \exp[i\omega_o t - i\beta(\omega_o)z]\exp[-\Gamma(z)(t-\beta_1 z)^2]$$

$$= \exp\left[i\omega_o(t - \frac{\beta(\omega_o)}{\omega_o}z\right]\exp\left[-\Gamma(z)\left(t - \frac{z}{1/\beta_1}\right)^2\right] \qquad (4.27)$$

The first exponent tells us that the carrier frequency harmonic ω_o propagating from 0 to z distance has a midband phase delay

$$t_p = \frac{z}{\omega_o/\beta(\omega_o)} = \frac{z}{v_\phi(\omega_o)} \qquad (4.28)$$

where

$$v_\phi(\omega_o) = \frac{z}{t_\phi} = \frac{\omega_o}{\beta(\omega_o)} \qquad (4.29)$$

is the midband *phase velocity*. The second exponent tells us that the pulse envelope at z is delayed by the time

$$t_g = \frac{z}{1/\beta_1} = \frac{z}{v_g(\omega_o)} \qquad (4.30)$$

where

$$v_g(\omega_o) = \frac{1}{\beta_1} = \frac{1}{(d\beta/d\omega)_{\omega=\omega_o}} \qquad (4.31)$$

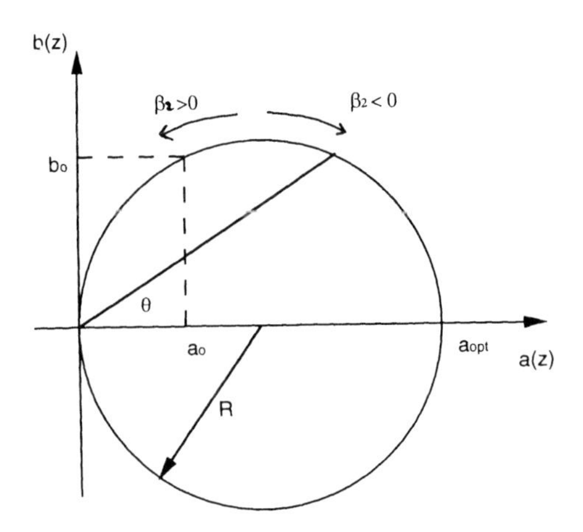

Fig. 4.1 Circle in the Γ plane illustrating the trajectory of a pulse

is the *midband group velocity*. Changes occur to the pulse while it is traveling in the +z direction. These changes are brought about by the changes of $\Gamma(z)$ with z.

2. The Evolution of the Pulse. Let us now restate the fact that at z=0 the pulse has a parameter

$$\Gamma_o = a_o - ib_o \tag{4.22}$$

where b_o = measure of chirp. At distance z, according to (4.26)

$$\frac{1}{\Gamma(z)} = \frac{1}{a(z) - ib(z)} = \frac{1}{\Gamma_o} + i2\beta_2 z = \frac{a_o}{a_o^2 + b_o^2} + i\left(\frac{b_o}{a_o^2 + b_o^2} + 2\beta_2 z\right) \tag{4.32}$$

Then

$$\begin{cases} \dfrac{a(z)}{a(z)^2 + b(z)^2} = \dfrac{a_o}{a_o^2 + b_o^2} \\ \dfrac{b(z)}{a(z)^2 + b(z)^2} = \dfrac{b_o}{a_o^2 + b_o^2} + 2\beta_2 z \end{cases} \tag{4.33}$$

After some algebra, we get

$$\begin{cases} a(z) = \dfrac{a_o}{(1 + b_o x)^2 + a_o^2 x^2} \\ b(z) = \dfrac{b_o(1 + b_o x) + a_o^2 x}{(1 + b_o x)^2 + a_o^2 x^2} \end{cases} \tag{4.34}$$

where

$$x = 2\beta_2 z \tag{4.35}$$

Recall that for a Gaussian pulse

$$\tau_p = FWHM \text{ for intensity} = \sqrt{\frac{2\ln 2}{a}} \tag{4.36}$$

$$\Delta\omega_p = FWHM \text{ for spectral band} = 2\sqrt{2\ln 2}\sqrt{a\left(1+\frac{b^2}{a^2}\right)} \tag{4.37}$$

We can now make the following observations:

(1) At each position z the pulse length is given by

$$\tau_p(z) = \sqrt{\frac{2\ln 2}{a(z)}} \tag{4.38}$$

A large $a(z)$ means a short pulse.

(2) Since, according to (4.33) the quantity

$$\frac{a(z)}{a(z)^2 + b(z)^2}$$

is constant, the bandwidth of the pulse does not change with z.

(3) With some algebra it is possible to show that

$$\left[a(z) - R^2\right] + b(z)^2 = R^2 \tag{4.39}$$

where

$$R = \frac{1}{2}\frac{a_o^{\,2} + b_o^{\,2}}{a_o} \tag{4.40}$$

The two variables $a(z)$ and $b(z)$ describe a circle in the (a,b) plane of radius R. This circle is shown in Figure 4.1.

If θ is the angle in Figure 4.1, then

$$\tan\theta = \frac{b(z)}{a(z)} = \frac{b_o(1+b_o x) + a_o^{\,2}x}{a_o} = \frac{b_o}{a_o} + \frac{a_o^{\,2} + b_o^{\,2}}{a_o^{\,2}}x$$

and

$$\theta = \tan^{-1}\left[\frac{b_o}{a_o} + \frac{a_o^{\,2} + b_o^{\,2}}{a_o^{\,2}}x\right]$$

But

$$\frac{d}{dx}\tan^{-1}[f(x)] = \frac{1}{1+f(x)^2}\frac{df(x)}{dx}$$

Then

$$f(x) = \frac{b_o}{a_o} + \frac{a_o^{\,2} + b_o^{\,2}}{a_o^{\,2}}x$$

and

$$\frac{d\theta}{dx} = \frac{\dfrac{a_o^{\,2} + b_o^{\,2}}{a_o^{\,2}}}{1+f(x)^2} \quad \rangle 0$$

This is true always. Therefore

$$\frac{d\theta}{dz} = \frac{d\theta}{dx}\frac{dx}{dz} = 2\beta_2 \frac{d\theta}{dx}$$

$$2\beta_2 \frac{d\theta}{dx} \rangle 0 \qquad if \quad \beta_2 \rangle 0$$

$$2\beta_2 \frac{d\theta}{dx} \langle 0 \qquad if \quad \beta_2 \langle 0$$

and

$\beta_2 \rangle 0 \quad \rightarrow$ counterclockwise direction

$\beta_2 \langle 0 \quad \rightarrow$ clockwise direction

Let us now consider a few examples.

(1) $b_o = 0$ (no chirp)

Transform-limited pulses at $z=0$ $(x = 2\beta_2 z)$

$$\begin{cases} a(z) = \dfrac{a_o}{1 + a_o^2 x^2} \\ b(z) = \dfrac{a_o^2 x}{1 + a_o^2 x^2} \end{cases} \tag{4.41}$$

$a(z)$ can only decrease with increase z, regardless of the sign of β_2: the pulse can only widen. $b(z)$ increases if $\beta_2 \rangle 0$, and becomes increasingly negative if $\beta_2 \langle 0$.

(2) $b_o \rangle 0, \quad \beta_2 \langle 0$ (chirp present at $z=0$)

Propagation in the case of figure will first increase the chirp, up to the highest point in the circle, where $b=a=R$, and then will compress the pulse up to point

$$a_{opt} = \frac{a_o^2 + b_o^2}{a_o}$$

At this point $b(z)=0$ and the pulse has become *transform limited*. The value of x that makes $a = a_{opt}$ can be calculated by setting $b(z)=0$:

$$b(z) = \frac{b_o(1 + b_o x) + a_o^2 x}{(1 + b_o x)^2 + a_o^2 x^2} = 0$$

or

$$b_o + b_o^2 x + a_o^2 x = 0$$

Then we obtain

$$x_{opt} = \frac{-b_o}{a_o^2 + b_o^2} \xrightarrow[a_o \langle\langle b_o]{} -\frac{1}{b_o} \tag{4.42}$$

and

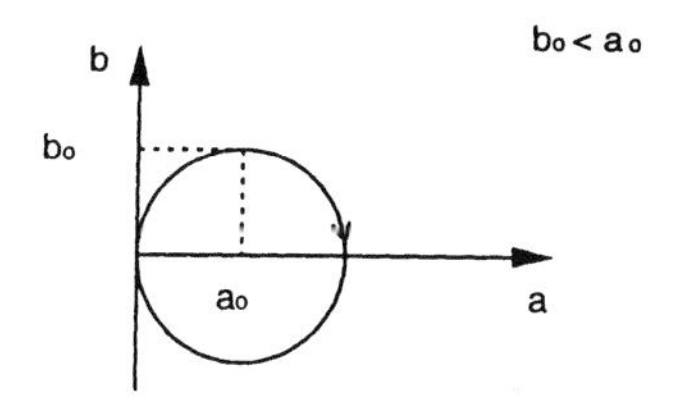

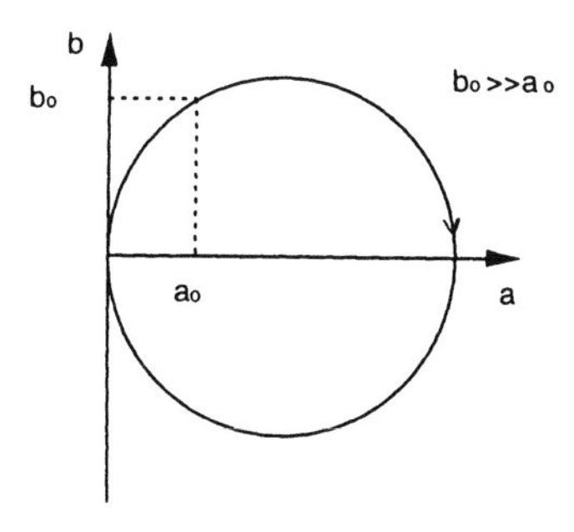

Fig. 4.2 Trajectories of pulse with equal initial spectral width; but different initial chirps

$$a_{opt} = \frac{a_o}{(1+b_o x_{opt})^2 1 + a_o^2 x_{opt}^2}$$

$$= \frac{a_o}{\left[1 + b_o \frac{-b_o}{a_o^2 + b_o^2}\right]^2 + a_o^2 \frac{b_o^2}{\left(a_o^2 + b_o^2\right)^2}} = \frac{a_o^2 + b_o^2}{a_o} = 2R \qquad (4.43)$$

z_{opt} is found from the following relation

$$x_{opt} = (2\beta_2 z)_{opt} = \frac{-b_o}{a_o^2 + b_o^2} \qquad (4.44)$$

The value of z_{opt} is called the *optimum compression length.* Since

$$\tau_{p\min} = \sqrt{\frac{2\ln 2}{a_{opt}}} = \sqrt{\frac{(2\ln 2)a_o}{a_o^2 + b_o^2}} \qquad (4.45)$$

$$\tau_{po} = \sqrt{\frac{2\ln 2}{a_o}} \qquad \text{(in the absense of chirp)} \qquad (4.46)$$

we obtain

$$\frac{\tau_{p\min}}{\tau_{po}} = \sqrt{\frac{a_o}{2\ln 2}}\sqrt{\frac{2\ln 2 a_o}{a_o^2 + b_o^2}} = \frac{a_o}{\sqrt{a_o^2 + b_o^2}}$$

$$= \frac{1}{\sqrt{1 + \frac{b_o^2}{a_o^2}}} \xrightarrow[a_o \ll b_o]{} \left|\frac{a_o}{b_o}\right| \qquad (4.47)$$

A large initial chirp $(b_o \rangle\rangle a_o)$, which implies a large initial $\sigma_t \sigma_\omega = \frac{1}{2}\sqrt{1+\frac{b_o^2}{a_o^2}}$, i.e., time length-bandwidth product, leads to the possibility of a large pulsewidth compression. This is also illustrated by Figure 4.2.

Summary

In this section we study the propagation of Gaussian pulse in media characterized by **phase velocity dispersion (PVD)** and **group velocity dispersion (GVD).**

If the pulse has initially a chirp, its time evolution may bring about a time compression, which is the greater the bigger the chirp. On the other hand, if we start with a transform-limited pulse, its width can only increase.

IV.C. Propagation of Transform-Limited Pulses

For a transform-limited pulse $b_o = 0$, and

$$a(z) = \frac{a_o}{1 + a_o^2 x^2} = \frac{a_o}{1 + a_o^2 4\beta_2^2 z^2} \tag{4.48}$$

The width of the pulse at $z=0$ is given by

$$\tau_p(0) = \sqrt{\frac{2\ln 2}{a_o}} \tag{4.49}$$

We know from (3.36) that

$$\sigma_t(0) = \frac{1}{2\sqrt{2\ln 2}} \qquad \tau_p(0) = \frac{1}{2\sqrt{a_o}} \tag{4.50}$$

Then

$$\sigma_t(z) = \frac{1}{2\sqrt{a(z)}} = \frac{1}{2\sqrt{\frac{a_o}{1 + a_o^2 4\beta_2^2 z^2}}}$$
$$= \frac{1}{2}\sqrt{\frac{1 + a_o^2 4\beta_2^2 z^2}{a_o}} = \sigma(0)\left[1 + a_o^2 4\beta_2^2 z^2\right]^{1/2} \tag{4.51}$$

Summary

In this section we relate the inevitable widening of transform-limited pulses, as they travel through a medium, to the GVD of the medium.

IV.D. Propagation in an Absorbing Medium

A medium which absorbs is often represented by a complex susceptibility

$$\chi = \chi_r + i\chi_i \tag{4.52}$$

The dielectric constant of the medium is also complex

$$K = 1 + 4\pi\chi = (1 + 4\pi\chi_r) + i(4\pi\chi_i) \tag{4.53}$$

The index of refraction is given by

$$n = \sqrt{K} = (1 + 4\pi\chi_r + i4\pi\chi_i)^{1/2}$$
$$\approx (1 + 2\pi\chi_r) + i(2\pi\chi_i) = n_r + in_i \qquad (4.54)$$

where we have assumed $\chi_r \langle\langle 1, \quad \chi_i \langle\langle 1$. Then

$$\begin{cases} n_r = 1 + 2\pi\chi_r \\ n_i = 2\pi\chi_i \end{cases} \qquad (4.55)$$

Consider a plane wave traveling in the $+z$ direction:

$$Ee^{i(\omega t - \vec{k}\cdot\vec{r})} = Ee^{i(\omega t - kz)} \qquad (4.56)$$

where $\vec{k}$ is the wave vector and $k = |\vec{k}|$. We can write

$$k = \frac{\omega}{c}n = \frac{\omega}{c}(n_r + in_i) = \frac{\omega}{c}n_r + \frac{i\omega n_i}{c} = \beta - i\frac{1}{2}\mu \qquad (4.57)$$

where c=velocity of electromagnetic waves in vacuum

$$\mu = -\frac{2\omega n_i}{c}, \qquad \beta = \frac{\omega}{c}n_r \qquad (4.58)$$

We find

$$e^{i\omega t - ikz} = e^{i\omega\left(t - \frac{k}{\omega}z\right)} = e^{i\omega\left[t - \frac{n_r + in_i}{c}z\right]}$$
$$= e^{i\omega\left(t - \frac{n_r z}{c}\right)} e^{\left(\frac{\omega n_i}{c}z\right)} = e^{i\omega(t - \beta z)} e^{\left(-\frac{1}{2}\mu z\right)} \qquad (4.59)$$

β is the propagation constant of the wave and it represents the rate of phase change with z. The parameter μ is called the *absorption coefficient*, because the intensity of the wave decreases as $e^{-\mu z}$.

From (4.58) and (4.55), we obtain

$$\mu = -\frac{2\omega n_i}{c} = -\frac{2\omega}{c}2\pi\chi_i = -\frac{4\pi\omega}{c}\chi_i \qquad (4.60)$$

Summary

An absorbing medium is modeled phenomenologically by a **complex susceptibility**.

When this is done, the dielectric constant, the index of refraction of the medium, and the wave vector are also complex. All these parameters can be expressed in terms of the real part χ_r and of the imaginary part χ_i of the susceptibility.

If both χ_r and $\chi_i \langle\langle 1$, the real parts of the parameters cited above are related to χ_r and their imaginary parts to χ_i.

IV.E. Propagation in a Resonant Medium

A resonant medium under the action of an electric field can be modeled by an ensemble of bound charges undergoing forced harmonic oscillations, a system already considered in Section II.B.

Each oscillator is characterized by a mass m, a charge e and a natural oscillation frequency $\omega_o = \sqrt{\frac{K}{m}}$ about its equilibrium position. A field is applied in the z-direction:

$$E(t) = E_o(\mathbf{x})e^{i\omega t} \tag{4.61}$$

Note that the sign of ω is opposite to the one used in Section II.B.

The equation of motion is

$$m\ddot{z} = -Kz + eE_o(\mathbf{x})e^{i\omega t} - m\gamma\dot{z} \tag{4.62}$$

where we have introduced a damping term $m\gamma\dot{z}$. We make the assumptions:

$$\begin{aligned} z &= \text{displacement} \ll \lambda_{rad} \\ \gamma &\ \text{independent of } \omega. \end{aligned} \tag{4.63}$$

The steady state solution is

$$z(\mathbf{x},t) = z_o(\mathbf{x})e^{i\omega t} \tag{4.64}$$

where

$$z_o(\mathbf{x})\frac{e/m}{\omega_o - \omega^2 + i\gamma\omega}E_o(\mathbf{x}) \tag{4.65}$$

A dipole moment is induced by the field

$$ez(\mathbf{x},t) = \frac{e^2/m}{\omega_o^2 - \omega^2 + i\gamma\omega}E_o(\mathbf{x})e^{i\omega t} = \alpha(\omega)E_o(\mathbf{x})e^{i\omega t} \tag{4.66}$$

where

$$\alpha(\omega) \quad = \quad \text{polarizability} \quad = \quad \frac{e^2/m}{\omega_o^2 - \omega^2 + i\gamma\omega} \tag{4.67}$$

The susceptibility is given by

$$\chi(\omega) = n_o\alpha(\omega) = \frac{n_o e^2/m}{\omega_o^2 - \omega^2 + i\gamma\omega} \tag{4.68}$$

where n_o = density of molecules. But

$$\omega_o^2 - \omega^2 = (\omega_o + \omega)(\omega_o - \omega) \approx 2\omega_o\Delta\omega \tag{4.69}$$

where

$$\Delta\omega = \omega_o - \omega$$

Then

$$\chi(\omega) = \frac{n_o e^2/m}{\omega_o^2 - \omega^2 + i\gamma\omega} \approx \frac{n_o e^2/m}{2\omega_o\Delta\omega + i\gamma\omega_o}$$

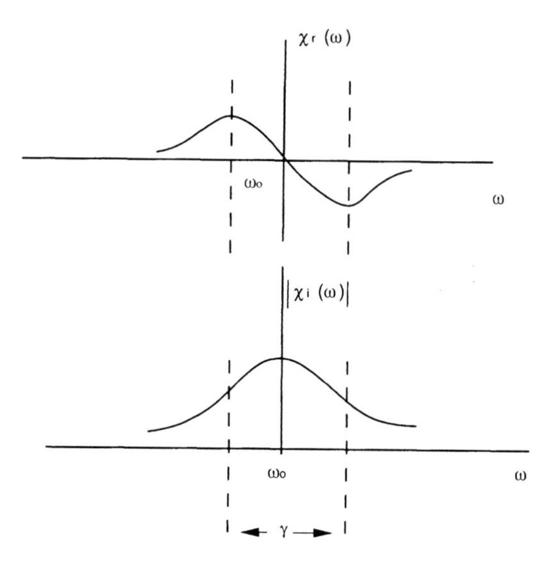

Fig. 4.3 The real and imaginary parts of the susceptibility

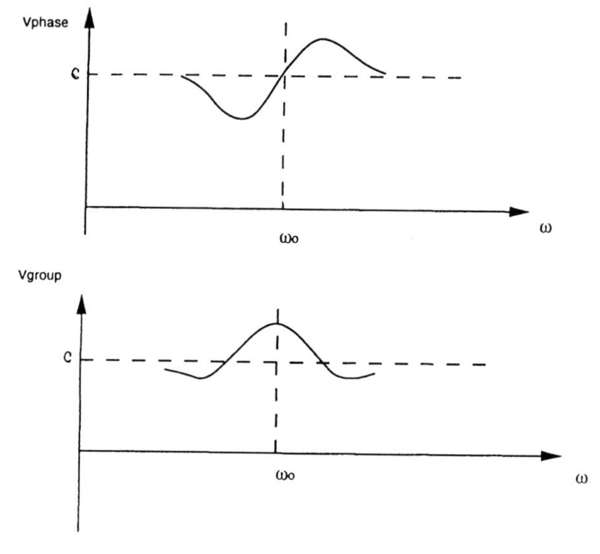

Fig. 4.4. Phase velocity and group velocity at frequencies close to a sharp resonance

$$\approx \frac{n_o e^2 / 2m\omega_o}{\Delta\omega + i\frac{\gamma}{2}} = \frac{n_o e^2}{2m\omega_o}\left[\frac{\Delta\omega}{(\Delta\omega)^2 + \left(\frac{\gamma}{2}\right)^2} - i\frac{\gamma/2}{(\Delta\omega)^2 + \left(\frac{\gamma}{2}\right)^2}\right] \tag{4.70}$$

and

$$\begin{cases} \chi_r = \dfrac{n_o e^2}{2m\omega_o}\dfrac{\Delta\omega}{(\Delta\omega)^2 + \left(\frac{\gamma}{2}\right)^2} = -\dfrac{2\Delta\omega}{\gamma}\chi_i \\ \chi_i = -\dfrac{n_o e^2}{2m\omega_o}\dfrac{\gamma/2}{(\Delta\omega)^2 + \left(\frac{\gamma}{2}\right)^2} \end{cases} \tag{4.71}$$

Plots of these functions are given in Figure 4.3. Note that for $\omega = \omega_o$, $\Delta\omega = 0$, we obtain

$$\chi_r(\omega_o) = 0, \quad \chi_i(\omega_o) = -\frac{n_o e^2}{m\omega_o}\frac{1}{\gamma} \tag{4.72}$$

Summary

The mathematical expression of the complex susceptibility depends on the microscopic model used to represent the absorbing medium. We consider a model which consists of an ensemble of bound charges undergoing forced harmonic oscillations under the action of an oscillatory electric field.

Two parameters define the model: the natural frequency of oscillations ω_o and the damping parameter γ.

The expression for $\chi_r(\omega)$ and $\chi_i(\omega)$ are found in terms of these parameters.

IV.F. Propagation at Frequencies Near Atomic Transitions

Assume that a wave transverses a distance L of a medium whose index of refraction is

$$n = n_r = 1 + 2\pi\chi_r(\omega) \tag{4.73}$$

The total phase shift experienced by the wave is

$$\frac{\omega}{c}nL = \frac{\omega}{c}[1 + 2\pi\chi_r(\omega)]L \tag{4.74}$$

The phase velocity is given by

$$v_\phi(\omega) = \frac{\omega}{\frac{\omega}{c}n} = \frac{c}{n} = \frac{c}{1 + 2\pi\chi_r(\omega)} \tag{4.75}$$

The group velocity is given by (see (4.4))

$$v_g(\omega) = \frac{c}{n + \omega\frac{dn}{d\omega}} \tag{4.76}$$

On the other hand

$$\frac{v_\phi}{1-\dfrac{\omega}{v_\phi}\dfrac{dv_\phi}{d\omega}} = \frac{c/n}{1+\dfrac{\omega}{c/n}\dfrac{c}{n^2}\dfrac{dn}{d\omega}} = \frac{c/n}{1+\dfrac{\omega}{n}\dfrac{dn}{d\omega}} = \frac{c}{n+\omega\dfrac{dn}{d\omega}} \tag{4.77}$$

Then

$$v_g = \frac{v_\phi}{n-\dfrac{\omega}{v_\phi}\dfrac{dv_\phi}{d\omega}} \tag{4.78}$$

Possible plots of v_ϕ and v_g are given in Figure 4.4.

We can make the following observations:

(1) The expressions for the real and imaginary parts of the susceptibility found in the previous section allow us to describe the variations of the phase velocity and of the group velocity at frequencies near an atomic transition. Sharp transitions may enhance considerable these changes and may bring about interesting new phenomena.

(2) For a strongly absorbing atomic transition the group velocity at the center of the transition can become faster than the velocity of light in the medium, or even in vacuum. A signal passing through such a medium could experience a strong attenuation and a considerable distortion.

(3) The results of experiments [7] and calculations [8] indicate that when a smooth pulse travels through a strongly absorbing medium, its envelope appears to travel at velocity close to v_g with little distortion of its shape, even when $v_g \rangle c$, or $v_g \langle 0$.

(4) These experiments and calculations use pulses with long and weak tails.
When these signals pass through the medium, different parts of the pulse spectrum are attenuated and phase-shifted differently, in such a way that the peak of the pulse envelope appears to move at velocity greater than c. What really happens is that the pulse envelope is greatly modified, but no part of the pulse is actually moving at velocity greater than c.
When calculations are done using pulses with a sharp leading edge, no output emerge at the output face of the sample at a time $\langle L/c$.

(5) The dispersion in the wings of a strong and narrow transition can produce a group velocity $\langle\langle$ velocity in the medium, and, of course $\langle\langle$ c in the wings, with a very small absorption [9].
If the entire spectrum of the pulse falls in the wings, far away from the resonance frequency, the pulse may be slowed down considerably, with a small net absorption.

Summary

The expressions for the real and imaginary parts of the susceptibility found in the previous section IV.E. allow us to describe the variations of the phase velocity and group velocity at frequencies near atomic transitions. Sharp transitions make these changes very relevant and may bring about interesting new phenomena, especially those related to the possibility that the group velocity may be greater than the velocity of light in the medium, or even in vacuum, and it may also acquire negative values.

IV.G. The Effect of Gain Dispersion

The primary effect of a passage of a pulse through a laser amplifies is the broadening due to the finite bandwidth of the system. Other effects are due to the gain dispersion, i.e. the variation of gain with frequency.

We shall consider a medium whose gain has a quadratic frequency dependence

$$\alpha(\omega) = \alpha_o - \frac{1}{2}\alpha_2(\omega - \omega_o)^2 \tag{4.79}$$

where

$$\begin{cases} \alpha_o = \alpha(\omega_o) \\ \alpha_2 = \left(\dfrac{d^2\alpha(\omega)}{d\omega^2}\right)_{\omega=\omega_o} \end{cases} \tag{4.80}$$

The first term in the expansion (4.79) amplifies uniformly all the components of a pulse, the second term leads to a change in the Gaussian pulse parameter. [See for analogy the formula (4.32)]:

$$\begin{aligned} \frac{1}{\Gamma(z)} &= \frac{1}{\Gamma_o} + 2\alpha_2 z = \frac{1}{a_o - ib_o} + 2\alpha_2 z = \frac{a_o}{a_o^2 + b_o^2} + i\frac{b_o}{a_o^2 + b_o^2} + 2\alpha_2 z \\ &= \left[\frac{a_o}{a_o^2 + b_o^2} + 2\alpha_2 z\right] + i\left[\frac{b_o}{a_o^2 + b_o^2}\right] = \frac{a(z)}{a(z)^2 + b(z)^2} + i\frac{b(z)}{a(z)^2 + b(z)^2} \end{aligned} \tag{4.81}$$

Then

$$\begin{cases} \dfrac{a(z)}{a(z)^2 + b(z)^2} = \dfrac{a_o}{a_o^2 + b_o^2} + 2\alpha_2 z \\ \dfrac{b(z)}{a(z)^2 + b(z)^2} = \dfrac{b_o}{a_o^2 + b_o^2} \end{cases} \tag{4.82}$$

We note that the imaginary part of $\frac{1}{\Gamma}$ does not change with z, but the real part increases with z. This means that the propagation of the pulse in the z-direction decreases the bandwidth of the pulse. The solutions of equations (4.82) are

$$\begin{cases} a(z) = \dfrac{a_o(1 + 2\alpha_2 a_o z) + 2\alpha_2 b_o^2 z}{(1 + 2\alpha_2 a_o z)^2 + (2\alpha_2 b_o z)^2} \\ b(z) = \dfrac{b_o}{(1 + 2\alpha_2 a_o z)^2 + (2\alpha_2 b_o z)^2} \end{cases} \tag{4.83}$$

1. Pulse-Broadening in Amplifiers. Consider a gain medium with a Lorentzian profile:

$$\alpha(\omega) = \frac{\alpha_o}{1 + \left[\dfrac{2(\omega - \omega_o)}{\gamma}\right]^2} \tag{4.84}$$

At $\omega = \omega_o$, $\alpha = \alpha_o$; at $\omega = \omega_o \pm \frac{\gamma}{2}$, $\alpha = \frac{\alpha_o}{2}$. We can expand $\alpha(\omega)$ as follows:

$$\alpha(\omega) \approx \alpha_o\left[1-\left(\frac{2}{\gamma}\right)^2(\omega-\omega_o)^2\right]$$
$$= \alpha_o - \frac{4\alpha_o}{\gamma^2}(\omega-\omega_o)^2 = \alpha_o - \frac{1}{2}\alpha_2(\omega-\omega_o)^2 \tag{4.85}$$

Then

$$\frac{1}{2}\alpha_2 = \frac{4\alpha_o}{\gamma^2}, \quad 2\alpha_2 = \frac{16\alpha_o}{\gamma^2} \tag{4.86}$$

$$\frac{1}{\Gamma(z)} = \frac{1}{\Gamma_o} + \frac{16\alpha_o}{\gamma^2} \tag{4.87}$$

Example

We have initially:

$$\Gamma_o = a_o, \quad b_o = 0$$
$$\tau_{po}{}^2 = \frac{2\ln 2}{a_o}$$

At distance z

$$a(z) = \frac{a_o}{1+2\alpha_2 a_o z}, \quad 2\alpha_2 = \frac{16\alpha_o}{\gamma^2}$$
$$\frac{1}{a(z)} = \frac{1}{a_o} + 2\alpha_2 z = \frac{1}{a_o} + \frac{16\alpha_o}{\gamma^2}z \tag{4.88}$$

and

$$\tau_p{}^2(z) = \frac{2\ln 2}{a(z)} = \frac{2\ln 2}{a_o} + \frac{(16\ln 2)2\alpha_o z}{\gamma^2} = \tau_{po}{}^2 + \frac{(16\ln 2)\ln G_o}{\gamma^2} \tag{4.89}$$

where

$$G_o = e^{2\alpha_o z} \tag{4.90}$$

Let us assume

$$\frac{\gamma}{2\pi} = 120GHz = 120\times 10^9\,Hz = 1.2\times 10^{11}\,Hz$$
$$G_o = 10^5$$

Then

$$\frac{(16\ln 2)(\ln G_o)}{\gamma^2} = \frac{11\times 11.51}{(2\pi\times 1.2\times 10^{11})^2} \approx (15\times 10^{-12}) = (15ps)^2$$

A delta-function signal pulse would be transformed into a 15 ps pulse. A 50 ps pulse would be transformed into a pulse of length

$$\tau_p = \sqrt{50^2+15^2} = 52.2ps$$

Note

Consider the quantity $y = 2\alpha_2 z$

$$2\alpha_2 z = \frac{16\alpha_o}{\gamma^2} z = 8\frac{2\alpha_o z}{\gamma^2}$$

Then

$$y = \frac{8\times 11.51}{(2\pi\times 1.2\times 10^{11})^2} = 1.6\times 10^{-22} \langle\langle 1$$

2. Possible Pulse-Narrowing in Amplifiers. We rewrite the expression for *a(z)* as follows

$$a(z) = \frac{a_o(1+2\alpha_2 a_o z)+2\alpha_2 b_o^2 z}{(1+2\alpha_2 a_o z)^2+(2\alpha_2 b_o z)^2} = \frac{a_o + a_o^2 y + b_o^2 y}{(1+a_o y)^2 + b_o^2 y^2} \tag{4.91}$$

where

$$y = 2\alpha_2 z \tag{4.92}$$

Assuming y small, we can write

$$\left(\frac{\partial a(z)}{\partial y}\right)_{y=0} = \left.\frac{(a_o^2+b_o^2)\left[(1+a_o y)^2+b_o y\right]-\left[2(1+a_o y)a_o+2b_o^2 y\right]\left[a_o+a_o^2 y+b_o^2 y\right]}{\left[(1+a_o y)^2+b_o^2 y^2\right]}\right|_{y=0}$$

$$= (a_o^2+b_o^2)-2a_o^2 = b_o^2 - a_o^2 \tag{4.93}$$

Then

$$a(z) \approx a_o + (b_o^2 - a_o^2)y = a_o + b_o^2 y - a_o^2 y = a_o\left(1 - a_o y + \frac{b_o^2}{a_o} y\right) \tag{4.94}$$

Neglecting for a moment the third term in the parentheses above,

$$a(z) = a_o(1 - a_o y)$$

we obtain for a Lorentzian profile

$$\begin{aligned}\tau_p^2(z) &= \frac{2\ln 2}{a(z)} = \frac{2\ln 2}{a_o(1-a_o y)} \approx \frac{2\ln 2}{a_o}(1+a_o y)\\ &= \frac{2\ln 2}{a_o} + (2\ln 2)y = \tau_{po}^2 + (2\ln 2)(2\alpha_2 z)\\ &= \tau_{po}^2 + (2\ln 2)\frac{16\alpha_o}{\gamma^2} z\end{aligned} \tag{4.95}$$

[see (4.89) and recall that $2\alpha_2 = \frac{16\alpha_o}{\gamma^2}$]

If the pulse has no chirp ($b_o = 0$), then it will broaden as indicated above. If the pulse has a large chirp at the beginning ($b_o \rangle\rangle a_o$), then the third term in the parentheses in (4.94) cannot be neglected: the pulse may then narrow. This effect can be explained physically as due to the fact that the pulse frequency sweeps across the gain profile: the central frequency harmonics are amplified by the gain at its peak value, whereas the lower or higher frequency are not amplified as much because they are in the wings of the gain profile.

Summary

This section considers the changes experienced by a pulse traveling in a medium with a frequency-dependent gain coefficient. An expected effect is the pulse broadening in time due to the finite bandwidth of the amplifying medium. Assuming a quadratic frequency dependence of the gain we verify that such a broadening takes place in general.

We then consider the case of medium with a Lorentzian gain shape and find formulae for the time dependence of the pulse width in terms of the initial width, the peak gain and the width of the Lorentzian profile. We then return to the more general case of a medium with a quadratic frequency dependence of the gain, in order to investigate the effects of the pulse chirp. We find that a pulse without chirp will broaden, whereas a pulse with a large chirp may narrow. This latter effect is due to the fact that the pulse frequency sweeps across the gain profile: the central frequency harmonics are amplified by the gain at its peak value, whereas the lower and higher frequency harmonics are not amplified as much, being in the wings of the gain profile.

V. PROPAGATION OF PULSES IN NONLINEAR MEDIA

The true worth of an experimenter consists in his pursuing not only what he seeks in his experiment, but also what he did not seek.
Claude Bernard

V.A. Propagation of Pulses in Laser Amplifiers

The basic constituent of a laser amplifier is a medium in which a population inversion has been established. A very narrow pulse that passes through such a medium experiences an exponential growth. A pulse with an appreciable width undergoes a more complex process of amplification since the leading part of the pulse affects the population inversion seen by the tail part of the pulse; in this case the time evolution of the pulse must be described by nonlinear time-dependent equations. The present treatment is based on the theory by Frantz and Nodvik [10].

Let a beam of monochromatic light transverse the amplifier medium along the x-direction as in Figure 5.1. The photon transport equation is given by

$$\frac{\partial n}{\partial t} + c\frac{\partial n}{\partial x} = \sigma c n\left(N_2 - N_1\right) \tag{5.1}$$

where

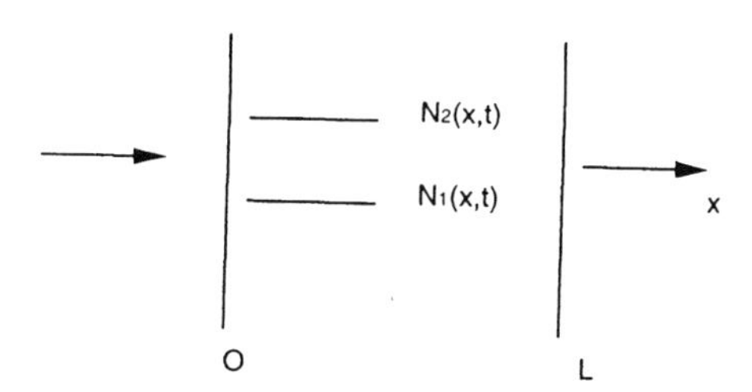

Fig. 5.1. A beam of monochromatic light transversing a laser amplifier

$n(x,t)$ =	photon density
$N_1(x,t)$ =	density of ground state atoms
$N_2(x,t)$=	density of excited state atoms
$N = N_1 + N_2$ =	total density of atoms (fixed)
σ =	resonance absorption cross section
$\sigma cn(N_2 - N_1)$ =	source term = difference between rate of stimulated emission and rate of absorption

In the above equation we have neglected the finite width of the amplifying transition and spontaneous emission. The latter can be neglected because the time of the passage of the pulse through the medium is much shorter than the lifetime of the excited state.

We note that the rate of photon creation is equal to the rate of growth of the ground state atoms:

$$\begin{cases} \dfrac{\partial N_1}{\partial t} = \sigma cn(N_2 - N_1) \\ \dfrac{\partial N_2}{\partial t} = -\sigma cn(N_2 - N_1) = -\dfrac{\partial N_1}{\partial t} \end{cases} \tag{5.2}$$

The equations to solve are then

$$\begin{cases} \dfrac{\partial n}{\partial t} + c\dfrac{\partial n}{\partial x} = \sigma cn(N_2 - N_1) \\ \dfrac{\partial N_1}{\partial t} = \sigma cn(N_2 - N_1) \\ \dfrac{\partial N_2}{\partial t} = -\sigma cn(N_2 - N_1) \end{cases} \tag{5.3}$$

or

$$\begin{cases} \dfrac{\partial n}{\partial t} + c\dfrac{\partial n}{\partial x} = \sigma cn(N_2 - N_1) \\ \dfrac{\partial (N_2 - N_1)}{\partial t} = -2\sigma cn(N_2 - N_1) \end{cases} \tag{5.4}$$

Let

$$\Delta = N_2 - N_1 \tag{5.5}$$

Then

$$\begin{cases} \dfrac{\partial n}{\partial t} + c\dfrac{\partial n}{\partial x} = \sigma cn\Delta \\ \dfrac{\partial \Delta}{\partial t} = -2\sigma cn\Delta \end{cases} \tag{5.6}$$

We shall impose the following boundary conditions

$$\begin{cases} N_2(x,-\infty) - N_1(x,-\infty) = \Delta_o(x) \qquad 0 \le x \le L \\ n(0,t) = n_o(t) \end{cases} \tag{5.7}$$

From the first of the equation (5.6), we obtain

$$\Delta = \frac{1}{\sigma cn}\frac{\partial n}{\partial t} + \frac{1}{\sigma n}\frac{\partial n}{\partial x}$$

Replacing this expression in the second equation, we obtain

$$\frac{\partial}{\partial t}\left[\frac{1}{\sigma cn}\frac{\partial n}{\partial t}+\frac{1}{\sigma n}\frac{\partial n}{\partial x}\right]=-2\sigma cn\left[\frac{1}{\sigma cn}\frac{\partial}{\partial t}+\frac{1}{\sigma n}\frac{\partial n}{\partial x}\right]$$

or

$$\frac{\partial}{\partial t}\left[\frac{1}{n}\frac{\partial n}{\partial t}+\frac{c}{n}\frac{\partial n}{\partial x}\right]=-2\sigma cn\left[\frac{1}{n}\frac{\partial n}{\partial t}+\frac{c}{n}\frac{\partial n}{\partial x}\right] \tag{5.8}$$

The next step is a change of variables:

$$\begin{cases} \xi=\dfrac{x}{c} & x=c\xi \\ \rho=t-\dfrac{x}{c}=t-\xi & t=\rho+\dfrac{x}{c}=\rho+\xi \end{cases} \tag{5.9}$$

Then

$$\begin{cases} \dfrac{\partial}{\partial t}=\dfrac{\partial}{\partial\rho} \\ \dfrac{\partial}{\partial x}=\dfrac{1}{c}\dfrac{\partial}{\partial\xi}-\dfrac{1}{c}\dfrac{\partial}{\partial\rho}=\dfrac{1}{c}\left(\dfrac{\partial}{\partial\xi}-\dfrac{\partial}{\partial\rho}\right) \end{cases} \tag{5.10}$$

On the other hand

$$\frac{\partial n}{\partial t}+c\frac{\partial n}{\partial x}=\frac{\partial n}{\partial\rho}+c\frac{1}{c}\left(\frac{\partial}{\partial\xi}-\frac{\partial}{\partial\rho}\right)=\frac{\partial n}{\partial\xi}$$

and

$$\frac{1}{n}\frac{\partial n}{\partial t}+\frac{c}{n}\frac{\partial n}{\partial x}=\frac{1}{n}\frac{\partial n}{\partial\xi}$$

The equation (5.8) becomes

$$\frac{\partial}{\partial\rho}\left[\frac{1}{n}\frac{\partial n}{\partial\xi}\right]=-2\sigma c\frac{\partial n}{\partial\xi}$$

or

$$\frac{\partial}{\partial\xi}\left[\frac{\partial\ln n}{\partial\rho}\right]=-2\sigma c\frac{\partial n}{\partial\xi}$$

or

$$\frac{\partial}{\partial\xi}\left[\frac{\partial\ln n}{\partial\rho}+2\sigma cn\right]=0$$

Therefore

$$\frac{\partial\ln n}{\partial\rho}+2\sigma cn=\frac{1}{n}\frac{\partial n}{\partial\rho}+2\sigma cn=f(\rho) \tag{5.11}$$

where $f(\rho)$ is independent of ξ. Let

$$a=\frac{1}{n},\quad n=\frac{1}{a} \tag{5.12}$$

Then

$$a\frac{\partial\frac{1}{a}}{\partial\rho}+\frac{2\sigma cn}{a}=f(\rho)$$

$$-\frac{1}{a}\frac{\partial a}{\partial\rho}+\frac{2\sigma c}{a}=f(\rho)$$

and

$$\frac{\partial a}{\partial\rho}+f(\rho)a=2\sigma c \tag{5.13}$$

<u>Claim</u>

The general solution of (5.13) is

$$a=e^{-\int f(\rho)d\rho}\left\{e^{\int f(\rho)d\rho}2\sigma cd\rho+g(\xi)\right\}$$

where $g(\xi)$ is an arbitrary function of ξ.

<u>Proof</u>

$$\frac{da}{d\rho}=-f(\rho)e^{-\int f(\rho)d\rho}\left\{\int e^{\int f(\rho)d\rho}2\sigma cd\rho+g(\xi)\right\}$$

$$+e^{-\int f(\rho)d\rho}2\sigma ce^{\int f(\rho)d\rho}=-f(\rho)a+2\sigma c$$

QED

Let

$$e^{\int f(\rho)d\rho}=\frac{dh}{d\rho}=h'(\rho)$$

Then

$$a=e^{-\int f(\rho)d\rho}\left\{\int e^{\int f(\rho)d\rho}2\sigma cd\rho+g(\xi)\right\}$$

$$=\frac{1}{e^{\int f(\rho)d\rho}}\left[2\sigma c\int e^{\int f(\rho)d\rho}d\rho+g(\xi)\right]$$

$$=\frac{1}{h'(\rho)}[2\sigma ch(\rho)+g(\xi)]$$

and

$$n=\frac{1}{a}=\frac{h'(\rho)}{2\sigma ch(\rho)+g(\xi)}$$

or

$$n(x,t)=\frac{\partial}{\partial t}h\left(t-\frac{x}{c}\right)\Bigg/\left[2\sigma ch\left(t-\frac{x}{c}\right)+g\left(\frac{x}{c}\right)\right] \tag{5.14}$$

One of the two boundary conditions requires

$$n(0,t) = n_o(t)$$

Then, from (5.14), setting $x=0$:

$$n_o(t) = \frac{\frac{\partial}{\partial t}h(t)}{2\sigma c h(t) + g(0)} = \frac{1}{2\sigma c}\frac{\partial}{\partial t}\ln[2\sigma c h(t) + g(0)]$$

or

$$\frac{\partial}{\partial t}\ln[2\sigma c h(t) + g(0)] = 2\sigma c n_o(t)$$

<u>Claim</u>

$$h(t) = -\frac{g(0)}{2\sigma c} + \lambda e^{2\sigma c \int_{-\infty}^{t} n_o(t')dt'} \tag{5.15}$$

where λ = arbitrary constant of integration.

<u>Proof</u>

$$2\sigma c h(t) + g(0) = 2\sigma c \lambda e^{2\sigma c \int_{-\infty}^{t} n_o(t')dt'}$$

$$\ln(2\sigma c h(t) + g(0)) = \ln(2\sigma c \lambda) - 2\sigma c \int_{-\infty}^{t} n_o(t')dt'$$

$$\frac{\partial}{\partial t}\ln[2\sigma c h(t) + g(0)] = 2\sigma c n_o(t)$$

QED

We have assumed that, as $t \to \infty$, $n_o(t) \to 0$ rapidly enough to allow the integral $\int_{-\infty}^{t} n_o(t')dt'$ to converge. This means that the total number of photons per unit area $c\int_{-\infty}^{t} n_o(t')dt'$ that have entered the medium was zero at some time. We have found

$$n(x,t) = \frac{\partial}{\partial t}h\left(t - \frac{x}{c}\right) \Bigg/ \left[2\sigma c h\left(t - \frac{x}{c}\right) + g\left(\frac{x}{c}\right)\right] \tag{5.16}$$

$$h(t) = -\frac{g(0)}{2\sigma c} + \lambda \exp\left(2\sigma c \int_{-\infty}^{t} n_o(t')dt'\right) \tag{5.17}$$

Then

$$2\sigma c h\left(t - \frac{x}{c}\right) + g\left(\frac{x}{c}\right) = g\left(\frac{x}{c}\right) - g(0) + 2\sigma c \lambda \exp\left(2\sigma c \int_{-\infty}^{t-\frac{x}{c}} n_o(t')dt'\right)$$

$$\frac{\partial}{\partial t}h\left(t-\frac{x}{c}\right)=\lambda 2\sigma c\lambda n_o\left(t-\frac{x}{c}\right)\exp\left(2\sigma c\int_{-\infty}^{t-\frac{x}{c}}n_o(t')dt'\right)$$

and

$$n(x,t)=\frac{\lambda 2\sigma c\lambda n_o\left(t-\frac{x}{c}\right)\exp\left(2\sigma c\int_{-\infty}^{t-\frac{x}{c}}n_o(t')dt'\right)}{g\left(\frac{x}{c}\right)-g(0)+2\sigma c\lambda\exp\left(2\sigma c\int_{-\infty}^{t-\frac{x}{c}}n_o(t')dt'\right)} \tag{5.18}$$

$$=\frac{n_o\left(t-\frac{x}{c}\right)}{\frac{g\left(\frac{x}{c}\right)-g(0)}{2\sigma c\lambda}\exp\left(-2\sigma c\int_{-\infty}^{t-\frac{x}{c}}n_o(t')dt'\right)+1}$$

Let

$$b(x)=\frac{g\left(\frac{x}{c}\right)-g(0)}{2\sigma c\lambda} \tag{5.19}$$

where λ is an arbitrary constant to be determined. Then

$$n(x,t)=\frac{n_o\left(t-\frac{x}{c}\right)}{1+b(x)\exp\left(-2\sigma c\int_{-\infty}^{t-\frac{x}{c}}n_o(t')dt'\right)} \tag{5.20}$$

From the second of (5.6)

$$\frac{\partial\Delta}{\partial t}=-2\sigma cn\Delta$$

<u>Claim</u>

$$\Delta=-\frac{\frac{1}{\sigma}\frac{\partial b}{\partial x}}{b+e^{2\sigma c\int\ldots}}$$

<u>Proof</u>

$$n\Delta=-\frac{\frac{1}{\sigma}\frac{\partial b}{\partial x}n_o\left(t-\frac{x}{c}\right)}{\left[1+be^{-2\sigma c\int\ldots}\right]\left[b+e^{2\sigma c\int\ldots}\right]}$$

$$-2\sigma c(n\Delta) = \frac{2c\dfrac{\partial b}{\partial x}n_o\left(t-\dfrac{x}{c}\right)}{b+e^{2\sigma c\int\dots}+b^2 e^{-2\sigma c\int\dots}+b}$$

On the other hand

$$\frac{\partial\Delta}{\partial t} = \frac{\dfrac{1}{\sigma}\dfrac{\partial b}{\partial x}2\sigma c n_o\left(t-\dfrac{x}{c}\right)e^{2\sigma c\int\dots}}{\left[b+e^{2\sigma c\int\dots}\right]^2}$$

$$= \frac{2c\dfrac{\partial b}{\partial x}n_o\left(t-\dfrac{x}{c}\right)e^{2\sigma c\int\dots}}{b^2+2be^{2\sigma c\int\dots}+e^{4\sigma c\int\dots}}$$

$$= \frac{2c\dfrac{\partial b}{\partial x}n_o\left(t-\dfrac{x}{c}\right)}{b^2 e^{2\sigma c\int\dots}+2b+e^{2\sigma c\int\dots}}$$

QED

At this point

$$\begin{cases} n(x,t) = \dfrac{n_o\left(t-\dfrac{x}{c}\right)}{1+b(x)e^{-2\sigma c\int\dots}} \\ \Delta(x,t) = \dfrac{-\dfrac{1}{\sigma}\dfrac{\partial b(x)}{\partial x}}{b(x)+e^{2\sigma c\int\dots}} \end{cases} \tag{5.21}$$

where

$$\int\dots = \int_{-\infty}^{t-\frac{x}{c}} n_o(t')dt'$$

Recall the boundary conditions

$$\begin{cases} n(0,t) = n_o(t) \\ \Delta(x,-\infty) = \Delta_o(x) \qquad 0 \le x \le L \end{cases} \tag{5.7'}$$

We have used the first one; now we shall use the second one:

$$\Delta_o(x) = \frac{-\dfrac{1}{\sigma}\dfrac{\partial b(x)}{\partial x}}{b(x)+1} = -\frac{1}{\sigma}\frac{d}{dx}\ln[b(x)+1]$$

$$\frac{d}{dx}\ln[b(x)+1] = -\sigma\Delta_o(x)$$

Integrating

$$\ln[b(x)+1] = -\sigma\int_0^x \Delta_o(x')dx' + c$$

$$b(x)+1 = \mu e^{-\sigma\int_0^x \Delta_o(x')dx'}$$

We recall that

$$b(x) = \frac{g\left(\frac{x}{c}\right) - g(0)}{2\sigma c\lambda}$$

Then

$$b(0) = 0 \qquad \mu = 1$$

and

$$b(x) = \exp\left[-\sigma\int_0^x \Delta_o(x')dx'\right] - 1 \tag{5.22}$$

We can write:

$$\begin{cases} n(x,t) = \dfrac{n_o\left(t-\dfrac{x}{c}\right)}{1-\left\{1-\exp\left[-\sigma\int_0^x \Delta_o(x')dx'\right]\right\}\exp\left[-2\sigma c \int_{-\infty}^{t-\frac{x}{c}} n_o(t')dt'\right]} \\ \\ \Delta(x,t) = \dfrac{\Delta_o(x)\left[-\sigma\int_0^x \Delta_o(x')dx'\right]}{\exp\left[-\sigma\int_0^x \Delta_o(x')dx'\right] - 1 + \exp\left[2\sigma c \int_{-\infty}^{t-\frac{x}{c}} n_o(t')dt'\right]} \end{cases} \tag{5.23}$$

Light which enters the amplifier medium at time t leaves it at time $t+\frac{L}{c}$. For an input impulse $n_o(t) = n(0,t)$, the output impulse is $n_L(t) = n\left(L, t+\frac{L}{c}\right)$.

We shall consider now a few examples.

1. Rectangular Pulse. For a rectangular pulse going through an amplifying medium with a uniform population inversion:

$$n_o(t) = n_o \qquad 0 \le t \le \tau$$
$$\Delta_o(x) = \Delta_o = const \qquad 0 \le x \le L$$

The first formula (5.23) now becomes

$$
\begin{cases}
n(x,t) = \dfrac{n_o}{1-\{1-\exp[-\sigma\Delta_o x]\}\exp\left[-2\sigma c n_o\left(t-\frac{x}{c}\right)\right]} \\
\quad = \dfrac{n_o}{1-\{1-\exp[-\sigma\Delta_o x]\}\exp\left[-2\sigma c\eta\left(t-\frac{x}{c}\right)\right]} \\
\quad\quad for \quad 0 \le t-\frac{x}{c} \le \tau, \quad \frac{x}{c} \le t \le \frac{x}{c}+\tau \\
=0 \quad\quad otherwise
\end{cases}
\tag{5.24}
$$

where $\eta = n_o c\tau$ = number of photons per unit area in the pulse. Note that

$$
n(x,t) \xrightarrow[x\to 0]{} n_o \qquad 0 \le t \le \tau
$$

Also for $x=ct$:

$$
n(x=ct,t) = n_o e^{\sigma\Delta_o x} \qquad t=\frac{x}{c} \tag{5.25}
$$

and for $x=L$:

$$
n\left(L,\frac{L}{c}\right) = n_o e^{\sigma\Delta_o L} \qquad t=\frac{L}{c} \tag{5.26}
$$

The leading edge of the pulse will be amplified as indicated above. The tail of the pulse will be amplified differently

$$
n\left(L,\tau+\frac{L}{c}\right) = \frac{n_o}{1-\left[1-e^{-\sigma\Delta_o L}\right]e^{-2\sigma\eta}} \tag{5.27}
$$

<u>Example</u>

$\sigma = 2.5\times 10^{-20} cm^2$

$\Delta_o = 8\times 10^{18} cm^{-3}$

$L = 10cm$

$\eta = 4\times 10^{18} cm^{-2}$

$\tau = 6\times 10^{-9}$ sec

$\sigma\Delta_o L = 2.5\times 10^{-20}\times 8\times 10^{18}\times 10 = 2$

$2\sigma\eta = 2\times 2.5\times 10^{-20}\times 4\times 10^{18} = 0.2$

The gain at the front of the pulse is

$$
e^{\sigma\Delta_o L} = e^2 = 7.389
$$

The gain at the end of the pulse is

$$
\frac{1}{1-\left[1-e^{-\sigma\Delta_o L}\right]e^{-2\sigma\eta}} = \frac{1}{1-\left[1-e^{-2}\right]e^{-0.2}}
$$

$$
= \frac{1}{1-[0.8647]0.8187} = \frac{1}{1-0.7079} = 3.42
$$

Note that

$$\frac{2\sigma\eta L}{c\tau} = 2\sigma\eta\frac{L}{c\tau} = \frac{0.2\times 10}{3\times 10^{10}\times 6\times 10^{-9}} = 0.011$$

so that

$$e^{\frac{2\sigma\eta L}{c\tau}} \approx 1$$

We can make the following observations:

(1) For any value of η, the photon density at $x=ct$, is amplified according to the exponential law

$$n\left(x,\frac{x}{c}\right) = n_o e^{\sigma\Delta_o x}$$

The gain at the front of the pulse is independent of intensity.

(2) At all later points it decreases, as shown in Figure 5.2 which pertains to the present example.

(3) We can find the half-width of the output pulse, i.e. the time t_o at which

$$\frac{n_L(t_o)}{n_L(0)} = \frac{1}{2}$$

Under the condition $e^{\frac{2\sigma\eta L}{c\tau}} \approx 1$, we find for $\eta = 4\times 10^{18}$

$$\frac{t_o}{\tau} = \frac{1}{2\sigma\eta}\ln\left[\frac{1-e^{-\sigma\Delta_o L}}{1-2e^{-\sigma\Delta_o L}}\right] = \frac{1}{2\sigma\eta}\ln\left[\frac{e^{\sigma\Delta_o L}-1}{e^{\sigma\Delta_o L}-2}\right]$$

$$= \frac{1}{0.2}\ln\left[\frac{7.389-1}{7.389-2}\right] = \frac{0.1702}{0.2} = 0.85$$

If η is increased to 4×10^{19}, the pulse-width is reduced to 0.085.

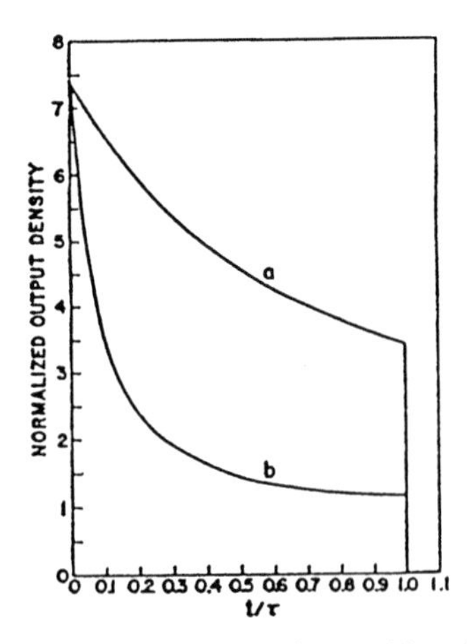

Fig. 5.2 Ratio of output photon density $n_L(t)$ to input photon density $n_o(t)$ for a rectangular pulse. (a) $\eta = 4.8\times 10^{10} cm^{-2}$, (b) $\eta = 4\times 10^{19} cm^{-2}$

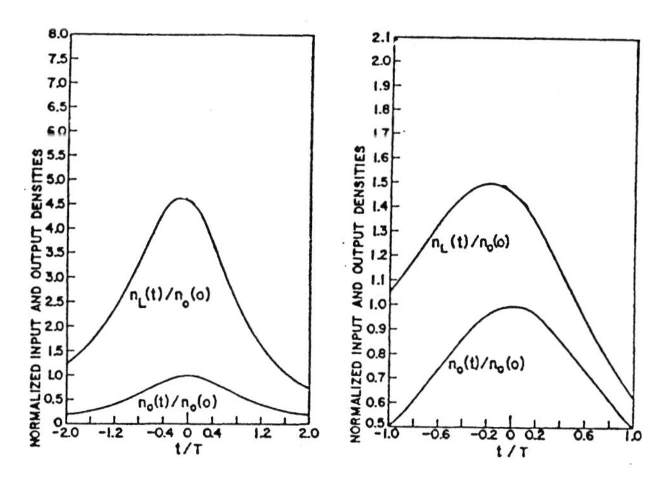

Fig. 5.3. Ratio of output photon density $n_L(t)$ to input photon density $n_o(t)$ for a Lorentzian pulse. The data for σ, Δ_o and L are the same as for the previous example.

2. Lorentzian Pulse. For a Lorentzian pulse

$$n_o(t) = \frac{\eta T/\pi c}{t^2 + T^2} \tag{5.28}$$

where T = half-width of pulse.

For a uniform population inversion Δ_o, we obtain

$$\begin{aligned} n(x,t) &= \frac{\eta T}{\pi c} \frac{1}{\left(t - \frac{x}{c}\right)^2 + T^2} \\ &\times \left\{ 1 - \left[1 - \exp(-\sigma \Delta_o x)\right] \left[\exp\left(-2\sigma c \int_{-\infty}^{t - \frac{x}{c}} \frac{\eta T}{\pi c} \frac{1}{t^2 + T^2} dt \right) \right] \right\}^{-1} \\ &= \frac{\eta T}{\pi c} \frac{1}{\left(t - \frac{x}{c}\right)^2 + T^2} \\ &\times \left\{ 1 - \left[1 - \exp(-\sigma \Delta_o x)\right] \left[\exp(-\sigma\eta) \left[1 + \frac{2}{\pi} \tan^{-1}\left(t - \frac{x}{c}\right)\Big/ T \right] \right] \right\}^{-1} \end{aligned} \tag{5.29}$$

This example is illustrated in Figure 5.3

We note that the width of the pulse does not change much, but the pulse appears to move forward in time, the more so, the higher is the pulse intensity. The same effect is found for Gaussian pulses.

Summary

The laser amplifier is a medium with inverted population. If a pulse of infinitesimal width transverses this medium its amplitude will grow exponentially. If the width of the pulse is appreciable, the tail of the pulses sees a population inversion that has been altered by the leading part of the pulse: the amplification process must be described by a nonlinear time-dependent equation. The treatment consists of the following:

(i) Initial conditions are established

The medium of thickness L, is ready to receive the incoming pulse. It has been prepared with a population inversion

$$N_2(x,-\infty) - N_1(x,-\infty) = \Delta_o(x) \qquad (0 \leq x \leq L)$$

The incoming signal is a pulse that, at the entrance *(x=0)*, is described by the photon density

$$n(0,t) = n_o(t)$$

(ii) Equations are set up for $n(x,t)$ and $\Delta(x,t)$

A **photon transport equation** in which a **source term** enters. This term represents the difference between the rate of stimulated emission and the rate of absorption; it is proportional to the resonance absorption cross section, the photon density, and the population inversion.
An equation that describes the time evolution of the **population inversion** in the medium: it contains a term that, similarly to the source term, is proportional to the resonance absorption cross section, the photon density, and the population inversion.

(iii) Calculations

Detailed calculations lead to general expressions for the photon density $n(x,t)$ and the population inversion $\Delta(x,t)$ as a functions of position x and time t.

(iv) Examples

Finally, different types of pulses (rectangular, Lorentzian, Gaussian) are considered, traveling through an amplifier with an initial population inversion uniform throughout the medium:
For the rectangular pulse the gain at the front of the pulse is independent of the intensity of the pulse, but it becomes smaller and smaller for the parts of the pulse farther and farther away in time from the front, giving the outgoing pulse a spike-like shape; the more intense is the pulse, the sharper is the spike.

For both the Lorentzian and the Gaussian pulses, the width of the output signal does not change much, but the pulse appears to move forward in time, an effect that gets enhanced at high pulse intensities.

V.B. Linearity and Nonlinearity of Media

Historically, until recently, all optical media were thought to be linear. The following was assumed:

(I) Optical properties such as refractive index and absorption coefficient are *independent of light intensity.*

(II) The *principle of superposition* holds:

An optical wave is described by a real function of $\mathbf{r}$ and t, a *wavefunction* $u(\mathbf{r},t)$, solution of the *wave equation*

$$\nabla^2 u - \frac{1}{c^2}\frac{\partial^2 u}{\partial t^2} = 0 \qquad (5.30)$$

where c = velocity of the waves. If $u_1(\mathbf{r},t)$ and $u_2(\mathbf{r},t)$ represent optical waves, then

$$u(\mathbf{r},t) = u_1(\mathbf{r},t) + u_2(\mathbf{r},t)$$

also represents a possible optical wave.

(III) The frequency of light cannot be changed by its passage through a medium.

(IV) Light cannot interact with light. Of course, light cannot control light.

The situation changed with the advent of the laser, a device that provided higher intensity of light in optical materials. The consequences are the following:

(I) The refractive index, and the speed of light in an optical medium, change with the light intensity.

(II) The principle of superposition is violated.

(III) The frequency of light is altered as the light passes through a nonlinear medium.

(IV) Light can interact with light: light can control light. (Light interacts with light via the medium.)

Nonlinear effects become appreciable when the electric field approaches the value of the interatomic fields. In order to get an order of magnitude for such fields, we calculate the electric field that acts on the electron in the atom of hydrogen:

$$r = 0.5 \times 10^{-8} cm = 0.5A$$

$$E = \frac{q}{r^2} = \frac{4.8 \times 10^{-10}}{(.5 \times 10^{-8})^2} = 19.2 \times 10^6 \frac{statvolt}{cm}$$

$$= 19.26 \times 3 \times 10^4 = 5.8 \times 10^{11} \frac{V}{m}$$

Summary

This section restates some of the basic concepts presented in the beginning of these lectures.

The optical properties of linear media such as refractive index and absorption coefficient are independent of the intensity of light. The principle of superposition holds. The frequency of light cannot be changed by its passage through a medium. Light cannot interact with light and light cannot control light.

The optical properties of a nonlinear medium change with light intensity. The principle of superposition is violated. The frequency of light is changed by its passage through a medium. Light can interact with light and light can control light.

How intense must light be in order to produce a nonlinear effect in a medium? In order to produce such effects the electric field has to come close to the value of the interatomic fields $\left(\approx 10^{11} \frac{V}{m}\right)$.

V.C. Propagation of Pulses in Nonlinear Dispersive Media

1. PVD and GVD. Consider a medium with a given relation between the propagation constant β and the frequency ω, as in Figure 5.4. The electric field of a plane polarized and monochromatic wave will propagate along the z-direction as follows

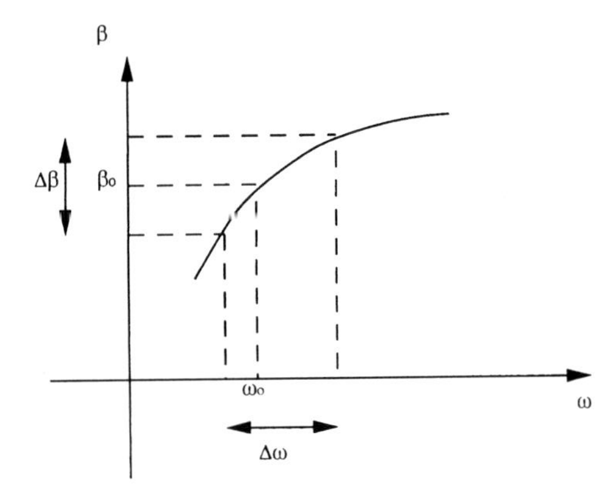

Fig. 5.4. Propagation constant versus ω

$$E \propto e^{i(\omega t-\beta z)} \tag{5.31}$$

where $\beta = \beta(\omega)$. The phase of the wave is

$$\phi = \omega t - \beta z \tag{5.32}$$

The velocity of the given phase front will be such that changes dt and dz must satisfy the condition

$$d\phi = \omega dt - \beta dz = 0 \tag{5.33}$$

This means that the phase front moves with a velocity

$$v_\phi = \frac{dz}{dt} = \frac{\omega}{\beta} = \text{phase velocity of the wave} \tag{5.34}$$

Consider now a pulse with

ω_o = central frequency
$\Delta\omega_o$ = bandwidth of the pulse

Assume

$$\beta = \beta_o + \left(\frac{d\beta}{d\omega}\right)_{\omega=\omega_o} (\omega - \omega_o) \tag{5.35}$$

The electric field can be expressed as follows

$$\begin{aligned} E(t,z) &= \int_{\omega_o-\frac{\Delta\omega_o}{2}}^{\omega_o+\frac{\Delta\omega_o}{2}} A_\omega e^{i(\omega t-\beta z)} d\omega \\ &= \int A_\omega e^{i\left\{(\omega_o-\Delta\omega_o)t-\left[\beta_o+\left(\frac{d\beta}{d\omega}\right)_{\omega_o}\Delta\omega\right]z\right\}} d\omega \\ &= e^{i(\omega_o t-\beta_o z)} \int_{\omega_o-\frac{\Delta\omega_o}{2}}^{\omega_o+\frac{\Delta\omega_o}{2}} A_\omega e^{i\left[\Delta\omega t-\left(\frac{d\beta}{d\omega}\right)_{\omega_o}\Delta\omega z\right]} d\omega \end{aligned} \tag{5.36}$$

Since

$$\omega = \omega_o + \Delta\omega$$
$$d\omega = d(\Delta\omega)$$

we can write

$$E(t,z) = e^{i(\omega_o t - \beta_o z)} \int_{-\frac{\Delta\omega_o}{2}}^{\frac{\Delta\omega_o}{2}} A_\omega \exp\left\{ i\Delta\omega t \left[t - \left(\frac{d\beta}{d\omega}\right)_{\omega_o} z \right] \right\} d(\Delta\omega)$$
$$= e^{i(\omega_o t - \beta_o z)} A\left[t - \frac{z}{v_g} \right] \tag{5.37}$$

In this relation $e^{i(\omega_o t - \beta_o z)}$ represents the carrier wave and $A\left(t - \frac{z}{v_g}\right)$ a function of $\left(t - \frac{z}{v_g}\right)$ the amplitude of the envelope. v_g is the velocity with which the envelope moves without changing its shape:

$$v_g = \left(\frac{d\omega}{d\beta}\right)_{\omega=\omega_o} \tag{5.38}$$

Consider now two pulses with central frequencies ω_1 and ω_2 and bandwidths $\Delta\omega_1$ and $\Delta\omega_2$, respectively, as shown in Figure 5.5, and of course, group velocities given by

$$\frac{1}{v_{g1}} = \left(\frac{d\beta}{d\omega}\right)_{\omega_1} \quad and \quad \frac{1}{v_{g2}} = \left(\frac{d\beta}{d\omega}\right)_{\omega_2}$$

If the peaks of the two pulses enter the medium at the same time, then, after traveling a distance L in the medium, they will be separated by a time interval

$$\Delta\tau = \frac{L}{v_{g2}} - \frac{L}{v_{g1}} = L\left[\left(\frac{d\beta}{d\omega}\right)_2 - \left(\frac{d\beta}{d\omega}\right)_1\right] \tag{5.39}$$

We shall assume that the dispersion relation $\beta = \beta(\omega)$ in the range between ω_1 and ω_2 can be approximated by a parabolic ω-dependence:

$$\left(\frac{d\beta}{d\omega}\right)_2 = \left(\frac{d\beta}{d\omega}\right)_1 + \left(\frac{d^2\beta}{d\omega^2}\right)_1 (\omega_2 - \omega_1) \tag{5.40}$$

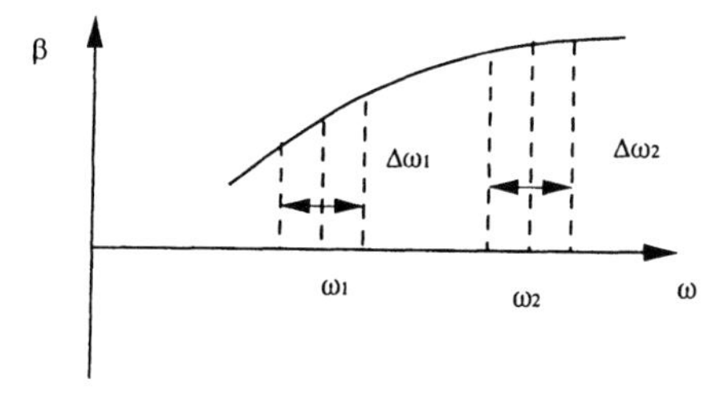

Fig. 5.5. Central frequencies and bandwidths of two traveling pulses

Consider now a pulse with central frequency ω_o and bandwidth $\Delta\omega_o$ so large that the dispersion within the pulse bandwidth is not linear. In this case, different spectral sections of the pulse will travel with different group velocities and the pulse-shape will change during the propagation. Two adjoining infinitesimal sections of the pulse spectrum, both $d\omega$ wide, and around the frequency ω will be delayed in time differently by the amount

$$d\tau = L\left(\frac{d^2\beta}{d\omega^2}\right)d\omega$$

We define *GVD* = *group velocity dispersion* as follows,

$$GVD = \frac{d^2\beta}{d\omega^2} \tag{5.41}$$

This quantity was called β_2 in section IV.A. Note that

$$GVD = \frac{d\left(\frac{1}{v_g}\right)}{d\omega} = -\frac{1}{v_g{}^2}\frac{dv_g}{d\omega} \tag{5.42}$$

2. <u>Passage of a Pulse through a Medium with Positive GVD</u> [11-14]. Let us consider a sharp pulse $(\tau_p \approx 6ps)$ traveling in a single-mode optical fiber of appropriate length and let the wavelength of the pulse fall in the region of positive GVD of the fiber (see Figure 5.6). A positive GVD means that the group velocity decreases with increasing ω.

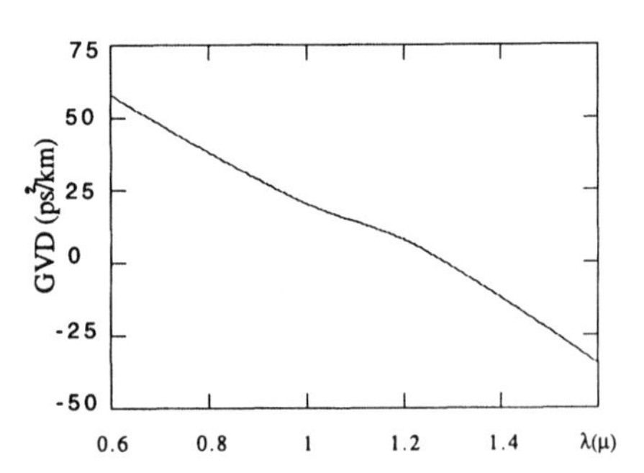

Fig. 5.6. *GVD* for fused silica =0 near $\lambda = 1.27\mu$ $GVD\rangle 0$ usually in the optical region *GVD* becomes $\langle 0$ somewhere in the near infrared region.

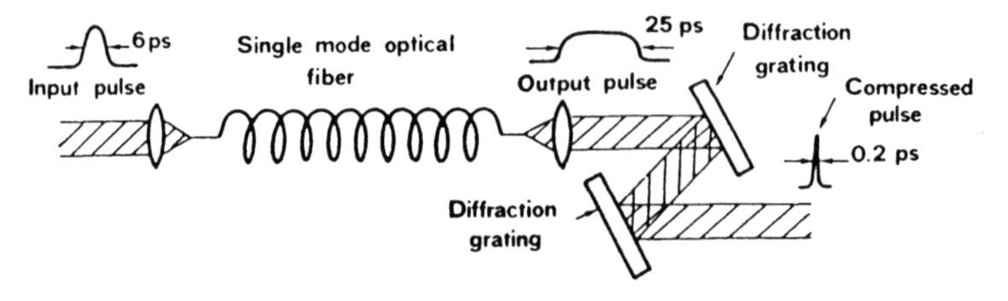

Fig. 5.7. Apparatus for pulse compression

We shall follow the pulse as it travels through the apparatus in Figure 5.7.

(I) The pulse propagates in the optical fiber. Given the small core diameter of the single-mode fiber ($\approx 4\mu$), the pulse produces a very high light intensity in the fiber core.

(II) The light field produces a change δn of the refractive index of refraction of the fiber, given by

$$\delta n = n_{2I} I(t) \tag{5.43}$$

where $I(t)$ = intensity of light, $n_{2I} \approx 10^{-16} cm^2 / W$ for fused quartz. This effect is called *optical Kerr effect.*

(III) The index of refraction of the material

$$n = n_o + n_{2I} I(t) \tag{5.44}$$

will be a function of time. Since the material has small losses, the intensity of the pulse field will remain high over the entire fiber length.

(IV) The pulse, *6ps* wide, propagates along the fiber a distance z. We shall assume for the moment that this material is without dispersion in order to single out the effects of nonlinearity. The pulse undergoes a phase change given by

$$\begin{aligned} \phi &= \omega_o t - \beta z = \omega_o t - \frac{\omega_o}{c_o}(n + \delta n)z \\ &= \omega_o t - \frac{\omega_o}{c_o} n_o z - \frac{\omega_o(\delta n)}{c_o} z \end{aligned} \tag{5.45}$$

where ω_o = carrier frequency of input pulse and c_o = velocity of light in vacuum.

This phase change induced by the variation of the light intensity at one particular place is called *self-phase modulation* or *SPM*. The instantaneous frequency of the light pulse at distance z is given by

$$\omega' = \frac{\partial \phi}{\partial t} = \frac{\partial}{\partial t}(\omega_o t - \beta z) = \omega_o - \frac{\omega_o z n_{2I}}{c_o} \frac{\partial I}{\partial t} \tag{5.46}$$

(V) The instantaneous carrier frequency $\omega'(t)$ is linearly dependent on the *negative time-derivative of the corresponding light intensity.* The effect of self-phase modulation is illustrated in Figure 5.8. Near the peak of the pulse, where the time-dependence of $I(t)$ is approximately parabolic, the instantaneous carrier frequency increases linearly with time: the pulse shows a *positive* chirp. In the outer wings ($t \langle t_A$ and $t \rangle t_B$) the pulse shows a *negative* chirp.

(VI) So far we have neglected the effect of *GVD*. This parameter is positive, a condition that corresponds to lower v_g's for higher frequency components. Different parts of the pulse will be affected differently.

(VII) Because of the broadening produced by *GVD* the peak intensity of the pulse is lower. The parabolic part of $I(T)$ is now extended to a larger region and the positive frequency chirp extends to a larger part of the pulse. These effects can be seen in Figure 5.9.

(VIII) Figure 5.9 shows that the pulse has broadened from an initial shape with width *6ps* to a rectangular pulse of $\approx 24ps$. The pulse has also acquired a nearly linear frequency chirp over the full pulse duration.

(IX) The interplay of *SPM* and *GVD* has introduced modifications into the pulse. As shown in Figure 5.10 the pulse spectrum has broadened from the initial transform-limited band of $0.5/6ps \approx 2.5cm^{-1}$ to a characteristic spectrum $\approx 50cm^{-1}$wide. The self-chirped pulse may be compressed externally by using a *linear* dispersive element with $GVD\langle 0$. Such elements may be provided by a pair of diffraction gratings [15,16] or by a sequence of prisms [17,18].

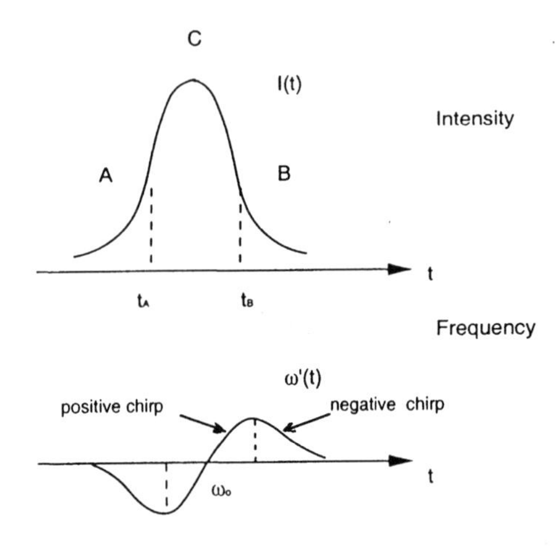

Fig. 5.8. The effect of self-phase modulation

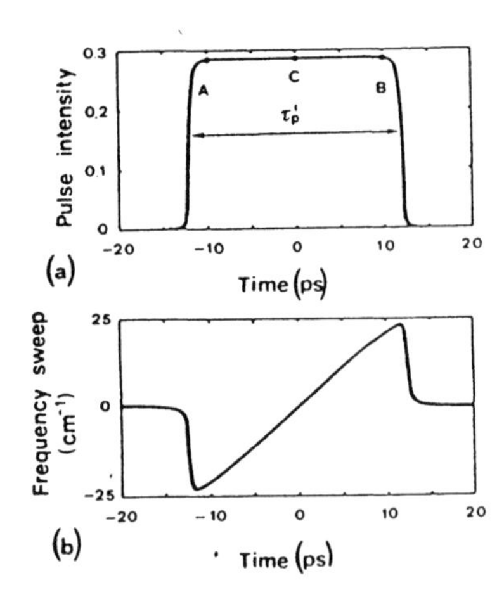

Fig. 5.9. Self-broadening of an initial *6ps* pulse after propagating through a single-mode fiber. (a) output pulse intensity, (b) output frequency chirp [12].

3. <u>Passage of a Pulse Through a Medium with negative *GVD*</u>. The interplay of *SPM* and *GVD* produces changes in the shape of a pulse, by an increasing amount with increasing distance.

Three different types of behavior may arise, depending on circumstances:

(I) Pulse broadening and enhanced frequency chirping (*GVD*⟩0),

(II) severe pulse distortion and break-up (*GVD*⟨0), and

(III) soliton formation and propagation (*GVD*⟨0).

We have examined case (I); we shall now concentrate on the latter two cases which are related to the *GVD*⟨0 condition.

Let us assume that the chirped pulse in Figure 5.11 travels through a nonlinear Kerr medium with a negative *GVD*. The follow scenario enfolds:

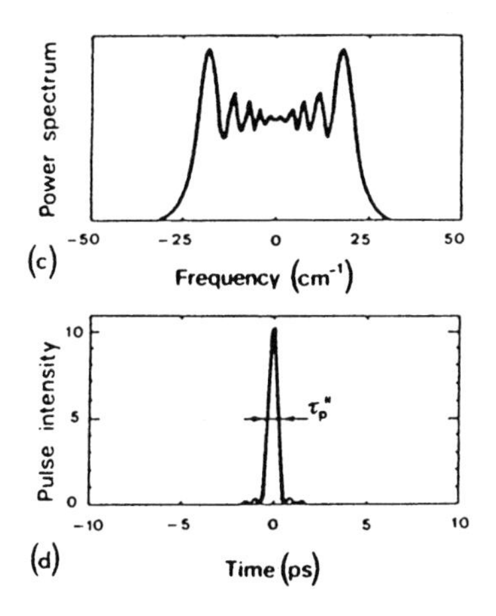

Fig. 5.10. (a) Output pulse spectrum,
(b) compressed pulse produced by a device with *GVD*⟨0 [12].

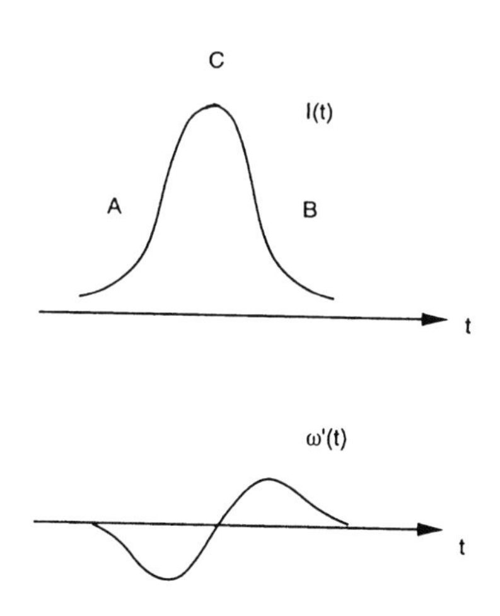

Fig. 5.11. Chirped pulse

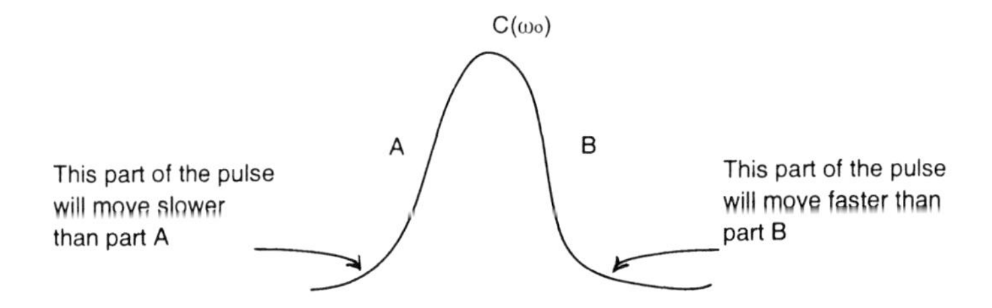

The central part of the pulse, while traveling in the fiber will be stretched.

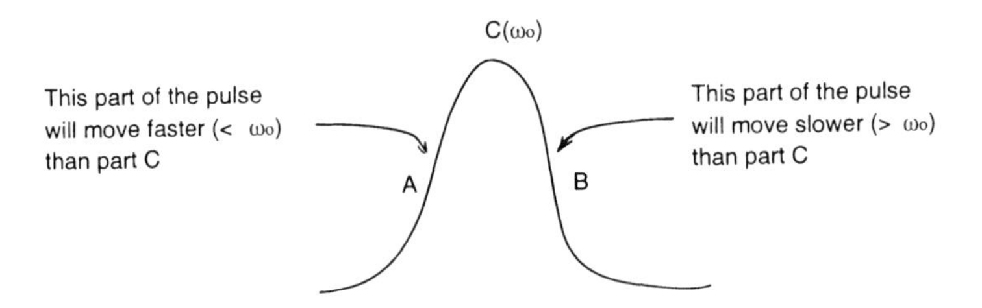

The outer parts of the pulse will be sharpened.

Pulse break-up

(I) The leading edge A of the chirped pulse begins to travel more slowly and to fall back against the main part of the pulse. The trailing edge B of the chirped pulse will begin to travel faster and to catch up with the main part of the pulse.

(II) The pulse will generally become compressed as its propagates, as a consequence of the *SPM*. As the pulse becomes more compressed its peak intensity will increase, and its rise and fall times will become shorter. Both effects will combine to greatly increase the self-chirping effects on the A and B parts of the pulse and this will increase the pulse compression, in a *runaway process*.

(III) The result will often be that the envelope of a high-power laser beam will begin to break up into complex subpulses within the main pulse envelope.

(IV) *Self-focusing* of the optical beam may complicate the picture.
Assume that an optical beam with moderate intensity I and a smooth transverse profile passes through a medium with a positive Kerr coefficient n_{2I} (see Figure 5.12). The higher intensity in the center of the beam will cause an increase of the index of refraction seen by the center, as compared to that seen by the wings of the beam.

(V) Pulse breakup and spectral-broadening effects, accompanied and *intensified* by self-focusing effects, can be a source of difficulty in many high-power lasers, and particularly in mode-locked lasers, where the peak power can be very high even if the average power or total energy may be low.
Self-phase modulation and self-focusing are especially strong runaway effects in such devices as multistage Nd:glass laser amplifier chains and in mode-locked Nd:glass lasers.

Soliton Waves [19-23]

(I) At wavelengths $\rangle 1.35\mu$, *GVD* in quartz-single-mode filbers is negative. The chirp produced by *SPM* through the Kerr effect will cause, at first, *time-compression* (and *not* broadening; broadening is for $GVD\rangle 0$).

(II) A smooth pulse will be steadily compressed and progressively distorted in shape, at least until it becomes sufficiently short that higher-order nonlinear effects begin to compete with the dispersive pulse compression.

(III) Such a pulse can approach a limiting pulse-shape which does not change further with distance and which represents the lowest-order soliton solution to the nonlinear equation (called *nonlinear Schroedinger equation*) that governs the wave propagation in the fiber.

(IV) Higher order soliton solutions represent pulses which do not propagate with constant shapes, but which, if initiated with a proper initial shape and amplitude, will return to that same initial shapes at periodic distances along the fiber [24]. Such solutions are often studied by means of numerical simulations. These higher-order solutions generally require higher amplitudes and energies than the lowest-order soliton, and the soliton period generally decreases with increasing pulse amplitude.

Summary

1. *PVD* and *GVD*

In this section, we start by considering a medium with a certain dispersion relation, i.e. a relation between the propagation constant β and the frequency ω, in order to restate the concepts of phase velocity and group velocity of a train of waves.

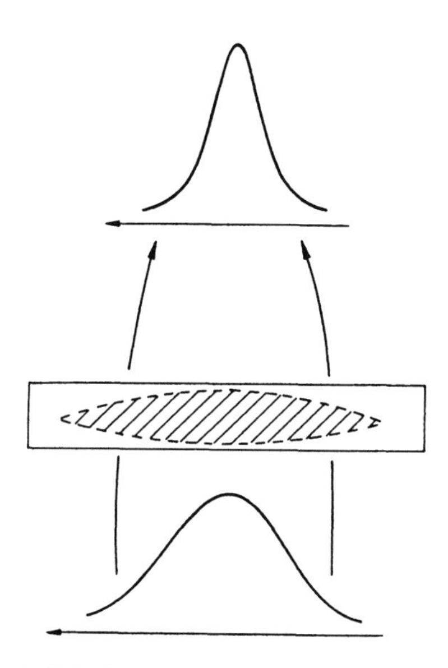

Fig. 5.12. A light beam inducing by itself a focusing lens

We then consider a pulse with a carrier frequency ω_o and with a bandwidth $\Delta\omega_o$, so large that the dispersion is not linear: different spectral regions of the pulse spectrum will travel with different (group) velocities and, of course, the shape of the pulse will change during propagation.

2. Passage of a Pulse through a Medium with Positive *GVD*

In order to study the effects of nonlinearity and positive *GVD* on a pulse we consider a pulse out of a mode-locked laser. The peak power of the pulse is $2kW$, the length of the pulse *6ps* and the carrier wavelength $5900\,\mathring{A}$.

The pulse passes through a single mode quartz optical fiber. The wavelength $5900\,\mathring{A}$ falls in the region of positive *GVD* of the fiber. For didactical reasons the effects of nonlinearity and *GVD* are examined separately.

Nonlinearity. The intense light field produces a change of the index of refraction of the fiber (optical Kerr effect). The result is a modulation of the phase induced by the variation of the intensity at one particular place during the passage of the pulse (**self-phase modulation, or *SPM***). The frequency varies across the pulse, with a positive chirp near the peak of the pulse and a negative chirp in the outer wings.

GVD. The leading edge of the chirped pulse travels faster than the main part of the pulse; the trailing edge travels slower than the main part of the pulse. The outer wing of the leading edge travels slower than the main part of the pulse and the outer wing of the trailing edge travels faster. The result is the stretching of the central part of the pulse and the sharpening of the outer wing. The peak intensity of the pulse is lower, and the positive chirp extends to a larger fraction of the pulse. The interplay of *SPM* and *GVD* introduces modifications in the pulse which consist of the following:

(I) The pulse of the example broadens from an initial shape with width *6ps* to a rectangular shape of $\approx 24ps$.

(II) The pulse acquires a nearly linear frequency chirp over its full duration.

(III) The pulse spectrum broadens from an initial transform-limited band of $\approx 2.5cm^{-1}$ to a characteristic phase-modulation spectrum of $50cm^{-1}$.

Compression of the pulse. The self-chirped pulse can be compressed by using a linearly dispersive device with a negative *GVD*.

3. Passage of a Pulse through a Medium with a Negative *GVD*

Pulse Break-up. When passing through a medium of negative *GVD*, a pulse experiences the following effects: the leading edge of the chirped pulse travels slower than the main part of the pulse; the trailing edge travels faster than the main part of the pulse.

SPM makes the leading edge of the chirped pulse travel more slowly and fall back against the main part of the pulse; it also makes the trailing edge travel faster and catch up with the main part of the pulse.

A **runaway process** sets in. The pulse becomes more compressed as it travels; its peak intensity increases and its rise and fall times become shorter. These effects increase the chirping of the pulse and the pulse is compressed more and more. The result is often the break-up of the pulse into complex subpulses.

Self-focusing (SF) is the effect present when a beam of smooth transverse profile passes through a nonlinear Kerr medium: the difference in intensity between the center of the beam and the wings produces an effective lens that focuses the beam. Spectral-broadening and pulse break-up are enhanced by self-focusing.

SPM and *SF* are strong runaway effects in Nd:glass laser amplifier chains, and in mode-locked Nd:glass lasers.

Soliton Waves. In a medium with negative *GVD* the chirp produced by *SPM* through the Kerr effect causes initially a time compression (and *not* a broadening; broadening is for $GVD \rangle 0$).

A smooth pulse is steadily compressed and progressively distorted in shape until it becomes sufficiently short that higher-order nonlinear effects start to affect the process. Such a pulse can approach a limiting pulse-shape which does not change further with distance and which represents the lowest order soliton solution of the nonlinear equation (called **nonlinear Schroedinger equation**) that governs the wave propagation in the fiber.

Higher order soliton waves represent pulses which do not propagate with constant shapes, but which, if started with a proper shape and amplitude, return to the same initial shapes at periodic distances along the fiber. These higher-order solutions generally require higher amplitudes and energies than the lowest-order solitons, and the soliton period generally decreases with increasing pulse amplitude.

DEDICATION

The author dedicates this article to his friend Renato Vinciguerra with gratitude and fond memories of the many scholarly discussions on the subject of Communication Theory he had with him at his house in Naples. The things the author learned from those discussions have a large bearing on this article. Grazie, Renato!

ACKNOWLEDGMENTS

The author wishes to thank the graduate students Giulio Gambarota, Yi Zhou and Chunlai Yang and Professor John Collins for their helpful criticism of this work and Kashawna Harling for the masterly typing of this paper. He wishes also to acknowledge the benefit of reading and consulting the following books which he would like to recommend to the readers of this article for additional reading.

(1) A.E. Siegman, *Lasers*, University Science Books, Mill Valley, California, 1986.

(2) B.E.A. Saleh and M.C. Teich, *Photonics*, Wiley Interscience Publications, New York, 1991.

(3) O. Svelto, *Principle of Lasers*, Plenum Press, New York, 1989.

REFERENCES

1. R. Loudon, *The Quantum Theory of Light*, Oxford University Press, New York, 1973, p. 64.

2. F.B. Hildebrand, *Advanced Calculus for Engineers*, Prentice-Hall, Inc., Englewood Cliffs, N.J., 1949, p. 523.

3. J.S. Toll, *Phys. Rev. 104*, 1760 (1956): J. Hamilton, *in Progress in Nuclear Physics*, Vol. 8, edited by O.R. Frisch, Pergamon, Elmsford, N.Y., 1960, p. 145.

4. P. Debye, *Polar Molecules*, Dover Publications, Inc., New York, 1929, pp. 84-85.

5. Ibidem, p. 107.

6. F.B. Hildebrand, *Advanced Calculus for Engineers*, Prentice-Hall, Inc., Englewood Cliffs, N.J., 1949, p. 529.

7. S. Chu and S. Wong, *Phys. Rev. Lett. 48*, 738 (1982).

8. C.G.B. Garrett and D.E. McCumber, *Phys. Rev. A 1*, 305 (1970).

9. D. Grischkowsky, *Phys. Rev. A. 7*, 2096 (1973).

10. L.M. Frantz and J.S. Nodvik, *J. Appl. Phys. 34*, 2346 (1963).

11. H. Nakatsuka, D. Grischkowsky, and A.C. Balant, *Phys. Rev. Lett. 47*, 910 (1981).

12. D. Grischkowsky, and A.C. Balant, *Appl. Phys. Lett. 41*, 1 (1982).

13. B. Nikolaus and D. Grischkowsky, and *Appl. Phys. Lett. 42*, 1 (1983); ibid. *43*, 228 (1983).

14. C.V. Shank et al., *Applied Physics Letters 40*, 761 (1982).

15. E.B. Treacy, *IEEE J. Quantum Electron. QE-5*, 454 (1969).

16. J. Desbois, F. Gires, and P. Tournois, *IEEE J. Quantum Electron. QE-9*, 213 (1973).

17. R.L. Fork, O.E. Martinez, and J.P. Gordon, *Opt. Lett. 9*, 150 (1984).

18. J.P. Gordon and R.L. Fork, *Opt. Lett. 9*, 153 (1984).

19. A.C. Scott, F.Y.F. Chu, and D.W. McLaughlin, *Proc. IEEE 61*, 1443 (1973). This is an extensive review of solitons with many references.

20. R.H. Stolen, L.F. Mollenhauer, and W.J. Tomlinson, *Opt. Lett. 8*, 168 (1983).

21. L.F. Mollenhauer and R.H. Stolen, *Opt. Lett. 9*, 13 (1984). This article reports on the *Soliton Laser.*

22. *Solitons in Action*, edited by K. Lonngren and A. Scott, Academic Press, New York, 1978.

23. R.K. Dodd, J.C. Eibeck, J.D. Gibbon and H.C. Morris, *Solitons and Nonlinear Equations*, Academic Press, London and New York, 1982.

24. *The Collected Papers of Enrico Fermi*, University of Chicago Press, 1962, II, 978.

TIME RESOLVED MOLECULAR SPECTROSCOPY AND MOLECULAR DYNAMICS

W. Demtröder, A. Kamal, E. Mehdizadeh,
G. Persch, Th. Weyh, and D. Zevgolis

Fachbereich Physik
Universität Kaiserslautern
D-67653 Kaiserslautern

ABSTRACT

With time resolved spectroscopic techniques radiative lifetimes, quenching cross sections and molecular dynamical processes in excited states can be studied. While for lifetime measurements a time resolution in the pico to nanosecond range is generally sufficient, many fast dynamical processes, such as molecular vibrations, breaking of chemical bonds and isomerization requires femtosecond time resolution. Different experimental techniques are explained and the information obtained with them on molecular dynamics is illustrated by several examples.

I. INTRODUCTION

There are two main goals of molecular spectroscopy. These are:

1. The elucidation of molecular structure
2. The detailed investigation of dynamical processes in excited molecules.

While the first goal is reached by determining the stationary energy states (eigenstates of the total Hamiltonian) of the molecule, the second is related with time-dependent processes, which take place, if nonstationary levels are excited. The time evolution of these levels might be due to radiative decay or to interactions with other levels. There might be either interactions between different energy levels of the same molecule resulting in energy redistribution within the free molecule and in a change of molecular structure, (intramolecular processes) or on the other hand collision-induced transitions where the excitation energy of a molecule is partly or completely quenched by transfer into internal energy of the collision partner or into translation energy (intermolecular relaxation).

Ultrafast Dynamics of Quantum Systems: Physical Processes and Spectroscopic Techniques, Edited by Di Bartolo and Gambarota, Plenum Press, New York, 1998

Both possibilities play an important role in the primary processes of chemical reactions [1,2]. The wide research field of photochemistry, for example, is concerned with the different pathways on which the excitation energy pumped into a molecule by absorption of one or several photons can flow within the molecule, or in case of dissociation, into the reaction products.

The time scale of these dynamical processes can range from milliseconds to femtoseconds. The longer times are observed for the radiative decay of long living excited states, for instance high vibrational levels in the electronic ground state.

The short time range is the domain of forming and breaking molecular bonds or of rearrangements of electron distributions in excited molecules [3]. With femtosecond time resolved spectroscopy these first steps of chemical reactions can be directly observed. A famous example of such an extremely fast process is the isomerization of rhodopsin molecules after absorption of a visible photon which forms the first step in a long cascade of chemical reactions governing the visual process in our eye [4].

The experimental techniques suitable for time resolved investigations of dynamical processes in molecules must be adapted to these different time scales. In the present lectures we will discuss various techniques of time-resolved spectroscopy, their advantages and limitations, and we will illustrate by several examples what we can learn from their results about molecular dynamics. Since these lectures are intended to give a survey on this field, the examples will not be restricted to our own work at the University of Kaiserslautern, but will also include specific experiments from other laboratories.

II. LIFETIME MEASUREMENTS OF SELECTIVELY EXCITED MOLECULAR LEVELS

One important information about the dynamics of excited molecular levels is provided by measurements of their lifetimes and their dependence on external parameters, such as pressure, temperature and the kind of collision partners. During the last decades several experimental techniques for measuring lifetimes have been developed. A very versatile method is the single photon counting delayed coincidence technique [5] which works as follows (Fig. 1):

The molecular sample is irradiated by a periodic sequence of short light pulses (typically with laser pulses of about 0,1 - 1 ns pulsewidth). The excited molecules decay either radiatively by emitting fluorescence photons, or nonradiatively by collisional energy transfer (quenching collisions). The decay of N_i excited molecules in level $|i\rangle$ after the end of the excitation pulse at $t = 0$ is then given by where A_i is the Einstein coefficient of spontaneous emission of level $|i\rangle$, N_B is the number density of collision partners B, σ_i (v,B) is the quenching cross section due to collisions with species B, and v_r is the relative velocity of the collision pair with the distribution function f(v) ($\int f(v)dv = 1$. Often the integral in (1) is approximated by its average value $\overline{\sigma_i(v,B) \cdot v}$

$$\frac{dN_i}{dt} = -A_i \cdot N_i - N_i \cdot N_B \cdot \int_0^\infty \sigma_i(v_r, B) f(v) dv \qquad (1)$$

The fluorescence of the excited molecules is collected by a lens, imaged onto the entrance slit of a monochromator, which selects a specific wavelength and detected by a fast photomultiplier.

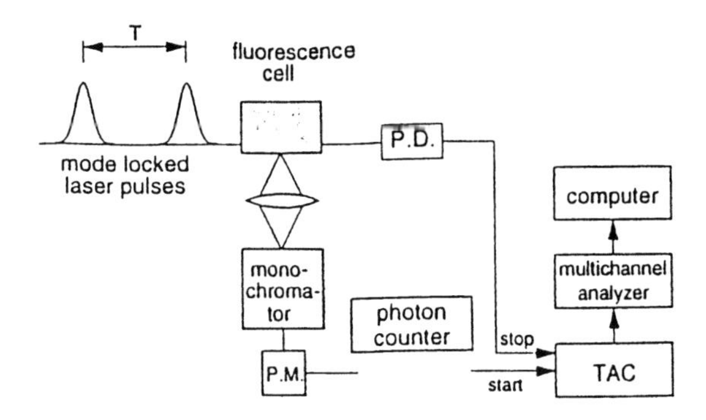

Fig. 1 Experimental arrangement for lifetime measurements with the single photon counting technique.

The output pulse generated by the first fluorescence photon, emitted after the laser pulse, stops a fast voltage ramp which had been started by the preceding laser pulse. The output voltage of this „time-to amplitude converter (TAC)“ is proportional to the time delay between excitation pulse and fluorescence photon emission (Fig. 2). If the number of collected fluorescence photons is sufficiently small, such that the probability of detecting a single photon per excitation pulse is much smaller than one, the pulse height distribution obtained after many excitation pulses represents the time distribution of emitted photons (Fig. 2). Because the TAC has a deadtime of a few microseconds after each ramp, it is advantageous to invert the time sequence, which means to start the TAC with the first detected fluorescence photon emitted at a time t after an excitation pulse and to stop it with the next excitation pulse. If the pulses have a constant repetition rate $f = 1/T$, this implies that the ramp time is $(T - t)$ instead of t. Since the rate dN/dt of detected fluorescence photons is small compared to f, this mode of operation allows the TAC to recover before the next ramp is started.

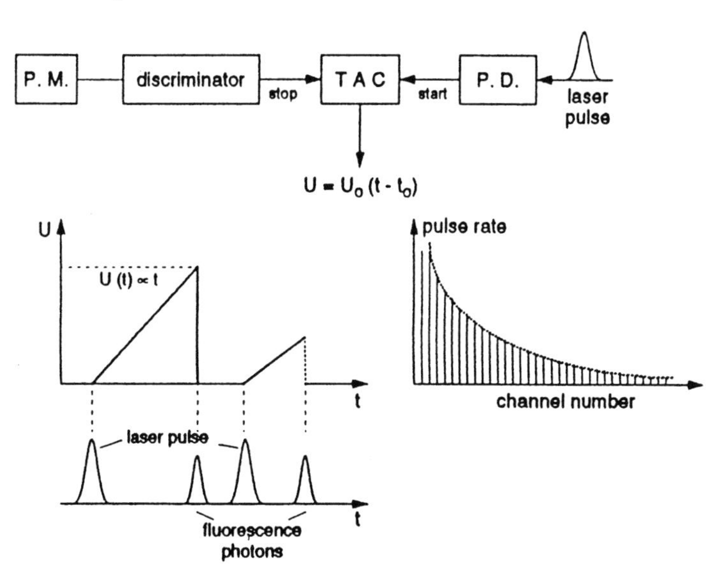

Fig. 2 Principle of single photon counting technique

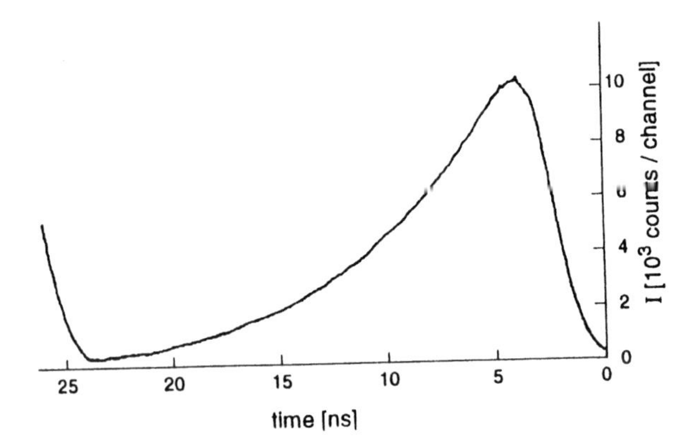

Fig. 3 Fluorescence decay curve of the excited level (v' = 43J' = 86) in the $D^1\Sigma_u$-state of the CS_2 molecule

The integration of eq. (1) yields

$$N_i(t) = N_i(0) \cdot e^{-t/\tau_{eff}} \tag{2}$$

where the effective lifetime τ_{eff} is given by

$$\frac{1}{\tau_{eff}} = A_i + N_B \cdot \bar{v} \cdot \overline{\sigma_i(v,B)} \tag{3}$$

If the molecules are excited in a vapor cell or a heatpipe [2] at a temperature T and a pressure p, the measured exponential decay (2) yields the effective lifetime of the excited molecular level under these conditions. Ideal pulsed light sources for this technique are cavity dumped mode-locked cw dye lasers, which provide pulse trains with a pulsewidth adjustable between a few picoseconds up to several nanoseconds, depending on the wanted spectral bandwidth of the laser pulses [7]. The pulse repetition rate can be controlled by the cavity dumping frequency and can be selected between 1 Hz to about 4 MHz. A typical example of such a measured decay curve is shown in fig. 3.

Often, the excitation energy of the molecules is above 3eV. Then UV-photons have to be used, which are generated by optical frequency doubling of the laser pulses in a nonlinear optical crystal, such as KDP (potassium-dihydrogen-phosphate) or BBO (beta-barium-borate). The achievable output energy of about 1nJ is sufficient to detect about 1 fluorescence photon per 100 excitation pulses, which gives an average counting rate of about $4 \cdot 10^4$/s at a pulse repetition rate of 4 MHz.

Plotting the inverse measured effective lifetime $^1/\tau_{eff}$ as a function of the density N_B of collision partners yields, according to eq. (3) the so-called Stern-Vollmer plot, illustrated in Fig. 4 for an excited state of the Na_2 molecules quenched by two different collision partners B. From the intersect of the sloped straight line with the axis p = 0 the radiative (or spontaneous) lifetime $\tau_{rad} = {}^1/A_1$ is obtained.

From the slope of the straight line

$$tg\alpha = N_B \cdot \bar{v} \cdot \sigma_i$$

of eq. (3), the quenching cross section can be derived, since the average velocity

$$\bar{v} = \sqrt{8kT/\pi \cdot \mu}$$

is determined by the temperature T and the reduced mass $\mu = m_A \cdot m_B/(m_A+m_B)$ of the collision

pair. Since the pressure p of collision partners B is related to the density N_B by

$$p = N_B \cdot k \cdot T,$$

the slope

$$tg\alpha = N_B \bar{v} \cdot \sigma_i = b \cdot \sigma_i \cdot p \quad \text{with} \quad b = \left(\pi\mu/8kT\right)^{1/2} \tag{4}$$

of the Stern Vollmer plot can be expressed by the measured quantities T and p and, therefore, allows one to determine the quenching cross section σ_i.

In order to assure that lifetime and quenching cross section of a single selected level are measured, the observed fluorescence should come solely from this level and should not be superimposed by fluorescence from other levels.

This can be achieved first of all by making the spectral bandwidth $\Delta\nu$ of the excitation pulse as small as possible. The lower limit $\Delta\nu \geq a/\Delta t$ is given by the Fourier-limit for a pulse with timewidth Δt where $a \approx 1$ is a constant which depends on the shape of the laser pulse [8].

If the pulse bandwidth $\Delta\nu$ is still larger than the average spacing between absorption lines of the molecule, several molecular levels are simultaneously excited. In this case the fluorescence light has to be spectrally dispersed by a monochromtor, which selects a single fluorescence line emitted from the wanted upper levels. This guarantees that the measured decay curve is that of a single molecular level.

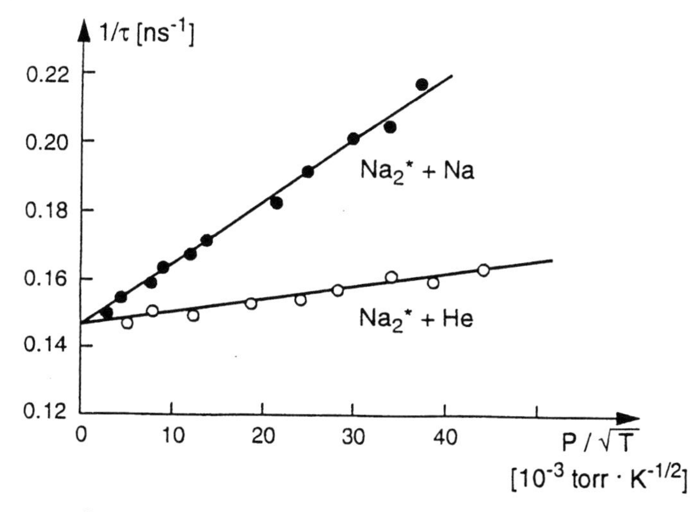

Fig. 4: Stern Vollmer plots of the inverse effective lifetime of the excited level ($v' = 6$, $J' = 27$) in the $D_1\Sigma_u$ state of Na_2 for two different collision partners

The observed quenching cross sections vary between $0{,}1 Å^2$ to $10^3 A^2$, depending on the excited molecule and the collision partners. Highly excited Rydberg levels have particularly large quenching cross sections but long radiative lifetimes. Their lifetimes must be, therefore, measured at sufficiently low pressures. Favorable conditions can be realized in a cold molecular beam, where the number density N_B is low and the relative velocity $\bar{v}_r$ is reduced, due to the low translational temperature.

These molecular beams have an additional advantage for measuring long lifetimes in the milli- to microsecond range. In cell experiments at low pressures the excited molecules may diffuse out of the observation region before they emit a fluorescence photon. This misleads the observer to assume a shorter lifetime than the actual one.

In a collimated molecular beam the molecules excited at the crossing point with the laser beam, all travel into the same direction (Fig. 5). When the fluorescence, emitted into a solid angle $\Delta\Omega$ around the molecular beam axis is observed, the detection probability can be made independent of the distance Δz traveled by the excited molecules, if the solid angle $\Delta\Omega$, covered by the detector is independent of Δz. This can be realized within a range $\Delta z = f \cdot D/d$ by an aperture in front

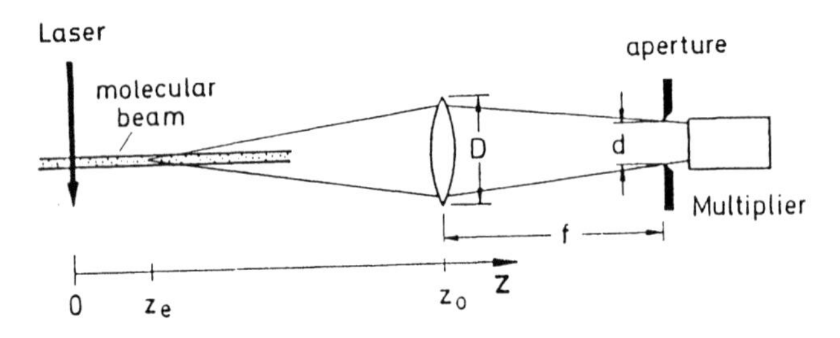

Fig. 5: Realization of constant detection probability for the fluorescence from long living excited molecules in a collimated molecular beam

of the detector, with a diameter d, which is smaller than the diameter D of the collimating lens with focal length f [9].

III. ANOMALOUS LIFETIMES

For many excited molecular levels the measured lifetimes differ considerably from those expected from calculated radiative transition probabilities or from measured integrated absorption cross sections, which are proportional to the oscillator strength of the corresponding transition. The reason for these „anomalous" lifetimes are various perturbations of the emitting level by other energetically close levels. These perturbations represent nonradiative couplings between the mutually perturbing levels, such as vibronic couplings, spin-orbit couplings or Coriolis-couplings.

They may shorten or lengthen the lifetime, depending on the perturbing state (Fig. 6). If, for example, the perturbing state has a repulsive potential, the coupling between the excited level to this state leads to predissociation and, therefore, to a shortening of the lifetime, because the excited level can now decay by radiation and by predissociation and the total decay probability is the sum of the radiative and the predissociative probabilities.

Often the dissociation energy of the molecule in its electronic ground state is larger than the excitation energy of higher electronic states. Examples are NO_2, SO_2, CS_2. In

such cases the excited electronic level can be coupled (for instance by vibronic couplings) to high vibrational levels of the electronic ground state which have long lifetimes. Then the coupling leads to an increase of the lifetime of the electronically excited level by up to three orders of magnitude [10].
For such perturbed molecular levels lifetime measurements provide information about the nature of the perturbing states. Every perturbation between two levels results in an energy shift of both levels. Furthermore, levels which cannot be reached by optical transitions from the ground state (dark levels) gain transition probability by their coupling to a „bright" level. The wavefunction of a perturbed level is a linear combination of the two mutually perturbing levels and, therefore, the dark level „borrows" oscillator strength from the bright level. In the absorption spectrum this results in a splitting of observable lines in two or more components, which represent transitions to these mutually perturbing states. For very strong perturbations the intensities of both components become comparable. If the lifetimes and the energy splittings of both components can be measured, the complete information about the coupling strength and the individual lifetimes of the „deperturbed" levels, (i.e. the lifetimes the levels would have without perturbation), can be derived [11].
Since the strength of the coupling is proportional to the square of the coupling matrix element, but inversely proportional to the energy separation of the two levels, the lifetimes of selectively excited short lived rovibronic levels within one electronic state can vary considerably with energy, if the excited electronic state is perturbed by another long lived state with different rotational or vibrational constants. This is illustrated by Fig. 7, where the lifetimes of several excited rovibronic levels in the $^2B_2$-state of NO_2 are plotted as a function of energy [12]. This illustrates that lifetime measurements provide an additional source of information on complex molecular structure and dynamics.

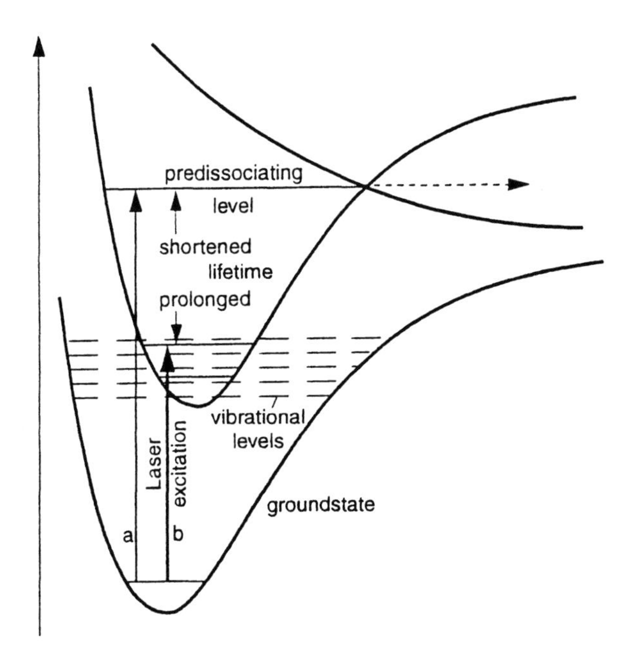

Fig. 6: Perturbations of excited levels by a dissociating state (excitation a) or by high vibrational levels of the electronic ground state (b)

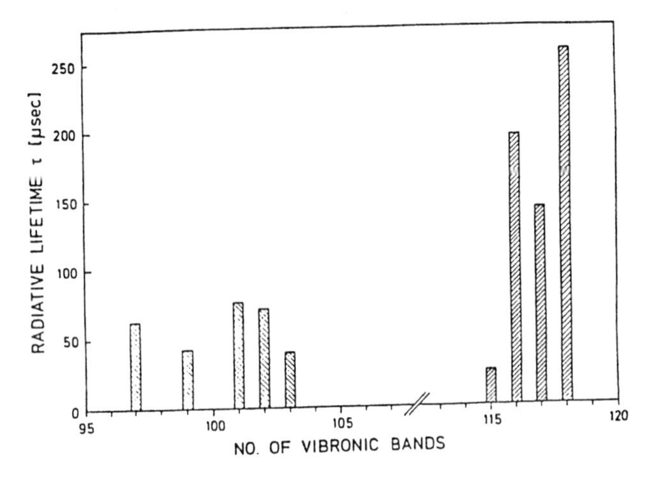

Fig. 7: Lifetimes of selectively excited rovibronic levels in the $^2B_2$-state of NO_2, which vary due to perturbations by high lying vibrational levels of the X^2A_1-ground state [12].

IV. MOLECULAR FEMTOSECOND SPECTROSCOPY

Vibrational motions of molecules as well as the forming and breaking of molecular bonds occur on a femtosecond to picosecond time scale, depending on the atomic masses and the depth of the potential. For instance, a nitrogen atom with a kinetic energy of 1 eV moves with a velocity of $4 \cdot 10^3$m/s. For 1 Å distance it therefore takes a time interval of 25 fs. Since vibrational motions cover a bond length change of the order of 1 - 3 Å, one sees, that the molecular dynamics proceeds on the femtosecond time scale. Also the rearrangement of the electron distribution, following optical excitation of a molecule proceeds within similar short time intervals.

With femtosecond laser pulses all these fast dynamical processes are now accessible to experimental investigations. The following examples, which represent only a small selection of work in many laboratories, illustrate some experimental techniques.

We start with experiments on Na_2, performed by G. Gerber and his group in Würzburg [13]. A pump laser pulse ($\Delta t \approx 65$ fs, $\lambda \approx 618$ nm) excites Na_2 molecules in a molecular beam from their ground state $X^1\Sigma_g^+(v'' = 0)$ into a coherent superposition of vibrational levels v' = 10 - 14 in the excited $A^1\Sigma_u^+$-state (Fig. 8). From the spectral width $\Delta\overline{\nu} \approx 1/(c \cdot \Delta t) \approx 500$ cm^{-1} and the vibrational spacings $\delta\overline{\nu} \approx 100$ cm^{-1} one can infer that about 5 vibrational levels are coherently populated. The superposition of their vibrational wavefunctions forms a wavepacket, which moves back and forth in time between the inner and outer potential well, with a group velocity that represents the velocity of the nuclei in the classical model of an unharmonic oscillator.

If the molecule in this coherent superposition of vibrational levels is further excited by a second laser pulse (called „probe"-pulse) into the $2^1\Pi_g$-state, it forms analogous wave packets in this state, which move, however, with another velocity because of the different potential. The excitation probability for this second step is much larger for the inner turning point in the A-state potential than for the outer turning point.

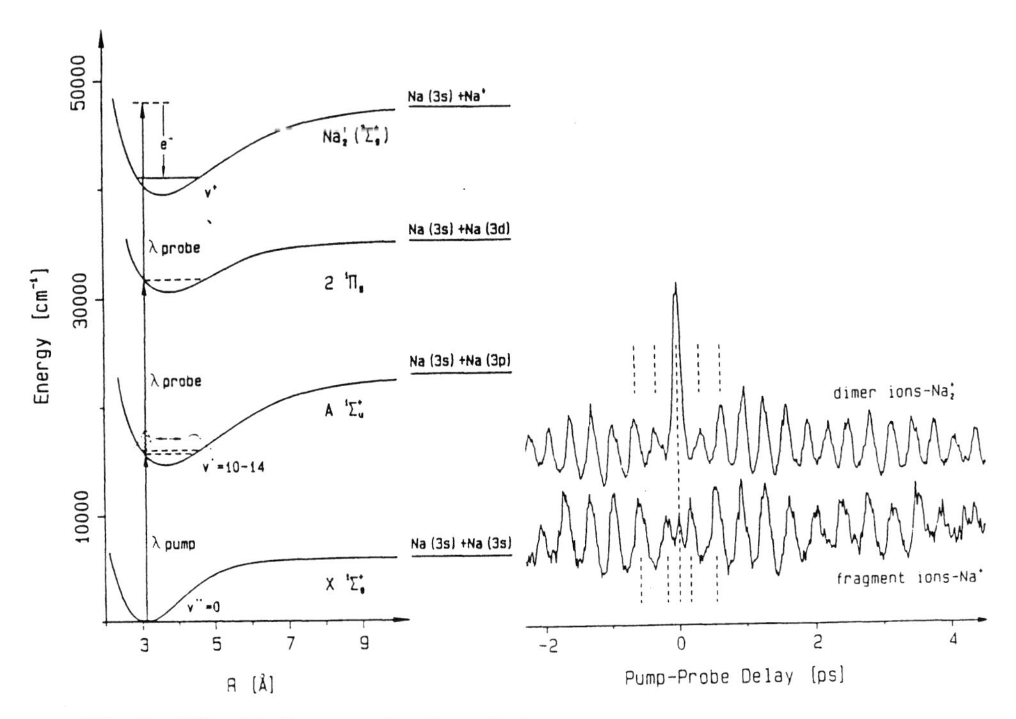

Fig. 8: Vibrational wavepacket dynamics in the Na_2-molecule
a) level scheme; b) observed ion intensity as a function of the delay time between pump and probe pulse [13]

of the population in the $2^2\Pi_g$ state) shows a periodic modulation with a period which equals the round trip vibrational period in the A-state.

Experimentally it is easier to observe the ions produced by further excitation of the $2^1\Pi_g$-state into the ionization continuum of the Na_2^+ ($^2\Sigma_g$)state. The time dependence of the ion rate is now influenced by the product of the two transition probabilities for the transitions $2^1\Pi_g \leftarrow A^1\Sigma_u$ and Na_2^+ ($^2\Sigma_g$) $\leftarrow$ Na_2 ($2^1\Pi_g$). While the first transition strongly depends on the nuclear distance R, the second does not. The wave packet dynamics therefore mainly depends on the movement in the $A^1\Sigma_u$ state. Because of the unharmonic potential the phase velocities of the different wavepacket-components differ and the wavepacket spreads out but is refocussed again after the recurrence time. The ion rate, measured as a function of time delay between pump and probe (Fig. 8) will show a periodic modulation behavior with a slowly varying amplitude, which reflects the movements of the Na-nuclei in the $A^1\Sigma_u$-potential.

With a proper selection of the laser wavelength it is even possible to control ionization or dissociation by choosing the correct time delay. This is illustrated in Fig. 9a where a high lying $^1\Sigma_u$-state with a double minimum potential is reached by two-photon excitation. At the inner turning point of the excited vibrational levels a third photon from the probe laser can reach the Na_2^+ ionization continuum following the scheme:

$$Na_2(^1\Pi_g, v') + h\nu \rightarrow Na_2^+(^2\Sigma_s^+) + e^-(E_{kin}^{(1)})$$

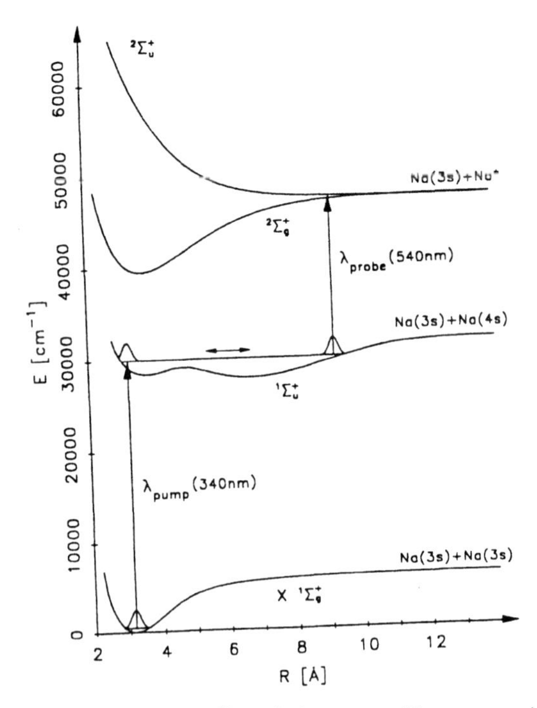

Fig. 9: Level scheme for photodissociation control by proper time delay [14]

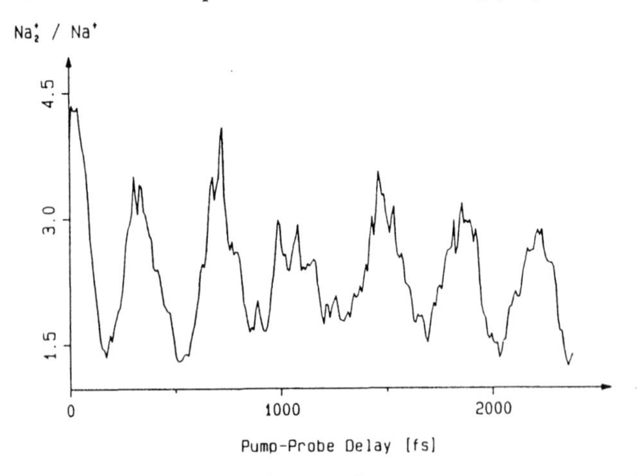

Fig. 10: Measured ratio of Na_2^+ and Na^+ ion rates as a function of the delay time between pump and probe laser [14]

but it misses the dissociative potential $^2\Sigma_u$, which can be only reached from the outer turning point in the $^1\Pi_g$ state (Fig. 9a), according to

$$Na_2(^1\Pi_g, v') + h\nu \rightarrow Na(3s) + Na^+ + e^-(E_{kin}^{(2)})$$

The measured ratio $N(Na^+)/N(Na_2^+)$ of atomic to molecular ion rate therefore strongly depends on the delay time between pump and probe pulse (Fig. 9b).

The kinetic energies E_{kin} of the photo-electrons are different for both cases. Their measurement can distinguish between both processes which may, of course, also be done by the mass spectrometric determination of the ion fragments. The combination of ion mass spectrometry and electron spectroscopy give very detailed information about the different pathways in this two or three photon ionization [14].

The Na_2 molecule serves only as a test ground to check various techniques of excitation and detection. Of more interest for the control of chemical reactions are larger molecules. Here more different pathways are possible and more vibrational degrees of freedom are present.

The first time-resolved experiments on molecular dissociation have been performed by A. Zewail and his group. We will illustrate the technique just by one example [14], which shows the wave packet motion within a quasi-bound potential of the NaI molecule and its partial reflection at the crossing point of two potentials (Fig. 11).

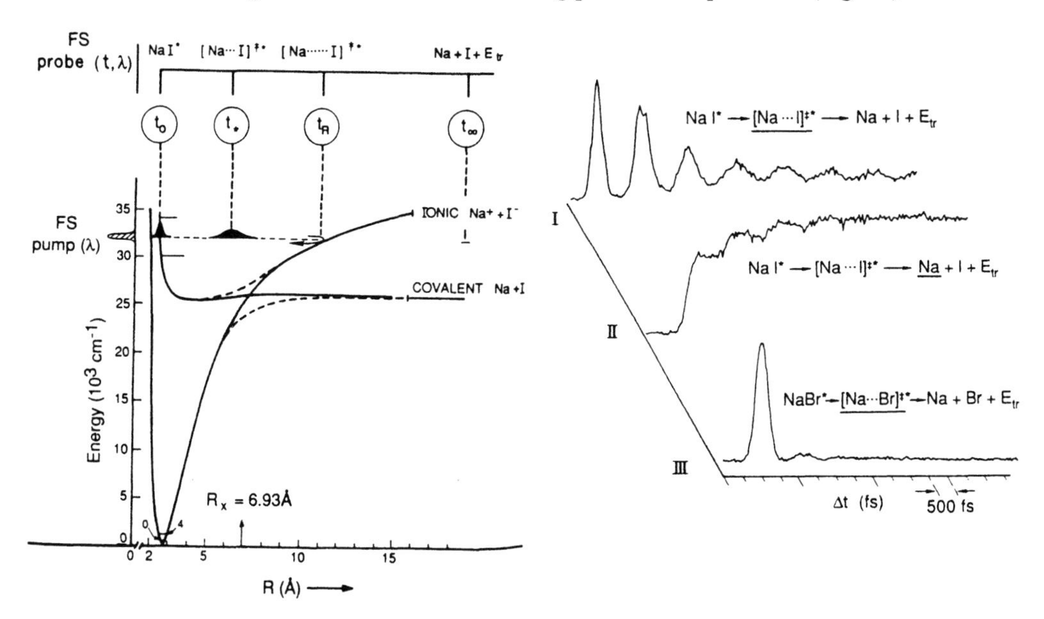

Fig. 11: a) Potential energy curves of NaI
b) Detection of (NaI)*(upper trace) and Na-Atoms (middle trace). For comparison the LIF-signal from (NaBr)* is shown in the lower trace.

At an internuclear separation of 0,7 nm there is an avoided crossing between the covalent curve dissociating into Na + I and the ionic curve $Na^+ + I^-$. If the molecule NaI is excited by a femtosecond pulse from the ground state into an energy range above the dissociation energy for Na + I but below that of $Na^+ + I^-$, a wave packet is formed at the inner slope of the potential of the activated complex $(NaI)^*$ which moves towards large internuclear separations. At the avoided crossing it may either tunnel through the barrier, leading to dissociation products Na + I, or it may be reflected.

The free Na atom can be detected by delayed probe laser excitation of the atomic 3P ← 3S transition, resulting in the emission of the yellow D-line while the (NaI)* can be monitored by excitation of higher states with subsequent emission of fluorescence at other wavelengths than the D-line. The time deendent rate of Na(a) and of the activated complex (NaI)*(b) in Fig. 11 clearly shows the oscillation of the wavepacket and the exponential decay of the signal amplitude, due to leakage through the barrier into the dissociation channel. For comparison the time sequence of the $(NaBr)^*$-fluorescence following the photodissociation of NaBr is shown, which illustrates a much faster decay.

There are many examples, where femtosecond techniques have unravelled details of the different steps of molecular structural change induced by absorption of photons [16, 17].

V. VIBRATIONAL RELAXATION BY COLLISIONS

If in a molecular ensemble conditions are created which represent a nonequilibrium energy distribution, collisions between the molecules will bring the system back to thermal equilibrium after the perturbation has ended.

The time scale of these relaxation processes depends on the number density of the collision partners and on the collisional cross sections. For investigations of fast relaxations the pump and probe technique is the most convenient experimental approach. The basic principle is explained in Fig. 12.

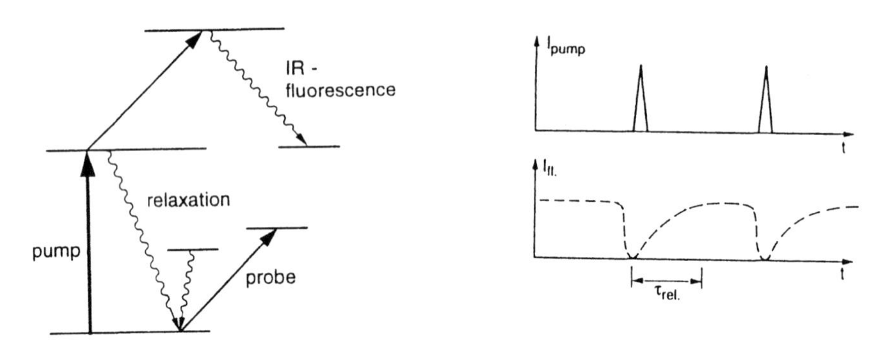

Fig 12: a) Level scheme for pump- and probe techniques
b) time dependent depletion and relaxation of the lower level

The pump pulse creates a sudden decrease of the lower level population and a corresponding increase of the upper level population. The time-delayed probe pulse probes the time dependent population change.

For vibrational excitation of molecules in liquids the collision-induced relaxation towards thermal equilibrium occurs on a picosecond time scale. Vibrational-translational energy transfer generally corresponds to the larger times on this scale while near resonant vibrational - vibrational transfer has the shortest relaxation times. A typical experimental arrangement is shown in Fig. 13. The pump pulse is provided by a mode-locked Nd-YAG-laser, which pumps a vibrational transition of molecules in liquids. Since these transitions are broadened, there are coincidences between the Nd-YAG-laser and vibrational transitions of several molecules. The probe pulse can be generated by Raman-shifting the pump pulse wavelength. Nowadays tunable Ti-sapphire lasers or optical parametric oscillators can be used which provide a larger tuning range.

Those investigations are of great importance for many problems of photochemistry of molecules in liquids. One example is the optical pumping of dye molecules in a solvent, such as ethanol, into the first excited singlet state S_1 where the excited rovibronic levels relax into the lowest vibrational level in the S_1 state from which spontaneous or, in case of dye lasers, induced emission takes place into high vibrational levels of the electronic ground state S_0 (Fig. 12). These levels relax by collisions within a few picoseconds into lower vibrational levels, thus depleting the terminal levels for laser emission and keeping the inversion between S_1 and S_0 sufficiently high for lasing conditions.

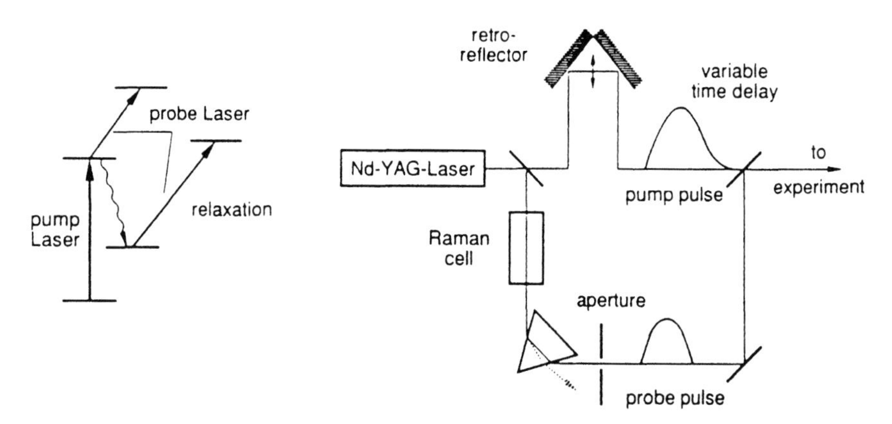

Fig. 13: Possible experimental arrangement for pump and probe spectroscopy of vibrational relaxation in liquids.

VI. CONCLUSIONS

The examples in this lecture only represent a small selection of numerous activities in many laboratories. They should illustrate the importance of time resolved spectroscopy for a more detailed insight into the dynamics of excited molecules and the many aspects of photochemistry as well as for the study of molecular collision processes which form the basis of chemical reactions. I want to thank the graduate students in my group, which provided the results for some of the examples presented here and I am also grateful to Prof. Gerber and Prof. Zewail who allowed me to use figures from their work.

VII. REFERENCES

1. M. Quack: Spectra and Dynamics of Coupled Vibrations in Polyatomic Molecules Ann. Rev. Phys. Chem. 41, 839 (1990)

2. R.D. Levine and R.B. Bernstein: Molecular Reaction Dynamics and Chemical Reactivity, (Oxford Univ. Press, Oxford 1987)

3. A.H. Zewail: Femtochemistry, World Scientific, Singapore 1994

4. R.A. Mathies, St. W. Lin, J.B. Ames, W.Th. Pollard: From femtoseconds to biology: Mechanisms of bacterion rhodopsin's light driven proton pump, Annu. Rev. Biophysics and Biophys. Chem., Vol. 20, (1991)

5. D.V. O'Connor, D. Phillips: Time Correlated Single Photon Counting, (Academic Press, New York 1984)

6. B. Steffes, X. Li, A. Mellinger, C.R. Vidal: Appl. Phys. B 62, 87 (1996)

7. W. Demtröder: Laser Spectroscopy, 2nd edition, (Springer, Heidelberg 1996)

8. J. Herrmann, B. Wilhelmi: Lasers for Ultrashort Light Pulses, (North Holland, Amsterdam 1987)

9. R. Kullmer, W. Demtröder: J. Chem. Phys. 84, 3672 (1986)

10. A.E. Douglas: J. Chem. Phys. 45, 1009 (1966)

11. Th. Weyh, W. Demtröder: J. Chem. Phys. 104, 6938 (1996)

12. G. Persch, H.J. Vedder, and W. Demtröder: Chem. Phys. 105, 471 (1986)

13. T. Baumert, M. Grosser, R. Thalweiser, and G. Gerber: Phys. Rev. Lett. 67, 3753 (1991)

14. T. Baumert, R. Thalweiser, V. Weiss, G. Gerber, in: Femtosecond Chemistry, ed. by J. Manz, L. Wöste, (VCH, Wiley, Weinheim 1995)

15. A.H. Zewail: Femtochemistry, Vol. I + II, (World Scientific, Singapore 1994)

16. T.S. Rose, M.J. Rosker, A. Zewail: J. Chem. Phys. 88, 6672 (1988)

17. W. Fuß, T. Schikarski, W.E. Schmid, S. Trushin, K.L. Kompa, P. Hering: J. Chem. Phys. 106, 2205 (1997)

FEMTOSECOND DYNAMICS OF MOLECULAR REACTIONS AT METAL SURFACES

Richard Finlay and Eric Mazur

Department of Physics
Harvard University
Cambridge, MA 02138

ABSTRACT

These lectures are an introduction to current research into photo-induced chemical reactions at metal surfaces. After an introduction to some qualitative quantum mechanics, we discuss the electronic and optical properties of metals, beginning from an introductory level. The Drude model is described in detail and then optical properties of matter are developed more completely by introducing band structures. The physics governing adsorption of reactants at a metal surface and other fundamental concepts in surface science are introduced. We describe the interaction with a subpicosecond laser pulse with a metal surface in preparation for discussion of some recent photochemistry experiments using subpicosecond laser pulses. The experiments address the nature of the photo-excited electrons that are responsible for chemical reaction of the adsorbates.

I. THE QUANTUM MECHANICS AND SPECTROSCOPY OF ATOMS

This section provides a brief overview of some fundamental concepts from quantum mechanics which are relevant to spectroscopy and molecular dynamics. The emphasis is on developing a physical interpretation of the Schrödinger equation. Introductions to quantum mechanics that approach the subject in more detail through mathematical formalism[1] or by following the historical development of the subject[2] are available.

Ultrafast Dynamics of Quantum Systems: Physical Processes and Spectroscopic Techniques, Edited by Di Bartolo and Gambarota, Plenum Press, New York, 1998

I. A. The Schrödinger equation

The justification for quantum mechanics is its agreement with experiment, and the consistency and elegance of the mathematical formalism. Nevertheless, it is instructive to provide a motivation based on classical mechanics. In classical mechanics the energy of a particle in an external potential $U(x)$ is

$$E = \frac{p^2}{2m} + U(x), \tag{1}$$

where p is the momentum and m is the mass of the particle. In quantum mechanics p and x are interpreted as quantum mechanical operators. Let's view x as proportional to an operator which means *multiply by the position* and p as proportional to an operator which means *take the derivative with respect to x*. These definitions are consistent with our intuition that the momentum specifies the rate of change of position. When the transformation

$$p \rightarrow -i\hbar\frac{d}{dx} \tag{2}$$

is applied to (1) the classical expression for the total energy is transformed into an operator expression. This operator is called the Hamiltonian, H:

$$\frac{p^2}{2m} + U(x) \rightarrow \frac{1}{2m}\left[-i\hbar\frac{d}{dx}\right]^2 + U(x) = -\frac{\hbar^2}{2m}\frac{d^2}{dx^2} + U(x) \equiv H. \tag{3}$$

The function that these operators act on is called the wavefunction, Ψ. The result is the time-independent Schrödinger equation:

$$H\Psi = E\Psi. \tag{4}$$

There is also a time-dependent Schrödinger equation. The following expression describes the time rate of change of the wavefunction in terms of the now time-dependent Hamiltonian[1]

$$H\Psi = i\hbar\frac{\partial\Psi}{\partial t}. \tag{5}$$

The time-dependent Schrödinger equation reduces to (4), the time-independent Schrödinger equation if the potential energy is independent of time: $U = U(x)$. To see this, write the wavefunction, Ψ, as a product of a time-dependent and a time-independent part:

$$\Psi(x,t) = \Psi(x)\theta(t). \tag{6}$$

Substituting (6) into (5), the equation separates into an expression with only spatial dependence on the left side, and only time dependence on the right side:

$$-\frac{\hbar^2}{2m}\left(\frac{1}{\Psi}\frac{d^2\Psi}{dx^2}\right) + U(x) = i\hbar\frac{1}{\theta}\left(\frac{d\theta}{dt}\right). \tag{7}$$

Because the right side is a constant independent of x, the left side must also be independent of x, and is therefore constant. Call the constant E.

$$-\frac{\hbar^2}{2m}\nabla^2\Psi + U(x)\Psi = E\Psi \tag{8}$$

$$i\hbar\frac{d\theta}{dt} = E\theta \tag{9}$$

The symbol ∇^2 represents the second derivative with respect to x. The first equation is simply the time-independent Schrödinger Equation 4 which justifies using the letter E for the constant. The second equation has a trivial solution:

$$\theta(t) = Ce^{-iEt/\hbar}. \tag{10}$$

What does this mean? In a time-independent potential, the time-dependence of the wavefunction is simply a rotation of the (complex) phase of the wavefunction. The frequency of this rotation is determined by the energy.

The product $\Psi(x,t)^*\Psi(x,t)$ is very important in quantum mechanics. Born interpreted it as a probability density; the probability of finding a particle described by wavefunction $\Psi(\vec{r},t)$ within a volume element $d\vec{r}$ of $\vec{r}$ is

$$\Psi(\vec{r},t)^*\Psi(\vec{r},t)d\vec{r}. \tag{11}$$

According to this interpretation, the wavefunction must be normalized to reflect the requirement that the particle is certainly somewhere in space:

$$\int \Psi(\vec{r},t)^*\Psi(\vec{r},t)d\vec{r} = 1. \tag{12}$$

The integral in Equation 12 is over all space. In the special case of a particle in a time-independent potential, Equation 10 contains all the time-dependence of the wavefunction. The product $\Psi(x,t)^*\Psi(x,t)$ is constant in time and so the wavefunction is called a stationary state.

The interpretation of $\Psi(x,t)^*\Psi(x,t)$ as a probability density places constraints on Ψ. It must be single-valued. It may not be infinite (unless it is infinite in only an infinitesimally small region of space). Furthermore, since the wavefunction is a solution of a second order differential equation (the Schrödinger equation) it must be continuous. If the potential is not too pathological then the first-derivative of Ψ is also continuous.

I. B. Origin of energy quantization

Rewriting the Shrödinger equation, we see that it specifies the curvature of the wavefunction:

$$\nabla^2\Psi = \frac{2m}{\hbar}[U(x) - E]\Psi. \tag{13}$$

Where $U(x) = E$ the curvature of the wavefunction is zero. In other regions, the sign of the curvature is determined by the signs of both Ψ and $U(x) - E$.

The meaning of the Schrödinger equation may be clarified by making qualitative sketches of possible solutions to Equation 13, using the equation to specify the sign of the curvature of Ψ. The left side of Figure 1 shows a potential energy surface and beneath it, two trial wavefunctions of energy E which have the same value at $x = x'$. Between x' and x'', both wavefunctions have downward curvature because Ψ is positive, and $E > U(x)$. Just beyond x'', Ψ is still positive but $E–U(x)$ changes sign. The curvature of both wavefunctions becomes upward. The upward curvature causes Ψ_A to diverge. Wave function Ψ_B soon passes below zero; Ψ_B changes sign, the curvature again becomes downward, and the wavefunction diverges downwards. Wavefunctions Ψ_A and Ψ_B are not acceptable because they are not normalizable. A wavefunction (not shown) with slope at x' between that of Ψ_A and Ψ_B would satisfy Equation 13 without diverging by tending to zero beyond x''.

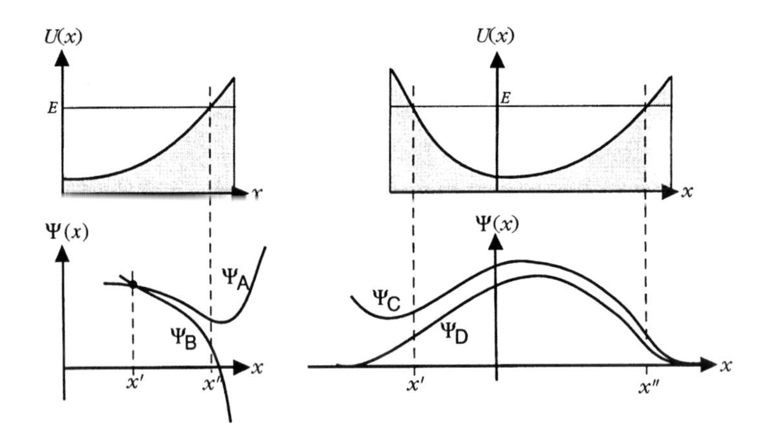

Figure 1 Qualitative trial wavefunctions of energy E for two potentials. Wavefunctions A, B and C are not acceptable because they diverge, and are not normalizable. D is a correct wavefunction. Starting with a trial energy, slope and value at x', the second derivative of the wavefunction is constrained by the Schrödinger equation. After [2].

The right side of Figure 1 demonstrates that a potential energy surface bounded on both sides further constrains the wavefunctions. Two wavefunctions are sketched. Both tend towards zero for positive x. However Ψ_C diverges on the left side. Wavefunction Ψ_D is normalizable. It has finite extent in the x-direction and is called a bound-state wavefunction. It turns out that normalizable wavefunctions exist only for special values of the energy: for most values of E there are no corresponding normalizable wavefunctions. This energy quantization is a consequence of the normalization condition. The values of the energy for which there is a corresponding wavefunction are labeled by quantum numbers.

Figure 2 shows another potential energy surface. Wavefunctions in the well region below E_1 have discrete, quantized energies. Above E_1, however, normalizable wavefunctions exist for all energies. We see in Section IV that molecules are bound to metal surfaces by interactions similar to that sketched in Figure 2.

I. C. The hydrogen atom

Study of the hydrogen atom was very important for the development of quantum mechanics because there was no classical explanation for the observed spectrum. It is still studied because it is relatively simple, and has exact analytic solutions. Furthermore, we will see that many other atoms (such as the alkali metals) have electronic levels similar to hydrogen.

The potential energy describing the interaction of the hydrogen nucleus (a proton) with an electron is:

$$U(r) = -\frac{1}{4\pi\varepsilon_o}\frac{e^2}{r} \tag{14}$$

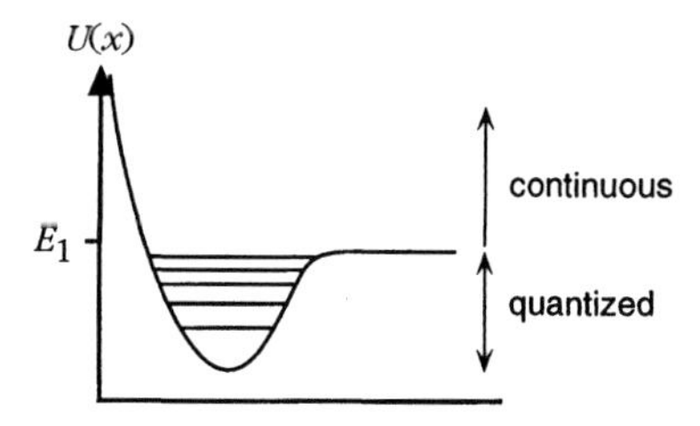

Figure 2 A potential energy surface with a region in which the energies are quantized (the bound-state well), and a region in which all energies are permitted. After [2].

where r is the radial distance from the nucleus to the electron. This potential produces a bounded well: we therefore expect quantized energies. Solution of the Schrödinger equation confirms that the energy of the electron depends on the quantum number n according to:

$$E_n = -E_I \frac{1}{n^2}. \tag{15}$$

The energy is defined in terms of the ionization energy, $E_I \approx 13.6$ eV. The energies of the electronic states of hydrogen are plotted in Figure 3.

The solutions of the Schrödinger equation for potential energy (14) and total energy (15) factor into radial and angular terms:

$$\Psi(r,\theta,\phi) = R_{n,l}(r) Y_{l,m}(\theta,\phi). \tag{16}$$

The angular dependence is entirely in the term $Y_{l,m}(\theta,\phi)$, called a spherical harmonic.[3] The radial dependence is contained in $R_{n,l}(r)$,[3] plotted in the left side of Figure 4 for various values of n = 0, 1, or 2. The radial distributions are usually expressed in terms of length scale

$$a = \frac{4\pi\varepsilon_o \hbar^2}{m_e e^2} \approx 0.54 \text{ Å}, \tag{17}$$

known as the Bohr radius.

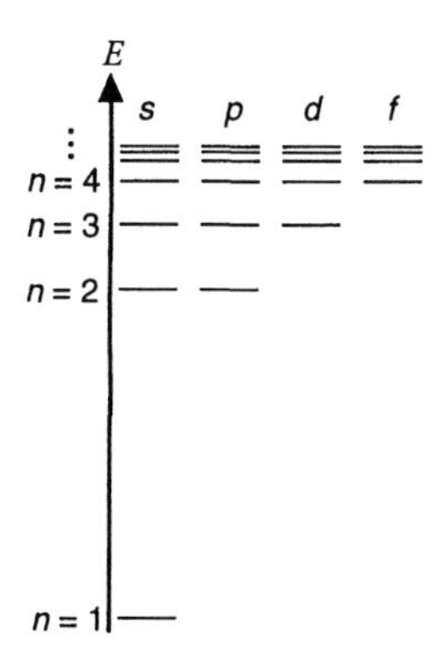

Figure 3 Energy levels of the hydrogen atom. The energies are independent of the values of l ($s,p,d,\ldots$). Each state may contain two electrons of opposite spin.

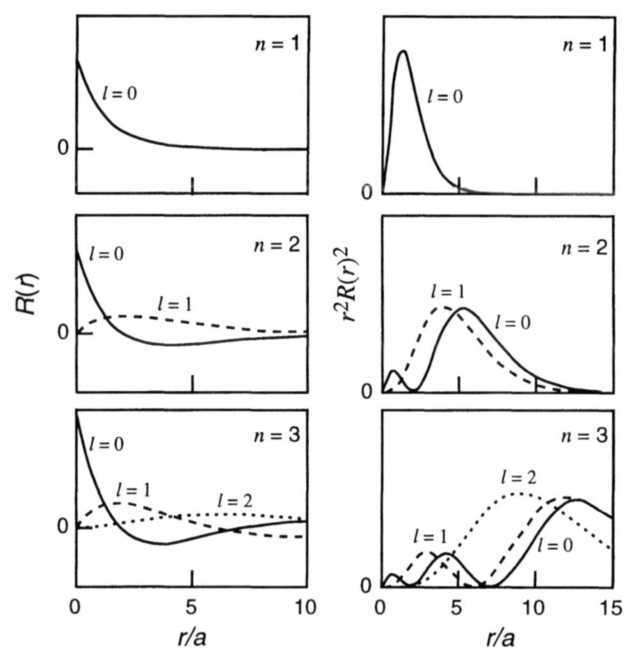

Figure 4 Radial dependence of the wavefunctions of the hydrogen atom (left). The l = 0 wavefunctions (the s-waves) are nonzero at the nucleus. The probability of finding the electron at radius r/a scales as r^2R^2 (right). The length scale a is the Bohr radius. After [2].

The values of the quantum number l in Equation 16 are integers in the range $0...(n-1)$. The values of l are represented with letters; s for $l = 0$, p for $l = 1$, d for $l = 2$, f for $l = 3$, with g and subsequent letters used for the higher values of l. The parameter m is called the magnetic quantum number. The values of m lie in the range $-l < m < l$.

The probability of finding the electron at radius r from the nucleus scales as $r^2 R_{n,l}(r)^2 dr$. The right side of Figure 4 shows how r^2R^2 depends on r. When $n = 1$, the electron is most likely found near $r/a = 1$, the Bohr radius. For $n = 2$ the radial distribution depends on l: the $l = 0$ probability distribution has a lobe which is closer to the nucleus than $l = 1$. A similar trend occurs for $n = 3$ where again the probability of finding the electron near the nucleus is greatest for lowest values of l.

According to Equation 15 and Figure 3, the energies of the electronic states of hydrogen do not depend on l. We will see below that in multi-electron atoms the energies of the states depend on l and their relative energies can be predicted from the spatial distributions of the wavefunctions.

It is possible to show using symmetry (and other) arguments that optical transitions between states of the hydrogen atom are constrained to satisfy selection rules. For example, a transition induced by an electric dipole interaction[3] between the atom and an external electric field must satisfy $\Delta l = \pm 1$. Transitions from s to p are allowed, but s to d is forbidden. Restrictions on Δm depend on the polarization of the light. There is no restriction on Δn.[2]

Figure 3 does not tell the whole story because it fails to represent the fine structure. The fine structure is a deviation of the electronic levels sketched in Figure 3 on a scale of about

10^4 times smaller than the ionization energy. Fine structure arises from several different interactions, including relativistic effects due to the high speed of the electron, effects related to the nonlocal distribution of the electron's wavefunction, and an interaction between the magnetic moment due to orbital motion of the electron around the nucleus with the magnetic moment associated with the spin of the electron. This last interaction is called spin-orbit coupling. When the magnetic moments of orbital and spin origin are parallel, the interaction energy is higher than when they are antiparallel. This energy difference is the spin-orbit coupling.[3]

A consequence of the spin-orbit coupling is that neither the orbital angular momentum ($\vec{l}$) nor the spin angular momentum ($\vec{s}$) is constant. (In quantum mechanical language we say that neither operator commutes with the Hamiltonian.) The sum, $\vec{j}$, however, is constant and has two possible magnitudes depending on the relative orientations of $\vec{l}$ and $\vec{s}$. Each value of $|\vec{l}|$ is called a level. For example, if $l = 2$, two levels are possible:

$$|\vec{j}| = |\vec{l} + \vec{s}| = 3/2 \text{ or } 1/2. \quad (18)$$

Each level has a different energy. The actual spectrum of hydrogen is also influenced by other contributions to the fine structure (listed above) which have effects on the same scale as the spin-orbit coupling.

There is also a hyperfine structure which is due to the spin of the proton. (The proton, like the electron, is a spin 1/2 particle.) One of the ways it influences the electronic levels is by interacting with the magnetic field caused by the motion of the electron. Hyperfine interactions are on the order of 10^3 times smaller than fine-structure interactions.[3]

I. D. Multi-electron atoms

It is not possible to find exact solutions for the energies and wavefunctions of atoms with two or more electrons. Interactions between electrons complicate the expression for the potential energy of any one electron. An approximate solution is obtained by considering one electron at a time and accounting for only an average interaction with the other electrons. The other electrons are pictured as forming a cloud around the nucleus. From outside the cloud, the rest of the atom appears to be a single positive charge because attraction to the Z charges in the nucleus (where Z is the atomic number) is partially offset by repulsion from $Z - 1$ electrons in the cloud. The effective potential outside the cloud varies as

$$V_{eff}(r) \sim \frac{e^2}{r}. \quad (19)$$

The electron outside the cloud is shielded from most of the nuclear charge. When the electron penetrates the cloud, however, it interacts with all of the nuclear charge, Ze, without shielding from the cloud. In this case the effective potential varies as

$$V_{eff}(r) \sim \frac{Ze^2}{r}. \quad (20)$$

According to Equations 19 and 20, electrons with wavefunctions that extend to the core experience a larger attraction to the nucleus and have a lower (more negative) energy than orbitals that do not extend inside the cloud.

Figure 5 shows two hydrogen atom wavefunctions. For fixed n, wavefunctions with low l penetrate towards the nucleus more than wavefunctions with higher l. The low l wavefunctions

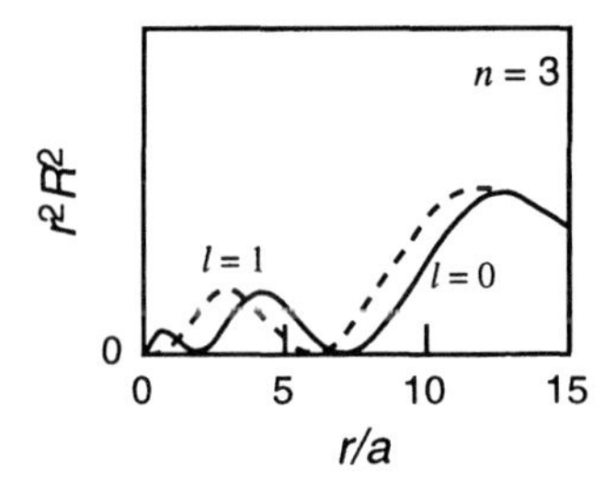

Figure 5 The $l = 0$ (s-orbital, bold) state extends closer to the nucleus ($r = 0$) than the $l = 1$ (p-orbital). When additional electrons form a shell around the nucleus, the penetration of the s-orbital towards the nucleus causes it to have lower energy than the p-orbital.

interact more with the core than the high l wavefunctions. Consequently, in a multi-electron atom, the $n = 3$, $l = 0$ energy, for example, is below the $n = 3$, $l = 1$ energy. The influence of the electron cloud is essential to this argument; in the single-electron atom the energy is independent of l.

The energy level diagrams for hydrogen, lithium, and sodium are indicated below. In lithium the shielding of the core causes the s, p, and d energies to be different. In sodium the difference is so great that the *3d* orbital is higher in energy than the *4s* orbital.

Shielding is particularly effective for alkali metals such as lithium, sodium, and potassium. Their outermost electron lies at a higher value of n than the rest of the electrons in the atom. The lower-lying electrons form a closed shell, which means that they occupy all states of lower n. A similar situation occurs for noble metals, such as copper, silver, and gold. These metals have a single electron in an s-orbital and 10 electrons in (fully occupied) d-orbitals. The fully-occupied d-orbitals tend to shield the outermost electron from the nucleus. The alkali metals and the noble metals are together called monovalent metals because they both have a single electron outside a shielded nucleus. Electrons that lie outside the closed shell are called valence electrons.

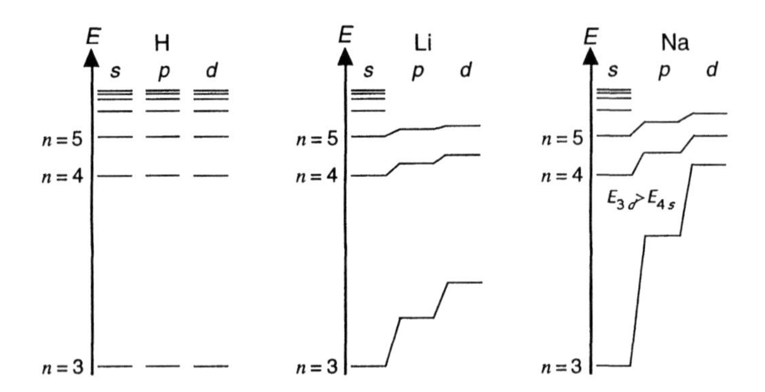

Figure 6 Energy level diagrams for the first few alkali metals. In hydrogen, the energy levels are independent of l. In atoms with closed shells, states of high l have relatively high energy compared with states of lower l and the same n. In sodium this effect is large enough that the energy of the 3d state is higher than the energy of the 4s state.

The approximate electronic state of an atom is stated by specifying its configuration. Configuration is represented by nl^i where i is the number of electrons occupying the orbital with the values n and l. For example, $1s$ and $1s^2$ refer to configurations in which the $n = 1$, $l = 0$ orbital is singly ($1s$, as in ground state hydrogen) or doubly ($1s^2$, as in ground state helium) occupied. Excited state configurations may be represented with this notation: the configuration of ground state oxygen, $1s^2 2s^2 2p^4$, becomes $1s^2 2s^2 2p^3 3s$ following excitation of an electron to $3s$. Each s-orbital ($l = 0$) may contain two electrons corresponding to the two spin states, and each p-orbital ($l = 1$) may contain six electrons corresponding to two electrons in each of the $m = -1, 0, 1$ states. The configuration only approximately gives the electronic state of the atom because the notation does not account for details in the electronic levels arising from the interactions between electrons in the atom. A more complete notation is described after a brief discussion of the helium atom.

I. E. Helium atom

The simplest atom in which there are electron-electron interactions is helium. Helium is not well described as a single electron bound to a nucleus with a screening cloud (as alkali metals are described) because both electrons occupy the same orbital. The energy of each electron is raised by Coulomb interactions with the other electron. Another complication is exchange interaction which is a purely quantum mechanical contribution to the energy. It arises from the laws of quantum mechanics applied to two indistinguishable objects. According to a law of quantum mechanics, if the electrons in a two-electron wavefunction are exchanged, there can be no change in the probability distribution, $|\Psi|^2$. For example, the wavefunctions representing one electron in an s-orbital and another in a p-orbital

$$\Psi_T = \Psi_s(1)\Psi_p(2) \text{ or, } \Psi_T = \Psi_p(1)\Psi_s(2), \tag{21}$$

have different values of $|\Psi|^2$, even though the only difference between them is an exchange of the roles of electrons labeled 1 and 2. Wavefunctions which are invariant under exchange of the electrons may be constructed as follows:

$$\Psi_\pm = \frac{1}{\sqrt{2}}\left[\Psi_s(1)\Psi_p(2) \pm \Psi_p(1)\Psi_s(2)\right]. \tag{22}$$

If the indices 1 and 2 are exchanged, representing exchange of the two electrons, the value of $|\Psi|^2$ is unchanged. The wavefunctions Ψ_+ and Ψ_- have different energies. The energy difference is called exchange splitting.

In classical mechanics, if two interacting bodies move under the influence of a central potential (such as two planets orbiting a sun) the total angular momentum of the two bodies is conserved. However, their mutual interaction exchanges angular momentum between them so that the angular momentum of either body alone is not conserved. Similarly, in the quantum mechanical description of an atom with two electrons, the angular momentum of a single electron, $\bar{l}_1$ or $\bar{l}_2$ is not constant, but the total orbital angular momentum,

$$\bar{L} = \bar{l}_1 + \bar{l}_2, \tag{23}$$

is constant. The quantum numbers corresponding to $\bar{L}$ are represented by the capital letters *S, P, D, etc.* The total spin angular momentum,

$$\bar{S} = \bar{s}_1 + \bar{s}_2, \tag{24}$$

$$^{2S+1}\{L\}_J$$

Figure 7 Standard atomic spectroscopy notation. The spectral term is represented inside the brackets with a letter. The level is specified in the lower right corner. $2S + 1$ in the upper left specifies the multiplicity of the electronic spin states.

may be either 0 or 1, depending on whether the spins of the electrons are parallel or anti-parallel. The values of L and S determine a spectral term.

If there were no interactions between the orbital angular momentum, $\vec{L}$, and the spin angular momentum, $\vec{S}$, then (23) and (24) would be good operators to describe the atom. However, as in the hydrogen atom, there are fine structure interactions which couple the angular and spin momenta. The total angular momentum

$$\vec{J} = \vec{L} + \vec{S}, \tag{25}$$

is conserved with fine structure interactions. There are $2S + 1$ possible values for J which together are called a multiplet. The quantity $2S + 1$ is called the multiplicity of the state. For example, when $S = 0$ there is one possible value for $\vec{J}$, determined by $\vec{L}$. When $S = 1$, the multiplicity is three. The $S = 0$ state is called a singlet, and the $S = 1$ state is called a triplet. A particular value of J determines the level. Figure 7 shows the standard spectroscopic notation. It specifies the term, the level, and the multiplicity of the atom. For example, the electronic ground state of helium is represented as $^1S_0$.

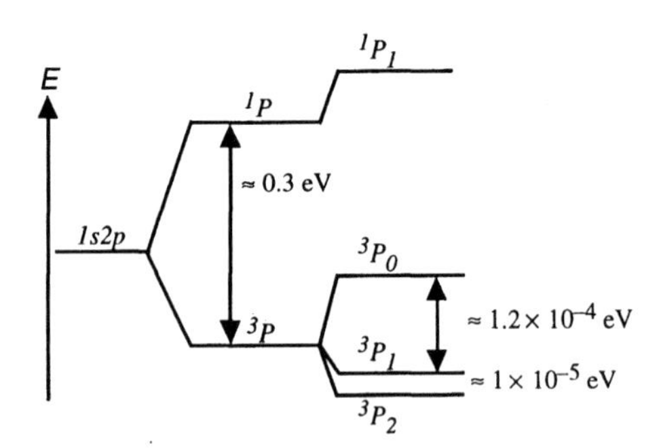

Figure 8 The energy level diagram for a two electron atom. Electrons in the $1s$ and $2p$ orbitals combine with total angular momentum $L = 1$, a P-state. Two possible terms are represented, corresponding to the two ways the spin angular momenta may combine: $S = 0$ or $S = 1$. Their energies are different because of exchange splitting. Fine-structure interactions cause the terms to split into levels corresponding to the $2S + 1$ possible values of J. After [3].

Figure 8 represents the energies of the possible levels in the $1s2p$ configuration. The energy difference between the levels has been exaggerated. Figure 8 does not tell the whole story. As in the hydrogen atom, an additional hyperfine splitting further complicates atomic spectra. Hyperfine splitting is described in many text books.[3]

I. F. Chemical bonds

When atoms are close together, the electrons from different atoms interact and may form chemical bonds. Attraction between the atoms causes the atoms in a molecule (or solid) to have lower total energy than separate atoms. At close range the atoms repel each other, preventing the molecule from collapsing. A potential energy surface which represents these qualitative features of the interaction is sketched in Figure 9.

The total energy of the molecule depends on the motion of the nuclei such as vibrations or rotations of the molecule and translation of the center of mass. While considering the internal excitation of the molecule, we typically ignore the translational energy. The various internal excitations evolve on very different timescales. Electronic motion occurs at about 10^{15} Hz. Vibrations occur at about 10^{13} Hz. Rotational motion is slower still at 10^{7}–10^{11} Hz. The electronic part of the Schrödinger equation may be solved while assuming the nuclei are at fixed locations because the electronic motion occurs so much more quickly than the nuclear motions. With this Born-Oppenheimer approximation we write the wavefunction as a separable function:

$$\Psi \approx \Psi_e(\vec{r}_e, \vec{R}_N)\chi_N(\vec{R}_N) \approx \Psi_e \Psi_v \Psi_R. \tag{26}$$

Here the nuclear wavefunction χ_N is written as a product of vibrational, Ψ_v, and rotational, Ψ_R, components.

The strongest bond is the covalent bond, which is formed by an overlap of charge distribution between neighboring atoms. The lowering of the total energy comes about because electrons are shared between nuclei. Figure 10 schematically illustrates covalent bonding between two atoms. If the two atoms get close enough for the atomic orbitals to overlap, then the wavefunctions of these orbitals can add with either the same or opposite phase. The two new orbitals have different energies. The lower level corresponds to the wavefunctions adding in phase while the upper level corresponds to adding with opposite phase. Electrons which occupy the symmetric state are concentrated primarily in between the two nuclei; they draw both nuclei towards the center by coulomb attraction to the positive charges on the nuclei. They are known as bonding orbitals. The electrons in the anti-symmetric wavefunctions, however, have a low probability of being between the nuclei. Electrons in these states tend to pull the molecule apart, and are called anti-bonding. If both of the original electronic levels in each atom were singly occupied, then in the ground state of the new system, the two electrons occupy the bonding orbital. Since this state is energetically

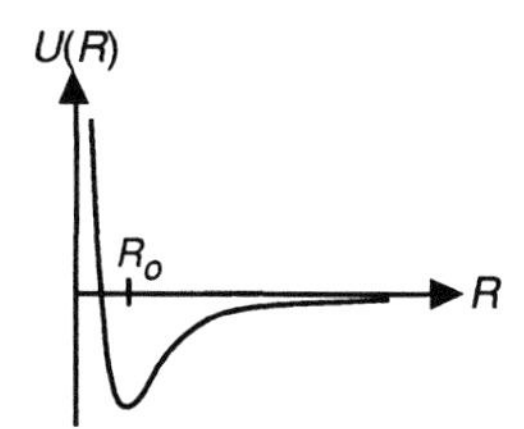

Figure 9 The qualitative form of the interaction potential between two nuclei in a diatomic molecule. The potential energy is plotted against the separation of the nuclei. The equilibrium separation of the nuclei is R_o.

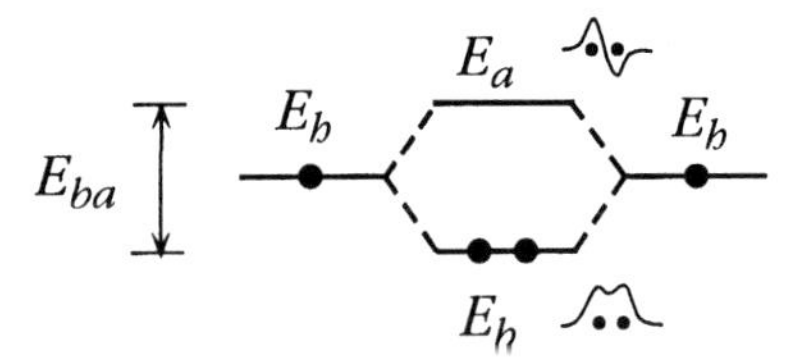

Figure 10 Energy level splitting in a covalent bond between two atoms.

more favorable than the state in which the two atoms are separated, the sharing of the two electrons results in a bond between the two atoms. In the case of a solid consisting of many atoms, the bonding and antibonding levels broaden into bands. We return to the topic of bands below.

The wavefunctions sketched in Figure 10 represent the combination of two s-orbitals. In general, the combination of the orbitals is more complicated. If the combined orbital is symmetric under rotation about the nuclear axis then the resulting bond is called a σ-bond.

The bonding of two atoms is determined by the balance between the electrons in bonding orbitals, and electrons in anti-bonding orbitals. For example, two hydrogen atoms (each in a ground-state $1s^1$ configuration) make H_2 because each atom contributes one electron to a σ bonding orbital. The orbital may contain two electrons without violating the Pauli exclusion principle because there are two available spins for each electron. Two He atoms, however, do not form a bond because the initial $1s^2$ configuration of each atom would result in a molecule with two electrons in a bonding orbital, and two electrons in an anti-bonding orbital, for no net bonding.

The formation of diatomic oxygen, O_2, is more complicated because each atom has 4 valence electrons. Three electrons from each atom go into bonding orbitals: 2 in each of two π orbitals, and 2 in the σ orbital. The remaining two electrons go into anti-bonding π^* orbitals. Overall there are 4 more electrons in bonding orbitals than anti-bonding. A full bond is formed when two electrons occupy a bonding orbital. Molecular oxygen has four such electrons, so O_2 has a double bond.

I. G. Vibrations of diatomic molecules

We now consider the vibrational motion of the nuclei in a diatomic molecule. Figure 9 qualitatively illustrates the interaction between two atoms comprising a molecule. The equilibrium separation of the atoms is R_o. The vibrations are usually small with respect to the internuclear separation, R. Writing the potential as a Taylor series expansion about the equilibrium position R_o in terms of $Q \equiv R - R_o$, we have

$$U(Q) = U(0) + \left(\frac{dU}{dQ}\right)_o Q + \frac{1}{2}\left(\frac{d^2U}{dQ^2}\right)_o Q^2 + \text{anharmonic terms} \tag{27}$$

The anharmonic terms contain additional nonvanishing derivatives of $U(Q)$. The first derivative of $U(Q)$ is zero at R_o. Neglecting anharmonic contributions, the potential is quadratic in Q:

$$U(Q) \approx \frac{1}{2}kQ^2,\ k = \left(\frac{d^2U}{dQ^2}\right)_o. \tag{28}$$

In classical mechanics, two atoms of mass m_a and m_b interacting with this potential vibrate sinusoidally at frequency

$$\omega = (k/\mu)^{1/2} \tag{29}$$

where μ is the reduced mass:

$$\frac{1}{\mu} \equiv \frac{1}{m_a} + \frac{1}{m_b}. \tag{30}$$

The Schrödinger equation describes the wavefunction for two particles interacting by the quadratic potential in Equation 28.

$$\nabla^2\Psi + \frac{2m}{\hbar}\left[E - \frac{1}{2}kQ^2\right]\Psi = 0 \tag{31}$$

The solution is an energy spectrum with equally-spaced energy levels:

$$E_v = \left(v + \frac{1}{2}\right)\hbar\omega, \quad v = 0,1,2,\ldots. \tag{32}$$

The wavefunctions are sketched in Figure 11. In the ground state the nuclei are most likely to be found at separation $Q=0$, *i.e.*, at their equilibrium separation, $R=R_o$. At higher levels of excitation the nuclei are most likely to be found at the limits of their oscillation (largest and smallest R). In classical mechanics a simple harmonic oscillator spends most of its time at the positions of greatest displacement where the velocity is lowest. The similarity of the quantum mechanical and classical behaviors for high quantum numbers is known as the correspondence principle.

The harmonic approximation is a good representation of the interaction potential at low levels of excitation where the Taylor series (27) remains a good approximation to the actual interaction potential. The harmonic approximation is particularly valuable because it has an exact solution.

Another expression for the interaction potential which has an exact solution is the Morse potential[4]:

$$U(Q) = D_e\left(1 - e^{-\alpha Q}\right)^2. \tag{33}$$

This is the function sketched in Figure 9; it increases rapidly at small separation and

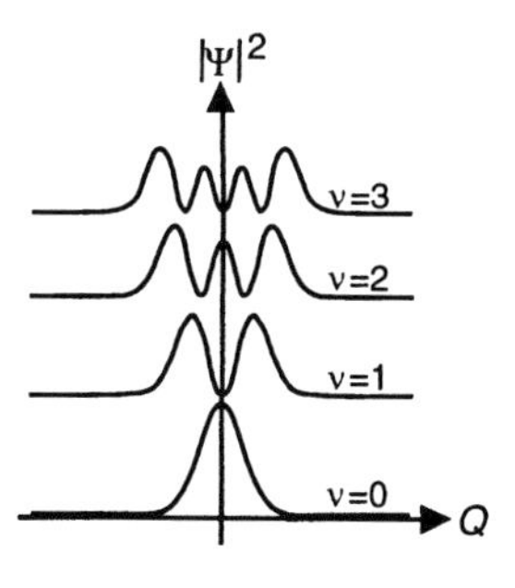

Figure 11 Squares of the wavefunctions of the harmonic oscillator (offset for clarity). At successively higher vibrational excitations (higher v), the nuclei are more and more likely to be found at large separation, in agreement with a classical harmonic oscillator.

disappears at large separation. The energies of the bound states in a Morse potential are

$$E_v = \left(v + \frac{1}{2}\right)\hbar\omega - \left(v + \frac{1}{2}\right)^2 \hbar\omega x. \tag{34}$$

The constant x depends on D_e and α. The second term is a correction to the energy of a harmonic oscillator (32). We stress that Equations 32 and 34 are the just the energies of states of the artificial potential energy surfaces (28, 33) used in these calculations. Real molecular spectra depend on the actual interaction potential between the atoms.

I. H. Electronic transitions in diatomic molecules

Electronic transitions are often depicted in potential energy *versus* nuclear separation diagrams such as Figure 12. The Figure shows an electronic ground state potential energy surface, X, and the potential energy surface corresponding to an excited electronic state, A. In this example, the excited state A has a minimum at larger R than the ground state. An arrow represents a transition from X→A. The transition is vertical in this diagram because electronic transitions occur much faster than nuclear motion, R.

The most probable transitions are between vibrational states with probability maxima at the same R. This is called the Frank-Condon principle. At high levels of vibrational excitation, the molecule is most likely to be found towards the limits of the oscillation, while in the ground state the probability distribution is concentrated near $R = R_o$ (Figure 11). The vertical transition in Figure 12 satisfies the Frank-Condon principle because it is a transition from the equilibrium position in the ground state to an extreme limit of motion in an excited (vibrational) state.

Spectra of molecules are further complicated by rotational excitations. Transitions between rotational states lie in the microwave region of the spectrum; transitions between vibrational states lie in the infrared region of the spectrum, and electronic transitions span the range from infrared through visible to the ultraviolet.

Additional details to this brief introduction are provided below as some of the concepts are applied to the study of chemical reaction of adsorbates at metal surfaces.

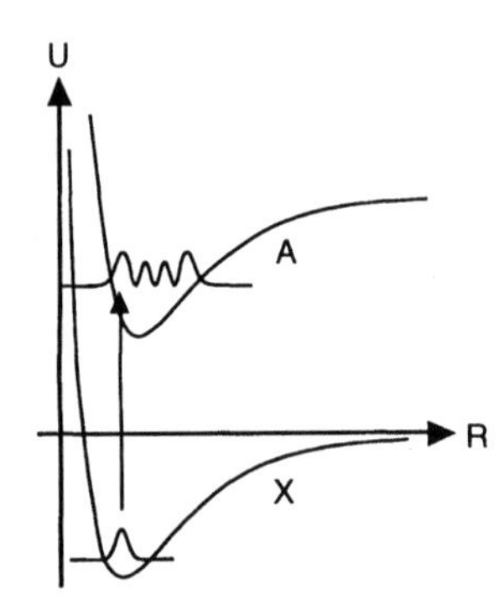

Figure 12 A transition between the electronic ground state of a molecule, X, and an excited electronic state, A. The displacement of these (hypothetical) potential energy surfaces ensures that the ground state wavefunction overlaps well with an excited vibrational state wavefunction. The transition shown satisfies the Frank-Condon principle.

II. OPTICAL PROPERTIES OF SOLIDS

Here we introduce some of the concepts that are relevant to understanding how light interacts with solids. A simple and very powerful model for the behavior of electrons in a solid is introduced, elucidating many optical properties of materials.

II. A. Propagation of electromagnetic waves in vacuum

In vacuum the frequency f and the wavelength λ of an electromagnetic wave are related by the speed c of light in vacuum,

$$f\lambda = c. \tag{35}$$

This yields a linear relation between the angular frequency $\omega \equiv 2\pi f$ and the wavevector $k \equiv 2\pi/\lambda$:

$$\omega = ck. \tag{36}$$

In a medium, the propagation of an electromagnetic wave is determined by the response of the material to electric and magnetic fields and is characterized by the dielectric constant ε, the magnetic permeability μ, and the electric conductivity σ. Except for the magnetic permeability, which is nearly frequency independent at optical frequencies, the response of the medium depends on the frequency of the incident wave, and so dispersion occurs: waves of different frequencies propagate at different speeds. The frequency-dependent speed v of light in a medium is given by

$$v = \frac{c}{\mathrm{Re}\sqrt{\varepsilon(\omega)}} \equiv \frac{c}{n(\omega)}, \tag{37}$$

where $\varepsilon(\omega)$ is the frequency-dependent dielectric function and $n(\omega)$ the index of refraction of the medium. This frequency dependence results in a nonlinear relation between the angular frequency and the wavevector of the electromagnetic wave:

$$\omega = \frac{c}{\mathrm{Re}\sqrt{\varepsilon(\omega)}}k. \tag{38}$$

Equation 38 is an example of a dispersion relation. It relates photon energy $E = \hbar\omega$ to wavevector. Figure 13 shows the dispersion relations, and the relationships between dielectric function and angular frequency, in vacuum and in a medium. Throughout this part of the lectures we use similar graphs to represent the optical properties of materials. The questions addressed are: why do different materials have different optical and electronic properties and what fundamental properties of solids are responsible for this behavior?

II. B. Propagation of electromagnetic waves through a medium

The electromagnetic response of a material varies over the frequency spectrum because the charges present in the material respond at widely different frequencies. Roughly speaking, we can subdivide the charges into the following categories: ionic cores (the nuclei and core electrons at each lattice site), valence electrons, and free electrons. The ionic cores can form dipoles that tend to orient themselves along the direction of the applied external fields. This motion is usually limited to low frequencies and therefore only contributes to the polarization — and hence the propagation of the wave — at frequencies in the microwave region and below (see Figure 14). At higher frequencies the dipoles can no longer follow the rapid

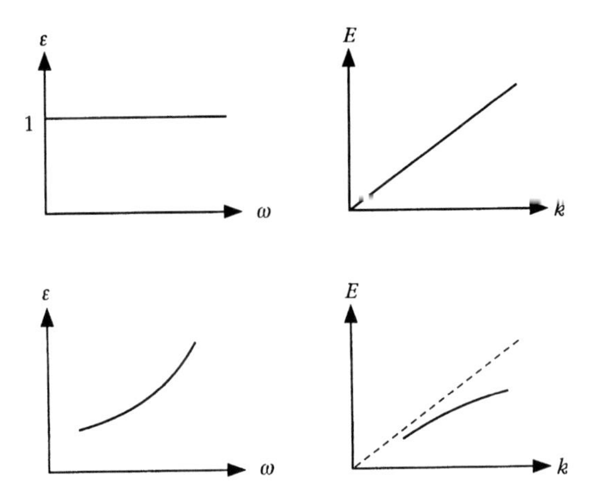

Figure 13 Dielectric function and energy-wavevector relationships in vacuum (top) and in a medium (bottom).

oscillation of the applied field and their contribution to the dielectric constant vanishes. Lattice vibrations (displacement of the ionic cores) induced by the applied field contribute at frequencies up to the infrared region of the electromagnetic spectrum. In the visible and ultraviolet regions only the response of the free and bound valence electrons remain. Core electrons contribute at high frequency (10–1000 eV), but unless absorptions occur their contributions to the polarizability are generally small and therefore the dielectric constant is close to 1 for x-rays. The dielectric function in Figure 14 is very schematic — real materials generally show more structure and, depending on the type and number of charges present, some of the features shown may not be there.

Let us begin by analyzing the motion of a bound valence electron in response to an external driving field. If the field oscillates at frequency ω, the electron oscillates at the same frequency with the phase and the amplitude of the oscillation determined by the binding and damping forces on the electron. The oscillation is described by the equation of motion of the electron:[5]

$$m\frac{d^2x}{dt^2} = -m\omega_0^2 x - m\gamma\frac{dx}{dt} - eE. \tag{39}$$

The first term on the right hand side of the equation represents a binding force with spring constant $k = m\omega_0^2$, where m and ω_0 are the mass and the resonant frequency of the bound electron, respectively. The second term is a velocity-dependent damping force and the third term is the driving force with E the applied field. Rearranging terms and assuming a sinusoidally varying applied field of amplitude E_o and frequency ω, we obtain an inhomogeneous second-order differential equation

$$m\frac{d^2x}{dt^2} + m\gamma\frac{dx}{dt} + m\omega_0^2 x = -eE_0 e^{-i\omega t}. \tag{40}$$

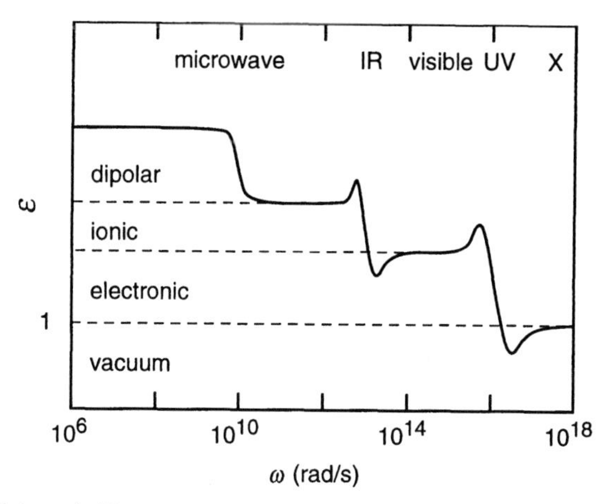

Figure 14 Schematic illustration of the various contributions to the dielectric constant across the electromagnetic spectrum.

The steady-state solution of this equation, representing the oscillating motion of the electron, must be of the form

$$x(t) = x_o e^{-i\omega t}. \tag{41}$$

Substituting this into Equation 40, we get for the amplitude of the motion

$$x_o = -\frac{e}{m}\frac{1}{(\omega_o^2 - \omega^2) - i\gamma\omega} E_o . \tag{42}$$

As is to be expected, the amplitude of the electrons is maximal when the driving frequency is equal to the resonance frequency. The motion of the electron results in an oscillating dipole moment

$$p(t) = -ex(t) = \left(\frac{e^2}{m}\right)\frac{1}{(\omega_o^2 - \omega^2) - i\gamma\omega} E_o e^{-i\omega t}. \tag{43}$$

In a sample with many bound electrons, the dipole moments of all the electrons contribute to a polarization

$$P(t) = \left(\frac{Ne^2}{m}\right)\sum_j \frac{f_j}{(\omega_j^2 - \omega^2) - i\gamma_j\omega} E_o e^{-i\omega t}, \tag{44}$$

where N is the total number of electrons, and f_j is the fraction of electrons having a resonant frequency ω_j and damping constant γ_j. In quantum mechanical terms, the factor Nf_j is the oscillator strength. This factor indicates how much each resonance contributes to the polarization.

The relation between the polarization P and the electric field E is usually written as $P(t) = \varepsilon_o \chi_e E(t)$, with χ_e the dielectric susceptibility, and $E(t) = E_0 e^{-i\omega t}$. The dielectric constant is given by $\varepsilon(\omega) = 1 + \chi_e$, so that

$$\varepsilon(\omega) = 1 + \frac{Ne^2}{\varepsilon_o m}\sum_j \frac{f_j}{(\omega_j^2 - \omega^2) - i\gamma_j\omega} = \varepsilon'(\omega) + i\varepsilon''(\omega). \tag{45}$$

Figure 15 shows the frequency dependence of the dielectric constant for a single resonance and the resulting $E(k)$-behavior. At resonance, dissipation of energy is maximized since the amplitude of the electron motion is maximized. This dissipation of electromagnetic energy is what we call absorption and is reflected by the peak in the imaginary part of the dielectric constant at the resonance frequency. Note, also, that the real part crosses through zero near the resonance.

Let us next turn to the response of free electrons to an oscillating electromagnetic wave. Setting the binding force in Equation 40 to zero we get

$$m\frac{d^2x}{dt^2} + m\gamma\frac{dx}{dt} = -eE_o e^{-i\omega t}. \tag{46}$$

Again, we obtain a solution of oscillating form,

$$x(t) = \frac{e}{m}\frac{1}{\omega^2 + i\gamma\omega}E_o e^{-i\omega t}, \tag{47}$$

but because there is no binding term, the motion does not exhibit any resonances. At low frequency, $\omega/\gamma << 1$, the applied electric field induces a time-varying current

$$J(t) \equiv \frac{dq}{dt} = -Ne\frac{dx}{dt} = \frac{Ne^2}{m}\frac{1}{\gamma - i\omega}E(t) \approx \frac{Ne^2}{m\gamma}E(t) \equiv \sigma E(t), \tag{48}$$

where N is the number of free electrons and σ the conductivity.

At high frequency the current can no longer keep up with the driving field and we may no longer ignore the imaginary part of the conductivity. Let us therefore again consider the oscillating dipole moment created by each electron:

$$p(t) = -ex(t) = -\left(\frac{e^2}{m}\right)\frac{1}{\omega^2 + i\gamma\omega}E_o e^{-i\omega t}. \tag{49}$$

For a sample with N free electrons, the polarization is

$$P(t) = -\left(\frac{Ne^2}{m}\right)\frac{1}{\omega^2 + i\gamma\omega}E(t) \equiv \varepsilon_o\chi_e E(t). \tag{50}$$

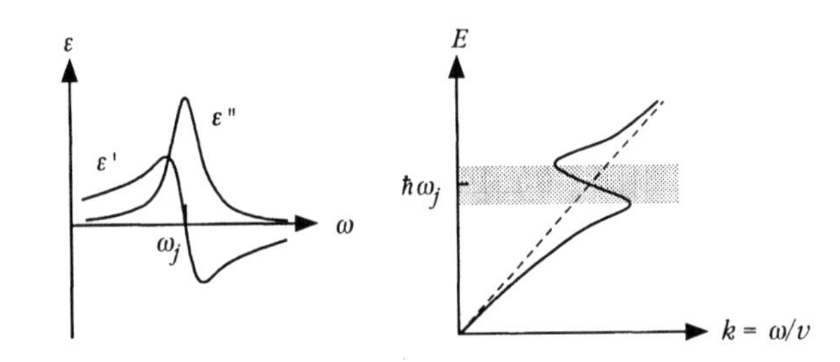

Figure 15 Frequency dependence of the dielectric function near a resonance (left) and the resulting relation between energy and wavevector (right). The shaded region indicates the range of values for which electromagnetic waves are strongly attenuated and do not propagate.

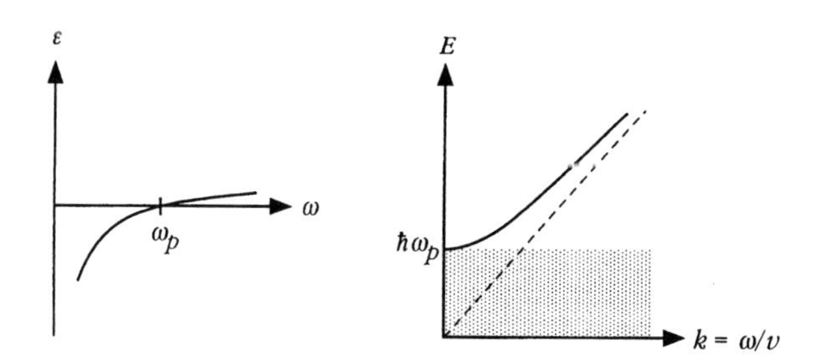

Figure 16 Dielectric function and $E(k)$ behavior for a plasma of free electrons with zero damping. The shaded region corresponds to a forbidden band of frequencies. Electromagnetic waves within this region are strongly attenuated.

From this expression we obtain the free-electron contribution to the dielectric constant:

$$\varepsilon(\omega) = 1 - \left(\frac{Ne^2}{m\varepsilon_o}\right)\frac{1}{\omega^2 + i\gamma\omega} = \varepsilon'(\omega) + i\varepsilon''(\omega). \tag{51}$$

If the damping is negligible, $\gamma << \omega$, the imaginary part of the free-electron contribution vanishes, and the real part becomes

$$\varepsilon'(\omega) = 1 - \frac{Ne^2}{m\varepsilon_o}\frac{1}{\omega^2} \equiv 1 - \frac{\omega_p^2}{\omega^2}, \tag{52}$$

where ω_p is called the plasma frequency. For frequencies below the plasma frequency, the dielectric constant is negative and so the index of refraction is purely imaginary resulting in strong attenuation of the electromagnetic wave (see Figure 16). In the $E(k)$-plot this attenuation gives rise to a range of 'forbidden frequencies', or frequency gap below the plasma frequency. In this regime, however, the reflectivity is nearly one and incident electromagnetic waves do not penetrate into the plasma. Above the frequency gap the electromagnetic wave propagates through the medium and at high frequency the dielectric function approaches unity. The free electrons thus act like a high-pass filter: below the plasma frequency reflection occurs; above the plasma frequency the free electrons are transparent. For intrinsic semiconductors, the plasma frequency lies in the microwave or infrared part of the electromagnetic spectrum. Metals reflect visible light because their plasma frequency is higher than visible optical frequencies.

The effect of small, but nonzero damping is illustrated in Figure 17. The damping results in a nonzero imaginary part at low frequency and a reduction of the frequency gap in the $E(k)$-plot. At high frequency the real part still approaches unity and the imaginary part vanishes.

II. C. Nonlinear optical interactions

Next we consider the nonlinear optical properties of materials. In the presence of an electric field $E(t)$, atoms in a solid become polarized giving rise to a polarization of the solid. For small electric fields, the induced polarization is linear in the applied field:

$$P(t) = \chi^{(1)}E(t), \tag{53}$$

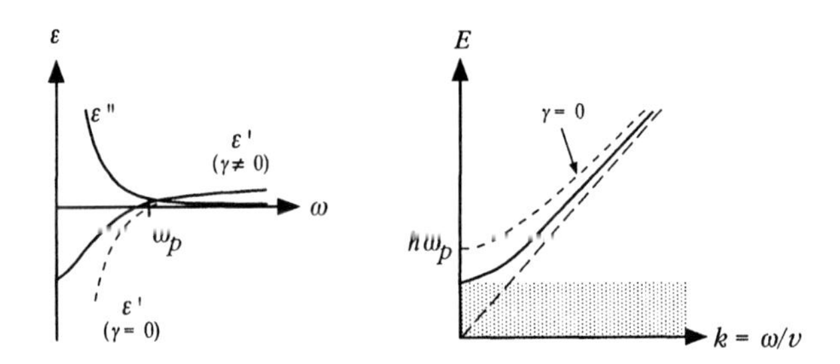

Figure 17 Dielectric function and $E(k)$ behavior for a plasma of free electrons with small nonzero damping. Damping reduces the forbidden band of frequencies.

with $\chi^{(1)}$ the linear susceptibility. For large applied fields, however, the induced polarization becomes nonlinear in the applied field[6]:

$$\begin{aligned} P(t) &= \chi^{(1)}E(t) + \chi^{(2)}E^2(t) + \chi^{(3)}E^3(t) + \ldots \\ &= P^{(1)}(t) + P^{(2)}(t) + P^{(3)}(t) + \ldots \end{aligned} \tag{54}$$

The first term represents the linear polarization, $P^{(n)}$ the n-th order polarization, and $\chi^{(n)}$ the n-th order nonlinear optical susceptibility. In general, the n-th order nonlinear susceptibility is not a scalar, but a tensor of rank $(n + 1)$. For typical materials the electric field has to be of the order of atomic field strengths E_{at} before the second-order term becomes comparable to the linear term:

$$\chi^{(n)} \approx \frac{\chi^{(1)}}{E_{at}^{n-1}}. \tag{55}$$

The nonlinear polarization can drive a new field. According to the wave equation that follows from the Maxwell equations, we have[6]

$$\nabla^2 E - \frac{n^2}{c^2}\frac{\partial^2 E}{\partial t^2} = \frac{4\pi}{c^2}\frac{\partial^2 P^{NL}}{\partial t^2}. \tag{56}$$

The second-order polarization, for instance, causes a driving term proportional to the square of the applied electric field, resulting in a new field at twice the applied frequency.

Consider an electron in the interstitial region between four atoms, as shown in Figure 18*a*. Physically, the doubling of the frequency comes about because the charges move along a curved potential plane. For an electric field oscillating in the plane of the drawing, the electron moves along the dotted arc; in addition to being displaced in the horizontal direction, the electron also undergoes a small vertical displacement. This vertical displacement gives rise to a small second-order dipole moment perpendicular to the horizontal first-order dipole moment. As Figure 18*b* shows, the period of oscillation of the second-order dipole moment is half that of the first-order dipole moment.

For systems with inversion symmetry, however, the second-order susceptibility vanishes. This can readily be seen by writing the second term of Equation 54 in tensorial form:

$$\mathbf{P}^{(2)} = \chi^{(2)} : \mathbf{EE}. \tag{57}$$

Applying inversion, we get

$$-\mathbf{P}^{(2)} = \chi^{(2)} : (-\mathbf{E})(-\mathbf{E}), \tag{58}$$

and from Equations 57 and 58 we see that $\chi^{(2)} = -\chi^{(2)}$, which can only be satisfied if $\chi^{(2)} = 0$.

For systems that do not have inversion symmetry, $\chi^{(2)}$ is not zero and an intense field causes a second-order polarization. Let the applied field be of the form

$$E(t) = \frac{1}{2} E e^{i\omega t} + \text{c.c.} \tag{59}$$

The second-order polarization is then

$$P^{(2)} = \chi^{(2)} E^2(t) = \frac{1}{2} \chi^{(2)} E E^* + \frac{1}{4}\left\{\chi^{(2)} E^2 e^{-i2\omega t} + \text{c.c.}\right\}. \tag{60}$$

The second term on the right-hand side oscillates at frequency 2ω and can drive a new electromagnetic wave at double the incident frequency. This process is called second-harmonic generation.

If two oscillating fields of different frequency are present,

$$E(t) = \frac{1}{2} E_1 e^{i\omega_1 t} + \frac{1}{2} E_2 e^{i\omega_2 t} + \text{c.c.}, \tag{61}$$

the second-order polarization contains terms at frequencies $2\omega_1$, $2\omega_2$, at the sum-frequency $\omega_1+\omega_2$, at the difference frequency $\omega_1-\omega_2$, and at zero frequency. Because of dispersion, the output beam at these new frequencies and the input beams at ω_1 and ω_2 travel at different velocities. To maximize output at any of the new frequencies it is therefore necessary to geometrically match the phases of the input and output beams.[6]

Let us next briefly turn to the third-order polarization which 'mixes' four electric fields — three input fields generate one new output field. When three different input frequencies are present, 13 new frequencies can be generated. An example of a third-order effect is coherent anti-Stokes Raman spectroscopy (CARS). Two of the input frequencies are chosen such that their difference matches a resonant frequency in the system:

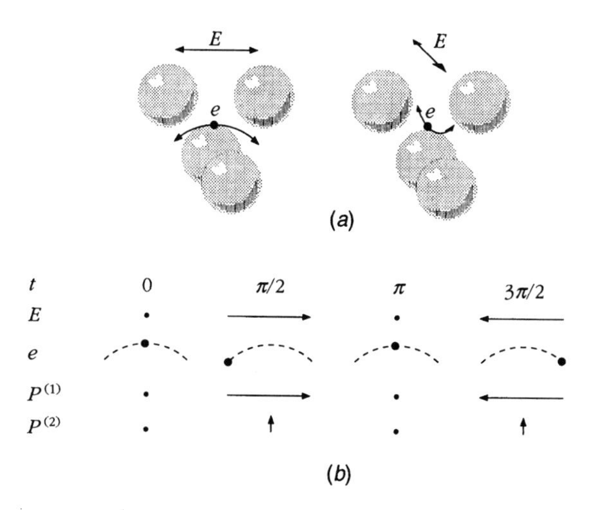

Figure 18 (*a*) An electron in the interstitial region between four atoms oscillates along a curved equipotential plane. (*b*) This oscillation causes a small second-order dipole in a direction perpendicular to the first-order dipole. The period of oscillation of the second-order dipole is one-half that of the first-order one.

$\omega_1 - \omega_2 = \omega_{res}$. The beating between the two input beams then coherently populates the upper level of the resonance. A third beam at frequency ω_3 then beats with the resonant oscillation in the system, generating a fourth beam at the anti-Stokes Raman frequency $\omega_a = \omega_1 - \omega_2 + \omega_3$. Note that this process is parametric, *i.e.*, the initial and final state is the same. The intensity of the coherent anti-Stokes beam is proportional to the population difference between the lower and upper levels of the resonance.[6] Hence, CARS can be used to measure population distributions.

Applications of nonlinear spectroscopy are summarized schematically in Figure 19.

III. ELECTRONIC AND VIBRATIONAL STATES OF SOLIDS

The atoms in a metal or semiconductor form an approximately regular three-dimensional pattern. Common arrangements of the atoms include face centered cubic (FCC) and body centered cubic (BCC), depicted in Figure 20, though many others are also common.[7] The energies of the states occupied by valence electrons depend on the arrangements of the atoms because valence electrons from neighboring atoms interact. The states of the valence electrons determine many properties of solids.

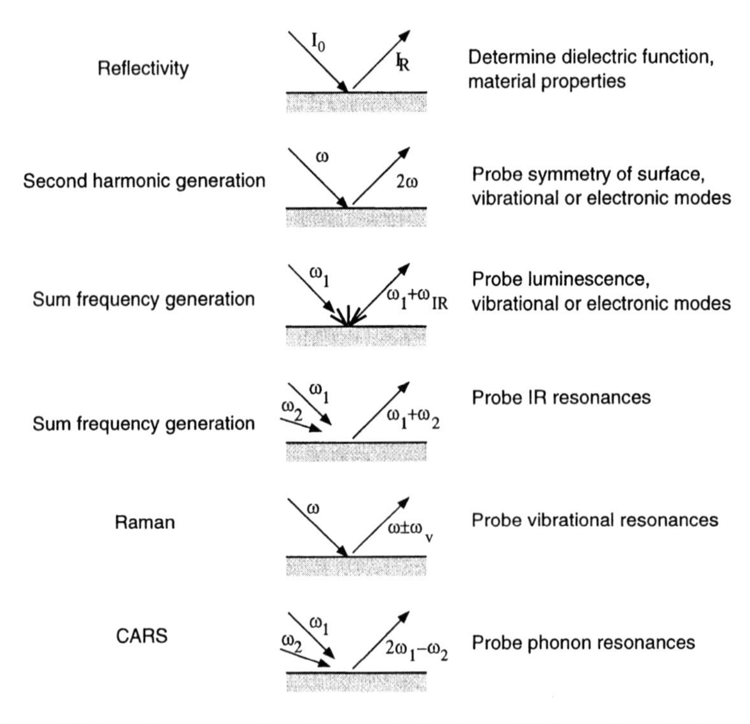

Figure 19 Schematic representation of some optical probes and their most important applications. Many phenomena can be studied with several different optical techniques. For example, vibrations can be studied with sum frequency generation, Raman scattering, and coherent anti-stokes Raman spectroscopy (CARS).

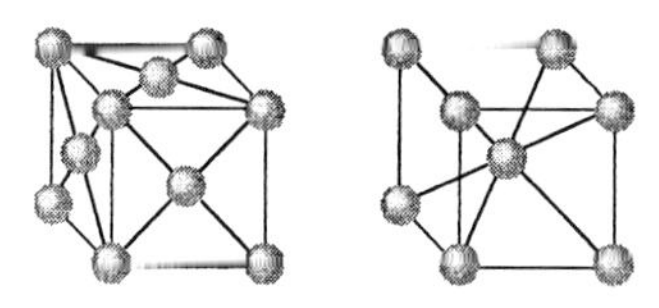

Figure 20 The face centered cubic (FCC, *left*) and body centered cubic (BCC, *right*) crystal structures. Not all the atoms are visible. The actual crystal is comprised of many of these cubes packed next to each other so that the atoms in the corners are shared by eight different cubes.

The wavefunction of an electron in a crystal is determined by the Schrödinger Equation 8 with $U(\vec{r})$ representing the potential of the charges in all the atoms. The equation cannot be solved analytically unless the real $U(\vec{r})$ is replaced with a simple function. Here we consider simple models to obtain a qualitative understanding of electronic states of solids.

III. A. Free electron states

We begin by considering free electrons for which $U(\vec{r}) = 0$. In the free electron model the Schrödinger equation yields a parabolic relationship between kinetic energy and momentum:

$$E = \frac{\langle p \rangle^2}{2m}. \qquad (62)$$

This equation is called a dispersion relation; it relates the energy of the electron to its momentum. The momentum of the free electron can take on any positive value, and hence the energy of the electron can be anywhere from zero to infinity. Equation 62 is represented in Figure 21 for a one dimensional crystal.

When $U(\vec{r})$ is not zero, a theorem known as Bloch's theorem constrains the electronic wavefunction. The theorem applies to electronic states in a periodic potential $U(\vec{r})$ where

$$U(\vec{r} + \vec{R}) = U(\vec{r}) \qquad (63)$$

for all $\vec{r}$. The vectors $\vec{R}$ which satisfy (63) are called lattice vectors. Bloch's theorem states that when (63) is satisfied there exists a vector $\vec{k}$ such that

$$\Psi(\vec{r} + \vec{R}) = e^{i\vec{k}\cdot\vec{R}}\Psi(\vec{r}). \qquad (64)$$

This equation says that between points in the crystal separated by a lattice vector $\vec{R}$, only the (complex) phase of the wavefunction may change. Wavefunctions in a periodic potential satisfy (64) and are called Bloch wavefunctions.[7, 8]

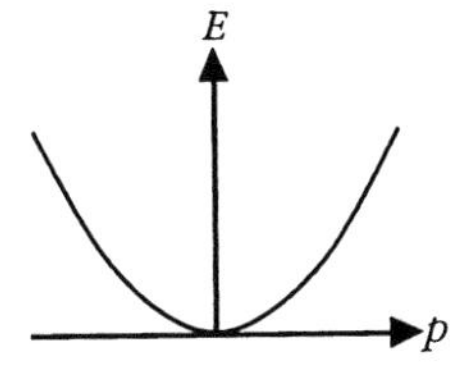

Figure 21 The parabolic dependence of energy on momentum for a free electron.

The quantity $\hbar\vec{k}$ is known as the crystal momentum. There are special values of the crystal momentum called reciprocal lattice vectors, $\vec{K}$, that satisfy

$$\vec{R}\cdot\vec{K} = 2\pi n, \tag{65}$$

where n is an integer. We may write a general $\vec{k}'$ in terms of $\vec{K}$:

$$\vec{k}' = \vec{k} + \vec{K}. \tag{66}$$

When this expression is substituted into (64), the definition (65) ensures that the factor containing $\vec{K}$ is 1:

$$\begin{aligned}\Psi(\vec{r}+\vec{R}) &= e^{i\vec{k}'\cdot\vec{R}}\Psi(\vec{r}) \\ &= e^{i(\vec{k}+\vec{K})\cdot\vec{R}}\Psi(\vec{r}) \\ &= e^{ik\cdot\vec{R}}\Psi(\vec{r})\end{aligned} \tag{67}$$

Equation 67 is the same as Equation 64. Therefore, two crystal momenta, $\vec{k}'$ and $\vec{k}$, which differ by a reciprocal lattice vector, $\vec{K}$, place the same constraints on the Bloch wavefunction. We therefore restrict attention to a small set of $\vec{k}$ values called the first Brillouin zone, comprised of those $\vec{k}$ whose magnitude cannot be made smaller by addition of a reciprocal lattice vector.

Bands are typically drawn only within the first Brillouin zone. Figure 22 shows an example in one dimension. The wave vectors $\vec{k}'$ and $\vec{k}$ differ by a reciprocal lattice vector of length $2\pi/a$. The horizontal arrows represent translations of the bands back into the first Brillouin zone. This representation is called the reduced zone scheme.

In a real crystal, the first Brillouin zone has a three-dimensional shape that reflects the symmetry of the real three-dimensional crystal lattice. Figure 23 depicts the first Brillouin zone of a FCC crystal. Conventional labels for points of high symmetry on the surface are indicated. There are many points on the surface that are equivalent to the labeled points because of the symmetry of the first Brillouin zone (and the crystal). The center of the polygon, $\vec{k} = 0$, is known as the *Γ-point.*

In a three-dimensional crystal, the $\vec{k}$ are three-dimensional and a graph of the band structure can no longer represent the states at all $\vec{k}$. Instead, the bands are shown along particular directions in $\vec{k}$-space. Figure 24 shows the band structure for free electrons in a FCC lattice in the directions of highest symmetry. The horizontal axis uses labels defined in Figure 23. Some locations in k-space (such as the Γ-point) are represented more than once.

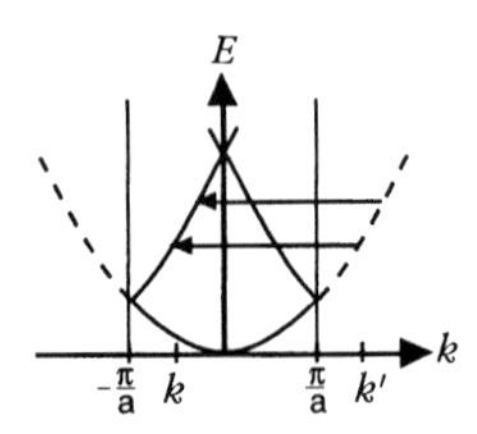

Figure 22 Wavevectors which differ by a reciprocal lattice vector (such as k and k') are physically equivalent. This motivates translating (represented by the horizontal arrows) the portions of the band that lie beyond $2\pi/a$ to produce a reduced zone representation.

Figure 23 The first Brillouin zone of a FCC crystal. This is a figure drawn in three-dimensional k-space. The origin, $\vec{k} = 0$, is located at the center of the polygon and is called the Γ-point. Other points of high symmetry on the outer surface of the figure are labeled with letters as shown.

For example, the band structure is represented along three different paths from the Γ-point to the edge of the Brillouin zone: Γ↔X, Γ↔L, and Γ↔K.

At the Γ-point, for energies near $E = 0$, the bands in Figure 24 have the quadratic dependence of Equation 62. When the bands reach the first Brillouin zone boundary (at X, L, or K for example), they are translated back into the first Brillouin zone, producing an apparent reflection of the parabolic shapes about the X, L, and K points. This reflection is analogous to that sketched in Figure 22 for a one-dimensional crystal.

The shape of the parabola depends on the particular path chosen in k-space. According to the dispersion relation (62), the energy of free-electron states depends on the square of the magnitude of $\vec{k}$. This fact explains the kink in the graph at L between Γ and W: the distance from the Γ-point increases more quickly between Γ and L than between L and W, as apparent in Figure 23. Throughout Figure 24, the bands reflect the dispersion relation for free electrons and the geometry of the first Brillouin zone.

The bands are not necessarily occupied with electrons. They cannot all be filled because the number of electrons in the crystal is finite, whereas there are infinite states available in the band structure. The probability that a state of energy E is occupied is given by the Fermi-Dirac function,[7]

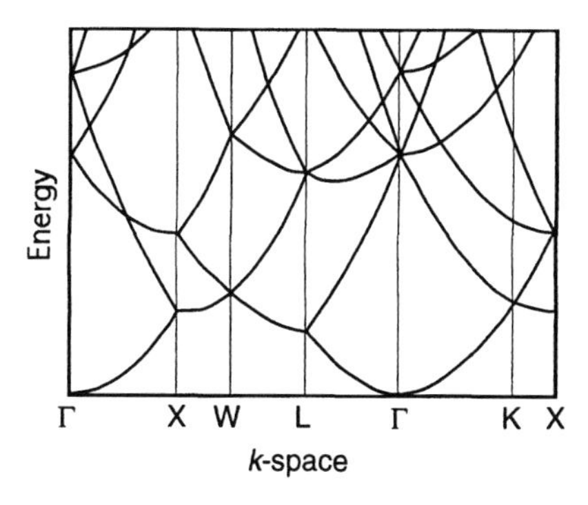

Figure 24 Free electron states for a FCC crystal. The states are plotted against representative directions in k-space. High symmetry positions in k-space are labeled with letters defined in Figure 23.

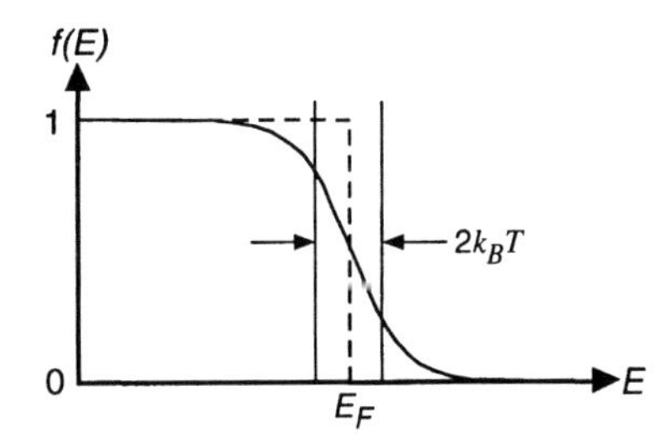

Figure 25 The Fermi-Dirac function at $T = 0$ (dashed line) and $T = 0.1E_F$ (solid line). The transition between $f(E) = 1$ and $f(E) = 0$ occurs over an energy range of approximately $2k_BT$.

$$f(E) = \frac{1}{e^{(E-E_F)/k_BT} + 1}, \tag{68}$$

where E_F is the Fermi energy, k_B is the Boltzmann constant, and T is the electron temperature. At $T = 0$, the Fermi-Dirac function is a step function: states below the Fermi energy are occupied and states above the Fermi energy are empty. At real temperatures the transition is smoother, as depicted in Figure 25. A Fermi-Dirac distribution is valid only when the electrons are in thermal equilibrium and temperature is well-defined.[8] Electrons out of thermal equilibrium are considered in the section on electron-electron scattering below.

In this section we have applied Bloch's theorem to determine how the symmetry of the crystal governs the electronic band structure. We have also assumed that the electrons are free, $U(\bar{r}) = 0$. These assumptions are, strictly speaking inconsistent. We see below, however, that when $U(\bar{r})$ is small but nonzero, the band structure resembles the free electron band structure. An excellent approximation of the band structure is obtained by applying only our knowledge of the symmetry of the crystal potential, (63) ignoring any interaction of electrons with the lattice.

III. B. Beyond free electron states

In a real crystal the electrons interact with the periodic potential of the lattice. The simplest band structures occur in metals where $U(\bar{r})$ is small. Recall that monovalent metals have a single electron outside a filled shell (or a filled sub-shell, such as Cu) and that this electron is shielded from the nucleus by its interactions with the other electrons. The shielding leads to relatively weak interactions with the lattice and small $U(\bar{r})$compared to other crystals. We expect the band structure of monovalent metals such as K and Cu to be approximately the band structure of free electrons.

Figure 26 is a graph of the solution to the Schrödinger equation obtained from perturbation theory[1] with the weak periodic potential $U(\bar{r})$. The influence of the interactions with the lattice is to segment and distort the free-particle solution (Figure 22). The distortion occurs near the Brillouin zone boundary; in other regions the bands remain parabolic. Alkali metals have bands which are very similar to Figure 26.

In other metals, there are often bands with free-electron character. For example, the bands of aluminum (configuration [Ne] $3s^23p^1$) depicted in Figure 27 are very similar to free electron bands. Copper, with configuration [Ar] $3d^{10}4s^1$, has a more complicated band

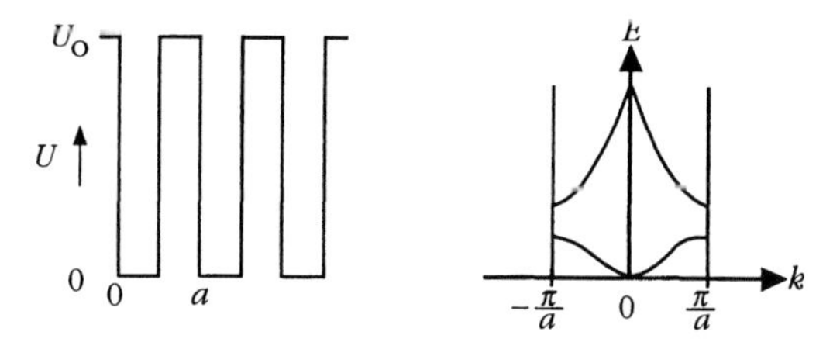

Figure 26 A periodic potential in one dimension (*left*) and the corresponding bands (*right*).

structure. The band structure of copper in Figure 28 has both horizontal and parabolic bands. The roughly horizontal bands arise from electrons in d-orbitals. The bands are horizontal because d-orbital electrons from one atom interact strongly with neighboring atoms, as expected from the graph of the spatial distribution of d-orbitals in Figures 3 and 5. The roughly parabolic bands in Figure 28 arise from the electrons in s-orbitals. The s-orbital electrons interact less with the lattice than the d-orbitals because of the small radial extent of s-orbitals. The parabolic bands in Figure 28 are similar to the free electron bands in Figure 24.

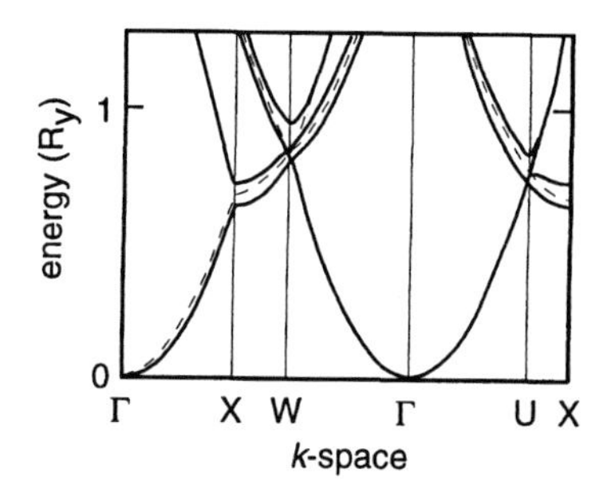

Figure 27 The band structure of aluminum has a lot of the parabolic character of free-electron bands. The dashed lines indicate the free electron bands. After [9]

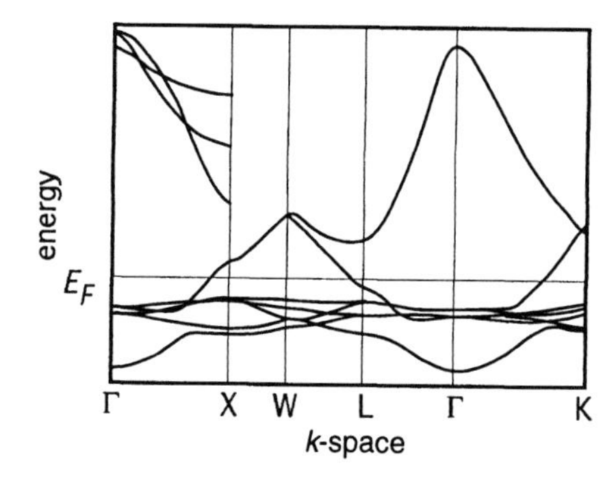

Figure 28 The calculated band structure of copper in a portion of k-space. Both parabolic free-electron-like bands and roughly horizontal d-bands are present. After [10].

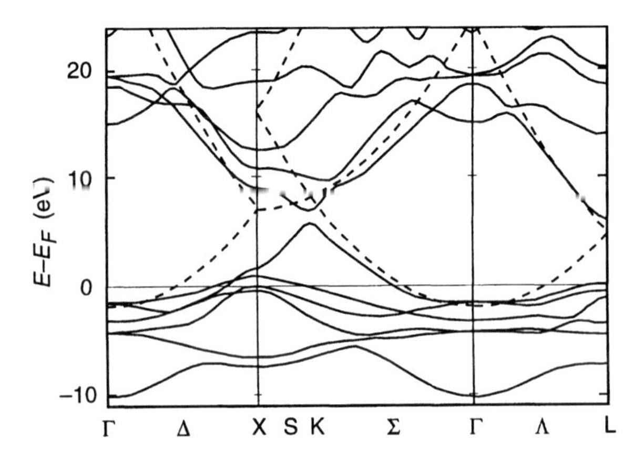

Figure 29 The calculated (solid lines) band structure of platinum. The dashed lines indicate the free electron bands for an FCC crystal. After [11]

The band structures of d-band metals share a number of features. Compare, for example, the band structure of copper in Figure 28 with the band structure of platinum shown in Figure 29. At low energy near the Γ-points, the quasi-free electron bands are similar because they are both FCC crystals. The Fermi levels of platinum and copper are at different locations with respect to the bands. In platinum (with configuration [Xe] $4f^{14}5d^{10}6s^{0}$) the Fermi level is in the horizontal d-bands, while in copper the Fermi level lies in the parabolic s-bands.

III. C. The dielectric function of metals

In section II.B, we studied the relationship between the energy (or frequency) of an electromagnetic wave propagating in a material, and the dielectric function, $\varepsilon(\omega)$ of the material. We derived the dielectric function for a material in which the electrons act as simple harmonic oscillators. Here we use that model to explain some features of the measured dielectric function of alkali metals, Figure 30. Additional features can be explained using the band structure of the alkali metals, Figure 31. More detailed structure of $\varepsilon(\omega)$ can be predicted from the band structure than from the simple harmonic oscillator model because the band structure is a more complete description of the electronic states in the crystal lattice.

The interaction of light with the electrons in a material is represented by transition of electrons from occupied to unoccupied states in the band structure. According to the Fermi-Dirac function, Equation 68, unless the temperature is very high, occupied states are below the Fermi level, and unoccupied states are mostly above the Fermi level. If the initial and final states of the electron are in the same band, the transition is intraband. If the transition is from one band to another, the transition is interband. In both cases, the transition is from below the Fermi level to above the Fermi level.

In both intraband and interband transitions, the crystal momentum is conserved. When the crystal momentum of the electron changes, as it must in an intraband transition, one or more other bodies must have an opposite change in crystal momentum. The photon does not contribute to conservation of crystal momentum because photons have negligible momentum. Phonons, on the other hand, do have crystal momentum, and intraband transitions are usually

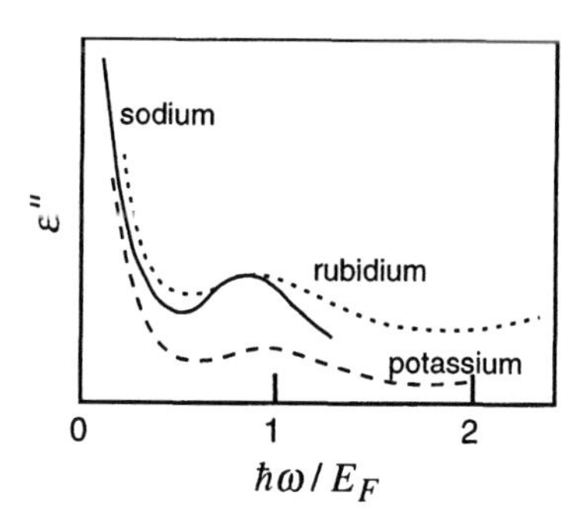

Figure 30 The experimentally determined imaginary parts of the dielectric function for the alkali metals. They are similar because the band structures are similar. After [8]

accompanied by excitation of a phonon. The total change in crystal momentum of the phonon and the electron is zero. Phonons are discussed in section III.E.

Intraband transitions are responsible for the decrease in $\varepsilon(\omega)''$ with increasing frequency in Figure 30. According to Figure 31, low energy excitation of electrons near the Fermi level occur in an approximately parabolic band. From Figure 21, we know that electrons in parabolic bands behave as free electrons, and from Figure 17 from section II.B, we know that in a material with free electrons the imaginary part of the dielectric function drops with increasing frequency. The dielectric function of alkali metals is similar to the dielectric function of free electrons because the band structure is nearly parabolic near the Fermi level.

The dielectric functions in Figure 30 do not monotonically decrease. The increase in $\varepsilon(\omega)''$ is attributed to the onset of interband transitions at energies sufficient to excited electrons across the gap between bands. The lowest energy interband transition (at constant crystal momentum) is shown in Figure 31. There are no transitions right at the Brillouin zone boundary because there are no electrons at the N-point: the Fermi level is below the energy at the N-point. The length of the arrow is about 0.65 E_F. The onset of interband transitions accounts for the rise in $\varepsilon(\omega)''$ in Figure 30 near 0.65 E_F.

Figure 32 shows the measured real and the imaginary parts of the dielectric function of platinum.[12] Overall, $\varepsilon(\omega)''$ falls with increasing energy—the expected free electron contribution to $\varepsilon(\omega)''$. The structure at about 0.8 eV, is attributed to the onset of interband transitions near the X point in the platinum band structure, Figure 29.[13]

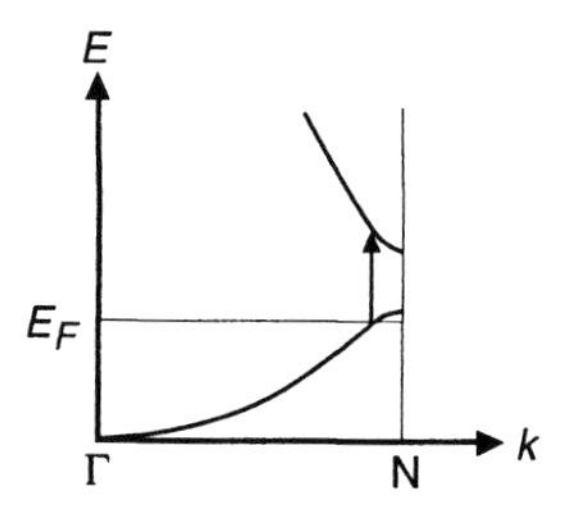

Figure 31 The band structure of an alkali metal, with an interband transition indicated. In an intraband transition (not shown) the initial and final states of the electron lie in the same band. After [8]

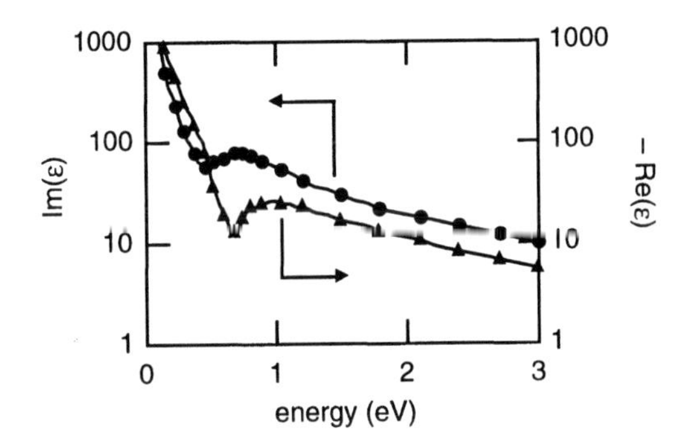

Figure 32 The dielectric function of platinum. The features at about 3/4 eV can be attributed to the band structure. Data from [12].

III. D. Electron-electron scattering

The electrons in a material do not necessarily have a distribution of energies described by the Fermi-Dirac equation. During absorption of light, for example, electrons acquire energies far in excess of k_BT. This energy is partitioned among all the electrons by collisions between electrons, until the distribution of electron energies is a Fermi-Dirac distribution. Fermi liquid theory predicts the rate at which a single electron excited above the Fermi level collides with another electron. One of the aims of the model is to predict the rate at which the excited electron scatters with other electrons.

Consider the collision of an excited electron of crystal momentum $\bar{k}_1$ with an electron of crystal momentum $\bar{k}_2$, depicted in Figure 33. The scattering rate depends on the probability of finding an electron of momentum $\bar{k}_2$, and also on the probability that there are empty states of momenta $\bar{k}_1'$ and $\bar{k}_2'$. In Fermi liquid theory it is assumed that, except for the single excited electron in question, the distribution of electron energies is described by the Fermi-Dirac function, Equation 68. With this assumption we can write the scattering rate as:

$$\frac{1}{\tau} \propto f(\bar{k}_2)\left[1 - f(\bar{k}_1')\right]\left[1 - f(\bar{k}_2')\right]. \tag{69}$$

This is the scattering rate for a particular $\bar{k}_1'$ and $\bar{k}_2'$. The scattering rate into all possible final states $\bar{k}_1'$ and $\bar{k}_2'$ depends on the total number of available states for $\bar{k}_1'$ and $\bar{k}_2'$.

We estimate the number of available states when the temperature is zero Kelvin. The calculation begins with some observations about the relative energies of the electrons before and after the collision. Figure 25 implies that the final states $\bar{k}_1'$ and $\bar{k}_2'$ must satisfy

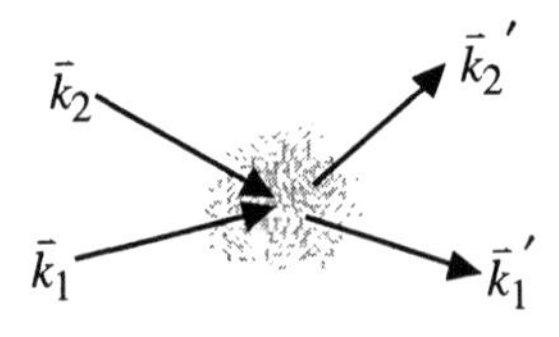

Figure 33 A binary collision between two electrons.

$E\left(\bar{k}_1'\right) > E_F$ and $E\left(\bar{k}_2'\right) > E_F$ because all the available (empty) states are above the Fermi level. The figure also implies that $E\left(\bar{k}_2\right) < E_F$ because the electron $\bar{k}_2$ must begin in an occupied state. Since the collision increases the energy of the electron $\bar{k}_2$, the energy of the electron k_1 must decrease to conserve energy: $E\left(\bar{k}_1'\right) < E\left(\bar{k}_1\right)$. By a similar argument, the final energy of the electron $\bar{k}_2$ must be less than the initial energy of the excited electron: $E\left(\bar{k}_2'\right) < E\left(\bar{k}_1\right)$. Both electrons end up with energies between E_F and $E\left(\bar{k}_1\right)$.

The number of available states between E_F and $E\left(\bar{k}_1\right)$ is proportional to $\left|\bar{k}_1\right| - \left|\bar{k}_F\right|$. Substituting this expression into (69) gives a scattering rate:

$$\frac{1}{\tau} \propto \left(\left|\bar{k}_1\right| - \left|\bar{k}_F\right|\right)^2. \tag{70}$$

This expression is valid when $T = 0$ and the initially excited electron $\bar{k}_1$ is near the Fermi level. Equation 70 shows that the lifetime of an electron becomes very large as the electron gets close to the Fermi level. Calculations of the scattering rate give electron-electron scattering times of 10 fs for an electron which lies 2 eV above the Fermi surface, and 1 ps for an electron which lies 0.2 eV above the Fermi surface. Equation 70 is often stated in terms of energy:

$$\frac{1}{\tau} \propto \left[E\left(\bar{k}_1\right) - E_F\right]^2. \tag{71}$$

Equations 70 and 71 are equivalent for energies close to the Fermi level.

At small (nonzero) temperatures the total number of states available for the electrons to scatter to depends on temperature. To calculate the scattering rate, we note that Figure 25 shows that the Fermi-Dirac distribution changes from 1 to 0 over a range of energies of order $k_B T$. If the target electron is to scatter into an available (empty) state, it must have an initial energy within the range $k_B T$ of E_F. Similarly, the number of available states scales as $k_B T$. Once the initial and final energies of the target electron are specified $E\left(\bar{k}_1'\right)$ is known because of energy conservation. The scattering rate depends the square of the temperature:

$$\frac{1}{\tau} \propto \left(k_B T\right)^2. \tag{72}$$

In practice, the scattering of electrons with energy close to E_F is dominated by scattering with phonons (see below) or impurities, and the quadratic dependence of scattering rate on temperature is not observed.[8]

III. E. Phonons

In the previous two sections we assumed the ionic lattice to be fixed and immobile. In this section we consider collective motions of the ions. The ions can be displaced from their equilibrium positions and such disturbances can travel through the solid in the form of phonons which play an important role in the electronic and optical properties of solids because they can interact directly with electromagnetic waves.

Let us begin by considering a linear chain of identical atoms separated by a spacing a as illustrated in Figure 34. The top of the drawing shows the atoms in their equilibrium position; at the bottom the atoms are displaced from their equilibrium position. Let us assume that only nearest neighbors exert forces on each other and that the interionic force obeys Hooke's law. The forces exerted on ion n by its two nearest neighbors are thus

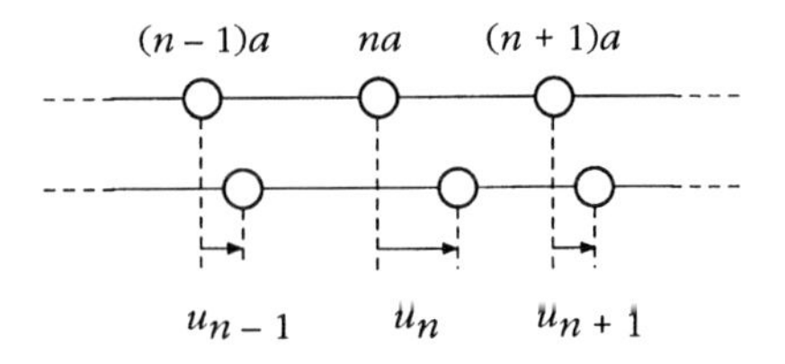

Figure 34 Vibrating linear chain of identical atoms spaced by a distance a.

$$
\begin{aligned}
F_{n-1,n} &= \gamma(u_{n-1} - u_n) \\
F_{n+1,n} &= \gamma(u_{n+1} - u_n)
\end{aligned}
\tag{73}
$$

where γ is the force constant. The equation of motion for the ion is then

$$m\frac{d^2u_n}{dt^2} = \gamma\left[u_{n-1} + u_{n+1} - 2u_n\right]. \tag{74}$$

We look for solutions in the form of a traveling harmonic displacement wave (called a normal mode)

$$u_n(t) = Ae^{i(qna-\omega t)}, \tag{75}$$

where A is the amplitude of the displacement wave, q the wavevector, and ω the angular frequency. Substituting this into Equation 74 we get

$$-m\omega^2 = \gamma[e^{-iqa} + e^{iqa} - 2] = -4\gamma\sin^2\left(\frac{qa}{2}\right), \tag{76}$$

so

$$\omega = \sqrt{\frac{4\gamma}{m}}\left|\sin\frac{qa}{2}\right|. \tag{77}$$

As Figure 35 shows, we only need to consider displacement waves of wavelength larger than $2a$ — due to the discreteness of the chain, all waves of shorter wavelengths are equivalent to certain waves of longer wavelengths. This means we can restrict our analysis to small wavevectors:

$$\lambda \geq 2a \quad \Rightarrow \quad q \leq \frac{\pi}{a}. \tag{78}$$

Figure 36 shows the dependence of the displacement frequency on wavevector q (77). For small wavevector, Equation 77 becomes linear in the wavevector

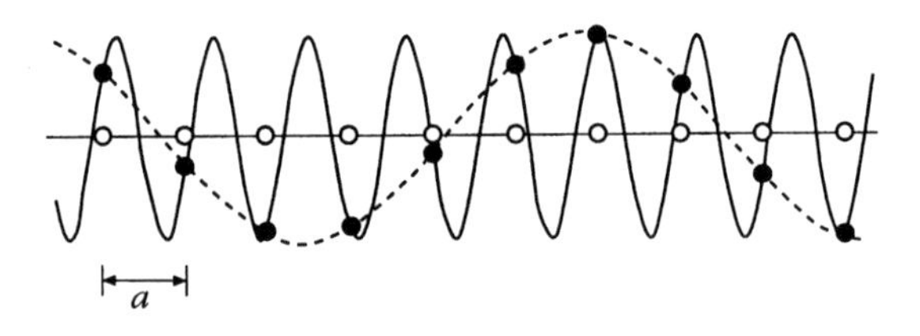

Figure 35 Oscillating chain of atoms showing instantaneous displacements. The solid curve conveys no information not given by the dashed one.

$$\omega = \sqrt{\frac{4\gamma}{m}}\frac{qa}{2} = \sqrt{\frac{\gamma a}{m/a}}q = v_s q\,, \tag{79}$$

with v_s the speed of sound. This is the relation one would obtain if the chain were continuous rather than discrete (when a approaches zero, π/a goes to infinity and the dispersion relation becomes linear throughout). The dispersion of waves near the edge of the Brillouin zone at π/a is therefore a direct consequence of the granularity of the chain.

In a two-atom linear chain the situation is more complicated because the atoms of different kind can either move in phase (such displacement waves are called acoustic phonons) or out of phase (optic phonons). Figure 37*a* illustrates the displacements that occur for transverse acoustic (TA) and optic (TO) phonons of small wavevector. While both displacements have the same large wavelength, the potential energy associated with the optic phonon is larger because the interatomic bonds are much more distorted. The dispersion relation now has two branches (see Figure 38*a*); for low wavevector the acoustic branch approaches zero, but because of the large distortion at low frequency, the corresponding energy for the optic branch is nonzero at zero wavevector.

Figure 37*b* shows the displacements for the acoustic and optic phonons of the shortest possible wavelength ($\lambda = 2a$). The corresponding energies (see Figure 38*a*) are slightly different. Figures 37*c* and 37*d* show how the cases illustrated in Figs. 37a and 37*b* relate to single-atom chain phonons: the optic branch vanishes as low wavevector optic phonons map onto large wavevector acoustic phonons. Note, in particular that the low-q TO phonon for the two-atom chain maps to a high-q TA phonon on the one-atom chain (*cf.* Figs. 37*a* and *c*). Similarly, the TO and TA phonon modes at the edge of the Brillouin zone for the two-atom chain, are identical on the one-atom chain (*cf.* Figs. 37*b* and *d*), but are now in the middle of a Brillouin zone that is twice as wide (Figure 38*b*).

III. F. Electron-phonon interaction

The Bloch wavefunctions (64) are solutions to the Schrödinger equation only when the

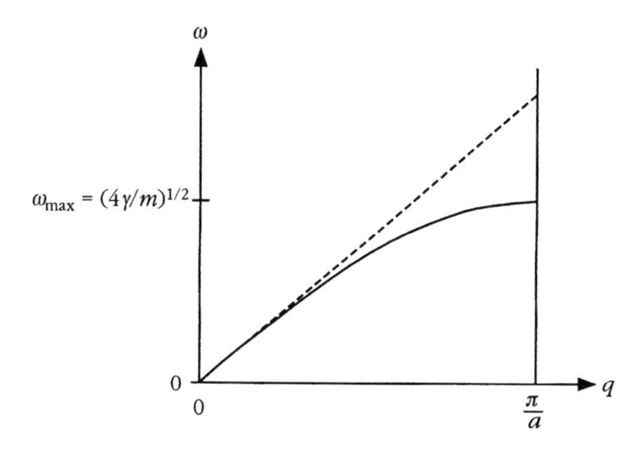

Figure 36 Dispersion of waves along a linear chain of atoms. The dashed line shows the result one would obtain for a continuous medium. The slope of the dashed line corresponds to the speed of sound waves in the medium.

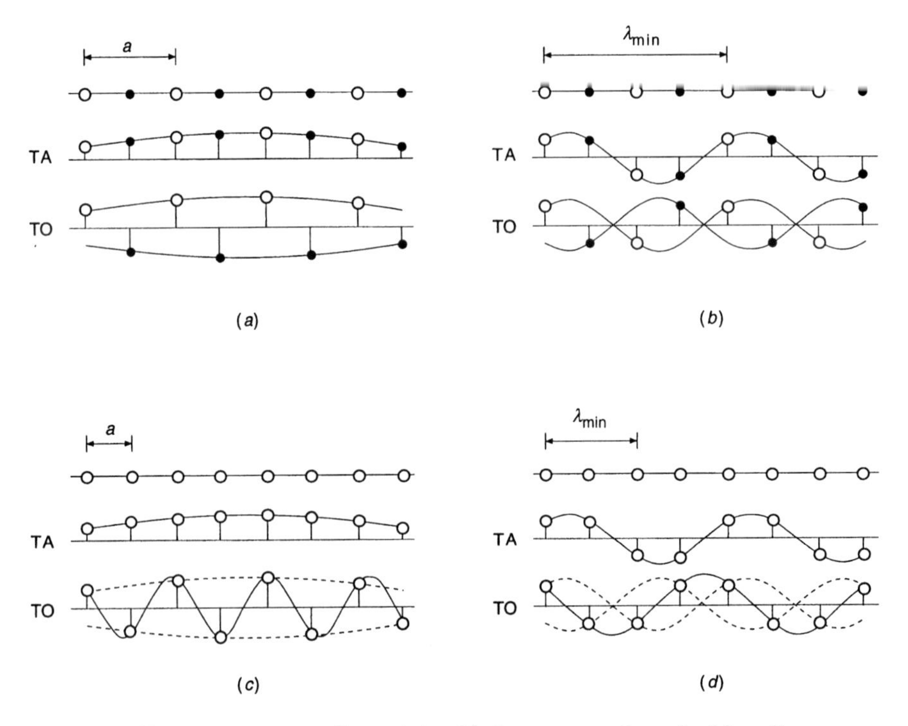

Figure 37 Waves on two-atom linear chains. Displacements are shown for (*a*) small and (*b*) large wavevector. The in- and out-of-phase waves correspond to acoustic and optic phonons, respectively. The bottom to graphs (*c* and *d*) show how the waves for a two-atom chain map onto waves of different wavevector for a one-atom chain.

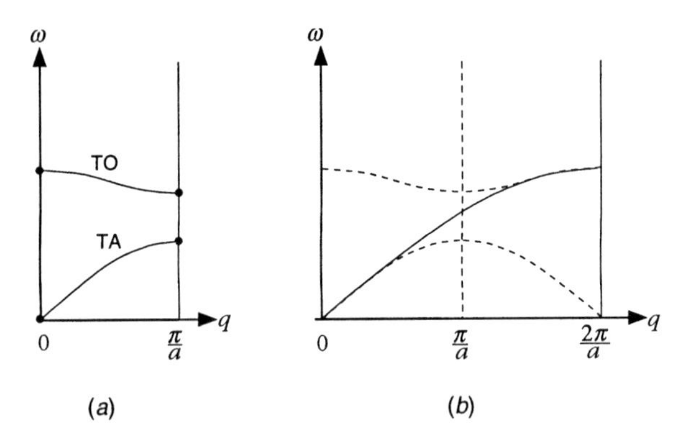

Figure 38 Dispersion relation for phonons (*a*) on a two-atom linear chain and (*b*) on a corresponding one-atom linear chain.

lattice is perfectly well ordered. In practice phonons cause a distortion of the lattice. The distortion allows electrons to make transitions between Bloch states. This process is described as a scattering of an electron with a phonon, and can either transfer energy to the phonon or to the electron.

We consider a phenomenological treatment of this scattering relevant when the electrons have been very highly excited by a laser pulse. For example, when a subpicosecond laser pulse with photon energy of about 2 eV strikes a metal surface, the energy is absorbed by the electrons. The electrons share this energy among themselves by electron-electron collisions (see above), reaching a Fermi-Dirac distribution. Simultaneously, the electrons scatter with phonons, effectively heating the metal. The spatial variation in the deposited laser energy creates a complication: the energy diffuses away from the surface towards the bulk.

These ideas are expressed in a model for the response of a metal to ultra-fast photo-excitation.[14, 15] This model states that the temperatures of the electrons, T_e, and phonons (the lattice), T_l, evolve according to:

$$\begin{aligned} C_e(T_e)\frac{\partial T_e}{\partial t} &= \kappa(T_e)\frac{\partial^2 T_e}{\partial x^2} - g(T_e - T_l) + A(x,t) \\ C_l(T_l)\frac{\partial T_l}{\partial t} &= g(T_e - T_l) \end{aligned} \tag{80}$$

where $A(x,t)$ is the energy deposited by the laser and C_e and C_l are the heat capacities of the electrons and the lattice. The constant g determines how quickly the electrons and phonons thermalize with each other. The first equation contains a term describing the diffusion of the electrons. The energy carried by phonon diffusion is small compared with that due to electron diffusion in a metal; phonon diffusion is neglected in this model. The constants are approximately known, and so the electron and lattice temperatures can be found by numerical solution of Equation 80.

Figure 39 shows the evolution of the electron and lattice temperatures at the surface of platinum following excitation by a 32 μJ/mm^2, 800-nm pulse. Initially the sample is in thermal equilibrium at 90 K. The laser pulse causes a large transient rise in the surface electron temperature followed by equilibration of the electron and lattice temperatures in a few picoseconds.

This model assumes that the electrons are always thermalized with each other—that they satisfy a Fermi-Dirac distribution at all times. In reality the laser pulse excites electrons far above the Fermi level and a finite time is required for the electrons to thermalize. For example, 800 nm photons have 1.6 eV energy, while at 90 K, $k_B T \approx 8$ meV. The relaxation of the photo-excited electrons to a Fermi-Dirac distribution has been the subject of numerous experiments.[16-19]

III. G. Photoemission spectroscopy

One of the most direct experimental methods for determining electronic states is photoemission spectroscopy, illustrated in Figure 40. Light stimulates an electron in a solid. If the electron is sufficiently excited, it escapes the material with energy

$$E_k = \hbar\omega - \Phi - E_i. \tag{81}$$

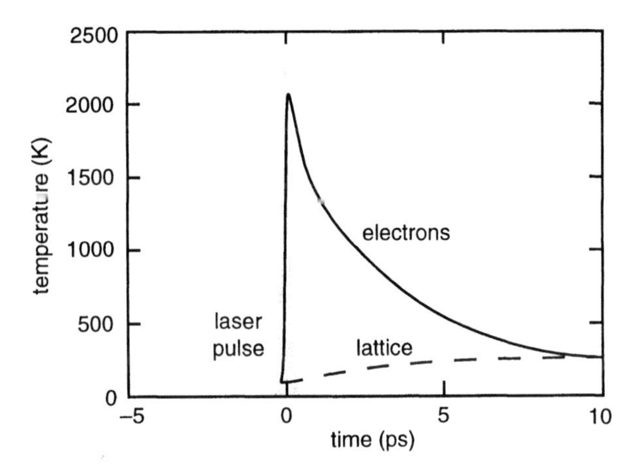

Figure 39 Calculated evolution of electron and lattice temperatures following excitation of platinum with a 100-fs, 32 μJ/mm^2, 800-nm laser pulse.

The energy of the photo-emitted electron depends on the initial photon energy, $\hbar\omega$, the initial state energy, E_i, and the energy required to extract the electron from the material, Φ, known as the work function. By measuring E_k, the energy of the initial state can be inferred. Generally electrons are emitted over a range of electron energies corresponding to the range of occupied states in the band structure, and the electrons comprise a photoelectron spectrum, as sketched in Figure 40.

The states that are probed in a photoemission experiment depend on the source of the initial excitation. Figure 41 depicts the some of the features which are observed with different excitation sources.

When the photon energy is low, the electrons may not receive enough energy from a single photon to overcome the work function of the material., but may escape if stimulated by two photons. This two-photon photoemission, or TPPE, is used to study the electronic states which lie between the Fermi level and the vacuum level. Because these states are above the Fermi level, the electrons rapidly scatter out of these states; TPPE using short pulses can therefore be used to study the dynamics of electron relaxation. Another way to study states between the Fermi level and the vacuum is to apply a strong DC electric field to the sample

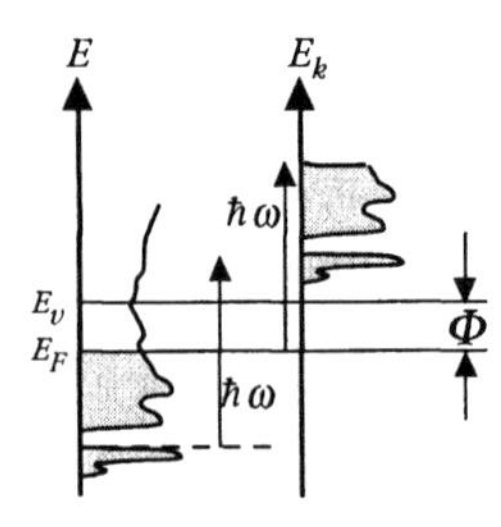

Figure 40 In photoemission spectroscopy the kinetic energy, E_k, of the emitted electron (right side of figure) depends on the initial state of the electron (left side of figure), the photon energy, and the work function. After [20].

surface. The potential barrier between an electron at the surface and a free electron in vacuum can be overcome if the applied field is sufficiently large.

At the other extreme, x-ray photons with energies in the 1000–eV range can eject core-level electrons. The resulting vacancy in the core level may be filled by an electron from a valence level. The energy released by this transition from the valence level to the core level may be imparted to another valence electron which is then ejected from the material with an energy depending on the energy levels of the states involved. This process is known as Auger recombination. The photoelectron spectrum following x-ray excitation has peaks which directly reflect the energies of the core levels, and peaks which arise from Auger recombination and reflect the energies of the valence and core levels. Spectroscopic techniques that rely on core-level ionization often use a monochromatic beam of electrons as the excitation source because electron-beam sources are conveniently produced in the laboratory from hot filaments.

The escape of an electron from the surface is not always as simple as depicted in Figure 40. The electrons may scatter and lose some of their energy. These collisions lead to a broad distribution of electron energies. When these electrons escape the surface, they are known as secondary emission. Secondary emission is observed as a broad feature at low E_k. The scattering of electrons accounts for the surface sensitivity of photoemission spectroscopy: only electrons emitted in the near surface region escape the material and are detected. The depth of the material that is probed with photoemission spectroscopy depends usually on the escape depth of the photoelectrons and not on the absorption length of the excitation source.

IV. A SURFACE SCIENCE PRIMER

This section is an introduction to the physics which governs chemical reactions at surfaces. Most of the examples are drawn from the chemistry of oxygen on platinum in preparation for section V of this chapter.

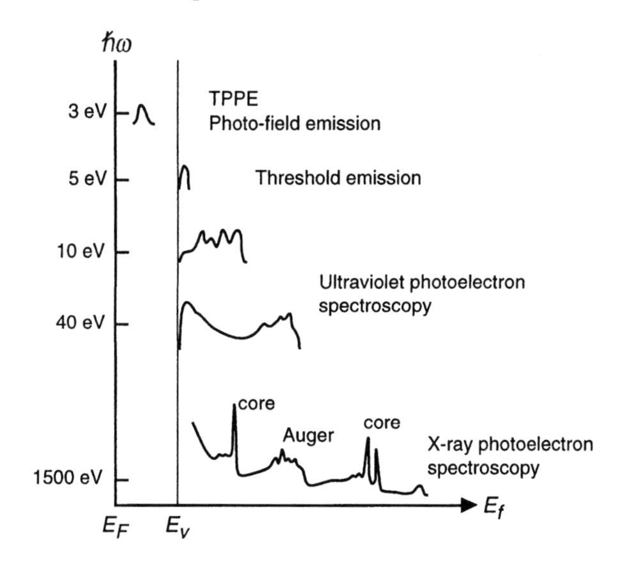

Figure 41 The photoelectron spectrum depends on the energy of the exciting photon. X-rays are able to induce emission of core electrons and Auger electrons. After [20]

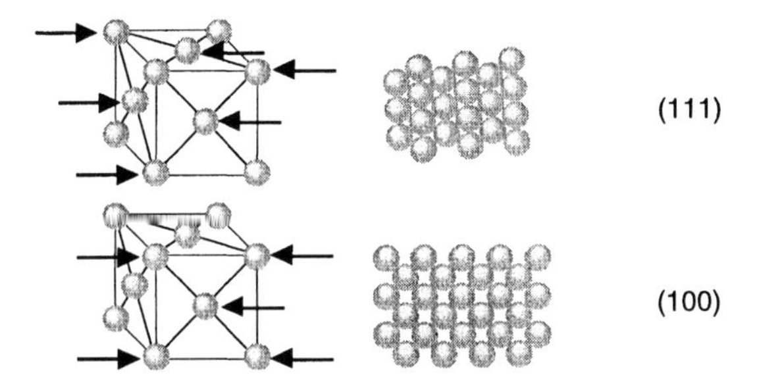

Figure 42 The (111) and (100) surfaces of an FCC crystal. The arrows indicate atoms in the crystal lattice which comprise the surface. In the diagrams of the crystals the spheres representing the atoms are not drawn to scale compared to the size of the cube.

Surface reactions are influenced by the chemical composition of the surface and the structure of the surface. To reduce the complexity, surface reactions are often studied on very clean, single crystal surfaces. A single crystal surface is cleaned in ultra-high vacuum (1×10^{-10} torr). Using the ideal gas law, one can calculate that at a pressure of 10^{-6} torr, approximately 1 s is required for every surface atom to be struck by a gas phase molecule.[21] Each such collision is an opportunity for a gas phase molecule to stick on the surface. Thus to maintain a surface clean for, say, 10^4 s a pressure near 10^{-10} torr is required.

The direction of the surface plane with respect to the lattice of a single crystal is specified by Miller indices.[7] Figure 42 shows examples of surfaces obtained from an FCC lattice. The (111) surface has a high density compared with the (100) surface. We concentrate our attention on the (111) surface of platinum, Pt(111).

IV. A. Sample preparation

Common procedures for cleaning a surface include annealing, and sputtering. The precise cleaning procedure is often determined by trial-and-error, using the diagnostic procedures described below to monitor the surface quality. As a starting point, there are compiled lists of the techniques that have been used to clean many crystals.[22]

When the sample is annealed in UHV (*i.e.* kept at elevated temperature for a certain time), some impurities simply desorb from the surface. For example, carbon monoxide desorbs from Pt(111) when the temperature exceeds about 300 K. Annealing may also provide the thermal energy required for surface atoms to rearrange themselves and correct small defects in the surface structure.

Sputtering is used to remove the first few layers of atoms from the surface. A noble gas, usually either neon or argon is admitted to the chamber. The gas is ionized and the ions are electrostatically accelerated to the surface. The ions dislodge material from the sample surface, including both the desired atoms and the impurities. This usually improves surface quality because impurities often cluster at the surface, particularly following annealing in vacuum or oxygen. Sputtering creates defects in the surface structure, so it is usually followed by annealing.

Pt(111) is usually cleaned by annealing in an oxygen atmosphere. During annealing, impurities such as silicon diffuse to the surface where they may bind with an oxygen atom, becoming trapped at the surface as an oxide. When these oxides are removed by sputtering, the density of impurities in the near-surface region is reduced. Annealing in oxygen also reduces carbon contamination because the carbon reacts to form carbon monoxide which desorbs from the surface.

IV. B. Adsorption of reactants

There are two classes of interaction between a molecule and a surface. Chemisorption refers to a molecule or atom attached on a surface by a chemical bond. Physisorption is a much weaker van der Waals interaction between a molecule or atom and a surface. The energy which binds a chemisorbed species to a surface is typically 0.4–10 eV while physisorption energies are typically 0.01–0.1 eV.

To understand physisorption, picture a metal surface interacting with a polarized molecule or atom as shown in Figure 43. Whenever a charged species is placed near a metal surface its electric field causes an image charge in the metal.[23] The image charge is positioned so that the electric field produced by a charge and its image charge is identical to the field which would be produced by the charge and the true distribution of surface charge on the metal. Figure 43 depicts a polarized adsorbate as two charges separated by a distance u. The corresponding image charges are shown.

The interaction of the adsorbate with the surface may be determined by finding the total electrostatic (Coulomb) interaction between the charges in the adsorbate and the image charges in the metal:

$$\begin{aligned} U(z) &= -\frac{e^2}{4\pi\varepsilon_0}\left[\frac{1}{2z}+\frac{1}{2(z-u)}-\frac{1}{(2z-u)}-\frac{1}{(2z-u)}\right] \\ &\approx -\frac{e^2}{4\pi\varepsilon_0}\frac{u^2}{z^3}. \end{aligned} \tag{82}$$

On the right side of this equation, the interaction potential $U(z)$ is expanded into a Taylor series for small u/z. The interaction potential, Equation 82, is zero if there is no polarization, $u = 0$. Otherwise the interaction varies as $1/z^3$.

When the adsorbate is very close to the metal, the image charge model shown in Figure 43 is not accurate because the interaction becomes repulsive. When the repulsive interaction is balanced against the attractive interaction of Equation 82, the adsorbate typically settles 0.3 – 1.0 nm from the surface in a shallow well. Physisorption is observed only at low temperatures. For example, O_2 will physisorb on Pt(111) at 45 K and form multiple physisorbed layers at 30 K.[24]

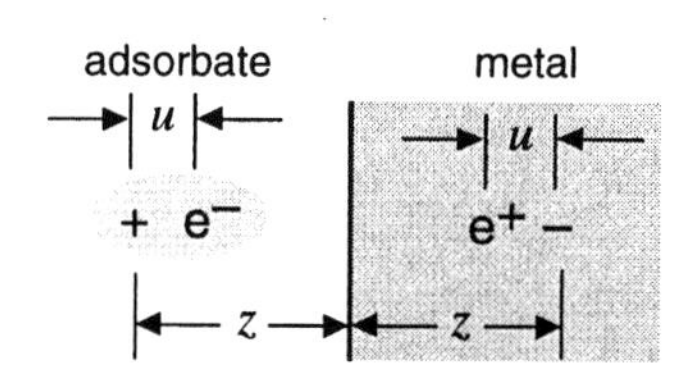

Figure 43 A polarized molecule or atom near a metal surface interacts with its image charges. After [21].

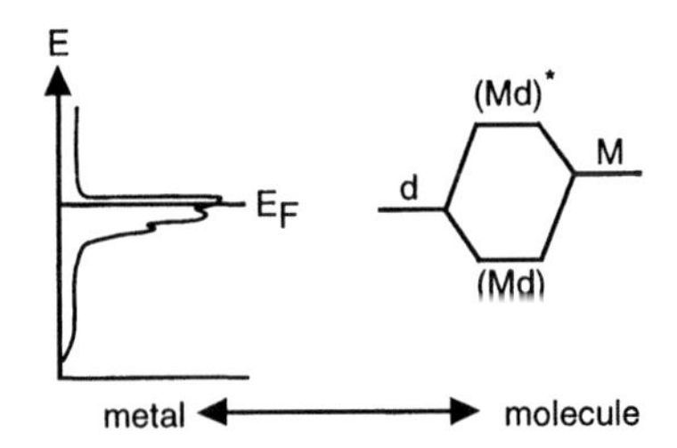

Figure 44 Chemisorption of a molecule on a metal surface is accomplished by mixing a *d*-orbital of the metal with an orbital of the molecule. Bonding (Md), and anti-bonding (Md)* orbitals are formed. After [21].

An adsorbate is chemisorbed if it forms a chemical bond with the substrate. The bond substantially changes the electronic states of the adsorbate. Figure 44 represents the energy levels of a molecule chemisorbed on a transition metal surface. The left side of the diagram represents the *d*-band of the transition metal. The right side represents a state of the free molecule. When the molecule chemisorbs on the metal surface the orbital of the molecule can mix with a *d*-orbital of the metal. The mixing is analogous to the formation of a bond between two atoms to make a molecule, Figure 9. The chemisorption of O_2 on Pt(111) is described in detail in section IV. F.

IV. C. Surface diagnostics

In the discussion of photoemission spectroscopy, we mentioned that electrons may be emitted from an atom by Auger recombination following removal of a core-level electron. The energy of the Auger electron depends on the energies of electronic states of the atom; since these energies are unique to each atom, the Auger spectrum identifies the atoms present in the sample. Auger spectroscopy is one of the most common means of determining chemical composition of a surface. Figure 45 shows the Auger spectrum obtained from a platinum surface. Interpretation of an Auger spectrum requires comparison of the observed spectrum with reference spectra from samples of known composition.[25]

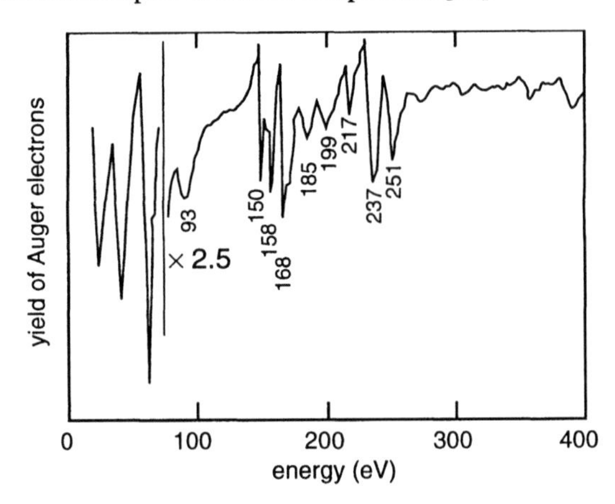

Figure 45 Auger spectrum of platinum. After [25].

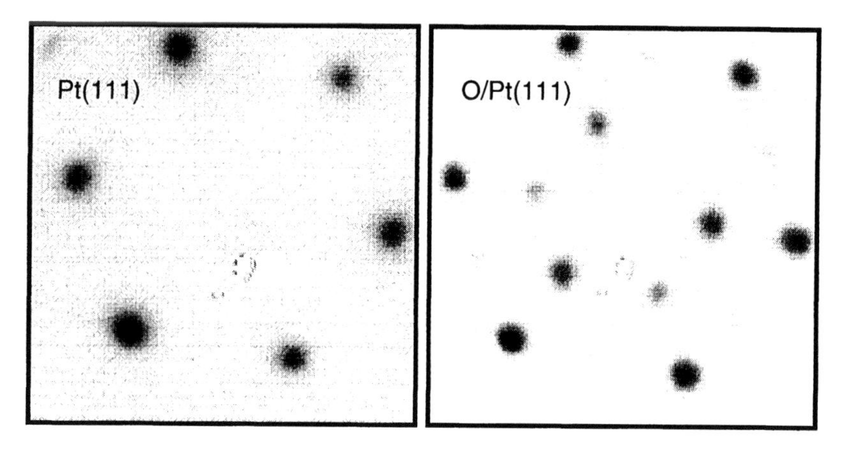

Figure 46 LEED patterns obtained from Pt(111), *left,* and O/Pt(111), *right*. The electron gun (not visible) obscures one of the spots due to oxygen.

Low energy electron diffraction (LEED) is used to determine the order of a surface. A monochromatic beam of electrons diffracts off the surface and is intercepted by a phosphor screen. The left side of Figure 46 shows the pattern formed by scattering of electrons from Pt(111). The hexagonal symmetry of the Pt(111) surface is reflected in the symmetry of the diffraction pattern. The right side of Figure 46 shows the diffraction pattern from a Pt(111) surface that has been exposed at 300 K to a few Langmuir of O_2: in addition to the diffraction spots due to the platinum atoms, there are spots attributed to diffraction of electrons from oxygen atoms. By comparing the positions of the spots due to oxygen to those due to platinum, it follows that the density of oxygen atoms is 1/4 the density of the platinum atoms. The oxygen atoms are arranged in a regular grid on top of the platinum surface; if their distributions were random they would not produce a sharp diffraction pattern.

The vibrational modes of an adsorbate can be probed by scattering a beam of monochromatic electrons off the sample. Some of the electrons lose energy by exciting discrete electronic or vibrational transitions in the adsorbates, so the distribution of adsorbate modes is reflected in the distribution of energies of the scattered electrons. The technique is called electron energy loss spectroscopy (EELS). Figure 47 shows the EEL spectrum of chemisorbed O_2 on Pt(111).[26-28] The signal at zero loss is 50 cm^{-1} wide; this is the resolution limit. The signals at 875 cm^{-1} and 700 cm^{-1} are attributed to stretching of the oxygen molecule along the O–O bond axis.[26] The signal at 380 cm^{-1} is assigned to vibration of the O_2 molecule perpendicular to the surface.[27] The signals at losses above 875 cm^{-1} are attributed to electrons that excited multiple quanta of the surface modes or coupled modes. The vibration of atomic oxygen with respect to the Pt(111) would scatter electrons with a loss of 480 cm^{-1}.[26, 27]

Infrared (IR) spectroscopy is another way to study the vibrational modes of adsorbates. Infrared light reflected off the surface is absorbed at frequencies that are resonant with vibrational frequencies of the adsorbates. IR spectroscopy using short laser pulses has been employed to study the decay of vibration of adsorbates due to loss of vibrational energy to the substrate. The decay is complete in roughly one picosecond for adsorbates on metal surfaces.[29, 30] Very short laser pulses are used to resolve these timescales. Their intensities are high enough for the nonlinear optical spectroscopies that were summarized in Figure 19.

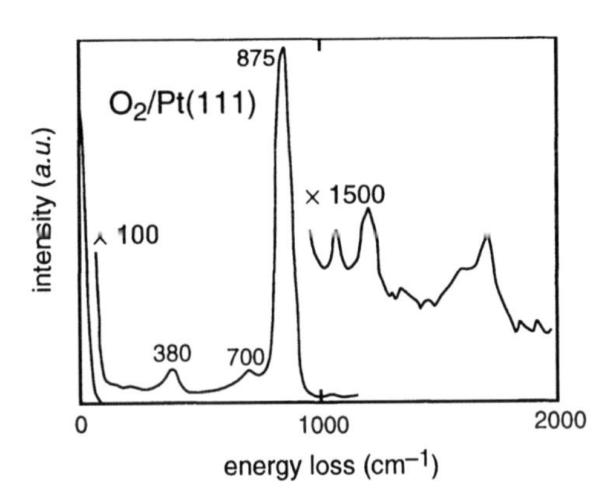

Figure 47 Electron energy loss spectroscopy of a saturated overlayer of O_2 on Pt(111) from a beam of 2 eV electrons incident at 75°. After [26]

The spatial distribution of surface electronic states can be imaged with atomic resolution by scanning tunneling microscopy (STM). The STM micrograph can be interpreted as a map of the surface because the electronic surface states are determined by the substrate and the adsorbates. In a recent study of the thermal dissociation of O_2 on Pt(111), an STM was used to determine that when O_2 thermally dissociates, the oxygen atoms break free of their mutual bond with enough kinetic energy to move approximately two platinum lattice constants.[31] This is experimental evidence that atoms can have ballistic motion on a surface. Below we summarize how the STM was used to determine the sites where O_2 chemisorbs on Pt(111).

IV. D. Thermal chemistry

We have seen that oxygen binds to platinum in several different forms: physisorbed molecular oxygen, chemisorbed molecular oxygen, and chemisorbed atomic oxygen. Because each form of oxygen is bound at a different site on the platinum with a different binding energy, these atoms become chemically active at different temperatures. A common way of studying adsorbate/substrate systems is to increase the sample temperature at a regular rate while monitoring the species desorbed from the surface. The experiment is called temperature programmed desorption (TPD) or temperature programmed reaction spectroscopy (TPRS) depending on whether the adsorbates simply desorb from the surface, or react to form new species. The yield in TPD or TPRS is plotted against the corresponding temperature, indicating the temperatures of thermally induced desorption and reaction.

Figure 48 shows the TPD of chemisorbed molecular oxygen, O_2/Pt(111), obtained with a 4 K/s heating rate. The signal at 36 atomic mass units (amu), is due to oxygen molecules comprised of the 18 amu isotope of oxygen, $^{18}O_2$. The 140 K signal, called α–O_2, is attributed to direct desorption of intact molecules. The O_2 desorption at 750 K, β–O_2, shows that not all the oxygen desorbs at 140 K. If the TPD is stopped between the α–O_2 and the β–O_2 desorption signals, an EEL spectrum has features attributed to atomic oxygen, but none attributed to molecular oxygen. The LEED pattern is the pattern produced by atomic oxygen on Pt(111) (Figure 46, right side). These observations indicate that there is no molecular oxygen on the surface, but there is atomic oxygen. The β–O_2 signal must be due to recombination of oxygen atoms, known as recombinative desorption.

So far we have discussed adsorption of a single species. The final section of this paper discusses photo-induced reaction between two adsorbed species, CO and O_2 on Pt(111). Production of CO_2 can be induced with light or heat. Figure 49 shows the signal at 48 amu detected during TPRS of CO/O_2/Pt(111). The surface was prepared with the isotopes $C^{18}O$ and $^{18}O_2$, and so the signal at 48 amu is due to CO_2. The different peaks correspond to the CO interacting with oxygen atoms in different states. The first desorption peak, α–CO_2 is not observed during TPRS of CO coadsorbed with atomic oxygen, CO/O/Pt(111).[32] The α-CO_2 is attributed to reaction of CO with an oxygen atom created by O_2 dissociation, before this atom equilibrates with the surface.[33] The other CO_2 peaks, β-CO_2, are due to reaction between CO and atomic oxygen after the atomic oxygen has equilibrated with the surface. The reaction occurs as the CO and O diffuse on the surface, forming islands; the multiple peaks arise from reaction as the CO and O collide during different stages of this diffusion.[34]

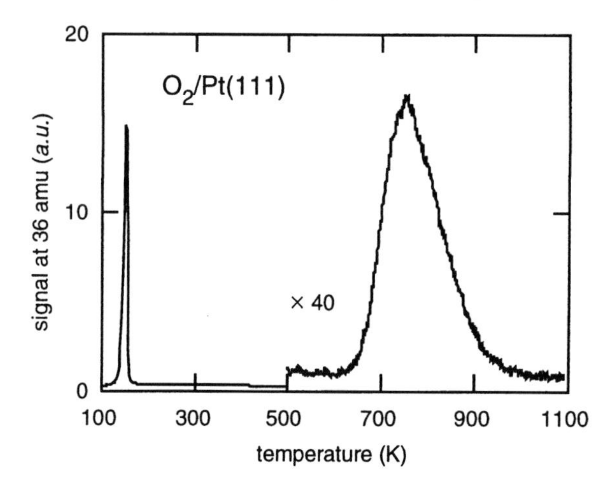

Figure 48 Temperature programmed desorption of O_2/Pt(111). The sample was dosed with isotopic oxygen, $^{18}O_2$, and the signal is detected at 36 amu with a mass spectrometer.

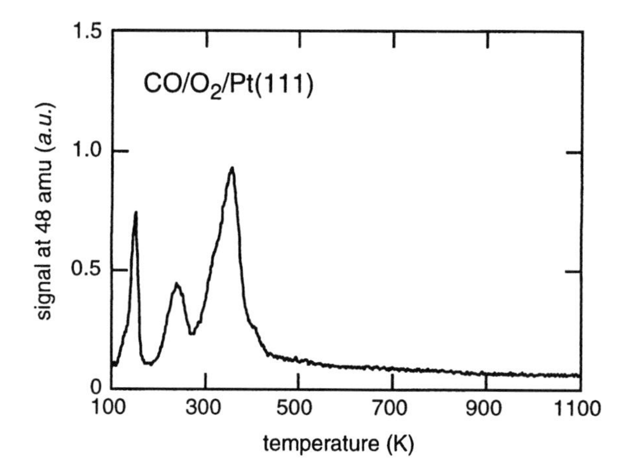

Figure 49 Temperature programmed reaction spectroscopy of CO/O_2/Pt(111). The signal at 48 amu is due to carbon dioxide, $C^{18}O_2$.

IV. E. Isotope exchange

Isotopic labeling of reactants is used to determine pathways in surface reactions. Different isotopes of a molecule have different vibrational frequencies and therefore have different EELS spectra. With a mass spectrometer of resolution better than 1 amu, different isotopes of the same product are distinguishable in TPD and TPRS.

Figure 50 shows the application of isotopic labeling to the study of thermal reaction in O_2/Pt(111). A mixture of 50% $^{16}O_2$ and 50% $^{18}O_2$ was prepared by mixing $^{16}O_2$ and $^{18}O_2$ outside the vacuum chamber. This gas was admitted to the sample surface to produce a chemisorbed overlayer denoted ($^{18}O_2$, $^{16}O_2$)/Pt(111). The left of Figure 50 shows the signal at 36 amu due to $^{18}O_2$ product. The α-O_2 signal is about half as large as in Figure 48, and the β-O_2 signal is about one quarter as large as in Figure 48. The signal at 32 amu due to desorption of $^{16}O_2$ looks very similar to the signal for $^{18}O_2$. The right side of Figure 50 shows the signal at 34 amu due to $^{16}O^{18}O$. There is no yield of α-$^{16}O^{18}O$, but comparison of the left and right sides of Figure 50 shows that the yield of β-$^{16}O^{18}O$ is twice as large as the yield of β-$^{18}O_2$. These observations are consistent with the evidence derived from EELS and LEED above. At 140 K, oxygen molecules desorb from the surface without opportunity to exchange isotopes of oxygen between molecules. Oxygen molecules also dissociate at 140 K. Near 700 K, the atoms recombine, but now there is opportunity for two different isotopes to form a molecule, $^{16}O^{18}O$. The probability of getting $^{18}O_2$ or $^{16}O_2$ at this stage is one half the probability of getting $^{16}O^{18}O$.

IV. F. Details of the bonding of oxygen to platinum

In the gas phase, molecular oxygen has 12 valence electrons in the $2\sigma_g$, $2\sigma_u^*$, $1\pi_u$, $3\sigma_g$, and $1\pi_g^*$ orbitals.[35] The right hand side of Figure 43 shows the electronic states of gas-phase O_2 which are most relevant to its bonding on platinum, including two π orbitals oriented normal to the molecular axis which have identical energies.

Experiments with scanning tunneling microscopes (STM) show that O_2 chemisorbs in at least three different configurations on Pt(111).[36] The O_2 is observed over a bridge site between two platinum atoms with the O–O bond axis parallel to the surface and aligned to

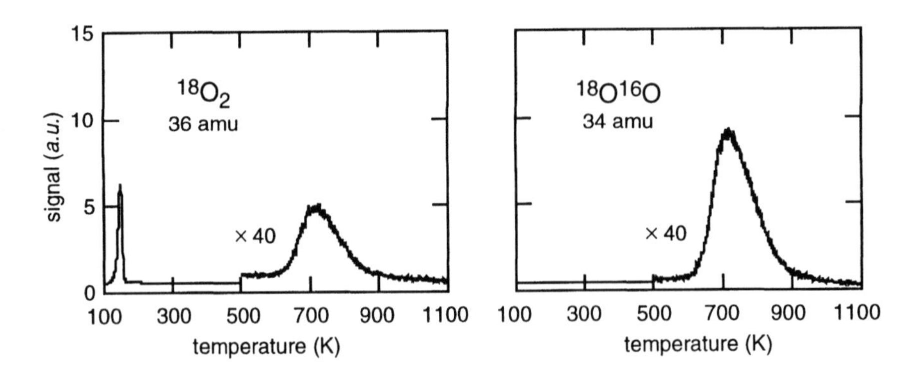

Figure 50 The TPD signal of (*left*) $^{18}O_2$ and (*right*) $^{16}O^{18}O$ from ($^{18}O_2$,$^{16}O_2$)/Pt(111). The yield of $^{16}O_2$ is very similar to the yield of $^{18}O_2$.

face the tops of platinum atoms. Calculations show that the O–O bond length in this configuration is 0.139 nm and the energy of vibration of the O–O bond is 850 cm^{-1}.[37] An O_2 species bound over FCC three-fold hollows is also observed with the STM.[36] Calculations show that this O_2 molecule is tilted 8° out of the plane of the surface. The O–O bond length is 0.143 nm and the energy of vibration is 690 cm^{-1}.[37] The calculated energies of vibration of the O–O bonds in the bridge and fcc configurations are consistent with observed vibrational energies from electron energy loss spectroscopy. The stretching of the O–O bond with respect to the 0.1207 nm length in the gas phase is consistent with the bond length inferred from x-ray spectroscopy.[38, 39] The stretching is due to transfer of electrons from the platinum into orbitals that are anti-bonding with respect to the O–O bond. The third O_2 species observed with STM is at O_2 adsorbed at step edges.[36, 40] Temperature programmed desorption experiments show that the binding energy of all the O_2 species is close to 0.4 eV.[27]

When the O_2 chemisorbs, some of the oxygen orbitals mix with platinum orbitals. The $1\pi^*$ orbitals perpendicular to the surface mix with the platinum *d*-band orbitals, producing π_b^* and π_b orbitals. The $1\pi^*$ orbitals of the oxygen parallel to the surface are not greatly perturbed and are denoted π_n.[39, 41] A shaded bar in Figure 51 represents the extent of the entire π_b and π_n region compiled from experimental[38, 42] and computational[41, 43, 44] sources. Overall, there is a net transfer of charge from the platinum to the O_2.[27]

A surface layer of atomic oxygen on Pt(111) can be obtained by exposing the surface to molecular oxygen at platinum temperatures above 145 K.[46] The surface coverage saturates at 0.25 ML with a p(2×2) LEED pattern.[27] The oxygen binds in fcc three-fold hollow sites,[47] with a 1.1 eV binding energy at 0.25 ML coverage,[27] and a 470-cm^{-1} Pt–O

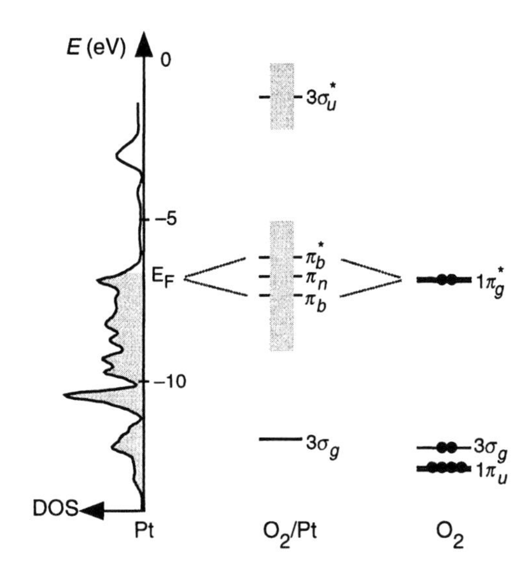

Figure 51 The calculated density of states of platinum[45] shown with the experimentally and theoretically determined states of O_2 chemisorbed on Pt(111). The vertical shaded bars indicate the approximate widths and locations of the orbitals of O_2 on the surface. Many of the O_2 orbitals are out of the energy range shown.[13]

vibration. Higher coverages of atomic oxygen can be attained by photodissociation of N_2O/Pt(111),[48] or by electron-beam dissociation of O_2/Pt(111).[26]

V. PHOTOCHEMISTRY OF OXYGEN ON PLATINUM

It has been well-established that high-intensity subpicosecond laser pulses induce reactions among adsorbates on a metal surface by photo-excitation of electrons in the metal substrate.[49-53] There is an active debate, however, about the energy distribution of the electrons responsible for the reaction between adsorbates on a metal surface. According to one proposal, the adsorbates interact with an essentially thermal substrate electron distribution.[54] According to another proposal the photon-energy-dependent electron distribution plays a significant role in the surface reactions.[52]

The yields in subpicosecond photo-induced reactions have a nonlinear dependence on fluence, a high quantum efficiency, highly-excited nonthermal internal-state distributions, and increasing translational energy with increasing laser fluence.[49, 51, 52, 54-59] Two-pulse correlation experiments show that the excitation has a lifetime of about 1 ps.[52, 58-61] Time-resolved surface second-harmonic generation indicates that desorption of CO from CO/Cu(111) is complete in less than 325 fs.[56]

We measured the desorption of O_2 and production of CO_2 from CO/O_2/Pt(111) induced with 0.3-ps laser pulses at 267, 400, and 800 nm to determine the dependence of the chemical reaction on the electron distribution. We find that the reaction is sensitive to photon energy and therefore we favor excitation of the adsorbates by nonthermal electrons. We argue that the two-pulse correlation experiments have been misinterpreted; the correlation time reflects the time required for relaxation of adsorbate vibrational modes, rather than the substrate electron-phonon coupling time.

When the photochemistry of CO/O_2/Pt(111) is induced with continuous light[28, 62] or nanosecond pulses,[51] the yields of O_2 and CO_2 scale linearly with fluence and depend strongly on photon energy. We report that the desorption induced by femtosecond laser pulses also scales linearly with fluence when the fluence is low. We observe the transition from linear to nonlinear dependence on fluence with 267- and 400-nm laser pulses.

V. A. Low-intensity photochemistry

Light from an arc lamp can induce desorption of O_2 from O_2/Pt(111). Figure 52 shows the dependence of the O_2 desorption yield on photon wavelength.[28] The yield increases with decreasing wavelength and there is no photodesorption at wavelengths longer than 550 nm. Figure 53 shows that the desorption rate is linear in fluence: $Y \propto F^1$. Irradiation also induces rearrangement of the O_2 molecules on the surface; temperature programmed desorption following irradiation of O_2/Pt(111) exhibits a broadened α-O_2 desorption peak compared to nonirradiated O_2/Pt(111).[62, 63]

A model called DIET, for desorption induced by electron transitions, explains the wavelength dependence of the data and the linear dependence of yield on fluence. Figure 54 shows how electronic excitation of an adsorbate can cause it to acquire the translational energy required for desorption. In the electronic ground state the interaction of the adsorbate with the surface is described by the potential energy surface labeled PES_1. The reaction coordinate could be, for example, the distance between the adsorbate and the surface, or the alignment of the adsorbate with respect to the surface. The interaction of the adsorbate with the substrate following electronic excitation of the adsorbate is represented by the potential energy surface labeled PES_2. In general, PES_2 does not have a minimum at the same reaction coordinate as PES_1. PES_2 could be purely repulsive, with no minimum at all.

Electronic excitation of the adsorbate is represented by the vertical arrow from the minimum of PES_1. This transition is an example of a Frank-Condon transition. The transition could be caused by a photon exciting an electron within the O_2/Pt(111) complex; such an excitation does not change the overall charge on the adsorbate. The transition could also be the transfer of an electron from the substrate to the adsorbate, changing the charge on the adsorbate. If an electron is transferred to the adsorbate, the shape of PES_2 reflects the interaction between the charged adsorbate and the image charge in the substrate.

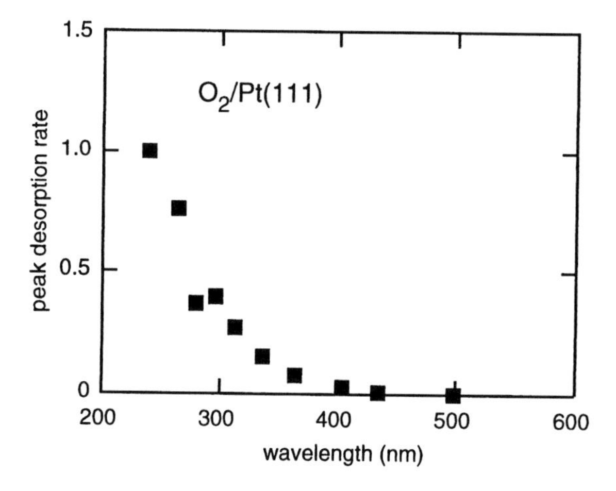

Figure 52 The rate of desorption of O_2 from O_2/Pt(111) under irradiation with continuous light from an arc lamp. After [28]

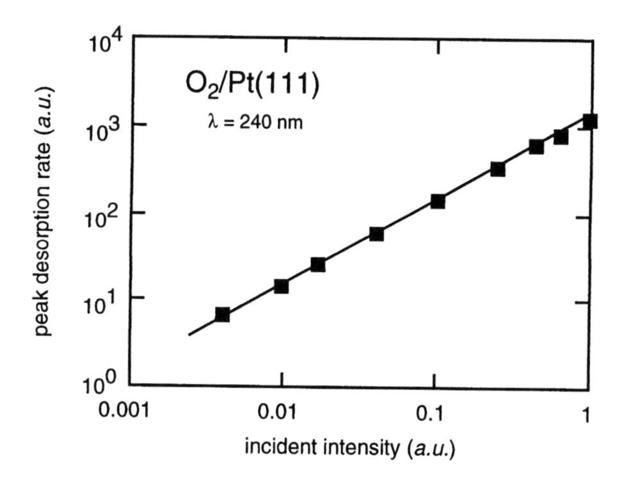

Figure 53 The rate of desorption of O_2 from O_2/Pt(111) induced with 240 nm irradiation, as a function of the incident intensity. The line has slope 1. After [28]

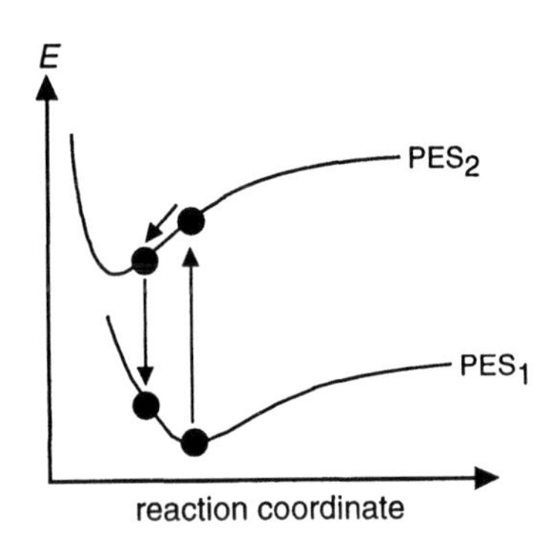

Figure 54 Desorption induced by electron transition (DIET) is explained in terms of two potential energy surfaces for the adsorbate-surface interaction.

An adsorbate excited to PES_2 accelerates towards the new potential energy minimum. When the electronic excitation relaxes, the adsorbate returns to PES_1, having acquired translational and potential energy, This energy accounts for desorption of the adsorbate.

Each photo-excitation of an electron and the possible subsequent excitation of an adsorbate acts independently of the other photo-induced excitations. This feature of the model accounts for the linear dependence of yield on fluence. The dependence of yield on wavelength reflects the need for photo-excited electrons to have energies appropriate to access available affinity levels of the adsorbates.

During irradiation of O_2/Pt(111) with an arc lamp, the fluence is kept low enough that the increase in surface temperature is only a few Kelvin: the surface temperature remains well below the 120 K temperature at which O_2 desorbs. According to the Fermi-Dirac function, Figure 25, the thermalized electron energies are near 20 meV. In contrast, the photon energy at, say, 400 nm, is about 3.2 eV. A photo-excited electron has an energy much greater than the thermal energy: it is these nonthermal electrons which govern the desorption. The nonthermal electrons have sufficient energy for the transition depicted in Figure 54. After thermalization of the electrons, the energy of an individual photon is distributed among all the electrons, and ultimately, all the surface modes. After thermalization, there is no longer sufficient energy in any individual electron to induce desorption.

The situation could be different when a subpicosecond laser pulses excites the material. According to Figure 38, the subpicosecond laser pulses creates a transient electron temperature far in excess of the temperature required for desorption of O_2 under equilibrium conditions. Could this hot, thermal distribution of electrons induce desorption? Our experiments are designed to address this question.

V. B. Subpicosecond laser desorption experiments

We studied the photochemistry of CO/O_2/Pt(111) using laser pulses from a 1-kHz regeneratively-amplified Ti:sapphire laser. The 100-fs, 800-nm pulses are frequency-doubled in a 1-mm thick lithium barium borate crystal and frequency-tripled in a 0.3-mm thick beta-barium borate crystal. The laser pulse energy is varied with a waveplate and a polarizing beam splitter. The 267- and 400-nm pulses have 0.26- and 0.3-ps duration, respectively. The 800-nm pulses are chirped to 0.3 ps so the pulse durations at all wavelengths are similar.

The energy of each laser pulse is measured with a photodiode that is calibrated with a power meter. The response of the power meter varies less than 3% over the range 267–800 nm. To ensure that there is no nonlinear absorption in the platinum, we measured the fraction of the laser pulse energy absorbed into platinum. The measured absorption of the platinum is constant over the range of fluences used in the experiments, and is in agreement with the reflectivity calculated from the published dielectric function of platinum.[12] We also verified that the absorption of the chamber window does not depend on fluence. These results confirm that the laser energy absorbed in the platinum is a constant fraction of the pulse energy measured outside the vacuum chamber.

The spatial profile of the laser pulses is measured with an ultraviolet-sensitive CCD camera. The profile captured by the camera is fit well by a Gaussian function. The fluence incident on the camera is reduced to a level where the camera response is linear by reflecting the beam off the front surfaces of two pieces of glass and is further attenuated with neutral-density filters. To confirm the accuracy of the camera-based spatial profile measurement, we measured the spatial profile of a Helium-Neon laser with the camera and compared it with the profile determined by scanning a pinhole through the beam while measuring the transmitted light with a photodiode. The camera and pinhole methods yield laser profiles that are identical to within 1%.

The absorbed laser fluence is determined from the energy absorbed in the platinum, and the spatial profile of the laser pulse, accounting for the 45° angle of incidence. The fluence varies over the profile of the laser spot; values quoted below refer to the absorbed fluence at the peak of the spatial profile. The tests described above confirm that there is no wavelength-dependent, nor any fluence-dependent systematic error in the calculation of absorbed fluence.

The experiments are conducted on a 12-mm diameter Pt(111) crystal in an ultrahigh vacuum chamber with a base pressure of 5×10^{-11} torr. All experiments are performed at a base temperature of 84 K. The crystal is cleaned using Ne ion sputtering at an ion energy of 500 eV, annealing in vacuum at 1100 K, and annealing in 10^{-8} torr oxygen at 500–1000 K.[22] Surface order is verified with low-energy electron diffraction and surface cleanliness is verified with Auger spectroscopy.[63]

After cleaning, molecular oxygen and carbon monoxide are adsorbed to saturation on the platinum surface.[27, 28] Molecular oxygen is deposited on the platinum surface as soon as the temperature has fallen below 94 K after a cleaning cycle. Carbon monoxide is deposited after the oxygen. To reduce background pressure, all adsorbates are deposited using a tube of 12-mm diameter brought to within 3 mm of the platinum surface.

The laser-induced O_2 desorption yield and CO_2 reaction yield are measured with a quadrupole mass spectrometer operating in pulse-counting mode. We alternate between detecting O_2 and CO_2 on successive laser shots. Between shots, we translate the sample to an unirradiated part of the sample. A potential difference of –90 V is applied between the sample and the ionizer of the mass spectrometer to prevent stray electrons from interacting with the sample. A tube of 4-mm inner diameter extends from the ionizer to the sample. This tube collects molecules desorbed from the surface within 14° of the surface normal. Using a high-speed mechanical shutter, we reduce the laser repetition rate to allow the gas-phase products to be pumped out of the chamber between successive laser shots.

The yield depends on the area of the sample preparation exposed to the laser pulses. To obtain an appreciable yield at low fluence, a large laser spatial profile of full-width-half-maximum up to 1 mm^2 is used. At high fluence, the spatial profile is decreased to as low as

0.05 mm^2 to reduce the absolute yield and to avoid saturating the pulse-counting electronics. The yields reported below are divided by the laser spot size to allow comparison between runs taken with different laser spot sizes. Below 20 μJ/mm^2, less than 1% of the adsorbates is depleted by a single laser pulse; to increase the signal in this regime, we admit up to 10 pulses at a 1-kHz rate to one spot on the sample and the mass spectrometer measures the total yield.

V. C. Results of the subpicosecond laser desorption experiments

Figure 55 shows the yield of oxygen molecules obtained from $CO/O_2/Pt(111)$ with 267-, 400-, and 800-nm laser pulses. Near 10 μJ/mm^2 there is a clear change in the dependence of the yield on absorbed laser fluence. Above 50 μJ/mm^2, the yield saturates because the pulse desorbs an appreciable fraction of the adsorbed oxygen.[52]

Below 10 μJ/mm^2 the yield from 267- and 400-nm pulses depends linearly on fluence. To determine the linear cross section in this regime, we measured the decreasing yield from a single spot on the sample as the surface coverage is depleted by 3,000 laser pulses. The linear cross sections thus obtained are $\sigma_{267} = (4 \pm 2) \times 10^{-23}$ m^2 and $\sigma_{400} = (4 \pm 2) \times 10^{-24}$ m^2 for 267- and 400-nm pulses, respectively. We do not observe any linear dependence of yield on fluence for 800-nm laser pulses; continuous light sources with wavelengths longer than 600 nm also do not induce reaction[28].

Between 10 and 50 μJ/mm^2 the yield depends nonlinearly on fluence. The data can be described by a simple power law, $Y \sim F^p$, where $p > 1$ and F is the fluence absorbed in the platinum. As Table 1 shows, the exponent p decreases with decreasing wavelength.

The wavelength dependence of the yields is also apparent in a comparison of the absolute yields at a particular fluence. Table 2 summarizes the wavelength dependence of the yields at 1 and 30 μJ/mm^2. At both fluences, the yield increases substantially as the wavelength decreases.

Figure 56 shows the yield of carbon dioxide from the same sample preparation as Figure

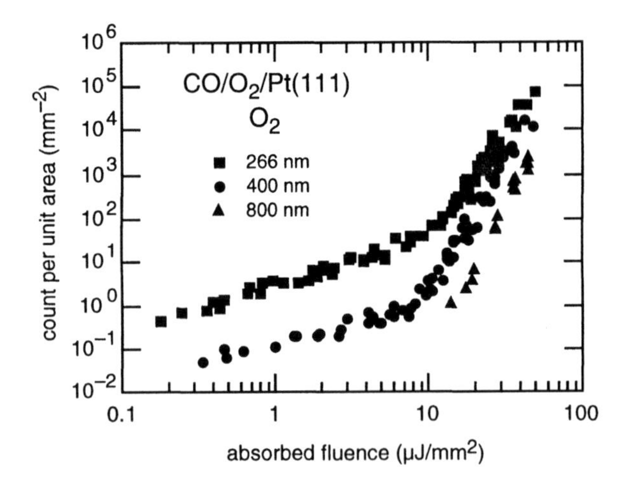

Figure 55 Yields of O_2 from $CO/O_2/Pt(111)$ obtained with laser pulses of 0.3-ps duration at ▲ 800-, ● 400-, and ■ 267-nm wavelengths.

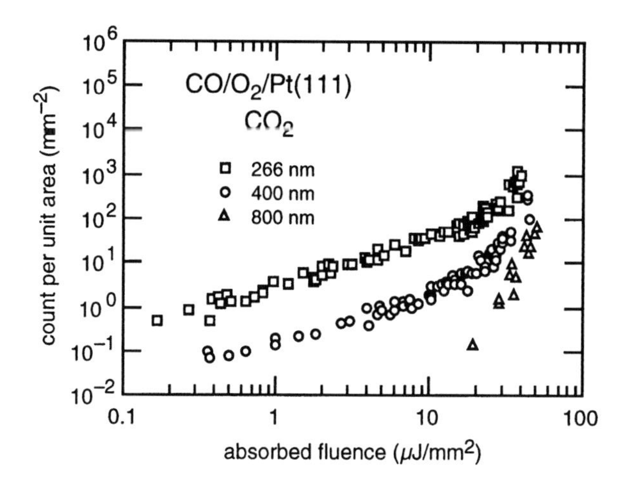

Figure 56 Yields of CO_2 from $CO/O_2/Pt(111)$ obtained with laser pulses of 0.3-ps duration at △ 800-, ○ 400-, and □ 267-nm wavelengths.

55. The dependence of the CO_2 yield on fluence is similar to that of O_2. Table 3 summarizes the ratio of yield of O_2 to yield of CO_2. When using 267- or 400-nm pulses at fluences below 10 $\mu J/mm^2$ (*i.e.*, in the linear regime), the yields of O_2 and CO_2 are the same. Above 20 $\mu J/mm^2$, the yield of O_2 is substantially more than the yield of CO_2, and the ratio is smaller at shorter wavelengths. The ratios shown in Table 3 are not corrected for the small dependence of the mass spectrometer detection efficiency on species.

To explore the time dependence of the desorption, we measured the total desorption yield from two 80-fs laser pulses as a function of the delay $t_1 - t_2$ between them. The pulses are orthogonally polarized to avoid interference. The resulting two-pulse correlation is shown in Figure 57. The dashed line shows the total yield when the two pulses act independently, *i.e.*, when $t_1 - t_2 \rightarrow \pm\infty$. The dependence of the signal on $t_1 - t_2$ reflects the evolution of the substrate and adsorbate excitations responsible for desorption. The data show a 1.8 ps wide peak centered at $t_1 - t_2 = 0$ on top of broad wings. The broad wings are approximately 0.1 ns wide.

V. D. Discussion

The desorption and reaction yields at fluences below 10 $\mu J/mm^2$, shown in Figures 55 and 56, scale linearly in fluence. The cross sections are about 10^{-23} m^2 and increase with decreasing wavelength. The yield of O_2 and CO_2 obtained from $CO/O_2/Pt(111)$ with continuous light[28] or nanosecond-pulses[51] also scales linearly in fluence. The cross sections measured with these low-intensity sources are about 10^{-23} m^2 and increase with decreasing wavelength. These similarities suggest that the linear surface femtochemistry is due to the same mechanism responsible for the surface photochemistry induced with continuous-wave or nanosecond-pulsed light sources.

The excitation of $CO/O_2/Pt(111)$ with these low-intensity sources has been attributed to electronic transitions into normally-vacant orbitals of the O_2.[28] As discussed above, this new electronic configuration causes the adsorbate atoms to move, accumulating vibrational or

Table 1 Wavelength dependence of the power law exponent. The yield is linear in fluence below 10 μJ/mm^2, but very nonlinear in fluence above 10 μJ/mm^2. There is no low-fluence yield with 800-nm laser pulses.

P	267 nm	400 nm	800 nm
$\leq 10\ \mu$J/mm^2	1.1 ± 0.1	0.9 ± 0.1	n.a.
$\geq 10\ \mu$J/mm^2	4.8 ± 0.5	6.0 ± 0.5	7.2 ± 0.5

Table 2 Wavelength dependence of the laser-induced yield. At all fluences studied, the yield increases with decreasing wavelength.

yield	267 nm	400 nm	800 nm
1 μJ/mm^2	3 ± 0.5	0.1 ± 0.02	0
30 μJ/mm^2	8000 ± 1500	2000 ± 500	120 ± 25

Table 3 Wavelength dependence of the ratio of O_2 to CO_2 yield. The ratio is one at fluences below 10 μJ/mm^2, but O_2 desorption is favored over production of CO_2 at fluences above 20 μJ/mm^2. At fluences above 20 μJ/mm^2, the ratio is strongly wavelength dependent.

O_2:CO_2	267 nm	400 nm	800 nm
$\leq 10\ \mu$J/mm^2	1 ± 0.1	1 ± 0.1	n.a.
$\geq 20\ \mu$J/mm^2	25 ± 5	55 ± 15	70 ± 10

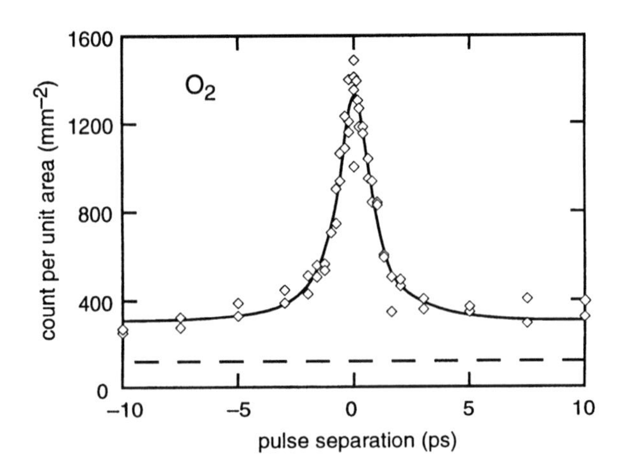

Figure 57 Desorption yield versus time delay $t_1 - t_2$ between two 80-fs excitation pulses at 800-nm for pulses of equal absorbed fluences. The dashed line denotes $Y(t_1 - t_2 = \pm\infty)$. The width of the central peak is 1.8 ps. The desorption yield is still enhanced after 75 ps.[52]

translational energy that may lead to desorption or reaction. Two mechanisms for the electronic transition have been proposed. The photon can stimulate a direct transition between orbitals in the O_2/Pt(111) complex.[28] Alternatively, the photon may excite an electron to a state above the Fermi level in the Pt(111) band structure, from which it crosses into an orbital of the O_2/Pt(111) complex.[64] Either way, the electron interacts with the O_2 while retaining a substantial portion of the initial photon energy. The surface photochemistry in the linear regime is therefore governed by electrons with a nonthermal distribution. The wavelength dependence of the cross section is due to the required matching of the energy of the photo-excited electron with the energies of the vacant O_2 orbitals.

Irradiation of a metal surface creates an electron distribution with thermal and nonthermal components. As indicated in Figure 39, after intense subpicosecond laser pulses excite a metal, the thermal distribution of electrons that can be several thousand Kelvin for a few picoseconds.[49] Several authors attribute the nonlinear dependence of yield on fluence, high desorption yields, and short excitation lifetimes to these transient hot electrons.[57, 65-67] The electron temperature depends on the pulse duration and the fluence absorbed in the platinum, but not on photon wavelength.[14]

Our experiments show three ways in which the nonlinear surface femtochemistry depends on wavelength. The power-law exponents summarized in Table 1 depend on wavelength, increasing from 4.8 ± 0.5 at 267 nm to 7.2 ± 0.5 at 800 nm. The desorption yields summarized in Table 2 are also wavelength dependent: at 30 $\mu J/mm^2$, the yield from 267-nm pulses is about 4 times that from 400-nm pulses, and about 65 times that from 800-nm pulses. The ratio between O_2 and CO_2 yields, summarized in Table 3, depends on wavelength, varying from 70 at 800 nm to 25 at 267 nm. The nonlinear femtochemistry of CO/O_2/Pt(111) depends on wavelength, and so the thermalized electron distribution cannot be solely responsible for exciting the adsorbates. We attribute the wavelength dependence of the nonlinear surface femtochemistry on CO/O_2/Pt(111) to interaction of the adsorbates with electrons from a nonthermal (non Fermi-Dirac) distribution.

The data from the two-pulse correlation experiments, Figure 57, show that the sample retains excitation from the first laser pulse for longer than 10 ps. This correlation time is longer than the electron-electron, electron-adsorbate,[55, 67] electron-lattice,[15, 49] and lattice-adsorbate[29, 30] relaxation times. The only remaining equilibration process is the cooling of the surface to the bulk, which occurs on roughly the same time scale as the decay of the wings. The correlation beyond 10 ps indicates that desorption is accomplished more easily from a pre-heated surface than from a cold surface.

We consider now the prominent 1 ps wide peak in Figure 57.[52, 61] Experiments show that in subpicosecond laser excitation of gold film, the nonthermal electron distribution persists for 0.5 ps.[16, 68] Though there are no published measurements of the electron-electron thermalization time for platinum, the thermalization time in platinum is likely less than the 1 ps correlation in Figure 57. It is therefore unlikely that the 1 ps correlation reflects the time for electrons to thermalize.

The 1 ps correlation in Figure 57 has been attributed to cooling of the thermalized electron distribution,[50, 59, 61] because, as Figure 39 shows, the electrons equilibrate with the lattice on a 1 ps time scale. Though the thermalized electron distribution may contribute to the laser-induced desorption and contribute to the 1 ps correlation, the wavelength dependence of our data indicates that thermalized electrons do not solely govern the desorption. The 1 ps correlation more likely reflects the time required for the adsorbates to dissipate the vibrational excitation induced by the first laser pulse. Indeed, the time scale for relaxation of vibrational excitation at a metal surface is approximately 1 ps.[29, 30]

V. E. Conclusions from the surface femtochemistry experiments

The absolute and relative cross sections for desorption and reaction depend on fluence. At high fluence, where desorption is more efficient than reaction, adsorbates can be desorbed with subpicosecond time-resolution for analysis. The efficient desorption could also be used to create empty reactive sites and increase reaction rate in a catalytic process in which site blocking inhibits the reaction. Finally, because the linear surface femtochemistry is caused by the same mechanism as the surface photochemistry induced with continuous light, the dynamics of the surface photochemistry induced with continuous light can be studied with femtosecond time-resolution by limiting the femtosecond pulses to low fluence.

The transition between nonlinear and linear surface femtochemistry is also of theoretical interest. At low fluence, the linear dependence of yield on fluence is due to desorption or reaction caused by a single electronic excitation. Above the transition fluence, the nonlinearity indicates that cooperative action of the photo-excited electrons dominates the linear process. These cooperative effects have been described as a frictional coupling between the substrate electrons and the adsorbates,[66] and as a repeated excitation of the adsorbate within the time required for cooling of the adsorbate vibration[67].

We describe the surface femtochemistry of O_2/CO/Pt(111) as follows. Linear surface femtochemistry, observed at fluences below 10 $\mu J/mm^2$, is due to the same mechanism as surface photochemistry induced with continuous-wave and nanosecond pulses. Above 10 $\mu J/mm^2$, another excitation mechanism dominates the reaction; yields are nonlinear in fluence and depend on wavelength.

It is not correct to completely attribute the desorption and reaction yields to the influence of a thermalized electron distribution, such as depicted in Figure 39. Models need to account for the nonthermal electrons to predict the wavelength dependence of our data. This result implies that previously-published two-pulse correlation data must be re-interpreted. The short, 10^{-12}-s correlation is due to relaxation of vibrational excitation of the adsorbates between laser pulses, and not due to cooling of the electrons to the bulk phonon temperature.

VI. REFERENCES

1. J. J. Sakurai, *Modern quantum mechanics*. S. F. Tuan, Ed. (Addison-Wesley, New York, 1985).
2. P. W. Atkins, *Molecular quantum mechanics* (Oxford University Press, Oxford, ed. 2nd, 1983).
3. C. Cohen-Tannoudji, B. Diu, F. Laloë, *Quantum mechanics* (John Wiley and Sons, New York, 1977), vol. 2.
4. M. Tinkham, *Group theory and quantum mechanics* (McGraw-Hill, New York, 1964).
5. J. D. Jackson, *Classical Electrodynamics* (John Wiley and Sons, New York, ed. Second Edition, 1962).
6. R. Shen, *The principles of nonlinear optics* (Wiley Interscience, New York, 1984).
7. C. Kittel, *Introduction to Solid State Physics* (John Wiley and Sons, New York, ed. 6th, 1986).

8. N. W. Ashcroft, N. D. Mermin, *Solid state physics.* D. G. Crane, Ed. (Saunders College, Philadelphia, 1976).
9. B. Segall, *Phys. Rev.* **124**, 1797 (1961).
10. G. A. Burdick, *Phys. Rev.* **129**, 138 (1963).
11. G. Leschik, R. Courths, H. Wern, S. Hüfner, H. Eckardt, J. Noffke, *Solid State Comm.* **52**, 221-225 (1984).
12. J. H. Weaver, C. Krafka, D. W. Lynch, E. E. Koch, *Optical Constants of Materials, Part 1, Physics Data* (1981).
13. P. N. Ray, J. Chowdhuri, S. Chatterjee, *J. Phys. F* **13**, 2569-2580 (1983).
14. S. I. Anisimov, B. L. Kapeliovich, T. L. Perel'man, *Sov. Phys. JETP* **39**, 375 (1974).
15. P. B. Corkum, F. Brunel, N. K. Sherman, T. Srinivasan-Rao, *Phys. Rev. Lett.* **61**, 2886-2889 (1988).
16. C. Suárez, W. E. Bron, T. Juhasz, *Phys. Rev. Lett.* **75**, 4536-4539 (1995).
17. W. S. Fann, R. Storz, H. W. K. Tom, J. Bokor, *Phys. Rev. Lett.* **68**, 2834-2837 (1992).
18. S. Ogawa, H. Petek, *Surf. Sci.* **357-358**, 585-594 (1996).
19. C.-K. Sun, F. Vallée, L. H. Acioli, E. P. Ippen, J. G. Fujimoto, *Phys. Rev. B* **50**, 15337-15347 (1994).
20. B. Feuerbacher, B. Fitton, R. F. Willis, Eds., *Photoemission and the electronic properties of surfaces* (Wiley, New York, 1978).
21. H. Lüth, *Surfaces and interfaces of solids.* G. Ertl, R. Gomer, D. L. Mills, Eds., Springer Series in Surface Science (Springer-Verlag, Berlin, 1993), vol. 15.
22. R. G. Musket, W. McLean, C. A. Colmenares, D. M. Makowiecki, W. J. Siekhaus, *Applications of Surf. Sci.* **10**, 143-207 (1982).
23. E. M. Purcell, *Electricity and magnetism*, Berkeley Physics Course (McGraw-Hill Book Company, New York, ed. 2, 1985).
24. J. Grimblot, A. C. Luntz, D. E. Fowler, *Journal of electron spectroscopy and related phenomena* **52**, 161-174 (1990).
25. L. E. Davis, N. C. MacDonald, P. W. Palmberg, G. E. Riach, R. E. Weber, *Handbook of auger electron spectroscopy* (Physical Electronics Industries, Inc., Eden Prairie, ed. 2nd, 1976).
26. H. Steininger, S. Lehwald, H. Ibach, *Surface Science* **123**, 1-17 (1982).
27. J. Gland, V. Secton, G. Fisher, *Surface Science* **95**, 587-602 (1980).
28. W. D. Mieher, W. Ho, *J. Chem. Phys.* **99**, 9279-9295 (1993).
29. T. A. Germer, J. C. Stephenson, E. J. Heilweil, R. R. Cavanagh, *Phys. Rev. Lett.* **71**, 3327 (1993).
30. T. A. Germer, J. C. Stephenson, E. J. Heilweil, R. R. Cavanagh, *J. Chem. Phys.* **101**, 1704 (1994).
31. J. Wintterlin, R. Schuster, G. Ertl, *Phys. Rev. Lett.* **77**, 123-126 (1996).
32. J. Gland, E. Kollin, *J. Chem. Phys.* **78**, 963-974 (1983).
33. T. Matsushima, *Surf. Sci.* **127**, 403-423 (1983).
34. K.-H. Allers, H. Pfnür, P. Feulner, D. Menzel, *J. Chem. Phys.* **100**, 3985-3998 (1994).
35. M. Orchin, H. H. Jaffé, *Symmetry, Orbitals, and Spectra (S.O.S.)* (Wiley-Interscience, New York, 1971).
36. B. C. Stipe, M. A. Rezaei, W. Ho, S. Gao, M. Persson, B. I. Lundqvist, *Phys. Rev. Lett.* **78**, 4410-4413 (1997).
37. A. Eichler, J. Hafner, *Phys. Rev. Lett.* **79**, 4481-4484 (1997).

38. W. Eberhardt, T. H. Upton, S. Cramm, L. Incoccia, *J. Vac. Sci. Tech. A* **6**, 876-877 (1988).
39. W. Wurth, et al., *Phys. Rev. Lett.* **65**, 2426-2429 (1990).
40. J. L. Gland, V. N. Korchak, *Surf. Sci.* **75**, 733 (1978).
41. A. W. E. Chan, R. Hoffmann, W. Ho, *Langmuir* **8**, 1111-1119 (1992).
42. D. A. Outka, J. Stöhr, W. Jark, P. Stevens, J. Solomon, R. J. Madix, *Phys. Rev. B* **35**, 4119-4122 (1987).
43. B. Hellsing, *Surf. Sci.* **282**, 216-228 (1993).
44. I. Panas, P. Siegbahn, *Chem. Phys. Lett.* **153**, 458-464 (1988).
45. N. V. Smith, *Phys. Rev. B* **9**, 1365 (1974).
46. N. Avery, *Chem. Phys. Lett.* **96**, 371-373 (1983).
47. K. Mortensen, C. Klink, F. Jensen, F. Besenbacher, I. Stensgaard, *Surface Science* **220**, L701-L708 (1989).
48. K. Sawabe, Y. Matsumoto, *Chem. Phys. Lett.* **194**, 45-50 (1992).
49. J. A. Prybyla, T. F. Heinz, J. A. Misewich, M. M. T. Loy, J. H. Glownia, *Phys. Rev. Lett.* **64**, 1537-1540 (1990).
50. F. Budde, T. F. Heinz, M. M. T. Loy, J. A. Misewich, F. de Rougemont, H. Zacharias, *Phys. Rev. Lett.* **66**, 3024-3027 (1991).
51. F.-J. Kao, D. G. Busch, D. G. da Costa, W. Ho, *Phys. Rev. Lett* **70**, 4098-4101 (1993).
52. S. Deliwala, R. J. Finlay, J. R. Goldman, T. H. Her, W. D. Mieher, E. Mazur, *Chem. Phys. Lett.* **242**, 617-622 (1995).
53. R. J. Finlay, T.-H. Her, C. Wu, E. Mazur, *Chem. Phys. Lett.* **274**, 499-504 (1997).
54. D. G. Busch, W. Ho, *Phys. Rev. Lett.* **77**, 1338-1341 (1996).
55. F. Budde, T. F. Heinz, A. Kalamarides, M. M. T. Loy, J. A. Misewich, *Surf. Sci.* **283**, 143-157 (1993).
56. J. A. Prybyla, H. W. K. Tom, G. D. Aumiller, *Phys. Rev. Lett.* **68**, 503-506 (1992).
57. L. M. Struck, L. J. Richter, S. A. Buntin, R. R. Cavanagh, J. C. Stephenson, *Phys. Rev. Lett.* **77**, 4576-4579 (1996).
58. R. J. Finlay, S. Deliwala, J. R. Goldman, T. H. Her, W. D. Mieher, C. Wu, E. Mazur, in *Laser Techniques for Surface Science II*. (1995), vol. 2547, pp. 218-226.
59. J. A. Misewich, A. Kalamarides, T. F. Heinz, U. Höfer, M. M. T. Loy, *J. Chem. Phys.* **100**, 736-739 (1994).
60. D. G. Busch, S. Gao, R. A. Pelak, M. F. Booth, W. Ho, *Phys. Rev. Lett.* **75**, 673-676 (1995).
61. F.-J. Kao, D. G. Busch, D. Cohen, D. G. da Costa, W. Ho, *Phys. Rev. Lett.* **71**, 2094-2097 (1993).
62. X.-Y. Zhu, S. R. Hatch, A. Campion, J. M. White, *J. Chem. Phys.* **91**, 5011-5020 (1989).
63. C. E. Tripa, C. R. Arumaninayagam, J. T. Yates, *J. Chem. Phys.* **105**, 1691-1696 (1996).
64. F. Weik, A. de Meijere, E. Hasselbrink, *J. Chem. Phys.* **99**, 682 (1993).
65. W. Ho, *Surf. Sci.* **363**, 166-178 (1996).
66. C. Springer, M. Head-Gordon, *Chem. Phys.* **205**, 73-89 (1996).
67. J. A. Misewich, T. F. Heinz, D. M. Newns, *Phys. Rev. Lett.* **68**, 3737-3740 (1992).
68. W. S. Fann, R. Storz, H. W. K. Tom, J. Bokor, *Surf. Sci.* **283**, 221-225 (1993).

SOME SELECTED ASPECTS OF LIGHT-MATTER INTERACTION IN SOLIDS

Claus F. Klingshirn

Institut für Angewandte Physik
der Universität Karlsruhe
Kaiserstraße 12
D-76128 Karlsruhe
Germany

ABSTRACT

The purpose of this contribution to the school is twofold. In the first part (chapters I to IV) some basic concepts of solid state physics are introduced in a tutorial way like the elementary excitations, the quasiparticles and interaction processes between them and with the electromagnetic light field. In the second part (chapters V to VII) some specific problems are treated in more detail like the concept of polaritons including systems of reduced dimensionality and disordered systems, the possibilities to determine the characteristic parameters of a resonance by linear and nonlinear spectroscopy in the frequency domain and finally some selected examples of incoherent population dynamics. Finally there will be a short conclusion and outlook.

I. INTRODUCTION

In this section we shall repeat some basic concepts of physics like the harmonic oscillator or harmonic waves. Later on we shall show how more complicated systems can be traced back to these simple concepts.

I.A. Harmonic oscillators and waves

The concepts of harmonic osciallators and waves are crucial for the understanding of physics. In principle a problem can be considered to be "solved" if it can be reformulated in the above categories. Therefore we repeat here shortly some aspects of their description in classical mechanics and in quantum mechanics.

Ultrafast Dynamics of Quantum Systems: Physical Processes and Spectroscopic Techniques, Edited by Di Bartolo and Gambarota, Plenum Press, New York, 1998

The equation of motion for a one-dimensional harmonic osciallator like a mass on a spring (Fig 1) without damping and external forces is given by eq. (1)

$$M \frac{\partial^2 u\ (t)}{\partial t^2} + D\ u(t) = 0 \tag{1}$$

where u(t) is the elongation from the equilibrium position, M is the mass and D the spring constant.

Equation (1) can be brought into the general form

$$\frac{\partial^2 u\ (t)}{\partial t^2} + \omega_0^2\ u(t) = 0 \tag{2}$$

with the (angular) eigenfrequency

$$\omega_0 = (D/M)^{1/2} \tag{3}$$

which is valid for all other harmonic oscillators like for pendula for small amplitudes with $\omega_0 = (g/l)^{1/2}$ where l is the effective length of the pendulum and g the acceleration at the surface of the earth ($g \approx 9.81\ ms^{-2}$) or for LC circuits with $\omega_0 = (LC)^{-1/2}$ where L is the self inductivity and C the capacity.

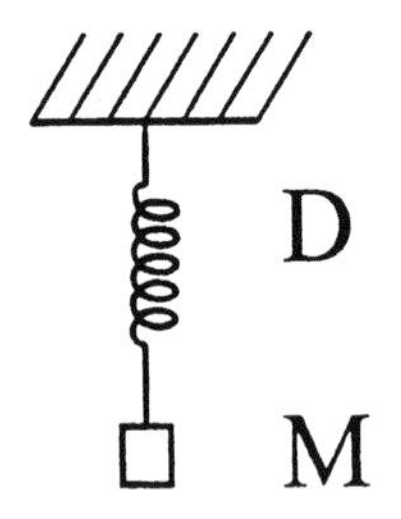

Figure 1 A harmonic oscillator consisting of a mass and spring.

The solution of (2) can be written e.g. as

$$u(t) = a \sin \omega_0 t + b \cos \omega_0 t \tag{4a}$$
$$u(t) = A \sin (\omega_0 t + \alpha) \tag{4b}$$
$$u(t) = A e^{i\alpha} e^{-i\omega_0 t}\ (+c.c.) \tag{4c}$$

If one wants to go from the complex notation of (4c) to reals variables u e.g. as in (4b) one has to add the complex conjugate (c.c.).

In any case there must be two free parameters in the solution to match some boundary conditions since (2) is a differential equation of second order.

It should be noted, that a constant external force i.e. a potential energy which is linear in the elongation u(t) chances the equilibrium position of the harmonic oscillator but not the eigenfrequency ω_0. This can be nicely demonstrated with the system of Fig 1 when gravitation is included as an external force. Equation (1) has to be rewritten in this case in the following form

$$M \frac{\partial^2 u(t)}{\partial t^2} - Mg + D\,u(t) = 0 \tag{5}$$

The substitution $u(t) = v(t) + Mg/D$ leads immediately to

$$\frac{\partial^2 v(t)}{\partial t^2} - \omega_0^2\,v(t) = 0 \tag{6}$$

with the same ω_0 as in (3).

Equation (1) can be extended by including damping (i.e. coupling to an ideal resevoir or heat both) and periodic external forces of frequency ω to give:

$$M \frac{\partial^2 u(t)}{\partial t^2} - \gamma M \frac{\partial u(t)}{\partial t} + D\,u(t) = A_0\,e^{-i\omega t} \tag{7}$$

Generally the damping is introduced as a term proportional to the velocity which simulates to some extend e.g. Stokes law or friction losses. This approach eliminates all memory effects or non-Markovian processes.

Equation (7) is the well-known formula to describe resonance phenomena as shown e.g. in Fig 2 for increasing damping.

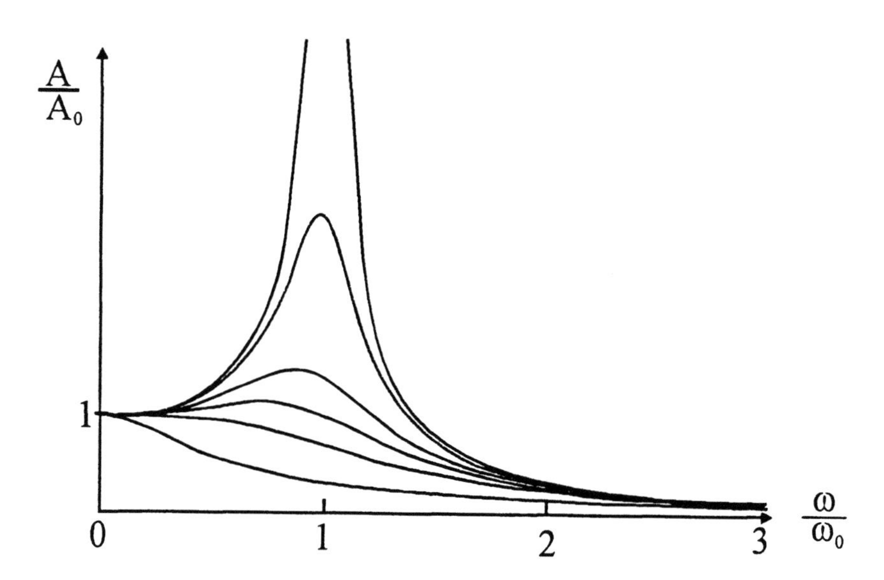

Figure 2 Resonance courves for a driven harmonic osciallator and various values of the damping.

Without damping there is a singularity at $\omega = \omega_0$. For increasing damping the maximum of the amplitude resonance curve decreases and shifts to lower frequencies. The curve which is essentially horizontal up to frequencies $\leq \omega_0$ corresponds to the aperiodic or critical damping which is very important in many technical applications like shock absorbers in cars or loudspeaker boxes in HiFi sets. For even larger damping there is no more osciallatory solution for the free system and the resonance curves start to decay rapidly for $\omega > 0$.

In quantum mechanics we have to consider the Schrödinger equation, which reads in its general form for a one dimensional system

$$H \varphi (x, t) = - \frac{i}{\hbar} \frac{\partial}{\partial t} \varphi (x, t) \tag{8}$$

To find the eigenstates, the time-independent Schrödinger equation has to be solved i.e.

$$H \varphi (x) = E \varphi (x) \tag{9}$$

$$\text{with } \varphi (x, t) = \varphi (x) e^{-i\frac{E}{\hbar}t} \tag{10}$$

In this case H is the sum of kinetic and potential energies. Including the momentum operator

$$p = \frac{\hbar}{i} \nabla \tag{11}$$

we obtain for the system of Fig 1

$$\left[-\frac{\hbar^2}{2M} \frac{\partial^2}{\partial x^2} + \frac{D}{2} x^2 \right] \varphi_n (x) = E_n \varphi_n (x) \tag{12}$$

As is well known, this system has eigenenergies E_n given by

$$E_n = (n + \frac{1}{2}) \hbar \omega_0 \tag{13}$$

The term $\frac{1}{2} \hbar \omega_0$ describes the zero point energy or -fluctuation. The harmonic oscillator can exchange energy with other systems only by integer multiples of $\hbar \omega_0$ the so-called energy quanta.

The wave functions φ_n are shown in Fig 3 for different quantum numbers together with the parabolic potential. For n = 1 the maximum probability to find the system is centered around x = 0. For increasing n the probability is pushed more and more towards the region of the classical turning points. In classical physics the probability to find the systems has actually a singularity at these turning points. This asymptotic coincidence is thus a nice example for the "correspondence" principle which states, that quantum mechanical systems tend to develop towards the classical behaviour for increasing quantum numbers.

When proceeding from the free, undamped harmonic oscillator of eg (2) to free, undamped harmonic waves we get in the simplest case in three dimensions a partial differential equation which contains second derivatives with respect to space and time

$$\frac{\partial^2}{\partial t^2} u(r,t) - C^2 \nabla^2 u(r,t) = 0 \tag{14}$$

where C is a constant and ∇ the Nabla operator, which reads in Cartesian coordinates

$$\nabla = \left(\frac{\partial}{\partial x}, \frac{\partial}{\partial y}, \frac{\partial}{\partial z} \right) \tag{15}$$

The general solution are all functions f (**r**, t) which have a well defined second derivative and which depend on **r** and t in the following way

$$u(\mathbf{r}, t) = u_0 f(\mathbf{kr} - \omega t) \tag{16}$$

where the wavevector **k** with $k = 2\pi\lambda^{1}$ and $\omega = 2\pi T^{-1}$ are connected in the way described below. T and λ are the temporal and spatial period (wavelength) of the wave.

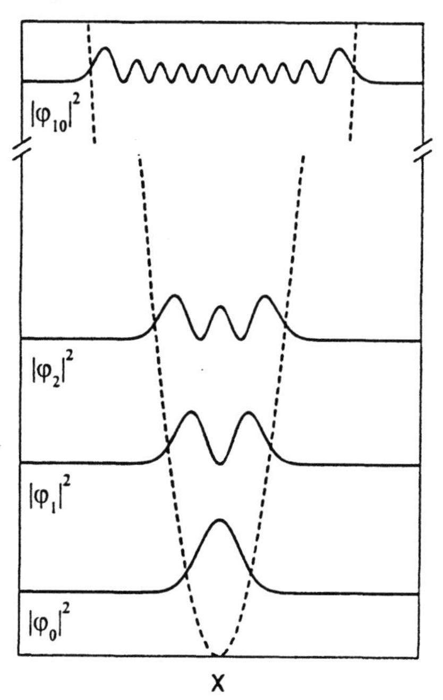

Figure 3 Eigenfunctions and eigenenergies of a one-dimensional harmonic oscillator.

From all the solutions comprised by (16) we select for the following the most simple ones, namely the so-called plane waves

$$u(\mathbf{r}, t) = u_0 \, e^{i(kr-\omega t+\alpha)} \; (+\text{c.c.}) \tag{17}$$

They got their name because the geometric places of constant phase at a fixed time are planes.

Inserting the ansatz (16) or (17) into (14) gives the relation

$$C^2k^2 = \omega^2 \tag{18}$$

This means that C is just the phase velocity

$$\upsilon_{ph} = \frac{\omega}{k} = C \tag{19a}$$

while the group velocity υ_g is given by

$$\upsilon_g = \frac{\partial\omega(k)}{\partial k} = \frac{1}{\hbar}\frac{\partial E(k)}{\partial k} = \frac{1}{\hbar}\,\text{grad}_{\mathbf{k}}\, E(\mathbf{k}) \tag{19b}$$

For the case of electromagnetic radiation in vacuum, C is evidently equal to the vacuum speed of light c.

The relation $\omega(\mathbf{k})$ is called dispersion relation. For photons $\upsilon_{ph} = c$ is constant and the dispersion relation is just a straight line as shown in Fig 4a. Fig 4b shows a quadratic dispersion relation, which is typical for massive particles, and Fig 4c a $|\sin ka|$ type dependence.

In quantum mechanics, the amplitude of the waves is quantized so that the energy of a certain mode $\omega(\mathbf{k})$ is given by

$$E_{n,\mathbf{k}} = (n + \frac{1}{2})\,\hbar\omega(\mathbf{k}) \tag{20}$$

for Bosonic particles.

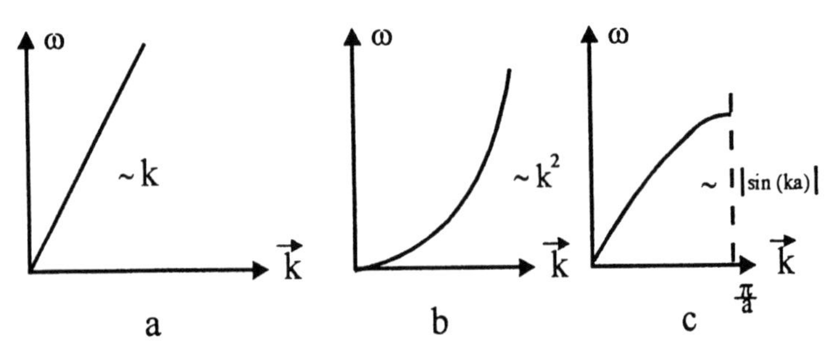

Figure 4 Some examples for dispersion relations.

Again $\frac{1}{2}\hbar\omega(\mathbf{k})$ describes the zero point energy and the wave can exchange energy with other systems only by integer numbers of energy quanta $\hbar\omega(\mathbf{k})$.

The reason why the two quantities ω and $\mathbf{k}$ have been chosen for the dispersion relation is, that there are conservation laws for the energy $\hbar\omega$ and for the quantity $\hbar\mathbf{k}$. For free particles in vacuum like photons, electrons, protons $\hbar\mathbf{k}$ is nothing but the momentum.

In crystals the quantity $\hbar\mathbf{k}$ is usually called quasi-momentum or psendo - or crystal momentum or, in the every day language in the lab simply "momentum" for two reasons. One is, that the eigenstates in a solid e.g. in a periodic lattice are often no eigenfunctions of the momentum operator given in (11) as for the case of Bloch-waves (see below) though they have a well defined $\mathbf{k}$ vector. The other reason is, that $\hbar\mathbf{k}$ is conserved in periodic structures with primitive translation vectors $\mathbf{a}_i$, i = 1, 2, 3 only modulo integer multiples of the so-called reciprocal lattice vectors **bi** given by

$$\mathbf{b}_i = 2\pi\frac{a_j \times a_k}{a_i\,(a_j \times a_k)} \tag{21}$$

i.e. states with $\mathbf{k}$ and $\mathbf{k} + \mathbf{G}$ are equivalent where

$$\mathbf{G} = \sum h_i\mathbf{b}_i;\ h_i = 0,\ \pm1, \pm2,... \tag{22}$$

As a consequence, the $\mathbf{k}$ space can be limited in periodic structures to the so-called first Brillouin zone which is defined as the cell around $\mathbf{k} = 0$ (the so-called Γ point) which contains all $\mathbf{k}$ which are closer to the origin than to any other point of the reciprocal lattice and which is obtained by constructing perpendicular planes in the middle of the vectors connecting the point $\mathbf{k} = 0$ with the (next) nearest points in $\mathbf{k}$-space. For a simple cubic lattice the first Brillouin zone extends from

$$-\frac{\pi}{a} \le k_i \le \frac{\pi}{a} \quad \text{with } i = x, y, z \tag{23}$$

Apart from defining the (quasi-)momentum of a (quasi-)particle by $\hbar\mathbf{k}$ there is the possibility to introduce it via the (19b) if the group velocity is multiplied by a suitable mass.

For further considerations of the topic "quasi momentum" see also the note by v. Baltz to this school.

Special attention should be paid to the fact that at plane surfaces or interfaces only the component of $\hbar\mathbf{k}$ parallel to the interface is conserved, due to the translational invariance of the problem in this direction (eventually modulo $\mathbf{b}_i$) while the normal component may change. Examples for this statement are e.g. the reflection and refraction of light described by Snellius law or a ball bouncing elastically off a wall.

To conclude this subsection we want to give some statements about the density of states. Since plane waves of the form (17) cannot be normalized in an infinite volume, one assumes a large but finite system e.g. a cube which contains all physically relevant parts of the system. The boundary conditions e.g. either nodes at the surface of the cube or periodic boundary conditions result in equidistantly spaced values of k_i on all three Karthesian coordinates in agreement with the uncertainly principle. If one wants to knwo the number of states N(k) in polar coordinates in a d-dimensional space contained in a shell betwen k and k+dk one finds

$$N(k)\,dk \sim k^{d-1}\,dk \tag{24}$$

This statement is correct for all plane-wave - like excitations or (quasi-)particles in vacuum and in solid.

If one wants to deduce from (24) the density of states (DOS) D(E) as a function of energy, the specific dispersion relution E(**k**) has to be known to come from N(k) dk to D(E) dE. For the dispersion relation of Fig 4a E ~ **k** one finds evidently

$$D(E)\,dE \sim E^{d-1}\,dE \tag{25a}$$

and for E ~ $\mathbf{k}^2$ corresponding to Fig 4b one gets

$$D(E)\,dE \sim E^{\frac{d}{2}-1}\,dE \tag{25b}$$

the relations of (25) are shown in Fig 5 for d = 3, 2 and 1. For quasi zero-dimensional systems the DOS would develop to a series of δ-functions.

I.B. Coupling of harmonic oscillators and waves

A system which has a lot of interest both in classical physics and in quantum mechanics are coupled harmonic oscillators.

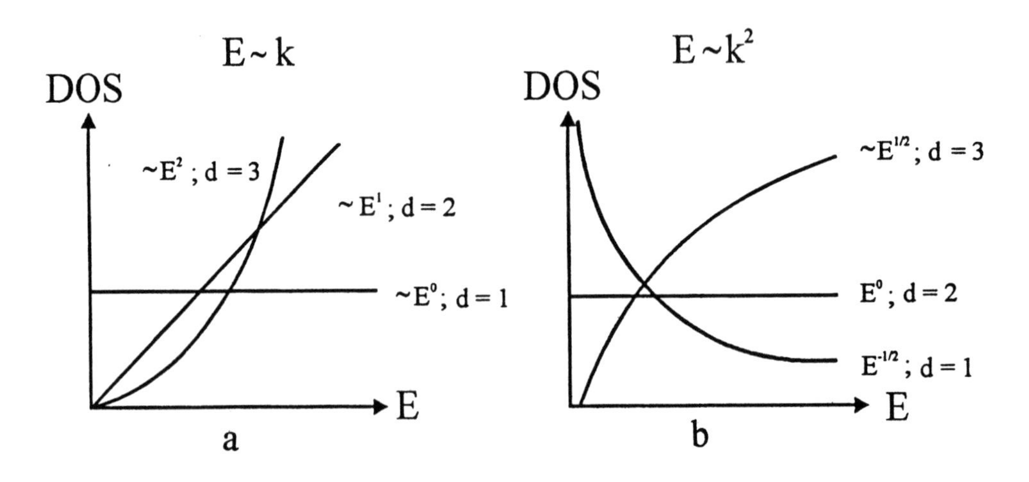

Figure 5 The density of states DOS for two different dispersion relations E(**k**) and three-, two- and one-dimensional systems.

As a simple system we consider two pendula with small amplitude as shown in Fig 6 coupled via a soft spring representing a coupling constant χ.

Their equation of motion read

$$\frac{\partial^2 u_1}{\partial t^2} + \omega_1^2 u_1 + \alpha(u_1 - u_2) = 0 \tag{28a}$$

$$\frac{\partial^2 u_2}{\partial t^2} + \omega_2^2 u_2 - \alpha(u_1 - u_2) = 0 \tag{28b}$$

with $\omega_i^2 = g/l_i$ $i = 1, 2$

The ansatz for the general solution is a superposition u of the two oscillations with in the moment unknown eigenfrequency ω_e

$$u = u_1 + u_2 = u_1^0 e^{-i\omega_e t} + u_2^0 e^{-i\omega_e t} \tag{29}$$

Inserting (29) into (28) gives

$$(\omega_1^2 - \omega_e^2 + \alpha) u_1^0 - \alpha u_2^0 = 0 \tag{30}$$

$$-\alpha u_1^0 + (\omega_2^2 - \omega_e^2 - \alpha) u_2^0 = 0 \tag{31}$$

In order to have a nontrivial solution for u_i^0 ($i = 1, 2$) the determinate of the coefficients has to vanish.

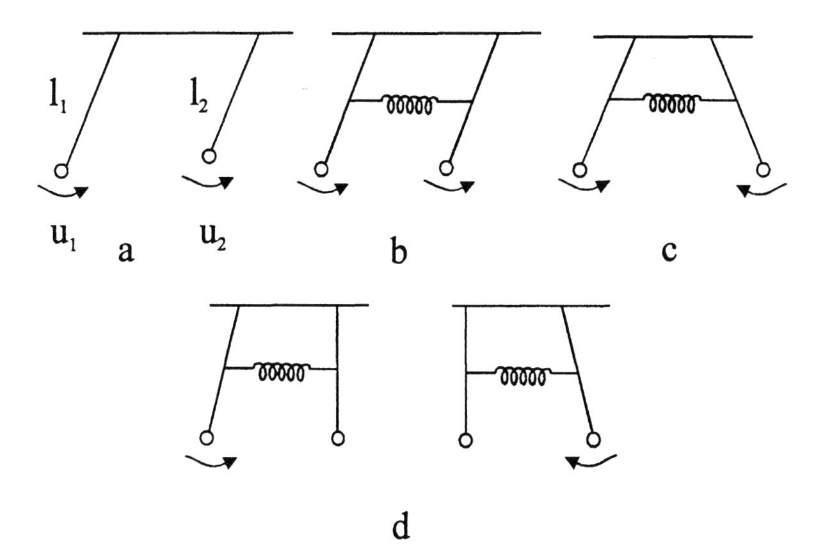

Figure 6 Two pendula which are uncoupled (a) or coupled by a weak spring and which are oscillating in phase (b) or in antiphase (c) or only one of the two pendula is oscillating and the other one is at rest (d).

This leads to a secular equation with the solutions

$$\omega_{e\,1,2}^2 = \frac{1}{2}\left[\omega_1^2 + \omega_2^2 + 2\alpha \pm \sqrt{(\omega_1^2 - \omega_2^2)^2 + 4\alpha^2}\right] \qquad (32)$$

For the two solutions of ω_e^2 we can find then the relative amplitudes of the osciallation modes e.g.

$$\frac{u_2^0}{u_1^0} = \frac{1}{\alpha}\left(\omega_1^2 + \alpha - \frac{1}{2}\left[\omega_1^2 + \omega_2^2 + 2\alpha \pm \sqrt{(\omega_1^2 - \omega_2^2)^2 + 4\alpha^2}\right.\right) \qquad (33)$$

We consider now the case of two pendula or generally of two harmonic oscillators which are degenerate without coupling i.e.

$$\omega_1 = \omega_2 = \omega_0 \qquad (34)$$

Then (32) simplifies to the expressions below with the resulting ratios of the amplitudes u_i

$$\omega_{e1}^2 = \omega_0^2 \rightarrow \frac{u_2^0}{u_1^0} = 1 \qquad (35)$$

$$\omega_{e2}^2 = \omega_0^2 + 2\alpha \rightarrow \frac{u_2^0}{u_1^0} = -1 \qquad (36)$$

This means a small but finite coupling α selects from all possible linear combinations of the oscillations of u_1 and u_2 as eigenmodes the symmetric and antisymative ones of Fig 6b and c. The splitting of the frequency is intuitively clear since the spring is not elongated in the symmetric mode of Fig 6b and produces consequently no additional restoring force while it does in the case of Fig 6c resulting in a larger eigenfrequency.

The general motion of the coupled pendula is a superposition of both modes and leads to the well known beating phenomena i.e. a periodic transition between the two configurations of Fig 6d.

If we start to make the eigenfrequencies $\omega_{1,2}$ different the eigenmodes develop continously from the ones of Fig 6b, c to the ones of Fig 6d, i.e. one of the oscillators has a very large amplitude which is in phase or antiphase with respect to the other which has only small amplitude. For the case $\omega_2^2 >> \omega_1^2, \alpha$ we find e.g.

$$\lim_{\omega_2 >> \omega_1, \alpha} u_2^0 / u_1^0 = -\omega_2^2 / \alpha \qquad (37)$$

The quantum mechanical description gives qualitatively the same results. If we have two uncoupled systems described by the Hamiltonious H_1^0 and H_2^0 we get in matrix notation the eigenenergies from the solution of

$$\begin{vmatrix} H_{11}^0 - E & 0 \\ 0 & H_{22}^0 - E \end{vmatrix} = 0 \tag{38}$$

i.e. both systems have their unpertubed eigenfrequencies

$$H_{ii}^0 = E_i \quad i = 1, 2 \tag{39}$$

If a perturbation H' is introduced so that $H = H^0 + H'$ we may get finite coupling matrix elements in the off diagonal resulting in (40)

$$\begin{vmatrix} H_{11}^0 - E & H_{12} \\ H_{21} & H_{22}^0 - E \end{vmatrix} = 0 \tag{40}$$

causing again a finite splitting. Though this phenomenon is known already in classical physics, it is often referred to as quantum-mechanical non-crossing rule or avoided crossing or level repulsion. This means if we have two classical oscillators or quantum-mechanical systems and we can vary e.g. the eigenfrequency or -energy of one of them by variation of a system parameter (e.g. the length of the pendulum in Fig 6 or, to address examples we shall see later the wave vector or the carrier density) we get the behaviour shown shematically in Fig. 7.

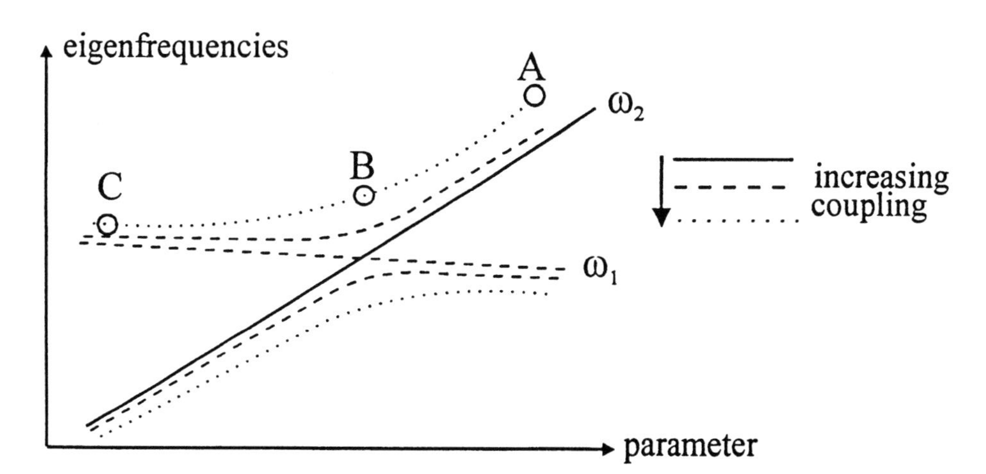

Figure 7 Schematic illustration of the non-crossing rule. The coupling between the two systems is supposed to be zero for the solid lines and to increase from the dashed to the dotted line.

Without coupling the eigenfrequencies of the two systems cross as a function of the parameter. With increasing coupling there is an increasing splitting of the former cross-over point. Furthermore we get with changing parameter a gradual transition from one state to the other. If we start e.g. at point A essentially the system 2 is excited. With decreasing parameter we come to point B where both systems have in the common eigenmodes approximately equal amplitudes. When we approach finally point C the system has essentially the character of system 1. The analog behaviour is found on the lower branch. If a certain perturbation can couple different states depends e.g. on the symmetry, if parity is a good quantum number, or more generally on group-theoretical considerations.

To conclude this subsection we want to outline what happens if we consider more than two coupled osciallators or quantum mechanical systems.

We have seen that the coupling of two identical systems leads to the splitting of an eigenvalue into two, a coupling of three identical systems will give three eigenfrequencies and the coupling of 10^{23} systems (which is roughly the number of atoms contained in 1 cm^3 of a solid) results in 10^{23} eigenfrequencies, which are usually so narrowly spaced that they form for all practical purposes a continous energy band. We come back to this aspect in subsection II.

The width of the band B is proportional to the coupling between the constituents. If we have no coupling the dispersion relation is flat as shown in Fig 8a. If we impose a wave-like excitation on the system the frequency ω_0 will be independent of the wavelength or of **k**. For finite coupling we get however a finite band width B as shown in Fig 8b.

If we produce a wave-paket in the system by exciting one or a few neighbouring osciallators it will stay where it is in the situation of Fig 8a but will propagate with a finite group velocity υ_g in the situation of Fig 8b since the group velocity is of a wave-paket centered around **k** given by the slope of the dispersion as stated with (19b).

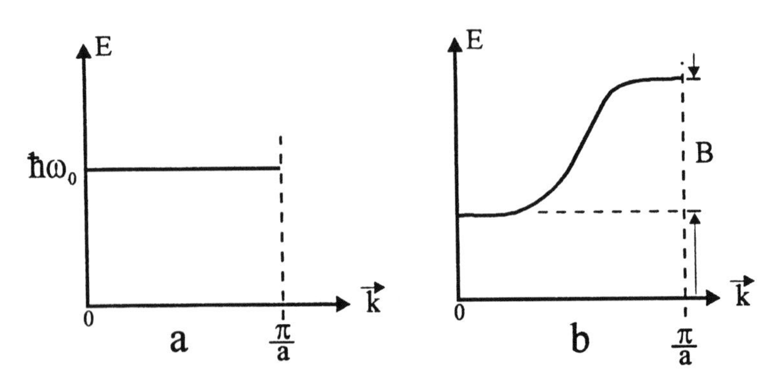

Figure 8 The dispersion relation of a system of uncoupled (a) and coupled (b) identical oscillators.

II. THE CONCEPT OF ELEMENTARYEXCITATIONS AND OF QUASIPARTICLES

In this section we introduce the concept of quasiparticles starting with some general considerations. The concepts developped in this section and partly also those of the preceeding one can be found in various degrees of sophistication in many text books on solid state physics (e.g. [1-6]) on semiconductor physics and optics (e.g. [7-11]) and last but not least in various contributions to the present and to preceeding courses of this school [12]. In order not to vaste too much place, the outline of the concepts developped in the following sections will be as simple and as short as possible. For more details the reader is referred e.g. to [1-12].

II.A. The basic concept

If we want to understand the properties of a solid, in principle we simply have to solve the Schrödinger equation

$$H\Psi = E\Psi \tag{41}$$

The Hamilton H and the eigenfunctions Ψ depend on the coordinates $\mathbf{R}_i$ of the ion cores with charge Z_ie and mass M_i and on the coordinates $\mathbf{r}_j$ of the electrons including spin. We get

$$H = \sum -\frac{\hbar^2}{2M_i}\nabla_i^2 + \sum -\frac{\hbar^2}{2m_e}\nabla_j^2 + \frac{e^2}{4\pi\epsilon_0}\left[\sum \frac{1}{|r_j' - r_j|} + \sum \frac{Z_i Z_i'}{|R_i - R_i'|} - \sum \frac{Z_i}{|r_j - R_i|}\right] \tag{42}$$

The problem is that the sums in (42) run over roughly 10^{23} particles per cm^3. Due to the Coulomb-coupling not even a separation is possible. Consequently there is no realistic chance to solve (41, 42) from first principles. The best what present day computers can do is to handle in this approximation molecules with up to about 10^2 atoms and electrons.

The solution is to consider only subsystems of the solid and, while doing so, not to care about the rest. The task is then, to find in this subsystem normal modes, the equation of motion of which looks - eventually after some approximations - like a harmonic oscillator or wave. Once this is accomplished the results of the preceeding section can be applied. A dispersion relation $\omega(\mathbf{k})$ can be established and the energy quanta $E(\mathbf{k})$ or $\hbar\omega(\mathbf{k})$ characterized by their energy and quasimonentum $\hbar\mathbf{k}$ can be introduced as the so-called quasiparticles. The name quasiparticles is given to account for the facts, that these quanta exist only in matter but not in vacuum and that the quantity $\hbar\mathbf{k}$ is a quasimomentum as explained earlier.

The price we have to pay for this approach is, that interaction processes between the various quasiparticles are not included.

The hope is that these interaction-processes are weak and can eventually be treated in perturbation theory and converge rapidly.

In the following we shall outline the above concept for phonons, plasmons, magnons and excitons. We shall start in all cases with volume or bulk bulk material but give at the end also a short outlock on systems of reduced dimensionality since a predominant fraction of solid state research is presently devoted to them.

II.B. Phonons

In the sense of the procedure above we consider first the collective motions of the ion cores. Bearing in mind, that

$$M_i >> m_e \tag{43}$$

in (42) we can state that the electrons can adiabatically follow the motion of the ion cores but not vice versa. This is the so-called adiabatic or Born-Oppenheimer approximation.

When we define the interaction potential V(r) between the ion-cores below, we do not bother about the electrons because they arrange themselves according to the position of the ions, i.e. they are simply included in V(r).

The interaction potential between atoms in solids (and molecules) looks in all cases schematically like in Fig 9.

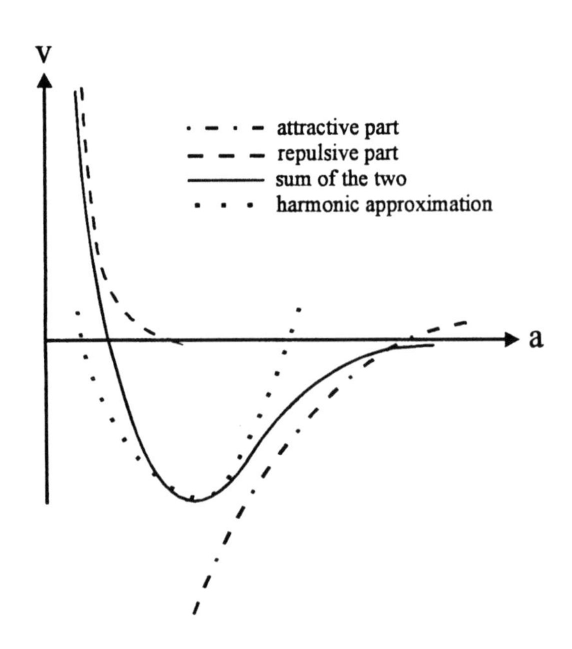

Figure 9 The interaction potential between neighbouring atoms in a solid as a function of their distance (i.e. lattice constant) a.

For all types of binding (ionic, covalent, metallic, etc.) there must be an attractive part otherwise the solid would not be formed. For small distances there must be in addition a repulsive part to prevent the solid from collapsing. This part is produced e.g. by the exchange interaction (or Pauli's exlusion principle) which starts to work, when the filled inneratomic shells start to overlap with decreasing distance. The sum has a minimum which gives the equilibrium distance. An approximation in the sense of IIA is now to approximate the interaction potential by a harmonic one as shown in Fig 9. This is a reasonable approximation for small amplitudes and will result in some interaction processes for larger amplitudes as outlined in IIIA.

The harmonic approximation allows us to represent the interaction between the atoms with mass M by springs with spring constant D. For a three-dimensional solid we could introduce different springs to represent next and next nearest neighbour interactions or the coupling between the outer electron shells and the inner ion core. Instead we use the simpliest possible model (see Fig 10a) which gives however already insight in the most important features of the collective motion of the ions.

The equation of motion for the atom at place n reads for a longitudinal wave, i.e. elongation $u_n(t)$ and direction of propagation along the chain

$$M \frac{\partial^2 u_n(t)}{\partial t^2} = D\left[(u_{n+1}(t)-u_n(t))-(u_n(t)-u_{n-1}(t))\right]$$

$$= D\left[u_{n+1}(t)-2u_n(t)+u_{n-1}(t)\right] \tag{44}$$

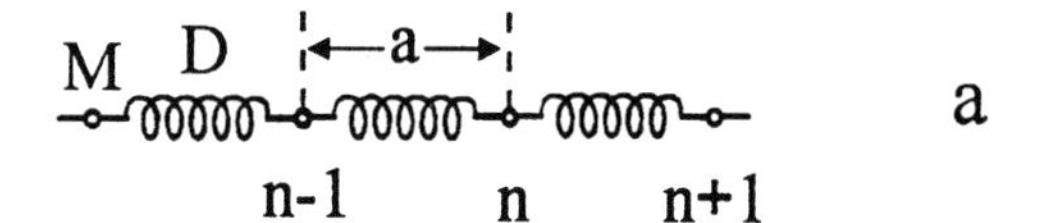

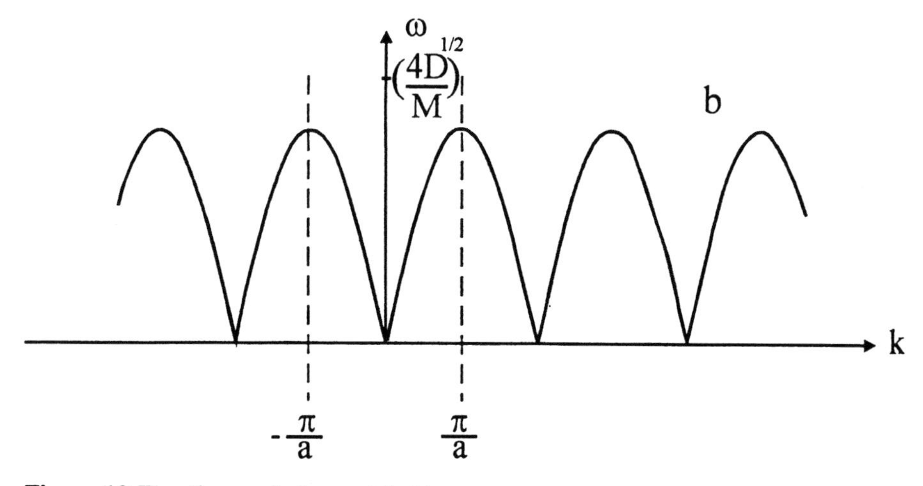

Figure 10 The linear chain model (a) and the resulting dispersion relation (b).

The second derivative with respect to space which appears in the treatment of a homogeneous string is replaced in (44) by a second order difference expression. Equation (44) looks almost like the one of a harmonic oscillator. We have now two ways to find a solution. One would be to define proper linear combinations of the $u_n(t)$, so-called normal modes (e.g. by closing the linear chain to a large ring), for which (44) is identical to a harmonic oscillator. The eigenmodes with frequency ω and wavevector $\mathbf{k}$ would then be quantized. These quanta of the collective excitation are then called quasi particles characterized by the energy $\hbar\omega$ and quasi-momentum $\hbar\mathbf{k}$. In the present case the name of these quasiparticles is phonons. The other way to treat (44) is to find an ansatz, which solves (44) directly and gives also the dispersion relation $\omega(\mathbf{k})$ and thus $\hbar\omega(\hbar\mathbf{k})$.

We choose here the second approach for simplificity and stress the fact, that the result for $\hbar\omega(\mathbf{k})$ must be the same for both ways, because the dispersion relation of phonons cannot depend on the way in which we calculated it.

The ansatz we choose is

$$u_n(t) = u_n^0 \exp\left[i(kna - \omega t)\right] \tag{45a}$$

$$u_{n+1}(t) = u_{n\pm 1}^0 \exp\left[i(k(n\pm 1)a - \omega t)\right] \tag{45b}$$

with the reasonable assumption that the amplitudes of all atoms are equal i.e.

$$u_n^0 = u_{n\pm 1}^0 \tag{46}$$

we find when inserting (45) into (46) the dispersion relation of longitudinal phonons

$$\hbar\omega(\mathbf{k}) = \left(\frac{4D}{m}\right)^{1/2} \left|\sin\frac{ka}{2}\right| \tag{47}$$

which is shown in Fig 10b. With the argument given with eqs. (21, 22) we can restrict ourselves to the interval $-\frac{\pi}{a} \le k \le +\frac{\pi}{a}$ i.e. the first Brillouin zone. If we go to a three-dimensional solid (Fig 11) we find, that we have apart from the longitudinal eigenmode for every $\mathbf{k}$ vector in addition two transverse eigenmodes. These two transverse modes can be degenerate or not, depending on the symmetry of the crystal and on the direction of $\mathbf{k}$ in the crystal.

In Fig 11 we show schematically the two cases for two different directions in $\mathbf{k}$-space. The three phonon branches starting at $\omega = 0$ for $\mathbf{k} = 0$ are always there and are called acoustic branches because they describe the propagation of sound in matter. If we have $s > 1$ atoms per unit cell then there are in addition

$$3s - 3 \tag{48}$$

so-called optic branches. They are called optic because they couple to the electromagnetic field if they carry an electric dipole moment.

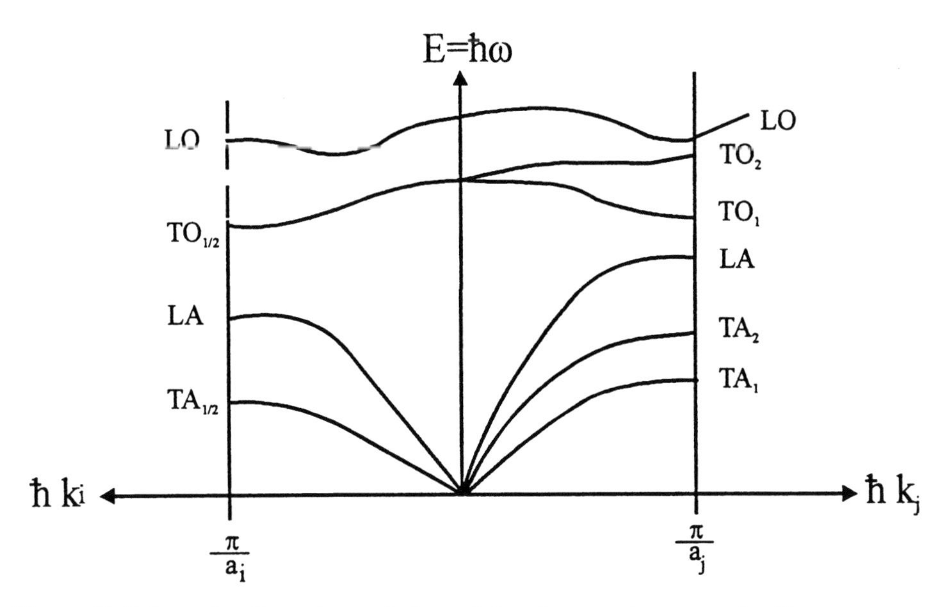

Figure 11 Schematic dispersion relation of the phonons in a crystal with two atoms per unit cell for different directions in **k**-space.

The existence of phonons and their disperison relation are very well established both experimentally and theoretically. The main experimental methods are inelastic neutron scattering. In this case a monochromatized beam of thermal neutrons is scattered from the sample. The conservation laws of energy and (quasi-) momentum read for the creation or annihilation of a phonon $\hbar\omega(\mathbf{k})$

$$\frac{\hbar k_f^2}{2m} = \frac{\hbar k_i^2}{2m} \pm \hbar\omega(\boldsymbol{k}) \tag{49a}$$

$$\boldsymbol{k}_f = \boldsymbol{k}_i \pm \boldsymbol{k} + \boldsymbol{G} \tag{49b}$$

where $\mathbf{k}_i$ and $\mathbf{k}_f$ are the wavevectors of the incident and scattered neutron. Since the wavevectors of thermal neutrons are comparable or larger than the first Brillouin zone, reciprocal lattice vectors **G** may come into play as indicated in (49b).

In optical spectroscopy one can see the optical phonon modes which are infrared active in the reflection spectrum as a stop-band extending roughly from $\hbar\omega_{T0}$ (k = 0) to $\hbar\omega_{L0}$ (k = 0). It should be noted that a finite coupling of light to a resonance in matter i.e. a finite (dipole) transition matrix element is necessarily connected with a finite splitting Δ_{LT} at k = 0 with the appearance of a "Reststrahl" or stop-band i.e. a region with R = 1 for vanishing damping between $\omega_T \leq \omega \leq \omega_L$, and as we shall see later, with the appearance of polaritons.

$$\langle f|H^D|i\rangle \neq 0 \;\leftrightarrow\; \Delta_{LT} \neq 0 \tag{50}$$

If phonons modulate the polarizability of matter, they show up in inelastic light scattering as Raman-lines for optic phonons and as Brillouin-lines for acoustic phonons. Since light has up to the far UV always wavevectors, which are much smaller than the first Brillouin zone, no reciprocal lattice vectors come into play in energy and momentum conservation in the crystal when only one phonon is created or annihilated i.e.

$$\hbar\omega_R = \hbar\omega_i \pm \hbar\omega(\mathbf{k}) \tag{51a}$$

$$\mathbf{k}_R = \mathbf{k}_i \pm \mathbf{k} \tag{51b}$$

where the indices i and R stand for the incident and scattered light wave. In addition refraction has to be considered for the in- and outgoing light beam at the surface. The frequency shift of the Raman lines corresponds to the eigenenergies of the Raman-active optical phonons, which are typically in the range from 10 to 100meV while the energy shift in Brillouin scattering is typically below or around 1meV due to the smallness of $\mathbf{k}_{i,R}$.

If a crystal has inversion symmetry like Si or NaCl so that parity is a good quantum number under the inversion operation $\mathbf{r} \to -\mathbf{r}$ odd optical phonons are IR active if the material has at least partial ionic binding while even phonons are Raman active. In crystals without inversion symmetry optical phonons may be both IR and Raman active.

Many examples of phonon dispersion curves determined by the various techniques mentioned above can be found e.g. in the textbooks [1-7, 10-12] including the contributions by Cardona and by Benedeck to this school or in [13].

If we go to systems of reduced dimensionality like a superlattice (Fig 12a) which consists of thin (typically several atomic layers thick) alternating layers of two different materials, the dispersion relations are modified with respect to those of the parent bulk materials.

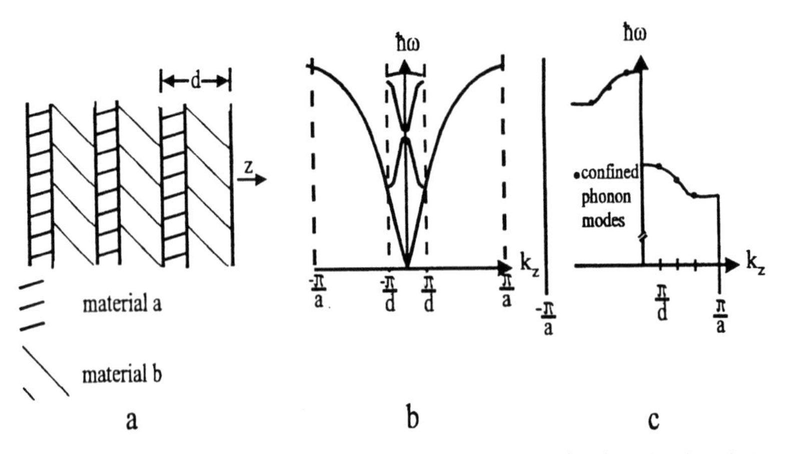

Figure 12 Schematic drawing of a superlattice (a) the folding back of the acoustic phonons into the new Brillouin zone (b) and confined optical phonon modes (c).

The most significant changes occur for wavevectors parallel to the growth (or z-) direction.

The artificial periodicity d reduces the Brillouin zone from

$$-\frac{\pi}{a} \leq k_z \leq \frac{\pi}{a} \text{ to } -\frac{\pi}{d} \leq k_2 \leq \frac{\pi}{d} \tag{52}$$

If the dispersion relation of the phonons in the two parent materials overlap, as is always the case for the beginning of the acoustic phonon branches, these branches are folded back into the new small Brillouin zone with reciprocal lattice vector $\frac{2\pi}{d}$ as shown schematically in Fig 12b. If they couple or if the bulk sound velocities in both materials are different small frequency gaps may open at k = 0 and $k_z = \pm\frac{\pi}{d}$. The states at k ≈ 0 of the backfolded modes can be seen in IR or Ramanspectroscopy depending on the selection rules. Experimental examples are found e.g. in [14].

If the phonon dispersions do not overlap energetically as is e.g. the case for the optic branches in GaAs/AlAs superlattices, the dispersion is not folded back. Instead confined (optical) phonons are formed, i.e. the lattice oscillates essentially only in one material or in the other. The wavelengths in z direction are then so, that an integer number of half wavelengths fits into the layer resulting in discrete points on the bulk-dispersion relation as indicated in Fig 12c. Experimental examples for their observation are found e.g. in [15].

II.C. Plasmons

As a next example we consider the collective motion of a gas of free carriers with density n, mass m and charge e with respect to a rigid background of oppositely charged ion cores, i.e. we do not consider for the moment phonons but come back to this aspect in subsection III.

If we elongate the ensemble of free electrons, e.g. in a metal, with respect to the positive background as shown schematically in Fig 13a by an amount u(t) we get surfaces charges ϱ_s and a resulting electric field E_x given by

$$\varrho_s = enu(t) \rightarrow E_x = \frac{neu(t)}{\epsilon_b \epsilon_0} \tag{53}$$

where ϵ_b is the background dielectric constant for frequencies above the plasma frequency defined below.

The equation of motion for the ensemble of free carriers reads then

$$m\frac{\partial^2 u(t)}{\partial t^2} = eE_x = \frac{ne^2}{\epsilon_0 \epsilon_b}u(t) \tag{54}$$

where ϵ_b is the background dielectric constant. This is directly the equation of a harmonic oscillator with an eigenfrequency generally called plasma frequency ω_{PL}^0

$$\omega_{PL}^0 = \left(\frac{e^2 n}{m \epsilon_0 \epsilon_b} \right)^{1/2} = \omega_L; \ \omega_T = 0 \tag{55}$$

This frequency is the longitudinal eigenfrequency at k = 0, the transverse one being zero, since an electron gas has, like all other gases, no static shear stiffness.

A more elaborate investigation (see e.g. [16]) shows a dispersion containing k^2 and k^4 terms i.e.

$$\omega_{PL}(k) = \omega_{PL}^0 \ (1 + \alpha k^2 + ß k^4 + ...) \tag{56}$$

This is shown in Fig 13b together with the range of possible two-particle excitations in a degenerate electron gas discussed in sub-section II.E. The energy quanta of the plasma oscillations are called plasmons. The numerial values are of the following order of magnitude

$$\text{metals: n} \approx 10^{23}\text{cm}^{-3}; \ 2 \text{ eV} \leq \hbar \, \omega_{PL}^0 \leq 15\text{eV} \tag{57a}$$

doped semiconductors:

$$0 \leq n \leq 10^{20} cm^{-3}; \ 0 \leq \hbar \, \omega_{Pl}^0 \leq 0.5eV \tag{57b}$$

The existence of plasmons can be verified by the observation of the Reststrahlbande which extends in this case from $\omega_T = 0$ to ω_{PL}^0 and explains the high reflectivity of metals in (large parts of) the visible and even the near UV, or by the energy losses of monochromatic electrons of about 100eV energy. In this so-called electron-energy loss spectroscopy (EELS) distinct peaks in the loss spectra are observed corresponding to the creation of one or more plasmons. The loss-function is given by

$$\frac{\partial^2 \sigma}{\partial \Omega \, \partial E} = \frac{\hbar}{(\pi e a_B)^2} \, \frac{1}{k^2} \, Im \left(\frac{1}{\epsilon(\omega, k)} \right) \tag{58}$$

where σ is the inelastic scattering cross section, a_B the Bohr radius and $\epsilon(\omega, \mathbf{k})$ the dielectric function. For examples of optical and EELS spectra of plasmons see e.g. [16, 17].

In systems of reduced dimensionality like quasi two dimensional quantum wells or one dimensional quantum wires, the restoring electric field is no longer constant i.e. independent of the wavelength as in the (capacitor-like) model of Fig 13a but decreases with increasing wavelength i.e. with decreasing k. As a consequence the dispersion relation of plasmons starts in systems of reduced dimensionality for k = 0 at $\omega = 0$ (see Fig 13c). The details of the dispersion relation depend on the details of the structure. A simple, quasi-two dimensional sheet of free carriers shows a dependence $\hbar \omega_{PL} \sim k^{1/2}$

while a stack of layers tends towards a relation $\hbar\omega_{PL} \sim k$ expecially in the transition range to three-dimensional behaviour. Examples for calculated dispersion relations and their verification by electronic Raman scattering are found e.g. in [18].

II.D. Magnons

In this subsection we discuss the spin system e.g. of an (anti-)ferromagnet without bothering about the other degrees of freedom. The Hamiltomian which describes magnetic ordering in a so-called Heisenberg (anti-)ferromagnet with localized spins reads

$$H = -2J \sum_{l \neq m} S_l \, S_m \tag{59}$$

S_l and S_m are the spins. Usually one considers only next neighbour interaction. J is the overlap integral describing the exchange interaction. J changes sign as a function of the

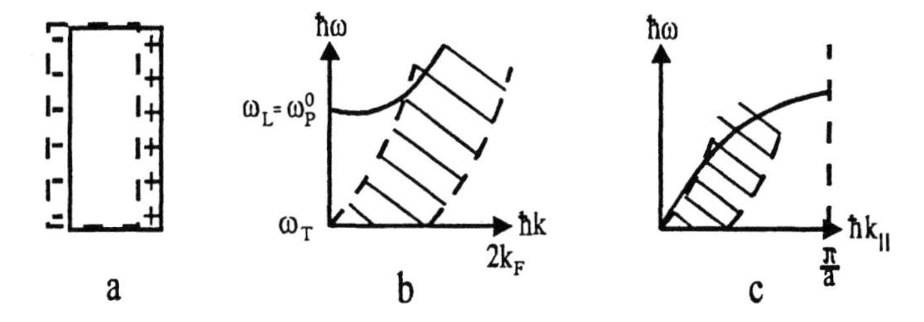

Figure 13 The set-up to deduce the equation of motion for plasmons (a) their dispersion relation together with the range of two-particle excitations (b) and the dispersion relation for a system of reduced dimensionality (c).

distance of the two electrons. This fact explains the appearance of ferro- and of antiferromagnetic ordering below a critical temperature in various materials. We concentrate here on Heisenberg (anit-)ferromagnets and refer the reader for itinerant - or band magnetism to the literature. Examples for the first and second case are e.g. EuO or MnTe and Fe, Co or Ni, respectively.

The saturation magnetisation of the Heisenberg ferromagnet at T = 0 is given by (see Fig 14a)

$$M_s(0) = g\mu_B \, N^S \tag{60}$$

where g is the Landé factor, μ_B the Bohr magneton, N the density of spins and s their angular momentum in the direction of magnetisation.

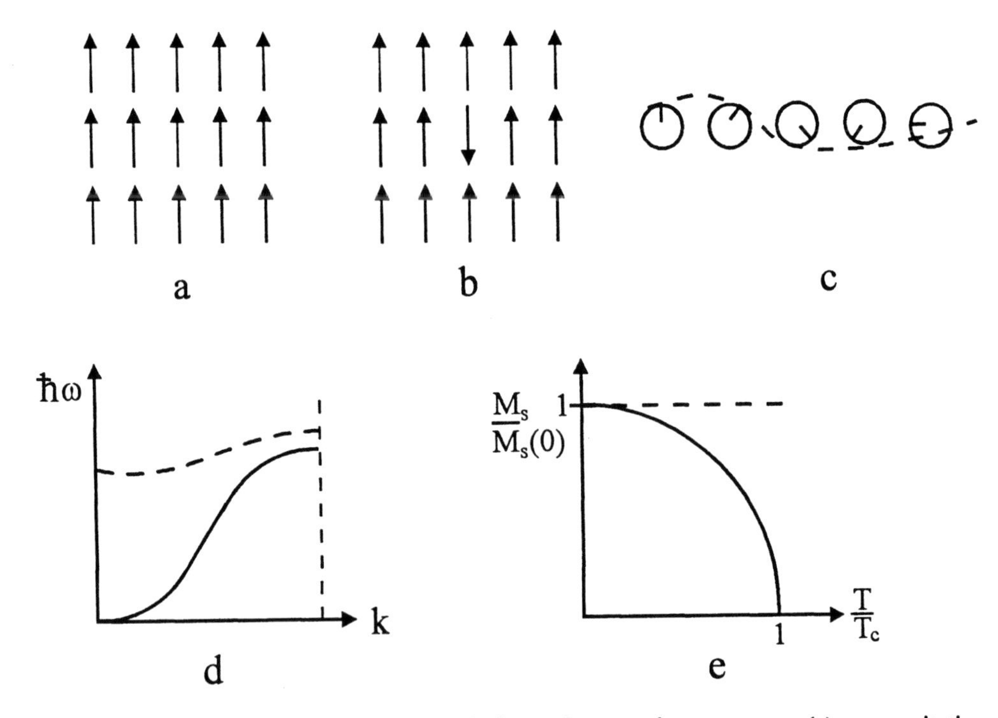

Figure 14 A perfect ferromagnetic ordering of magnetic moments (a) an excitation involving a spinflip of a single electron (b) a collective excitation of the spin-system (c) the dispersion relation of an "acoustic" magnon (–) in a ferromagnet and the additional "optic" branch in an antiferro or ferrimagnet (---) (d) and the saturation magnetisation of a ferromagnet as a function of temperature (e).

The simpliest excitation of the system would be to flip one spin (Fig 14b). This exciation contains a finite energy E_{flip} which depends on J in (59) and the coordination and reduces M_s by two units. As a consequence one would expect, that the saturation magnetisation remains constant with increasing temperature as long as $k_BT << E_{flip}$ holds and starts to decay only for $k_BT \leq E_{flip}$ until it vanishes at the Curie temperature T_C. Actually the experimentally observed behaviour (Fig. 14e) is different. $M_s(T)$ starts to decrease immediatly for $T \geq 0$ according to

$$\frac{M_s(0)-M_s(T)}{M_s(0)} = AT^{3/2} \tag{61}$$

where A is a material parameter. This behaviour can be understood only if one assumes, that there are low energy collective excitations in the spin system, the excitation of which reduces M_s. The picture for these excitations which are called magnons is sketched in Fig 14c. One assumes that every spin and magnetic moment is tilted slightly against the direction of magnetization in a wave-like way. It can be shown that the excitation of a magnon reduces M_s by one unit, and that the dispersion relation starts for $\mathbf{k} = 0$ at $\hbar\omega = 0$ with a parabolic dependence as shown schematically in Fig 14d.

The existence of magnons can be derived apart from $M_s(T)$ in a more direct way by inelastic neutron scattering through processes in which a magnon is created or anihilate in a similar way as in (49). Neutrons are necessary in this case since they interact with a crystal essentially via their magnetic moment. In the case of antiferromagnetic ordering the unit cell is twice as big, as confirmed by the difference between inelastic X-ray and neutron scattering. In this case additional branches appear which are sometimes called "optic" magnon branches (see Fig 14d). The names "acoustic" and "optic" magnons are chosen simply in analogy to phonons. This nomenclature does not imply that the optic mangons couple strongly to the light field. Indeed there are not many experiments showing conclusively the excitation of a single "optic" magnon by the absorption of a suitable IR photon. For some discussions of this point see e.g. [19] and references therein.

II.E. Electronic bandstructure and pair excitations

In this subsection we want to discuss pair excitations in the electronic system of metals, semiconductors and insulators.

Before we are able to treat this topic, we have to introduce shortly the one particle states i.e. the band-structure.

In the general Hamiltomian (42) we assume, that the ion cares are fixed at their equilibrium positions, and omit all terms which depend only on their coordinates $\mathbf{R}_i$ and on derivatives with respect to $\mathbf{R}_i$. The term with $|r_j - R_i|$ in the denominator produces then a periodic potential for the electrons. The Coulomb-term with $|r_{j'} - r_j|$ prevents then still a separation ansatz of the electron wavefunction in a product or more appropriately in a Slater determinate.

Instead of using such an ansatz and going through the Hartree-Fock procedure we proceed immediately to the so-called one electron approximation. In this approach one assumes that the interaction among all electrons produces on the overage also a lattice periodic potential which is combined with the one from the ion-cores to a $V(\mathbf{r})$ with

$$V(\mathbf{r}) = V(\mathbf{r} + \mathbf{R}) \tag{62}$$

where $\mathbf{R}$ is a translation vector of the lattice. The problem is then reduced to

$$H = -\frac{\hbar^2}{2m} \nabla^2 + V(r) \tag{63}$$

once the one particle states $\varphi(\mathbf{r})$ and their eigenergies are known one can populate these states according to Fermi-Dirac statistics until all electrons are accomodated.

It is well known that the solutions of a periodic potential fulfil the Bloch- or Ewald-Bloch theorem. The eigenstates are the so-called Bloch waves

$$\varphi_{i,k}(r) = e^{ikr}\, U_{i,k}(r) \text{ with } U_{i,k}(r) = U_{i,k}(r+R) \tag{64}$$

which consist of a lattice periodic part $U_{i,\mathbf{k}}(\mathbf{r})$ where i is the band index (see below) which is related to the parent atomic orbitals and a plane wave factor, showing that the Bloch waves can move as free particles through the whole crystal. The eigenenergies are periodic in **k** space according to

$$E_i(\mathbf{k}) = E_i(\mathbf{k} + \mathbf{G}) \tag{65}$$

where **G** is a vector of the reciprocal lattice. In Fig 15a the dashed line shows the disperion of free electrons in vacuum. Introducing a weak periodic potential causes some scattering of the otherwise plane electron waves.

The superposition of all scattered waves cancels essentially for a general wave-vector. But there are special wave-vectors, namely just those at the border of the Brillouin zone for which all waves scattered for an incident wave with wave-vectors **k** overlap constructively to form a counterpropagating wave with -**k**. The superposition of these two results in a standing wave, for which the spatial position of the nodes and anti-nodes is fixed. For the same wave-vector we can have two standing waves with different potential energy namely one which has its anti-nodes at the position of the lattice-periodic potential minima and another where they are situated between these minima. Consequently an energy gap opens at the borders of the Brillouin zones as shown by the solid line in Fig 15a. Applying (65) we can shift the various branches of the dispersion relation into the first Brillouin zone, resulting in the reduced bandscheme of Fig 15b. On the other hand we can start with the atomic terms of the parent atoms (Fig 15d) which split into bands according to the statements given at the end of section I, separated eventually still by gaps or forbidden regions as shown in Fig 15c. Making use of the translation invariance we can introduce the concept of **k** vector or of quasi momentum $\hbar\mathbf{k}$ resulting in a transition from Fig 15c to b.

Now we populate the bands with electrons according to Fermi-Dirac statistics, and we call all completely filled bands valence bands and all partly filled or empty bands conduction bands. If we end up at T = 0 K with some completely filled valence bands, then a forbidden gap and then completely empty bands we have an insulator for

$$E_g > 4 \text{ eV:insulator} \tag{66a}$$

and a semiconductor for

$$0 \le E_g \le 4eV\text{:semiconductor} \tag{66b}$$

since a completely empty band trivially does not carry any electric current and a completely filled one not due to Pauli's principle. If we end up at T = 0K with one (or more, energetically overlapping) partly filled band we have a metal. At finite temperatures we can introduce in a semiconductor some electrons in the lowest conduction band which is empty at T = 0K or remove some electrons from the highest valence band which is completely occupied at T = 0K either by thermal excitation of carriers accross the gap (intrinsic conductivity) or by doping with donors or acceptors, resulting in finite conductivity.

In regions, where the dispersion of the bands is parabolic, it is a useful concept to introduce an effective mass m_{eff}, which is generally a tensor and which is inversely proportional to the (constant) curvature of the band via

$$\left(\frac{1}{m_{eff}}\right)_{ij} = \frac{1}{\hbar^2}\frac{\partial^2 E}{\partial k_i \partial k_j} = \frac{1}{\hbar^2}\frac{\partial^2 E}{\partial k^2};\ i,j = x,y,z \tag{67}$$

where the third term is valid for isotropic media.

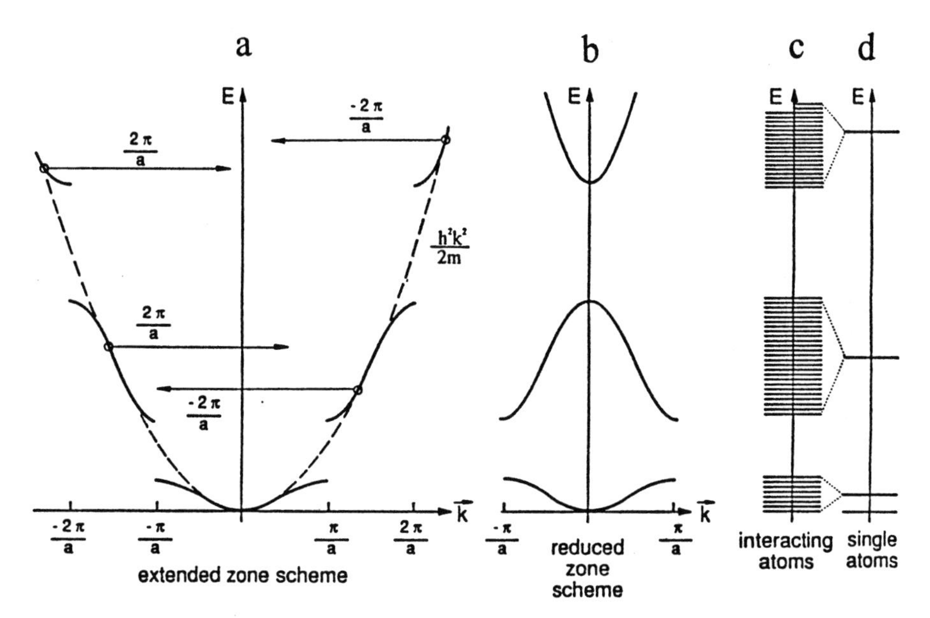

Figure 15 The appearance of reduced bandstructure scheme for electrons in a periodic crystal potential (b) starting from nearly free electrons (a) or from atomic orbitals (d) with some interaction (c).

Instead of considering a almost filled valence-band with some electrons missing, we can consider the few empty states and call them defect electrons or holes, which have charge, **k** vector, effective mass and spin opposite to the ones of the missing electron. The crystal electrons and holes are quasiparticles, which exist only in the periodic potential of the crystal, which are characterized by their dispersion relation E($\hbar$**k**) and which obey Fermi-Dirac statistics.

It should be noted that (crystal) electrons and holes are not yet excitations of the elctronic system but rather one particle states.

The excitations of the system involve always two particles namely an electron which is brought into a previously empty state and the hole, which is left behind just by the excitation of this electron.

In the following we elucidate this concept first for metals, then for semiconductors and insulators. In Fig 16 we show shematically the conduction band of a metall partly filled up to the Fermi-energy (or chemical potential) E_F and the cooresponding Fermi-Vector k_F.

We can now excite an electron from an occupied state in the Fermi sea below E_F to an unoccupied state above, creating simultaneously an empty state i.e. a hole in the Fermi sea. In contrast to insulators and semiconductors, the Coulomb interaction in the excited electron- hole pair is essentially screened by the many other free carriers. In three and two dimensional metals, we get a whole band of pair excitations, which extends for $E \to 0$ from $k = 0$ to $k = 2k_F$ and develops for increasing energy as schematically shown in Fig 13b. For a one-dimensional **k**-space the band reaches the value $E = 0$ only at $k = 0$ and $k = 2k_F$. For more details of this topic see e.g. [16, 17].

In semiconductors and insulators, the pair excitation involves the transfer of an electron from a valence-band across the gap into a conduction band as shown schematically for a direct gap semiconductor in Fig 17a.

Due to the Coulomb attraction between electron and hole, which is not screened by many other free carriers as in the case of a metal, a positronium- or hydrogen- like series of bound states is formed below the bandgap known as excitons. The dispersion of these two particle states cannot been drawn properly in the one-particle diagram of Fig 17a (compare Fig 16 and 13b). Instead one defines the unexcited crystal as ground - or vacuum state with $E = 0$ and $\mathbf{k} = 0$ where **k** is now the wave-vector of the center of mass motion. If the band to band transititon is dipole-allowed, excitons with s envelope function, in the notation of the hydrogen quantum numbers, are coupling strongly to the radiation field leading to peaks in the absorption spectrum $\alpha(\omega)$ at $\mathbf{k} = 0$ and an ionisation continuum above E_g. The tree-dimensional Sommerfeld-enhancement produces a rather flat $\alpha(\omega)$ curve in the ionsisation continuum instead of the square-root dependence expected from the combined density of states (Fig 17c). The peaks in the absorption spectrum show also up as resonance structures in the reflection spectra (Fig 17d). In the case of an indirect gap material as shown in Fig 17e excitons are formed, too, with minima in their dispersion at the respective **k**-values (Fig 17f). A transition from the ground state to the excitons is possible only with participation of a momentum-conserving phonon. This reduces the oscillator strength considerably and results in a square-root dependence of the absorption edges provided the transition-probability is **k** independent (Fig 17g). At higher energies direct gap excitons may appear, too (Fig 17f). Apart from excitons involving the upmost valence and the lowest conduction bands, there are also processes involving the excitation of an electron from or - in other words - the creation of a hole in deeper valence bands.

These so-called core excitons are often prominent at places in the dispersion relation where critical points occur in the combined density of states and extend in principle in energy up to the positions of the L or K X-ray absorption edges.

The simplest case of a quasi two-dimensional structure is a single quantum well in which electron and hole are confined in the same material (Fig 18a).

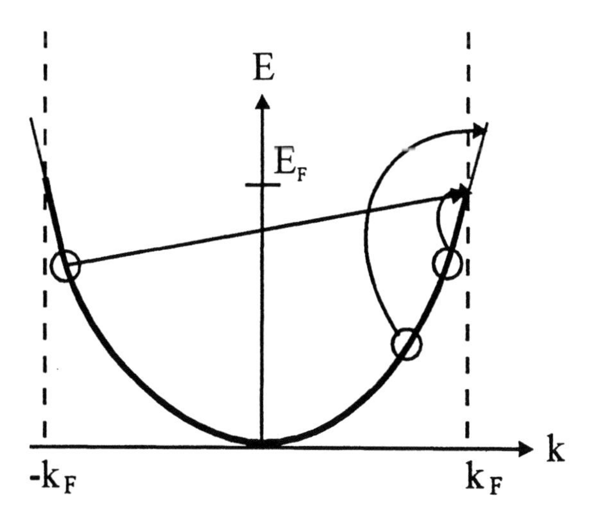

Figure 16 The partly filled conduction band of a metal and various excitation processes of electrons.

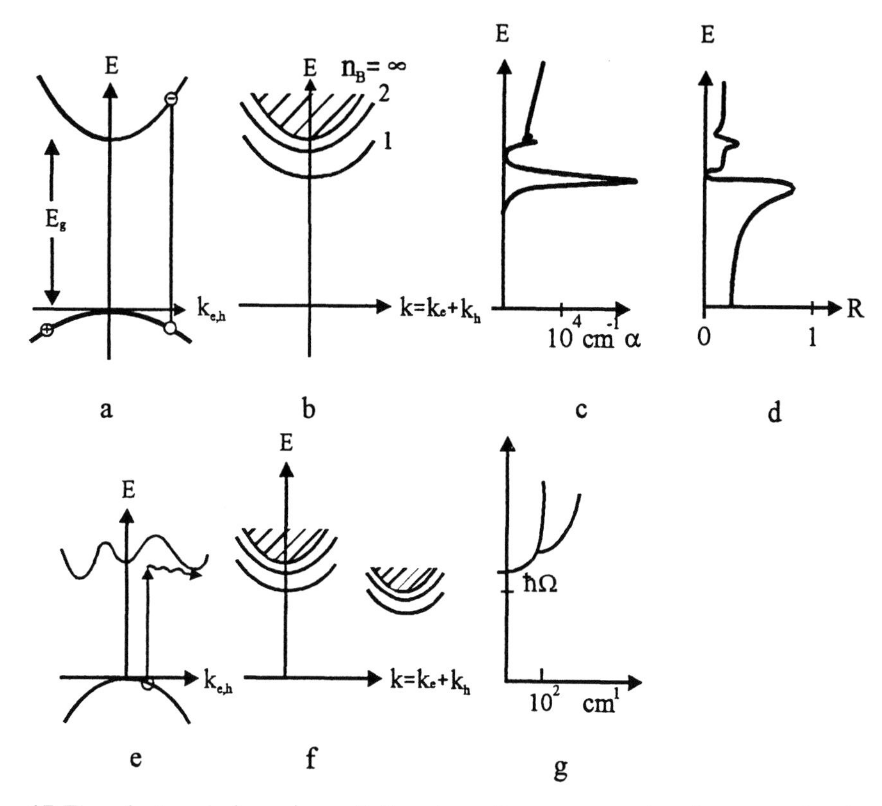

Figure 17 The electron-hole pair excitation in a direct gap semiconductor or insulator (a), the dispersion of the exciton states resulting from (a) via Coulomb attraction (b) and the corresponding absorption (c) and reflection spectra (d). The analog for an indirect gap material (e), (f) and (g).

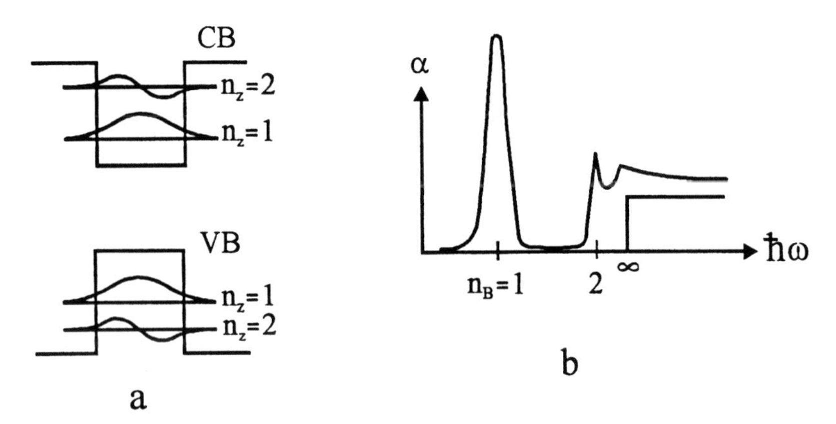

Figure 18 Schematic drawing of the alignment of conduction and valence bands of a type I quantum well (a) and the absorption spectrum of the exciton series connected with the first subband transitions (b).

Quasi two-dimensional subbands with quantum number n_z are formed. In the ideal case the selection rules for the transitions between CB and VB reads

$$\Delta n_z = n_{zVB} - n_{zCB} = 0 \tag{68a}$$

For every intersubband transition a series of excitons is formed with the dispersion of (68c) which is compared with the 3d case in (68b)

$$\text{3 dim: } E\,(n_B,\mathbf{k}) = E_g - Ry^* \frac{1}{n_B^2} + \frac{h^2 k^2}{2M} \tag{68b}$$

$$\text{2 dim: } E\,(n_B,n_z,K_{\parallel}) = E_g + E_Q(n_z) - Ry^* \frac{1}{\left(n_B - \frac{1}{2}\right)^2} + \frac{h^2 k_{\parallel}^2}{2M} \tag{68c}$$

with the effective Rydberg energy of the exciton given by

$$Ry^* = 13{,}6\ eV\ \frac{1}{\epsilon^2} \cdot \frac{\mu}{m_0} \tag{68d}$$

$$\text{with } \mu = \frac{m_e m_h}{m_e + m_h};\ M = m_e + m_h \tag{68e}$$

$$\text{and } \mathbf{k} = \mathbf{k}_e + \mathbf{k}_h \tag{68f}$$

The quantization energy is given by E_Q. The third term in (68c) on the r.h.s. is valid for a strictly 2d system i.e. infinite barriers and the wellwidth $l_z \to 0$. Consequently one reaches in real quasi 2d systems for the binding energy of the exciton ground state only approximately $3Ry^*$ as compared to $4Ry^*$ following from eq (68c). The excitonic absorption spectrum looks approximately as in Fig 18b. It is repeated for every subband n_z. The Sommerfeld factor is less pronounced in 2d systems and varies only by a factor two. In quasi one-dimensional quantum wires the "one over square-root" dependence of the density of states is reduced in the absorption spectra by a Sommerfeld-factor < 1.

In uncoupled quantum dots one has a confinement in all three directions of space and consequently a sharp energy level scheme as indicated in Fig 19a, b for one particle- and for pair-states.

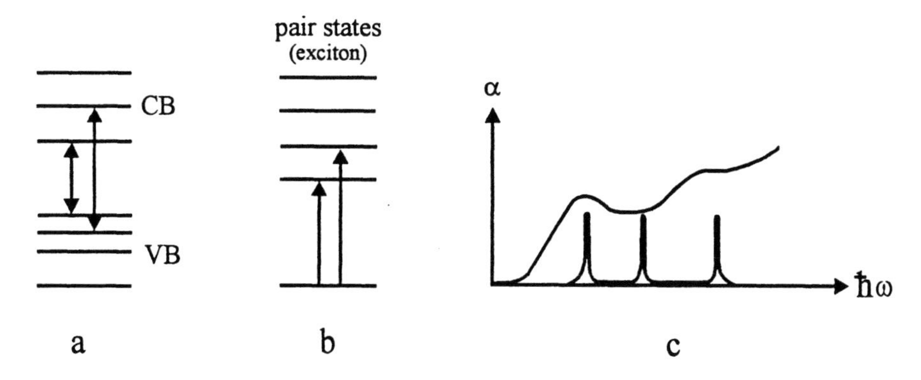

Figure 19 One particle (a) and two particle states (b) in quantum dots and the resulting absorption spectrum for one dot size and for a distribution of finite width (c).

From these Figures one would expect a series of rather sharp peaks in the absorption spectrum, the widths of which is determined by their phase-relaxation time e.g. through coupling to phonons. Actually one observes generally only a weakly structured absorption spectrum, which results from narrow features via inhomogeneous broadening especially through a finite distribution of the dot sizes.

III. INTERACTION PROCESSES

In this section we give some examples for interaction processes which we have neglected so far.

III.A. Phonon-phonon interaction

The anharmonicity in the interaction potential between neighbouring atoms, which we neglected in the introduction of phonons (Fig 9) causes rather strong interaction processes between phonons. In a strictly harmonic case, phonons could propagate through each other without interaction as is e.g. the case for light quanta in vacuum. The anharmonicity results in phonon phonon scattering influencing e.g. the transport of heat by so-called "Umklapp-processes". Furthermore the thermal expansion with increasing lattice temperature of most solids is a consequence of the anharmonicity. For a strictly harmonic potential, the "center of gravity" of the phonon modes coincides with the equilibrium position, i.e. with the minimum for all quantum numbers n. In the anharmonic case of Fig 9 it shifts with increasing quantum number i.e. for higher lattice temperatures to larger interatomic distances.

III.B. Electron-phonon interaction

When we introduced the band structure, we assumed that the lattice is "rigid" i.e. that all atoms are sitting in their equilibrium position. This is actually not the case and consequently there are carrier-phonon interaction processes, which are e.g. decisive for the electric conductivity at higher temperatures.

The interaction processes, which we discuss in the following are the
- deformation potential coupling
- piezo-acoustic coupling
- polar-optic or Fröhlich interaction.

The band structure a crystal depends on the lattice parameters $\mathbf{a}_i$ and on the arrangement of the atoms in the unit cell, i.e. on the basis. Consequently the bands shift and change their dispersion if a crystal is subject to hydrostatic or uniaxial strain. The shift of the energetic position of a band with strain is described by the corresponding hydrostatic or uniaxial deformation potential.

If a phonon propagates through a crystal it will periodically deform the arrangement of the atoms, resulting in a periodic modulation of the band-structure via the deformation potential which in turn is felt by the carrier. In principle it is necessary to give the deformation potential for every phonon branch because they are different. For many materials only a part of the various deformation potentials are known.

A material which has no center of inversion and at least partly ionic binding shows the piezoeffect, i.e. (uniaxial) strain produces a polarisation of the sample and a corresponding electric field and vice versa (piezo effect and electrostriction).

An (acoustic) phonon travelling through a crystal produces periodic changes of the lattice parameter. In piezo electric materials this change of the lattice parameters can be connected with the appearance of an electric field, depending on the type of phonon and the symmetry of the crystal structure. This electric field interacts with the carriers.

The polar optic or Fröhlich coupling describes the interaction of the electric field of LO phonons with free carriers. The coupling strength is given in this case by the dimensionless coupling constant α given by

$$\alpha = \frac{e^2}{8\pi \epsilon_0 \hbar \omega_{LO}} \left(\frac{2m_{r.e.}\omega_{LO}}{\hbar} \right)^{1/2} \left(\frac{1}{\epsilon_b} - \frac{1}{\epsilon_s} \right) \tag{69a}$$

here ω_{LO} is the angular frequency of the optical phonon, $m_{r.e.}$ the effective mass of the rigid lattice, ϵ_s and ϵ_b are finally the static and background values of the dielectric function well below and above the frequency of the optic phonons. The bigger their difference at $k = 0$ the stronger the coupling of the optic phonon to the electromagnetic radiation field, as espressed by the Lyddone-Sachs-Teller relation

$$\frac{\omega_L^2}{\omega_T^2} = \frac{\epsilon_s}{\epsilon_b} \tag{69b}$$

as a special case of the statement given in connection with Fig 7.

For most semiconductors we find $\alpha < 1$ while $\alpha > 1$ occurs in ionic insulators as shown in the table below. For element semiconductors like Si α is zero

$$\alpha_{InSb} = 0.015$$
$$\alpha_{CdS} = 0.65$$
$$\alpha_{LiI} = 2.4$$
$$\alpha_{RbBr} = 6.6$$

Take III.1 Fröhlich coupling constants of some semiconductors and insulators.

For weak coupling, i.e. for $\alpha < 1$ the Fröhlich interaction results not only in scattering processes of carriers i.e. in a renormalization of the damping or imaginary part of their eigenenergy, but also of the real part.

The bandgap shrinks with a shift ΔE of conduction and valence band given by

$$\Delta E_{c,V} = \alpha_{e,h} \cdot \hbar\omega_{LO} \tag{69c}$$

the effective masses increase, since a carrier in an ionic material is surrounded by a cloud of longitudinal optical phonons, according to

$$m_{Pol} = m_{r.l.} \left(1 - \frac{\alpha}{6} \right)^{-1} \tag{69d}$$

This lattice distorion has a polaron radius r_{Pol} given by

$$r_{Pol} = \left(\frac{\hbar}{2m_{r.l.}\omega_{LO}} \right)^{1/2} \tag{69e}$$

In most experiments carried out to determine the band parameters one measures the polaron mass and the polaron gap like in cyclotron resonance absorption

$$\omega_c = \frac{eB}{m_{Pol}} \tag{70}$$

or from the series limit of excitons ($n_B \to \infty$) or from the temperature dependence of the intrinsic conductivity of semicondcutors.

For strong coupling i.e. $\alpha \geq 1$ the approximations in (69) are no longer valid and self-localization of carriers may occur.

III.C. Plasmon - phonon coupling

Plasmons and LO phonons can interact via their longitudinal electric fields. While $\omega^0_{PL} \sim n^{1/2}$ (see (55)) the optical phonon frequencies are not much changed in semiconductors up to $n \leq 10^{20} cm^{-3}$. As a consequence the phonon and plasmon frequencies cross as a function of n without coupling as shown in Fig 20a. The above mentioned coupling validates the concept of avoided crossing introduced in subsection IB. and results for small, optically accessible k-vectors in the situation depicted in Fig 20b.

The transverse and longitudinal eigenfrequencies can be deduced from the poles and zeros of a dielectric function $\epsilon(\omega)$ for negligited damping δ which consists of the sum of a Lorentz and Drude term and read for k ≈ 0

$$\epsilon(\omega) = \epsilon_b + \frac{f_{phonon}}{\omega^2_{TO} - \omega^2 - i\omega\gamma_{Ph}} - \frac{f_{Plasmon}}{-\omega^2 - i\omega\gamma_{Pl}} \tag{71}$$

ϵ_b is here the background dielectric constant for $\omega > \omega_{LO}, \omega^{(0)}_{PL}$. The resulting reflection spectra are shown in Fig 20c, d for two different densities.

The above concept ist experimentally very well verified, e.g. recently for highly n-doped ZnO:Ga [20].

For large wave-vectors **k**, i.e. wave-vectors larger than the inverse, density dependent screening length, the situation changes, since the plasmon "disappears" in the continuum of the two particle states (see Fig 13b) and the damped optical phonon frequencies reemerge.

III.D. Polaritons - coupling of quasiparticles with the electro-magnetic field

As a last example we discuss the coupling of the elementary excitations to the electromagnetic radiation field, i.e. to photons.

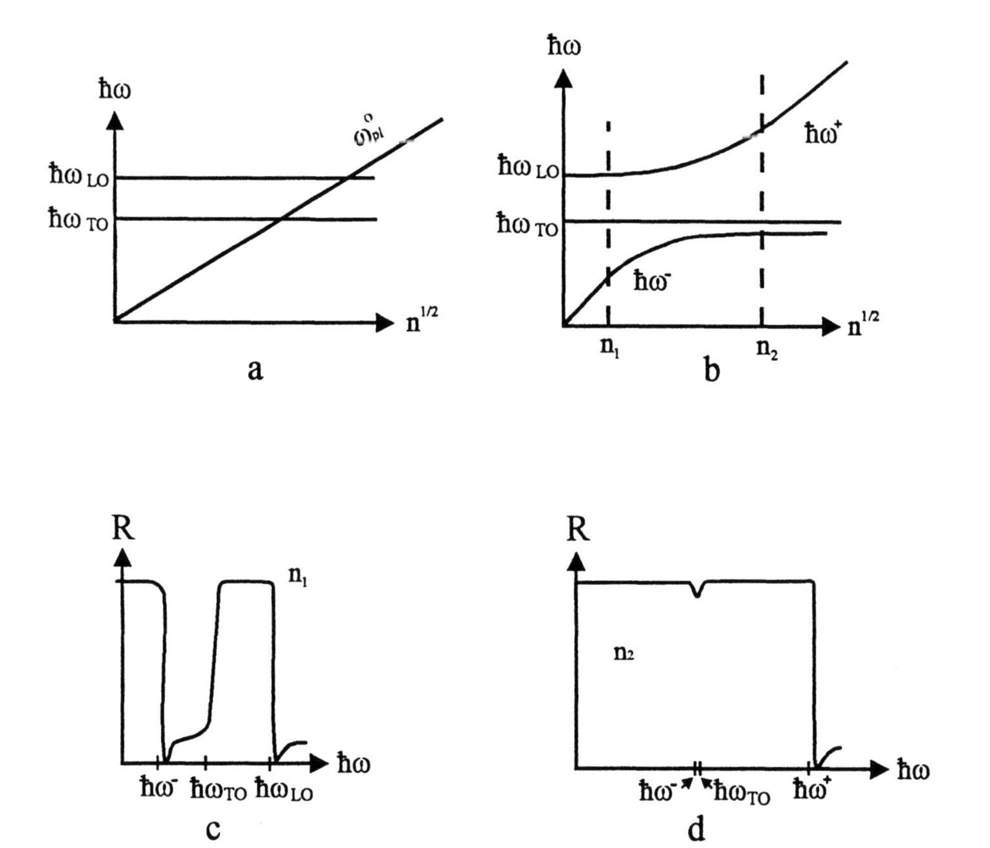

Figure 20 The dependence of the plasmon and the TO and LO phonons on the carrier density $n^{1/2}$ without coupling (a) and the avoided crossing resulting for a finite coupling (b) and the reflection spectra for two different densities n_1 and n_2 (c, d) indicated in (b).

In the so-called weak or perturbative treatment one plots the dispersion of an elementary excitation, e.g. a phonon and the one of photons. At the intersection point they can interact under conservation of energy and (quasi-) momentum. Usually one considers in dipole approximation simply the point $\mathbf{k} = 0$ and calculates the coupling strength or transition rates in perturbation theory.

On the other hand it is obvious that the interacting system between photons and quasiparticle in a solid is a perfect candidate for the application of the avoided crossing rule behaviour of subsection IB. Applying this concept leads to the strong coupling approach or to the polariton concept. Two modifications appear compared to Fig 21a as shown in Fig 21b. A finite coupling leads necessarily to a finite longitudinal-transverse splitting Δ_{LT} and the crossing between the disperions of photon and elementary excitation is avoided. The dispersion relation of Fig 21b follows from a combination of the dielectric function $\epsilon(\omega, \mathbf{k})$ (72a) where we introduced for generality also a $\mathbf{k}$ dependence of the characteristic parameters eigenfrequency ω_T, damping δ and oscillator or coupling strength f and the so-called polariton equation (72b) which follows from Maxwell's equations.

$$\epsilon(\omega,\boldsymbol{k}) = \epsilon_b + \frac{f(\boldsymbol{k})}{\omega_0^2(\boldsymbol{k}) - \omega^2 - i\omega\gamma(\boldsymbol{k})} \tag{72a}$$

$$\epsilon(\omega,\boldsymbol{k}) = \frac{c^2k^2}{\omega^2} \tag{72b}$$

The quanta of light in matter propagating according to the dispersion relation of Fig 21b are quanta of a mixed state of polarisation wave and electromagnetic wave. They are called polaritons with a prefix indicating the type of quasiparticles involved.

The lower polariton branch (LPB) starts at k = 0, $\omega = 0$ "photonline" i.e. linear but the slope, $c(\epsilon_s)^{-1/2}$ which is different compared to the one in vacuum (namely c) indicates that a mixture of a electromagnetic and a polarization wave propagates. The LPB bends over to a phonon or exciton or magnon-like behaviour. There is a finite Δ_{LT}, at $\hbar\omega_L$ eventually a longitudinal eigenmode starts characterized by $\epsilon(\omega_L) = 0$ and the upper polariton branch which bends over to a "photonlike" i.e. linear dispersion, now with a slope $c(\epsilon_b)^{-1/2}$.

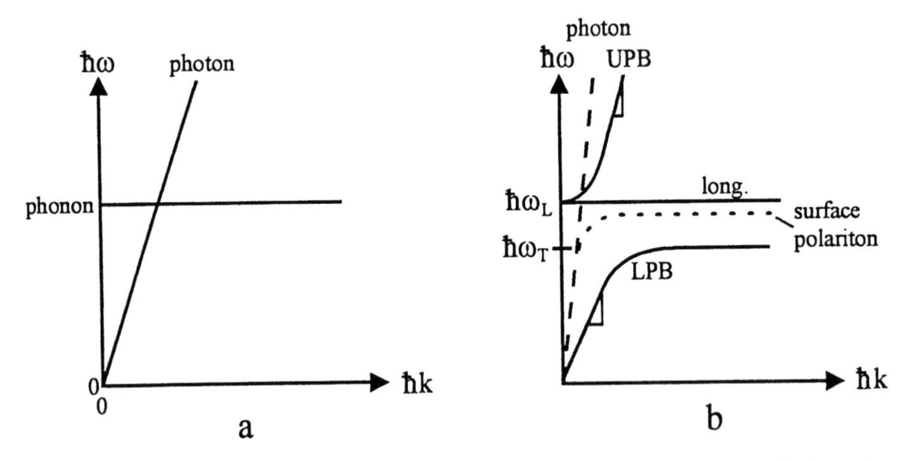

Figure 21 The interaction of light (photons) with an elementary excitation (quasi-particle) in matter in the weak (a) and strong (b) coupling limit.

In Fig 22 we summarize the dispersion relations for phonons, plasmons and excitons in the weak coupling limit (l.h.s.) and of the corresponding polaritons (r.h.s). Note that the transverse eigenfrequency of plasmons is zero. In addition we give in Fig 21b and 22 r.h.s. the dispersion of photons in vacuum and of surface plaritons (see below).

The problem of polaritons in systems or reduced dimensionality will be treated for the example of exciton polaritons in section V.

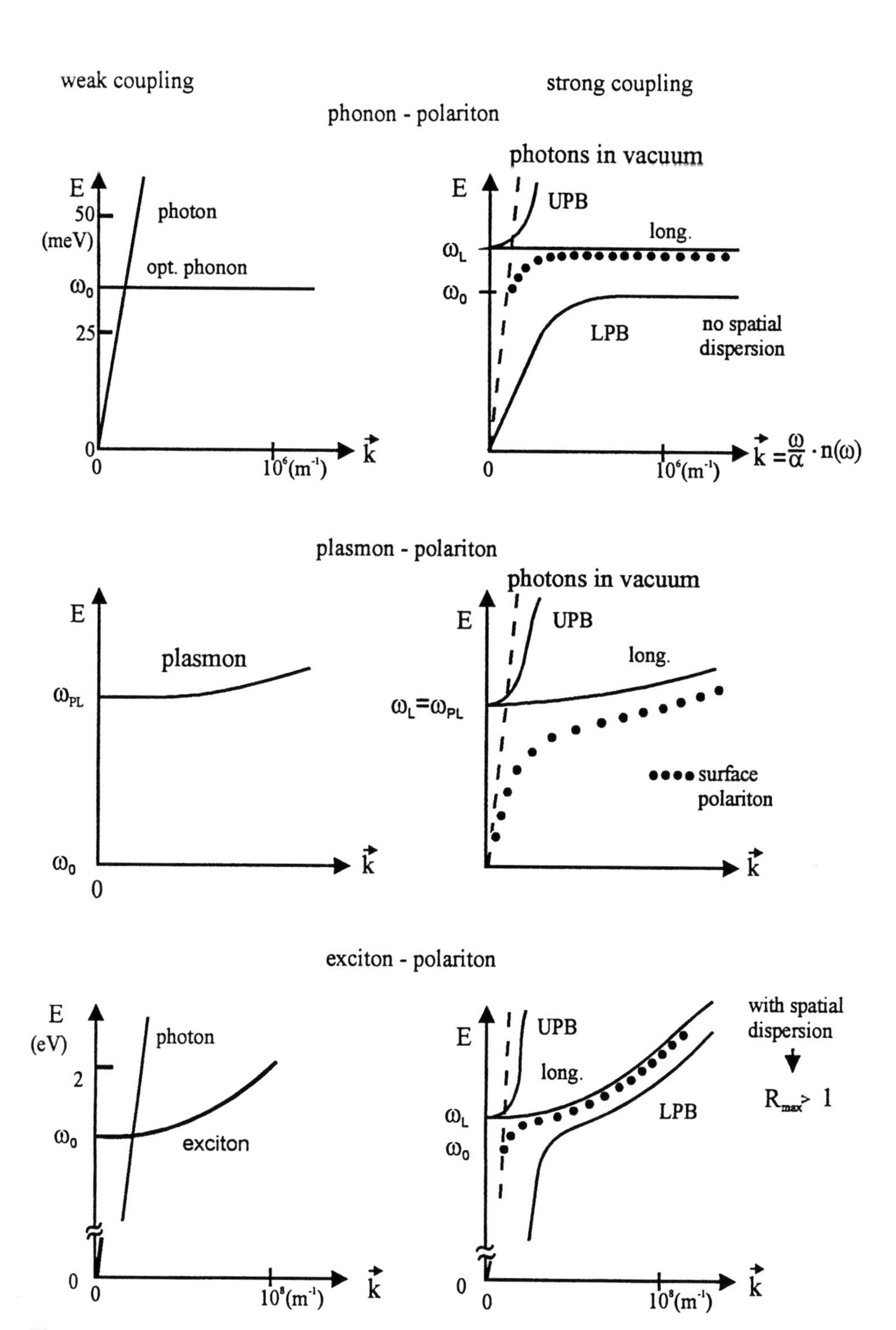

Figure 22 The dispersion relation of phonons, plasmons and excitons in the weak coupling limit (l.h.s.) and the resulting polariton dispersion (r.h.s.). The dashed lines give for comparison the dispersion of photons in vacuum and the dotted ones of surface polaritons.

IV. SURFACE POLARITONS

The inspection of the surface of a lake tells us, that waves cannot propagate only in the volume of some medium but also in a limited layer along the surface of a medium or more generally along an interface.

Surface phonons are treated in various contributions to this school [21]. We concentrate in this short section on surface excitations which involve also the electromagnetic field, i.e. on surface or interface polaritons. To make things easy we always assume, that one half-space is vacuum (air) i.e. n = 1 and $k = \frac{\omega}{c}$. The other half-space consists of a medium with a refractive index n(ω) resulting in $k = \frac{\omega}{c}n(\omega)$.

Since the topic of this section is generally not so well known to a broader audience, we start as an introduction with a simpler, but related phenomenon, namely total internal reflection.

We assume that a light beam falls from a medium with n(ω) >1 on the plane interface to vacuum as shown in Fig 23a.

Usually the situation is described by Snellius laws of refraction

$$\frac{\sin\alpha_i}{\sin\beta} = n(\omega) \tag{73a}$$

an reflection

$$\alpha_i = \alpha_r \tag{73b}$$

where all beams and the normal to the interface are in one plane, neglecting birefringence for the moment.

Since we have sin ß≤ 1 and n(ω) > 1 there is a limiting or critical angle α_c so that for $\alpha \geq \alpha_c$ there is no more refracted beam but only a reflected one, which carries all the energy flux, justifying so the name total (internal) reflection. We find for α_c

$$\sin\alpha_c = \underset{= 1}{\sin\beta} / n(\omega) = 1 / n(\omega) \tag{73c}$$

We consider the problem now from another point of view which, we need later in connection with quantum well polaritons. At a plane interface one has translational invariance only parallel, but not normal to the interface and therefore conservation of (quasi-)momentum also only in the plane of the interface. For medium I and II we have

$$K_{\mathrm{I}} = n(\omega)\,\frac{\omega}{c};\; K_{\mathrm{II}} = \frac{\omega}{c} \tag{74a}$$

and

$$K_{\mathrm{I}}^{\mathrm{I}} = \frac{n(\omega)\omega}{c} \sin\alpha; \; K_{\mathrm{II}}^{\mathrm{I}} = \frac{\omega}{c} \sin\beta \tag{74b}$$

It is easily seen that the laws of (73a, b) follow directly from (74a, b). We can rewrite (74a, b) to give

$$K_{\mathrm{II}}^{\mathrm{I}} = \frac{\omega}{c} \sin\beta = \left(\frac{\omega^2}{c^2} - K_{\mathrm{II}}^{\perp^2}\right)^{1/2} \tag{74c}$$

For

$$\alpha \geq \alpha_c \rightarrow K_{\mathrm{II}}^{\perp^2} \leq 0 \rightarrow K_{\mathrm{II}}^{\perp} \text{ imaginary} \tag{74d}$$

This means we have a so-called evanescent wave in medium II which varies periodically in space parallel to the interface and decays in medium II exponentially with increasing distance as shown schematically in Fig 23c and in (74e) for the field A

$$A \sim e^{i(K^{\mathrm{I}} r^{\mathrm{I}} - \omega t)} \, e^{-K_{\mathrm{II}}^{\perp} r^{\perp}} \tag{74e}$$

The necessity for the evanescent wave follows also from the boundary conditions deduced from Maxwell's equations which state that across an interface the normal or parallel components of the following fields must be steady

$$E_{\text{parallel}}, \; B_{\text{normal}} \tag{75a}$$

and in the case of varnishing surface charges and currents also

$$D_{\text{normal}}; \; H_{\text{parallel}} \tag{75b}$$

This distance over which the evanescent wave decays normal to the interface is comparable to the wavelength as follows from (74c). If a medium with n(ω) is brought to the interface within a distance comparable to the wavelength (Fig 24) the evanescent wave can couple to the light field in the lower half space resulting again in a propagating wave. This effect evidently reduces the reflectivity to values below 1. This effect is known as optical tunneling or attenuated total reflection and is known actually longer than the quantum-mechanical tunnel effect [22].

A surface polariton is now a quantum of a mixed state between a polarisation wave in the medium and an electro-magnetic wave on both sides of the interface, which propagates along the interface and decays exponentially on both sides as shown schematically in Fig 25 and (76)

$$A \sim e^{i(k^{\mathrm{I}} r^{\mathrm{I}} - \omega t)} \, e^{-K_{\mathrm{I,II}}^{\perp} r_{\perp}} \tag{76}$$

where $K_{\mathrm{I}}^{\perp}$ and $K_{\mathrm{II}}^{\perp}$ may be different.

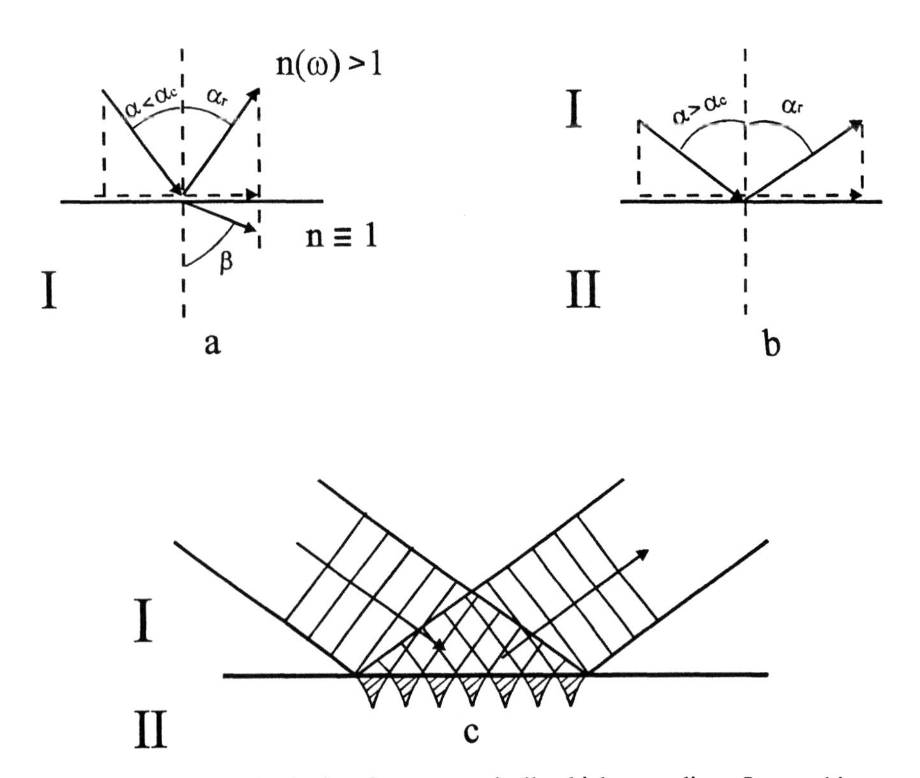

Figure 23 A light beam inpinging from an optically thicker medium I on a thinner one II (here vacuum) below (a) and above (b, c) the critical angle for total internal reflection. Wavevectors are given in (a), (b) and wavefronts in (c).

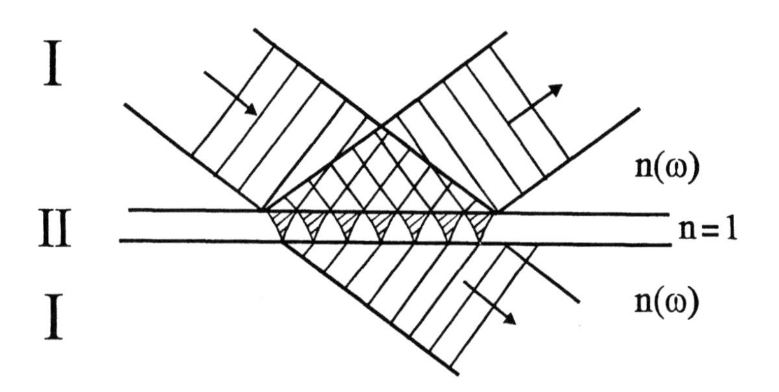

Figure 24 An arrangement to observe the optical tunnel effect.

In order to understand the spectral region, where we can expect surface polaritons we have to find arguments, why the excitation cannot decay to either side of the interface:

For the vacuum side we adopt what we learned above. If

$$K^{\parallel} > \frac{\omega}{c} \tag{77a}$$

the surface excitation cannot decay on the vacuum side. If in addition

$$\omega_T \leq \omega \leq \omega_L \tag{77b}$$

the surface excitation cannot decay into the medium, because there are no propagating modes in the region of the stop-band.

In order to define the dispersion relation of surface polaritons $\hbar\omega_{SP}(k^{\parallel})$ one has to use an ansatz like (76) and the boundary conditions (75). This has been done in great detail in [23, 24] distinguishing even the cases without and with a k-dependence of the eigenfrequency of the dielectric function, the so-called spatial dispersion.

The result of these calculations is that the dispersion of surface polaritons starts at the intersection of the transverse eigenfrequency with the dispersion of photons in vacuum i.e. at

$$\hbar\omega_T\left(K^{\parallel} = \frac{\omega}{c}\right) \tag{78a}$$

and reaches for large $k^{\parallel}$ asymptotically the condition where

$$\epsilon(\omega, k^{\parallel}) = -1 \tag{78b}$$

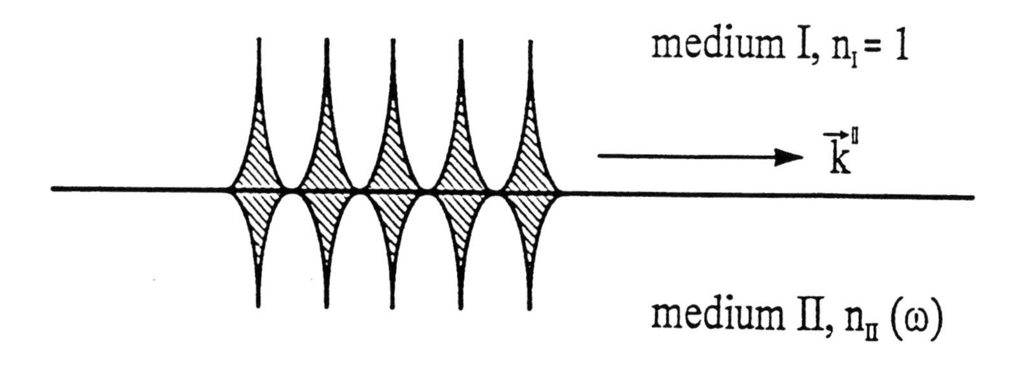

Figure 25 Schematic drawing of a surface polariton wave.

In Fig 21, 22 the dispersion relations of phonon-, plasmon- and exciton-surface polaritons are shown.

The concept of surface polaritons and their dispersion relation has been verified experimentally for phonons, plasmons and excitons see e.g. [11, 25 - 30] and references therein.

Since the surface polaritons do not decay to either side of the interface, it is also not possible to couple to them by simply sending a light beam on the interface. Instead some tricks have to be used. We outline here two different possibilities from the regime of linear optics.

If a light beam is sent into a prism or semi-zylinder of e.g. fused silica above the angle of total internal reflection at its base (Fig 26a) we get there an evanescent wave.

The frequency of this evanescent wave can be varied trivially and $K^{\parallel} > \frac{\omega}{c}$ within a certain range by changing α. If the medium under investigation is brought to the base within a distance $\leq \lambda$, the evanescent wave can couple to a surface polariton mode if both $\hbar\omega$ and $\hbar k^{\parallel}$ match. In this case part of the incident energy is transferred to the surface polariton and the reflectivity drops below unity (in a similar way as in the optical tunneling experiment in Fig 24). This so-called attenuated or frustrated total internal reflection allows thus to measure within a certain range of $k^{\parallel}$ the dispersion of surface polaritons.

In the other method, illustrated inf Fig 26b a grating is produced on the surface. The artificial periodicity Λ perpendicular to the grooves of the grating result in reciprocal lattice vectors

$$G^{\parallel} = n\frac{2\pi}{\Lambda};\ n = 0, \pm 1, \pm 2, ... \tag{79}$$

which can be added to $k^{\parallel}$ of the incident light field.

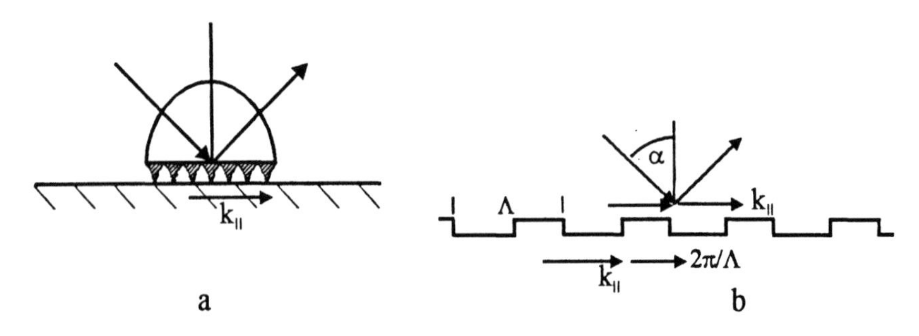

Figure 26 The determination of the dispersion relation of surface polaritons by attennated total reflexcion (a) or by a surface grating (b).

By tuning ω and the angle of incidence α of the incoming light beam one can tune

$$\hbar\omega,\ \hbar k^{\parallel} = \hbar\left(\frac{\omega}{c}\sin\alpha \pm n\frac{2\pi}{\Lambda}\right) \tag{80}$$

If both quantities match a point on the dispersion of the surface polariton, again energy is transferred to this mode and the reflectivity decreases. Examples for both techniques can be found e.g. in [24-30].

V. POLARITONS

In this section we want to have a closer look to the concept of polaritons choosing the exciton polaritons as an example. We start with three-dimensional or bulk material, proceed to exciton polaritons in quantum wells and in Fabry Perot resonators and finish with a generalized definition of the term polariton.

V.A.Bulk exciton polaritons

In Fig 27a we show schematically the dispersion of a simple exciton-polariton resonance including the lower and upper polariton branches, a longitudinal exciton branch and a triplet exciton which involves a spinflip in its excitation and which is therefore assumed to couple neither to the radiation field nor to the singlet exciton polariton.

As a consequence both dispersion curves cross. (In reality the triplet excitons can have a tiny oscillator strength resulting in very small coupling.) In Fig 27b-e we give a selection of some physical quantities, the spectra of which can be measured and allow to deduce more or less directly the dispersion relation, namely the spectra of reflection, of luminescence, of the group velocity and of the refractive index. The reflection spectra (Fig 27b) reach in the resonance region only values significantly below 1 even for vanishing damping, due the kinetic energy connected with the center of mass motion of the exciton. This term results in at least one propagating mode and one evanescent mode below $\hbar\omega_L$ and at least two propagating modes above. This phenomenon necessitates the introduction of so-called additional boundary conditions (additional to the ones given by Maxwell's equations see (75)). For details and realistic reflection spectra see e.g. [11, 13, 31] and the references given therein.

The luminescence spectra (Fig 27c) are determined by the distribution of the polaritons over their dispersion curve, by the probability that they reach the surface during their (diffusive) path through the sample, and by the probability that they are transmitted through the surface which in turn depends on the refractive index and on the angle under which they hit the surface. As discussed already in connection with total internal reflection, the exciton polaritons can be transmitted through the surface only if the parallel component of their wave-vector is smaller than ω/c i.e. the wave-vector in vacuum. As a consequence the internal solid angle in **k**-space from which exciton polaritons contribute to the luminescence decreases rapidly with increasing **k**. As we shall see later, the situation is quite different for polaritons in quantum wells or wires since there the parallel component of the wave-vector is identical to $\mathbf{k}_{total}$. In contrast to the zero-phonon line, all excitons can contribute to the LO phonon replica since the LO phonons have a rather flat dispersion and can accomodate the **k** vector of the polaritons in the initial state. For realistic examples of luminescence spectra see e.g. [11] and references given therein.

If a short ps pulse is send on the sample with a spectral width below Δ_{LT} one can measure its group velocity when propagating through the sample, e.g. by measuring the delay compared to a pulse propagating in vacuum (Fig 28b). The group velocity gives just the slope of the dispersion relation (19b) and looks schematically for the lower and upper polariton branches as shown in Fig 27d.

From the experimental data of $v_g(\hbar\omega)$ one can reconstruct the dispersion relation. It should be noted that the values of v_g can be as low as $c \cdot 10^{-5}$. For realistic data and further references see again [11, 31].

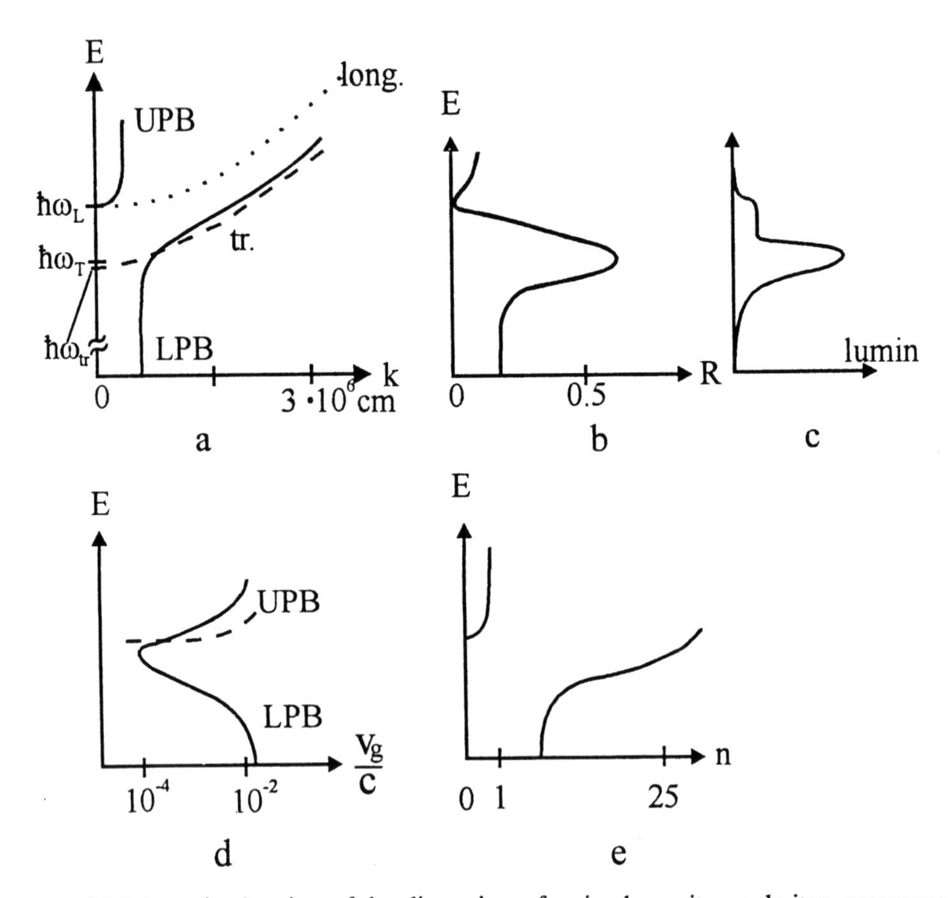

Figure 27 Schematic drawing of the dispersion of a simple exciton polariton resonance (a) and the resulting spectra of reflection (b) luminescence (c) group velocity (d) and refractive index (e).

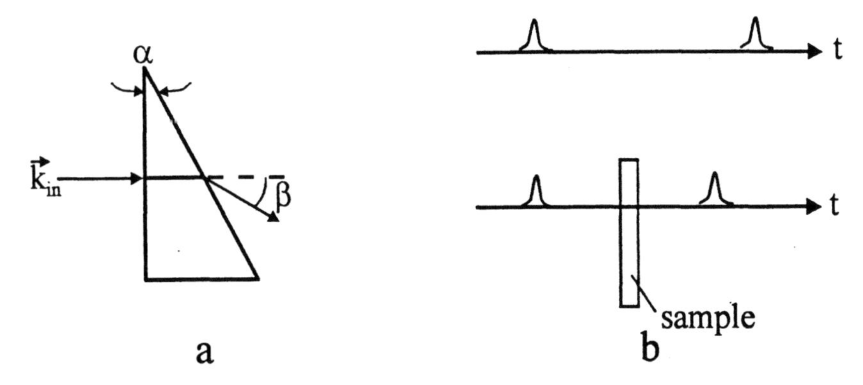

Figure 28 Schematic drawing of the setup to determine the dispersion of polaritons from the refraction of a (thin) prism (a) and from the time of flight of ps pulses (b).

The last method mentioned here is conceptually the most simple one: one measures the refraction of light by a prism with known angle α. From the spectra $\beta(\hbar\omega)$ the spectrum of the refractive index $n(\hbar\omega)$ can be deduced. In the geometrical arrangement of Fig 28a this relation reads simply

$$\frac{\sin[\alpha+\beta(\omega)]}{\sin\alpha} = n(\omega) \tag{81}$$

and the dispersion relation follows from

$$k(\omega) = \frac{\omega}{c} n(\omega) \tag{82}$$

The problem is, that the absorption coefficient reaches in the resonance region values up to $10^0 \dots 10^2 \mu m^{-1}$. This means that the prism base has to be in the μm range to get some light transmitted. Fortunately some CdS platelets grow directly in the desired prism shaped form. For realistic experimental data and further references see again [11, 31]. It should be noted that values of n up to 25 have been measured [32] resulting in $\epsilon_1 \geq 600$ and $k \geq 3\cdot10^6 cm^{-1}$.

Further methods to determine experimentally the dispersion of polaritons in bulk materials like two-photon Raman scattering or the evaluation of Fabry-Perot modes are reviewed also in [11, 13]. To summarize this subsection it can be stated that the concept of exciton polaritons is very well established both experimentally (see above) and theoretically [7-10].

V.B. Exciton Polaritons in structures of reduced dimensionality

The concept of exciton polaritons in structures of reduced dimensionality is just recently being developped, especially from the side of theory. A decade ago one could still hear statements, that polaritons do not exist in structures of reduced dimensionality due to the missing translational in variance in one or more directions.

We consider various cases shown in Fig 29. If we have a light beam incident normally on a single quantum well with thickness l_z (Fig 29a) we have for the case $l_z << \lambda$ a rather simple situation.

The light is reflected and transmitted through a thin layer containing oscillators, whith infinite tranlational effective mass in the direction of **k**. The optical density $l_z \cdot \alpha$ is for typical values of e.g. a GaAs QW (l_z = 10 nm, $\alpha = 10^4 cm^{-1}$) much smaller than unity and therefore difficult to detect both in transmission and reflection spectra. Instead of direct absorption spectroscopy one uses often photo-luminescence- or photo-current excitation spectroscopy. See e.g. [33 - 35] and references therein. The reflection spectra are often dominated by surfaces and interfaces other than the ones of the quantum well [11]. Only few examples are known showing clear resonances in the reflection spectra at normal incidence. Often an angle of incidence close to Brewsters angle is chosen [35a].

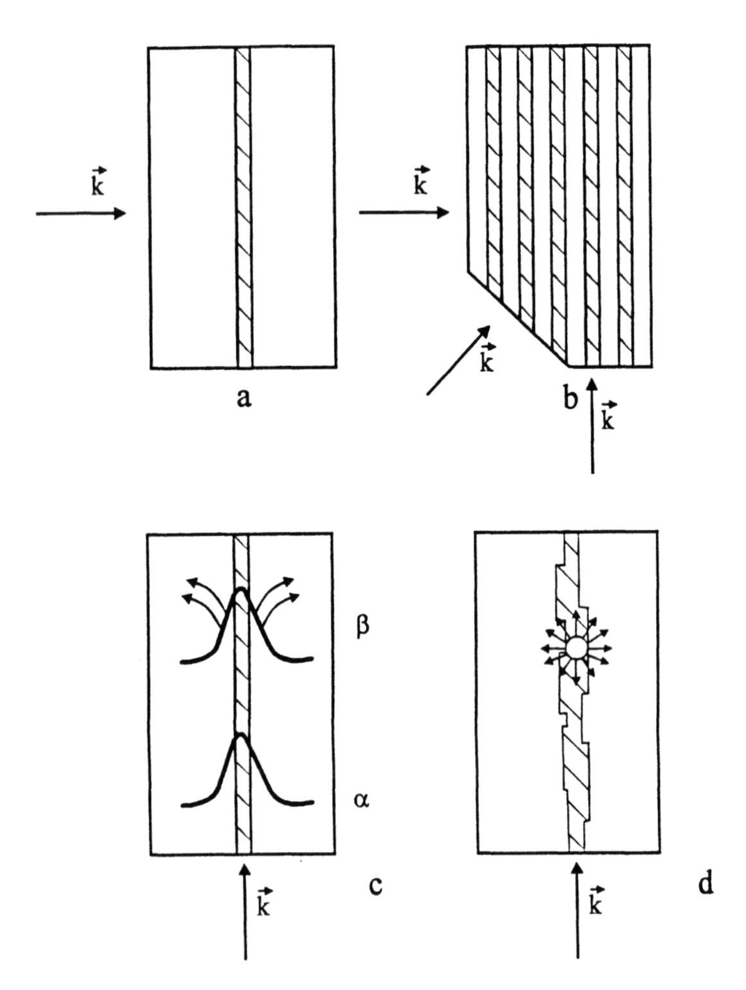

Figure 29 Idealized geometries to discuss the problem of polaritons in quantum-strucu-res: single quantum well and normal incidence (a) multiple quantum well or superlatti-ce and arbitrary incidence (b) a single quantum well and parallel incidence (c) and a realistic, disordered quantum well (d).

If the layer of the film has a thickness comparable to the wavelength, its own dispersion becomes important and various points of the dispersion curve of the bulk polaritons of the well material can be excited with **k** vectors in the layer given by [35b]

$$k_m = m\frac{\pi}{l_z} \tag{83}$$

In the case of a multiple quantum well structure or a superlattice with a period $<<\lambda$ and a total thickness $>>\lambda$ (Fig 29b) we can treat the problem as a "homogeneous" effective medium, since the light cannot "resolve" the fine internal structure for all directions of incidence. Consequently the concept of bulk polaritons can be applied to this situation but the optical properties will strongly depend on the orientation of **E** (and of **k**) relative to the internal structure, i.e. the material will show strong birefringence and dichroism. See also [36] and the selection rules mentioned below. As a first guess one would expect that the effective mass of the resonance is finite for **k** parallel to the layers and infinite perpendicular to it. A polarization coupling between the layers treated theoretically in [37] could even modify the second value.

Eventually the most interesting situation is the one of Fig 29c, where a light beam is sent parallel to one (or a few) quantum wells with $l_z << \lambda$. Since there is only tranlational invariance along the plane of the quantum well, **k** is only defined in this plane i.e. $\mathbf{k}_{total} = \mathbf{k}^{\|}$. The exciton dispersion is shown schematically in Fig 30a, which is based on calculations in [38-40].

For simplificity we consider only one resonance. In reality the $n_B = 1$ exciton in zincblende type structures (point group T_d) with symmetries $\Gamma_3+\Gamma_4+\Gamma_5$ splits in a QW grown on a (001) plane (point group D_{2d}) into heavy hole excitons with symmetries $\Gamma_1+\Gamma_2+\Gamma_5$ and light hole excitons with symmetries $\Gamma_3+\Gamma_4+\Gamma_5$ where the dipole allowed transitions are Γ_5 in T_d and Γ_5 for $E \perp z$ and Γ_4 for $E \parallel z$ in D_{2d} [38].

For the quantum well exciton Δ_{LT} varies both with $k^{\|}$ and l_z as [38]

$$\Delta_{LT} \sim K^{\|} \cdot l_z \tag{84}$$

This can be understood since for $l_z \to 0$ there is no quantum well and the light propagates entirely in the barrier material. The same holds for $K^{\|} \to 0$ i.e. $\lambda \to \infty$ since a beam can be decreasingly less focussed to and confined in a QW wave guide for increasing wavelength. Fig 30b shows the resulting polariton dispersion assuming that the background dielectric constant ϵ_b of the well material and the (non-resonant) dielectric constant of the barrier material are equal. (For the situation of different values see [38]). The dashed line gives the dispersion of light propagating in the barrier i.e.

$$k_{barrier}(\omega) = \frac{\omega}{c}\epsilon_b^{1/2} \tag{85}$$

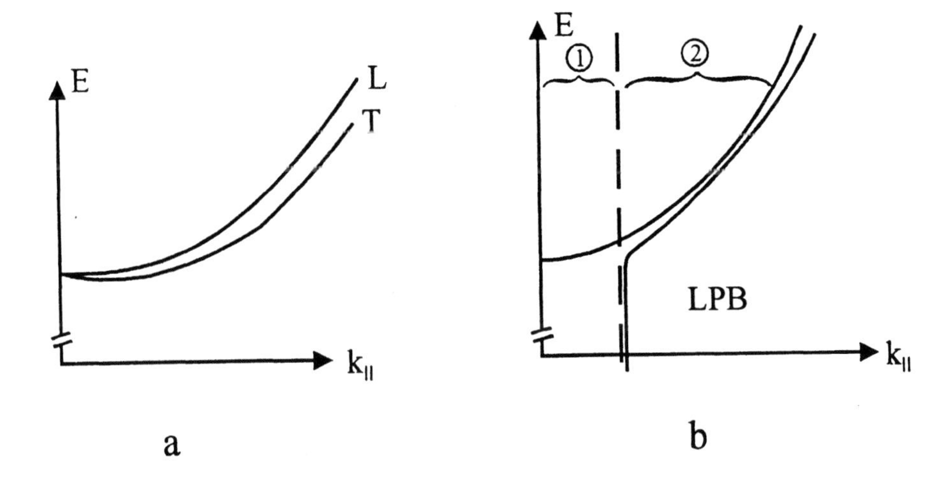

Figure 30 Schematic drawing of the inplane dispersion of dipole allowed exciton (a) and of an exciton polariton (b). region 1: radiant modes, region 2: guided or "localized" modes.

The polariton dispersion starts with a lower polariton branch and a longitudinal exciton branch above looking similar to the ones known for bulk material (Fig 27a). However, due to the facts, that $\mathbf{k}^{\mathbf{l}} = \mathbf{k}$ and that $k^{\mathbf{l}}(\omega) > k_{barrier}(\omega)$ (84) these states cannot radiate into the barrier material for arguments similar to the ones discussed already in connection with total internal reflection and surface polaritons. Actually these are guided (or localized) modes in the sense of a wave guide structure, as indicated schematically at point α in Fig 29c.

The upper polariton branch is missing in comparison to bulk material. It would have $k^{\mathbf{l}} < k_{barrier}$ and would correspond to an "antiguided" mode, which radiates rapidly into the barrier material due to diffraction with a typical time constants T_r in the 10ps regime [38-40] as shown schematically at point ß in Fig 29c. Since even at temperatures $T \leq 10K$ most of the exciton polaritons are occupying already states in region 2 in thermodynamic equilibrium the decay time T_d of the luminescence is much longer than T_r and is expected to increase linearly with temperature.

$$T_d = \alpha T_r \cdot T \tag{86}$$

This behaviour has experimentally been verified in GaAs and other QW materials [11, 41-43] at least for temperatures up to $T \approx 100K$. Since the proportionality constant α in (86) varies however considerably for samples with otherwise similar data concerning e.g. well width and barrier composition [11, 43] an alternative and eventually more realistic model has been put forward, that assumes that the luminescence arises essentially from localized tail states formed by fluctuations of the well width and of the barrier composition as indicated in Fig 29d and from defects. This model is realistic whenever the width of the excitonic luminescence and absorption features exceeds at low temperatures values around 0.1 meV. This value results from the phaserelaxation times of ≥ 10ps and the dispersion.

Actually most samples exceed this criterion. Good QW samples show a width ≤ 1 meV, others (partly used in the experiments in [41, 42]) even values of 5 to 10 meV, indicating clearly the dominance of localization effects and consequently the limited applicability of the model outlined above and developped for the idealized structure of Fig 29c. The analysis of the data following still (86) results in T_r values of (a few) hundred ps, a value typical for bound or localized exciton states in many materials.

The attempts to verify the polariton dispersion according to Fig 30b experimentally are rather limited. Measurements, which relay on the influence of the polariton concept on the shape of the emission spectra [44], give only rather indirecte evidence. There is an example of time of flight measurements [45] in which ps pulses propagated along the guided modes over a distance of 250 μm. A clear decrease of the group velocity has been observed at the spectral positions of various resonances expected for the heavy and light hole exciton polaritons for different polarizations (see above). Values down to $\approx 0.1c$ have been obtained which are evidently still considerably higher than then ones mentioned above for bulk polaritons. This has two reasons, namely the facts that the guided wave propagates only to a small central fraction in the well while the wings extend into the barrier (point α in Fig 29c) and that the resonance itself is in [45] spectrally washed out to a width ≥ 19 meV by the disorder processes mentioned above.

Concerning the dispersion relation of exciton polaritons in quantum wires, we found in the literature at the time of the school essentially theoretical considerations [46] but only experimental verifications via luminescence spectroscopy [44]. We expect for the near future enhanced activities to verify the concept of exciton polaritons in QW and other structures of reduced dimensionality. Experimental techniques other than the time of flight methode mentioned already above with [45] and which are proposed here for this purpose to our knowledge for the first time include the prism methode and the investigation of the spacing of Fabry-Perot modes, which proved successful already for bulk materials [11]. The geometries are shown schematically in Fig 31.

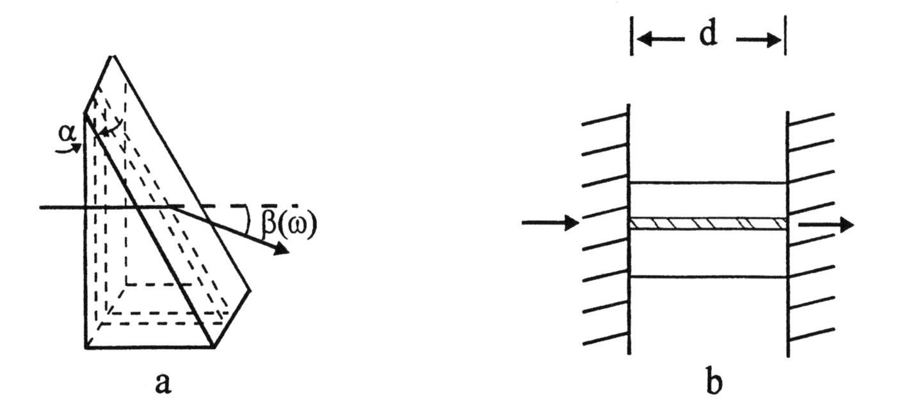

Figure 31 Two possibilities to measure the polariton dispersions for quantum well (or -wire) structures using the prism (a) and the Fabry-Perot resonator method (b).

In both cases light is coupled into the guided modes. In Fig 31a it is refracted from the prism. The measurements of $\beta(\omega)$ allows to deduce the effective refractive index (81) and from this the $\mathbf{k}^{\mathbf{l}}$ vector of the guided polariton mode as a function of photon energy. For the Fabry-Perot resonator (FPR) method the quantum-well wave guide is inserted between two highly reflecting (dielectric) mirrors forming a FPR in the way shown in Fig 31b. Since the transmission maxima are equally spaced in $\mathbf{k}^{\mathbf{l}}$ by $\Delta k^{l} = \pi/d$, where d is the geometrical thickness, one can reconstruct from the knowledge of the photon energies of the transmission maxima the polariton dispersion. Since the absorption coefficients for the guided quantum well polaritons are only in the $10^2\,cm^{-1}$ regime [45], sample lengths in the prism and Fabry-Perot method can be of the order of 100 μm.

It should be noted, that the above proposed techniques should work not only for exciton polaritons but also for phonon polaritons or for plasmon polaritons using e.g. modulation doping to create a sufficiently high carrier concentration in the well.

V.C. Cavity polaritons

An aspect of the polariton concept which obtains presently a lot of attention are the so-called cavity-polaritons. We will approach the problem in various steps, but mention already here, that it can be traced back (once again!) to the problem of level repulsion between coupled oscillators , here between an electronic transition and the eigenmodes of a Fabry Perot resonator (FPR), which is enhanced by the enhancement of the electric field strength in high finesse FPR.

In the picture of weak or perturbative coupling between photons and an electronic transition in atoms, molecules or solids (Fig 32) there is a quantitiy called Rabi-Frequency which is given by the product of the transition dipole moment and the amplitude of the incident electric field strength E_0 [8, 9].

$$\hbar\omega_{Rabi} = \langle f|er|g\rangle E_0 \tag{86}$$

and which describes the following phenomenon. An incoming photon which is close to, but not exactly in resonance with an electronic transition virtually excites the transition. The dipole moment of the virtually excited state radiates again a photon which in turn causes a virtual excitation etc. The frequency with which the system oscillates back and forth between the two states of electromagnetic field and virtually excited state is just the Rabi frequency. Since we have two energetically close lying systems namely the photon field and the electronic excitation, which are coupled via the dipole transition matrix element, we expect also to see the effect of level repulsion. Since the frequency of the incident field is fixed from outside, it cannot shift, but the energy of the electronic transition can, and this is what it does in the presence of a photon field. This frequency shift called ΔE_{OSE} in Fig 32 is known as optical (or ac) Stark effect. The amount of this splitting is proportional to the square of the Rabi frequency devided by the detuning.

$$\Delta E_{OSE} \sim \frac{|\hbar\omega_R|^2}{\Delta E_{gf} - \hbar\omega} \tag{87}$$

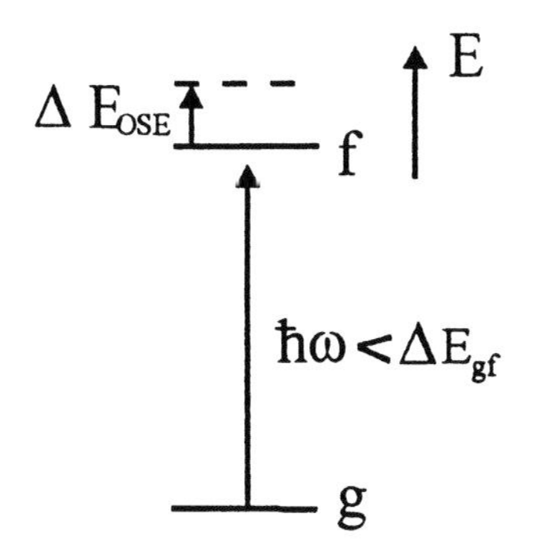

Figure 32 Level scheme to explain the concepts of Raby frequency and optical Stark effect.

An atom in an almost resonant photon field is also known as a "dressed" atom [47]. More recently this phenomenon has been treated theoretically and observed experimentally also for excitons in various semiconductors. See e.g. [48, 49] and references therein.

For the interested reader the above introduction of the Rabi frequency and of the optical Stark effect might introduce a conceptual problem, since we introduced the polariton concept already in a similar way namely as mixed state of - or an oscillation between - the photon field and a (collective) excitation in the medium. The coupling strength has been characterized, however, by the longitudinal -transverse splitting Δ_{LT} which in turn is proportional to the transition matrix element squared but does not depend on the field strength as ω_R does (86).

The way out of this apparent contradiction is the following. The polariton concept including finite values of Δ_{LT} can be considered as an effect of "Rabi-splitting" caused by the always present zero-point vibrations of the electro-magnetic field. If in addition to these vacuum fluctuations a light beam is send into the sample it causes an additional shift of the transition energies, which is however only observable once ΔE_{OSE} becomes comparable to Δ_{LT}.

The next ingredient to the problem of cavity polaritons is the FPR (Fig 33). Whenever an integer number of half waves fits into the FPR all partial waves, which are reflected or transmitted from the two mirrors forming the FPR, interfere in it constructively.

This constructive interference results in an intracavity amplitude, which can be significantly higher than the one of the incident field, if the relfectivity of both mirrors forming the FPR is close to one. Furthermore the transmission and reflection of the FPR reach one and zero, respectively. In between, the interference of the partial waves is mainly destructive and the reflectivity is high, while the transmission and the intracavity amplitude are low. For formula, including also absorption, see e.g. [11].

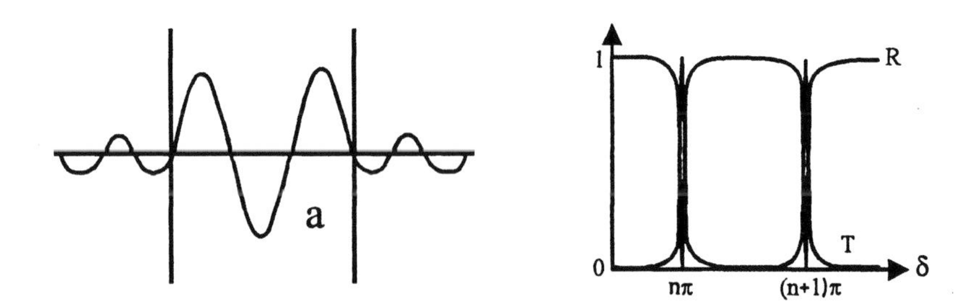

Figure 33 The filed amplitude inside and outside a lossless FPR (a) and its transmission and reflectionspectra as a function of the phase shift δ of one half round trip.

In the simplest case the mirrors of the FPR can be formed by evaporating thin metal films on a flat substrate. This approach is however only of limited use, since very thin metal films have only moderate reflectivity, while thicker ones absorbe and introduce unwanted losses in the FPR. A better way is to use so-called dielectric mirrors. The reflectivity of a single surface of a non-absorping, dielectric material is rather small and given by

$$R = r^* r = \frac{(n-1)^2}{(n+1)^2} \tag{88}$$

If, however, an alternating stack of quarter-wave layers of materials with high and low reflective index is deposited the whole stack may have a reflectivity ≥ 0.95 over a certain range of frequencies, which is essentially determined by the dispersion of the materials. See Fig 34, 35a.

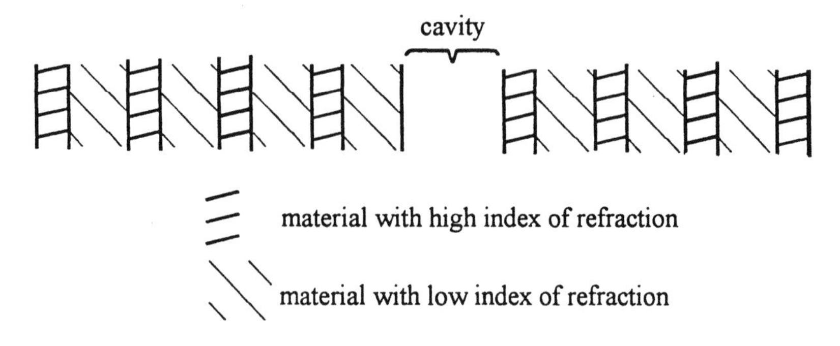

Figure 34 Schematic drawing of a FPR with stacks of dielectric mirrors on both sides of the cavity.

If a FPR is formed by two of these dielectric mirrors with a spacing of $m\lambda/2$ ($m = 1, 2, 3, \ldots$) then an extremely narrow FPR mode appears in reflection (Fig 35b) and transmission. This mode is one of the two oscillators considered in the following. To introduce the electronic transition of Fig 32 as the second oscillator we place a thin layer of semiconductor at the position of an antinode in the FPR using e.g. the lowest exciton resonance. The layer should be thin, not to introduce too much absorption losses in the FPR which would decrease the Q-value or finesse of the cavity. Typically one or a few QW are used.

If the FPR mode at $\hbar\omega_R$ and the exciton resonance at $\hbar\omega_{ex}$ are spectrally well separated, they appear at their characteristic energies, $\hbar\omega_{ex}$ showing up e.g. as a weak dip in the reflection due to the damping introduced by the absorption peak, or as luminescence peak if the FPR is excited above the reflection plateau (Fig 35b) e.g. in the band-to-band transition region of the semiconductor. One has now the possibility to tune $\hbar\omega_R$ e.g. by changing the width of the cavity or even simplier, by tilting the FPR (Fig 36).

The crossing of $\hbar\omega_R$ and $\hbar\omega_{ex}$ as a function of the (external) angle of incidence is transformed in an avoided crossing in the way shown schematically in Fig 36. Real data are found e.g. in [50-52] and references therein. The splitting at the former crossing point is a measure for the coupling between exciton and FPR resonances. This coupling is enhanced compared to an idential semiconductor sample in vacuum, since at the resonance of the FPR not only the internal field resulting from a light beam sent on the FPR is enhanced, but also the zero point fluctuations, which are responsible for Δ_{LT} as outlined above.

The fact that the modes shown by the dots in Fig 36 are actually mixed states of the FP and the exciton resonances in the sense of a polariton follows among others from the fact, that luminescence appears at both frequencies if the system is excited above the high reflection band. A FPR alone would never luminescence at $\hbar\omega_R$ if excited above. Since the system has translational invariance only in the plane of the FPR **k** is only a good quantum number in this plane. By changing the angle of incidence $\mathbf{k} = \mathbf{k}^{\|}$ can be tuned and the dots in Fig 36 can thus be considered as the dispersion of the cavity polaritons.

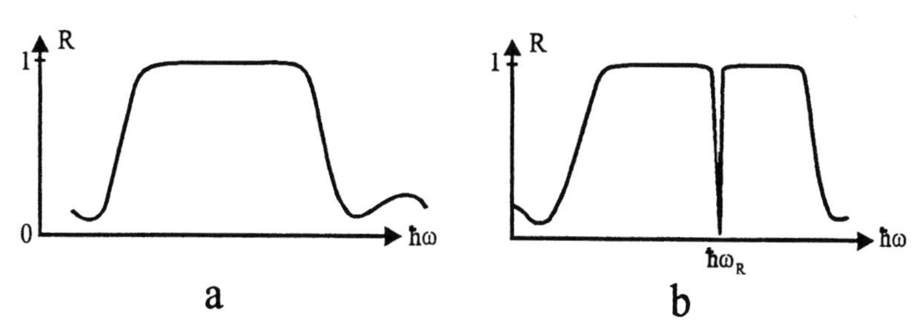

Figure 35 The reflectivity spectrum of a dielectric mirror (a) and of two mirrors with a $m\lambda/2$ cavity (b).

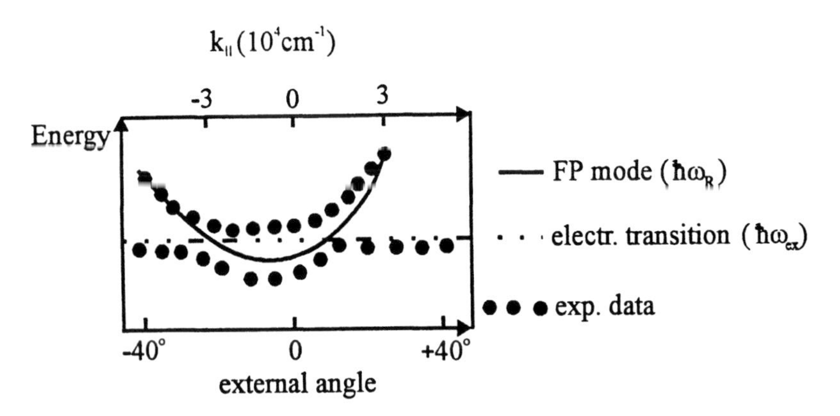

Figure 36 The tunning of the Fabry Perot mode with the angle of incidence ( ) and the electronic transition energy (-·-·) without coupling and the experiment data (.....) (schematic drawing).

To conclude this subsection we want to have a short alternative look on a FPR. It is well known, that a periodic, real potential, e.g. a periodic rectangular Kronig-Penney potentail for electrons, produces allowed bands, which are separated by forbidden gaps, in which no electronic eigenstates exist. In a similar way the periodic modulation of the refractive index in a dielectric mirror produces spectral regions, in which a light field can propagate separated by forbidden "photonic" band gaps. The region of high reflectivity in Fig 35a corresponds to such a one-dimensional photonic band gap. If a defect is introduced in the periodic potential for electrons e.g. by a potential which is different from the others by width and / or depth, this may result in an allowed defect state in the forbidden gap. The cavity in Fig 34 is such a "defect" in the optical analogon. If properly designed it produces allowed "defect" state namely just the FP resonance at $\hbar\omega_R$. More details about the fascinating new field of photonic band gaps and the possibilities to suppress or enhance even spontaneous emission are found e.g. in [53].

V.D. A generalisation of the polariton concept

We have considered so far the polariton concept for bulk materials, for quantum-wells and -wires including the enhancement of the coupling in a cavity. We want to consider now the situation of a random arrangement of resonators or oscillators which can couple to the electromagnetic field, ending with a proposal for a generalized definition of the term "polariton".

The oscillators which we are considering are of rather general nature. They must have one (or more) transitions with transition energies $\hbar\omega_0{}^i$ which couple to the light field. They can be realized by atoms, molecules or - corresponding to the topic of this contribution - by semiconductor nanocrystals (also called quantum-dots or artifical atoms) which are formed e.g. as tiny semiconductor precipitates in glasses or organic matrices, by nanostructuring of quantum wells or in a "self-assembled" growth mode. For details see e.g. [54-56] and references therein.

If we place a single one of these oscillators in an electromagnetic radiation field, we do not have to consider polaritons. Instead we have a rather simple scattering problem. The incident light field drives the oscillator and will be scattered off or on resonance from it as shown schematically in Fig 37.

If we have several such oscillators in the coherence volume of the incident field, the scattered waves interfere destructively or constructively depending on their relative phases, i.e. we have more complieated situation.

The situation gets simplier again, if we have many oscillators within the coherence volume of light. The coherence volume depends on the properties of the light source, but a lower limit is given by the wavelength λ

$$V_{coh} \geq \lambda^3 \tag{89a}$$

If n is the density of oscillators, "many" means that the number N per coherence volume is large compared to 1

$$N >> 1 \tag{89b}$$

with

$$N = V_{coh} \cdot n \geq \lambda^3 n \tag{89c}$$

so that the fluctuations given e.g. by $(N)^{1/2}/N$ are much smaller than 1. In this case, the scattered waves cancel essentially to zero (apart from a small Rayleigh - scattering) except for the forward direction for which we obtain a constructive interference of all partial waves. This is exactly the point, where the laws of streight propagation of light beams in homogeneous media appear or the laws of reflection and refraction.

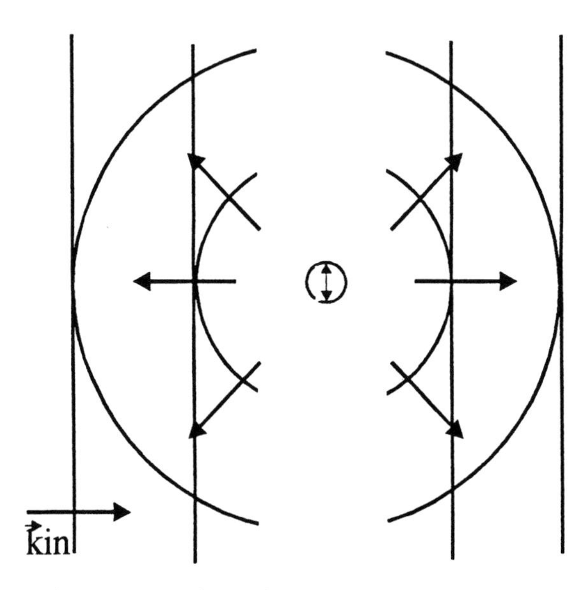

Figure 37 Schematic drawing of the light scattered by a single oscillator in an incident electromagnetic filed.

As we have seen, these laws are based on the conservation laws of **k** (or of (quasi-)momentum). Many every day experiences like looking through a window, into the water of a clear lake or even through air, demonstrate the law of **k**-conservation is not bound to periodic (or cristalline) media or vacuum, but holds also for disordered systems like amorphous materials, liquids and gases, provided the criterion (89b) is fulfilled.

We can formulate this point also in another way. If (89) holds, the average distance d between the oscillators is much smaller than λ

$$d \ll \lambda \qquad (89d)$$

This means, that the light field cannot "resolve" the fine structure but propagates through an effective medium. See also [60] and references therein.

However, there are some differences compared to vacuum or a crystal. In vacuum **k** is a good quantum number for all wavelength. In a perfect crystal **k** is conserved only modulo reciprocal lattice vectors **G** resulting in the definition of Brillouin zones and the restriction to the first one.

In a disordered system **k** is a good quantum number for long wavelength, fulfilling (89). With decreasing λ or increasing $|k|$ this criterion is less and less satisfied and **k** conservation is more and more relaxed. This idea can be illustrated by comparing e.g. a quartz crystal (c-SiO_2) and amorphous fused silica (a-SiO_2). In the visible or near UV, **k** conservation and all the above mentioned laws connected to it hold both for c-SiO_2 and a-SiO_2. If we go to shorter wavelength in the X ray region c-SiO_2 shows well defined diffraction spots resulting from **k**-conservation modulo reciprocal lattices vectors (Ewald's construction) while a-SiO_2 shows only blurred ring structures indicating a substantial relaxation of **k**-conservation at these short wavelength.

In the following we assume that the conditons (89) for the applicability of the effective medium theory are fulfilled. The dielectric function $\epsilon(\omega)$ can then be written as

$$\frac{c^2k^2}{\omega^2} = \epsilon(\omega) = 1 + \sum\!\!\!\!\!\!\int \frac{f_i}{\omega_{0i}^2 - \omega^2 - i\omega\gamma_i} \qquad (90)$$

The sign $\sum\!\!\!\!\!\!\int$ on the right hand side tells us, that we have to sum over discrete resonances and to integrate over continously distributed ones which occur e.g. if the eigenfrequencies ω_0 of the oscillators are distributed over a certain energy range (inhomogeneous broadening) or if we have a continous spectrum of transitions like in the ionization continuum of excitons. The oscillator strength f_i contains the density of oscillators n and the transition matrix element.

We consider in the following two examples for disordered systems. The first one are semiconductor quantum dots dispersed in a silicate glass [54, 55].

TEM pictures of a glass samples containing CdSe dots show e.g. an average radius of about 2.5 nm [58]. This radius is smaller than the excitonic Bohr radius in bulk CdSe (a_B = 5.4 nm), this means, that excitons are strongly confined in all three dimensions.

Furthermore, such figures show that the criterion (89) is fulfilled, even if a certain inhomogeneous size distribution is considered. We can therefore apply the effective medium theory, confirmed by the fact, that such "semiconductor doped glasses" are used since several decades as edge filters in optics.

Compared to bulk CdSe the exciton resonances are modified in various ways. Due to the three-dimensional confinement the bands are split up into a discrete level scheme, similar to atoms justifying the synonym "artifical atoms" for quantum dots. Due to the strong confinement of electrons and holes the overlap of their wave function and thus the transition matrix element e.g. of the lowest allowed (singlet) exciton transition in a dot is enhanced. This effect is however overcompensated in the oscillator strength of (90) due to the low density n of osciallators or in other words by the low filling factor of the glass, which is typically in the 10^{-2} to 10^{-3} regime. Consequently the longitudinal-transverse splitting Δ_{LT} following from (90) for a single transition energy and vanishing damping is reduced in the CdSe doped glass compared to bulk CdSe. On the other hand, the singulet-triplet splitting, resulting from the short range part of the exchange interaction between conduction and valence band is enhanced in the quantum dot by the strong confinement, resulting in a level scheme for the 1s exciton in bulk CdSe and in a CdSe doped glass as shown in Fig 38.

Due to the inhomogeneous broadening of the transitions caused by the size distribution of the dots no sharp spectral features are expected in the dispersion of light in a semiconductor doped glass. Instead one expects a behaviour as shown schematically in Fig 39.

Though the difference of the dispersion between undoped and doped glass might be small, it should be measurable by the prism method or by determining Brewster's angle.

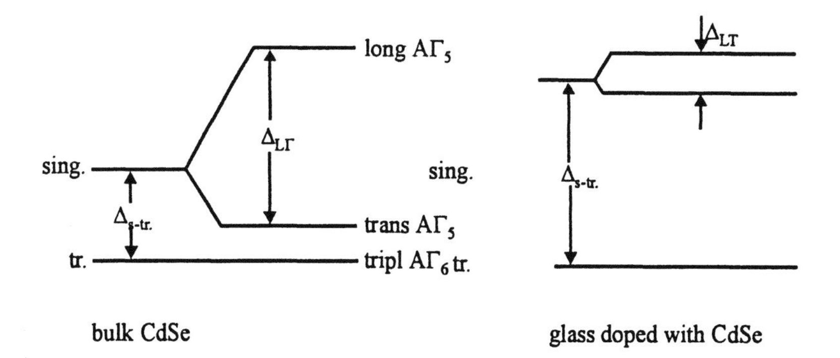

Figure 38 Schematic level scheme of the lowest exction in bulk CdSe and in CdSe doped glass containing the singulet-triplet splitting Δ_{s-tr} and the longitudinal transverse splitting of the singlet Δ_{LT}.

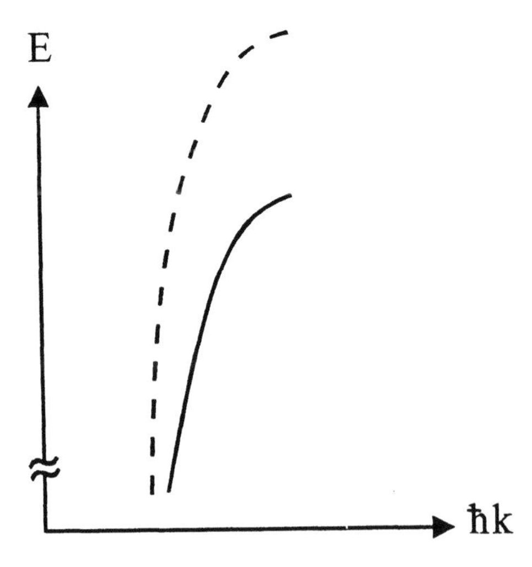

Figure 39 Schematic drawing of the dispersion of light in a glass without (---) and with semiconductor nanocrystals ( ).

In order to give an example with a more pronounced effect we consider a disordered system with a much smaller inhomogeneous broadening namely Na vapor. Towards the end of the last and the beginning of this century the dispersion of light in metal vapours has been measured, using a modification of the prism method of Fig 28. Instead of keeping the metal vapour in a prism shaped cell, one used parallel windows in an established oven in which the metal was thermally evaporated but a concentration gradient was perpendicular to the propagation direction of the measuring light beam. The concentration gradient in turn was the consequence of a transverse temperature gradient.

In Fig 40 we show an example from [59] for Na vapour. The refractive index shows a beautiful resonance structure around the yellow Na-line. If $n(\omega)$ is multiplied by $\frac{\omega}{c}$ and if the axes are then interchanged we get the dispersion of light in Na-vapour. It is obvious that the resulting data or those of Fig 40 are beautiful examples of a polariton dispersion in (disordered) matter, here in a gas.

Our scientific "grandfathers" who carried out these experiments had of course no idea of the polariton concept, and - at the end of the last century - not even of the notation of a photon.

From our present point of view, we can however reinterpret the old data in terms of the polariton concept. We know that the quanta of light carry the energy $\hbar\omega$ both in vacuum and in matter e.g. from the external and internal photo effects. Whenever the momentum $\hbar\mathbf{k}$ of these quanta is given by $\hbar\omega/c$ we have a pure electro-magnetic wave i.e. photons. Whenever $\hbar\mathbf{k}$ deviates from this value, i.e. whenever the refractive index $n(\omega)$ or the dielectric function $\epsilon(\omega)$ are different from one, we know that the electromagnetic wave is accompanied by a polarisation wave. We propose to call the quanta of this mixed state "polaritons" independent if they propagate in a crystal, in an amorphous solid, a liquid or a gas.

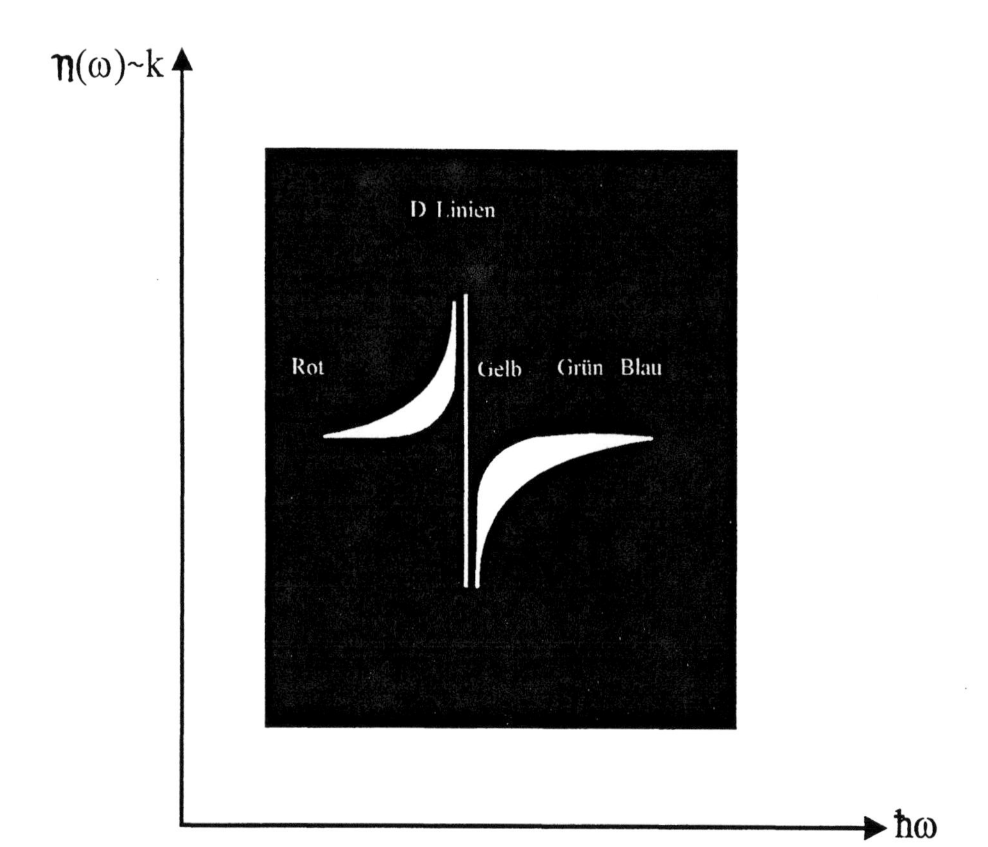

Figure 40 The spectrum of the refractive index of Na-vapour in the visible part of the spectrum. According to [59].

VI. WHAT CAN WE LEARN ABOUT A RESONANCE FROM LINEAR SPECTROSCOPY?

A resonance is characterized by its oscillator strength, its eigenfrequency and its damping (90). We have seen in this and previous schools [61], how these parameters and especially the damping can be determined from coherent, time resolved spectroscopy especially from four wave mixing experiments including photon-echos, quantum beats etc. We do not want to go into this field here. Instead we want to answer the questions: What can we learn about the parameters of a resonance from stationary, linear spectroscopy alone and what additional methods are available other than coherent, time resolved spectroscopy.

If we measure only the reflection and transmission spectra with standart techniques as shown e.g. in Fig 41a we can eventually deduce the real and imaginary parts of the complex index of refraction or of the dielectric function by using Kramer-Kronig relations or ellipsometry, but otherwise we do not learn much, apart from the fact that there are obviously some transitions.

If on the other hand, we have additional information about the resonance, e.g. if it is homogeneously or inhomogeneously broadened or if we have a **k** dependence of the eigenfrequency ω_0 (**k**) or not, the so-called spatial dispersion, we can deduce a lot of details from the spectra. For example, the analysis of low temperature excitonic reflection spectra has been cultivated in the seventies and eighties to a degree, that the transverse eigenenergy, the longitudinal transverse splitting i.e., the oscillator strength, the damping and even the translational mass could be deduced. For details see e.g. [11, 31] and references therein.

To give an example, it has been found that the damping parameter $\hbar\gamma$ of the $n_B = 1$ AΓ_5 exciton is situated for high quality CdS samples at low temperatures ($T_l \leq 5$ K) in the range between 50μeV and 1 meV [62] much later, the phase relaxation time $T_2 = (2\gamma)^{-1}$ has been determined from time resolved FWM experiments to be for CdS, CdSe or CuCl in the range between 1ps and 20ps [63, 64] in agreement with the older data.

The distinction between homogeneous and inhomogeneous broadening, which is crucial for the further interpretation and evaluation of spectra like in Fig 42, can be done with slightly more elaborate methods of linear spectroscopy like photoluminescence (excitation) spectroscopy (PL and PLE) or, which is essentially the same, with site selective spectroscopy.

In the case of PL one excites resonantly in an inhomogeneously broadened transition and measures the resulting luminescence. For PLE one detects the intensity of a spectrally narrow part of an inhomogeneously broadened luminescence band and tunes the spectrally also narrow excitation source. Fine-structures in the inhomogenously broadened transitions in quantum dots like the singlet-triplet splitting (see Fig 39) have been detected in this way. See e.g. [55, 65] and references therein.

Apart from the coherent spectroscopy in the time domain, which is, as mentioned above, not a topic of this contribution, there are also nonlinear optical techniques in the frequency domain, which may give additonal information. Two of them will be shortly mentioned namely spectral hole burning (SHB) and nondegenerate four wave mixing (NDFWM).

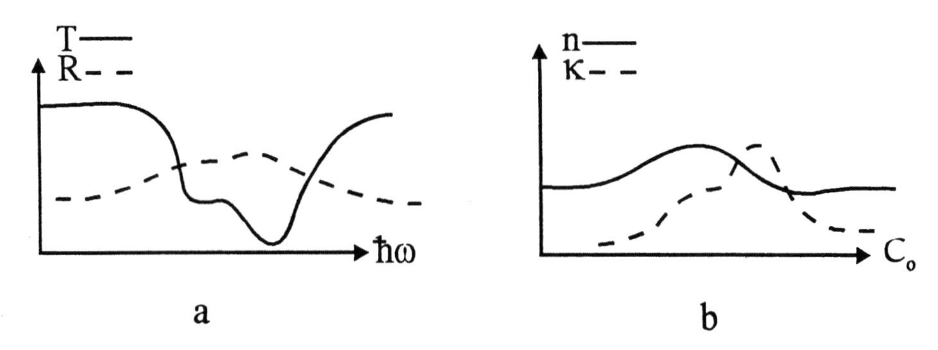

Figure 41 Schematic drawing of the transmission and reflection spectra of a resonance (a) and the real and imaginary parts of the complex index of refraction (b).

If one has a homogenously broadened transition, which can be saturated, one can pump in the absorption band at any photonenergy and the whole band will decrease as shown schematically in Fig 42a. In the case of an inhomogeneously broadened absorption band, which arises from a superposition of many independent oscillators with slightly different eigenfrequencies, the pumping at $\hbar\omega_{pump}$ reduces the absorption at this frequency by the creation of a so-called spectral hole (Fig 42b) and leaves the rest of the absorption band essentially unaffected.

If care is taken to use a sufficiently low pump power and spectrally narrow source, the homogeneous width of the individual transition can be deduced from the spectral hole. Examples for spectral hole burning in quantum-dot ensembles can be found in [55] for other examples see e.g. contributions to previous schools [66, 67] and references therein.

The basic idea of the spectroscopy with four wave mixing or laser-induced gratings is the following: one has to bring two coherent laserbeams (generally of equal frequencies, so-called degenerate FWM) to interference on the sample under a small angle. The interference results in a periodic modulation of the intensity impinging on the sample (Fig 43a). If there is any optical nonlinearity, one gets consequently a periodic modulation of the optical properties i.e. an amplitude and / or a phase grating.

From this grating, light can be diffracted. In the simplest case the two beams which create the grating, are diffracted themselves (self-diffraction), but one can also diffract a third, independent beam which can have another frequency or can arrive at a later time in pulsed experiments. A review of different variants of this method is found in [11].

In the case of NDFWM one creates the interference pattern with two independent lasers of frequencies ω_1 and ω_2 and wave vector $\mathbf{k}_1$ and $\mathbf{k}_2$. The lasers must be spectrally very narrow (e.g. $\Delta\omega < 0.1$ meV) to ensure a sufficiently long coherent overlap-time. If the lasers are detuned with respect to each other, i.e. if $\omega_1 \neq \omega_2$, the interference pattern moves laterally with a speed v_{lat} given by

$$v_{lat} = \frac{\omega_1 - \omega_2}{|k_1 - k_2|} \tag{91}$$

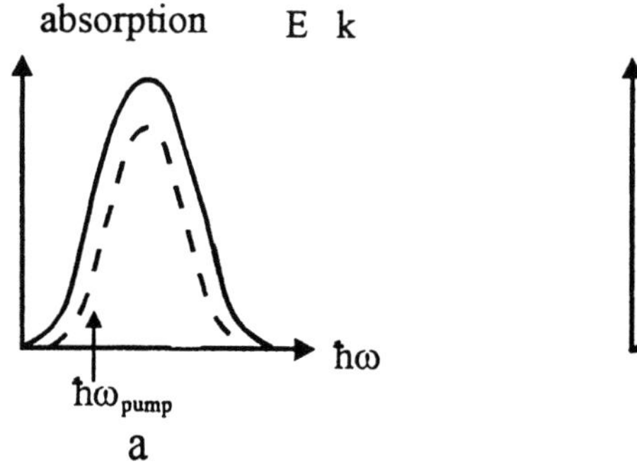

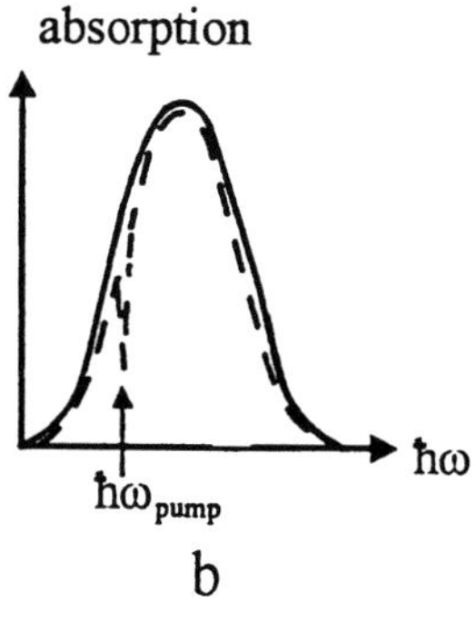

Figure 42 The bleaching of a homogeneously (a) and of an inhomogeneously broadened absorption band (b).

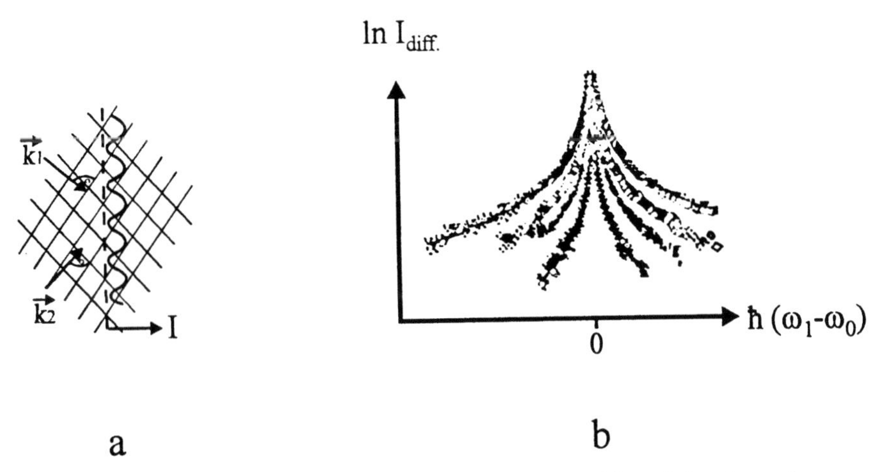

Figure 43 The basic set-up four wave mixing experiments (a) and the dependence of the diffracted intensity as a function of detuning in the case of nondegenerate four wave mixing (NDFWM) (b) according to [55] and references therein.

We assume that the species, which are responsible for the optical nonlinearity, have a characteristic decay time T_{ch}. This is e.g. the lifetime T_1 in the case of an incoherent population grating or the phase relaxation time T_2 in the case of a coherent polarisation grating. If the grating moves over one grating period Λ in a time much longer than T_{ch}, the spatial modulation can follow the moving grating and the diffraction efficiency is constant. In the oposite case, the grating is washed out and the diffraction efficiency (or the intensity of the first diffracted order) drop with increasing detuning $\omega_1 - \omega_2$ as shown schematically in Fig 43b. From a fit to experimental data it is possible to deduce T_{ch}. An example is the measurement of the T_2 time of excitons in quantum dots as a function of temperature and excitation density. Values of T_2 as short as 35fs have been determined by this method [54, 66c] a value which is hardly accessible in time-domain experiments. For more details and other aspects of (N)DFWM see also [11, 68].

VIII. INCOHERENT POPULATION DYNAMICS

In this section we outline some examples of incoherent population dynamics, referring the reader for the coherent processes e.g. to [8, 9, 11, 61] and the references given therein. We start with a rather old example, namely the relaxation of polaritons in bulk CdS on their dispersion curve at low temperature [69]. In Fig 44 the upper and lower polariton branches are shown.

We assume that electron-hole pairs have been created by a short pulse with some excess energy. The first relaxation process occurs in typically $\leq$ 1p by the emission of LO phonons as indicated by the arrow in Fig 44. If the kinetic energy of the carries is too low, to emit another LO phonon, the relaxation continues via the less efficient emission of acoustic phonons.

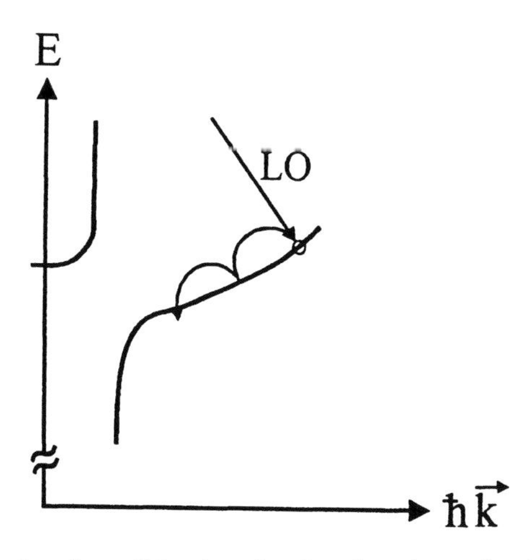

Figure 44 Schematic drawing of the intraband relaxation of polaritons in bulk samples.

Since the acoustic phonons have for small **k** a linear dispersion relation it gets for the polaritons with decreasing **k** increasingly difficult to relax further on the parabolic part of the exciton like dispersion curve by emitting acoustic phonons under energy and (quasi-) momentum conservation. As a consequence they accumulate in the so-called bottle neck, i.e. the transition region from the exction -to the photon - like part of the dispersion curve. This accumulation is enhanced by the fact that a scattering onto the photonlike part has a low probability due to the much smaller density of final states. If the polaritons hit the surface, they can be transmitted as luminescence as discussed already in more detail above. The temporal evolution of the polariton luminescence allows therefore to monitor the population dynamics. In [69] it has been found that the luminescence from the higher **k**-states decays with a time constant below 1ns, while values up to 3ns have been observed in the bottle-neck region. It should be noted, that this time is not simply a radiative decay time, but depends in a rather complex way on the excitation properties and on the diffusion of the exciton - like polaritons to the surface.

As we have discussed earlier, the emission process in QW is different from bulk. Excitons or exciton - like polaritons with large **k** - cannot decay radiatively at all. A very nice experiment of hot luminescence has been performed recently, which allows to monitor the in plane **k** distribution of excitons in a ZnSe QW [71]. In Fig 45 we show on the right hand side, here in the weak coupling picture the dispersion of photons (in the barrier material) and of excitons in the QW. For simplicity only the 1s hh exciton and the onset of the continuum states are shown together with a tail of localized states caused by disorder and symbolized in Fig 45 by the set of horizontal lines.

If the excitation is chosen slightly more than one LO above the exciton energy at **k** = 0, the first relaxation process will be the emission of an LO phonon, which brings the exciton in a guided-mode state (compare Fig 30b). After this fast relaxation process the excitons have two further relaxation processes: They can slowly relax on the 1s dispersion relation by emission of acoustic phonons until they are trapped in the localized states (see Fig 30) and contribute to the zero phonon line.

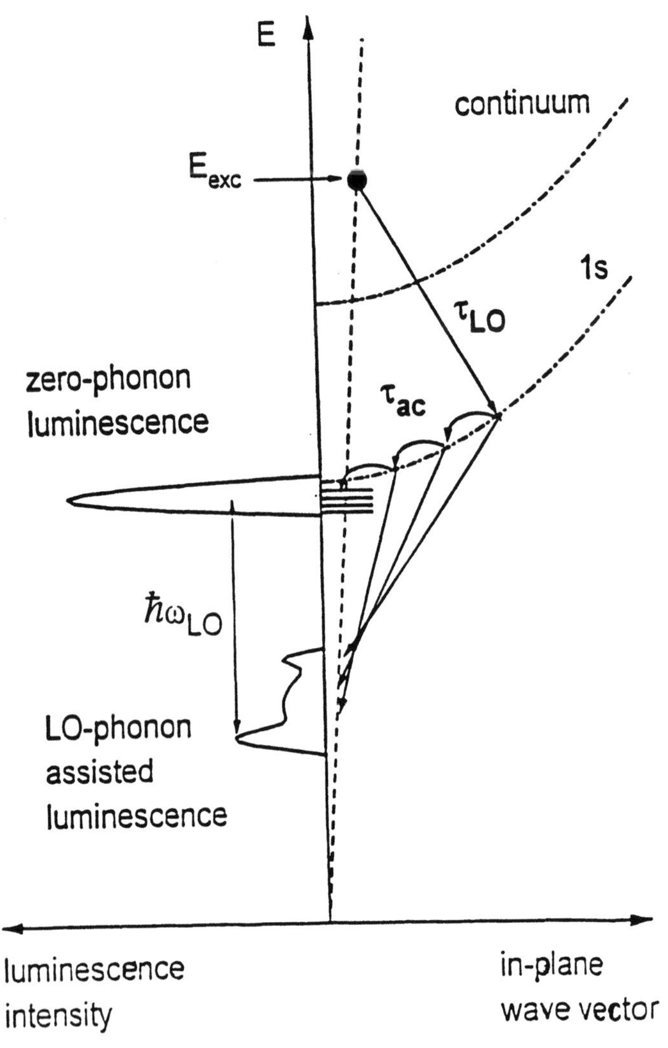

Figure 45 The temporal evolution of the exciton population after ps excitation (r.h.s.) and the luminescence spectra (l.h.s.) schematic after [71].

Alternatively they can emit another LO phonon which scatters the excitons on radiative states, contributing thus to the LO phonon assisted luminescence band. In the ZnSe QW the branching ratio for both processes is such, that the temporal evolution of the population on the 1s exciton dispersion can be monitored in the first LO phonon satellite. It has been found in [70] that it takes approximately 100ps until the excitons reach a thermal distribution, however still with an effective temperature above lattice temperature. The above model was shown to allow also an almost quantitative interpretation of the PLE spectra obtained under stationary excitation conditions [70, 71].

Further examples of hot luminescence in III-V materials are found e.g. in [72] and references given therein.

Some comments on the luminescence decay dynamics in GaAs QW have been given already above. Therefore we mention here only one more example where the intraband relaxation in a GaAs MQW has been monitored by fs time resolved pump and probe spectroscopy [73]. Electron-hole pairs have been excited in this experiment with a 70fs pulse about 50meV above the exciton resonances. Measuring the differential transmission at time delay increments of 50fs allowed to monitor how the bunch of excited carriers relaxes through the continuum states and reaches after about 200fs the exction levels. Since the excess energy used in this experiment was about 50meV, at least the high energy fraction of the carriers might relax by LO phonon emission.

In [74] the temporal develepment of vertical cavity surface emitting lasers (VCSEL) has been investigated after optical pumping by short pulses. The active media were either GaAs QW's or a thin $In_{1-y}Ga_yAs$ bulk layer. After a first short excitation pulse with some excess energy, stimulated emission sets in after some time delay which depends on the number of carrier pairs which has been created and which is in the 10ps regime. This is evidently the time it takes the carriers to relax down in their bands to form a degenerate electron-hole plasma population, with allows lasing. If the sample is excited again by second pulse after lasing has set in, one would eventually expect that the stimulated emission increases. This is the case only if one excites directly above the chemical potential of the electron-hole pair system. If again some excess energy is used, the laser is switched off by the second pulse for a time interval of the order of 5ps and then resumes lasing. The reason for the transient switch off is the heating of all carriers by the additional ones via carrier-carrier scattering, which rises the carrier temperature to such high values that the population is no longer degenerate. The remarkable point is, apart from the observation of the effect itself, the fact that the dynamics of the emission and of the carriers in their bands could be modeled using the semiconductor Bloch equations in close agreement with the experimental data. The importance of this effect for optical data handling is obvious.

Our next example concerns the intraband relaxation dynamics of excitons in bulk, crystalline $CdS_{1-x}Se_x$. This material is a prototype for localization, which is due to composition-fluctuation of the atoms on the anion sublattice. For recent reviews of this topic see e.g. [75]. In Fig 46 we show as a dashed line the low-temperature luminescence under weak stationary excitation. The highest peak is the zero phonon luminescence out of the localized states (the mobility edge is situated in this sample around 2.5eV) followed by two LO phonon sattelites of the CdS sublattice. If the sample is excited with a 100ps pulse above the mobility edge, the localized tail states and some of the extended states above are populated.

The luminescence spectrum monitors the temporal evolution of the population. Localized excitons recombine, in this sample predominantly radiatively [77]. Consequently excitons in extended or shallow localized states can either recombine or relax to neighbouring deeper states by acoustic phonon assisted tunneling. This process explains the gradual red shift of the emission maximum which has even after 30ns not completely reached the position it has under cw excitation (see inset).

The population of the localized states can be rather easily inverted at low temperatures [78] resulting in stimulated emission. If again 100ps pump pulses are used the stimulated emission, the bleaching of absorption and the resulting blue shift of Fabry-Perot modes follows first essentially the excitation pulse. When the stimulated emission stops, the slower spontaneous decay dynamics are observed [79, 80].

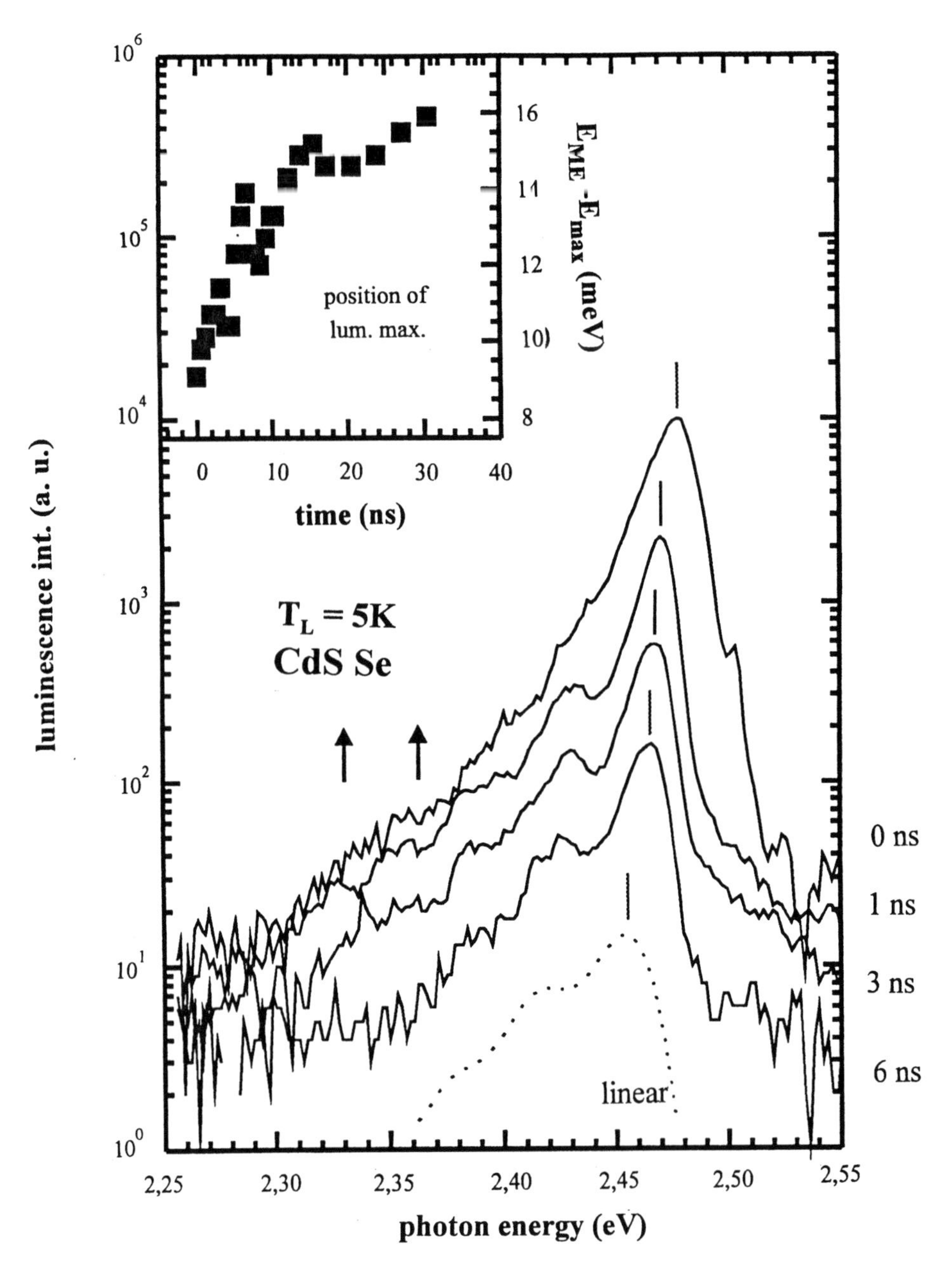

Figure 46 The temporal evolution of the localized exction luminescence of a $CdS_{1}Se_{-}$ sample.

Under special conditions, however, a selforganized pulsation of the stimulated emission occurs, the period of which is wavelength dependent and determined by the group velocity. For more details of this topic see [79, 80].

As a last example we want to address some recent experiments on the spin dynamics of bulk GaAs and GaAs QW [81].

The samples are excited with σ^+ circularly polarized light and the direction of obervation of the luminescence is parallel to the excitation $\mathbf{k}_{exc}$. The selection rules in this case are such, that essentially electrons with $\sigma_z = +\frac{1}{2}\hbar$ are created, which recombine giving again preferentially σ^+ polarized light as long as the spin orientation did not decay or is otherwise influenced. Exactly this can be done by applying a magnetic field normal to $\mathbf{k}_{exc}$. The electron spin start then a Larmor-precission, while the hole spins are assumed to randomize rapidly presumably due to spin-orbit interaction [82]. After half a Larmor period the electrons have $\sigma_z = -\frac{1}{2}\hbar$ resulting in σ^- polarized luminescence. After one period, the light has again σ^+ polarisation and so on. The Larmor precission can be seen in spontaneous and even in stimulated emission and can be followed in the first case over several hundred ps at low temperatures. This means that the spin relaxation is decoupled from the relaxation of the electronic polarisation, which has typically only decay times of a few ps up to the ten ps range [43]. For more details see [81].

VII. CONCLUSION AND OUTLOOK

The contribution of numerous colleagues and of the author to this and proceeding schools of this series [12a-j] demonstrate, that the spectroscopy of semiconductors is since several decades a lively, stimulating and fruitful field of research, which comprises both aspects of basic research and of application. It does not take much of a prophetic gift to predict, that semiconductor optics will continue to be a major topic in solid state physics well into the next century.

Acknowledgements: The author wants to thank his coworkers for their fruitful work, many colleagues for stimulating discussions who are both too numerous to mention them here by name. With special pleasure I should like to mention the discussions on symmetries and on the concept of (quasi-)momentum which we had during the school in Erice with Prof. Cardona (Stuttgart) and Prof. v. Baltz (Karlsruhe). Last, but not least I should like to thank Ms M. Brenkmann for the careful and patient typing of the manuscript and Ms U. Bolz for the artwork. The work has been supported by the Deutsche Forschungsgemeinschaft through various projects and by the European Community in the frame of INTAS.

REFERENCES

[1] O. Madelung, Festkörpertheorie, Heidelberger Taschenbücher Vols. 104, 109, 126, (1972/73)
O. Madelung, Introduction to Solid State Theory, Springer series in Solid State Sciences, Berlin (1978)

[2] Ch. Kittel, Einführung in die Festkörperphysik, 10th ed. R. Oldenbourg, München (1993)
Ch. Kittel, Introduction to Solid State Physics, 6th ed, J. Wiley and Sons, New York (1986)

[3] N.W. Ashcroft, N.D. Mermin, Solid State Physics, Holt-Souders, New York (1976)
[4] H. Ibach, H. Lüth, Festkörperphysik, 4th ed. Springer, Berlin (1995), english version available
[5] K. Kopitski, Einführung in die Festkörperphysik 2nd ed, Teubner, Stuttgart (1989)
[6] J.R. Christman, Festkörperhsyik, Oldenbourg, München (1995)
J.R. Christman, Fundamentals of Solid State Physics, J. Wiley and Sons, New York (1988)
[7] O. Madelung, Grundlagen der Halbleiterphysik, Heidelberger Taschenbücher 71, Springer, Berlin (1970)
[8] H. Haug, S.W. Koch, Quantum theory of the optical and electronic properties of semicondcutors, 3rd ed., World Scientific, Singapore (1994)
S.W. Koch, Microscopic Theory of Semiconductors, World Scientific, Singapore (1996)
N. Peyghambarian, S.W. Koch, A. Mysyrowicz, Introduction to Semiconductor Optics, Prentice Hall, Englewood Cliffs (1993)
[9] F. Henneberger, S. Schmitt-Rink, E.O. Göbel, Optics of Semiconductor Nano-structures, Akademie Verlag, Berlin (1993)
[10] P.Y. Yu, M. Cardona, Fundamentals of Semiconductors, Springer, Berlin (1996)
[11] C. Klingshirn, Semiconductor Optics 2nd print, Springer, Berlin (1997)
[12] B. DiBartolo ed, Previous Proceedings of this school, edited by Plenum Press, New York
a Optical Properties of Ions in Solids, NATO ASI series B 8 (1975)
b Spectroscopy of the Excited State; ibid. B 12 (1976)
c Radiationless Processes, ibid. B 62 (1980)
d Collective Excitations in Solids, ibid. B 88 (1983)
e Energy Transfer Processes in Condensed Matter, ibid. B 114 (1984)
f Spectroscopy of Solid-State Laser Type Materials, Ettore Majorana Intern. Science Series, Physics Vol. 33
g Disordered Solids, ibid. 46
h Advances in Nonradiative Processes in Solids, NATO ASI Series, B 249 (1989)
i Optical Properties of Excited States in Solids, ibid. B 301
j Nonlinear Spectroscopy of Solids, ibid. B 339
k Spectroscopy and Dynamics of Collective Excitation in Solids, ibid. B 356
l Ultrafast Dynamics of Quantum Systems, in press
[13] Landolt Börnstein, New Series, Group III, Vol. 17 a-i and Vol. 22a, b, Springer, Berlin, (1981-1995)
[14] M. Cardonna and G. Güntherodt, Light Scattering in Solids V, Topics in Appl. Phys. 66, Springer, Berlin (1989)
[15] B. Lou, Sol. State Commun. 76, 12 (1990)
M. Vogelsang et al., J. of Raman Spectr. 27, 239 (1996)
[16] v. Baltz, Ref. 12 j, p 303
[17] H. Raether, Excitation of Plasmons and Interband Transitions by Electrons, Springer Tracts in Mod. Phys. 88 (1980)
H. Raether, Surface Plasmons on Smooth and Rough Surfaces and on Gratings, ibid. 111 (1988)
[18] D. Olego et al., Phys. Rev. B 26, 7867 (1982)
G. Fasal et al., Phys. Rev. Lett. 56, 2517 (1986)
L.C.O. Súilleabhaín et al. Sol. State Commun. 87, 517 (1993)

[19] M. Grüninger et al., Europhys. Lett. 35, 55 (1996)
D. Reznik et al., Phys. Rev. B 53, R14741 (1996)
M.R.F. Jensen et al., Phys. Rev. B 55, 2745 (1997)
[20] M. Göppert et al., J. Luminesc. 72-74, 430 (1997)
[21] S. Benedeck, Ref. [12d] p 523 and this volume
[22] R.W. Pohl, Optik und Atomphysik, 13th ed, Springer, Berlin (1976)
[23] J. Lagois, Ph. D., Stuttgart (1976)
[24] J. Lagois, Habilitation Thesis, Erlangen (1981)
[25] Modern Problems in Cond. Matter Sciences Vol. 1, Surface Polaritons and Vol. 9 Surface Excitation, North Holland, Amsterdam
[26] N. Marshall and B. Fischer, Phys. Rev. Lett. 28, 811 (1972)
[27] N. Marshall, B. Fischer and H.-J. Queisser, Phys. Rev. Lett. 27, 95 (1971)
[28] J. Lagois and B. Fischer, Festkörperprobleme / Adv. in Sol. State Phys. 18, 197 (1978)
[29] J. Lagois, Sol. State Commun. 39, 563 (1981)
[30] T. Zettler et al., Phys. Rev. B 39, 3931 (1989)
[31] C. Klingshirn and H. Haug, Phys. Reports 70, 315 (1981)
B. Hönerlage et al., ibid. 124, 161 (1985)
[32] I. Broser et al., Sol. State Commun. 39, 1209 (1981)
M.V. Lebedev et al., JETP Lett. 39, 366 (1984)
[33] Landolt Börnstein, New Series, Group III, Vol. 34C (in preparation)
[34] R. Cingolani and K. Ploog, Adv. in Phys. 40, 535 (1991)
[35] a) E.L. Ivchenko et al., phys. stat. sol. b 161, 217 (1990)
a) E.L. Ivchenko, A.V. Karokin, Sov. Phys. Sol. State 34, 968 (1992)
b) Schultheis and K. Ploog, Phys. Rev. B 29, 7058 (1984)
b) U. Neukirch et al., JOSA B 13, 1256 (1996)
[36] E. Tsitsishvili, Appl. Phys. A 62, 255 (1996)
[37] T. Stroucken, A. Knorr, P. Thomas, S.W. Koch, Phys. Rev. B 53, 2026 (1996)
D.S. Citrin, Pys. Rev. B 51, 14361 (1995)
[38] S. Jorda, U. Rössler, D. Broido, Phys. Rev. B 48, 1669 (1993)
S. Jorda, ibid. 50, 2283 (1994)
[39] D.S. Citrin, phys. stat. sol. b 188, 43 (1995)
D.S. Citrin, Phys. Rev. B 49, 1943 (1994) and B 50, 5497 (1994)
[40] L.C. Andreani, F. Bassani, Phys. Rev. B 41, 7536 (1990)
F. Tassone, F. Bassani, L.C. Andreani, ibid. 45, 6023 (1992)
R. Atanasoo, F. Bassani, V.M. Agranovich, ibid. 49, 2658 (1994)
[41] J. Feldmann et al., Phys. Rev. Lett. 59, 2337 (1987)
R. Eccleston et al., Phys. Rev. B 44, 1395 (1991)
and ibid. 45, 11403 (1992)
[42] L.C. Andreani, Sol. State. Commun. 77, 641 (1991)
M. Gurioli et al., Phys. Rev. B 44, 3115 (1991)
[43] D. Oberhauser et al., phys. stat. sol. 173, 53 (1992)
[44] L. Schultheis, K. Ploog, Phys. Rev. B 29, 7058 (1984)
U. Neukirch et al., phys. stat. sol. b 196, 473 (1996)
and Phys. Rev. B 55, 15408 (1997)
M. Kohl, D. Heitmann, P. Grambow, K. Ploog, Superl. and Microstructures 5, 235 (1989)
[45] K. Ogawa, T. Katsuyama, H. Nakumara, Phys. Rev. Lett. 64, 796 (1990)
[46] S. Jorda, Sol. State Commun. 87, 439 (1993)
D.S. Citrin, Phys. Rev. B 48, 2535 (1993)

[47] W. Demtröder, Laser Spectroskopie 2nd ed, Springer Berlin (1991) and his contribution to this book
[48] R. Zimmermann, Festkörperprobleme / Adv. in Sol. State Phys. 30, 295 (1990)
[49] U. Neukirch, K. Wundke, Sol. State. Commun. 99, 607 (1996)
[50] C. Weisbuch, M. Nishioka, A. Ishikawa, Y. Arakawa, Phys. Rev. Lett. 69, 3314 (1992)
[51] R. Houdré et al., Phys. Rev. Lett 73, 2043 (1994)
[52] P. Kelkar et al., Phys. Rev. B 52, R5491 (1995)
F. Jahnke et al., Festkörperprobleme / Adv. in Sol. State Physics 37 (1997) in press
H.M. Gibbs, ibid.
[53] C.M. Soukoulis, Photonic Band Gaps and Localization, ed. NATO ASI Series 308, Plenum Press, New York (1993)
[54] a) L. Bányai and S.W. Koch, Semicondcutor Quantum Dots, World Scientific, Singapore (1993)
[55] U. Woggon, Optical Properties of Semiconductor Quantum Dots, Springer Tracts in Mod. Physics 136, Springer, Berlin (1997)
[56] M. Grundmann, D. Bimberg, Phys. Blätter 53, 517 (1997)
[57] C. Klingshirn, phys. stat. sol. b, in press (1997)
[58] F. Gindele, Diploma Thesis, Karlsruhe (1996)
[59] Müller-Pouillet's Lehrbuch der Physik und Metreologie, 2. Band, 3. Buch, Vieweg, Braunschweig (1909) and Ref. [22]
A. Kundt, Pogg. Ann. 144, 128 (1872)
O. Lummer, E. Pringsheim, Phys. Z. S. 4, 430 (1903)
[60] A.J. Sievers in Ref [12k] p 227
F. Comas, C. Trallero-Giner and M. Cardona, Phys. Rev. B, in press
see also ref [55], chapter 4
[61] J.M. Hvam, Refs. [12j] p 91, [12k] p 147 and the contribution to this volume
R. Zimmermann in Ref [12k] p 123
R. v. Baltz, contribution to this volume
[62] I. Broser and M. Rosenzweig, Phys. Rev. B 22, 2000 (1980)
G. Blattner et al., Phys. Rev. B 25, 7413 (1982)
K.-H. Pantke, I. Broser, Phys. Rev. B 48, 11752 (1993)
[63] K-H. Pantke, et al., phys. stat. sol. b 173, 69, 91 (1992)
D. Weber et al., Phys. Rev. B 55, 12848 (1997)
[64] J.Y. Masumoto, S. Shionoya and T. Takagahara, Phys. Rev. Lett. 51, 923 (1983)
F. Valée, F. Bogani and C. Flytzanis ibid. 66, 1509 (1991)
[65] U. Woggon et al., Phys. Rev. B 54, 1506 (1996)
[66] a) M.W. Berz et al., Sol. State Commun. 80, 553 (1991)
b) U. Woggon, phys. stat. sol. b, in press
c) U. Woggon and M. Portuné, Phys. Rev. B 51, 4719 (1995)
[67] R.M. Macfarlane, in Ref. [12i] p 399, in Ref. [12j] p 151, in Ref. [12k] p 471 and references given therein
[68] D.G. Steel, H. Wang, S.T. Cundiff, in Ref [9] p 75
[69] P. Wiesner, U. Heim, Phys. Rev. B 11, 3071 (1975)
[70] H. Kalt et al., Proc. Intern. Conf. Phys. II-VI Compounds, Grenoble 1997, in press
[71] M. Umlauff, Ph. D. Thesis, Karlsruhe (1997)
[72] D.N. Mirlin, V.I. Perel', Semicond. Sci. Technol. 7, 1221 (1992)
[73] D.S. Chemla et al., in Optical Nonlin. and Instabilities in Semiconductors, H. Haug ed, Academy Press, New York (1988), p 83

[74] S.G. Hense, M. Elsässer, M. Wegener, Festkörperprobleme / Adv. in Sol. State Physics (1997), in press
[75] S. Permogorov, A. Reznitsky, J. Luminesc. 52, 201 (1992)
H. Schwab et al., phys. stat. sol b 172, 479 (1992)
U. Siegner et al., Phys. Rev. B 46, 4564 (1992)
[76] T. Breitkopf et al., J. Crystal Growth 159, 788 (1996)
and J. Opt. Soc. Am. B 13, 1251 (1996)
[77] R. Westphäling et al., J. Luminesc. 72-74, 980 (1997)
R. Westphäling, Ph. D. Thesis, Karlsruhe (1998)
[78] F.A. Majumder et al., Z. Physik B 66, 409 (1987)
[79] T. Breitkopf et al., Proc. 23. Intern. Conf. Phys. Semicond., M. Scheffler and R. Zimmermann eds., World Scientific, Singapore (1996), p 349
T. Breitkopf, Ph. D. Thesis, Karlsruhe (1998)
[80] C. Klingshirn, phys. stat. sol. b 202, 857 (1997)
[81] M. Oestereich et al., Festkörperprobleme, Adv. in Sol. State Physics 37 (1997), in press
[82] R. Ferreira and G. Bastard, Europhys. Lett. 23, 439 (1993)

FEMTOSECOND PULSE GENERATION: PRINCIPLES AND FIBER APPLICATIONS

Erich P. Ippen

Department of Electrical Engineering and Computer Science
Department of Physics
Research Laboratory of Electronics
Massachusetts Institute of Technology
Cambridge, MA 02139

ABSTRACT

The principles of ultrashort pulse generation by passive modelocking are outlined. Particular attention is paid to the fast saturable absorber case as exemplified by Kerr-lens modelocking (KLM) and polarization additive-pulse-modelocking (P-APM). Steady-state behavior is described in the context of the master equation, and limitations of theory and practice are discussed. Application of these principles to ultrashort-pulse fiber lasers is demonstrated by soliton laser and stretched-pulse laser examples.

I. ULTRASHORT PULSE GENERATION

I. A. Introduction

Since lasers first made it possible to generate pulses shorter than those achievable with electronic flash [1], we have witnessed steady progress in the technology of ultrashort pulse generation and the development of increasingly widespread applications of ultrafast optics in physics, chemistry, biology and engineering. Particularly rapid advances occurred following demonstration of the first sub-picosecond pulses in 1974 [2]. The continuous wave (CW) dye laser that made this possible remained at the center of developments for the next two decades. Femtosecond measurement techniques became increasingly sophisticated, scientific applications developed rapidly [3] and, with the generation of pulses as short as 6 fs [4], the temporal resolution limits of optics began to be tested. Figure 1 illustrates this advance with a plot of the pulse duration versus year. The early dominance of the dye laser, advancing from the first CW achievements to the invention of

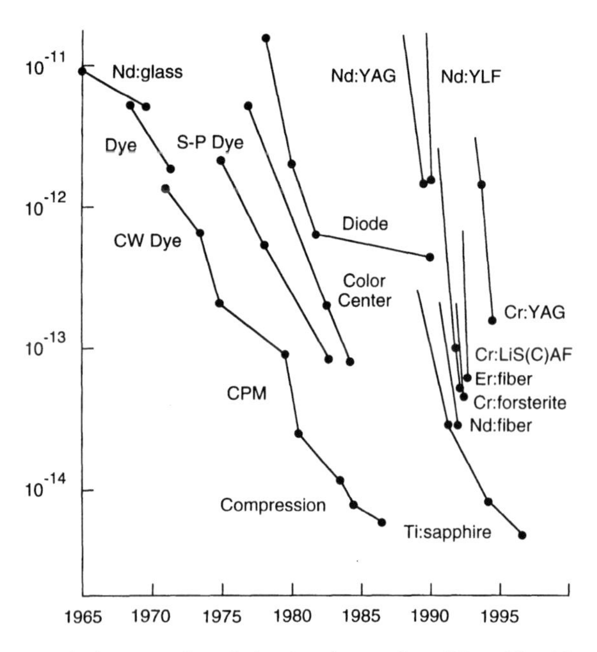

Figure 1: Progress in the generation of ultrashort laser pulses. The achievable pulse duration versus year for a number of different laser technologies.

the colliding-pulse-modelocked (CPM) version to amplified systems that permitted further compression to the ultimate limits, is clear from this plot. Similar techniques were applied to semiconductor diode laser and color-center lasers to achieve ultrashort pulses at higher repetition rates and different wavelengths. All of these systems had an important characteristic in common: they were able to generate ultrashort pulses without any high-speed modulation or nonlinearity. Their pulse shortening mechanism, a surprise at first, relied on dynamic laser gain saturation to repetitively sharpen the trailing edge of a pulse, while a saturable absorber elsewhere in the cavity sharpened the leading edge. An analytic description of this cooperative mechanism was developed and dubbed slow-saturable-absorber passive modelocking [5]. Solid-state lasers, more attractive for many applications because of their robustness and reliability, could not be modelocked in this fashion because their low emission cross-sections and long-lived upper states prevented any significant dynamic gain saturation. They had to rely completely on rapid recovery of the saturable absorber for pulse shortening. That kept them from producing pulses much shorter than 5 ps for many years. Evident from Figure 1, however, is the fact that there was a revolutionary breakthrough in ultrashort-pulse solid-state lasers in the late 1980's. New, broadband laser media were developed, and new techniques were invented for shaping femtosecond pulses without gain dynamics. Quite naturally, the general method is known as fast-saturable-absorber modelocking [6], even though in most cases no real absorption is employed.

With the advent of potentially compact and practical all-solid-state femtosecond lasers, the speed of the ultrafast optics revolution has accelerated. Pulses as short as 5 fs, about 2 optical cycles in duration, are being generated with femtosecond Ti:sapphire lasers [7,8]. An array of other solid-state lasers [9] make pulses of a few 10's of femtoseconds available at other wavelengths. Many of these, including optical fiber lasers, can be pumped by semiconductor diodes for more practical application. As illustrated in Figure 2,

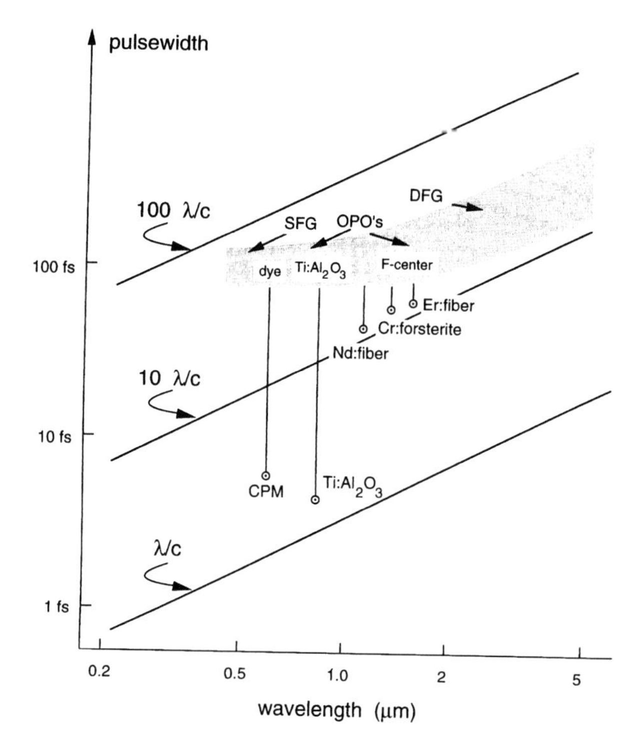

Figure 2: The range of wavelengths over which femtosecond optical pulses are available at high repetition rate. Single points indicate fixed wavelength records; the shaded area indicates continuous coverage. The straight lines show the pulse duration of a specific number of cycles versus wavelength.

the relatively high powers of these lasers, or their amplified extensions, permit wavelength shifting of the femtosecond pulses by nonlinear optics to the full spectrum of wavelengths from the visible to mid-infrared [10]. Ultrahigh-power amplified systems extend this capability into the extreme ultraviolet and even x-ray regime, and produce peak optical intensities that can be used to accelerated electrons to relativistic velocities in less than a femtosecond.

I. B. Pulse Basics - Time Domain Propagation and Filtering

We now consider how a short optical pulse propagates inside a laser and how it is shaped by the nonlinear and time-varying dynamics that produce modelocking. The basic elements, shown in Figure 3 are: gain, loss, bandwidth limiting, self-phase modulation and self-amplitude modulation. Loss is simply represented by a linear factor l. Gain is also assumed to have no dynamic or shaping effect on the pulse, but depends ultimately to the average power I inside the resonator.

$$g = g_o/(1 + I/I_s) \qquad (1)$$

where I_s is the saturation intensity. The other two linear and time-independent effects are (a) pulse broadening by bandwidth limiting in the laser, and (b) pulse broadening and chirping by group velocity dispersion (GVD). In the frequency domain, a single pass through the gain changes an incident spectral amplitude $E(\omega)$ into $E'(\omega)$ where

$$E'(\omega) = \{1 + g\,[1 - (\omega - \omega_o)^2/\,\omega_g^{\,2}]\,\}E(\omega) \qquad (2)$$

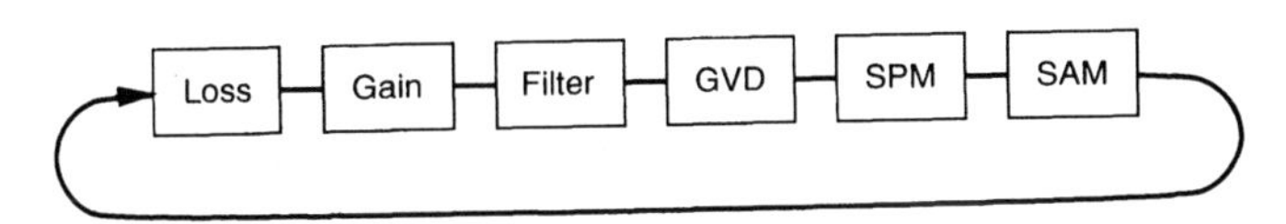

Figure 3: The elements in a passively modelocked laser: loss, gain, spectral filtering, self-phase modulation and self-amplitude.

if we assume that the gain is small and that deviation from the center of the gain is small compared to the gain bandwidth. Then, by Fourier transformation we find the output pulse in terms of the input.

$$E'(t) = \{1 + g\,[1 + (1 / \omega_g^2)\, d^2/dt^2]\,\}E(t) \tag{3}$$

Similarly, passage through an element with GVD (a quadratic phase variation with frequency) produces

$$E'(\omega) = [1 - jD(\omega - \omega_o)^2]\, E(\omega) \tag{4}$$

where, again, we assume a small change over the frequency range. The new pulse in the time domain is broadened and chirped. It becomes

$$E'(t) = [1 + jD\, d^2/dt^2]\, E(t) \tag{5}$$

Thus, we have two dispersion operators for use in the time domain, one real (bandwidth limiting) and one imaginary (frequency dispersive). It can be shown [11] that the bandwidth limiting produces a fractional increase in pulsewidth per pass of $\Delta\tau/\tau = 2g/\omega_g^2\tau^2$. The second, GVD, term stretches (chirps) the pulse more rapidly for short pulses producing a $\Delta\tau/\tau = 2D^2/\tau^4$; but it does not change the spectral bandwidth. So, to produce short pulses, some sort of pulse shortening modulation or nonlinearity is needed in the laser to compensate for these pulse lengthening effects. As indicated above, femtosecond pulse shortening in solid-state lasers requires a fast saturable absorber.

I. C. The Fast Saturable Absorber

A fast saturable absorber is an element that responds essentially instantaneously to changes in light intensity. That is, it can recover its initial absorption in a time short compared to the optical pulse duration. Thus, it can produce pulses in a laser without any help from gain saturation dynamics. Figure 4 shows how this works in the steady state. The absorber shapes the pulse on both leading and trailing edges and discriminates against background light between pulses. Rate equation analysis for the loss, in the limit of absorber recovery time $\tau_a << \tau_{pulse}$, yields

$$l_a(t) = l_o - \gamma\, |E(t)|^2 \tag{6}$$

where $|E(t)|^2$ is defined as the photon flux density, γ is the self-amplitude-modulation (SAM) coefficient, and we have assumed $\gamma|E(t)|^2 << l_o$. The gain is assumed to be approximately constant during the pulse, and equal to its saturated level determined by the steady-state average power. This is the case for media with small gain cross-sections and long upper-state lifetimes. It is, in a practical sense, the defining condition for the fast-absorber modelocking model.

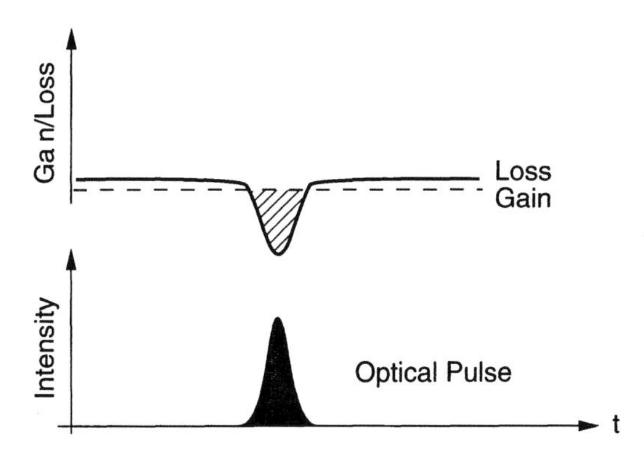

Figure 4: Illustration of the fast saturable absorber dynamic. Loss before and after the pulse is greater than the gain; but the pulse intensity decreases (saturates) the loss to create a window of net gain for the pulse.

The fact that fast-saturable absorber modelocking ceases to become effective when the pulse duration becomes comparable to the absorber recovery time (generally an excited state lifetime in the absorber) limited its use for many years to high-power picosecond systems. Only recently, with the advent of broadband cw systems has it become possible to modelock solid-state lasers passively, in a controlled way. Semiconductor saturable absorbers have been used with success in color-center lasers [12], in coupled cavity systems [13], in anti-resonant Fabry Perot devices [14] and in fiber lasers [15]. Real absorbers have the advantage of simplicity. Because of their real lifetimes, however, they have not, by themselves, produced the shortest pulses. That has been accomplished with artificial fast saturable absorbers as we now describe.

The fastest optical nonlinearities are reactive and nonresonant. The index of refraction nonlinearity in glass, for example, has response on the order of a few femtoseconds [16]; and it may be utilized over a wide range of wavelengths. Because they are fast, such nonlinearities are also relatively weak. Their potential applicability to modelocking has been recognized for some time in the context of pulsed systems [17], but it was not until the emergence of fiber and high-power cw solid-state lasers that their utility has been fully appreciated. A recent review documents this renaissance [18]. Several different realizations of artificial saturable absorbers have been described in the literature [19-21]. In each case, changes of refractive index with intensity are converted into amplitude modulation. They all have clear advantages over real absorbers in that they need not dissipate power (since they deflect power out of the laser instead of absorbing it) and their operational parameters can be varied experimentally. The effective small signal losses they introduce and their effective SAM coefficients can be optimized by proper choice of lenses, beam splitters and cavity dimensions. We will focus our attention here on a technique using dynamic self focusing, called Kerr-lens modelocking (KLM), because it has led to the shortest pulses and is widely used in bulk element solid-state laser systems. Later, in the context of fiber lasers, we will discuss nonlinear polarization rotation, or polarization-additive-pulse-modelocking (P-APM).

I. D. Kerr Lens Modelocking

The importance of dynamic self-focusing [22] became apparent shortly after the observation of self-modelocking in a cw Ti:sapphire laser [23] which revealed different

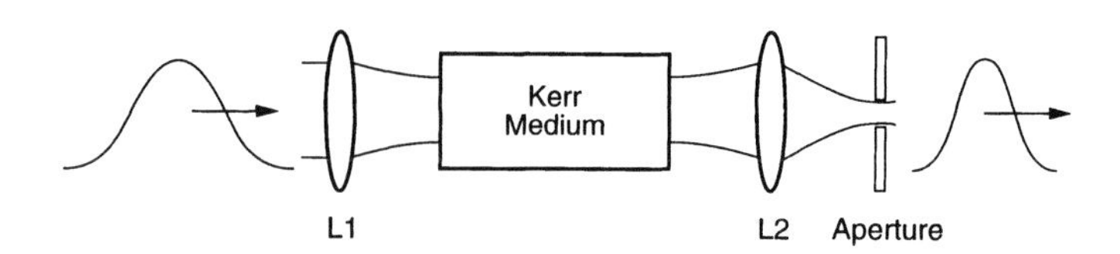

Figure 5: Self-focusing change the optical beam parameters inside a laser dynamically. With proper cavity design and some sort of aperturing, cavity loss is less for the high pulse intensity than for the low level background.

spatial mode parameters under short pulse operation than under cw. Piche [24] suggested that self-focusing might be responsible. Cavity designs were developed to enhance the lensing effects and utilize them for effective saturable absorber action as illustrated in Figure 5. The technique was dubbed Kerr lens modelocking (KLM) by Spinelli et.al.[25] who demonstrated improved modelocking with an aperture appropriately placed in the cavity. Excellent results were quickly achieved by a variety of groups in both Ti:sapphire and other systems [9].

Several numerical [26-28] and analytical studies [20,29] provide insight into various characteristics of the process. Since self-focusing is produced by index of refraction changes, KLM is linked fundamentally to the presence of self-phase modulation (SPM). This phase modulation may be written in a form similar to that of equation (6):

$$\phi(t) = \phi_o + \delta\,|E(t)|^2 \qquad (7)$$

A relationship between the SPM coefficient δ and SAM coefficient γ for KLM has been determined analytically[20] and the figure of merit, $M = \gamma/\delta I_o$, under proper design, is found to be about the same as that for other artificial saturable absorber systems. KLM Ti:sapphire lasers have produced the shortest modelocked laser pulses to date [30,31].

Since the peak amplitude of the pulse shortening modulation, m_f, produced by a fast absorber is proportional to the peak intensity of the pulse, it depends inversely on the pulse duration for a given average power in the laser: $m_f = \gamma W/2\tau$, where W is the pulse energy. The parabolic temporal curvature of the modulation also becomes stronger as the pulse gets shorter. Together they produce a modulation (transmission) function that can be written $T = 1 - (\gamma W/2\tau)(t^2/\tau^2)$. The resultant decrease in pulsewidth per pass is obtained from $1/\tau'^2 = 1/\tau^2 + \gamma W/2\tau^3$, which gives

$$\Delta\tau/\tau = \gamma W/2\tau \qquad (8)$$

Thus, fast-absorber dynamics produce an ever increasing pulse shortening rate as the pulse gets shorter. This is the true strength of such systems. It differs from the slow-absorber case in which the shortening rate is independent of pulsewidth [32]. A limit is reached, of course, when the absorber is completely saturated or when the pulse duration approaches its response time. The downside of such strong shortening for ultrashort pulses is that the shortening is very weak initially when the pulse is long. This latter aspect has led to problems with self-starting in fast-absorber systems. More about that below; first we are now in a position to evaluate the steady-state pulse characteristics of a KLM, or any other fast-absorber system.

I. E. The Master Equation

Each of the pulse shaping elements described above are introduced as terms in the modelocking master equation [6,20]. To ensure a steady-state solution one new term $t_D(d/dt)$ is also added to account for the possibility of a net timing shift of the pulse envelope per round trip.

$$\{ g - l + j\phi_o + g/\omega_g^2(d^2/dt^2) + jD(d^2/dt^2) + t_D(d/dt) + (\gamma - j\delta)|E(t)|^2 \} E(t) = 0 \qquad (9)$$

This master equation has a remarkably elegant analytical solution [33]:

$$E(t) = E_o \operatorname{sech}(t/\tau) \cdot \exp\{j\beta \ln \operatorname{sech}(t/\tau)\} \qquad (10)$$

which is expressed in terms of a chirp parameter as well as the pulsewidth.

Equations for τ, β and W are obtained by plugging (10) into (9). Figures 6a and 6b show plots of pulsewidth τ and chirp β as a function of GVD for a fixed energy, W, and gain bandwidth, ω_g, but for different values of SPM, δ. Here it is assumed that δ is positive (positive change in index with intensity) as is the case for most femtosecond nonlinearities in solid state materials. For zero SPM the minimum pulsewidth occurs at zero GVD and has a value of [34]

$$\tau_o = 4g/\gamma W \omega_g^2 \qquad (11)$$

consistent with a simple balance of the SAM pulse shortening and filter broadening as described above. As SPM increases, the point of minimum pulsewidth moves to negative GVD where the chirp is compensated. The GVD value required for $\beta = 0$ is given analytically by [34]

$$|D| = (\delta/\gamma)(g/\omega_g^2) \qquad (12)$$

and at this point again

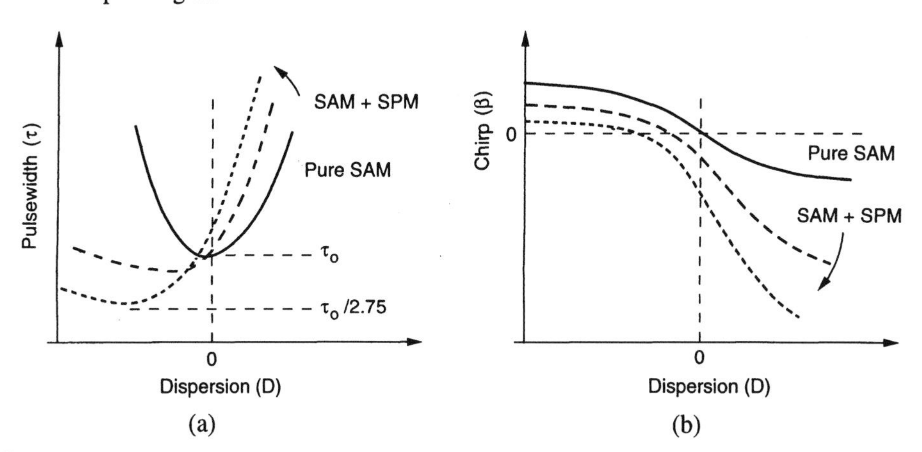

Figure 6: Plots of solutions to the master equation. (a) Pulsewidth versus dispersion for different values of SPM. Proper balance of GVD and SPM reduces the pulse duration by a factor of 2.75. (b) corresponding plots of the chirp parameter versus dispersion.

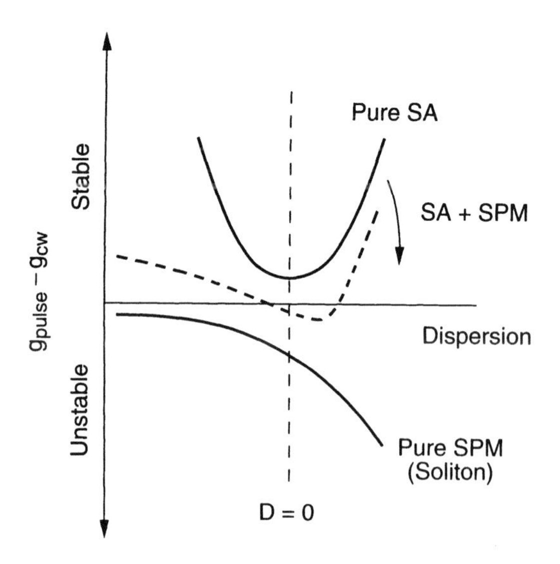

Figure 7: A plot of the stability parameter $\beta = g_{pulse}$-g_{cw} versus dispersion for different values of SPM. For $\beta < 0$ the pulse is not stable with respect to the growth of noise.

$$\tau = 4|D|/\delta W \tag{13}$$

It is interesting to note that shorter pulses with chirp can be obtained at higher values of |D|; but, within the approximations of this theory, they cannot be more than a factor of 2.75 shorter than they would be with SAM alone. We should note further, however, that such a comparison is based on the assumption that the laser output would be limited by the cavity bandwidth in the absence of such soliton shortening. In some systems the steady state pulsewidth may be limited by higher order effects (saturation of the SAM and higher order GVD) before the bandwidth limit is reached.

The solutions thus predicted by the master equation are not always stable. Stability requires not just that a steady-state pulse shape be produced but that it have greater gain than competing cw oscillation. To elucidate this condition we plot g_{pulse} - g_{cw} in Figure 7. It reveals an interesting condition: a pure soliton laser (i.e. one without any amplitude modulation) is always unstable). Because the soliton requires more bandwidth than cw oscillation at the gain peak, the latter will always take over. Even with SAM, SPM can drive the system unstable unless sufficient GVD is introduced at the same time.

I. F. Limitations

The master equation also incorporates several approximations that are certainly violated by some practical laser systems. At high powers the saturable absorber saturates fully, so that its modulating effect is no longer linear with intensity. Soliton shaping can keep the pulse somewhat shorter than the width of the flattening amplitude shaping window, but the steady-state pulse duration will be longer than that predicted by the analytical theory alone. If the peak power of the pulse is limited, for example, by the particular saturation characteristics of an artificial saturable absorber, the single-pulse soliton energy is also limited. In fiber lasers this, along with periodic perturbations, can cause pulse break-up into multiple solitons.[35]. One method for increasing the energy obtainable in a single pulse is the stretched-pulse APM technique used in fiber lasers[36]. Finally, in soliton systems, when the roundtrip SPM phase shift becomes large, so that the cavity length is no longer much shorter than the soliton period, roundtrip-periodic perturbations produce pulsewidth-limiting instabilities and spectral sidebands [36,38].

At very short pulsewidths, higher order dispersion can be the dominant factor. In the master equation formalism it introduces higher-order time-derivative terms for which no simple analytical solution has been found. Numerical simulations as well as experimental evaluations indicate that third-order dispersion limits pulse shortening and produces asymmetries in shape and spectrum. In some systems, this has been corrected for and fourth order dispersion is thought to be the limiting factor in ultrashort pulse generation [39]. Higher order dispersion compensation by proper dielectric multi-layer mirror design is being actively pursued as a means for getting to pulses of a few cycles or less.

Finally, when the changes produced by individual elements are large, the pulse may have significantly different durations and spectra in different parts of the cavity. The ordering of the elements can then be important, particularly in the case of strong SPM. A laser operating under these conditions has been called a solitary laser [40]. The pulse duration may be approximated by

$$\tau = 3.53|D|/\delta W + \alpha \delta W \tag{14}$$

where α is an empirical constant. The first term, like the master equation result, is consistent with the prediction of adiabatic soliton shaping.; the second term gives the increase due to the large periodic changes. Equation (14) has shown good correspondence with results obtained with some lumped-element solid-state lasers[40] and fiber lasers[41].

I. G. Self-starting

Finally, we note that self-starting is a problem inherent to fast saturable absorber systems. In theory, if you wait long enough, a short pulse should develop out of any initial fluctuation. In practice, this does not always happen. One reason is competition from pulse dispersing processes. If the pulse does not shorten significantly within a cavity mode coherence time [42], it will be dispersed. This sets a power threshold for self-starting. Mode pulling due to spurious reflections [43] can cause a short coherence time, as can spatial hole-burning[44]. It has been shown[45] that unidirectional ring operation, which reduces both spatial hole-burning and the effects of multiple reflection, greatly facilitates self-starting. Still, there is not yet any completely satisfying way to ensure starting in all systems. Current methods (other than banging on the table) include tilting plates and moving mirrors [46-48], adding a real saturable absorber either intracavity [49] or in RPM mode[50], synch-pumping [51-53], regenerative initiation [54] or active modulation [55]. Each method brings with it trade-offs. Which is actually used will depend upon the need for low starting threshold and cost.

I. H. Summary of pulse generation

The generation of ultrashort pulses by passive modelocking is now an advanced and highly sophisticated art. A large number of passively modelocked lasers has been developed for different wavelength regimes, different power levels, and different applications. All of them have somewhat different components, specific design needs and operating characteristics. The purpose of this paper has not been to discuss all of these systems and to document their characteristics. Instead we have tried to illustrate some of the common principles by which they operate. With a master-equation, perturbational time-domain analysis we have arrived at an analytical solution of steady state pulse parameters and their stability. Although more detailed, numerical analysis is becoming increasingly necessary as these systems are refined and pushed to their limits, the classic analytical analyses of fast-absorber modelocking continues to provide valuable insight.

II. SHORT PULSE FIBER LASERS

II. A. Introduction

Fiber lasers were first demonstrated in the early 1960's [56] but it was really the advent of erbium-doped fibers in the 1980's [57] sparked the rapid development of this technology. Erbium doped fibers not only provide amplification near the important 1.55 μm optical communication band, they provide that gain over a wide bandwidth, sufficient for either multiple wavelength channels or for ultrashort pulse generation. Among their many other advantages are ruggedness, compactness, amenability to diode pumping, and operation in a single spatial-mode. The latter property, of course, also eliminates the self-focusing described in the previous section as a possible modelocking mechanism. So, different techniques have had to be developed. In the sections below we will focus on the method of polarization-additive-pulse-modelocking (P-APM) which has been exploited successfully for the generation of pJ-energy sub-picosecond solitons [58,59,60] and for the generation of nJ-energy pulses shorter than 100 fs [61,62]. We will discuss how soliton effects in these lasers act to control and define pulse characteristics and then how they can be circumvented to achieve higher powers and shorter pulses.

II. B. Polarization Additive Pulse Modelocking (P-APM)

Nonlinear polarization rotation can occur in an optical fiber when the initial polarization state is elliptical. The ellipse can be resolved into right- and left-handed polarization components of different intensities. These two circular components then accumulate different nonlinear phase shifts related to the intensity dependence of the index of refraction [63,64]:

$$n = n_o + n_2 I \qquad (15)$$

The polarization ellipse rotates while maintaining its ellipticity and handedness. An optical fiber is particularly well-suited for nonlinear polarization rotation because they maintain high intensity over long interaction lengths. Figure 8 illustrates how nonlinear polarization rotation is used in conjunction with a birefringent wave plate and a polarizer to obtain artificial absorber action. (Higher intensity rotates the polarization to more effectively pass through the polarizer.) The technique is called polarization additive pulse modelocking (P-APM) since the coherent addition of the right- and left handed components at the final polarizer determines the modulation.

Since the nonlinear index of refraction, and therefore the polarization rotation angle, is linearly proportional to the optical intensity, P-APM is equivalent to the fast absorber modelocking described above. The master equation governing the steady state is the same as that given in equation (9). Thus, there are basic similarities between P-APM fiber lasers and the bulk component KLM lasers discussed previously. The same factors that inhibit self-starting in KLM lasers are also at work in fiber lasers. Spatial hole burning, produced by reflections and standing waves, is a particular problem in fiber lasers since the gain

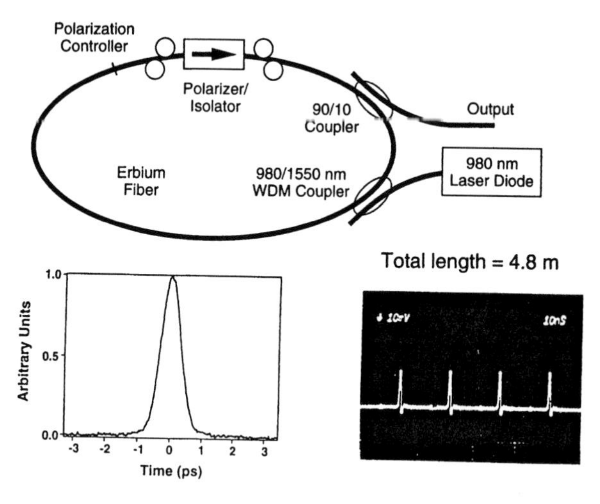

Figure 8: Diagram of the elements required for P-APM. Nonlinear polarization rotation occurs in the Kerr medium and produces power dependent transmission through the analyzer.

medium (the erbium doped fiber) is distributed over relatively long lengths (meters). Thus, lasers constructed in unidirectional ring configurations have been shown to self-start more easily.

A typical fiber ring laser [65] is shown in Figure 9. The erbium-doped fiber is pumped at 980 nm via a dichroic coupler. Output is extracted with a 90/10 fiber beam-splitting coupler. A Faraday isolator element ensures unidirectional oscillation and also acts as the polarizer-analyzer needed for the P-APM action. (Such an isolator would make KLM Ti:sapphire lasers self-starting, but the dispersion it introduces prevents those systems from reaching their shorter pulse durations.) "Rabbit ear" polarization controllers (acting as birefringent wave plates) are positioned before and after the isolator/polarizer to compensate for residual fiber birefringence and create the optimum elliptically polarized state for P-APM. Limits on the density with which a fiber can be doped with erbium, in addition to the lengths required for the fiber pigtailing of isolator and couplers, require the overall cavity to be several meters in length. Typical output from such a ring in shown in the lower part of the figure. Single pulses per round trip are emitted at a repetition rate of about 40 MHz. The pulse duration, determined by nonlinear (SHG) intensity autocorrelation is about 450 fs. Why is it not shorter? As we discuss below, the pulsewidth in this laser is not limited by the bandwidth of the gain medium (which could according to the master equation could easily produce pulses of less than 100 fs) but by soliton effects.

II. C. Soliton limiting

A pulse traveling around the fiber ring is shaped by the balance between SPM and GVD which produces a soliton-like pulse.

$$A(T,t) = A_o \mathrm{sech}\,(t/\tau)\exp\{-(j\delta|A_o|^2/2)(T/T_R)\} \tag{16}$$

with

$$2|D|/\tau^2 = \delta|A_o|^2 \tag{17}$$

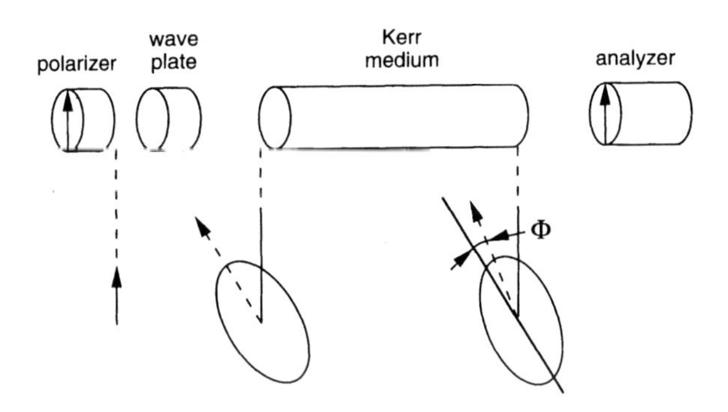

Figure 9: An all-fiber soliton ring laser using P-APM. Lower insets are the SHG autocorrelation of the pulse and an oscilloscope race of the pulse train.

This constraint (eqn.17) is the so-called area theorem that fixes the product $|A_o|\tau$ be a constant characteristic of the system. If the amplitude of the soliton is increased, its pulsewidth should shorten.

A practical limit to this shortening is evidenced in the data of Figure 10. The pulse spectrum displays the narrow sidebands characteristic of periodic soliton modulation first described by Kelly [38] and analyzed theoretically by Gordon [66]. (A log-scale temporal autocorrelation trace shown in the inset shows how these sidebands correspond to a spreading pulse background.) Figure 11 illustrates the mechanism for this sideband generation. The soliton has a nonlinear phase shift per pass of $\delta|A_o|^2/2$ relative to the phase shift of linear waves at its center frequency. Linear waves experience dispersion which causes phase shifts equal to $-|D|\Delta\omega^2$ about this center frequency. When the sum of the two phase shifts becomes equal to a multiple of 2π, scattering of the soliton into the continuum becomes phase matched. Power is then scattered resonantly at the specific phase matched frequencies into the dispersive continuum.

This can also be described in terms to the "soliton period" $Z_{o,}$ the distance over which a soliton accumulates a nonlinear phase shift of $\pi/4$. When the soliton shortens, becomes more intense and increases its nonlinear phase shift to the point (2π) where $8Z_o$ becomes equal to the cavity length, resonant scattering would occur at the center frequency of the soliton. Before the pulse gets this short, however, increasing loss to the encroaching sidebands limits the energy and stops the shortening process. In practice, this occurs already when $8Z_o$ reaches $4L_{cavity}$. Typical limiting values for a soliton fiber laser with length 5 m are $\tau > 200$ fs and internal pulse energies < 60 pJ. Shorter cavity lengths would enable shorter pulses and higher energies ($W \propto (1/\tau) \propto L^{1/2}$) but restrictions on erbium doping densities have made that impractical to date.

It is possible to eliminate the spectral sidebands, and associated energy background, by introducing a filter into the cavity [67]. A resulting spectrum and autocorrelation measurement are shown in Figure 12. It turns out that the filter need not increase the pulse duration. In fact, with decreasing filter bandwidth, the pulse actually shortens a bit over the sideband-limited case at first, since the sidbands are a parasitic drain on the pulse energy. Of course, when the filter is narrowed further the output pulse duration lengths proportionately. The spectrum of the shortest pulses is still narrower than that of the erbium gain bandwidth, so that it is possible to tune the center wavelength over a range as large as 50 nm [68].

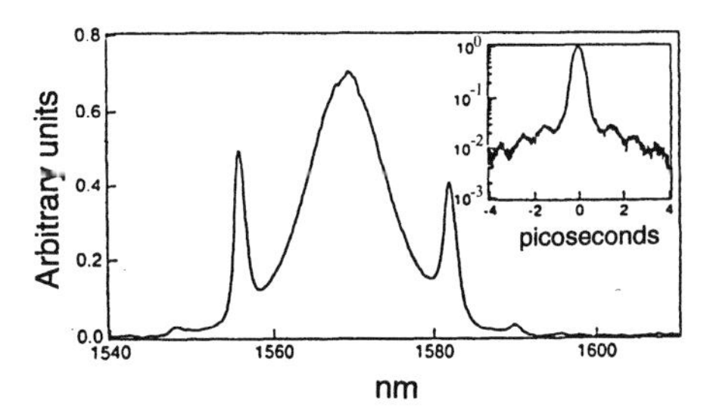

Figure 10: The spectrum of the soliton laser output. Inset is the SHG auto-correlation on a log scale.

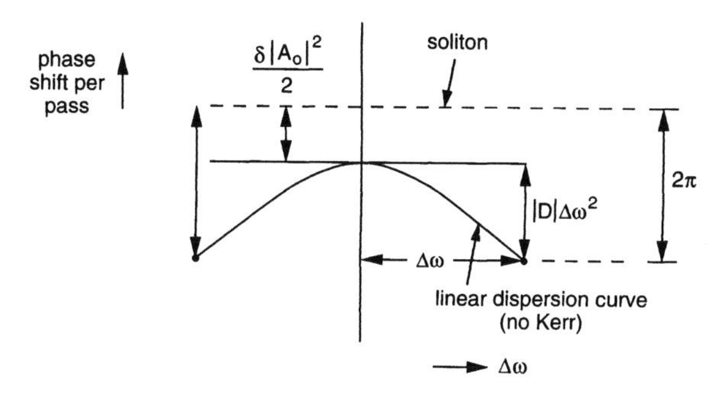

Figure 11: Illustration of the phasematching condition between the soliton and the background continuum.

II. D. Harmonic modelocking

In order to obtain the high (GHZ) pulse repetition rates required by applications such as high data rate optical communication systems and optical sampling, fiber lasers must be induced to emit more than one pulse per round trip. To do this in a controlled manner, an intracavity electro-optic modulator is generally employed. The idea is to drive the modulator at a high (N'th) harmonic of the cavity roundtrip rate and permit multiple (N) pulses to circulate within the cavity. When such harmonic modelocking was first demonstrated, however, large pulse-to-pulse fluctuations were observed [69,70]. This occurs because the slow ($\approx$10 ms) relaxation time of the erbium gain medium does not permit it to stabilize, by dynamic saturation, energy fluctuations on the timescale of the pulse period ($\approx$ 1 ns). One method for stabilizing repetition rate and pulse energy is to introduce an optical sub-cavity into the loop in such away that roundtrip reflections in the sub-cavity couple subsequent pulses to each other [71,72]. A disadvantage of this method of course is that it requires interferometric stabilization of the sub-cavity to loop. More appealing are methods that use passive nonlinear means to limit and stabilize pulse energy. One of these makes use of the nonlinear polarization rotation phenomenon discussed above. With a construction similar to that used for P-APM one can adjust the polarization biasing wave plates such that higher pulse intensity produces more, rather that less, loss. That technique has been dubbed APL (additive pulse limiting) [73]. Another method that does not rely on polarization rotation, and can therefore be used in systems built with polarization maintaining fiber, takes advantage of soliton properties. Increased energy in a

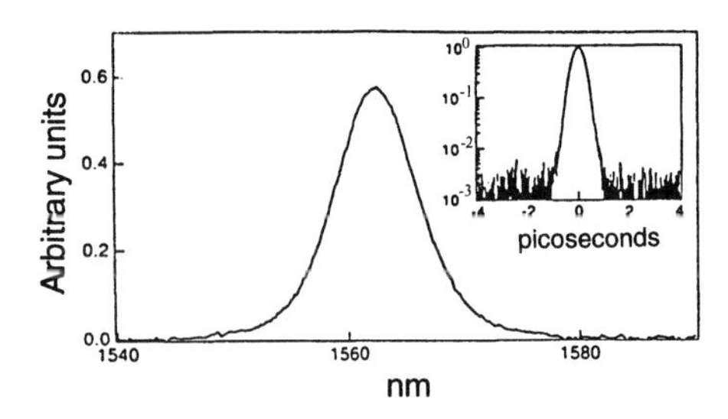

Figure 12: The spectrum of the soliton laser with an intracavity filter. Inset is the SHG autocorrelation on a log scale.

soliton leads to pulse shortening and therefore spectral broadening. A spectral filter in the laser that prohibits such broadening provides a fast damping mechanism [74,75].

In order for stabilization by soliton effects to be effective, the pulsewidth in the laser must be controlled by the soliton mechanism. This implies that the steady-state pulsewidth must be shorter than that which the active modulator would produce by itself. Master-equation modelocking theory similar to that outlined above [76] predicts that stable modelocking is possible with soliton-like shortening by a factor of about three over that produced by active modelocking alone [77]. Kaertner et al [78] have extended the theory to show that with sufficiently large GVD (and compensating SPM) even further shortening can be stable.

Two conditions must be met for stable operation of a harmonically modelocked soliton laser: the modulator must be able to maintain synchronization, and the soliton state must be stable with respect to the longer pulse actively modelocked state. The former condition is a practical matter of keeping the cavity (roundtrip) harmonic close enough to the modulation frequency so that the modulator can make up for any difference by retiming the pulses. We simply note that this gets more difficult as the pulses get shorter since retiming by the modulator decreases proportionally to τ^2. The second condition we can analyze by comparing the roundtrip loss encountered by the filtered soliton with the modulation loss of the longer pulse mode [79]. For stability the loss for the longer pulse must be greater. Thus,

$$\pi^2 M\Omega_m{}^2\tau^2/24 + 1/3L\Omega_f{}^2\tau^2 < \mathrm{Re}\{(M\Omega_m{}^2/2)[(1/L\Omega_f{}^2 - jD/L]\} \qquad (18)$$

where M is the modulation index divided by the cavity length L, Ω_m is the modulation frequency, and Ω_f is the filter bandwidth. The left-hand side is the loss experienced by the soliton and the right-hand side expresses the modulation loss of the longer pulse. In addition, the modulator must not drive the soliton unstable. According to soliton perturbation theory [80] energy fluctuations are damped if

$$\pi^2 M\Omega_m{}^2\tau^2/24 < 1/L\Omega_f{}^2\tau^2 \qquad (19)$$

The energy of the soliton is then given by equation (13). These stability criteria were investigated by Jones et al [81] using a polarization maintaining ring laser about 45 m in length and driven at 5 GHz. By varying modulation depth and filter bandwidth they investigated the region proscribed by equation (18) and found good agreement. Well inside this region stable pulses were obtained; outside, the pulses deteriorated. For a filter bandwidth they obtained pulses 634 fs in duration, a factor of 4.4 shorter than that predicted by active modelocking theory and the shortest harmonically modelocked pulses achieved to date from a fiber laser employing soliton compression.

Harmonic modulation and soliton stability have important consequences also for optical memories [82]. One realization of a memory is an optical storage loop in which a pattern of pulses (data) circulates, with zeros as well as ones in the harmonically defined time slots. It has been shown that with proper adjust of nonlinearity and filtering, as well as gain in the loop, such a pattern, can be maintained indefinitely. Experiments have demonstrated this with 2 kbits of 20 Gbit/s data in a loop [83].

II. E. The Stretched-Pulse Laser

As we have discussed to this point, soliton lasers have inherent practical limits on their pulse width and energy due to spectral sideband generation and saturation of the P-APM. To circumvent this difficult Tamura et al [36] constructed a ring laser comprised of two segments of oppositely dispersive fiber. A schematic of this laser is shown in Figure 13. Modelocking is achieved by P-APM as in the soliton laser, but as the pulse gets shorter it is alternately dispersed, positively and negatively, by the two different fiber segments. The shorter it gets, the more strongly it stretches and compresses, the shorter the distance over which it remains short. This has the effect of reducing the effective nonlinearity in such a way as to prevent the peak power from being clamped. It also breaks up the phasematching to sidebands. As a result, both higher energies and shorter pulses are produced. Figure 14 illustrates extraction of the pulses from the stretched-pulse laser at a point where they are highly chirped. Pulse autocorrelation measurements taken after different degrees of dispersion compensation in an external fiber are also shown in Figure 14. They demonstrate the linearity of the chirp produced by the combination of SPM and positive (normal) dispersion in the segment prior to extraction. With this laser pulses as short as 76 fs were produced [84]; with subsequent versions [85] pulses as short as 63 fs were achieved. An analytical theory for stretched-pulse P-APM has also been formulated [86]. It predicts the observed positions of the pulse minima inside the laser, the stability of highly chirped solutions, the Gaussian-shaped spectra and the lack of sideband radiation.

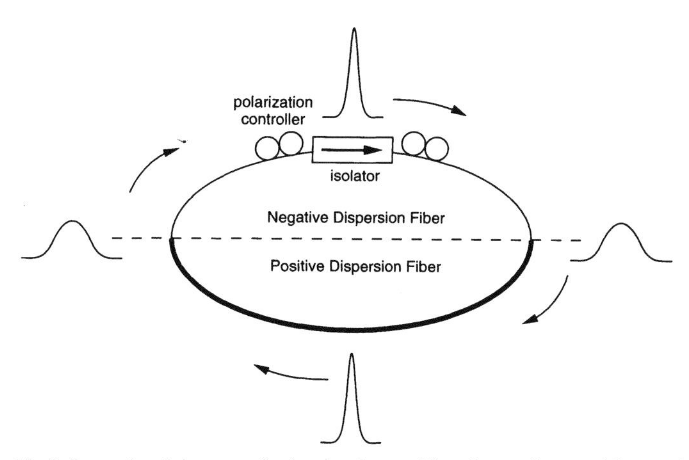

Figure 13: Schematic of the stretched pulse laser. The alternating positive and negative dispersion causes the pulse to stretch and compress as indicated.

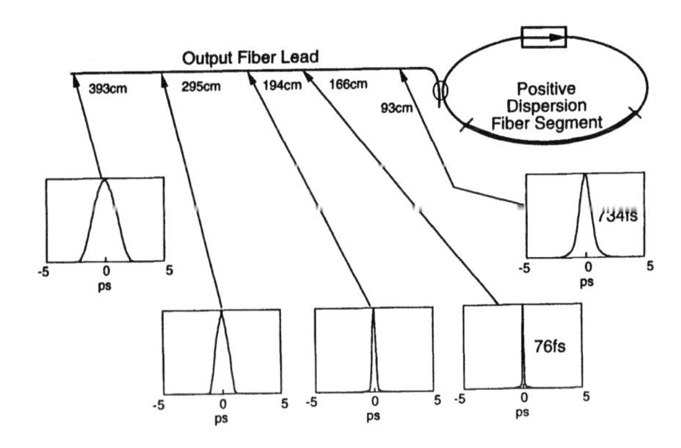

Figure 14: Autocorrelation traces of the output of a stretched-pulse laser after different distances of external fiber dispersion compensation.

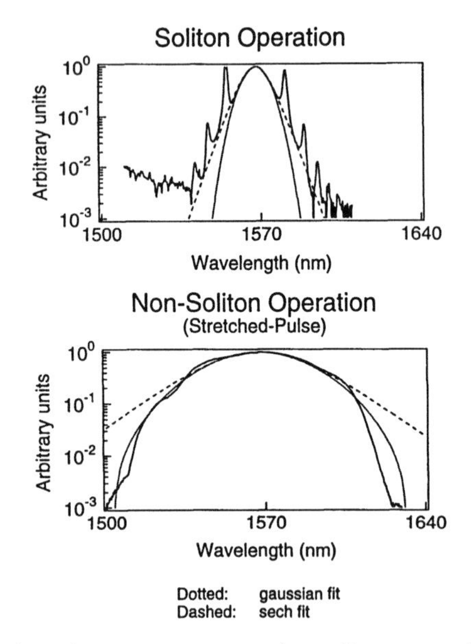

Figure 15: Comparison between spectra under soliton operation (with resonant sidebands) and nonsoliton operation. Dashed lines indicate fit for sech pulse shape; solid lines for gaussian.

An experimental comparison of the spectra produced under soliton and nonsoliton conditions are shown in Figure 15. The soliton case exhibit sidebands and a shape corresponding to a sech pulse (dashed theoretical curve); the stretched pulse case more closely follows the Gaussian shape (solid theoretical curve.)

Advanced, and now commercialized, versions of the stretched-pulse laser have a slightly modified design to permit extraction of higher power. It was observed experimentally [61] that higher pulse energy could be obtained from the polarizing beam splitter that serves as the rejection port of the P-APM. Although one might expect the pulses exiting from this port to have poor quality, surprisingly high quality pulses, albeit somewhat longer than those inside the ring, can be achieved. Pulse durations shorter than 100 fs are routine; and, with high-power diode-laser pumping [62,87], pulse energies are in excess of 2 nJ. These latter capabilities have opened interesting application opportunities for this system. Nanojoule pulse energies are easily sufficient for a variety of scientific studies using femtosecond pump-probe measurement; and the stretched-pulse laser produces the shortest pulses available in the 1.55 μm wavelength regime. The high peak powers permit efficient frequency doubling to 780 nm, a wavelength where femtosecond pulse amplification to high power is possible using Ti-sapphire amplifiers [87]. The frequency-doubled pulses are shorter than and of even higher quality than those at the fundamental since the nonlinear conversion process discriminates against the wings and eliminates excess spectrum. The 270 pJ energies at 780 nm are more than sufficient to seed a high power amplifier [88]; so the stretched-pulse laser can be a compact, less expensive replacement for argon-laser-pumped Ti:sapphire oscillators in these applications. Finally, just the broad spectral width of this laser alone makes it a potentially attractive transmitter for pulse trains of many different wavelengths simultaneously in a dense wavelength-division-multiplexed (WDM) fiber communication system [89].

II. F. Summary of short-pulse fiber lasers

We have described the application of polarization additive-pulse modelocking (P-APM) to erbium-doped fiber lasers for the generation of ultrashort pulses. The importance of soliton effects produced by the combination of self-phase modulation (SPM) and group-velocity dispersion (GVD) was discussed. They limit pulse durations and cause energy shedding to the continuum, but they can also be used for pulse stabilization and quantization in harmonically modelocked systems. Development of the stretched-pulse laser, to circumvent the deleterious soliton effects and permit the generation of higher pulse energies and shorter pulses, was described. It now provides femtosecond pulses for a variety of applications including broadband optical communication networks as well as the seeding of high-power femtosecond amplifiers for scientific studies.

III. ACKNOWLEDGEMENT

Many of the experiments described in these notes were performed by graduate students in the optics group of the Research Laboratory of Electronics at MIT. Kohichi Tamura, Chris Doerr, Lynn Nelson and David Jones deserve the credit for the work. My colleague Hermann Haus inspired, and collaborated on, all of it. And, it could not have been accomplished without support from AFOSR and DARPA.

IV. REFERENCES

1. A. J. DeMaria, D.A. Stetser and H. Heynau, Appl. Phys. Lett. **8**, 174 (1966).
2. C.V. Shank and E.P. Ippen, Appl. Phys. Lett. **24**, 373 (1974).
3. Ultrafast Phenomena, Springer Series in Chemical Physics, Springer Verlag, 1978-1996.
4. R.L. Fork, C.H. Brito Cruz, P.C. Becker and C.V. Shank, Opt. Lett. **12**, 483 (1987).
5. H.A. Haus, IEEE J. Quant. Electron., **QE-11**, 736 (1975).
6. H.A. Haus, J. Appl. Phys. **46**, 3059 (1975).
7. A. Boltuska, Z. Wei, M. Pshenichnikov and D. Wiersma, Opt. Lett. **22**, 102 (1997).
8. M. Nisoli, S. deSilvestri, O. Svelto, R. Szipocs, K. Frenz, S. Sartonia, Ch. Spielman and K. Frausz, Opt. Lett. **22**, 522 (1997).
9. U. Keller, K. J. Weingarten, F. X. Kaertner, D. Kopf, B. Braun, I. D. Jung, R. Fluck, C. Hoenneger, N. Matuschek, J. Aus der Au, IEEE J. Selected Topics in Quant. Electron. **2**, 435 (1996)

10. P.E. Powers, R.J. Ellingson, W.S. Pelouch and C.L. Tang, JOSA B, **10**, 2162 (1993).
11. H.A. Haus, "Waves and Fields in Optoelectronics", Prentice Hall, 1984.
12. C.L. Cesar, M.N. Islam, C.E. Soccolich, R.D. Feldman, R.F. Austin and K.R. German, Opt. Lett. **15**, 1147 (1990).
13. E.A. DeSouza, C.E. Soccolich, N. Pleibel, R.H. Stolen, M.N.Islam, J.R. Simpson and D.J. DiGiovanni, Electon. Lett. **29**, 447 (1993).
14. U. Keller, D.A.B. Miller, G.D. Boyd, T.H. Chiu, J.F. Ferguson and M.T. Asom, Opt. Lett. **17**, 505 (1992.
15. M.H. Ober, M. Hofer, U. Keller, T.H. Chiu, Opt. Lett. **18**, 1532 (1993).
16. A. Owyoung, R.W. Hellwarth and N. George, Phys. Rev. **B5**, 628 (1972).
17. L. Dahlstrom, Opt. Commun. **3**, 399 (1971).
18. I.D. Jung, F.X. Kärtner, N. Matuschek, D.H. Sutter, F. Morier-Genoud, Z. Shi, V. Scheuer, M. Tilsch, T. Tschudi, and U. Keller, Appl. Phys. B. **65**, 137, (1997).
19. E.P. Ippen, H.A. Haus and L.Y. Liu, J. Opt. Soc. Am. B **6**, 1736 (1989).
20. H.A. Haus, J.G. Fujimoto and E.P. Ippen, IEEE J. Quant. Electron. **28**, 2086 (1992).
21. M.L. Dennis and I.N. Duling III, Appl. Phys. Lett. **62**, 2911 (1993).
22. M.C. Marconi, O.E. Martinez and F.P. Diodati, Opt. Commun. **63**, 211 (1987).
23. D.E. Spence, P.N. Kean and W. Sibbett, Opt. Lett. **16**, 42 (1991).
24. M. Piche, Opt. Commun. **86**, 156 (1991).
25. L. Spinelli, B. Couillaud, N. Goldblatt and D.K. Negus, Digest of Conf. Lasers Electro-Opt. (CLE), Opt. Soc. Amer. 1991, paper (PDP7).
26. F. Salin, J. Squier and M. Pichë, Opt. Lett. **16**, 1674 (1991).
27. O.E. Martinez and J.L.A. Chilla, Opt. Lett. **17**, 1210 (1992).
28. I.P. Christov, H.C. Kapteyn, M.M. Murnane, C.-P. Huang and J. Zhou, Opt. Lett. **20**, 309 (1994).
29. T. Brabec, P.F. Curley, Ch. Spielmann, E. Wintner and A.J. Schmidt, J. Opt. Soc. Amer. B **10**, 1029 (1993).
30. I.D. Jung, F.X. Kärtner, N. Matuschek, D. Sutter, F. Morier-Genoud, U. Keller, V. Scheuer, M. Tilsch, and T. Tschudi, Opt. Lett. **22**, 1009 (1997).
31. L. Xu, Ch. Spielmann, F. Krausz and R. Scipocz, Opt. Lett. **21**, 1259 (1996).
32. E. P. Ippen, Appl. Phys. B **58**, 159 (1994)
33. O.E. Martinez, R.L. Fork and J.P. Gordon, J.O.S.A. B **2**, 753 (1985).
34. H.A. Haus, J.G. Fujimoto and E.P. Ippen, J. Opt. Soc. Amer. B **8**, 2068 (1991).
35. D.J. Richardson, R.I. Laming, D.N. Payne, M.W. Phillips and V.J. Matsos, Electron. Lett. **27**, 730 (1991).
36. K. Tamura, E.P. Ippen, H.A. Haus and L.E. Nelson, Opt. Lett. **18**, 1080 (1993).
37. N. Pandit, D.U. Noske, S.M.J. Kelley and J.R. Taylor, Electron. Lett. **28**, 455 (1992).
38. S.M.J. Kelley, Electron. Lett. **28**, 802 (1992).
39. I. Christov, M. Murnane, H. C. Kapteyn, J. Zhou and C. P. Huang, Opt. Lett. 19, 1149 (1994)
40. T. Brabec, Ch. Spielmann and F. Krausz, Opt. Lett. **16**, 1961 (1991).
41. T. Brabec, C. Spielmann and F. Krausz, Opt. Lett. **17**, 748 (1992).
42. K. Krausz, T. Brabec and Ch. Spielmann, Opt. Lett. **16**, 135 (1991).
43. H.A. Haus and E.P. Ippen, Opt. Lett. **16**, 1331 (1991).
44. F. Krausz and T. Brabec, Opt. Lett. **18**, 888 (1993).
45. K. Tamura, J.M. Jacobson, E.P. Ippen, H.A. Haus and J.G. Fujimoto, Opt. Lett. **18**, 220 (1993).
46. N.H. Rizvi, P.M.W. French and J.R. Taylor, Opt. Lett. **17**, 279 (1992).
47. Y.M. Liu, K.W. Sun, P.R. Prucnal and S.A. Lyon, Opt. Lett. **17**, 1219 (1992).
48. M.H. Ober, M. Hofer and M.E. Fermann, Opt. Lett. **18**, 367 (1993).
49. N. Sarukara, Y. Ishida and H. Nakano, Opt. Lett. **16**, 153 (1991).
50. U. Keller, G.W. t'Hooft, W.H. Knox and J.E. Cunningham, Opt. Lett. **16**, 1022 (1991).
51. L.F. Mollenauer and R.H. Stolen, Opt. Lett. **9**, 13 (1984).

52. F. Krausz, Ch. Spielman, T. Brabec, E. Wintner and A.J. Schmidt, Opt. Lett. **17**, 204 (1992).
53. A. Seas, V. Petricevic and R.R. Alfano, Opt. Lett. **16**, 1668 (1991).
54. D.E. Spence, J.M. Evans, W.E. Sleat and W. Sibbett, Opt. Lett. **16**, 1762 (1991).
55. J.D. Kafka, M.L. Watts and J.-W.J. Pieterse, IEEE J. Quant. Electron. **QE-28**, 2151 (1992).
56. E. Snitzer, Phys. Rev. Lett. **7**, 444 (1961).
57. S.B. Poole, D.N. Payne, R.J. Mears, L. Reekie, M.E. Fermann and R.J. Laming, L. Lightwave Technol. **22**, 159 (1986).
58. M. Hofer, M.E. Fermann, F. Haberl, M.H. Ober, A.J. Schmidt, Opt. Lett. **16**, 502 (1991).
59. V.J. Matsas, T.P. Newson, D.J. Richardson, D.N. Payne, Electron. Lett. **28**, 1391 (1992).
60. K. Tamura, H.A. Haus, E.P. Ippen, Electron. Lett. **28**, 1391 (1992).
61. K. Tamura, C.R. Doerr, L.E. Nelson, E.P. Ippen, H.A. Haus, Opt. Lett. **19**, 46 (1994).
62. G. Lenz, K. Tamura, H.A. Haus, E.P. Ippen, Opt. Lett. **20**, 1289 (1995).
63. P.D. Maker, R.W. Terhune, Phys Rev. Lett. **12**, 507 (1964).
64. R.H. Stolen, J. Botineau, A. Ashkin, Opt. Lett. **7,** 512 (1982).
65. K. Tamura, H.A. Haus, E.P. Ippen, Electron. Lett. **28**, 2226 (1992).
66. J. P. Gordon, J. Opt. Soc. Am. B **9**, 91 (1992).
67. K. Tamura, C.R. Doerr, H.A. Haus, E.P. Ippen, IEEE Photon. Tech. Lett. **6**, 697 (1994).
68. K. Tamura, E.P. Ippen, H.A. Haus, IEEE Photon. Tech. Lett. **6**, 1433 (1994).
69. J.D. Kafka, T.Baer, D.W. Hall, Opt. Lett. **14**, 1269 (1989).
70. A. Takada, H. Miyazawa, Electon. Lett. **26**, 216 (1990).
71. E. Yoshida, Y, Kimura, M. Nakazawa, Appl. Phys. Lett. **60**, 932 (1992).
72. G.T. Harvey, L.F. Mollenauer, Opt. Lett. **18**, 107 (1993).
73. C.R. Doerr, H.A. Haus, E.P. Ippen, M. Shirasaki, T. Tamura, Opt. Lett. **19**, 31 (1994).
74. M. Nakazawa, K. Tamura, E. Yoshida, Electron. Lett. **32**, 461 (1996).
75. K. Tamura, M. Nakazawa, Opt. Lett. **21**, 1930 (1996).
76. H.A. Haus, Y. Silberberg, IEEE J. Quantun Electron. **QE-22**, 325 (1996).
77. D.J. Kuizenga, A.E. Siegman, IEEE J. Quantum Electron. **6**, 694 (1970).
78. F.X. Kärtner, D. Kopf, U. Keller, J. Opt. Soc. Am. B **12**, 486 (1995).
79. H.A. Haus, A. Mecozzi, Opt. Lett. **17**, 1500 (1992).
80. H.A. Haus, A. Mecozzi, IEEE J. Quantum Electron. **QE-29**, 983 (1993).
81. D.J. Jones, H.A. Haus, E.P. Ippen, Opt. Lett. **21**, 1818 (1996).
82. C.R. Doerr, W.S. Wong, H.A. Haus and E.P. Ippen, Opt. Lett. 19. 1958 (1994)
83. J.D. Moores, K.L. Hall, S.M. LePage,. K.A. Rauschenbach, W.S. Wong, H.A. Haus and E.P. Ippen, IEEE Photon. Technol. Lett. 7, 1096 (1995)
84. K. Tamura, L.E. Nelson, H.A. Haus, E.P. Ippen, Appl. Phys. Lett. **64**, 149, (1994).
85. K. Tamura, E.P. Ippen and H.A. Haus, Appl. Phys. Lett. **67**, 158 (1995)
86. H.A. Haus, K. Tamura, L.E. Nelson, E.P. Ippen, IEEE J. Quantum Electron. **QE-31**, 591 (1995).
87. L.E. Nelson, S.B. Fleischer, G.Lenz, E.P. Ippen, Opt. Lett. **21**, 1759 (1996).
88. T.B. Norris, Opt. Lett. **17**, 1009 (1992).
89. M. Nuss, W.H. Knox, U. Koren, Electron. Lett. **32**, 1311 (1996).

LASER SOURCES FOR ULTRAFAST SPECTROSCOPY

Allister I Ferguson

Department of Physics and Applied Physics
University of Strathclyde
Glasgow
Scotland, UK

ABSTRACT

New methods for generating ultrashort light pulses from solid-state-lasers are opening up many new areas of application. I discuss methods for the generation of femtosecond pulses from diode-pumped solid-state-lasers and optical parametric oscillators. The application of these light sources in imaging systems is also described.

1. INTRODUCTION

Progress in the generation of ultrashort light pulses over the past few years has been spectacular. When this is coupled with progress made in diode pumped solid state lasers, new areas of science and technology are being opened up, for the first time.

In this chapter I will describe some of the highlights in the new methods of ultrashort pulse generation and describe how these techniques have been rapidly incorporated into diode pumped solid state laser technology. This technology provides compact, efficient and reliable laser light sources which can be used in a wide range of applications. In order to give a flavour of the application areas to which these new sources can be applied I will describe three imaging techniques which have benefited greatly by the developments. These are, optical coherence topography, terahertz imaging and multiphoton induced fluorescence imaging.

2. MODE-LOCKING

2.1 Principles of Mode-Locking

Soon after the invention of the laser it was realised, that in addition to the phenomenal collimation and focusability of a laser beam, the temporal characteristics of lasers are very rich. A laser, of course, is made up of an excited medium, which provides gain over a certain bandwidth $\Delta\nu$, surrounded by mirrors which provide feedback. The mirrors form an optical cavity which can sustain oscillation at only certain allowed frequencies. The frequencies which can be sustained must lie within the gain bandwidth of the medium and must correspond to fitting an integer number of half wavelengths into the length of the cavity. The allowed frequencies are given by

Ultrafast Dynamics of Quantum Systems: Physical Processes and Spectroscopic Techniques, Edited by Di Bartolo and Gambarota, Plenum Press, New York, 1998

$$\nu = nc / 2\ell(f) \quad (1)$$

where n is the number of half wavelengths in the cavity, c is the velocity of light and $\ell(f)$ is the effective length of the cavity for light at frequency ν. In the absence of dispersion $\ell(f)$ is just the physical length of the cavity ℓ. The frequency separation between adjacent modes is given by

$$f = c / 2\ell \quad (2)$$

where we have ignored dispersion.

If the condition $f << \Delta\nu$ is satisfied the laser will be capable of oscillating on many modes. In general, these modes will oscillate independently and give an output which, when averaged over time, is constant but which has massive fluctuation on timescales of less than the cavity round trip time.

This is perhaps best illustrated by the figure 1(a) [1,2]. This is the output of a multimode laser in frequency and time. It has been assumed that a large number of modes oscillate with a Gaussian distribution of amplitude but with random relative phase. The output intensity, as a function of time, is also shown. The timescale shown in this figure represents one transit in the cavity. It can be clearly seen that the power of the laser fluctuates randomly in time. The width of the fluctuations τ is approximately given by $\tau = 1/\Delta\nu$. It is important to stress here that the mode amplitudes and phases are constant in time, under the assumptions made here, but that this still gives an apparently randomly fluctuating output. A consequence of the assumption of fixed amplitude and phases is that the fluctuating power output in time repeats every cavity round trip time.

We now look at what happens when we make the assumption that the phases of each of the modes is related to the phase of the neighbouring mode. Let us assume that all modes have the same phase but that they still have the same amplitudes as before. The resulting situation is shown in figure 1(b). The output in time is now a single spike and the phase of the output is constant in time. Indeed the height of the peak has had to be greatly reduced

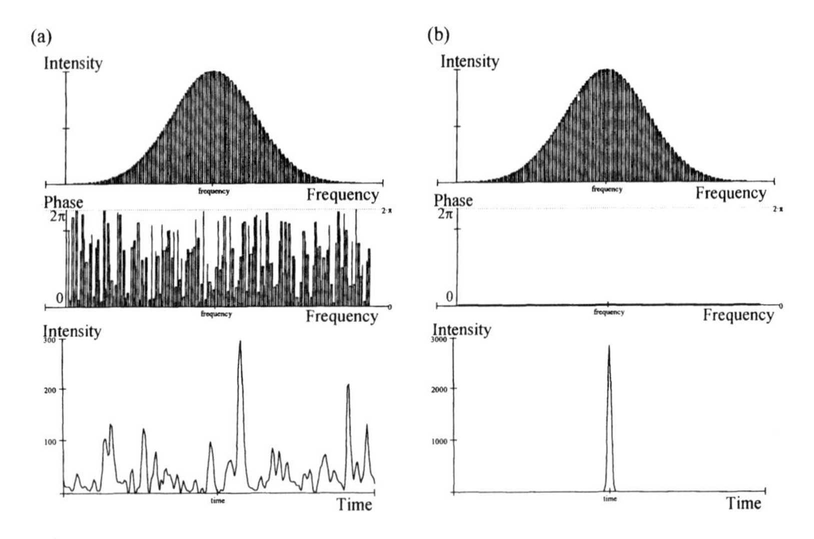

Figure 1. (a) Frequency domain traces of the intensity and phase of a multimode laser. We have assumed 150 modes with random phases. The output intensity is calculated using a Fourier transform and the resulting trace is shown as a function of time. Noise spikes with a duration comparable to the inverse of the spectral width can be seen. (b) All modes are assumed to have the same phase (zero). The output intensity is now a single spike of high intensity.

to fit into the figure. This spike will appear at each roundtrip of the cavity. The width of the spike is similar to the width of the fluctuation in figure 1(a) and is given approximately by $\tau = 1/\Delta\nu$. This rather special mode of operation of a laser is called mode-locking. Laser mode-locking is a truly remarkable phenomenon and is the technique used to generate the very shortest pulses of light. Light of just a few optical cycles has been readily generated by mode-locking.

Let us now look at mode-locking in a little more detail. Why does phase relationship give rise to ultrashort pulses? If we think of what we mean by setting the phases of all the modes to be equal what we are saying is that at some point in time all of the mode amplitudes will add coherently giving a giant enhancement in peak power. Each of the modes will then evolve in time, the amplitudes will no longer add constructively and the output power will drop to a very small value. At some point later in time the amplitudes will again add constructively giving another giant pulse. This will occur at one cavity round trip time later. The output will thus be in the form of a train of pulses separated by the cavity round trip time. We can think of a mode-locked laser as a series of oscillators of fixed amplitude oscillating in phase. In principle, if we had a narrow enough spectral filter we could take the output from a mode-locked laser and filter it down to just a single mode. This would appear as a single frequency continuous-wave oscillator.

Let us consider what is happening in the time domain. In this case the multimode laser produces randomly fluctuating power which fills the cavity. When this system is mode-locked, all of the power is compressed in time to give a single pulse which propagates back and forth inside the cavity. Each time the pulse reaches the output coupler a pulse emerges giving rise to a train of pulses.

The art of laser mode-locking is therefore concerned with the control of the phases of the modes of a multimode laser such as to compress the output of the laser in time to produce a single pulse within the cavity.

2.2 Methods of Mode-Locking

Mode-locking methods can be classified into two broad categories. These methods are called active and passive mode-locking. In the case of active mode-locking an external source of some kind determines the timing of the system while in passive mode-locking the laser itself provides the timing.

2.2.1 Active mode-locking. A schematic diagram of an actively mode-locked laser is shown in figure 2. In this case a modulator is placed in the cavity. This modulator can be an amplitude or phase modulator. The case of an amplitude modulator is, perhaps, easier to understand. The amplitude modulator is driven at a frequency such as to provide loss modulation at the cavity mode spacing $f = c/2\ell$. Light which passes through the modulator at a loss minimum will experience less loss than light passing through the modulator at other times. The timing of loss modulation and transit around the cavity is arranged such that the light experiencing low loss will continue to pass through the modulator at low loss on each transit while other light will continue to suffer loss. On many successive round trips all of the light will eventually be compressed to pass through the modulator at low loss. This thus gives rise to mode-locking.

We can also think of active mode-locking in the frequency domain where the modulation puts sidebands onto the oscillating modes. If the cavity mode spacing and the modulation frequency coincide the oscillating modes will be coupled together, thereby giving the required phase relationship.

A good feature of active mode-locking is that it is virtually a universal mode-locking technique. Provided a loss or phase modulator can be made to operate at the oscillating wavelength, and can be driven to provide modulation at the cavity mode spacing, it is possible to get mode-locked operation. The main drawback with active mode-locking is that the pulse compressing mechanism is not very strong and, as the pulses become shorter and shorter in time, the compression mechanism becomes weaker. A related problem with active mode-locking is that the matching of the cavity round trip time and the modulation

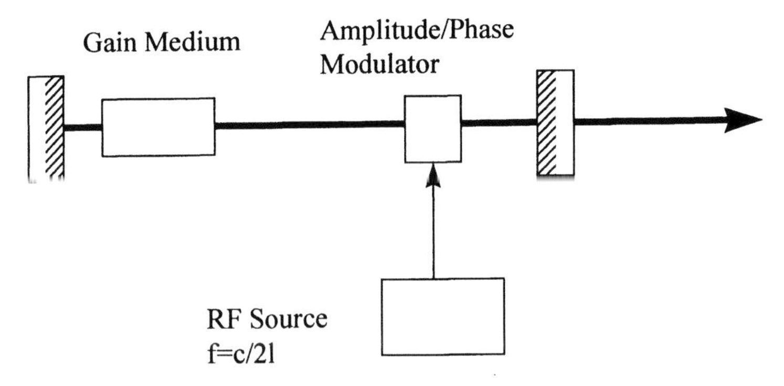

Figure 2. A schematic diagram of an actively mode-locked laser.

frequency becomes more and more critical as the pulses become shorter. The active mode-locking technique becomes impractical at pulse duration of typically around 10psec when operating at about 100MHz repetition rate.

2.2.2 Passive mode-locking. Passive mode-locking was one of the first mode-locking techniques to be exploited. The passive method has many practical implementations but in its most general form we can think of a saturable absorber in a cavity. The transmission of a saturable absorber is shown schematically in figure 3. In this figure we plot the transmission of the absorber as a function of the light intensity falling upon it. At low intensity the light is strongly absorbed whereas at high intensity the light transmission is very high.

If a saturable absorber is placed in a laser cavity, and if the gain and linear absorption are properly balanced, the laser will oscillate. If it oscillates in a large number of modes, as we have seen, there will be massive fluctuations in the intensity. The low intensity light will be absorbed and experience loss whereas the high intensity light will be transmitted. The high intensity light will therefore dominate and become even more intense, and the

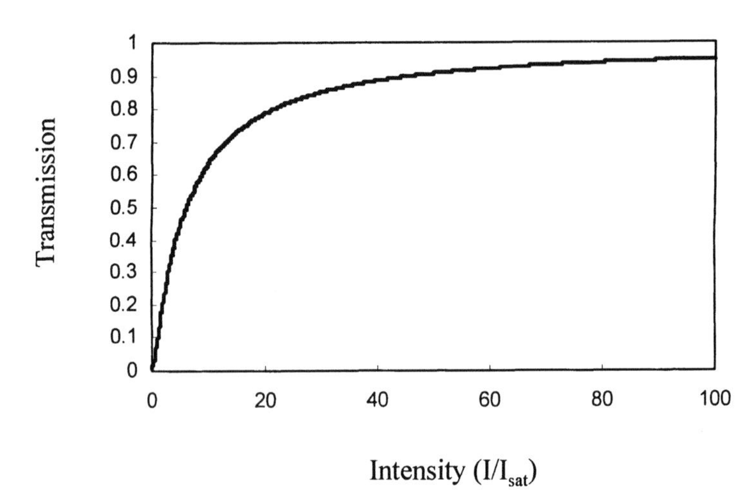

Figure 3. The transmission of a saturable absorber as a function of the incident intensity. At low intensity the transmission is negligible whereas at high intensity the transmission approaches unity and the material becomes transparent.

absorber will become even more transparent. Eventually, a single pulse containing all of the energy will dominate and propagate around the cavity, giving rise to mode-locked operation.

The attractive feature of passive mode-locking is that the pulse compression mechanism gets stronger as the pulses become shorter. The shortest light pulses ever generated have been accomplished by use of passive mode-locking techniques.

Traditionally, passive mode-locking has been performed using saturable absorber dyes. These dyes tend to degrade with time and often have to be circulated to keep them fresh. There has been a revival of passive mode-locking methods due to the new generation of mode-locking techniques. These techniques are based on actual and virtual saturable absorption in solid state media and are very reliable. Some of these methods will be described in more detail later. The main disadvantage of passive mode-locking is that it is quite system specific. The right characteristics are needed from the saturable absorber, including the appropriate level of linear absorption, the saturation intensity and the recovery time. These parameters have been worked out for most combinations of gain medium and saturable absorber.

3. THE OPTICAL KERR EFFECT

The new generation of mode-locking techniques depends, to some degree, on the optical Kerr effect. The optical Kerr effect is the high frequency extension of a phenomenon observed by John Kerr in Glasgow in 1875. Kerr was interested in the connection between light and electricity. He observed that if an electric field was applied to a transparent medium, the polarisation of light could be modified. He found that this effect was proportional to the square of the applied field and that it occurred in all transparent media whether they be anisotropic or isotropic. The Kerr effect is an extremely rapid phenomenon and applies up to optical frequencies. The square of the electric field at optical frequency is proportional to the intensity of light (power per unit area). The optical Kerr effect (or ac Kerr effect) can be expressed as a modification to the refractive index of a medium given by

$$n(r,t) = n_o + n_2 I(r,t) \tag{3}$$

where n_o is the normal linear refractive index and n_2 nonlinear refractive index coefficient. We have explicitly allowed for a spatial variation in intensity $I(r,t)$ as well as a temporal variation. The nonlinear refractive index coefficient n_2 can be a positive or negative quantity but under most circumstances is a positive quantity. The optical Kerr effect manifests itself in two phenomena associated with the spatial and temporal variation of intensity, namely self-focussing and self-phase modulation, respectively.

3.1 Kerr Lens

A Kerr lens occurs when there is a spatial variation in the intensity of a light source passing through a medium exhibiting the Kerr effect. If we assume that the nonlinear index is positive we see that, for example, a laser beam with a Gaussian intensity distribution would give rise to a positive lens and hence self-focussing. This is because the centre of the beam has higher intensity than the wings and so, on passing through the medium, the optical path length will be greater at the centre of the beam than in the wings. The resulting medium will appear to be a positive lens and the beam will be focussed as it passes through the medium.

3.2 Self-Phase Modulation

Self-phase modulation is the change in phase of an optical pulse as it passes through a medium exhibiting the Kerr effect. In this case we are only concerned with the temporal properties of the light and can write the refractive index in the presence of the light as

$$n(t) = n_o + n_2 I(t). \tag{4}$$

The Kerr effect can be taken to be almost instantaneous. The change in phase of the transmitted pulse after having travelled a distance L through the medium is

$$\phi(t) = -\delta n L \omega_o / c \tag{5}$$

where ω_o is the carrier (or centre) frequency of the light pulse and $\delta n = n_2 I(t)$ is the nonlinear change in refractive index. We now consider the time varying change of phase by considering the instantaneous frequency $\omega(t)$ which is given by

$$\omega(t) = \omega_o + \delta\omega(t) \tag{6}$$

and

$$\delta\omega(t) = d\phi(t)/dt. \tag{7}$$

The instantaneous frequency shift $\delta\omega(t)$ is shown in figure 4 for a Gaussian beam in time and assuming that n_2 is a positive quantity. The leading edge of the pulse is shifted to lower frequency and the trailing edge is shifted to higher frequency. Several important points are illustrated by this figure. The first is that new frequencies which were not present in the original pulse have now been generated and so the spectrum of the emitted light is broader than when it entered the medium. The second point to note is that, at the centre of the pulse, the instantaneous frequency of the light returns to its original carrier frequency value of ω_o. The third point to note is that in this central region the frequency is swept from low frequency to high frequency in an approximately linear manner. This is called a frequency chirp.

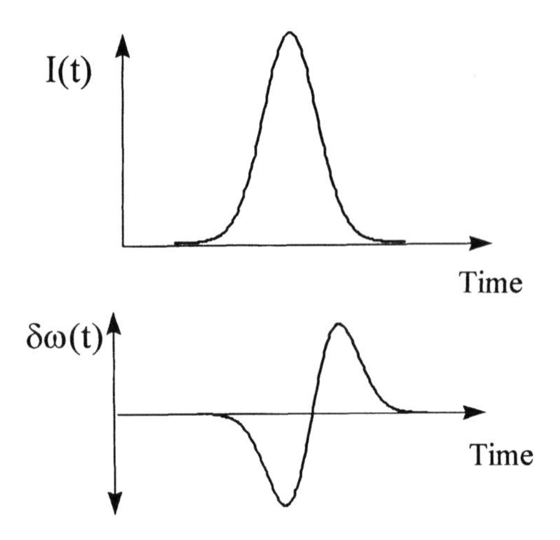

Figure 4. The intensity of a Gaussian pulse as a function of time together with the change in frequency of the pulse relative to the carrier frequency. The leading edge of the pulse (left-hand side) is down shifted whereas the trailing edge is up-shifted due to self-phase modulation.

Under most circumstances it is impossible to disentangle the effects of self-focussing and self-phase modulation because they both take place together as light passes through the medium. One circumstance where it is possible to ignore self-focussing is the case where the light passes along a single mode-optical fibre. The index guiding in the fibre dominates any self-focussing effect and so we can isolate self-phase modulation.

4. ADDITIVE PULSE MODE-LOCKING

One of the remarkable modern mode-locking effects which relies on the Kerr nonlinearity is a technique known as additive pulse mode-locking (APM). A schematic diagram of an idealised APM system is shown in figure 5. The laser itself is very simple with a gain medium and a cavity. All, or part, of the output of the laser passes through a medium exhibiting the Kerr nonlinearity, onto a mirror, and back into the laser cavity. In order to neglect self-focussing effects, and also to provide a large self-phase modulation at modest power, the nonlinear medium would normally be a single mode optical fibre. The length of the nonlinear cavity is adjusted to be equal to the laser cavity length or an integer multiple of that length.

The mode-locking mechanism can be thought of in the time or frequency domain. For the sake of simplicity we will just consider the time domain. We know that a multimode laser oscillator will exhibit strong amplitude fluctuations. Let us consider one of these pulses. On exiting the laser this pulse will pass through the nonlinear medium, onto the mirror and back again. The pulse will experience self-phase modulation in passing through the nonlinear medium twice. It will therefore be frequency chirped, the centre frequency equalling the carrier frequency of the original pulse, the leading edge down shifted and the trailing edge upshifted in frequency. Since the nonlinear cavity is matched to the laser cavity, the chirped pulse will meet the remnants of the original pulse still in the cavity. These two pulses will then interfere. If the phase of the returning chirped pulse is appropriately adjusted, it will constructively interfere with the original pulse and enhance it. This can only occur at the centre where the two pulses have the same frequency. At the wings the pulses will have different frequency and so will tend to cancel each other. The consequence is that the original pulse is now shorter and higher in power than the original. When this pulse now propagates through the nonlinear medium it will experience even more self-phase modulation and so will produce an even shorter and more intense pulse when it returns to the laser cavity. As the pulse becomes shorter and shorter the pulse compression mechanism becomes stronger and stronger, and all of the energy in the cavity compresses to a single pulse. The pulse duration is normally limited by either the bandwidth of the gain medium or dispersion in the cavity. The main advantage of this technique is this very powerful compression mechanism but the main disadvantage is the need for interferometric alignment of two cavities.

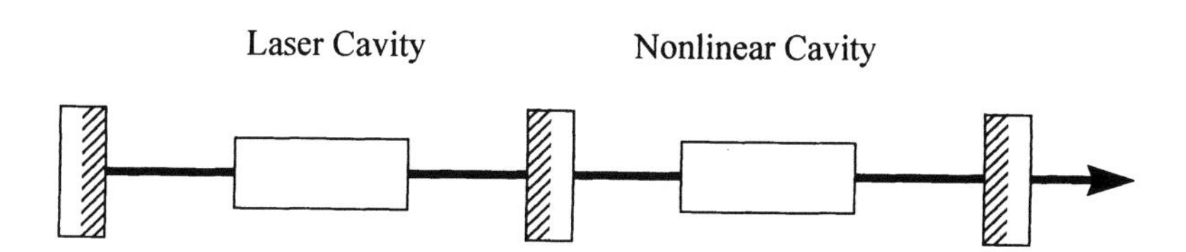

Figure 5. A schematic diagram of an additive pulse mode-locked laser also known as a coupled cavity mode-locked laser.

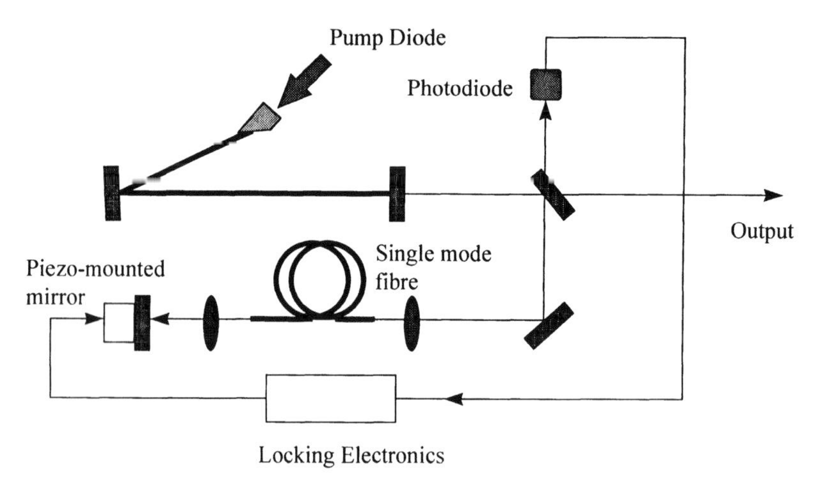

Figure 6. A diagram of a diode pumped Nd:YLF mode-locked laser based on additive pulse mode-locking.

A practical implementation of a diode pumped solid state laser version of an APM system is shown in figure 6. In this case the active gain medium is Nd:YLF operating on the 1047nm line. The gain medium is placed at the end of the cavity so that it can be easily pumped by laser diodes. One end of the rod is high reflection coated at 1047nm and antireflection coated at the pump wavelength of 792nm. The other end of the rod is at Brewster's angle. This ensures operation at 1047nm and also prevents any internal reflection from surfaces. The pump is made up of two 4W laser diodes tuned to the absorption in Nd:YLF. The output power from a system of this type is in the region of 1.5W. A few hundred milliwatts are diverted into a single mode optical fibre. The light from the fibre goes on a mirror and is then retroreflected back through the fibre and into the laser cavity. The length of the external cavity is adjusted to match the laser cavity. When the matching occurs, interference fringes are observed on the photodiode. This can be used to provide an error signal for a servo system which holds the phase of the external cavity at an appropriate level to induce mode-locking. This system typically produces pulse durations of 2-3psec at 1047nm with about 1.2W of output at about 100MHz repetition rate. This gives a massive amount of peak power from an all solid state system and is presently being used in our laboratories for nonlinear optical processes such as synchronous pumping of optical parametric oscillators and the generation of squeezed light with sub-Poissonian noise characteristics.

In some applications it is highly desirable to have even shorter pulses than a few psec or even to have adjustable pulse duration. An elegant scheme for achieving this is shown in figure 7. The output from the laser is passed through a single mode fibre. The fibre generates new frequencies through self-phase modulation as was described earlier. As these new frequencies are generated, the pulse will broaden spectrally and temporally. The temporal broadening is due to the normal dispersion of the fibre. The resulting output is a chirped pulse with nearly all the energy contained in a region where the frequency is linearly swept. The linear sweep in frequency can be undone by passing the light through a delay line with, effectively, negative dispersion. The slow parts of the pulse will catch up with the fast parts to produce a pulse which is significantly shorter than the starting pulse. Compressions of the order of thirty or more are quite routine. The compression is normally accomplished by use of a grating delay line, shown schematically as a pair of gratings but is usually a single grating multipassed, in practice. Almost 50% throughput has been achieved

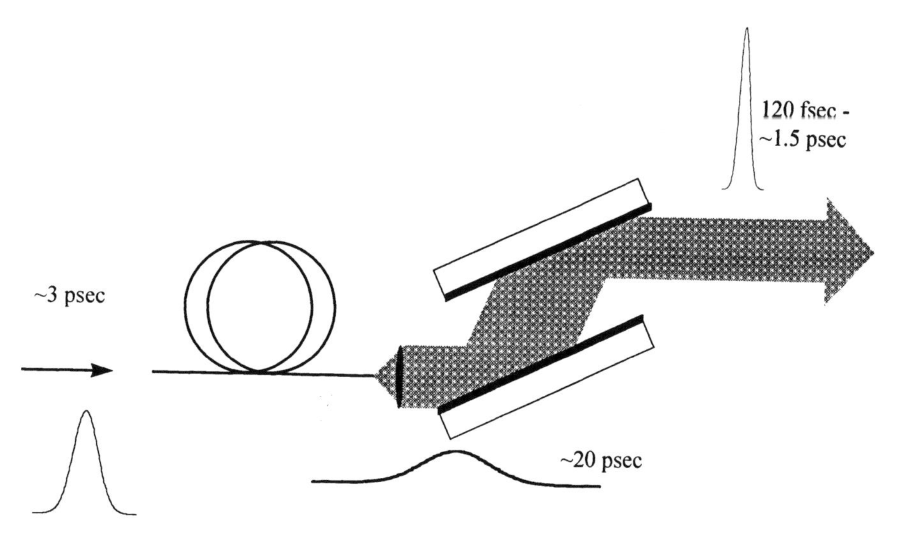

Figure 7. A schematic diagram of a fibre-grating pulse compressor. The initial pulse is stretched and chirped by the single mode fibre and then compressed using a grating (shown here as a pair of gratings).

with this device at pulse durations as short as 120fsec and average powers of 500mW. Systems of this type are now routinely used for multiphoton imaging to be discussed in section 10.3.

5. KERR LENS MODE-LOCKING

The second of the manifestations of the Kerr nonlinearity has led to a revolution in the way in which we look at mode-locking. Although the concept and practical implication of self-focussing of laser beams has been around almost as long as the laser itself, and the concept of using self-focussing as a mode-locking mechanism had been discussed in the literature, it was not until the advent of the Ti:$A\ell_2O_3$ laser, and the problems of mode-locking raised by it, that the concept of Kerr lens mode-locking was practically implemented.

A typical Kerr lens mode-locked Ti:$A\ell_2O_3$ laser is shown in figure 8. The Ti:$A\ell_2O_3$ gain medium is excited by an argon ion laser operating in the green region of the spectrum. The cavity is formed by four mirrors, two plane end mirrors and two concave focussing mirrors. A slit is placed in one arm of the resonator. If the separation between the concave mirrors and the position of the gain medium is carefully adjusted, it is possible to ensure that the intensity fluctuations present in all multimode lasers, will give rise to some self-focussing in the gain medium. A carefully adjusted cavity will lead to this high intensity spike being focussed and passing through the slit more readily. The high intensity light will experience reduced losses. This is precisely the requirement needed for passive mode-locking by saturable absorption. We can thus regard the Kerr lens as a very fast saturable absorber, discriminating against low intensity light and exhibiting low loss at high intensity. As is the case with many mode-locking techniques based on the Kerr effect, the mode-locking mechanism gets stronger as the pulses become shorter.

A key advantage of the Kerr lens techniques over saturable absorbers is that we no longer have to have an actual absorbing medium to discriminate against low intensity light as this is provided by the slit. The second important advantage is that the speed of the process, and its recovery, are limited only by the Kerr nonlinearity, which is known to be virtually instantaneous. A feature of Kerr lens mode-locking is that for most systems the natural fluctuations of the multimode laser are not sufficient to initiate mode-locking and a number of starting mechanisms have been devised. Examples of starting mechanisms have been a judicious tap on one of the mirror mounts, an intracavity tipping Brewster plate and active mode-locking.

The technique described above, with an intracavity slit, is sometimes referred to as a 'hard aperture' method. It has been observed that Kerr lens mode-locking can also be accomplished without a slit. In this case the mechanism is thought to be the mode overlap between the pump and laser mode. The laser is set up such that the Kerr lens enhances the mode overlap between pump and laser mode. This kind of Kerr lens mode-locking is referred to as 'soft aperture' mode-locking. Soft aperture mode-locking clearly requires a good quality pump beam to be effective.

The second arm of the $Ti:A\ell_2O_3$ laser shown in figure 8 contains a pair of prisms. The primary purpose of these prisms is to provide dispersion compensation. As with all Kerr lens, and related mode-locking methods, the pulse duration is limited either by the gain bandwidth of the active medium or by dispersion. Control of the dispersion is one of the most important tools in generating short pulses in $Ti:A\ell_2O_3$ since the gain bandwidth of the medium is so enormous.

A prism pair deployed in the way shown in figure 8 has an effective negative (or anomalous) dispersion. Most materials have positive dispersion, which is to say that the refractive index increases as the wavelength gets shorter. The dispersion of the prism pair can be adjusted by the choice of the material of the prisms and their separation. The separation is usually adjusted to give negative overall dispersion for the complete system, including the other components in the cavity. By translating one of the prisms into or out of the beam the amount of material dispersion provided by the prisms can be adjusted from negative, through zero to positive, to provide a range of operational regimes in the laser.

$Ti:A\ell_2O_3$ is by far the most fully investigated solid state gain medium for the generation of femtosecond optical pulses. The crucial issue in generating the shortest possible pulse is control of dispersion. At the very shortest pulse durations, which have associated with them extremely broad spectral width, higher order terms in dispersion become important. It is not possible to correct higher order terms with simple prism pairs. The aim then becomes that of minimising material in the laser cavity. In practice, wavelength dependent phase shifts in dielectric mirrors can be the limiting factor in the production of the shortest pulses. Ideally these mirrors should have minimum dispersion over the bandwidth of the laser or, at least, have a well controlled phase shift.

A very compact system design can be produced by deliberately modifying the dispersion characteristics of dielectric laser mirrors. By modifying the normal quarter wave stack such that different wavelengths penetrate different distances into the dielectric layers it is possible to provide dispersion control through the mirrors. These are often referred to as 'chirped' mirrors. This approach has led to the generation of pulses of duration less than 10fsec, corresponding to just a few optical cycles.

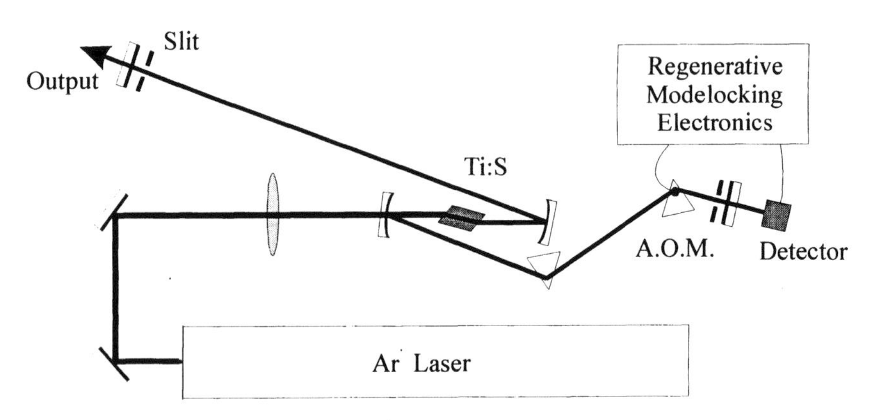

Figure 8. A diagram of a $Ti:A\ell_2O_3$ laser set up for Kerr lens mode-locking.

6. DIODE PUMPED FEMTOSECOND LASERS

Although the Ti:Aℓ_2O_3 system has revolutionised the whole of ultrafast laser physics it does still have a major drawback. Ti:Aℓ_2O_3 absorbs primarily in the green and has to be pumped by a laser operating at around 500nm. Furthermore, the laser system can only reach its full spectral potential when pumped with several watts of green laser power. Until recently, the argon ion laser has been the laser of choice for pumping Ti:Aℓ_2O_3. This has restricted the widespread adoption of femtosecond laser technology outside of the research laboratory. Progress has been made, recently, in the commercial production of diode pumped Nd:YVO_4 lasers, intracavity frequency doubled to 532nm. Output powers of up to 10W are now available from these diode pumped systems. The costs may still prohibit their widespread adoption.

A material system which has attracted considerable attention in recent years has been Cr:LiSAF (Cr^{3+}:LiSrAℓF_6). This material has a similar emission band to Ti:Aℓ_2O_3 in the region of about 800nm to 950nm and so is capable of sustaining short pulses. The main reason for interest in Cr:LiSAF has been its strong absorption around 670nm. Laser diodes operating in this spectral region are now readily available and can be used to directly pump Cr:LiSAF. Furthermore, the threshold for oscillation in Cr:LiSAF has been found to be just a few 10s of mW. For the first time, a really compact, reliable and efficient diode pumped femtosecond system is in prospect.

Despite the low threshold and good slope efficiency of diode pumped Cr:LiSAF the intracavity power associated with this system is much less than can readily be generated in Ti:Aℓ_2O_3. This has therefore given rise to a study of the detailed performance of Kerr lens cavities for low threshold operation. In these models the gain medium, which also acts as a Kerr lens, is split into a large number of segments. A matrix is used to model the intensity dependent refractive index at each segment to give an overall Kerr lens for the rod. The rod is then incorporated into a four mirror cavity with two plane end mirrors and two concave focussing mirrors. The spot size w at one of the end mirrors is then examined as a function of separation between concave mirrors, length of arms and position of gain medium between the concave mirrors. A parameter which measures the change in this spot size as a function of intracavity power p extrapolated to zero power is then calculated. This parameter δ is given by

$$\delta = \left(\frac{1}{w}\frac{dw}{dp}\right)_{p \to o} . \tag{8}$$

If this parameter is negative, in the plane where the aperture is located, the high intensity peaks will pass through the aperture more easily and will experience less loss than the low intensity light. This is the condition for Kerr lens mode-locking.

A very useful configuration for low power Kerr lens mode-locking threshold is when the two arms are matched in length and the two concave mirrors are pulled together a few millimetres closer than the centre of the stability region. If the centre of the gain medium is placed a few millimetres towards the slit, the conditions will be optimised for Kerr lens mode-locking. Using considerations of this kind it has been possible to design laser systems with very low Kerr lens threshold and indeed produce diode pumped Cr:LiSAF producing femtosecond pulses [3].

A typical diode pumped Cr:LiSAF laser system is shown in figure 9. The system is similar to Ti:Aℓ_2O_3 lasers with the exception of the diode pump and the careful adjustment of cavity mirror positions. Another feature to note about this system is that mode-locking is initiated and maintained by a loss modulator. The scheme used is called regenerative mode-locking and consists of a photodiode which detects the pulse train. The fundamental beat, at the repetition rate of the laser, is detected and used to generate phase shifted radiofrequency power to drive the amplitude modulator. As the cavity length drifts the radiofrequency oscillator follows. This provides trouble-free hands-off mode-locking

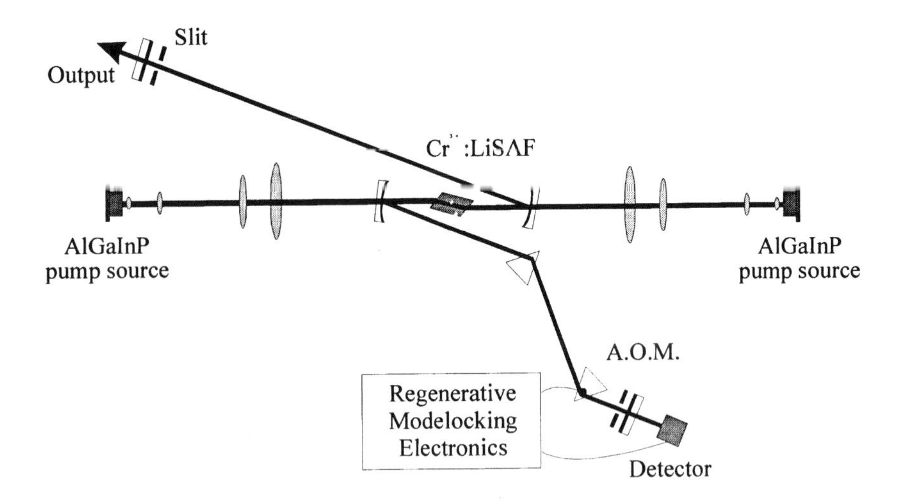

Figure 9. Schematic diagram of a Kerr lens mode-locked Cr:LiSAF laser pumped by a laser diode.

indefinitely. A suitably designed set of cavity mirrors and well-chosen material system for the dispersion compensating prisms has enabled pulses of 18fsec to be generated with this system. Unless special efforts are taken with thermal management in the Cr:LiSAF, output powers of up to about 100mW are as much as can be expected. There are designs in which the thermal load has been carefully spread in Cr:LiSAF to produce almost 1W of output power.

7. SEMICONDUCTOR SATURABLE ABSORBERS

The early days of mode-locking were dominated by passive techniques based on saturable absorber dyes in contact with one of the end mirrors. A modern-day version of this technique is the semiconductor saturable absorber [4]. There are many approaches made possible by the flexibility of semiconductor preparation. An elegant approach is based on the idea of a saturable absorbing quantum well embedded into a multilayer

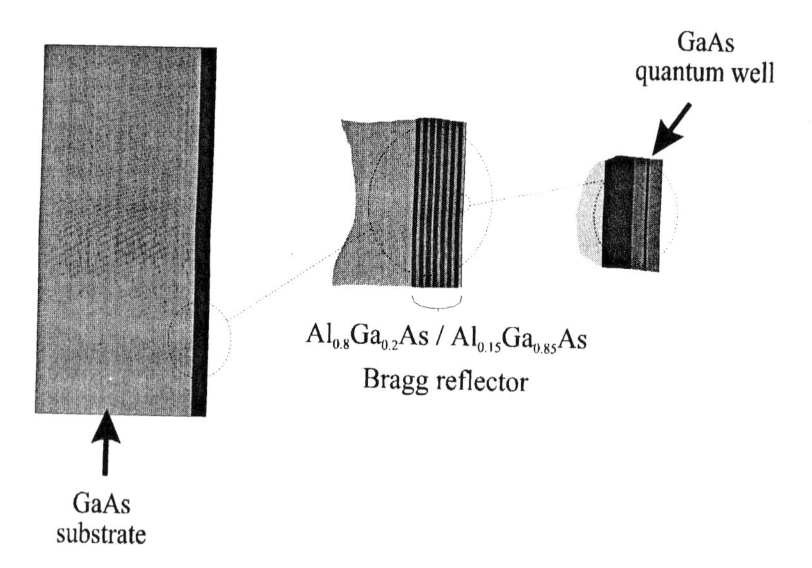

Figure 10. The layer structure of a saturable Bragg reflector used for mode-locking the Cr:LiSAF laser.

reflecting structure. The basic design of such a device is shown in figure 10. This mirror is grown on a GaAs substrate. Successive layers of Aℓ Ga As with different compositions are used to build up a reflecting structure similar to the layers on a dielectric laser mirror. This is called a Bragg reflector. A GaAs quantum well is placed within the top layer. The position of the well within the layer determines the field strength experienced by the well, whereas the thickness of the well determines the wavelength of absorption. By appropriately choosing the Bragg stack and the quantum well thickness, it is possible to produce a mirror which is highly reflecting for high intensity light and absorbing for low intensity light. This is called a saturable Bragg reflector (SBR) and provides a very reliable mode-locking element [5].

A simple and compact SBR mode-locked Cr:LiSAF laser is illustrated in figure 11. In this case only a single pump diode is used. The cavity is arranged such that the SBR is at a small waist where the intensity is high. The normal dispersion compensating prisms are also present in the other arm of the resonator. This laser is self-starting and produces pulses of about 80fsec output powers of about 40mW with a pump power of 400mW. The centre wavelength is, to a large extent, determined by the SBR quantum well thickness and for this system is 860nm.

The SBR approach is quite general and can be applied to most laser systems. We have demonstrated the use of a specially designed SBR centred at 1047nm for mode-locking of Nd:YLF. Using a 5W fibre coupled laser diode as a pump source up to 1W of mode-locked output has been obtained at pulse durations of about 5psec.

8. PARAMETRIC PROCESSES

The laser systems which have been described in the previous sections have either been fixed wavelength or have been tunable over a limited spectral region. The spectral performance of a laser system, of course, is limited by the gain bandwidth of the active medium. This clearly puts restrictions on just how far lasers can be tuned. The spectral restrictions on laser systems can be greatly reduced if a fixed wavelength, or laser with restricted tunability, is used to excite a parametric process.

Parametric processes occur where a transparent medium is illuminated by an intense laser beam, normally called the pump beam, with frequency ω_p. If the material symmetry is such that it lacks a centre of inversion, this pump beam will produce on polarisation of the medium. The medium will exhibit optical gain on two frequencies normally called the signal and the idler at frequency ω_s and ω_i, respectively. The energy matching condition is that $\omega_p = \omega_s + \omega_i$. In this way no net energy is transferred to the medium and the medium returns to the ground state. The parametric gain will be maximised when the wave-vector matching condition $\underline{k}_p = \underline{k}_s + \underline{k}_i$ is satisfied. This condition is called phase-matching and when the condition is satisfied the interaction is said to be phase-matched. Satisfaction of the phase-matching condition is primarily what determines the signal and idler frequencies and hence wavelengths. The wavevector in a medium is given by $k = 2\pi n/\lambda$ where n is

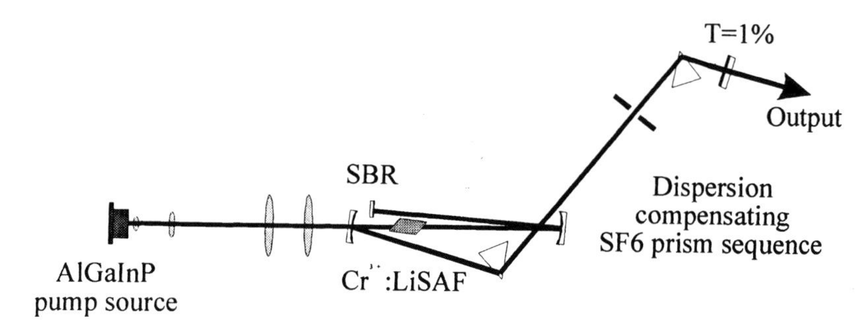

Figure 11. A schematic diagram of a Cr:LiSAF laser mode-locked using a saturable Bragg reflector.

the reflective index and λ is the wavelength. It is possible to tune the wavelength of the signal and idler by modifying the refractive index such that the phase-matching condition is satisfied. The refractive index can be modified either by changing the temperature of the material or modifying the propagation direction in a birefringent material. These are called temperature tuning and angle tuning, respectively.

We now have a situation where parametric gain can be generated at the signal and idler wavelength. Optical gain can be converted into oscillation by means of optical feedback; it is thus possible to make an optical parametric oscillator (OPO). An OPO is very similar to a laser, having a gain medium and a cavity. It is possible to resonate either the signal or the idler or both. For stability and ease of use it is desirable to resonate only the signal or the idler. This is called a singly resonant OPO.

8.1 Optical Parametric Oscillator

Optical parametric oscillators come in many different configurations. The configuration most appropriate for generation of a train of ultrashort pulses is called synchronously pumped. In a synchronously pumped OPO the length of the OPO cavity is matched to the repetition rate of the train of pump pulses. In this way, parametric gain will be experienced by light passing through the nonlinear medium synchronously with a pump pulse. The amplified light will propagate around the cavity and return to the nonlinear crystal just in time to meet the next pump pulse, where it will be further amplified. If the net gain experienced by the light in the OPO cavity exceeds the losses in going around the cavity, the OPO will oscillate. Just as with a laser, the OPO will have a well-defined threshold. When the resonated beam (either signal or idler) oscillates, the other beam (idler or signal) will also be generated. The OPO will therefore give two outputs at quite different wavelengths such that the sum of the energy of the signal and idler photon is equal to the energy of the pump photon. The condition where the signal and idler wavelengths are the same is called degeneracy.

A remarkable OPO material to have emerged in recent years, is called periodically poled lithium niobate (PPLN). The idea behind the use of this material goes right back to the earliest days of nonlinear optics. This concept is that of quasi-phase-matching. Let us first consider what happens when a nonlinear interaction is not phase-matched. We will take, for the sake of simplicity, second harmonic generation. In second harmonic generation the fundamental beam at frequency ω propagates through the medium. The intensity of the fundamental is such as to create a polarisation in the medium at frequency 2ω. This polarisation will thus generate light at frequency 2ω. The phase of the polarisation at 2ω will reflect the phase of the fundamental beam and will have a certain phase velocity. The generated light at 2ω will also have a phase velocity. If the phase velocity at ω and 2ω can be made the same we have a phase-matched interaction and the second harmonic will grow quadratically at the expense of the fundamental, as it propagates through the medium. If the phase velocities are different at ω and 2ω the polarisation and the generated wave will get out of step with each other. When they are sufficiently out of step the second harmonic will be reconverted back into fundamental. As the beams propagate through the medium the second harmonic will appear and disappear. The distance over which the second harmonic increases is called the coherence length. Clearly the amount of second harmonic generated under non-phase-matched conditions is small. The idea behind quasi-phase-matching is that the non-phase-matched interaction takes place and, just as the second harmonic starts to be reconverted into fundamental, the sign of the nonlinear coefficient is changed and instead of getting reconverted, the second harmonic is enhanced over the next coherence length where, again, the sign of nonlinear interaction is reversed. The reversal of this nonlinear interaction can be brought about by the poling of the material. In quasi-phase-matching the nonlinear coefficient is modulated with a period twice the coherence length of the interaction, to offset the accumulated phase mismatch. The coherence length ℓ_c is given by

$$\ell_c = 2 / \Delta k$$

where $$\Delta k = k_p - k_s - k_i. \tag{9}$$

The vectorial notation on the wavevectors has been dropped since the waves co-propagate in this case.

Quasi-phase-matching, by periodic poling, has several major advantages. The first is that all beams co-propagate and so overlap over a long length. This contrasts with angle phase-matching where, in general, the birefringence causes the propagation directions in crystal to be different. The second feature is that the propagation direction in the crystal can be chosen to maximise the nonlinear coefficient. In birefringent phase-matching this direction is constrained by the need to phase-match. These advantages combine to make quasi-phase-matched lithium niobate one of the most efficient nonlinear materials currently available.

Quasi-phase matching in lithium niobate has attracted a great deal of attention since it is a mature, readily available material with a broad transparency range [6]. Quasi-phase-matching can be accomplished in ferroelectric materials such as lithium niobate by periodic reversal of the ferroelectric domains since anti-parallel domains correspond to sign reversal of the nonlinear coefficient. A very successful way of accomplishing this reversal of the domains is by use of an applied electric field to a precisely defined mask structure produced by lithographic techniques. Large samples of periodically poled lithium niobate (PPLN) have been produced in this way.

The tuning range of a synchronously pumped PPLN OPO pumped by a 1047nm mode-locked Nd:YLF laser is shown in figure 12. The period of the PPLN was varied by translating the crystal through the pump beam. For a particular period, the wavelength of the signal and idler could be varied by changing the temperature. The temperature variation is also shown in the diagram. When pumped at 1047nm the PPLN can produce output wavelengths from 1.33μm out to 4.85μm. Beyond 4.85μm the absorption of the idler inhibits the oscillation of the OPO.

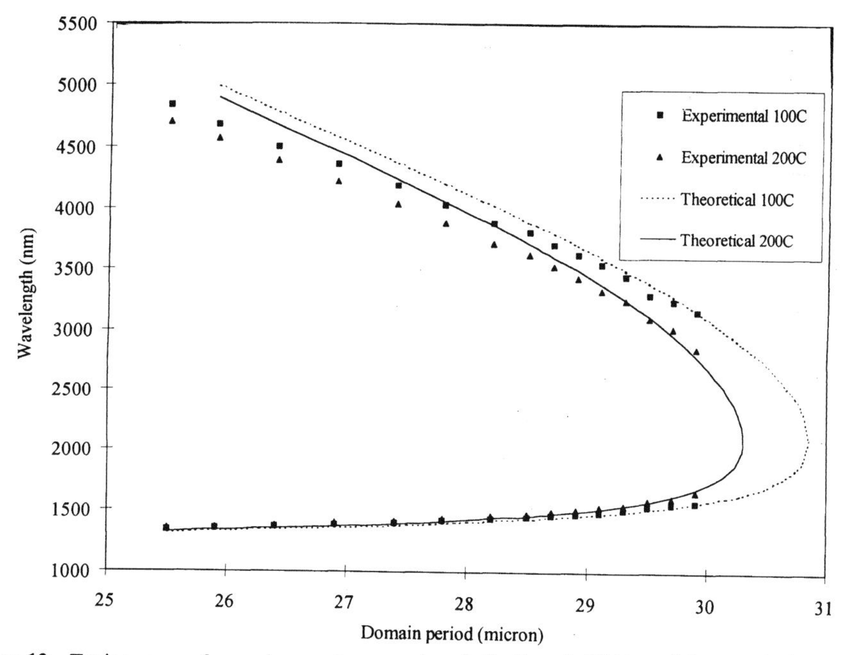

Figure 12. Tuning range of a synchronously pumped, periodically poled lithium niobate, optical parametric oscillator pumped at 1047nm.

8.2 Optical Parametric Amplification

Many applications in time resolved spectroscopy require very broad spectral coverage at relatively modest repetition rate (few 10s of kHz). An approach to developing light sources which operate from the ultraviolet through to the mid-infrared is the use of parametric amplification. The main difference between an OPO and an optical parametric amplifier (OPA) is that the OPA usually has a very substantial single pass gain such that a very weak seed beam, or even noise, can be amplified up to a significant level (as a small fraction of the pump power) in a single pass. In these circumstances there is no need for a resonator.

The pump power required to get the kind of gain needed to make an OPA is usually in the region of 100MW or more, and so the energy in a 100fsec pulse should be more than, say, 10μJ. This kind of energy is normally obtained by amplification of the mode-locked laser beam. A popular approach to the kind of amplification needed is to adopt the technique of regenerative amplification. In this method a small seed beam is injected into a cavity containing a gain medium which is usually pumped by a pulsed laser of high energy. The seed pulse is repetitively amplified as it propagates around the cavity. When the energy of the amplified pulse has built up to a point where it has depleted the stored energy in the gain medium, the amplified pulse is kicked out of the cavity. Single femtosecond pulses of between 1μJ and 1mJ at repetition rates of between 1kHz and 300kHz have typically been produced in this way.

The peak power in a regeneratively amplified pulse is more than adequate to excite parametric amplification processes. One of the most remarkable of these processes is that of white light continuum generation. In this scheme, the laser pulse is focussed onto a transparent medium (glass, quartz, sapphire etc.). Many non-phase-matched parametric, and other, processes are induced and, as the name suggests, a very broadband spectral output it produced. The output virtually covers the transparency region of the material in question. This light source itself can be quite useful as a femtosecond white light source. It can also be used as a seed for an OPA. In this scheme, a small fraction of the white-light continuum is passed through the parametric medium. The spectral component which is phase-matched will be amplified and so produce a high energy at a specific wavelength. The system can be broadly tuned by angle or temperature tuning the nonlinear crystal. The energies of the OPA output are such that the amplified beams can subsequently be used for second harmonic generation, together with sum and difference frequency generation. In this way an extremely widely tunable source of femtosecond pulses is now available for use in a wide variety of applications.

9. ULTRASHORT PULSE MEASUREMENT

The measurement and characterisation of ultrashort pulses is a very important part of the processes of pulse generation and also of characterisation of the operation of the laser source. Spectral and temporal measurements of ultrashort pulses are essential diagnostic tools. Very short pulses provide few real problems in measuring their spectral content since the spectrum of a broad pulse is so short, however one can only easily measure the time averaged spectrum. Pulse duration measurements present a few more problems. Real-time detectors and analysis equipment are unable to resolve pulses of a few picoseconds or shorter and so indirect measurement techniques have to be deployed. The most common pulse characterisation technique is that of the second order autocorrelation. This technique allows the characterisation of the very shortest possible pulses and, when used carefully, can give accurate numbers for duration and a good idea of pulse shape.

A schematic diagram of a particular type of autocorrelator is shown in figure 13. The device is very similar to a Michelson interferometer. The source is split into two approximately equal components. Each part goes through a delay line and the two components are recombined on a beamsplitter. At least one of the arms of the interfero-

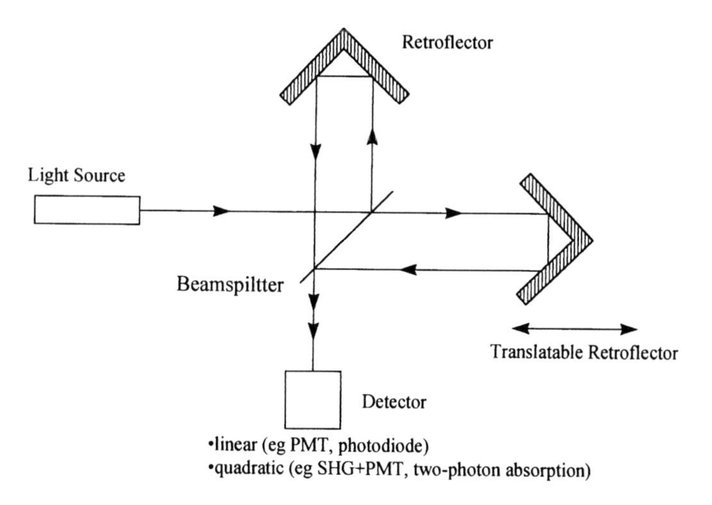

Figure 13. A schematic diagram of a Michelson interferometer set up to measure the autocorrelation function of laser pulses.

meter is movable so that a variable delay can be accomplished. The combined beams fall on a detector system. In a standard Michelson arrangement the detector is a linear detector such as a photodiode or a photomultiplier. As one of the arms is scanned, an interference pattern is generated which reflects the spectrum of the light source. The actual spectrum can be obtained by taking the Fourier transform of the interference pattern. This arrangement is often called a Fourier transform spectrometer. If the spectrum of the source is extremely narrow, the interference pattern will be sinusoidal and will extend over a large path length difference between the arms. This situation is shown in figure 14(a) where we have assumed monochromatic light. Each fringe maximum corresponds to a path length difference of one half the wavelength of the light source. If the source has a broad spectrum an interference pattern will only be observed over a small path length difference. This is illustrated in figure 14(b). In this case we still see fringes corresponding to half-wavelength path length differences in the arms but after a certain distance the fringes get washed out. The distance over which the fringes are observed is called the coherence length. The broader the spectrum, the shorter the coherence length. Of course, a broad spectrum and short coherence length is no guarantee of a short pulse.

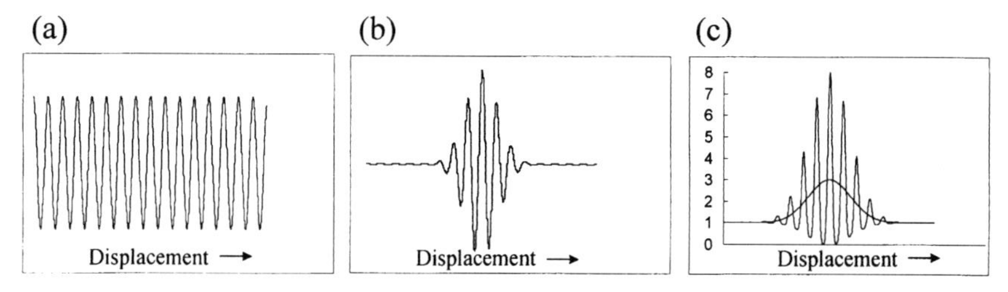

Figure 14 **(a)** First order correlation of a laser with very narrow linewidth.
(b) Correlation of a broad bandwidth light source showing the short coherence length.
(c) Second order autocorrelation of a short laser pulse. The interferometric trace shows an 8:1 ratio. When these fringes are averaged we get a 3:1 ratio.

In order to gain information about the pulse duration, a nonlinear detector system is needed. The normal nonlinear detector is a second harmonic crystal which produces frequency doubled light. This light is then detected with a photomultiplier. The photomultiplier output gives the second order autocorrelation function $g^2(\tau)$ which is given by

$$g^2(\tau)=\frac{\int_{-\infty}^{\infty}\left(E_1(t)+E_2(t+\tau)\right)^4 dt}{\int_{-\infty}^{\infty}E_1(t)^4+E_2(t)^4 dt} \tag{10}$$

where $E_1(t)$ and $E_2(t)$ represents the real electric field amplitude in the two arms of the interferometer given by

$$E_i(t)=A(t)\cos\left(\omega_o t+\phi_i(t)\right) \qquad i=1,2. \tag{11}$$

In this expression A(t) represents the slowly varying field amplitude, ω_o is the carrier frequency and $\phi_i(t)$ is the phase of the electric field. Note that $\phi_i(t)$ can be time-varying which allows this expression to represent frequency chirps.

A schematic second order autocorrelation is shown in figure 14(c). We have assumed a Gaussian amplitude distribution A(t) and that there is no chirp, and hence $\phi_i(t)$ is a constant. There are two traces in this graph. The trace with the rapid oscillations is called the interferometric autocorrelation. The fringes correspond to interference in a similar way to the linear detection case. The important point to note is that the peak to background ratio on this graph is 8:1. In practice the 8:1 ratio is a good indicator that an autocorrelator is set up well and that the pulses are well-behaved. The width of the trace can be used to deduce the pulse duration. Although this technique does not give direct evidence for the shape of the pulse, the second order autocorrelation has been calculated for a number of pulse shapes. Detailed fitting of experimental autocorrelation with the calculated functions can be used to deduce the pulse shape and hence the pulse duration.

Figure 14(c) also shows a smooth autocorrelation trace. In this case the rapid oscillations have been averaged out. This non-interferrometric trace has a peak to background ratio of 3:1. Again, this ratio is an indicator of good performance of the laser and autocorrelator.

In many practical applications there is a need for rapid assessment of the quality and duration of pulses from a mode-locked laser. A very appealing extension of the autocorrelator technique has been the recent rediscovery that $g^{(2)}(\tau)$ is not only given by second harmonic generation but is also given by two-photon absorption. Both effects are proportional to the square of the intensity of the light. A number of schemes have emerged where two-photon absorption in a semiconductor diode has been used as the detection mechanism. Provided the sum of the energy of two photons exceeds the band gap, electron-hole pairs will be created. These can then be detected as a current. This makes for an elegant and inexpensive solution to generation on second order autocorrelation. There is no need for a doubling crystal and photomultiplier. Apart from the obvious convenience factors there are a couple of other important advantages of the two-photon technique. The first is that the method is spectrally insensitive. The two-photon energy just has to be above the bandgap. This compares with second harmonic generation which has to be phase-matched and has a relatively narrow spectral window. Secondly, the method has a huge dynamic range. It has been shown to be linear over six orders of magnitude in signal. This kind of performance is difficult, if not impossible, in a photomultiplier. A final, and important advantage is that the sensitivity is high and incident powers of a few microwatts have been used to produce autocorrelations with good signal to noise. We have shown that the waveguide provided by a laser diode is also a good medium for the detection by two-photon absorption. The guide ensures confined beams over a long distance and hence increased signals. An interesting variation on this idea is that the laser diode can be forward biased to produce a laser output beam which can be used for alignment purposes. Once alignment has been achieved the laser diode then becomes the detector.

10. APPLICATIONS

The applications of solid state femtosecond light sources are immense. Most applications are concerned with the observation of rapid dynamical processes in physics, chemistry and biology. There is an increasing use of ultrashort pulses in materials processing, where the very high peak power can be used to initiate a number of processes which would otherwise be difficult to obtain. We will deal here only with the use of ultrashort pulses in forming images. Three methods which make use of the unique properties of ultrashort pulses will be described. These are optical coherence tomography, terahertz imaging and multiphoton imaging.

10.1 Optical Coherence Tomography

In many imaging applications it is desirable not only to form a two dimensional image of an object (in the x-y plane) but also to gain information into the depth of the object (the z-plane). Optical coherence tomography makes use of the fact that a broad bandwidth light source has a short coherence length. In section 9, we described a Michelson interferometer and pointed out that interference fringes could only be observed over a small distance for broad bandwidth pulses. A schematic diagram of a set-up for optical coherence tomography is shown in figure 15. Light goes down one arm of the interferometer and is returned by a mirror in the normal way. The light going down the other arm falls onto the sample and part of it is reflected. The reflection light then goes back along the arm and is combined on a beamsplitter. When the distance travelled from the beamsplitter to the mirror, and the beamsplitter to the reflecting part of the sample, are matched, interference will occur. By moving the mirror at a particular speed, the interference pattern will have a well-defined frequency which will enable it to be detected against the noise. The amplitude of the signal and position of the mirror give information about the reflectivity of the sample at particular depths. The x-y image in a particular plane can be obtained by scanning the beam over the sample. It is normal, in optical coherence tomography, to scan in just the x or y plane and plot this against the z-plane. In this way information can be obtained about the properties of a line deep into the sample.

The depth resolution of the technique is limited by the bandwidth of the light source and the spatial resolution by the wavelength of the source. The technique works well with incoherent sources such as super luminescent light emitting diodes, however, as the spatial solution becomes higher, less light gets through the system. An attractive feature of the use of lasers for this purpose is that the spatial resolution can be high without the loss in detection efficiency.

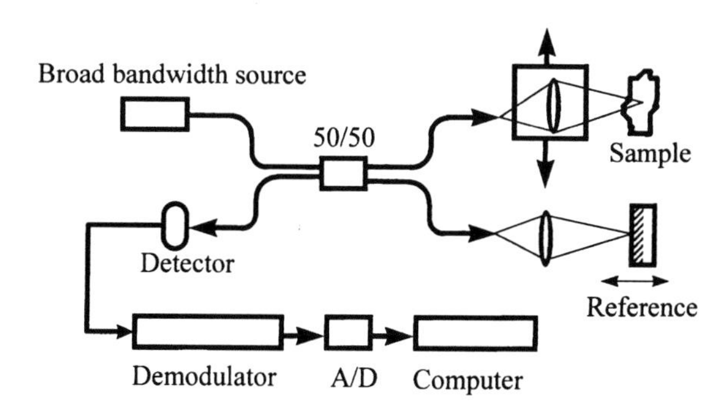

Figure 15. A schematic diagram of an optical coherence tomography set-up.

In a recent experiment, a mode-locked Cr:forsterite laser operating in the 1.3μm region has been used for OCT [7]. The laser was first launched down a single mode fibre where self-phase modulation broadened the spectrum, thereby reducing the coherence length and improving the z-resolution. The single spatial mode nature of the laser beam ensures that it is possible to get diffraction limited spatial resolution with relatively small loss of signal.

The OCT technique is beginning to become widely used. It is particularly useful in penetrating into tissue and has been used in conjunction with a fundus camera to identify lesions within the retina. Although ultrashort pulse lasers are not an essential element in an OCT set-up they do have a role in improving the performance of the systems and making them more widely applicable.

10.2 Terahertz Imaging

Terahertz imaging is a technique which has been available for some time but has only come to the fore in recent years with the advent in reliable sources of very short pulses. A terahertz wave is simply an electromagnetic disturbance with frequency components extending into the terahertz region (1THz = 10^{12}Hz). This is a spectral region where the properties of materials are very different from their properties in the visible region of the spectrum. Terahertz rays are particularly strongly attenuated by water vapour and so can be used as a method of tracing the presence of water in samples. The technique has the potential to produce images beneath the surface of materials and, for example, could be used to replace x-rays in some applications.

A schematic diagram of a terahertz imaging system is shown in figure 16. The system consists of a femtosecond laser source such as a Ti:Aℓ_2O_3 laser or Cr:LiSAF laser which is focussed down onto the gap in a stripline deposited on a semiconducting sample. A bias voltage is applied across the stripline. When the semiconductor absorbs the light it will create a carrier in the semiconductor and allow the current to flow while the pulse is on. This rapid burst of current gives rise to an electromagnetic wave with frequency components up to the inverse of the pulse duration of the pulse; a picosecond pulse (10^{12}sec) can therefore produce a terahertz (10^{12}Hz) wave. The terahertz wave will emanate from the point source and can be collected with some focussing optics. In the scheme shown in the figure, the collecting optics are a hemispherical silicon lens and off-axis paraboloids. These optics focus the terahertz radiation onto the sample. The radiation is then collected by a similar off-axis parabola and hemispherical lens combination and focussed onto a detector which is similar to the terahertz generator. The electric field waveform of the terahertz radiation can be detected by a heterodyne scheme.

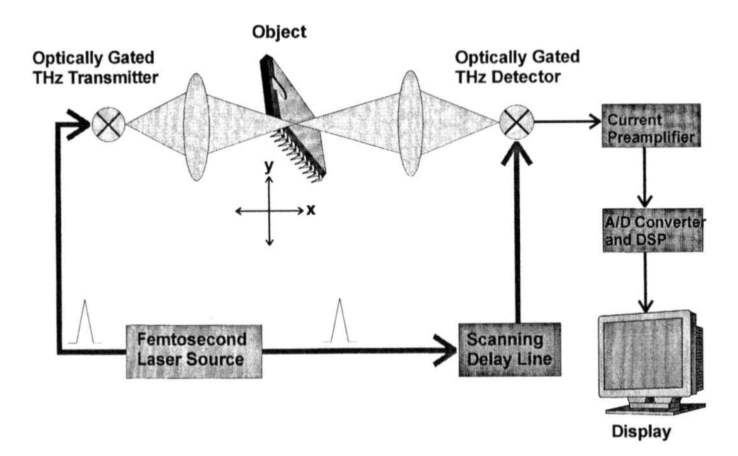

Figure 16. A schematic diagram of an arrangement for building up the image of a sample using terahertz radiation.

An image of the sample is obtained by scanning the sample in the x-y plane while detecting the amplitude of the terahertz radiation. The nature of the sample can also modify the waveform of the transmitted wave. Analysis of this wave can give information on the nature of the sample. The set-up can also be adapted to look at reflection from the sample to form a tomographic image in a similar way to OCT.

Terahertz imaging has been used to quantify the amount of water in living materials such as leaves of plants. There is also the possibility of using the technique to look at burns in skin where the moisture content can give valuable information on whether a skin graft is needed in a particular area where burns have occurred. It is still early days in the science and technology of terahertz imaging but we can expect to see some widespread adoption of this technique as the instrumentation and analysis of data develop.

10.3 Multiphoton Imaging

Femtosecond laser technology has triggered an enormous activity in the use of multiphoton induced fluorescence as an imaging technique. The main advantages of this technique are that it enables live biological samples to develop with minimum disturbance, it allows deep penetration and detection in the sample and reduces bleaching effects.

The multiphoton imaging technique builds on the widely adopted technique of confocal imaging. A diagram showing the principles of confocal imaging system is shown in figure 17. In this scheme an illuminating source, usually an argon ion laser beam at around 500nm, is reflected by a dichronic beamsplitter and through a microscope objective into a simple which is stained with a fluorescent dye, or is naturally fluorescent. The resulting fluorescence is collected by the objective, passes through the dichronic beamsplitter, and falls on a photomultiplier. An image of the system can then be built up by scanning the laser in the x-y plane. A key feature of the confocal imaging system is a pinhole located in front of the photomultiplier. This pinhole is set up such a way the out-of-plane fluorescence is rejected. The remarkable depth sectioning that this permits has led to instruments based on this principle being used throughout the world in biomedical instrumentation. Depth sectioning also means that three dimensional images of biological samples can be built up by taking successive x-y plane scans at different z-depths.

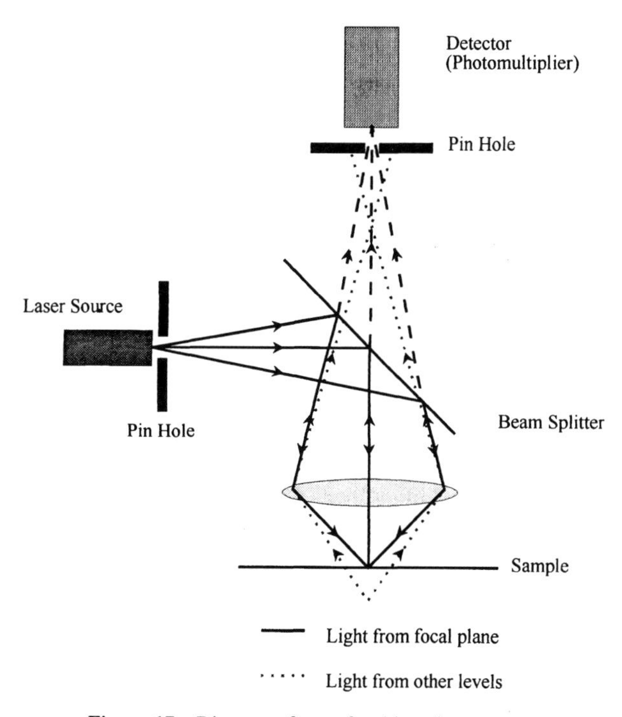

Figure 17. Diagram of a confocal imaging system.

One of the main difficulties of the confocal imaging system is that fluorescence is excited throughout the cone of illumination of the incident laser beam. Since fluorescence is collected only at the focus, the bulk of the excitation is wasted and leads to the generation of toxic products which modify, and eventually kill, the sample under observation. Furthermore, the fluorescent dyes only undergo a limited number of excitation and emission cycles before bleaching and giving rise to reduced signals. The unnecessary out of focus excitation can be avoided if multiphoton excitation of the dye is used. If we consider the case of, say, two-photon absorption, the excitation wavelength is shifted to the region of 1000nm (compared to 500nm for one-photon) and the excitation probability goes as the square of the intensity. The probability of excitation is therefore much greater at the focus than elsewhere. We thus have a localised excitation which gives rise to reduced toxicity and bleaching. In addition, depth sectioning is automatically provided by the localised excitation and so the confocal pinhole can be dispensed with. This has many advantages, including the fact that even if the fluorescence is scattered, it will still register on the photomultiplier and give a signal. This means that scattering samples can be looked at and even samples with low scattering can be observed far deeper into the material.

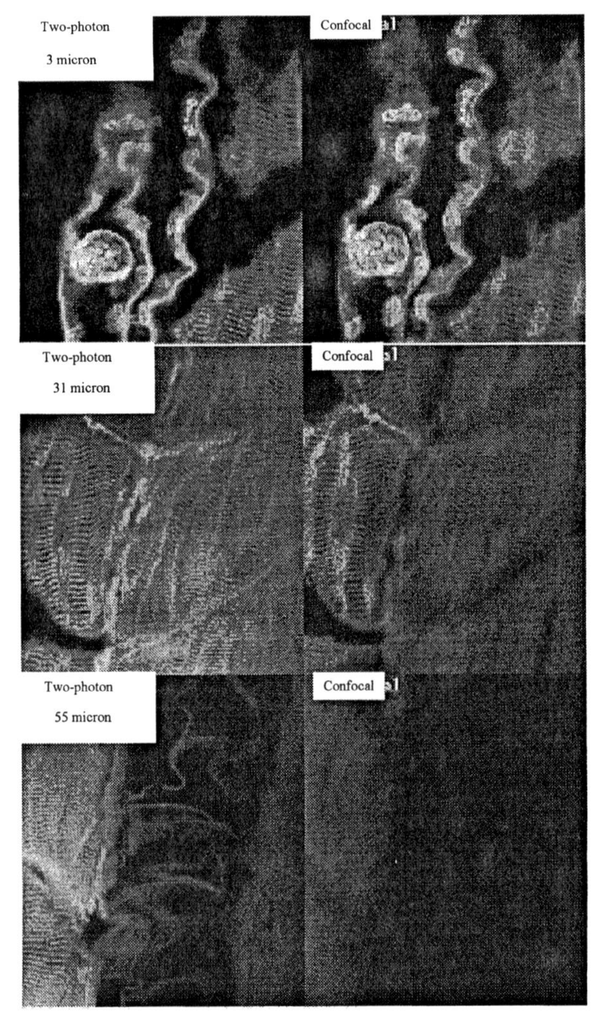

Figure 18. Comparison of two-photon and confocal imaging for different depths into the sample as shown in the figure. The loss of image quality for deeper sections into the sample can be clearly seen.

The probability of two-photon excitation is normally very small but does depend on the square of the intensity. In confocal microscopy the excitation beam is already very tightly focussed and so the intensity is quite high. However, if the technique is to be practical the excitation rate must approach that of the one photon excitation rate. The average power to the sample is limited because of thermal heating caused by partially absorbing materials. In order to get the two-photon probability up to reasonable levels, the peak power of the laser must be increased while maintaining low average power. Of course, this is accomplished by use of a mode-locked laser. At typical repetition rates of about 100MHz, average powers of about 10mW and pulse durations of about 100fsec produce enough peak power to get excitation rates which compete with the one-photon case. At pulse durations lower than about 100fsec dispersion in optics and samples begin to limit the duration of the pulses at the sample. Compact femtosecond lasers based on Ti:$A\ell_2O_3$, Cr:LiSAF and Nd:YLF are now being used in two-photon imaging systems.

In figure 18 we show a comparison of one-photon (confocal) and two-photon imaging at different depths into the sample. By the time the excitation is 55 microns into the sample scattering of fluorescence reduces the image quality in the confocal system and much information is lost. This is typical of many systems that have been studied in the recent past.

Two-photon microscopy has already had a massive impact in the biological sciences and is beginning to become widely adopted. It has only been shown recently that the peak power available from the femtosecond sources now available are capable of producing three-photon and higher order processes. This gives access to energies equivalent to ultraviolet energies while only using visible or near infrared excitation wavelengths. The technique has also recently been applied to the investigation of microcircuitry in integrated circuits by use of multi-photon excitation of currents in the silicon substrate of the circuit.

REFERENCES

1. G.H.C. New, The Generation of Ultrashort Laser Pulses, Reports in Progress in Physics, Vol. 46, No. 8, pp 877-971 (1983).
2. P.M.W. French, The Generation of Ultrashort Laser Pulses, Reports in Progress in Physics, Vol. 58, No. 2, pp 169-262 (1995).
3. M.J.P. Dymott and A.I. Ferguson, Pulse Duration Limitations in a Diode-Pumped Femtosecond Kerr-Lens Mode-Locked Cr:LiSAF Laser, Applied Physics B, Vol. 65, pp 227-234 (1997).
4. W.J. Weingarten, F.X. Kärtner, D. Kopf, B. Braun, I.D. Jung, R. Fluck, C. Hönninger, N. Matuschek, Aus der Au, Semiconductor Saturable Absorber Mirrors (SSAMs) for Femtosecond to Nanosecond Pulse Generation in Solid-State Lasers, IEEE Journal of Selected Topics in Quantum Electronics, Vol. 2, No. 3, pp 435-453 (1996).
5. S. Tsuda, W.H. Knox, S.T. Cundiff, W.Y. Jan, J.E. Cunningham, Mode-Locking Ultrafast Solid-State Lasers with Saturable Bragg Reflectors, IEEE Journal of Selected Topics in Quantum Electronics, Vol. 2, No. 3, pp 465-472 (1996).
6. L.E. Myers, R.C. Eckardt, M.M. Fejer, R.L. Byer, Quasi-Phase-Matched Optical Parametric Oscillators in Bulk Periodically Poled $LiNbO_3$, Journal of the Optical Society of America B-Optical Physics, Vol. 12, No. 11, pp 2102-2116 (1995).
7. J. Fujimoto, J.A. Izatt, M.D. Kulkorni, H-W. Wang, K. Kobayashi, M.V. Sivak Jr., Optical Coherence Tomography and Microscopy in Gastrointestinal Tissues, IEEE Selected Topics in Quantum Electronics, Vol. 2, No. 4, pp 1017-1028 (1997)
8. D.L. Wokosin, V. Centonze, J.G. White, D. Armstrong, G. Robertson and A. I. Ferguson, All Solid-State Ultrafast Lasers Facilitate Multiphoton Excitation Fluorescence Imaging, IEEE Journal of Selected Topics in Quantum Electronics, Vol. 2, No. 4, pp 1051-1065 (1996).

PHONONS, ELECTRONS, AND ELECTRON-PHONON INTERACTION: SEMICONDUCTORS AND HIGH-T_c SUPERCONDUCTORS

Manuel Cardona

Max-Planck-Institut für Festkörperforschung
Heisenbergstr. 1, D-70569 Stuttgart
Federal Republic of Germany, European Union

ABSTRACT

These lectures discuss the information which can be obtained about electronic structure, phonons, and electron-phonon interaction in crystals by means of laser Raman spectroscopy. As prototype materials we choose conventional semiconductors (Ge, Si, GaAs) for which considerable amount of knowledge has accumulated during the past 40 years, and high-T_c superconductors which have been in the headlines for the past ten years. The Raman technique yields information about low frequency excitations (bosons), of energies typically below 0.5 eV. These excitations can be either of a vibronic (phonons, localized vibrational modes) or an electronic nature(electronic interband transitions, plasmons, magnons). Examples of excitations of a mixed electronic and vibronic nature will also be discussed.

I. INTRODUCTION

The Raman effect was discovered by C.V. Raman in Calcutta (India) in 1928.[a] It had been predicted by Smekal in 1923 [1]. The first observations took place for organic liquids. The observed spectra consisted of sharp bands which were correctly assigned to molecular vibrations. Work on crystalline and amorphous solids followed. Raman observed in his first report to the Indian Academy of Sciences in March 1928 [2] that while liquids and crystalline solids displayed sharp inelastic scattering spectra, amorphous solids (e.g. glasses) showed broad bands:

> *It is also of great interest to remark that the solid crystal ice also shows the sharp modified lines in the scattered spectrum in approximately the same position as pure water. The only observations made with amorphous solids are with optical glass. Here the modified scattered spectrum consists of diffuse bands and not sharp lines. Whether this is generally true for all amorphous solids, and whether*

[a]For this discovery C.V. Raman was rewarded with the Nobel Prize in Physics in 1930. It is the only Physics Nobel Prize awarded for work performed in a Third World country (under colonial rule!).

any changes occur at low and high temperatures remains to be determined by experiment.

The paragraph above, taken verbatim from Ref. [2], exemplifies the keen experimental insight of Raman. Today we know that the sharp inelastic scattering lines of crystalline solids are a consequence of $\boldsymbol{k}$-vector conservation: δ-function-like peaks (usually Lorentzian) are seen for scattering by one phonon while the bands with sharp structure observed for scattering by two phonons correspond to van Hove singularities in the dispersion relations. For amorphous solids $\boldsymbol{k}$ is not a good quantum number and thus need not be conserved: broad bands are obtained. The first convincing evidence of this interpretation had to wait till 1971 when the first-order Raman spectra of amorphous silicon and germanium were measured and interpreted as corresponding to the whole density of phonon states, instead of sharp phonons at the center of the Brillouin zone [3,4].

The *cross section* for Raman scattering is very small. In a solid, instead of cross section one usually talks about *scattering efficiency* which is the ratio of scattered to incident power per unit path length L and unit solid angle Ω, integrated over the width of the spectral structure being observed in the case of sharp structures such as those which correspond to scattering by one phonon. In the case of broad structures, one usually refers to the efficiency per unit spectral width:

$$\frac{\partial^2 S}{L \partial \omega_R \partial \Omega} \qquad (\text{in units of } \omega_R^{-1} \text{ cm}^{-1} \text{ sterad}^{-1}) \tag{1}$$

It is customary to express ω_R in cm^{-1}, i.e., in inverse wavelengths (wavenumbers). In the case of sharp structures we use:

$$\frac{\partial S}{L \partial \Omega} \qquad (\text{in cm}^{-1} \text{ sterad}^{-1})\ . \tag{2}$$

Typical values for the differential scattering efficiency of Eq. (2) are 10^{-7} cm^{-1} sterad^{-1}. The cross sections corresponding to Eqs. (1) and (2) are obtained by multiplying these equations by a reference volume which can be either that of a unit cell, the whole scattering volume or the unit of volume. This is the reason why the term "cross section" in a solid leads to ambiguities and should be avoided.

The pathlength L corresponding to the scattering light accepted into a Raman spectrometer is either the depth of focus, the sample thickness, or the absorption length in the case of strongly absorbing samples, whatever is smaller. The absorption length is $(\alpha_L + \alpha_S)^{-1}$ where α_L and α_S are the absorption coefficients for the laser and scattered light, respectively. Typical depths of focus are on the order of 2 mm, while values of $(\alpha_L + \alpha_S)^{-1}$ for metallic samples are about 100 nm. Hence, the ratio of scattered to incident light intensity is usually between 10^{-8} and 10^{-12}. This requires very good spectrometers and often long integration times. Because of these small signals, the powerful spectroscopic technique discovered by Raman remained, for three decades after its discovery, privy to a few laboratories throughout the world. As a source of monochromatic incident radiation, mercury arcs were used [5] until S.P.S. Porto and coworkers realized, around 1963, that the gas lasers were ideal sources for Raman spectroscopy [6]. Since the mid sixties an explosion in the field of Raman spectroscopy has taken place, involving from instrumental development and applications to fundamental studies and materials characterization. A wide range of high performance laser sources, spectrometers and detectors are nowadays commercially available.

Probably the most recent important developments concern scattered-light detectors. During the early years photographic plates were used. They allowed for simultaneous recording of a wide spectral range. Their main disadvantages were nonlinearity and poor sensitivity. Since the mid sixties photographic plates have given way to optoelectronic devices. First, single channel photomultipliers were used. While they allowed single photon counting, the limitation to only one frequency channel, thus rejecting most of the available scattered signal, resulted in long measurement times. In more recent years single channel photomultipliers have been replaced (except for applications requiring very high resolution) by multichannel systems operating on principles similar to those of highly sensitive TV-camera tubes. Nowadays the most commonly used ones are cooled charge coupled devices (CCD) [7]. High aperture instruments in which diffraction gratings are largely replaced by notch filters are commercially offered (only one grating is kept as a dispersive element to generate the scattered spectrum). The high collection efficiency of these systems, in connection with CCD detectors, allows fast gathering of data. Under those conditions it is possible to obtain spatial and temporal resolution: the topography of the Raman spectrum over a sample surface can be measured (resolution $\sim$1 μm) while time dependent spectra can be collected as a function of time (resolution $\sim$1 msec).

As mentioned above, in the case of transparent solids macroscopic depths are probed. The spectra are then representative of the sample bulk. In the case of strongly absorbing samples (e.g. metals) the probing depths can be as small as 10 nm. In this case bulk features may still appear but features characteristic of the surface may also show up. The sampling depths of various commonly used types of spectroscopies are listed in Table 1 for the case of metallic samples, in order to give a feeling for their surface vs. bulk sensitivities.

Table 1: Sampling depths of several kinds of spectroscopies and diffraction techniques as applied to semiconductors and metals (e.g. high T_c superconductors).

UPS (He lamp as source)	0.5 nm
XPS (ESCA, AlK_α source)	5 nm
Raman, ellipsometry in the visible	50 nm
Infrared spectroscopy	500 nm
X-rays diffraction	5 μm
Neutron scattering	whole sample

The Raman sampling depth of $\sim$50 nm listed above applies to metals and also to semiconductors in the visible and/or near uv. This depth makes Raman spectroscopy ideal for the investigation of semiconductor nanostructures. In particular, since the proposal by Burstein, Pinczuk, and Mills [8] the Raman technique has proven to be invaluable for the investigation of two-dimensional electron gases such as found in semiconductor nanostructures. As an example Fig. 1 shows the scattering spectra for incident and scattered parallel and crossed polarizations of a GaAs quantum well in an $Al_{0.18}Ga_{0.82}As$ barrier [9]. The scattering configuration is given in Porto's notation as $z(x',x')\bar{z}$ and $z(x',y')\bar{z}$ where the various letters represent (from left to right) the direction of propagation of the laser, its $\boldsymbol{E}$-polarization, the $\boldsymbol{E}$-polarization of the scattered light, and its direction. x' and y' represent the $2^{-1/2}(x+y)$ and $2^{-1/2}(x-y)$ directions. The crossed-polarized spectrum shows a peak which corresponds

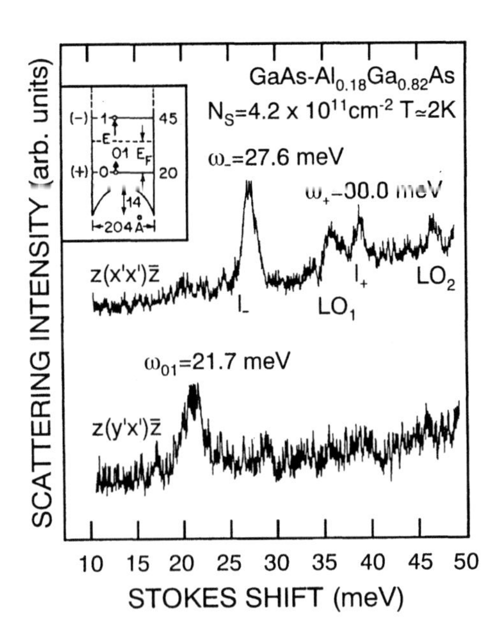

Figure 1: Light scattering spectra of a modulation-doped multiple quantum well GaAs($Al_{0.18}Ga_{0.82}$)As heterostructure. The inset shows calculated energies of the lowest quantum well states, the band bending and the Fermi energy. The GaAs well is 204 Å. Only the energy range of the lowest intersubband excitation is shown. The LO_1 and LO_2 peaks are from the ($Al_{0.18}Ga_{0.82}$)As barriers [9].

to transitions from the partly occupied (through doping) ground state of the quantum well to the first excited state. The main peaks in the parallel polarized spectrum (labeled I_- and I_+ in Fig. 1) represent such transitions mixed with LO-phonon excitations through Coulomb interaction. Note that the transition energies are given in meV, a unit conventional in electronic spectra. Energies of frequencies in vibronic spectra are usually given in wavenumbers (cm^{-1}). The conversion factors between these and other energy-related units are:
$1 \text{ eV} = 1.60 \times 10^{-12} \text{ erg} = 8.07 \times 10^3 \text{ cm}^{-1} = 1.16 \times 10^4$ Kelvin.
The interested reader can choose among a large number of reference books in the field of light scattering. We mention here only as an example Refs. [10–14]. Reference [10] contains a collection of 7 books covering most aspects of light scattering in solids as they have developed since 1975.

II. SPONTANEOUS RAMAN SCATTERING BY PHONONS

II.A. Conservation Laws and Feynman Diagrams

One distinguishes between Stokes and anti-Stokes Raman scattering. In the former, frequency down-conversion occurs as an elementary excitation is created while in the latter an excitation is absorbed by the scattered photon which is then up-shifted. Stokes scattering takes place down to the lowest temperatures ($T = 0$) while for anti-Stokes scattering the temperature must be high enough so that the excitations to be absorbed are present. The process of Raman scattering by a phonon (or any other boson) can be represented by one of the two diagrams in Fig.2 where the zig-zag lines represent photons, the curly line the

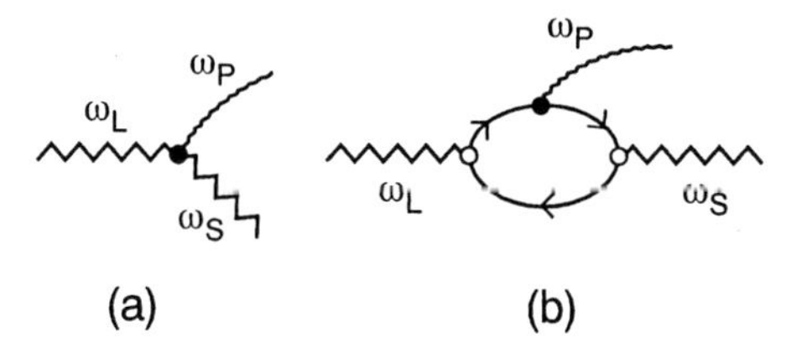

Figure 2: Feynman diagrams for (a) ionic Raman scattering, (b) Raman scattering by bosons (e.g. phonons, frequency ω_p) mediated by electron-hole excitations. ω_L (ω_S) represents the laser (scattered) frequency.

phonon of frequency ω_p (i.e., energy $\hbar\omega_p$) and the straight lines represent either excited electrons ($\longrightarrow$) or holes ($\longleftarrow$). The diagram in Fig. 2(a) corresponds to the so-called ionic Raman scattering: the three bosons interact "simultaneously" at one and the same vertex. A Raman process of this type (the so-called ionic Raman process) has yet to be identified. The usual type of Raman scattering by one phonon is represented in Fig. 2(b) and takes place through the mediation of an electronic excitation. The incident photon produces such an excitation (an electron-hole pair), the phonon scatters the excited electron (or the hole) and the remaining electron-hole pair recombines while creating the scattered photon. The *Feynman*[b] diagrams of Fig. 2 are not simply "cartoons" illustrating the scattering process: They can be used to obtain the scattering efficiencies by replacing the various lines by so-called propagators or Green's functions which are algebraic expressions depending on the various frequencies and other quantum numbers involved. Integrating (or summing) over all "internal frequencies" yields the desired efficiencies as functions of ω_L. The rules to perform these calculations can be seen in Ref. [15]. Note that the three lines labeled ω_L, ω_p and ω_S can be permuted, giving rise to nonequivalent diagrams which must also be included in the calculation of the efficiency. Such calculation, as described, yields the *scattering amplitude*: The amplitudes corresponding to the various processes must be added. The squared magnitude of this sum yields the scattering efficiency. Note that in the process of adding and squaring different diagrams, interference effects may result [16,17].

Conservation laws restrict the possible diagrams in Fig. 2 which can contribute to the scattering efficiency. We mention first *energy conservation* which dictates that:

$$\omega_S = \omega_L \mp \omega_p \tag{3}$$

with the $-$ $(+)$ corresponding to the Stokes (anti-Stokes) case. While Eq. (3)is quite general and applies to any scattering system such as gases, liquids, glasses and crystals, it only holds for the "external" frequencies (we often use frequency as synonymous of energy since they only differ by the factor $\hbar$.). Energy need not be conserved in intermediate states (such as the electronic excitations in Fig. 2(b)) which are short lived and thus called virtual.

In crystalline solids the wavevector $\boldsymbol{k}$ must be conserved at each vertex of Eqs. (2) (to a reciprocal lattice vector, but this restriction is irrelevant in light scattering). If we designate the three "external" wavevectors by $\boldsymbol{k}_L$, $\boldsymbol{k}_p$, and $\boldsymbol{k}_S$ they must obey in a crystal:

[b] R.P. Feynman was awarded the Physics Nobel Prize in 1965 for his diagram techniques as applied to Quantum Electrodynamics. The present lectures were held at the Feynman Room in Erice, where the walls are covered with beautiful Feynman diagrams made out of wrought iron.

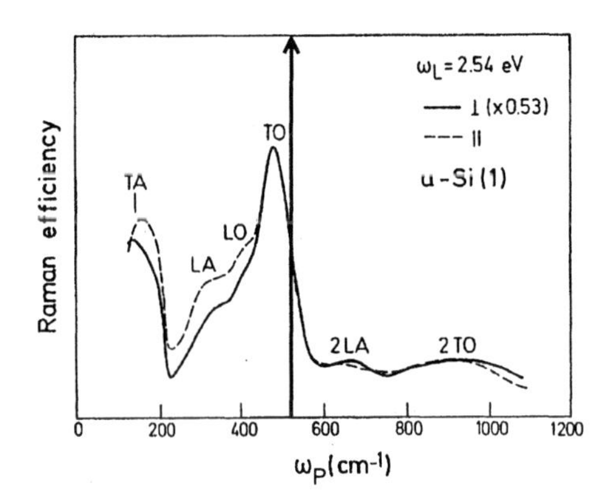

Figure 3: First and second order phonon Raman spectra of amorphous Si (a-Si) for parallel (∥) and crossed (⊥) polarized scattering configurations. The vertical line at 520 cm^{-1} represents the sharp first order spectrum of the $\boldsymbol{k} = 0$ phonons in crystalline Si (c-Si) [3,19].

$$\boldsymbol{k}_S = \boldsymbol{k}_L \mp \boldsymbol{k}_p \qquad (4)$$

where the $-$ $(+)$ sign corresponds to Stokes (anti-Stokes) scattering. Note that a generic $\boldsymbol{k}_p$ is an arbitrary point within the first Brillouin zone (BZ) and should have a magnitude on the order of π/a_0 (a_0 = lattice constant). This value of $\boldsymbol{k}$ is very large compared with that of visible or near-visible photons and therefore Eq. (4) implies that only excitations with $\boldsymbol{k}_p$ near the center of the BZ can be investigated in Raman scattering ($k_p \simeq 0$). In order to investigate other excitations with $k_p \neq 0$ one can use inelastic neutron scattering[c]: thermal neutrons have wavevectors with a typical magnitude several times the size of the Brillouin zone and therefore, according to Eq. (4), the whole Brillouin zone can be investigated when varying the relative direction of the incident and scattered neutrons. Recently, inelastic x-rays scattering (IXS) (using the x-rays emitted in an electron storage ring) has achieved sufficient resolution to measure excitations of a few meV [18]. Because of the shorter wavelength it can also cover most of the Brillouin zone. However, the resolution of this technique is still about two orders of magnitude worse than that which can be achieved with laser Raman spectroscopy.

As already mentioned, Eq. (3) holds for all scattering systems, but the energy only needs to be conserved between the initial and final states. On the other hand Eq. (4) holds at each vertex but only for crystalline systems since it is a direct consequence of the existence of *translational invariance.* In glassy (amorphous) solids it need not to be obeyed (except for the long wavelength acoustic phonons: after averaging over their wavelengths even glasses become translationally invariant [18]). This fact had already been inferred by Raman in 1928 [2] (see also italicized text in page 1 of this review). As an illustration of the effect of $\boldsymbol{k}$-nonconservation Fig. 3 shows the Stokes spectrum of amorphous Si for both, parallel and crossed incident and scattering polarization [3,19]. The broad bands labeled TA, LA, LO, and TO correspond roughly to the four phonon bands of c-Si labeled in the same manner. The integrated strength of the sharp line (resulting from $\boldsymbol{k}$ conservation) which represents the $\boldsymbol{k} = 0$ phonons of c-Si (at 520 cm^{-1}) is roughly the same as that of the sum of the a-Si bands below 520 cm^{-1}. Notice that above 520 cm^{-1} two weaker bands appear in a-Si (also

[c] For the development of inelastic neutron scattering spectroscopy B.N. Brockhouse was awarded the Physics Nobel Prize in 1994.

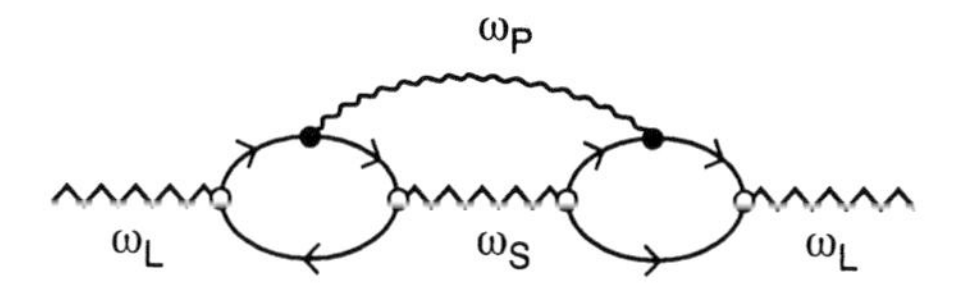

Figure 4: Diagram for the self-energy of ω_L photons induced by the Raman process of Fig. 2(b). The imaginary part of the corresponding SE yields the scattering efficiency.

in c-Si). Each corresponds to scattering by two phonons (1 and 2) for which ω_p in Eq. (3) must be replaced by $\omega_1 + \omega_2$ (in c-Si we would also have to replace $\boldsymbol{k}_p$ by $\boldsymbol{k}_1 + \boldsymbol{k}_2$ in Eq. (4)).

The diagram in Fig. 2(b) represents the amplitude of the scattering corresponding to the phonon ω_p. As already mentioned, the corresponding analytical expression must be squared so as to obtain the scattering efficiency. The latter can also be obtained directly as the imaginary part of the analytic expression associated with the diagram of Fig. 4. The fact that Fig. 4 contains twice the diagram of Fig. 2(b) already suggests that the former corresponds to the square of the latter, i.e., to the scattering efficiency.

Figure 4 only has the laser photons as *external lines*. Thus, energy is automatically conserved. This diagram can be viewed as a so-called self-energy (SE) diagram for the ω_L photon as induced by the Raman process. This SE has an imaginary part which describes the lifetime of the ω_L photon induced by the decay into the ω_S photon plus an ω_p phonon. The inverse lifetime is, of course, directly proportional to the scattering efficiency. The real part of the SE represents a negligibly small renormalization of the photon velocity which is irrelevant for our purposes. The ω_p and ω_S lines in the middle of the diagram, when evaluated according to the rules, lead to an imaginary part which is a δ-function of (ω_l-ω_S-ω_p) (or a Lorentzian if ω_p has a width). In this manner, the energy conservation which we had to impose in Fig. 2(b) is automatically recovered.

The representation of Fig. 4 has the advantage of being able to exhibit explicitly the interference of different possible scattering channels. For this purpose one must take for the left bubble one of the channels and for the right a different channel. Such a diagram corresponds to the interference contribution which is also obtained by squaring the sum of two different diagrams of the type shown in Fig. 2(b). The interference diagram of the type of Fig. 4 must, of course, be added to the standard (squared) diagrams in which the same process appears for the right as for the left bubble [17].

II.B. Quasistatic Approximation: Raman Tensors

While the evaluation of the diagrams in Figs. 2 and 4 yields the exact quantum-mechanical expressions for the scattering efficiencies, in many cases it is possible to use a semiclassical approach which is more intuitive and physically transparent. The approach is called the *quasistatic approximation* and it requires that the phonon frequency be small compared with the characteristic frequencies (usually related to those of electronic transitions ω_e) responsible for the electric susceptibility at the laser frequency $\chi(\omega_L)$. This condition can be written more precisely as:

$$\omega_p < |\omega_e - \omega_L + i\Gamma_e| \ , \qquad (5)$$

where Γ_e represents the lifetime broadening of the electronic excitations of frequency ω_e contributing to $\chi(\omega_L)$. If Eq. (5) is fulfilled, we can expand $\chi(\omega_L)$ in power series of the phonon displacement u and keep only first order terms (note that an optical phonon with $\boldsymbol{k} = 0$ does not change the translational symmetry of the crystal):

$$\begin{aligned} \chi(\omega_L, u(t)) &= \chi(\omega_L) + \frac{d\chi}{du} \cdot u(t) + \ldots \\ u(t) &= u_0 e^{\pm i\omega_p t} \end{aligned} \tag{6}$$

where we have explicitely displayed the fact that u depends sinusoidally on the time t. Under the action of a light beam of frequency ω_L Eq. (6) yields a polarization P equal to:

$$P = \chi(\omega_L) E_L e^{-i\omega_L t} + \frac{d\chi(\omega_L)}{du} E_L e^{-(\omega_L \mp \omega_p)t} u_0 \tag{7}$$

The first term in the r.h.s of Eq. (7) is responsible for the standard linear propagation of the laser beam in the crystal while the second term represents a polarization varying sinusoidally with t with the frequency $\omega_L \pm \omega_p$. This polarization represents an oscillating dipole which will generate dipole radiation at the Stokes ($\omega_S = \omega_L + \omega_p$) and anti-Stokes ($\omega_S = \omega_L - \omega_p$) frequencies. The radiated power P is given by;

$$P \propto \frac{\omega_S^4}{c^3} \left| \frac{d\chi}{du} \right|_{\omega_L}^2 \langle u^2 \rangle E_L^2 \tag{8}$$

(the interested reader should make an effort to insert in Eq. (8) the appropriate numerical and physical constants so as to replace $\propto$ by $=$. The treatment of dipole radiation can be found in any standard classical electrodynamics textbook).

In order to obtain from (8) the scattering efficiency we must find an explicit expression for $\langle u^2 \rangle$ at the temperature T. Using the quantum-mechanical completeness relation, plus the fact that the operator u only connects the n-th state of a harmonic oscillator with either $n+1$ or $n-1$, we find

$$\begin{aligned} \langle u^2 \rangle_T &= \frac{\hbar}{4M\omega_p N} [\langle n |u| n+1 \rangle_T \langle n+1 |u| n \rangle_T] + \langle n |u| n-1 \rangle_T \langle n-1 |u| n \rangle_T = \\ &= \frac{\hbar}{4M\omega_p N} [(n_B + 1) + n_B] \end{aligned} \tag{9}$$

where N is the total number of unit cells in the scattering volume, M is a reduced mass (here, on purpose, only loosely defined) and n_B the Bose-Einstein[d] statistical factor:

---

[d] Albert Einstein was awarded the Physics Nobel prize in 1921 for his quantum theory of the photoelectric effect.

$$n_B = \frac{1}{e^{\frac{\hbar\omega_p}{k_B T}} - 1}, \tag{10}$$

with k_B the Boltzmann constant. The $n_B + 1$ term in Eq. (9) corresponds to Stokes scattering while n_B (which vanishes for $T \to 0$, as expected) corresponds to anti-Stokes. Note that in Eqs. (7) and (8) χ and $d\chi/du$ have been assumed to be scalars while, in general, they will be tensors. Even if χ is a scalar for a cubic or an isotropic material (such as the a-Si of Fig. 3), $(d\chi/du) = \boldsymbol{R}$ will, in general, be a second rank tensor (the so-called Raman tensor) which must be symmetric since χ is symmetric. In order to obtain the scattering efficiency this tensor must be contracted with the unit vectors $\hat{\boldsymbol{e}}_L$ ($\hat{\boldsymbol{e}}_S$) directed along the incident (scattered) electric fields. The scattering efficiency thus becomes:

$$\frac{\partial S}{L \cdot \partial\Omega} \propto \frac{\omega_S^4}{c^3} \left|\hat{\boldsymbol{e}}_L \cdot \boldsymbol{R} \cdot \hat{\boldsymbol{e}}_S\right|^2 \frac{\hbar}{4M\omega_p} \begin{cases} n_B + 1 & \text{Stokes} \\ n_B & \text{anti-Stokes} \end{cases} \tag{11}$$

As already mentioned, the Raman tensor $\boldsymbol{R}$ must be a symmetric tensor within the quasistatic approximation. Its components are complex in the case of a metal while for a semiconductor or insulator they are only complex above the absorption edge. If the quasistatic approximation does not apply, $\boldsymbol{R}$ can become asymmetric. As an example see Ref. [20].

It follows from Eq. (11) that it is possible to determine the temperature at *the scattering volume* by measuring the ratio of the anti-Stokes (I_{AS}) to the Stokes (I_S) signal. Assuming that $\boldsymbol{R}$ is the same for both signals, and neglecting the difference in ω_S, we find:

$$\frac{I_{AS}}{I_S} = e^{-\frac{\hbar\omega_p}{k_B T}}. \tag{12}$$

The restriction of equal values of ω_S and $\boldsymbol{R}$ can be lifted by measuring I_{AS} and I_S at the same *scattered* frequency (although this requires a tunable laser).

The tensor $\boldsymbol{R}$ contains the polarization selection rules for the phonons under consideration. As an example we treat a diamond-type crystal (diamond, Si, Ge, gray tin). The three-fold-degenerate optical phonons at $\boldsymbol{k} \simeq 0$ have the symmetry of xy, yz, zx, i.e. belong to the irreducible representation $\Gamma_{25'}$ of the O_h crystallographic point group [21] (for these phonons the two atoms in a primitive cell vibrate against each other. The interested reader should check the symmetry of xy, yz, zx with respect to the operations of the full cubic group). The Raman tensor must have the same symmetry properties as the corresponding phonon, it therefore must have the form:

$$\boldsymbol{R}_x = R\begin{pmatrix} 0 & 0 & 0 \\ 0 & 0 & 1 \\ 0 & 1 & 0 \end{pmatrix}; \ \boldsymbol{R}_y = R\begin{pmatrix} 0 & 0 & 1 \\ 0 & 0 & 0 \\ 1 & 0 & 0 \end{pmatrix}; \ \boldsymbol{R}_z = R\begin{pmatrix} 0 & 1 & 0 \\ 1 & 0 & 0 \\ 0 & 0 & 0 \end{pmatrix}. \tag{13}$$

Each one of the three components of $\boldsymbol{R}$ in Eq. (13) correspond to vibrations polarized along x, y, and z, respectively. The Raman selection rules are obtained by contracting each of these tensors with a chosen set of $\hat{\boldsymbol{e}}_L$ and $\hat{\boldsymbol{e}}_S$, squaring each of the contracted results and adding these squares so as to take into account the threefold degeneracy. For backscattering (a configuration usually employed with opaque samples) on a (100) surface we only get scattered light for $\hat{\boldsymbol{e}}_L \parallel [010]$ and $\hat{\boldsymbol{e}}_S \parallel [001]$ or vice versa (also for $\hat{\boldsymbol{e}}_L \parallel \hat{\boldsymbol{e}}_S \parallel [011]$). The interested reader should derive the corresponding selection rules for backscattering on a [111] or [110] surface.

We mention at this point a rather general selection rule which applies to all materials with a *center of inversion*. In this case parity is a good quantum number and the excitations must be of either odd or even symmetry with respect to the inversion. In (*dipole allowed*) Raman scattering two *odd* electric fields ($\boldsymbol{E}_L$, $\boldsymbol{E}_S$) participate. Therefore only *even* excitations can scatter (since odd $\times$ even $\times$ odd $=$ even). For (dipole allowed) ir-absorption, on the contrary, only one odd electric field is involved: the corresponding excitations must therefore be *odd* so as to preserve parity. We have thus derived an *exclusion* rule valid, of course, only for materials with a center of inversion: Raman and ir activities are mutually exclusive. In the case of the diamond structure the center of inversion is located between the two atoms of the basis (primitive cell). The optical phonons are invariant (even) upon this inversion. They therefore are Raman active but ir-inactive. In the case of the NaCl structure, centers of inversion are located at the atomic sites. The corresponding optical vibrations reverse their sign upon inversion (i.e. they are odd). Therefore they are ir-active but Raman-inactive.

We have discussed the Raman tensor of Ge. We discuss next the case of zincblende-type materials such as GaAs, with a tetrahedral structure similar to that of Ge except that the center of inversion has been lifted by making the two atoms in the primitive cell different (the point group is that of the tetrahedron, T_d). In this case, the (yz, zx, xy) symmetry is the same as that of (x, y, z), i.e., that of an electric dipole. The $\boldsymbol{k} \simeq 0$ optical phonons are thus not only Raman active but also *ir-active*. At the small but finite values of $\boldsymbol{k}$ which apply to the backscattering configuration the threefold phonon degeneracy is split: the vibrations along the direction of $\boldsymbol{k}$ induce a longitudinal field proportional to $div\, \boldsymbol{u} = i\boldsymbol{k} \cdot \boldsymbol{u}$ which strengthens the restoring force constant. This field does not appear for the transverse vibrations for which $\boldsymbol{k} \cdot \boldsymbol{u} = 0$. The phonon frequencies thus split into longitudinal (ω_L) and transverse (ω_T) by an amount:

$$\omega_L^2 - \omega_T^2 = \frac{4\pi N (e_T^*)^2}{\epsilon_\infty \mu} \tag{14}$$

where N is the number of primitive cells per unit volume, e_T^* the transverse effective charge (responsible also for the ir absorption), ϵ_∞ the so-called ir dielectric constant and μ the reduced ionic mass (for GaAs $\mu^{-1} = M_{\mathrm{Ga}}^{-1} + M_{\mathrm{As}}^{-1}$). The Raman selection rules must be derived separately for longitudinal and transverse phonons using the tensor components of Eq. (13):

$$\begin{aligned} \text{For } \omega_L &: \quad \boldsymbol{R}_L = \boldsymbol{R}_x \hat{q}_x + \boldsymbol{R}_y \hat{q}_y + \boldsymbol{R}_z \hat{q}_z \\ \omega_T &: \quad \boldsymbol{R}_T = \boldsymbol{R}_x q_x^T + \boldsymbol{R}_y q_y^T + \boldsymbol{R}_z q_z^T \end{aligned} \tag{15}$$

where $\boldsymbol{q}^T$ is a unit vector perpendicular to $\hat{\boldsymbol{q}}$. It is now easy to see that for backscattering on a GaAs (001) surface it is only possible to observe LO phonons, with the same polarization selection rules as given above for the diamond structure. The interested reader should be able to show that for backscattering on a (111) surface both LO and TO phonons can be observed while for a (110) surface only TO phonons are Raman active within the approximations being considered. We note, however, that very close to *a resonance* forbidden LO scattering can appear for $\hat{e}_L \parallel \hat{e}_S$. This phenomenon can be explained as induced by the long range electrostatic *Fröhlich interaction* characteristic of LO phonons [19].

II.C. Resonance Raman Scattering and Modulation Spectroscopy

Since the scattering efficiency and its dependence on ω_L is determined by $d\boldsymbol{\chi}/du$, we briefly discuss next the structure of the susceptibility $\boldsymbol{\chi}(\omega_L)$ which is determined by electronic transitions between filled (valence) and occupied (conduction) states. We consider only so-called direct transitions, i.e., interband transitions in which $\boldsymbol{k}$ is conserved. Indirect (e.g. phonon mediated) transitions are too weak to be of interest here. In a nonmetalic crystal there is a minimum energy at which such transitions can occur, the so-called fundamental energy gap. At this energy, ω_o, $\boldsymbol{\chi}(\omega)$ has some kind of a singularity (sometimes called a van Hove critical point), possibly of the square root type:

$$\chi(\omega) \simeq A(\omega - \omega_0)^{1/2} + \text{constant} \tag{16}$$

Similar critical points can occur at higher frequencies. In a metal there is no fundamental gap but van Hove singularities, often located at high symmetry points of the Brillouin zone, can also be found. They appear as a consequence of the translational invariance and, for reasons related to Raman's remarks reproduced in page 1, they are absent in amorphous materials.

The function $d\boldsymbol{\chi}/du$, needed to calculate $\partial S/L\partial\Omega$ according to Eq. (11), becomes particularly interesting near a van Hove singularity (Eq. (16)). Its derivative with respect to u has two contributions, one arising from the prefactor A (which contains matrix elements for the optical transitions) and the other from the dependence on u of the critical energy ω_g (we have replaced the lowest gap ω_0 by a generic critical energy ω_g):

$$\frac{d\chi(\omega)}{du} \simeq \frac{dA}{du}(\omega - \omega_g)^{1/2} - \frac{1}{2}(\omega - \omega_g)^{-1/2} \cdot \frac{d\omega_g}{du} \ . \tag{17}$$

Note that the lineshape of $\partial S/L\partial\Omega$ is likely to be dominated near ω_g by its most singular contribution:

$$\frac{\partial S}{L\partial\Omega} \sim |\omega - \omega_g|^{-1} \cdot \left|\frac{d\omega_g}{du}\right|^2 \ . \tag{18}$$

The term $d\omega_g/du$ can be replaced by the difference in the derivatives of ω_c, the final state, and ω_v, the initial state, with respect to u. These two derivatives, which we write as D_c and D_v, represent the effect of the phonon distortion on the corresponding electronic state, e.g., they are a measure of the electron-phonon interaction.

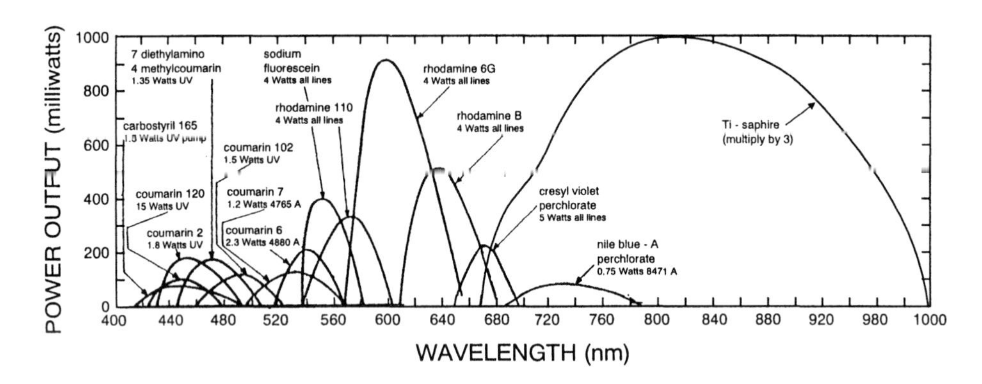

Figure 5: Typical powers obtained with cw dye lasers, and with the titanium-sapphire laser pumped with an argon laser, vs. wavelength.

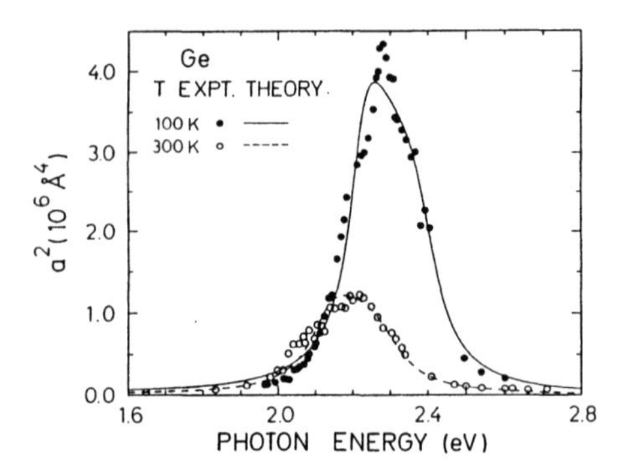

Figure 6: Raman efficiency of germanium resonating near the E_1 and $E_1 + \Delta_1$ interband critical points measured vs. laser photon energy at two temperatures. From Ref. [22].

The discussion above illustrates the power of Raman spectroscopy. It is usually employed to measure the energy of low frequency excitations such as phonons. However, if we measure the scattering efficiency around an electronic interband critical point vs. laser frequency (Eq. (18)) we also obtain the energy of the critical point ω_g, at which the scattering efficiency exhibits a maximum. This technique, like the phenomenon on which it is based, is called *resonant* Raman scattering. If we measure the scattering efficiency in *absolute units* [19] it also becomes possible to determine $D_c - D_v$, i.e., to obtain information about electron-phonon interaction.

Resonant Raman experiments ideally require continuously tunable lasers. The most commonly used ones are dye lasers and, more recently, the titanium-sapphire laser. We display in Fig. 5 the typical power vs. wavelength emitted by those devices.

A typical resonance Raman spectrum of the phonon scattering efficiency vs. laser photon energy measured for germanium (the so-called resonance profile) is shown in Fig. 6. It corresponds to the critical points which are labeled E_1 and $E_1 + \Delta_1$. The lines through the experimental points have been obtained using standard theory [19,22]. In this figure, the scattering efficiency is equivalently represented by the independent element of the Raman tensor squared, labeled $|a^2|$. The element a of the Raman tensor is given in Å^2 as corresponds to the derivative of χ with respect to u multiplied by the volume of the primitive cell (hence the dimensions of Å^2). We recall that E_1 and $E_1 + \Delta_1$ are gaps along the $\langle 111 \rangle$ directions close to the L-point of the BZ. Because of the small spin-orbit splitting, the spectra in Fig. 6 are dominated by changes in matrix elements (the prefactor of the singularity in Eq. (16)) and not by $d\omega_g/du$. From the fit to the experimental data of the theoretical curves two electron-phonon coupling constants (*deformation potentials*) are obtained [22].

II.D. Modulation Spectroscopy

The resonance Raman method as applied to the determination of electronic interband critical points bears some resemblance to a family of spectroscopic techniques called reflectance modulation or, more generally, modulation spectroscopies [23]. In these techniques one applies to a crystal an external sinusoidal perturbation (such as a stress, a sinusoidally varying temperature or electric field) which modulates its reflectivity. Upon reflection on the modulated sample, the light beam acquires a small periodic intensity modulation which can be detected with high sensitivity by means of a *lock-in amplifier.* The modulated signal has an amplitude which corresponds to the derivative of the reflectivity with respect to the modulating agent. The procedure just described, like resonant Raman scattering, enhances critical point structure and yields precise information about interband transition energies. Resonant Raman scattering can be considered as a form of modulation spectroscopy in which the modulation of the electronic transition is produced internally in the sample by one of the Raman phonons [24].

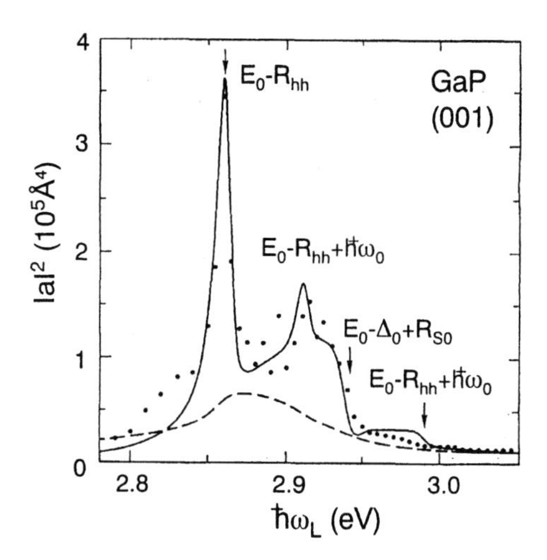

Figure 7: Resonant Raman scattering vs. ω_L corresponding to the creation of LO phonons on a (100) surface of GaP at low temperatures. The dashed line represents the theoretical prediction for uncorrelated electron-hole pairs as intermediate states [25].

So far we have attributed structure in resonant Raman scattering to electronic interband transitions. Implicitly this assumes the creation by the laser photons of uncorrelated electron-hole pairs. While this is often (such as in the case of Fig. 6) a reasonable approximation, sometimes, especially near the fundamental absorption edge, one must take into account the Coulomb attraction between the electron and hole created by the ω_L photon (excitonic interaction). This interaction enhances resonances calculated on the basis of uncorrelated pairs by as much as one order of magnitude (see Fig. 7).

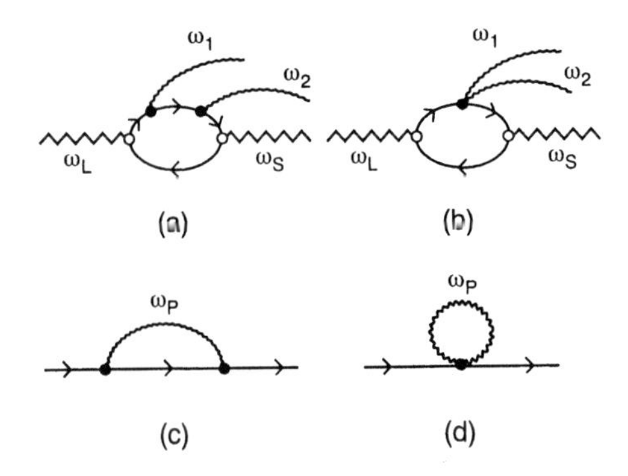

Figure 8: (a,b) Prototypes of Feynman diagrams involved in the Raman scattering by two phonons: (a) iterated electron one-phonon coupling, (b) coupling of an electron to two phonons simultaneously at one vertex. For comparison we have included (c,d) the closely related processes which determine the self energy of an electron state resulting from the interaction with phonons.

II.E. Second Order Raman Scattering

We have already mentioned in page 6 that while only $\boldsymbol{k} = 0$ phonons can be observed in first order Raman scattering (and ir-absorption), $\boldsymbol{k}$ conservation requires only $\boldsymbol{k}_1 + \boldsymbol{k}_2 = 0$ in second order scattering in which two phonons (1 and 2) are created. Hence, all phonons can in principle participate. The two prototypes of Feynman diagrams for two-phonon scattering are depicted in Fig. 8. In this figure we have included the two diagrams which contribute to the self energy of an electron state induced by coupling to phonons (Fig. 8(c) and (d)). This self energy is responsible for the dependence of the gaps on temperature [26] and on isotopic mass [27]. It is therefore possible to estimate the temperature dependence of a gap from the corresponding resonant Raman efficiency for scattering by two-phonon overtone processes and vice versa [19].

While in principle all phonons can contribute to the second order Raman spectra, very close to a resonance phonons with $\boldsymbol{k} = 0$ predominate. The reason is that for these phonons one can achieve conditions close to triple resonance in which all energy dominators which correspond to the three intermediate states in Fig. 8(a) nearly vanish. The resulting resonance profiles are rather sharp, as illustrated in Fig. 9.

II.F. Phonons in High T_c Superconductors

The so-called 214 superconductors (e.g. $La_{2-x}Ba_xCuO_4$) reach critical temperatures T_c close to 40 K[e] while the 123 materials ($YBa_2Cu_3O_{7-\delta}$) have broken the "liquid nitrogen barrier" with $T_c \simeq 93$ K. We shall use the latter for our illustration of Raman spectroscopy as applied to high T_c superconductors.

The primitive cell of $YBa_2Cu_3O_7$ is shown in Fig. 10. The Y atom, one Cu (Cu1) and one oxygen (OI) are at centers of inversion: their vibrations are odd and hence not Raman active. There are 5 pairs of atoms separated by centers of inversion (Ba, Cu2, OII, OIII,

[e] For the discovery of the 214 ($La_{2-x}Ba_xCuO_4$) superconductors G. Bednorz and K.A. Müller were awarded the Nobel Prize for Physics in 1987.

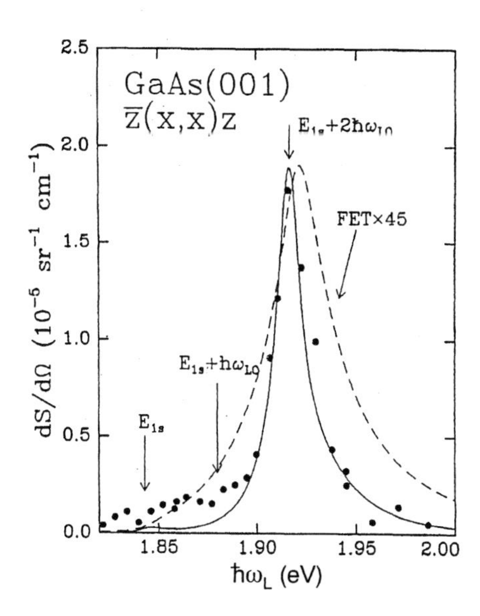

Figure 9: Resonance profile for Raman scattering by 2LO phonons in GaAs. The solid line represents a calculation including excitonic interaction which agrees quantitatively with the experimental points. The dashed line represents the calculation for uncorrelated e-h pairs which gives a maximum scattering efficiency 45 times smaller than observed [28].

OIV). For each of the three cartesian directions x, y, z one can construct odd and even vibrations of these atoms,the former being ir-active, the latter Raman active. Through annealing at about 900 C the OI atoms (chain oxygens) abandon the material which then becomes semiconducting (and nonsuperconducting) with the composition $YBa_2Cu_3O_6$. The new crystal structure is tetragonal (D_{4h} point group) as opposed to the orthorhombic $YBa_2Cu_3O_7$ (D_{2h}) superconductor. Since the in-plane lattice parameters a and b are nearly equal, and the chain OI does not contribute to Raman active modes, we shall assume (with the proper *caveats*) that the $YBa_2Cu_3O_7$ is also tetragonal. The Raman spectra are dominated by vibrations along the z-axis: As shown in Fig. 11, one observes 5 main modes corresponding to vibrations of the Ba (115 cm^{-1}), Cu2 (150 cm^{-1}), OII-OII (340 cm^{-1}), OII-OIII (430 cm^{-1}) and the so-called apical oxygen OIV (500 cm^{-1}) [29]. Under the assumption made above of tetragonal D_{4h} point group symmetry (instead of the exact orthorhombic D_{2h}) four of the vibrations along z should be completely symmetric upon the D_{4h} operations (A_{1g} symmetry). These four are the vibrations of Ba, Cu2, of the apical OIV, and a combination of the OII-OIII oxygens of the CuO_2 planes (displacements of both OII and OIII equal and along the same direction). These vibrational modes can mix since they have the same symmetry, but a number of experimental facts indicate that the admixture is small. The remaining Raman active vibrational mode along z corresponds to OII and OIII vibrating in opposite directions by the same amount. The reader should check that upon the operations of the tetragonal point group D_{4h} this OII-OIII mode behaves like $x^2 - y^2$, a behavior which is referred to as B_{1g} symmetry (the interested reader should check this behavior e.g. by applying a 90° rotation around the z-axis). The Raman tensors of A_{1g} and B_{1g} symmetry have the general form (in the D_{4h} point group):

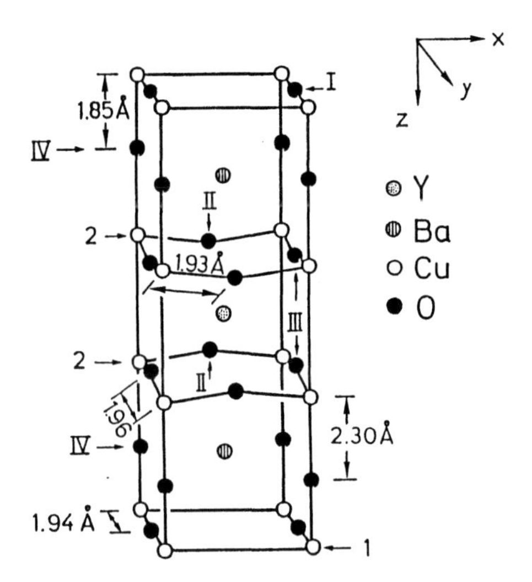

Figure 10: Primitive cell of $YBa_2Cu_3O_7$. Note the orthorhombic D_{2h} symmetry (tetragonal D_{4h} if we remove the OI oxygens of the chains). The two CuO_2 planes around the Y (which can be substituted by almost any rare earth atom) are believed to be the site of the quasi-two-dimensional high T_c superconductivity.

$$A_{1g}: \begin{pmatrix} a & 0 & 0 \\ 0 & a & 0 \\ 0 & 0 & c \end{pmatrix} \; ; \; B_{1g}: \begin{pmatrix} b & & \\ & -b & \\ & & 0 \end{pmatrix} . \tag{19}$$

Note that the B_{1g} symmetry is lowered to A_g in the D_{2h} orthorhombic point group: $x^2 - y^2$ is invariant with respect to all D_{4h} operations which preserve the CuO chains of Fig. 10. The Raman tensor of the five Raman active phonons polarized along z has then the general form:

$$A_g: \begin{pmatrix} a & 0 & 0 \\ 0 & b & 0 \\ 0 & 0 & c \end{pmatrix} \tag{20}$$

which is simply obtained as a linear combination of the A_{1g} and B_{1g} of Eq. (19). As already mentioned, the tensors of Eq. (19) are exact for the semiconductor $YBa_2Cu_3O_6$ and a good approximation for $YBa_2Cu_3O_7$, especially for *twinned* crystals which contain microscopic domains with the chain along y and along x, thus not allowing to distinguish between the a and the b directions of Fig. 10 even if they are different.

It is easy to find out which of the five observed z-polarized Raman phonons is that of B_{1g} (D_{4h} point group) symmetry. The reader should find the selection rules for backscattering on an x-y surface which follow from Eq. (19). He will realize that for in-plane polarizations the A_{1g} phonons are only observed for parallel incident and scattered polarizations while the B_{1g} phonon should appear for parallel polarizations along either x or y ($z(x,x)\bar{z}$ and $z(y,y)\bar{z}$

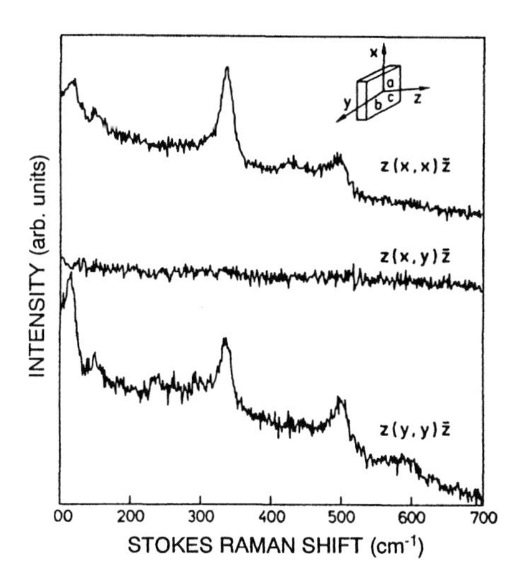

Figure 11: Phonon Raman spectra of untwinned single crystals of $YBa_2Cu_3O_7$ for three backscattering configurations (given in Porto's notation). Note the differences between the $z(x,x)\bar{z}$ and the $\bar{z}(y,y)\bar{z}$ spectra which reflect the orthorhombicity of the crystal [30].

in Porto's notation) but not for both polarizations parallel to either of $2^{-1/2}(x \pm y)$. B_{1g} is allowed, however, for the crossed polarizations $e_L \parallel (011)$; $e_S \parallel (0\bar{1}1)$. Using these rules it was early discovered that the 340 cm^{-1} OII-OIII phonon has the B_{1g} symmetry [30]. This makes of this phonon, with a frequency higher than the Ba and the Cu2 phonons but lower than the OII+OIII and the apical OIV, the "keystone" in the assignment of the observed Raman phonons. A definite assignment of the 115 and 150 cm^{-1} phonons to *nearly unmixed* Ba and Cu2 vibration has been made recently by means of isotopic substitution: when replacing Ba isotopes the frequency of the 115 cm^{-1} phonon changes nearly like $M_{Ba}^{-1/2}$ while that of the 150 cm^{-1} does not budge. The opposite happens upon isotopic substitution involving the copper atoms [31].

II.G. Resonant Raman Scattering by High T_c Superconductors

Figures 7 and 8 indicate that it is possible to calculate from the electronic band structure the resonant profiles of Raman scattering by phonons, at least in the simplest possible crystals with two atoms per primitive cell (note that in order to have Raman active phonons, i.e., optical phonons at $\boldsymbol{k} = 0$, we must have at least two atoms per primitive cell). Until the discovery of high T_c superconductivity this was the state of the art, no calculations were available for more complex crystals. The importance of the high T_c materials produced a "quantum jump" in that state of the art, going from 2 atoms per primitive cell to the 13 atoms of $YBa_2Cu_3O_7$ as soon as one-electron band structure calculations became available [32]. The high T_c superconductors should have, except very near the Fermi surface, electronic state widths $\Gamma_e \simeq 0.1$ eV. Hence, Eq. (5) should hold and one can estimate $\partial S/Ld\Omega$ using the calculated spectrum of $\partial\chi/\partial u$. The latter can be obtained by evaluating $\chi(\omega_L)$ at two different positions of the vibrating atoms and dividing the difference by the corresponding atomic displacement so as to find the derivative. This is particularly easy for the A_g phonons (those which interest us anyhow) since they do not break the symmetry of the crystal and therefore the band structure code does not have to be modified in order to accommodate

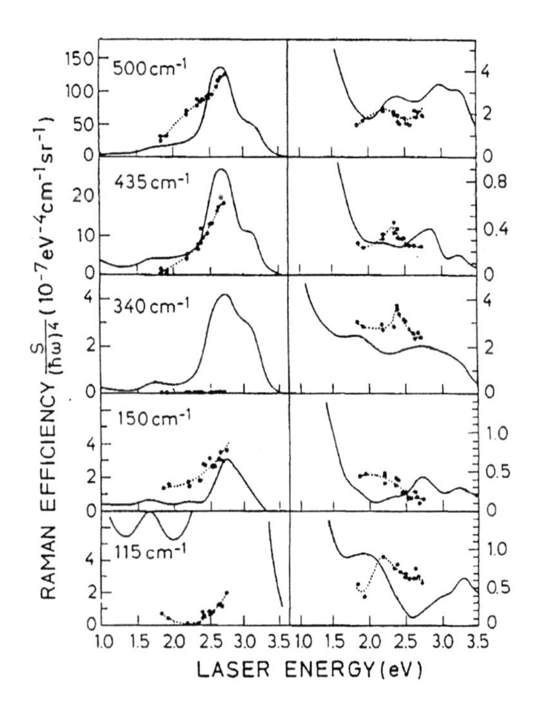

Figure 12: Resonant Raman spectra (in absolute efficiency units) of the four A_{1g} and the B_{1g} (340 cm^{-1}) phonons of $YBa_2Cu_3O_7$. The points, and the dotted lines connecting them are experimental. The solid lines represent the theoretical calculations based on the LMTO electronic band structure [32]. Left panels: (z, z) polarizations; right panels: (x, x), (y, y) average polarizations.

a phonon displacement. If displacements of more than one pair of atoms contribute to the phonon under consideration this can also be easily taken into account provided the relative contributions (i.e., the eigenvectors) are known.

Figure 12 shows the results of such calculations of *absolute* scattering efficiencies for the five A_g phonons of $YBa_2Cu_3O_7$ compared with experimental data [32]. An average of the $z(x, x)\bar{z}$ and $z(y, y)\bar{z}$ is given since the crystals used for the measurements were twinned. The electronic band structure employed to calculate $\chi(\omega_L, u)$ was obtained with the LMTO-ASA method (linear muffin tin orbital, atomic sphere approximation). The agreement between measured and calculated results is rather startling, especially when one considers that both are given in *absolute efficiency units.*

Recently, similar calculations and measurements have been performed for $YBa_2Cu_4O_8$ (also called 124), a superconductor which has a structure similar to that of Fig. 10 except that there are two instead of one CuO chains, staggered on top of each other, with the Cu of one of them bonding with the oxygen of the other [33]. The strong Cu-O bonds make the structure stoichiometric (contrary to the case of $YBa_2Cu_3O_{7-\delta}$ where δ can vary from 0 to 1). Moreover, 124 crystals are single domain and the $z(x, x)\bar{z}$ and $z(y, y)\bar{z}$ differ considerably. Although the eigenvectors of 124 are more complicated and less well known than those of 123 (there are in total 7 phonons of A_g symmetry which leads to more admixture of atomic modes) reasonable agreement is found between calculated and measured resonant profiles, including also the absolute magnitude of the scattering efficiency.

II.H. Scattering Interference between Orthogonal Polarization Configurations: High T_c Superconductors

The Raman tensors shown in Eqs. (19) and (20) generate interesting interference phenomena observable when the scattering polarizations are not along the crystal axes. Let us consider the tetragonal Raman tensor of Eq. (19) and assume that both $\hat{\mathbf{e}}_L$ and $\hat{\mathbf{e}}_S$ are parallel to each other and to the x, z plane. We designate as ϑ the angle they subtend with the z axis. A straightforward calculation leads to the scattering efficiency as a function of the angle ϑ [34]

$$\frac{\partial S(\vartheta)}{L\partial\Omega} \sim |a|^2 \sin^4(\vartheta) + |c|^2 \cos^4(\vartheta) + 2|a||c| \cos(\varphi_{ac}) \sin^2(\vartheta) \cos^2(\vartheta) \tag{21}$$

where φ_{ac} represents the phase difference between the a and the c components of the Raman tensor.

The 123 high T_c superconductors (and their semiconducting counterparts) are ideal materials for the observation of the interference phenomena between the xx and the zz channels represented by Eq. (21). In the case of twinned 123 superconductors one must use for a in Eqs. (2), (19), and (21) the average of the a and b of Eq. (20). Conversely, such interference phenomena can be used for the characterization of thin films and single crystals of these materials.

Interference phenomena of the type described above have been recently observed in $SmBa_2Cu_3O_{7-\delta}$ thin films oriented with the z-axis in-plane [34]. We display the data obtained for the four A_{1g}-like phonons of a thin film of $SmBa_2Cu_3O_{7-\delta}$ (a 123 superconductor) in Fig. 13. The film was grown with the z-axis in-plane and therefore, according to Eq. (19), no interferences are expected for the B_{1g} phonon. The observed interference between (x, x) and (z, z) scattering enables us to determine the relative values of $|a|$ and $|c|$ and also the differences in their phases. These values have been compared with *ab initio* calculations in Ref. [34]. Similar effects should also be observed in untwinned crystals (remember that $YBa_2Cu_4O_8$ is always untwinned) for incident and scattered polarizations both in the x-y plane.

II.I Reflectance Modulation by Coherent Phonons

During the past few years a Raman related technique has been at the center of attention of the short pulse (fs), time-dependent spectroscopy community [35]. It involves the excitation at the surface of *coherent* phonons with $\boldsymbol{k} = 0$ by a short laser pulse. These phonons ring near the surface at the corresponding Raman frequency and consequently modulate the reflectivity r vs. time at this frequency with a strength:

$$\Delta r = \frac{dr}{d\chi} \cdot \frac{d\chi}{du} \cdot u(t) = \frac{dr}{d\chi} \cdot \boldsymbol{R} \cdot u(t) \tag{22}$$

Thus the strength of the reflectivity modulation is determined by the corresponding Raman tensor $\boldsymbol{R}$.

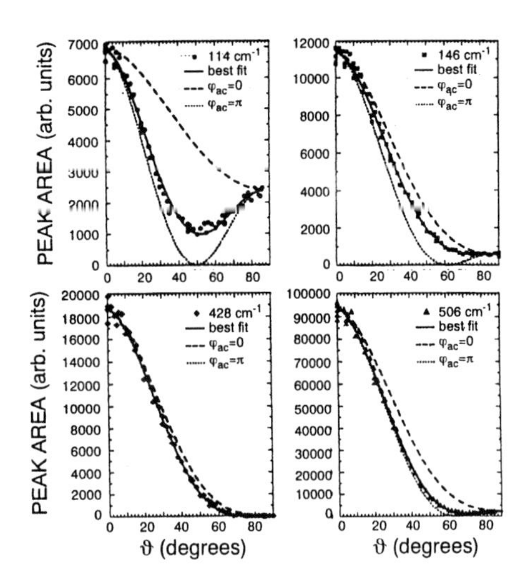

Figure 13: Raman efficiencies of the four A_{1g}-like phonons of $SmBa_2Cu_3O_{7-\delta}$ observed at room temperature on a (110)-surface in parallel polarization as function of the angle ϑ. Full symbols correspond to experimental values determined from the spectra by numerical line fits. Full lines represent the results of a least squares fit of Eq. (21) to the data. The best fit values for φ_{ac} are 2.08, 1.45, 2.58, and 2.21 rad. for the Ba, Cu2, OII+OIII, and OIV phonon, respectively. The dashed and dotted lines show intensity functions calculated using the same values of $|a|$ and $|c|$ as determined for the best fit, but phase shifts of $\varphi_{ac} = 0$ and $\varphi_{ac} = \pi$, respectively [34].

Figure 14 displays the time dependence of the reflectivity measured for $YBa_2Cu_3O_7$. The oscillations superimposed on the continuously decaying background represent the sum of the modulations of Eq. (22) for all possible phonons. The fast Fourier transforms of these oscillations are shown in the inserts of Fig. 14: two peaks are seen at ~115 and 150 cm^{-1} which corresponds to the A_g Raman modes of Ba and Cu, respectively [36]. Because of the limited time resolution, Raman phonons of higher frequencies do not appear in the Fourier transform of Fig. 14.

Of particular interest is the change in the relative intensities of the 115 and the 150 cm^{-1} peaks between 300 ($T > T_c$) and 40 K ($T < T_c$). This change is likely to be related to electron-phonon renormalizations below T_c of the type which has been recently observed for $HgBa_2Ca_3Cu_4O_{10+\delta}$ [37,38].

III. EFFECT OF PERTURBATIONS ON THE PHONONS OF A CRYSTAL

We have already mentioned that the most powerful tool available nowadays for the investigation of the full dispersion relations of all phonons in the whole Brillouin zone is inelastic neutron scattering (INS). However, this technique is cumbersome, slow, and costly (last but not least there is the ecological onus of nuclear waste disposal!). Moreover the resolution and frequency accuracy is typically one to two orders of magnitude less than it can be reached by Raman spectroscopy (it is particularly difficult to measure phonon linewidths by INS since the resolution is not much better than 5 cm^{-1}). Progress in enhancing INS accuracy and resolution is possible but slow.

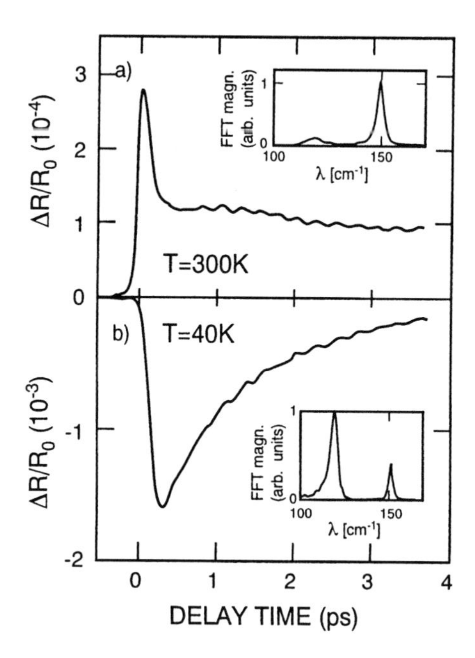

Figure 14: Time resolved reflectivity of an $YBa_2Cu_3O_{7-\delta}$ film on $SrTiO_3$ at 300 K (a) and 40 K (b, below T_c). The inserts display the fast Fourier transforms of these signals which roughly correspond to the standard Raman spectra measured directly in the frequency domain [36].

III.A. Temperature Dependence of Phonon Linewidths and Frequencies

Raman spectroscopy is a particularly useful technique to accurately determine phonon linewidths Γ_p and frequencies ω_p. Concerning linewidths one can usually distinguish between inhomogeneous linewidths, corresponding to a mixture of different pieces of material with a spread of frequencies, and homogeneous ones, which correspond to a finite lifetime τ_p of the phonon according to the "frequency-lifetime" *uncertainty principle*:

$$4\pi c\Gamma_p \cdot \tau_p \simeq 1 \ , \tag{23}$$

where Γ_p has been chosen to be the *half width at half maximum* (HWHM) (in cm^{-1}).

As an example we show in Fig. 15 the Raman spectra at 77 and 334 K of the $\boldsymbol{k} = 0$ phonons of gray tin (α-tin), a rather interesting material with diamond structure obtained from white tin by cooling below 13°C or, more recently, by molecular beam epitaxy (MBE) on a zincblende-type substrate (InSb or CdTe are usually chosen because their lattice constant nearly matches that of α-Sn). Note the broadening (which can be shown to be homogeneous) and the shift of the spectral lines upon raising the temperature. Such phenomena should not take place for a perfectly harmonic crystal. However, such crystal does not exist in nature: anharmonic components of the interatomic potentials ($\sim r^3$ and higher powers of r, where r represents changes in interatomic distances). The spectra of Fig. 15 illustrate vividly the fact that phonons are not *particles* (with infinite lifetime) but only quasiparticles with a

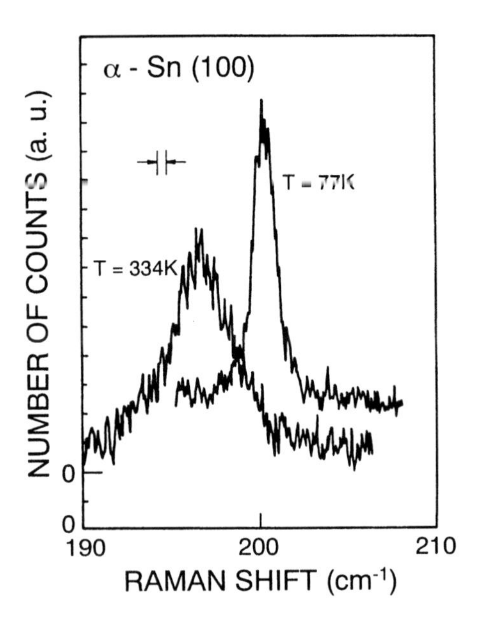

Figure 15: Raman spectra of gray tin measured at 77 and 334 K [39]. The shift and broadening with increasing temperature are a measure of the real and the imaginary parts of the self energy represented by the diagrams of Fig. 16.

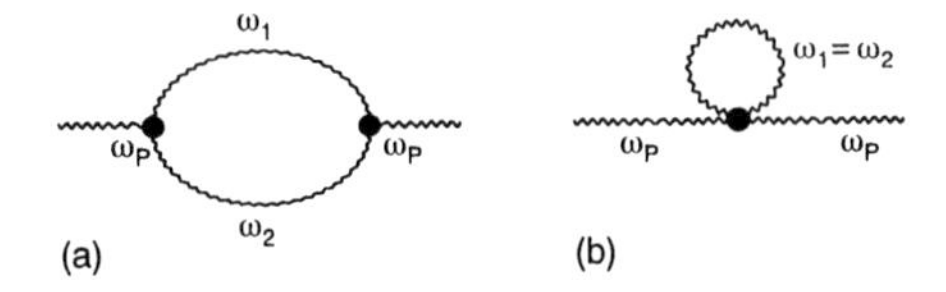

Figure 16: (a) Self-energy diagram which determines, to lowest order, the real and imaginary parts of the anharmonic self-energy contribution to the phonon frequency and linewidth. (b) Contribution to the phonon frequency of the fourth order anharmonicity in first order perturbation theory. Note that (b) does not have any imaginary contribution (i.e., any contribution to the linewidth). Compare these diagrams with those of Fig. 8 (c) and (d).

lifetime which can be estimated with Eq. (23), and $2\Gamma \simeq 1$ cm^{-1} at 77 K found in Fig. 15, to be $\simeq 6$ psec.

The temperature dependence of the phonon frequency and linewidth is represented, to the lowest order of approximation, by the diagrams of Fig. 6. The imaginary part of the self-energy results from terms in Fig. 16(a) (similar to those in Fig. 8(c)) in which energy is conserved in the intermediate state:

$$2\Gamma_p = \frac{2\pi}{\hbar^2} |M_{an}|^2 [1 + n_B(\omega_1) + n_B(\omega_2)] \, N_d(\omega_1 + \omega_2 = \omega_p) \tag{24}$$

where N_d is the density of two-phonon states (ω_1, ω_2) which fulfill the energy conservation condition $\omega_p = \omega_1 + \omega_2$ and M_{an} is the matrix element of the anharmonic hamiltonian corresponding to the vertices of Fig. 16(a). Equation (24) follows from to the so-called

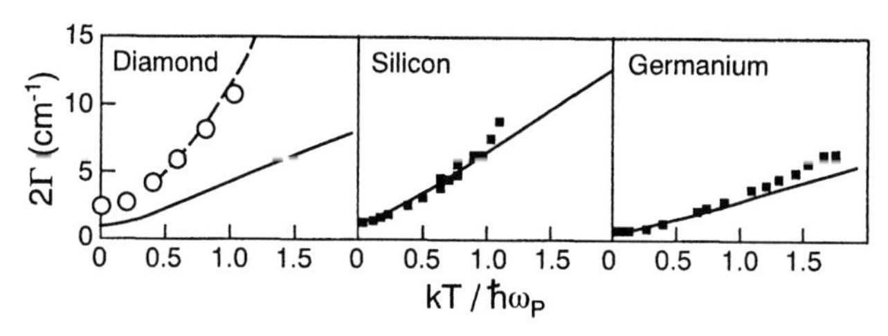

Figure 17: Temperature dependence of the full width at half maximum, 2Γ, of the LTO phonon in C, Si, and Ge. Solid lines are the result of an *ab initio* calculation [40]: the points represent experimental results [39,41] while the dashed line, in the case of diamond, represents the results of molecular dynamics calculations [42].

Fermi Golden Rule.[f]

In the Klemens approximation one assumes that $\omega_1 = \omega_2 = \omega_p/2$. In this case the square brackets of Eq. (24), which determine the temperature dependence of Γ_p, reduce to $[1 + 2n_B(\omega_p/2)]$, i.e.:

$$2\Gamma_p(T) = \frac{2\pi}{\hbar^2} |M_{an}|^2 N_d(\omega_p) \left[1 + 2n_B\left(\frac{\omega_p}{2}\right)\right] \tag{25}$$

Note that Eq. (25) enables us to predict the temperature dependence of Γ_p without adjustable parameters if $\Gamma_p(T = 0)$ is known. An adjustable parameter can be introduced if we assume that $\omega_1 \neq \omega_2$: in this case the temperature dependence of Γ_p becomes steeper than that predicted for $\omega_1 = \omega_2$. Fitting the temperature dependence of Γ_p measured for Si, Ge, and α-Sn led to the conjecture that $2\omega_1 \simeq \omega_2 = \frac{2\omega_p}{3}$ [39]. Recent *ab initio* calculations based on a pseudopotential electronic band structure and LDA (local density approximation) perturbation theory have confirmed this conjecture [40]: ω_1 corresponds to TA and ω_2 to LA phonons near the edge of the Brillouin zone. The calculation presented in Ref. [40] are displayed in Fig. 17 together with experimental data for Ge and Si [39]. In the case of diamond, the calculations of Ref. [40] give values of 2Γ which are throughout a factor of two smaller than the experimental ones [41]. Note that for diamond $\omega_1 \simeq \omega_2 \simeq \omega_p/2$ should be a good approximation. The semiempirical molecular dynamics calculations [42] represented by the dashed curve in Fig. 17 give an excellent representation of the experimental data.

The real part of the diagram of Fig. 16(a) plus the diagram of Fig. 16(b) represent, to a first approximation, the anharmonic correction to the phonon frequency. In most cases a decrease of frequency (frequency softening) is observed with increasing temperature [39, 41,42]. Notice that if a good calibration is available a, Raman measurement of a phonon frequency can be used for the determination of the temperature of the scattering volume.

Equation (25) yields a finite value of $2\Gamma_p$ even at $T = 0$. This finite low temperature width results from the anharmonicity related to the *zero point vibrational amplitude.* In materials with only one type of element (e.g. C, Ge, Si) the zero point amplitude is inversely proportional to $M^{-1/2}$ (M = atomic mass which can be varied if the element has different stable isotopes). Note that at high T Eq. (25) yields Γ_p proportional to T and independent of M.) A similar effect is observed for the frequency renormalization at $T = 0$ [27,43]. In materials with more then one element (ZnSe, CuCl, $YBa_2Cu_3O_7$) each element, if available with different isotopic masses, produces a different isotopic effect on each of the phonon

[f] E. Fermi was awarded the Physics Nobel Prize in 1938 for his work on nuclear reactions.

frequencies and linewidths under consideration. For examples see Ref. [43].

III.B. Effects of Pressure on the Phonon Frequencies and Linewidths

The use of pressure as a parameter to modify the physical properties of solids was pioneered by P.W. Bridgman[g] and his successor W. Paul[h] at Harvard University. Truly hydrostatic pressure was limited those days to about 3 Gigapascal (GPa), i.e., 30 kbar. Nowadays pressures as high as 100 GPa can be obtained with a small, simple and inexpensive device called the diamond anvil cell (DAC).

The DAC operates under the principle that this is possible to obtain very high hydrostatic pressures by pressing with the flat faces of two diamonds (anvils) onto a special metal gasket in the center of which a hole of about 100 μm diameter has been drilled. As pressure transmitting fluid one used, when the technique was developed around 1975, a mixture of methanol and ethanol [44,45]. Nowadays it is advisable to use helium if pure hydrostatic pressures higher than 10 GPa are to be reached [46]. The crux of the DAC shift with pressure in order to measure the pressure through a suitable calibration procedure [44]. The small size of the sample required makes of Raman spectroscopy an ideal technique for the investigation of elementary excitations under hydrostatic pressure [45]. There has also been interest in Raman measurements under uniaxial stress (because of sample breaking these measurements seldom achieve 5 GPa) [46]. Measurements of the effect of uniaxial stress on Raman phonons can be used to determine the strain state of epitaxially grown semiconductor, micro-, and nanostructures induced by lattice mismatch [46–48].

We display in Fig. 18 the pressure dependence of the Raman frequencies of Ge and Si as measured with the DAC. The dashed-dotted lines represent *ab initio* calculations based on a pseudopotential-LDA electronic band structure. The excellent agreement found for the absolute frequencies of Ge is to be viewed as fortuitous since a deviation of 1.5% (8 cm^{-1}) is found for Si [49]. The measured pressure dependence is reproduced extremely well by the calculations, a fact that experience with other materials shows not to be fortuitous.

The deviation between calculated (513 cm^{-1}) and experimental (520 cm^{-1}) zero pressure frequencies of silicon represents the state of the art of the so-called *ab initio* calculations and the possibility of such errors must be taken very seriously. As an example of the problems which may arise we show in Fig. 19 the effect of pressure on the linewidth (FWHM) of the Raman phonons of Ge and Si [49]. In spite of minor deviations between the measured and the calculated absolute values of the FWHM, the linear measured pressure dependence is represented rather well by the calculations, except for an upturn observed in the calculated points above 7 GPa. An analysis of the *ab initio* calculations shows that such upturn, which is not observed experimentally, results from the fact that ω_p approaches a sharp singularity in $N_d(\omega_p)$ with increasing pressure. This approach can be understood by considering that ω_p *increases* with pressure while ω_1 ($\approx$ the TA frequency at the edge of the BZ) *decreases* [46]. If one corrects "by hand" the error in the calculations of the absolute value of ω_p shown in Fig. 18 the encounter of ω_p with $\omega_{\mathrm{TA}} + \omega_{\mathrm{LA}}$ does no longer occur below 10 GPa. The experimental FWHM vs. pressure can then be satisfactorily reproduced by the theory. This should be a warning to theorists not to make unwarranted claims till their calculations have been contrasted with measurements (Karl Popper would say "either falsified or verified" by experiment). At this point a sobering thought may be the following quotation from a great

[g] For this work P.W. Bridgman was awarded the Nobel prize in Physics in 1946

[h] The author's Ph.D. thesis advisor.

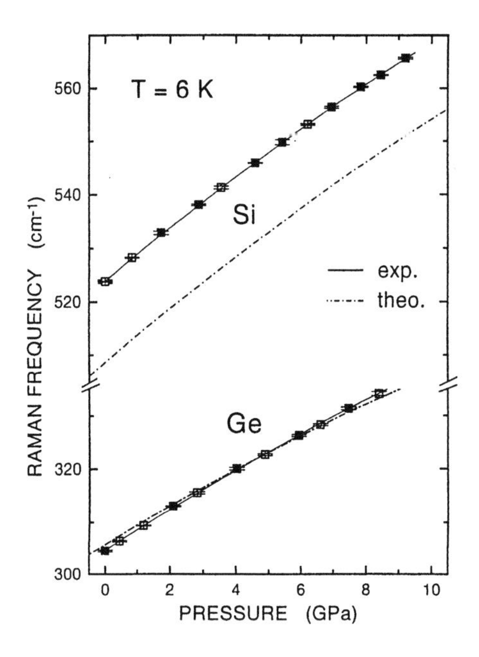

Figure 18: Raman phonon frequencies for Si and Ge, measured for increasing (filled squares) and decreasing (open squares) pressure. The solid lines correspond to fitted quadratic relations. The dashed-dotted lines represent the results of *ab initio* calculations [49].

theorist, Max Planck[i]:

> *Experiments are the only means of knowledge at our disposal. The rest is poetry, imagination.*

We close this section by mentioning that most crystals experience phase transitions under pressure. The pressures at which the transitions occur can also be "predicted" by *ab initio* calculations. Raman spectroscopy in connection with the DAC has proven to be an excellent technique for the observation of these transitions. For a review see Ref. [46] and references therein.

III.C. Contribution of Isotopic Disorder to the Phonon Linewidth

Equation (25) represents the FWHM of a phonon in a crystal which is perfect except for the vibrations of the lattice (which can be considered as a form of dynamical intrinsic imperfections). Most crystals, however, contain static, nonintrinsic imperfections such as impurities, vacancies, dislocations, and even the surface is to be regarded as an imperfection from the point of view of the three-dimensional translational symmetry. In this case a T-independent term (an additional adjustable parameter) must be added to Eq. (25) in order to represent the scattering of the phonons by the defects.

The simplest possible defects appear in crystals composed of elements which have several stable isotopes. When natural constituents are used for the crystal growth one expects to find in the crystal atomic mass distributions corresponding to the natural abundance of

[i] Max Planck was awarded the Physics Nobel Prize in 1918 for his discovery of energy quanta (i.e, $\hbar\omega$).

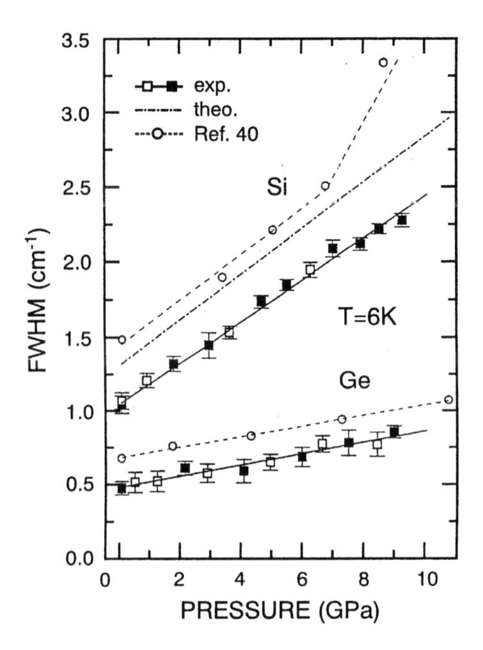

Figure 19: Pressure dependence of the full width at half maximum of the Raman phonons in Si and Ge. Filled and open squares refer to measurements under increasing and decreasing pressure, respectively. The solid lines represent fits to the experimental points. The dashed lines are calculated results from Ref. [40]. The dashed-dotted line represents the results of corrected calculations for Si (see text) [49].

the various isotopes (diamond: $^{12}C_{0.99}{}^{13}C_{0.01}$; $^{28}Si_{0.92}{}^{29}Si_{0.05}{}^{30}Si_{0.03}$; $^{70}Ge_{0.21}{}^{72}Ge_{0.27}{}^{73}Ge_{0.08}$ $^{74}Ge_{0.08}{}^{74}Ge_{0.36}{}^{76}Ge_{0.08}$. The case of polyatomic crystals (ZnSe, $YBa_2Cu_3O_7$) can be particularly complicated. Here we confine ourselves to crystals with only one kind of atom (diamond, germanium, α-Sn) since the simpler treatment displays already the basic physics involved.

The contribution of isotopic disorder to the phonon linewidth $2\Gamma_i$ of a monoatomic crystal can be obtained in the simplest possible way by perturbation theory[27]:

$$2\Gamma_i = \frac{\pi \omega_p^2 g}{12} N_d^{(1)}(\omega_p) \tag{26}$$

where $N_d^{(1)}$ is the density of one-phonon states, normalized to six per unit cell, and g is the mass variance:

$$g = \frac{\langle M_i^2 \rangle}{\langle M_i \rangle^2} - 1 \tag{27}$$

In Eq. (27) M_i represent the mass of each isotope present and $\langle \ldots \rangle$ the average of M_i weighted by its abundance. Concerning the contribution of g to the FWHM of the Raman phonon we must distinguish two cases: (a) the case of Ge, Si, and α-Sn, and (b) the case of diamond. The origin of the difference is illustrated in Fig. 20. In case (a) the maximum of the optical

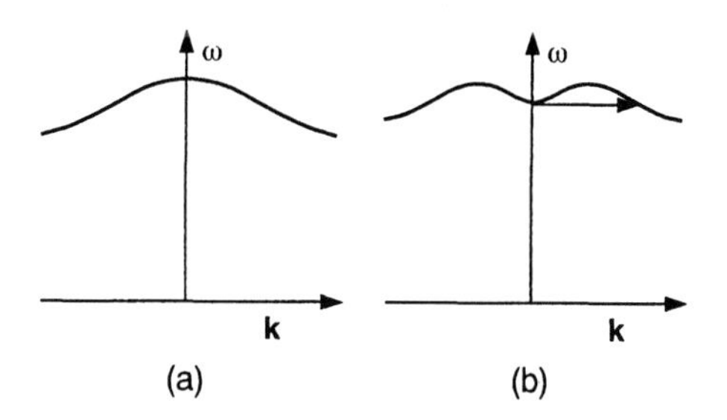

Figure 20: Diagram illustrating the scattering of Raman phonons ($\boldsymbol{k} = 0$) by isotopic mass fluctuations. Case (a) refers to Ge, Si, and α-Sn. At $\boldsymbol{k} = 0$ the density of states $N_d^{(1)}(\omega_p)$ vanishes and, according to Eq. (26), no broadening is induced by isotopic fluctuations. Broadening will take place, however, for phonons with $\boldsymbol{k} \neq 0$. In case (b), which applies to diamond, the reentrant dispersion relation leads to a finite density of states, and considerable broadening, even for $\boldsymbol{k} = 0$. The horizontal arrow represents an elastic scattering process not possible in case (a).

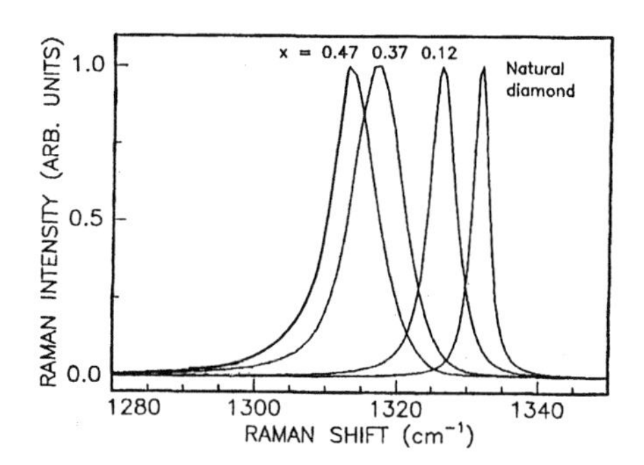

Figure 21: Spectra of the Raman phonons of diamond at 300 K with several isotopic compositions. The concentration of ^{13}C is denoted by x. With increasing x the linewidths are considerably enhanced up to $x = 0.5$; between $x = 0.5$ and $x = 1$ (not shown here) the linewidth decreases again. From Ref. [50].

branches occurs exactly at $\boldsymbol{k} = 0$. At this point $N_d^{(1)} = 0$ and, to the order of Eq. (26), no broadening occurs: the scattering at the static spatial isotopic mass fluctuations is elastic and no states are available to scatter into ($N_d^{(1)} = 0$). In the case of diamond, however, the maximum occurs for $\boldsymbol{k} \neq 0$ and $N_d^{(1)} \neq 0$. In this case considerable broadening is expected. This is illustrated in Fig. 21 for several diamonds grown with tailor-made mixtures of ^{12}C and ^{13}C. The linewidths observed for 47% ^{13}C are between two and three times wider than those observed for pure ^{13}C and ^{12}C [50].

In the case of Ge (we are in the process of investigating these isotope effects in Si also) the low temperatures FWHM are about 0.7 cm^{-1} for a pure isotope. This linewidth, of purely anharmonic origin, varies with isotopic mass like M^{-1} (as expected from Eq. (24)). A very small increment $2\Gamma_i = 0.035$ cm^{-1} appears for natural Ge ($g = 5.9 \times 10^{-4}$) and 0.55 cm^{-1} for the isotopic mixture with the largest possible $g = 1.5 \times 10^{-3}$. $2\Gamma_i$ should vanish in the case of Ge according to Eq. (26). The value of $2\Gamma_i$ measured is indeed very small but nonnegligible. It has been attributed to a nonvanishing $N_d^{(1)}(\omega_p)$ generated by the anharmonic broadening [27,51,52], i.e., by a higher order perturbation involving anharmonicity and isotopic disorder.

The width represented by Eq. (26) corresponds to the imaginary part of the self-energy

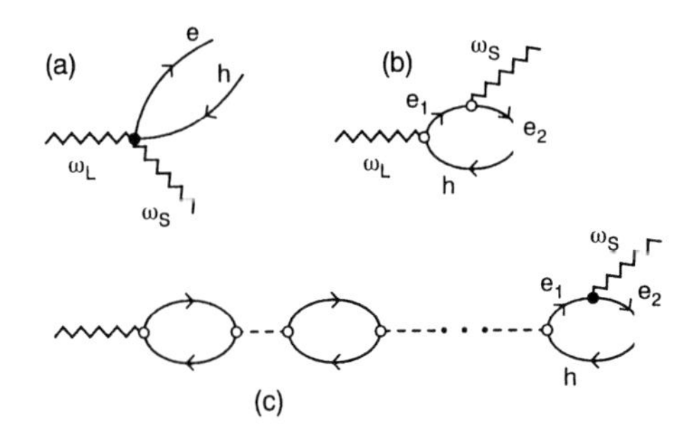

Figure 22: Basic diagrams for light scattering by intraband electronic excitations in metals. (a) Thomson scattering by free electrons. (b) Scattering mediated by interband transitions. (c) Bubble diagrams which contribute to the screening of (a) and (b). Here the dashed lines represent the Coulomb interaction while the dots stand for an arbitrary number of screening bubbles. The scattering amplitude is obtained by adding the corresponding expressions for (a), (b), and (c) with all possible numbers of bubbles.

induced by the isotopic disorder. This suggests the presence of a real part of this self energy which should manifest itself in a shift (renormalization) of the observed ω_p with respect to the value expected for the average isotopic mass. The real part of a self-energy can be written as a sum (or an integral) of squared matrix elements of the perturbation divided by energy denominators to an intermediate state. In the case of Ge and α-Sn, the intermediate states lie all below ω_p. The measured self-energy shift must therefore correspond to a hardening of ω_p. For $g = 1.5 \times 10^{-3}$ a positive self-energy shift of $+1.06$ cm^{-1} has been observed in the case of Ge [52].

Very recently similar work has been published concerning α-Sn with different isotopic compositions [53].

IV. PHONON RENORMALIZATION BY ELECTRONIC TRANSITIONS IN METALS AND DOPED SEMICONDUCTORS

IV.A. Intraband Transitions in Metals and Doped Semiconductors

The reason why metals conduct is the existence of low energy intraband excitations which can carry both electricity and heat. Semiconductors heavily doped with electrically active impurities (either donors or acceptors) can also be considered as metals with adjustable (usually low) carrier concentrations. We shall refer to them as *metals* unless otherwise required. The basic diagrams for the amplitude of light scattering by intraband electronic excitations are shown in Fig. 22.

The coupling vertex in (a) is simply:

$$\frac{1}{m}\frac{e^2}{c^2}\left(\boldsymbol{A}_L \cdot \boldsymbol{A}_S\right) \tag{28}$$

where $\boldsymbol{A}_L$ and $\boldsymbol{A}_S$ are the vector potentials of the incident and the scattered radiation. Diagram (b) can be conveniently combined with (a) whenever the $e_1 - h$ excitations are not close to ω_L and ω_S (non-resonant scattering). Using the well known $\boldsymbol{k}\cdot\boldsymbol{p}$ theory [21] one finds that the combined contributions of (a) and (b) can be represented by an expression similar to Eq. (28) with the free electron mass m replaced by the free carrier effective mass tensor $\boldsymbol{\mu}$.

$$\frac{e^2}{c^2}\left(\boldsymbol{A}_L\cdot\frac{1}{\boldsymbol{\mu}}\cdot\boldsymbol{A}_S\right) . \tag{29}$$

The screening diagrams of Fig. 22(c) contribute to a reduction of the scattering efficiency.

The scattering efficiency per unit frequency interval, including screening, can be written as:

$$\frac{\partial S}{L\partial\Omega\partial\omega} \simeq \frac{\hbar r_0^2 k_R^2 \epsilon_\infty^2}{4\pi^2 e^2}\left|\left(\hat{\mathbf{e}}_L\cdot\frac{m}{\boldsymbol{\mu}}\cdot\hat{\mathbf{e}}_S\right)\right|^2 \mathrm{Im}\frac{-1}{\epsilon(\boldsymbol{k}_R,\omega_R)} \tag{30}$$

where ϵ_∞ is the interband contribution to the long wavelength dielectric function k_R (ω_R) the scattering vector (frequency), $r_0 = \frac{e^2}{mc^2}$ the Thomson[j] radius and $\epsilon(k_R,\omega_R)$ the dielectric function including the free carriers (the so-called Lindhard[k] dielectric function). Equation (30) must be multiplied either by the Bose-Einstein factor $[1+n(\omega_R)]$ for Stokes- or $n(\omega_R)$ for anti-Stokes scattering.

The dielectric function $\epsilon(k_R,\omega_R)$ diverges for $\omega_R \to 0$, a fact which produces a strong decrease in scattering efficiency at low frequencies. The low frequency scattering strength is shifted to the region where $\epsilon_r = 0$. (This happens near the frequency at which $-\mathrm{Im}\epsilon^{-1}(k,\omega)$ has a maximum, a fact which results in a peak in the Raman spectrum.) The frequency where this happens is called the *screened plasma frequency.*

We shall only mention here a few additional interesting effects related to the intraband scattering in metals. The interested reader will find examples and details in Refs. [9], [54], and [55]. The first interesting point concerns the possible observation of *unscreened* scattering (i.e., situations in which the effect of the bubbles of Fig. 22(c) cancels or, at least, does not dominate for $\omega_R \to 0$). This happens in two typical situations:

1. in the presence of spin-orbit coupling one can flip the electron spin in the scattering process. In this case diagram (b) can lead to the creation of a pure spin-density excitation without charge fluctuations which does not couple via Coulomb interaction to the ensemble of intraband excitations and the screening diagrams of Fig. 22(c) vanish.

2. In the case of a multicomponent electron gas (i.e., electrons with different effective mass tensor $\boldsymbol{\mu}$, such as for n-type Si and Ge) it is possible to excite the various components in such a way that the resulting density waves have no net charge (although each one of the component does). This situation also obtains for metals with a Fermi surface whose various points do not all have the same mass. This situation applies to the normal state of high T_c superconductors [55]. The scattering is then related to the variance of the mass around the Fermi surface [54].

[j] J.J. Thomson received the Physics Nobel Prize in 1906 for his investigations of conduction processes in gases.

[k] J. Lindhard, a professor at the University of Aarhus (Denmark), passed away in October 1997 while the author was writing this article.

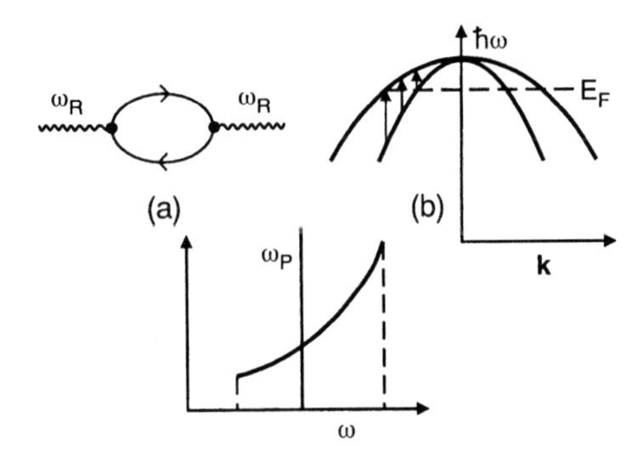

Figure 23: (a) Feynman diagram for the self-energy of a phonon ω_R induced by electronic excitations. (b) Electronic excitations which contribute to (a) in p-type Ge or Si. Note that strictly speaking the transitions depicted by arrows in (b) are *interband* (from light to heavy hole bands). (c) Density of electronic transitions corresponding to the band diagram in (b). The vertical line labeled ω_p corresponds to the Raman phonon. If it overlaps with the electronic continuum an imaginary part Σ_i results. Otherwise only the real part Σ_r is finite.

The discussion above applies to the normal state of metals. We shall close this section by mentioning that at the transition from the normal to the superconducting state a gap opens in the excitation spectrum represented by either $\mathrm{Im}\varepsilon(k,\omega)$ or $-\mathrm{Im}\epsilon^{-1}(k,\omega)$. This gap can be seen by Raman spectroscopy, in fact in the case of high T_c superconductors Raman measurements provided some of the first evidence for the existence of a superconducting gap [55,56]. The observed dependence of the gap scattering on polarization configuration also led to the conclusion that the gap is anisotropic, in agreement with the presently accepted wisdom that the order parameters in these materials has d-like $k_x^2 - k_y^2$ symmetry (i.e., B_{1g} for the tetragonal D_{4h} point group) [55,57].

IV.B. Renormalization of Phonons via Electron-Phonon Interaction: Semiconductors

In standard metals the phonons interact with the intraband electronic excitations. The resulting self-energy $\Sigma = \Sigma_r + i\Sigma_i$ effects a renormalization of the phonon frequencies (Σ_r) and yields a contribution to the homogeneous quasiparticle phonon widths ($-\Sigma_i$ = HWHM). Unfortunately, since in a metal it is usually not possible to switch on and off or to vary Σ, it is difficult to extract information about Σ from experimental spectra. Some exceptions exist, such as the so-called Kohn anomalies which appear for phonons whose $\boldsymbol{k}$ vector connects *nesting* parts of the Fermi surface, a phenomenon closely related to the so-called *Peierls instabilities* which occur in one-dimensional metals. In the case of heavily doped semiconductors, it is possible to change the carrier concentration through doping and thus to observe changes in the corresponding phonon self-energies. Particularly interesting are the cases of p-type Ge and p-type Si. A diagram of the valence band structure of these materials is shown in Fig. 23. This figure also displays the corresponding self-energy diagram and a plot of the density of electronic excitations and their position in the frequency scale as compared with the Raman phonon.

We show in Fig. 24 the effect of doping with acceptors (gallium) on the Raman spectrum of the $\boldsymbol{k} = 0$ phonons of germanium [58]. The measured spectra depend only on

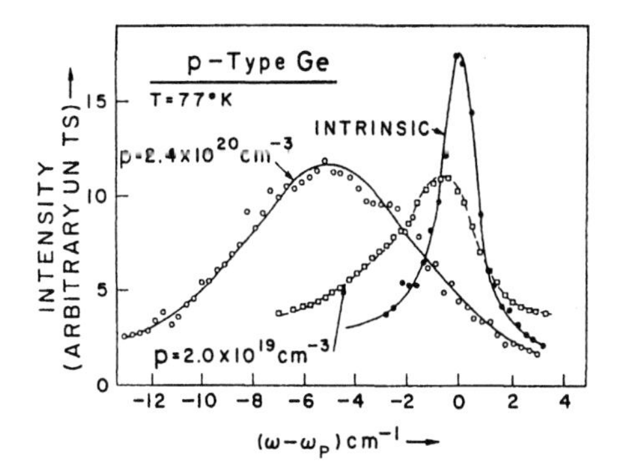

Figure 24: Effect of doping with acceptors on the spectra of the Raman phonon of germanium at 77 K. Note that for a hole doping $P = 2.4 \times 10^{20}$ cm^{-3}, 0.1% of the valence electron concentration produces a shift of 6 cm^{-1}, i.e., 2%. The enhancement factor of $\approx$20 reflects the small energy denominators which corresponds to Fig. 23(c) [58].

the hole concentration, not on the type of dopant. Also, if the acceptor-doped material is compensated with donors, the spectra revert to the shape found for undoped material. It is therefore logical to attribute the observed softening and down-shift to the diagrams of Fig. 23. From the broadenings and shifts observed in Fig. 24 values of the corresponding electron (or hole)-phonon coupling constant (i.e. deformation potential) $d_0 \simeq 40$ eV can be derived [21].

The large renormalization of the Raman phonons of p-Ge just discussed poses the question of what may be the renormalization of the other phonons with $\boldsymbol{k} \neq 0$: acoustic as well as optic, longitudinal as well as transverse. Inelastic neutron scattering measurements have been performed for the acoustic phonons at n- and p-type silicon [59]. Attempts to measure optic phonons have not been successful: the large phonon amplitudes of the acoustic phonons (not of the optic phonons!) at room temperature (see Eqs. (10) and (11)) makes such measurements possible.

IV.C. Fano Lineshapes

Sometimes the renormalization of the parameters of phonons which interact with electronic continua leads to interference effects as represented by the four diagrams of Fig. 25.

The simplest case of Fano (also called Breit-Wigner[l]) lineshape is obtained for a flat continuum of electronic excitations overlapping a discrete phonon, with all vertices assumed to be frequency independent. Note that the electron-phonon coupling constants D_1 and D_2 need not be the same: D_1 describes the standard phonon Raman scattering while D_2 represents the coupling of the phonons to the intraband electronic excitations. In the simple case just mentioned one finds the lineshape:

$$\frac{\partial^2 S}{L \partial\omega \partial\Omega} \propto \frac{(\epsilon + q)^2}{1 + \epsilon^2} \tag{31}$$

[l] For his work on fundamental symmetry principles and their application to nuclear and particle physics E.P. Wigner was awarded the Physics Nobel Prize in 1963.

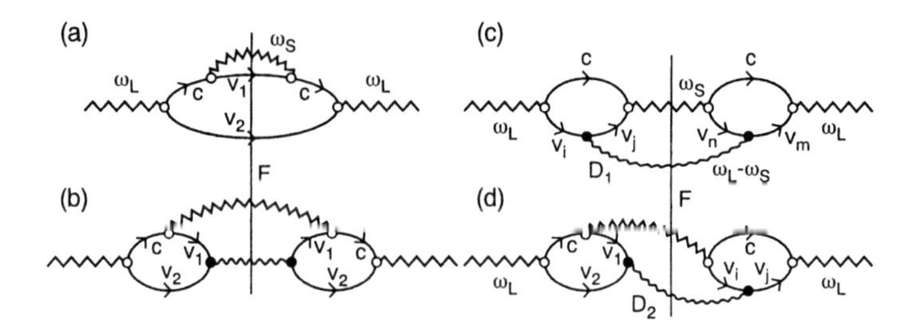

Figure 25: Four scattering efficiency diagrams of the type of Fig. 4 which, when added, give rise to Fano profiles [17]. (a) Diagram for electronic excitations like those of Fig. 24. (b) and (c) Diagrams for the scattering by (renormalized) phonons. (d) Diagram which represents the scattering interference between the phonon and the electronic continuum. From Ref. [17]

where $\epsilon = (\omega - \omega_R)/\Gamma$ (ω_R is the renormalized phonon frequency). The Fano parameter q is given by:

$$q = \left[V\left(T_p/T_e\right) + V^2 R(\omega_p)\right] / \pi V^2 N_e(\omega_p) \ , \tag{32}$$

where V is the matrix element of the electron-phonon hamiltonian, T_p (T_e) the scattering amplitude for the Raman scattering by the phonons (electronic continuum), N_e is the density of electronic excitations and R represents the real part of the self energy given by:

$$R(\omega_p) = P \int_{-\infty}^{+\infty} \frac{N_e(\omega)}{\omega_p - \omega} d\omega \tag{33}$$

where one must insert $N_e(-\omega) = -N_e(\omega)$ and P stands for the principle part of the integral.

According to Eq. (31) the sign of the dimensionless parameter q specifies whether the destructive interference takes place for $\omega < \omega_R$ ($q > 0$) or for $\omega > \omega_R$ ($q < 0$). The sign of q is determined by a number of parameters according to Eq. (32). If $R(\omega_p)$ is small compared with $T_p/T_e V$, the sign of q is determined by the product of the signs of T_p, T_e, and V.[m]

We show in Fig. 26 the Fano profiles measured for the $k = 0$ phonon of heavily doped p-type Si ($P = 1.6 \times 10^{20}$ cm^{-3}) by means of Raman spectroscopy. The solid lines represent calculations based on the diagrams of Fig. 25 under the reasonable assumption that $D_1 \simeq D_2$. The calculations take into account resonant energy denominators in Fig. 26 so that the only adjustable parameter in the fit is Γ (the same for all curves). The agreement with experiment is rather remarkable.

IV.D. Phonon Self-Energies in High T_c Superconductors

It was early realized by Zeyher and Zwicknagl [60] that phonons whose frequencies are close to the maximum value of the energy gap (i.e., the order parameter) $2\Delta_0$ of a high T_c

[m] We have implicitly assumed that q is real. If q had an imaginary part, it could be transformed into a real q plus a non-interfering background [9].

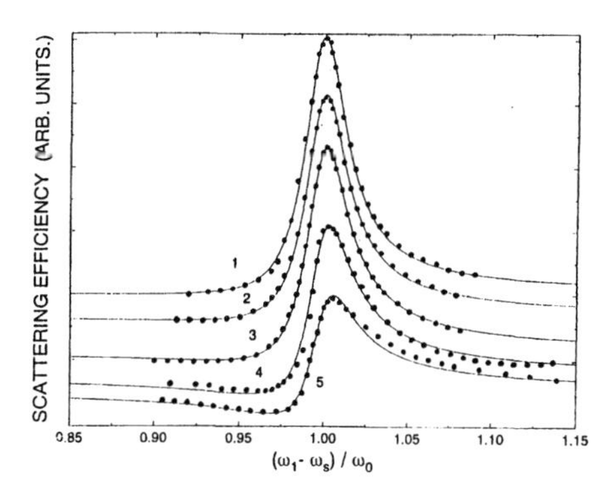

Figure 26: Measured (dots) and calculated (solid lines) Fano profiles for the Raman phonon of p-type Si ($P = 1.6 \times 10^{20}$ cm^{-3}) obtained for different laser frequencies [17].

superconductor should suffer strong changes in their electronic renormalization when going from the normal to the superconducting states. Such effects were detected soon after the discovery of high T_c materials [29,56]. Three types of phenomena are observed when lowering the temperature from above to below T_c:

a) Frequency shifts at T_c: the canonical effect is a down-shift for $\omega_p < 2\Delta_0$ and an up-shift for $\omega_p > 2\Delta_0$.

b) A decrease in linewidth $2\Gamma_p$ if ω_p is well below $2\Delta_0$; an increase for $\omega_p \simeq 2\Delta_0$.

c) A change in the integrated strength of the phonon peak. This change can be obtained from the expression:

$$\frac{\partial S_p}{L \partial \Omega} \propto \pi q^2 \Gamma \tag{34}$$

where q is the Fano parameter of Eq. (33). This equation indicates that a change in q (and therefore in Eq. (34)) can result from the change in the density of states of electronic excitations $N_e(\omega_p)$ which takes place in the superconducting state as the gap $2\Delta_0$ opens. T_p and T_e depend, of course, on ω_L. Using a laser frequency ω_L such that $VT_p/T_e \simeq 0$, a strong increase in the intensity of the Raman phonon can take place below T_c if $\omega_p \simeq 2\Delta_0$ as a result of the increase in $|R(\omega_p)|$. These changes in intensity have been observed for $YBa_2Cu_4O_8$ [61], in $YBa_2Cu_3O_7$ [62], and, more recently, in the mercury-based high T_c superconductors (e.g. $HgBa_2Ca_3Cu_4O_{10+\delta}$ [37,38], also referred to as Hg-1234). These effects, if interpreted with reasonable models, enable one to estimate the electron-phonon coupling constant of the phonons under consideration.

As an illustration of the strong renormalization effects observed in Hg-1234 (D_{4h} point group) we show in Fig. 27 the evolution of one A_{1g} phonon ($\omega_p \simeq 375$ cm^{-1}) related to vibrations of the four CuO_2 planes with some contribution of the three Ca planes. The phonon frequency drops by nearly 60 cm^{-1} below T_c while the FWHM increases from 10 to 90 cm^{-1} immediately below T_c (as $2\Delta_0(T)$ crosses ω_p) and falls down to 20 cm^{-1} at lower T's. Maybe the most interesting effect is the growth of the integrated Raman intensity (I_p in Fig. 27).

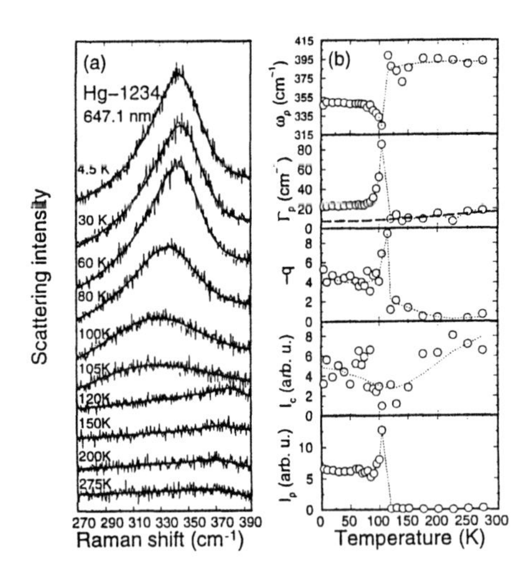

Figure 27: Raman spectra covering the 390 cm^{-1} A_{1g} mode measured in $x'x'$ polarization with the 647.1 nm laser line at different temperatures, and the corresponding fits with Fano profiles (a). The fitted frequencies, linewidths Γ_p (HWHM), lineshapes parameters q and the phonon intensities are represented in (b) by open circles. Smooth dotted lines joining the experimental points are given as a guide to the eye. The dashed line in the linewidth panel represents a fit to the widths found above T_c with the function of Eq. (25), taking $\omega_p = 390$ cm^{-1}. This fit yields $\Gamma_p(T=0) = 7.3$ cm^{-1}.

In most of the conventional superconductors the superconductivity results from a coherent state involving all electrons held together by the electron-phonon interaction. This state is composed of pairs of electrons (the so-called Cooper[n] pairs) with opposite spin and opposite $\boldsymbol{k}$-vectors held together by the exchange of phonons. The dimensionless effective electron-phonon interaction constant λ is less than 0.5 for most conventional superconductors [63]. From the data of Fig. 27 it is possible to estimate a value of $\lambda \simeq 0.1$ for the 375 cm^{-1} $\boldsymbol{k} = 0$ phonon. Hg-1234 has 20 atoms per unit cell which correspond to 60 vibrational modes. Hence we can estimate a total value of $\lambda \simeq 0.1 \times 60 = 6$ as the dimensionless electron-phonon coupling constant that one would have *if all phonons would couple to electrons with the same strength.* This is a huge coupling constant which makes it hard to believe that phonons play no role whatsoever, neither positive nor negative, in the phenomenon of high T_c superconductivity.

ACKNOWLEDGMENTS

The author is indebted to a multitude of colleagues and coworkers with whom he has interacted during his 30 years adventure in the field of Raman spectroscopy. They are too numerous to be mentioned; many appear in the by far not exhaustive literature references. Thanks are also due to Sabine Birtel for her skillful help in the preparation of the manuscript. Financial support from INTAS (a European agency for the support of scientific collaboration with Eastern Europe), the German Fonds der Chemischen Industrie and a Max-Planck research award is greatfully acknowledged.

[n] L.N. Cooper shared the Physics Nobel Prize with his coauthors J. Bardeen and J.R. Schrieffer in 1972. The prize was awarded for their development of the BCS theory of superconductivity.

REFERENCES

1. A. Smekal, Naturwiss. **11**, 873 (1923).
2. C.V. Raman, Indian J. Phys., 387 (1928).
3. J.E. Smith, Jr., M.H. Brodsky, B.L. Crowder, M.I. Nathan, A. Pinczuk, Phys. Rev. Lett. **26**, 642 (1971).
4. M. Cardona, J. of Molecular Structure **141**, 93 (1986).
5. H.L. Welsh, M.F. Crawford, T.R. Thomas, and G.R. Love, Canad. J. Phys. **30**, 577 (1952). B. Stoicheff, Canad. J. Phys. **32**, 330 (1954). These papers desscribe the so-called Toronto Arc, the strongest light source for Raman spectroscopy used before the advent of the laser.
6. H. Kogelnik and S.P.S. Porto, JOSA **53**, 1446 (1963). In this paper the use of the He-Ne laser was introduced.
7. J.C. Tsay, in Ref. [10], Vol. IV, p. 232.
8. E. Burstein, A. Pinczuk, and D.L. Mills, Surface Science **98**, 451 (1980).
9. G. Abstreiter, M. Cardona, and A. Pinczuk, in Ref. [10], Vol. IV, p. 80.
10. *Light Scattering in Solids*, Vols. I – VII (1975 – 1988), ed. by M. Cardona and G. Güntherodt, Springer Verlag, New York, Heidelberg.
11. *Scattering of Light by Crystals*, by W. Hayes and R. Loudon, Wiley-Interscience, (1986).
12. *The Raman Effect*, Vol. I and II, ed. by A. Anderson, M. Dekker, New York (1971).
13. *Light Scattering in Magnetic Solids*, by M.G. Cottam and D.J. Lockwood, Wiley-Interscience, (1986).
14. *Practical Raman Spectroscopy*, by J.J. Gardner and P.R. Graves, Springer, Heidelberg, 1989.
15. V.I. Belitsky, M. Cardona, I.G. Lang, S.T. Pavlov, Phys. Rev. B **46**, 15767 (1992).
16. A. Cantarero, C. Trallero-Giner, and M. Cardona, Phys. Rev. B **39**, 8388 (1989).
17. V.I. Belitsky, A. Cantarero, M. Cardona, C. Trallero-Giner, and S.T. Pavlov, J. Phys.: Condens. Matter **9**, 5965 (1997).
18. C. Masciovecchio, G. Ruocco, F. Sette, P. Benassi, A. Cunsolo, M. Krisch, V. Mazzacurati, A. Mermet, G. Monaco, and R. Verbeni, Phys. Rev. B **55**, 8049 (1997).
19. M. Cardona, in Ref. [10], Vol. II.
20. F. Cerdeira, E. Anastassakis, W. Kauschke, and M. Cardona, Phys. Rev. Lett. **57**, 3209 (1986).
21. *Fundamentals of Semiconductors*, by P.Y. Yu and M. Cardona, Springer, Heidelberg, 1996.

22. M.I. Alonso and M. Cardona, Phys. Rev. B **33**, 10107 (1988).

23. *Modulation Spectroscopy*, by M. Cardona, Academic, New York, 1969.

24. M. Cardona, Surface Science **37**, 100 (1973).

25. A. Cantarero, C. Trallero-Giner, and M. Cardona, Solid State Commun. **69**, 1183 (1989).

26. M. Cardona and S. Gopalan, in *Progress in Electron Properties of Solids (Festschrift in Honor of Franco Bassani)*, ed. by R. Girlanda *et al.* (Kluwer Academic Publishers, 1989), p. 51.

27. M. Cardona, Festkörperprobleme/Advances in Solid State Physics **34**, ed. by R. Helbig (Vieweg, Braunschweig/Wiesbaden, 1994), p. 35.

28. A. García-Cristóbal, A. Cantarero, C. Trallero-Giner, and M. Cardona, Phys. Rev. B **49**, 13430 (1994).

29. C. Thomsen, in Ref. [10], Vol. VI (1991) and references therein.

30. R. Liu, C. Thomsen, W. Kress, M. Cardona, B. Gegenheimer, F.W. de Wette, J. Prade, A.D. Kulkarni, and U. Schröder, Phys.Rev. B **37**, 7971 (1988).

31. R. Henn, T. Strach, E. Schönherr, and M. Cardona, Phys. Rev. B **55**, 3285 (1997).

32. E.T. Heyen, S.N. Rashkeev, I.I. Mazin, O.K. Andersen, R. Liu, M. Cardona, and O. Jepsen, Phys. Rev. Lett. **65**, 3048 (1990).

33. B. Lederle, D.Sc. Thesis, University of Stuttgart, 1997.

34. T. Strach, J. Brunen, B. Lederle, J. Zegenhagen, and M. Cardona, Phys. Rev. B, in press.

35. For a review see: R. Merlin, Solid State Commun. **102**, 207 (1997).

36. W. Albrecht, T. Kruse, and H. Kurz, Phys. Rev. Lett. **69**, 1451 (1992).

37. Xingjiang Zhou, M. Cardona, D. Colson, and V. Viallet, phys. stat. sol. (b) **199**, R7 (1997).

38. V. G. Hadjiev, X.J. Zhou, T. Strohm, M. Cardona, Q.M. Lin, and C.W. Chu, Phys. Rev. B , submitted.

39. J. Menéndez and M. Cardona, Phys. Rev. B **29**, 2051 (1984).

40. A. Debernardi, S. Baroni, and E. Molinari, Phys. Rev. Lett. **75**, 1819 (1995).

41. H. Herchen and M.A. Capelli, Phys. Rev. B **43**, 11740 (1991); E.S. Zouboulis and M. Grimsditch, Phys. Rev. B **43**, 12492 (1991); note: we have not been able to reproduce the "Klemens" curve in the second reference. The calculated curve in Fig. 6 of the first reference seems to have been obtained by following with a spline each one of the points in Fig. 8 of Ref. [42]. We believe that this is unreasonable. The smooth curve of our Fig. 17 makes more sense to us.

42. C.Z. Wang, C.T. Chan, and K.M. Ho, Phys. Rev. B **42**, 11276 (1990).

43. A. Göbel, T. Ruf, Cheng-Tian Lin, M. Cardona, J.-C. Merle, and M. Joucla, Phys. Rev. B **56**, 210 (1997).

44. G.J. Piermarini, S. Block, J.O. Barnett, and R.A. Forman, J. Appl. Phys. **46**, 2774 (1975).

45. B. Weinstein and G.J. Piermarini, Phys. Rev. B **12**, 1172 (1975).

46. E. Anastassakis and M. Cardona, in *High Pressure Semiconductor Physics* Volume of *Semiconductors and Semimetals*, ed. by T. Suiski and W. Paul (Academic Press, San Diego, 1998).

47. M. Cardona, phys. stat. sol. (b) **185**, 5 (1996).

48. G.H. Loechelt, N.G. Cave, and J. Menéndez, Appl. Phys. Lett. **66**, 3639 (1995).

49. C. Ulrich, E. Anastassakis, K. Syassen, A. Debernardi, and M. Cardona, Phys. Rev. Lett. **78**, 1283 (1997).

50. K.C. Hass, M.A. Tamor, T.R. Anthony, and W.F. Banholzer, Phys. Rev. B **44**, 1246 (1991); J. Spitzer, P. Etchegoin, M. Cardona, T.R. Anthony, and W.F. Banholzer, Solid State Commun. **88**, 509 (1993).

51. H.D. Fuchs, C.H. Grein, R.I. Devlen, J. Kuhl, and M. Cardona, Phys. Rev. B **44**, 8633 (1991).

52. J.M. Zhang, M. Giehler, A. Göbel, T. Ruf, M. Cardona, E.E. Haller, and K. Itho, Phys. Rev. B, in press.

53. D.T. Wang, A. Göbel, J. Zegenhagen, and M. Cardona, Phys. Rev. B, in press

54. M. Cardona and I.P. Ipatova, in *Elementary Excitations in Solids*, ed. by J.L. Birman, C. Sébenne and R.F. Wallis (Elsevier Science Publ., Amsterdam, 1992), p. 237.

55. T. Strohm and M. Cardona, Phys. Rev. B **55**, 12725 (1997).

56. K.B. Lyons, S.H. Liou, M. Hong, H.S. Chen, I. Kwo, and T.J. Negran, Phys. Rev. B **36**, 5592 (1987).

57. T. Devereaux and D. Einzel, Phys. Rev. B **50**, 10287 (1994); ibid Phys. Rev. B **54**, 15547 (1996).

58. F. Cerdeira and M. Cardona, Phys. Rev. B **5**, 1140 (1972).

59. L. Pintschovius, J.A. Vergés, and M. Cardona, Phys. Rev. B **26**, 5658 (1982).

60. R. Zeyher and G. Zwicknagl, Solid State Commun. **66**, 617 (1988).

61. E.T. Heyen, M. Cardona, J. Karpinski, E. Kaldis, and S. Rusiecki, Phys. Rev. B **43**, 12958 (1991).

62. B. Friedl, C. Thomsen, H.-U. Habermeier, and M. Cardona, Solid State Commun. **81**, 989 (1992).

63. P.B. Allen and B. Mitrović, in *Theory of Superconducting T_c*, Solid State Physics, Vol. 37, ed. by H. Ehrenreich and D. Turnbull (Academic Press, 1982).

SURFACE PHONONS AND THEIR ROLE IN ULTRAFAST PHENOMENA

Giorgio Benedek

Istituto Nazionale per la Fisica della Materia
Dipartimento di Scienza dei Materiali dell'Universitá
Via Emanueli 15, I-20126 Milano, Italy

ABSTRACT

Many ultrafast optical phenomena studied with pump-and-probe techniques occur on or below the time scale of microscopic vibrations and can therefore involve or even generate high-frequency phonons. Since most semiconductor structures currently studied with ultrafast optical methods are obtained with planar technologies and retain a quasi-2D character, surface phonons are actually to be considered. In the first two lectures the basic theory and fundamental aspects of surface phonons are introduced, with a summary description of the surface phonon dispersion curves for some relevant semiconductor crystal surfaces. In the third lecture some recent pump-and-probe investigations of ultrafast phenomena involving surface phonons on semiconductor as well as metal surfaces are briefly discussed.

1. THE NATURE OF SURFACE PHONONS

More than a century ago Lord Rayleigh demonstrated mathematically [1] that the long-wave component of earthquakes is due to elastic waves travelling along the surface. The displacement amplitudes of these waves are localized in the surface region of a semiinfinite elastic continuum. They are polarized in a plane normal to the surface along the propagation direction called the sagittal plane as shown in Fig. 1 [2]. Rayleigh's discovery is commonly taken as the starting point of surface phonon physics. Soon afterwards the theory of Rayleigh waves and other types of surface waves in semiinfinite elastic continua was extended beyond the realm of geophysics. Solid state physicists and materials scientists became interested because of the potential of surface acoustic waves for applications in delay lines, optoacoustic systems and signal processing devices [3]. The theoretical research, initiated by Rayleigh's fundamental work, was soon extended to nanometric wavelengths where the discrete lattice structure of solids becomes effective and determines the dispersion relations.

Ultrafast Dynamics of Quantum Systems: Physical Processes and Spectroscopic Techniques, Edited by Di Bartolo and Gambarota, Plenum Press, New York, 1998

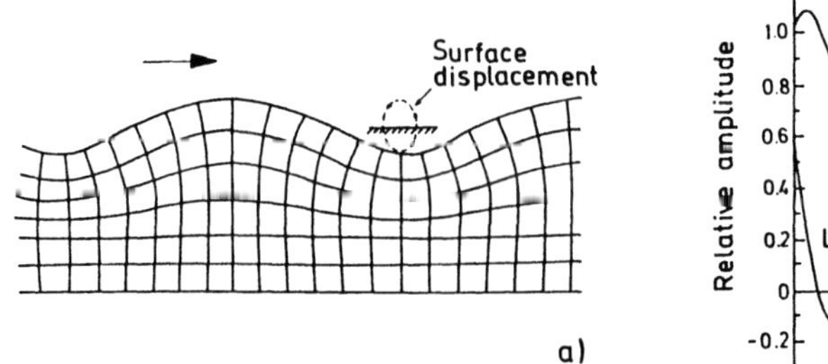

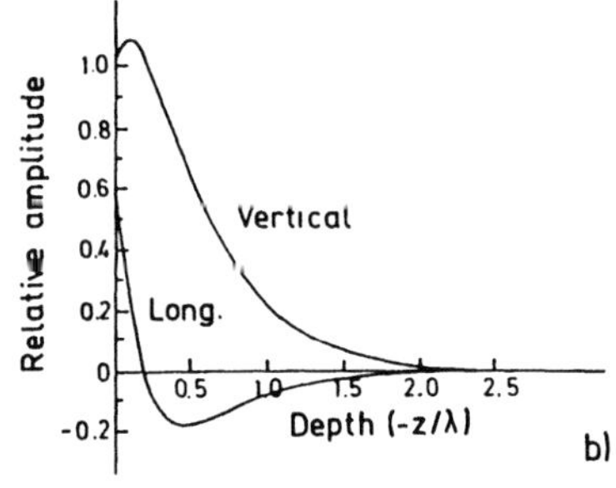

Fig. 1: (a) Schematic diagram showing the displacement pattern of a Rayleigh wave (RW) in the vicinity of the surface of a semiinfinite homogeneous elastic solid in the saggital plane (from G.W. Farnell [2]). The displacement field has a dominant component normal to the surface. The smaller longitudinal component has a $\pi/2$ phase shift with respect to the normal one which gives an elliptical polarization in the sagittal plane. (b) Both components decay exponentially inside the solid with a decay length proportional to the RW wavelength.

1.1 The Born-van Kármán approach

A phenomenological microscopic treatment of surface lattice dynamics is provided by the early Born-von Kàrmàn theory [4,5,6]. In this approach the equations of motion for the harmonic vibrations of any array of interacting atoms are derived from a phenomenological expression for the total potential energy

$$\phi = \frac{1}{2}\sum_{ij} \mathrm{v}(\mathrm{r}_{ij}) + \frac{1}{3!}\sum_{ijk} \mathrm{w}(\cos\vartheta_{ijk}), \tag{1}$$

where v is a two-body radial potential between atom pairs and w is a three-body term depending on the angle ϑ_{ijk} formed by the vectors $\mathbf{r}_{ij} = \mathbf{r}_i - \mathbf{r}_j$ and $\mathbf{r}_{kj} = \mathbf{r}_k - \mathbf{r}_j$ connecting atoms i and k to atom j [7] (Fig. 2). Three-body angle-bending potentials provide perhaps the simplest model for those many-body forces between atoms which are mediated by the surrounding valence electrons. Ab-initio theories based on the density functional approach in the local density approximation provide a detailed microscopic description of the internal electronic degrees of freedom and the effective electron-mediated interactions between atoms involved in surface dynamics [8]. In the Born-von Kàrmàn theory the equations of motion for the harmonic vibrations of a monoatomic lattice are then given by

$$- M_i \ddot{u}_\alpha(i) = \Sigma_{j\beta} R_{\alpha\beta}(i,j) u_\beta(j), \tag{2}$$

where the $u_\alpha(i)$ are the Cartesian components of the displacement $\mathbf{u}(i)$ from the equilibrium position $\mathbf{r}(i)$ of atom i, M_i is the atomic mass and

$$R_{\alpha\beta}(i,j) = \frac{\partial^2 \phi}{\partial r_{i\alpha} \partial r_{j\beta}} \tag{3}$$

is the atom-atom force constant matrix. When used as indices α,β denote Cartesian coordinates. For a two-body central interatomic potentials $v(\mathbf{r}_{ij})$, the force constant matrix can be expressed in terms of derivatives of the potential as

$$R_{\alpha\beta}(i,j) = \left(r_\alpha r_\beta / r^2\right)\beta_{ij} + \left(\delta_{\alpha\beta} - r_\alpha r_\beta / r^2\right)\alpha_{ij} \quad , \tag{4}$$

where $\mathbf{r} \equiv \mathbf{r}_{ij}$ and

$$\beta_{ij} \equiv \frac{\partial^2 v}{\partial r_{ij}^2}, \qquad \alpha_{ij} \equiv \frac{1}{r_{ij}} \frac{\partial v}{\partial r_{ij}}, \tag{5}$$

are the two-body radial and tangential force constants, respectively. Similarly the derivatives of the three-body potential provide a set of angle-bending force constants. This Born-von Kármán procedure has also been the basis of the preliminary analysis of most of the bulk phonon dispersion curves measured by neutrons [9].

We consider a three-dimensional (3D) lattice of basis vectors $(\mathbf{a}_1, \mathbf{a}_2, \mathbf{a}_3)$, with $N_1 \times N_2 \times N_3$ unit cells each of which contains s atoms of masses M_κ ($\kappa = 1,2, \ldots s$). The periodicity in three dimensions is established through the cyclic boundary conditions which require that the translations $N_1\mathbf{a}_1$, $N_2\mathbf{a}_2$, $N_3\mathbf{a}_3$ (or any combination thereof) brings any atom into itself. The same lattice, however, can be generated by two-dimensional (2D) translations in the $(\mathbf{a}_1, \mathbf{a}_2)$ plane of a larger unit cell, each containing sN_3 atoms. A lattice generated in this way is called a *slab*.

To solve the dynamical equations, Eq. (2), solutions are chosen in the form of Bloch waves with 2D wavevectors $\mathbf{Q}$ in the $(\mathbf{a}_1, \mathbf{a}_2)$ plane and angular frequencies $\omega_{\mathbf{Q}\nu}$ as

$$u_\alpha(i) \propto u_\alpha(l_3\kappa, \mathbf{Q}\nu) \exp\left[\mathrm{i}\mathbf{Q}\cdot(l_1\mathbf{a}_1 + l_2\mathbf{a}_2) - \mathrm{i}\omega_{\mathbf{Q}\nu}t\right] \quad , \tag{6}$$

where the atomic index i is now replaced either by the set of integers (l_1, l_2, l_3) and κ required to specify the unit cells in three dimensions and atoms inside a cell, respectively. Alternatively, the sets (l_1, l_2) and $(l_3\kappa)$ can be used to label the large cells in two dimensions and the atoms inside each large cell, respectively.

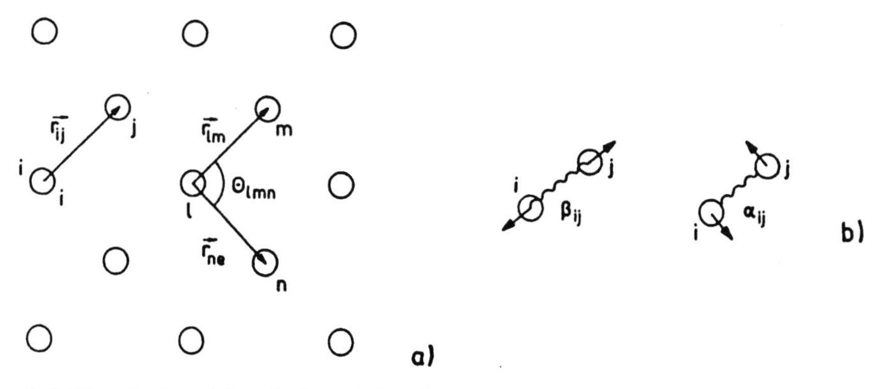

Fig. 2: (a) Two-body and three-body angle bending interactions between atoms in a lattice; (b) radial and tangential components of the interatomic two-body force constants.

Inserting Eq. (6) into Eq. (2) leads to a coupled set of equations for the slab which can be written as the following eigenvalue problem

$$M_\kappa \omega^2_{Q\nu} u_\alpha(l_3\kappa, Q\nu) = \sum_{\beta l_3'\kappa'} R_{\alpha\beta}(l_3\kappa, l_3'\kappa', Q)\, u_\beta(l_3'\kappa', Q\nu) \quad , \tag{7}$$

where $l_3 = 1,2,\ldots N_3$ labels the N_3 atomic layers forming the slab and $R_{\alpha\beta}(l_3\kappa, l_3'\kappa_3', \mathbf{Q})$ is the 2D Fourier transform of the force constant matrix. The index ν labels the $3sN_3$ vibrational modes corresponding to each wavevector $\mathbf{Q}$. The set of modes of given ν and all $\mathbf{Q}$ values forms the ν-th phonon branch and the function $\omega_{\mathbf{Q}\nu}$ is its dispersion relation. From the solution of Eq. (7) the eigenfrequencies $\omega_{\mathbf{Q}\nu}$ and the mass-weighted eigenvectors $M_\kappa^{1/2} u_\alpha(l_3\kappa,\mathbf{Q}\nu)$ are obtained. The eigenvectors $u_\alpha(l_3\kappa,\mathbf{Q}\nu)$ are normalized to unity and determine the polarization of the displacement field. The quantized physical displacements of the sN_3 atoms in the (l_1,l_2)-th big unit cell for the $(\mathbf{Q},\nu)$-th mode is thus given by

$$u_\alpha(l_1,l_2,l_3,\kappa;\mathbf{Q}\nu,t) = (\hbar / 2N_1N_2\omega_{\mathbf{Q}\nu})^{1/2} u_\alpha(l_3k,\mathbf{Q}\nu)\exp\left[i\mathbf{Q}\cdot(l_1\mathbf{a}_1 + l_2\mathbf{a}_2) - i\omega_{\mathbf{Q}\nu}t\right] \quad . \tag{8}$$

Pairs of such eigenwaves with different $(\mathbf{Q},\nu)$-labels are mutually orthogonal (normal modes). As long as the cyclic boundary condition is maintained also for the third translation $N_3\mathbf{a}_3$ the above solutions are just those of the 3D lattice which have been relabelled for a 2D representation. The eigenfrequencies $\omega_{\mathbf{Q}\nu}$, for each $\mathbf{Q}$ and for ν varying from 1 to $3sN_3$, form now 3s bands, each one containing N_3 modes. For $N_3 \to \infty$ the individual modes can no longer be resolved and the bands become continuous. They are referred to as bulk phonon bands projected onto the ($\mathbf{a}_1$, $\mathbf{a}_2$) surface.

The correspondence between three- and two-dimensional representations of the phonon branches in a solid with 3D cyclic boundary conditions is illustrated in Fig. 3, where the dispersion surface of an acoustic phonon mode in the 3D representation is converted into the typical "spaghetti" band of the 2D representation (*surface-projected* bulk band). The 3D dispersion surface represents the phonon energy $\hbar\omega$ as function of the 3D wavevector $\mathbf{q}$, which is split into the 2D component $\mathbf{Q}$ parallel to the slab and the normal component, $\mathbf{q} = (\mathbf{Q}, q_z)$. Each line of the 2D band represents the phonon energy as function of $\mathbf{Q}$ for a given value of q_z, and there are as many lines as the number of possible values of q_z , that is N_3. If there are s atoms in the unit cell, each atom having three degrees of freedom, the phonon spectrum is formed by $3s$ different, partially superimposed bands, i.e., by $3sN_3$ lines altogether, which are conveniently labelled by the single index ν.

The frequency distribution within the bands is given by the $\mathbf{Q}$-selected phonon density of states (DOS)

$$D(\mathbf{Q},\omega) = \frac{1}{N_3}\sum_\nu \delta(\omega - \omega_{\mathbf{Q}\nu}) . \tag{9}$$

The above 2D representation provides a convenient basis for the calculation of the vibrations at the surface of a crystal lattice. Let us take the macroscopic lattice with three-dimensional cyclic boundary conditions and cut all physical interactions across the interface between two adjacent layers, e.g., the layers at $l_3 = 1$ and N_3. In this way the third boundary condition is removed and two free surfaces are created at $l_3 = 1$ and N_3: the original cyclic lattice has been transformed into a slab. This operation determines a perturbation of the force constant matrix and a consequent change of the eigenfrequencies and eigenvectors of Eq. (7).

The eigenwaves corresponding to the perturbed dynamics can be obtained by either a Green's function (GF) method [10-23] or through a variational method based on trial eigenvectors [24-28] or by the direct diagonalization method [29-40]. The first two analytical methods, which require a reduced computational effort but a certain amount of algebraic work, are nowadays less popular than the direct diagonalization method (slab method), which takes advantage of the availability of fast computers.

In the slab method the dynamics of a crystal slab consisting of N_3 parallel atomic layers is solved by letting the number of layers N_3 increase until the spectral features of the surface layers are clearly distinguished from those of the bulk. In practice N_3 is bounded by the size of the force constant matrix which can be diagonalized with present computational methods. Typically N_3 lies somewheres between 10^1 and 10^2 [29,41]; however by using algebraic codes to set up the force constant matrix it has been possible to go to larger numbers of layers. This has been especially useful for calculating stepped surfaces [42], where, because of the substantially larger unit cells, N_3 must be increased correspondingly.

Figure 4 shows the calculated dispersion curves for an ideal two dimensional slab of a single layer of densely packed atoms of mass corresponding to that of nickel, arranged according to the close packed (111) surface of an fcc lattice [43]. The two-dimensional

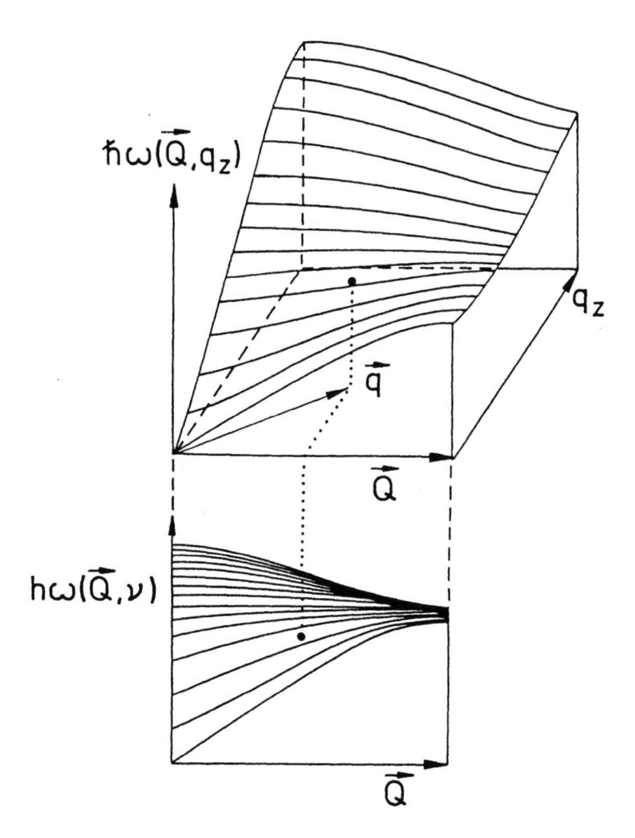

Fig. 3. The energy surface $\hbar\omega$ ($\mathbf{Q}$,q_z) generated by the dispersion curves of an acoustic phonon as function of parallel ($\mathbf{Q}$) and normal (q_z) wavevector components is converted into a two-dimensional (surface-projected) band $\hbar\omega$ ($\mathbf{Q}$,ν). The normal component q_z is replaced by an index ν labelling all modes of the band having the same parallel wavevector $\mathbf{Q}$. In this example $N_3 = 15$.

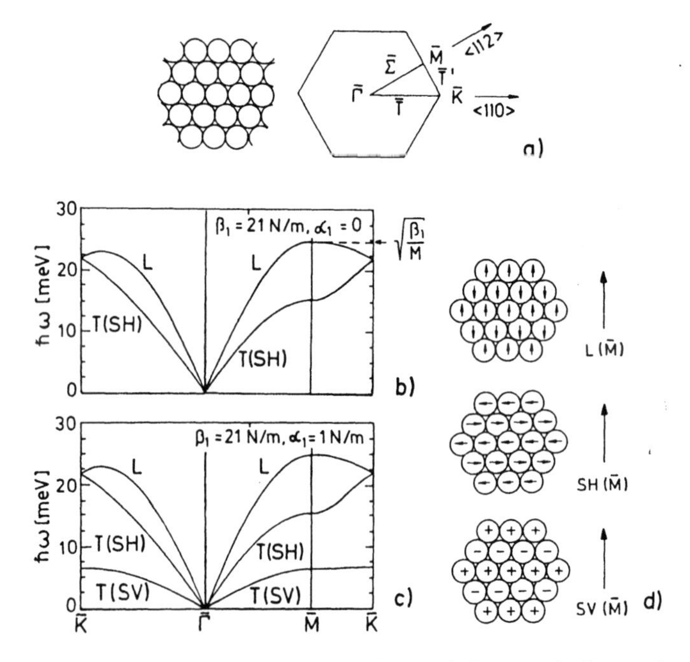

Fig. 4: Calculated dispersion curves for a single (111) atomic layer equivalent to the surface layer of an fcc crystal [43]. Part (a) shows the surface geometry of the closely packed hexagonal structure of atoms and the first Brillouin zone as well as the irreducible part of the zone boundary. Part (b) shows the dispersion curve for nearest neighbors interactions with only radial forces β_1. Part (c) shows the effect of adding the nearest neighbor tangential force constant α_1. L denotes longitudinal, T (SV) transverse shear vertical (out-of-plane) and T(SH) transverse shear horizontal (in-plane) polarizations. The displacement patterns for the three modes at the $\overline{M}$point are shown in (d).

Brillouin zone and the labelling of the symmetry directions is given in Fig. 4(a). The dispersion curves in Fig. 4(b) were calculated for a nearest neighbor (nn) pairwise interaction characterized by a single radial force constant β_1. A layer with no stress is simulated by setting the nn tangential force constant α_1=0. In this case only two non-zero modes with in-plane longitudinal (L) polarization and an in-plane transverse (T) polarization commonly referred to as a shear horizontal (SH) mode are obtained. With a finite value of α_1 also the third mode is non-zero and has a transverse polarization out of the plane (shear vertical (SV)) while the other modes are only weakly affected, as shown in Fig. 4(c). The displacement patterns for the three modes at the zone boundary point $\overline{M}$ are also shown in Fig. 4(a). This simple model the importance of the tangential force constant for out-of-plane vibrations.

The slab method provides direct insight into how some of the normal modes of a thin film gradually evolve into those of the bulk and others into surface modes as the thickness of the film increases layer by layer from one layer to macroscopic dimensions This is illustrated in Fig. 5(a) for the case in which the same two force constants β_1 and α_1 are used throughout, not only within the layers but between the layers [44]. The addition of a second layer leads to the appearance of three new modes which have dispersion curves which are qualitatively similar to those of the single layer except that they have finite frequencies at the zone origin $\overline{\Gamma}$. They can be attributed to vibrations of the two planes with respect to each other. As additional layers are added the number of modes increases accordingly. The zone origin modes evolve into a collective motion involving standing waves along a direction normal to the planes [45]. Beyond about 10 layers a new structure in the distribution of the dispersion curves becomes apparent with some modes well separated from dense bands. In the limit of $N_3 = \infty$ the former are the surface modes, while the bands are attributed to bulk modes projected onto the surface.

This behaviour can also be understood as resulting from the perturbation of the force constant matrix in the first few atomic layers of the surface region resulting from the removal of the layers above the surface layer. This perturbation leads to a peeling off of a few frequencies from either the lower or the upper edges of the bulk bands for each value of **Q** leading to surface phonon branches. According to Rayleigh's theorem [13] the number of localized frequencies from each band is actually given by the number of degrees of freedom which are affected by the surface perturbation. Most surface branches are actually below the respective bulk bands because the reduced coordination of the surface atoms yields a softening

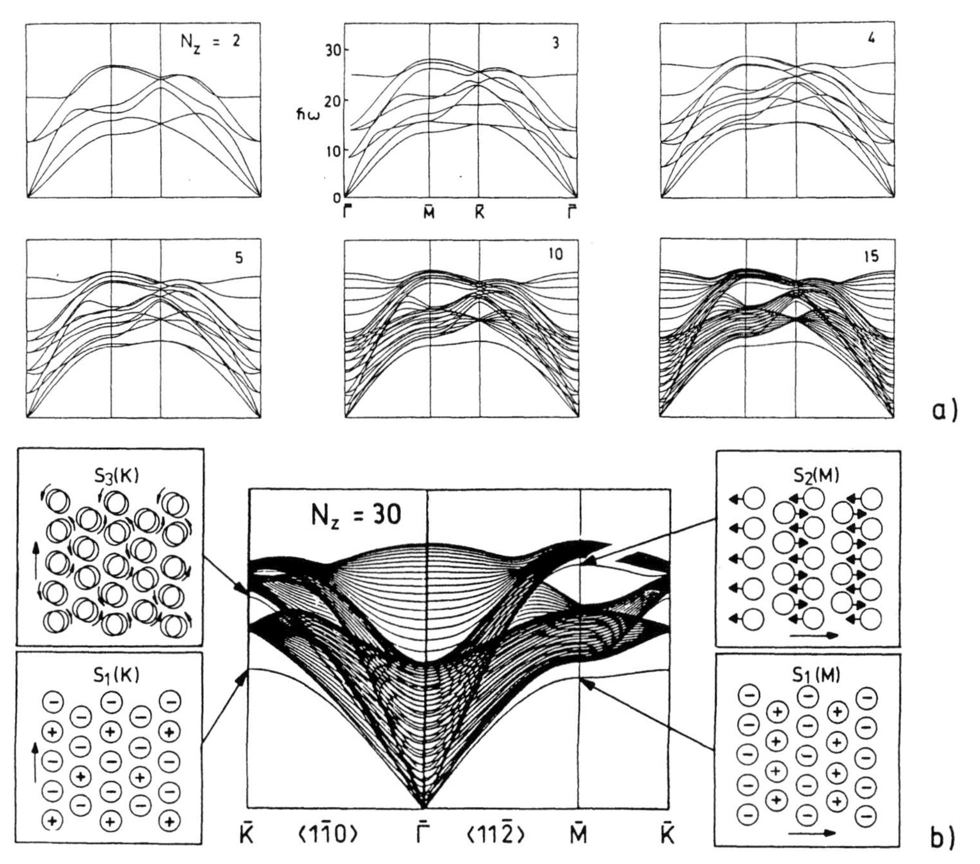

Fig. 5: Part (a) shows the evolution to the surface phonon dispersion curves of a fcc(111) slab as additional layers are added to the single layer shown in Fig. 1(c) [44]. Part (b) shows the dispersion curves for $N_z = 30$ where the behaviour for an infinitely thick slab is approached and the distinction between the isolated surface modes and the closely spaced bulk bands becomes apparent. For this case the displacement patterns of the surface atoms (top view) for the surface modes at the zone-boundary symmetry points are shown in the insets to the left and right. Small arrows indicate displacements parallel to the surface which can be either linearly ($S_2(\overline{M})$) or circularly ($S_3(\overline{K})$) polarized; ± indicate displacements perpendicular to the surface. The long arrow in each set indicates the direction of the wavevector.

of the phonon frequencies. In exceptional cases, however, the elastic relaxation of the surface can cause a stiffening of certain modes.

In this case the new modes appear above the respective bulk bands. For an ideal surface in which the surface atoms retain the same positions which they had in the bulk no more than one surface localized phonon branch is expected for each band per surface. Since in the slab model of a surface there are in fact usually two identical surfaces, the localized frequencies occur as pairs of nearly degenerate branches with a splitting which disappears with increasing thickness of the slab.

In this context the term localization has two different meanings. In the spectral sense, isolated δ-functions appear outside the continuous DOS of the bulk bands. In the spatial sense, surface phonons have a displacement field which decays exponentially with increasing distance from the surface, in the same fashion as Rayleigh waves do (Fig. 1). The spectral localization of the eigenfrequencies is related to the spatial localization of the eigenvectors in the surface region: the larger the separation of the surface phonon frequency away from the bulk band edge, the greater the localization of the corresponding displacement field at the surface of the crystal. Thus when the penetration is much less than the slab thickness the pair of modes arising at the two surfaces of the slab are practically degenerate. In a slab of infinite thickness the two surfaces are independent of each other and the surface mode frequencies are equal, which is equivalent to replacing the thick slab with a semiinfinite crystal with a single surface.

1.2 Classification of surface modes

The polarization of each surface branch reflects approximately that of the parent bulk band from which it originates, but also depends on the symmetry of the propagation direction on the surface. We shall refer to the *sagittal plane*, defined by the wavevector direction and the normal to the surface, and consider first the case when the sagittal plane coincides with (or is parallel to) a mirror-symmetry plane of the crystal. This case normally (but not always) occurs for a propagation in the surface symmetry directions. As illustrated in Fig. 6(a), in both the high symmetry directions of the fcc(100) surface the sagittal plane is a mirror-symmetry plane, whereas on the fcc (111) surface [Fig. 6(b)] this is true for only the <11$\bar{2}$> direction. For such mirror-symmetry directions the displacement fields of the surface modes have either even or odd symmetry with respect to the sagittal plane: the former are fully contained in the sagittal plane and are called *sagittal* ($\perp$), the latter are perpendicular and are called *shear horizontal* (SH). We note, however, that the plane of the surface is not a mirror-symmetry plane. This has a general consequence, common to all waves propagating along a surface: the shear vertical (SV) and longitudinal (L) components of a displacement in the sagittal plane are coupled together and out-of-phase with respect to each other, leading to an *elliptical* polarization. Depending on the dominant component of the sagittal polarization the two modes are designated as quasi-transverse (~SV) and quasi-longitudinal (~L).

Figure 7 shows the displacement fields for the three surface acoustic modes (a) and their location in the spectrum (b): the RW (which is ~SV) and the ~L modes have retrograde elliptical polarization in the sagittal plane, whereas SH is linearly polarized in the surface plane. It is important to note, however, that at those high symmetry points in the reciprocal space, either at the origin or the boundaries of the Brillouin zone, where the sagittal modes have a vanishing group velocity an exactly linear polarization is recovered, which is either SV or L. The polarization of surface modes becomes more complicated if the sagittal plane is not parallel to any mirror-symmetry plane, as occurs, e.g., for the <1$\underline{1}$0> direction of the fcc(111) surface [Fig. 6(b)] or other non-symmetry directions. In this case the polaritation is a mixture

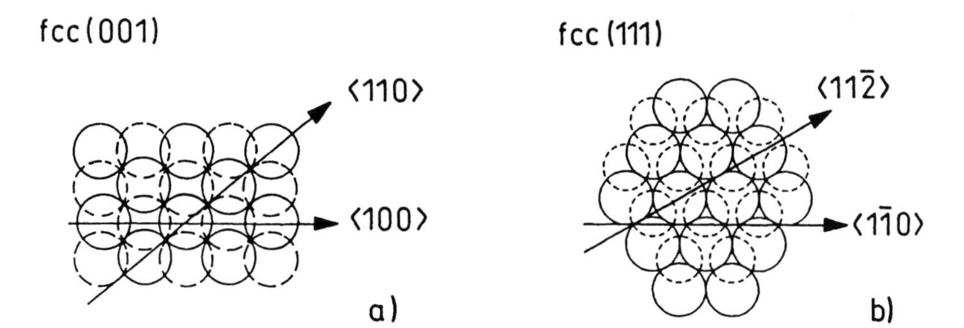

Fig. 6: The (001) and (111) surfaces of a face-centered cubic (fcc) crystals showing the atoms in the first (full circles) and second layer (broken circles) and the symmetry directions. For the (001) surface the sagittal plane, defined by the surface wavevector and normal to the surface, is a mirror plane along either the <110> or the <100> directions; for the (111) surface only the sagittal plane along <112> is a mirror plane, whereas along <110> it is not. The argument for fcc(111) also holds for the (0001) surface of a hexagonal monoatomic crystal. Surface phonons propagating in a mirror-symmetry plane have either pure sagittal or shear horizontal polarization.

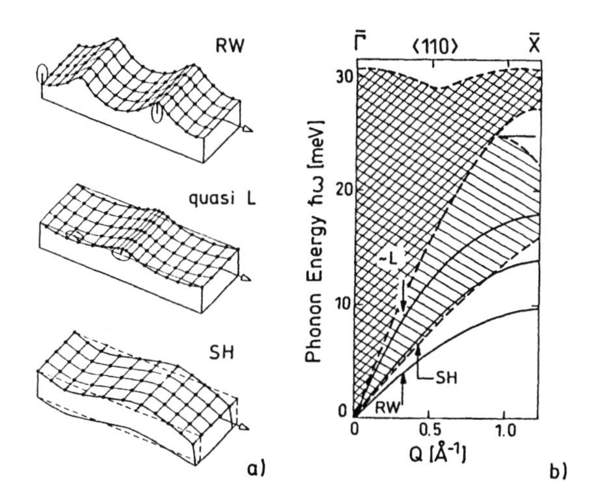

Fig. 7 (a): The three polarizations of surface modes in the topmost layer of a monoatomic crystal along a high symmetry direction (e.g., the <100> direction for the fcc(001) surface). Both the Rayleigh waves (RW), which are the quasi-shear vertical (~SV) and the quasi-longitudinal (~L) modes have an elliptical polarization in the sagittal plane with the major axis either normal or parallel to the surface, respectively. The shear-horizontal (SH) modes have a linear, purely transverse polarization in the surface plane. (b) The two sagittal modes have a wavelength equal to eight times the interatomic distance and are therefore located at one fourth of the SBZ, whereas the SH mode, with a wavelength six times the interatomic distance, falls at one third of the SBZ. At the zone boundary the three modes acquire a pure SV, L or SH linear polarization.

of sagittal and SH components, though the modes can still be designated from their dominant component as ~ SV, ~L and ~SH.

Since the bulk bands partially overlap each other, a surface mode originating from a given band may overlap with another band and as a result of hybridization may not be localized. In general the following three different situations are encountered:

(1) Localized surface modes have frequencies for a given wavevector which lie outside the spectrum of all the bulk modes with the same wavevector. As a result no mixing (hybridization) can occur their displacement fields must decay exponentially inside the crystal. They have δ-functions in the phonon DOS.

(2) Surface resonances have frequencies for a given wavevector which lie inside a band of bulk modes and a polarization which is not orthogonal to that of the bulk modes. As a result of mixing their displacement field is large in the surface region but does not decay to zero far inside the crystal, tending in an oscillatory fashion to some bulk mode of equal frequency and polarization. The resonance peaks in the phonon density of states have a Lorentzian profile, with a finite width proportional to the bulk DOS calculated at the resonance frequency.

(3) Pseudo-surface modes have frequencies for a given wavevector which lie inside a band of bulk modes of orthogonal polarization. Due to orthogonality no mixing occurs: these modes are still localized and decay exponentially inside the crystal, but any infinitesimal rotation of the propagation direction away from the symmetry direction leads to a mixing with the bulk bands and transforms these modes into resonances.

These different modes are illustrated by the phonon dispersion curves calculated for the surface of a diatomic crystal, NaF(001). The arrangement of Na and F ions in the (001) surface is shown in Fig. 8 together with one eigth of the fcc BZ of the three dimensional NaF crystal. The shaded triangle $\overline{\Gamma M X}$ in the xy plane is also one eight of the surface BZ. The phonon dispersion curves as calculated by the Green's function (GF) method [46,47] are displayed for both the sagittal and shear horizontal modes in Fig. 9 along the symmetry directions $\overline{\Gamma M}$ (<100>) $\overline{\Gamma X}$ (<110>) and along the line in reciprocal space connecting $\overline{X}$ with $\overline{M}$ which represents the Brillouin zone boundary. Since there are two atoms per unit cell, altogether six different bulk bands for the acoustical and optical modes exist for each symmetry direction.

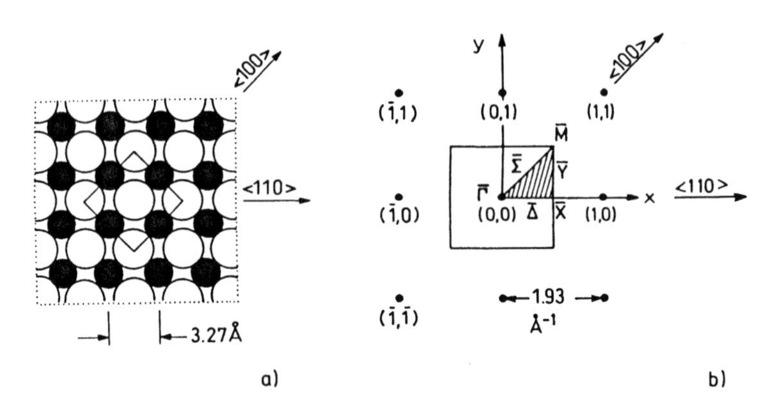

Fig. 8: (a) the arrangement of sodium (full circles) and fluorine (open circles) ions in the NaF(001) surface with indication of the symmetry directions and of the interionic distance r_o. (b): the corresponding surface Brillouin zone (BZ). Its irreducible part is represented by the shadowed area; Γ, $\overline{M}$, $\overline{X}$ label the symmetry points of the surface BZ.

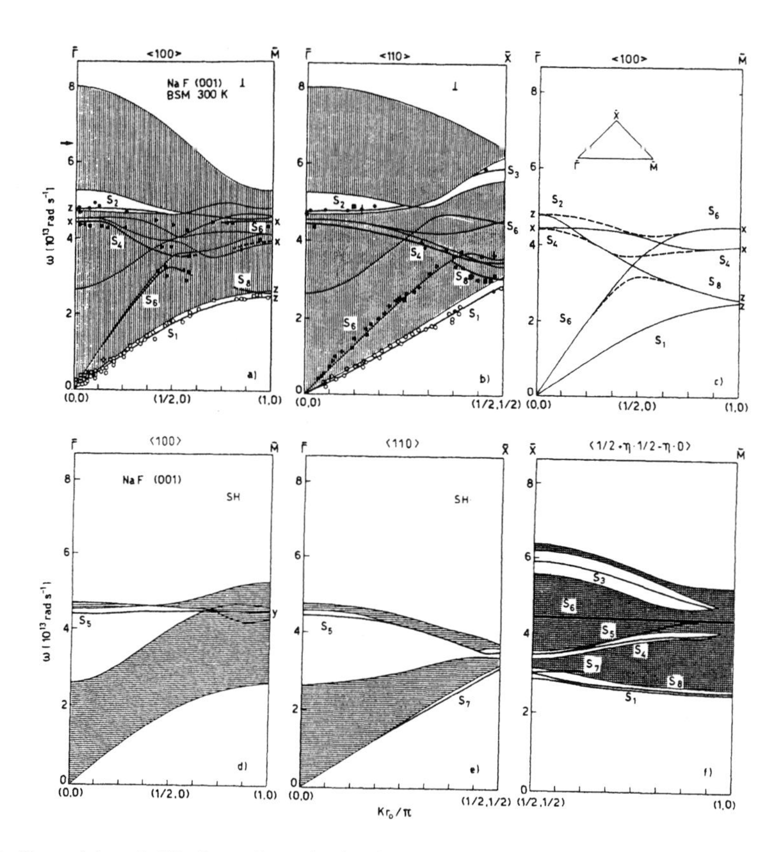

Fig. 9: Parts (a) and (b) show the calculated and HAS experimental surface phonon dispersion curves of NaF(001) along the two symmetry directions ΓX (<110>) and ΓM (<100>) for sagittal polarization (⊥). The open circles show the HAS measurements of Brusdeylins, Doak and Toennies [48] and the black dots are the HAS measurements of Brusdeylins *et al.* [49]. The heavy lines are surface modes and the shaded areas correspond to bulk bands projected on to the surface. Part (c) illustrates the complex hybridization scheme of the sagittal modes along <100>. Parts (d) ad (e) show the calculated dispersion curves for the shear horizontal (SH) polarized modes, while part (f) shows the total spectrum of surface modes and bulk bands along the zone edge XM. Calculations were made by Benedek et al. [46] with the GF method and the breathing shell model with room temperature parameters.

The dispersion curves calculated for the sagittal plane, together with the experimental points from helium atom scattering (HAS) measurements [48,49], are shown in Fig. 9(a) for the <110> direction and Fig. 9(b) for the <100> direction. Heavy lines are surface modes; thin lines are the edges of the bulk bands (shadowed areas). The complex pattern of avoided crossings appearing along <100> is explained in Fig. 9(c) as an effect of the mixing between SV and L polarizations and between acoustic and optical characters occurring away from symmetry points. The SH surface modes and bands are illustrated in Fig. 9(d) and (e). Along the zone boundary $\overline{MX}$ the sagittal and the SH components are mixed together and are shown superimposed in Fig. 9(f). The optical surface modes occurring in diatomic crystals are characterized by an anti-phase motion of the two ions of the unit cell. This is schematically corresponding to the Wallis mode S_2 [51], ~SV and SH, corresponding to the two Lucas modes S_4 and S_5, respectively [50,51]. The sagittal modes S_2 and S_4 have an elliptical polarization, which means that their displacement patterns of Fig. 10(a,b) are indeed coupled out of phase.

The surface-projected phonon DOS for sagittal polarization in NaF(001) obtained from the same GF calculation [46] are displayed in Fig. 11 for the $\overline{\Gamma}$ and $\overline{M}$ points. Unperturbed and perturbed spectra are plotted in the upper and lower parts of the Figure, respectively. Localized surface modes as well as pseudo-surface modes and surface resonances are clearly seen outside and inside the perturbed surface-projected phonon bands. The labels S_j (j = 1,2,...) relate the DOS spectral features to the dispersion curves shown in Fig. 9.

1.3 Surface versus Bulk Phonons

By way of introduction to the next Section, the important similarities and differences between the surface and bulk phonon dispersion curves are illustrated for four systems representing different types of materials in Fig. 10. For each system the dispersion curve of the lowest frequency mode measured in the bulk using neutrons is compared with the behavior of the corresponding surface phonon dispersion curves as illustrated by HAS measurements along the equivalent crystallographic directions. A basic goal of surface phonon studies is to obtain

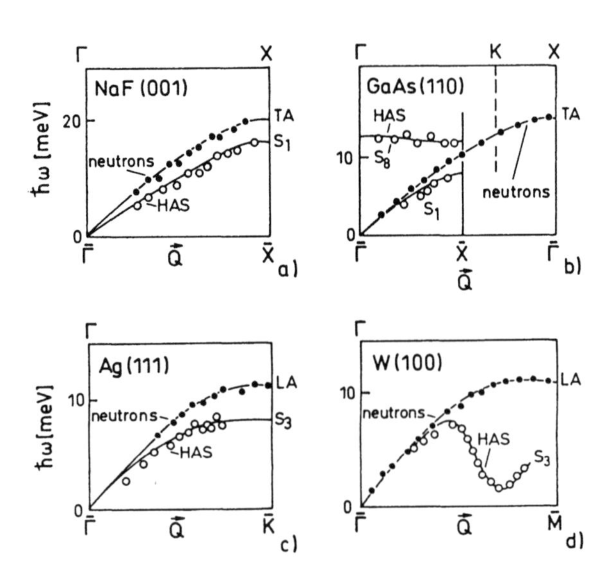

Fig. 10: Some acoustic phonon dispersion curves measured by helium atom scattering at the surface of various types of solids are compared to those obtained with neutron scattering along the same crystallographic directions in the bulk. For the closed-shell ionic NaF(001) [48,52] (a) and the semiconductor GaAs(110) [53,55] (b) the comparisons are between the Rayleigh wave (RW, S_1) and the transverse acoustic (TA) bulk mode. In GaAs(110) the surface Brillouin zone is only half that in the bulk with the consequence that the RW is folded back into an upper branch S_8 with a splitting at $\overline{X}$. For the fcc noble metal Ag(111) [59,60] (c) and the bcc transition metal W(001) [62,63] (d), the anomalous longitudinal (S_3) surface branch is compared with the longitudinal acoustic (LA) bulk mode.

detailed microscopic insight concerning the origin of these similarities and differences in the structure and interactions occuring at the surface with respect to the bulk.

As illustrated in Fig. 10(a) the surfaces of crystals made of closed-shell ions or atoms, such as the alkali halides, or rare gas solids etc., exhibit only modest changes with respect to the corresponding atomic planes inside the bulk. This is related to the fact that, apart from some hybridization induced by the lower point-group symmetry, the atoms or ions at the surface of a solid have the same electronic structure as in the bulk

The observed reduction in the frequency of the surface modes, as illustrated in Fig. 10(a) for NaF (001) [48,52], is thus largely a geometric effect resulting from the reduced number of nearest neighbors at the surface. The situation becomes much more complex when the surfaces of crystals not composed of closed shell atoms or ions are considered.

In covalent homopolar semiconductors, like diamond, Si and Ge, the equal share of the bond charge between the two newly created surfaces of the cleaved crystal initially leaves half-filled bands consisting of dangling bonds at the surfaces. This situation is unstable against a symmetry-breaking reconstruction and leads to a variety of new surface structures, provided the surface temperature is much smaller than the splitting energy between the bonding and antibonding surface states induced by reconstruction. A related reconstruction occurs in heteropolar tetrahedrally-coordinated semiconductors, e.g. GaAs(110), where the bonding charge is not equally shared by neighboring atoms. The charge transfer between surface ions is enough to stabilize the surface into a buckled, unreconstructed phase [53,54] with a surface BZ which is twice as small as the corresponding section of the bulk BZ. Thus, as shown in Fig. 10(b), the surface phonon dispersion curve of GaAs (110) along the $<1\bar{1}0>$ direction is folded at the surface BZ edge [53], whereas the bulk one extends beyond up to the bulk BZ edge [55]. The buckling opens a gap between the folded surface phonon branches [53,54,56]. Despite the complexity of the associated structures the bond lengths show only little variations as compared to the larger changes in bond angles resulting from the reduced coordination. Thus the buckling affects the force constants mostly through the rotational invariance condition referred to the new equilibrium configuration [56]. It is therefore not so surprising that it has been possible to discriminate between subtle structural differences from an analysis of the frequencies of zone boundary surface phonons. For example the buckling angles of π-bonded chains in Si(111) 2x1 [56] or GaAs(110) [53] can be determined in this way.

In metals the partitioning of the free electrons between the two newly created surfaces also generally results in an unstable situation leading to a rearrangement of electrons in real space and on the Fermi surface. This intimate connection between structure and electronic distortion is not surprising since in metals the interactions between ion cores is mediated by the interposed electrons. Even for unreconstructed surfaces the changes in surface force constants required to explain experimental surface phonon frequencies are so significant that they can only be described in terms of many-body effects due to the redistribution of the electronic charge. Moreover there is a connection between the redistribution of the electronic charge at the surface and the inward structural relaxation of the outermost layer [57], so that the change of surface force constants is directly related to the relaxation. Another related quantity is the surface stress, which is in turn connected with the occurrence of anomalies in surface phonons dispersion curves [58].

In Fig. 10(c) the anomalous longitudinal (AL) surface phonon branch S_3 observed in Ag(111) along <110> at room temperature [59] is compared with the bulk longitudinal acoustic (LA) branch [60]. In this case the surface phonon branch shows a broad, albeit sizeable, softening towards the zone edge. This has been interpreted in terms of surface electronic states. Surface bands crossing the Fermi level may lead to a large electron susceptibility over a narrow range of wave vectors. At these wave vectors a strong electron-phonon coupling can occur and the resulting kinks or sharp depressions in the surface phonon dispersion curves are generally referred to as Kohn anomalies [61]. This is illustrated for W(001) along <100> at 500 K [62] in Fig. 10(d), in comparison with the regular dispersion curve of bulk LA phonons [63]. In extreme cases, which have been seen in only a few systems such as W(110) + H(1x1), the anomalies are extremely sharp and deep so that phonon frequencies almost vanish [64,65]. Structural instabilities such as those leading to a reconstruction are thought to be induced by such sharp Kohn anomalies. Indeed the remarkable softening of the surface phonon branch of W(001) with respect the bulk longitudinal mode eventually drives a $\sqrt{2}\times\sqrt{2}$ surface reconstruction at temperatures below about 280 K, where the splitting energy becomes greater than the surface thermal energy.

2. SEMICONDUCTORS: SURFACE PHONONS VERSUS STRUCTURE

The lattice dynamics of the semiconductors are of considerable interest because of the strong covalent bonds in these substances [66-68]. The neutron measurements of bulk phonons stimulated the development of several different models of which the bond charge model [69-72] has been found to be particularly useful also for describing the surface phonons. The surface dynamics of elemental and compound semiconductors is rendered more complex than in the case of the alkali halides because of the ubiquitous spontaneous reconstruction of their surfaces. This is due to the intrinsic instability of the surface half-filled orbitals associated with the broken (dangling) bonds.

2.1 The (111) Surface of Silicon

The native (111) surface of silicon, as obtained from cleavage, reconstructs spontaneously to a (2x1) structure. At $T_s \geq 500$ K this surface then reconstructs irreversibly to produce the well-known 7x7 structure [73]. Unlike the bulk, which is non-polar and has no LO-TO splitting between the $\mathbf{q} = 0$ optical phonons, the surface hosts polar (optically active) phonons at $\mathbf{Q} = 0$ due to the lack of inversion symmetry. For example, on Si(111) (2x1) there is an optical mode at 55 meV strongly active in electron energy loss spectroscopy (EELS) [74]. Moreover in Si(111) (2x1) [75], like in GaAs(110) [Fig. 10(b)] there is a flat, nearly dispersionless surface phonon branch with frequencies of about 10 meV (Fig. 11). This originates from the folding of the Rayleigh branch produced by the doubling of the surface unit cell with respect to that of the bulk. For Si(111) 2x1 this is due to reconstruction; for GaAs(110) to the lower symmetry of the (110) surface.

Reconstruction in Si(111)(2x1) yields a feature which is not present in GaAs(110): a sequence of alternating five- and seven-atom rings (inset in Fig. 11) instead of the regular array of six-atom rings occuring in the unreconstructed case. Because of the large size of the rings the surface perturbation extends at least down to the fifth atomic layer, so that one can speak of a buried interface between the reconstructed and deeper regular bulk regions [56]. With respect to the ideal surface the more compact five-atom rings are expected to exhibit stiffer modes whereas the expanded seven-fold rings will have softer modes. Moreover, some modes are associated with the buried interface. This explains the large number of surface modes occurring in Si(111)(2x1): in practice one expects as many surface branches as many degrees of freedom per unit cell are involved in the reconstruction. The considerable depth of reconstruction in this case serves as a reminder that slab calculations should always be based on a sufficiently thick slab.

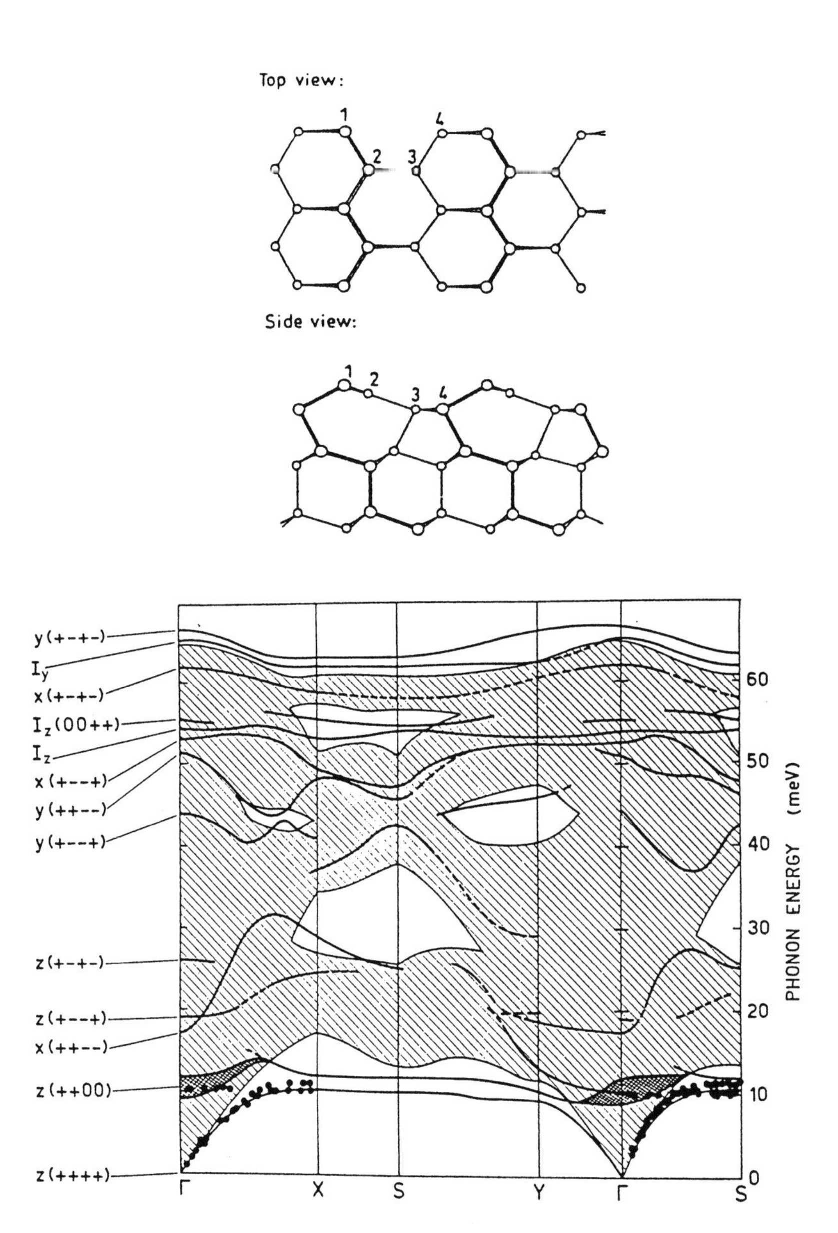

Fig. 11: Comparison of bond-charge model calculations of surface phonon dispersion curves for the Si(111) (2x1) reconstructed surface [56] with the results of Helium atom scattering experiments (•) [75] and EELS (o) [74]. The calculation predicts a broad resonance around 10 meV (shaded regions). The inset (top left) shows the sequence of 7-fold and 5-fold atomic rings according to Pandey's reconstruction model [76] as appears in a projection onto a $(1\bar{1}0)$ plane normal to the surface. Atoms 1 and 2 belong to the surface π-bonded chain and atoms 3 and 4 to the groove between two neighboring chains. The displacements of the surface atoms and polarizations α for the zone center surface modes are indicated by their orientation (+ or -) or their being at rest (o) with respect to the direction α, where α=x: parallel to the chain: α=y: on the surface normal to the chain: α=z to the surface. I_α are interface modes.

The experimental curves are compared with the results of a bond-charge-model (BCM) calculation for a 22 layer slab [56]. This calculation is in good agreement with both the 10 meV and the optical mode at 55 meV. The almost isotropic dipolar activity of the latter was found to arise from the *vertical* motion of atoms in the interface region. The BCM theory makes it possible to acquire structural information on the surface. It indicates that the small surface phonon gap at the zone boundary along the $\overline{\Gamma S}$ direction, where folding occurs, is a sensitive measure of the surface π-bonded chain buckling. Here buckling refers to the difference in the vertical positions of the two inequivalent chain atoms. The interpretation of the measurements of Harten et al. [75] suggests that the correct buckling is 0.20 Å in agreement with density functional calculations [77], but significantly smaller than the value of 0.30 Å previously deduced from fits to LEED data [78].

The structure and dynamics of the Si(111)(7x7) reconstructed surface represented a challenge to both experimentalists and theorists because of the structural complexity of the dimer-adatom-stacking fault model introduced by Takayanagi et al. [79] and the very large unit cell, which means a very small Brillouin zone. Only recently the resolution of inelastic HAS has been improved to the extent that the surface phonon dispersion curves could be measured in the acoustic region [80].

The experimental points (o) in the (ΔK, ΔE) plane corresponding to the phonon peaks in the HAS time-of-flight spectra are plotted in Fig. 12, after folding into the positive quadrant, for the two high-symmetry directions and two different incident energies. The experimental points are then compared to the set of curves which are obtained by translating the Rayleigh wave (RW) branch of the "ideal" (hydrogen-saturated) Si(111)-1×1 surface [81] by all possible reciprocal lattice vectors $\mathbf{G}_{(7\times7)}$ of the 7×7 superstructure. The Si(111)-1×1 RW branch is best fitted by a sum of sinusoidal components

$$\Delta E(\mathbf{Q}) = (7.5\ \mathrm{meV})\{\sum_{i=1}^{3} [\sin^2(\mathbf{A}_i\cdot\mathbf{Q}) + 0.55\sin^2(2\mathbf{A}_i\cdot\mathbf{Q}) + 0.75\sin^6(2\mathbf{A}_i\cdot\mathbf{Q})]\}^{1/2} \quad (10)$$

where $\mathbf{A}_1 = a(1,0)$, $\mathbf{A}_2 = a(-1/2, \sqrt{3}/2)$ and $\mathbf{A}_3 = a(1/2, \sqrt{3}/2)$, with $a = 3.84$ Å, are the two-dimensional basis vectors of the *unreconstructed* hexagonal (111) surface, and $\mathbf{Q} \equiv \Delta\mathbf{K} - \mathbf{G}_{(7\times7)}$. In Fig. 12 full and broken lines represent Eq. (10) for in-plane and out-of-plane $\mathbf{G}_{(7\times7)}$ vectors, respectively.

The experimental points follow quite well the expected phonon branches for many different $\mathbf{G}_{(7\times7)}$ vectors, though measurable inelastic processes only occur for $\mathbf{G}_{(7\times7)}$ vectors giving large diffraction intensities: in Fig. 12(b), for example, RW data are missing at those $\mathbf{G}_{(7\times7)}$ vectors whose diffraction intensities are comparatively weak. (Fig. 12). The best results are obtained with a lower incident energy [Fig. 12(a)] where, besides the replicas of the RW branch at six different in-plane $\mathbf{G}_{(7\times7)}$ vectors, one has a good number of events along the lowest out-of-plane branches. However the full extension of the spectrum up to the maximum of the RW branch is better probed in experiments at higher incident energy [Fig. 12(b,c)], though the spread of data points looks comparatively large on the present expanded ΔK scale.

The effects of surface reconstruction occurring in the acoustic region are better observed after folding the data of Fig. 12 into the small Brillouin zone of the 7x7 phase [Fig. 13(a)]. There is a depletion of phonons around 12 meV with a downward shift of branches into the gap at 9 meV and an upward shift of the highest branches above the RW maximum of the ideal surface [arrows in Fig. 13(a)]. These perturbations have the following explanation. Despite the complex reconstruction, the low frequency lattice dynamics is dominated by the collective

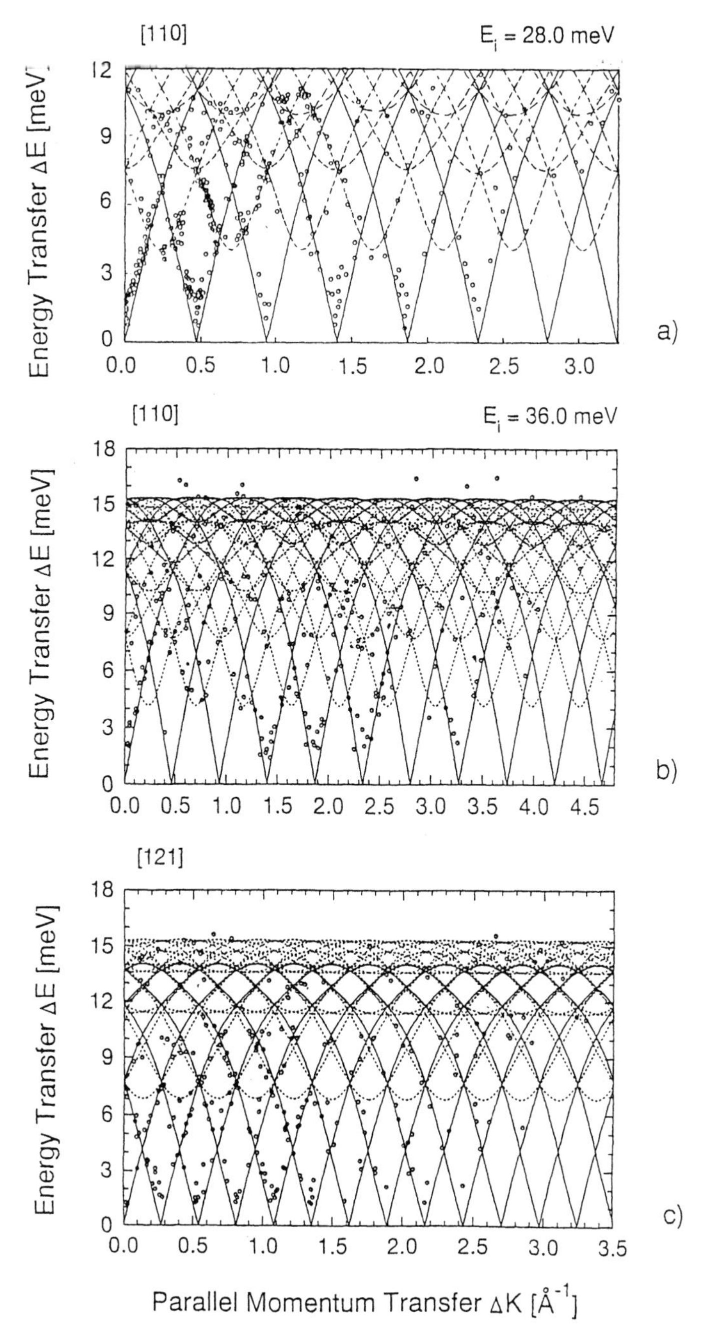

Fig. 12: Experimental (o) and theoretical (lines) dispersion curves of Si(111)(7×7) for the symmetry directions [110] (a,b) and [121] (c). The theoretical curves are obtained by reproducing the Si(111)(1×1) RW branch for all possible in-plane (full lines) and out-of-plane (broken lines) reciprocal lattice vectors of the (7×7) reconstructed surface.

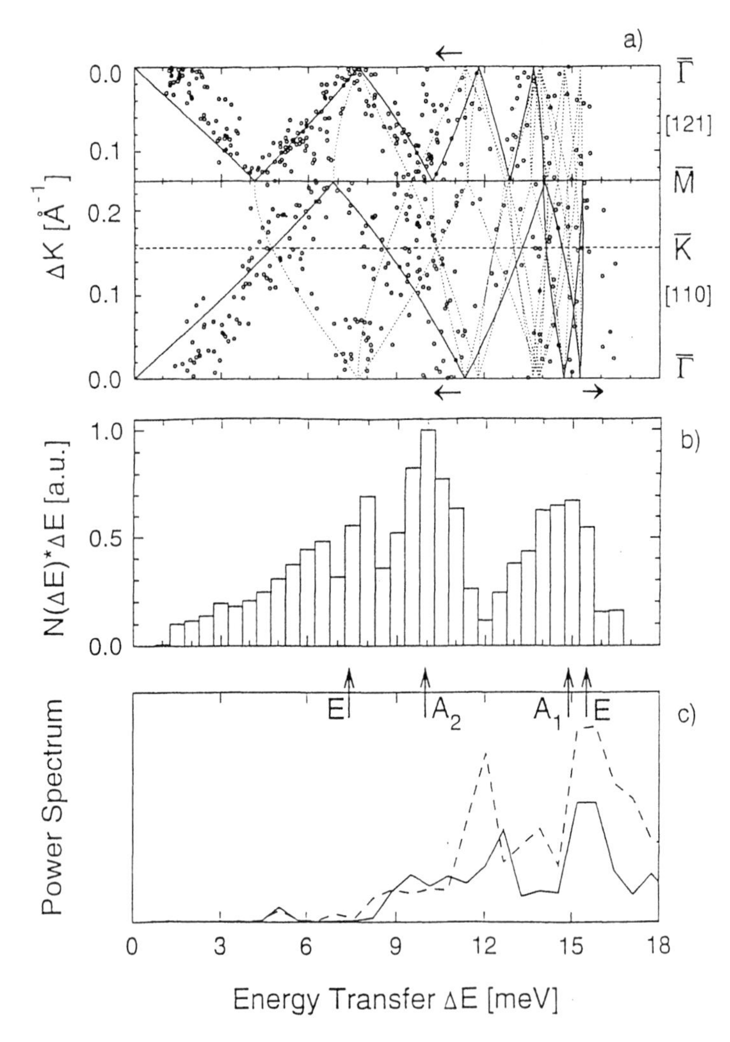

Fig. 13: Dispersion curves and data points for Si(111)(7×7) folded into the reduced surface Brillouin zone (a), and corresponding density of states (b). This is compared to the calculated low-frequency power spectrum (c) obtained by first-principles (Kim *et al* [82]). The solid line represents the power spectrum of polarization vectors projected along the [111] direction normal to the (111) plane; the dotted line is for projection in the plane. Arrows indicate the lowest Q = 0 frequencies obtained by Štich *et al* [83] with first-principle MD and the multiple signal classification algorithm. Symmetry labels refer to the irreducible representations of the C_{3v} group.

vibrations of a large number of atoms extending fairly deeply into the crystal and is little affected by the complicated surface structure. On the other hand the high-frequency portion of the folded acoustic branch is heavily affected by the complex reconstruction, whose dynamics in the low-frequency region is dominated by the motion of adatoms [82]. The phonon density as obtained by summing up the experimental points measured in all directions [Fig. 13(b)] shows a gap at about 12 meV separating the two above frequency regions. These features of the acoustic spectrum of Si(111)(7x7) are well reproduced by recent Car-Parrinello calculations by Štich *et al.* [83].

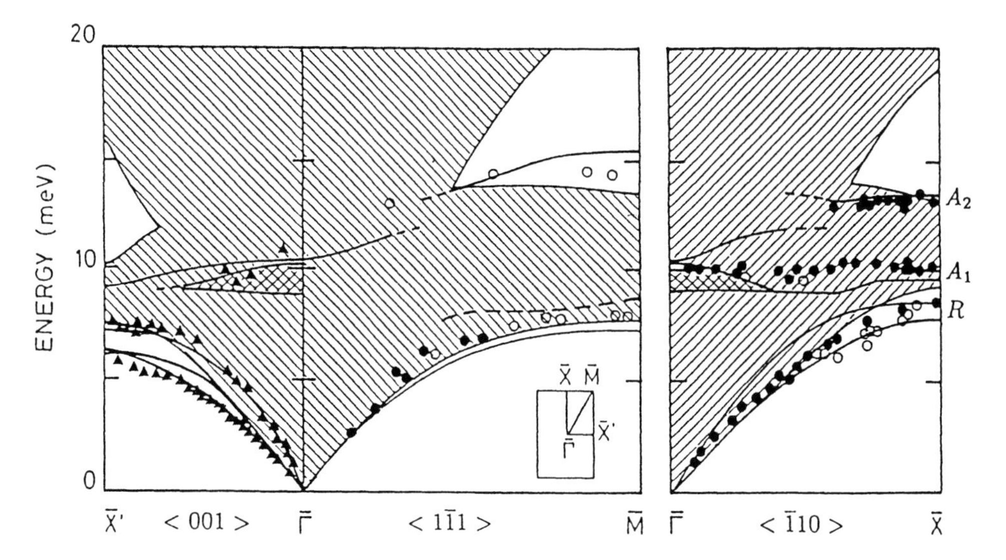

Fig. 14. (a) Comparison of a bond-charge model calculation of the surface phonons dispersion curves in the lower part of the spectrum with the experimental HAS data [92,93,53] for the unreconstructed GaAs (110) surface. Solid lines are the calculated surface phonon branches. The actual width of the strong resonance at ~ 10 meV is represented by the squared areas. Weaker resonances are indicated by broken lines. The surface Brillouin zone is shown in the inset. (b) The surface structure of GaAs(110); the side view shows the buckling of the topmost atomic rows and the consequent deformation of the hexagonal rings.

2.2 GaAs(110)

A more favorable situation occurs in GaAs(110) for which the six-atom rings are not broken at the surface as on Si(111) 2x1. The structure and reciprocal lattice of the GaAs(110) surface is shown in Fig. 14(b). It is now well established that the surface relaxation is mainly characterized by a bond length conserving rotation of the surface chains by a tilt angle of about 30° [84], with the As atoms shifted above the ideal (110) plane and the Ga atoms shifted towards the bulk. The high frequency modes were early on investigated by EELS which revealed a strong FK mode at about 36 meV [85,86]. The HAS dispersion curves are shown in Fig. 14(a).

Semiempirical tight-binding [54,87], BCM [53] and various first-principle calculations [88-90] agree in reproducing the experimental data both from HAS [53,91,92] and EELS [93-95] measurements and particularly the flat low-lying branch, which also occurs at 10 meV. Unlike the silicon case, this branch falls well above the zone-boundary Rayleigh wave (S_1) because no seven-fold ring softening occurs. The calculated surface phonon frequencies were also found to depend strongly on the geometrical arrangement of surface atoms, which allows for a determination of the surface crystallography from an accurate fit of the dispersion curves [53].

3. SURFACE PHONONS IN ULTRAFAST PHENOMENA

3.1 Pump & Probe Experiments with Phonons.

The application of surface acoustic waves (SAW) in integrated optoacoustic devices implies a direct conversion of photons into acoustic waves and, viceversa, of acoustic waves into modulated electromagnetic waves. The great perfection of monocrystalline surfaces and overlayers which can be achieved with present epitaxial growth techniques permits in principle to extend the SAW frequencies up to the THz range where the phonon wavelengths become comparable to the lattice distances. Since the amount of information carried by a signal rapidly increases with its frequency, there is an obvious advantage in working in the THz domain. In this domain, however, the dispersion of surface phonons can no longer be neglected, and so is for the microscopic interactions - notably the electron-phonon interaction – which are responsible for various, sometimes anomalous features of the dispersion relations. These effects, representing at first a serious difficulty, could support in principle new unconventional applications. To this purpose a detailed knowledge of the surface phonon dispersion curves is useful.

These perspectives have stimulated many studies on the direct generation and detection of surface phonons by means of photon beams. Since the phonon periods in the acoustic dispersive and optical regions are in the *ps* or sub-*ps* range, and also the phonon lifetimes can be in the *ps* range (especially in nonequilibrium conditions), only the advent of ultrashort laser pulses from *ps* down to the *fs* scale has permitted to investigate the phonon dynamics directly in the time domain.by means of pump-and-probe (P&P) experiments (Fig. 15) [96,97].

The first P&P studies of phonon dynamics have been carried out for bulk phonons and consisted in time-resolved Raman spectroscopy from the longitudinal optical (LO) phonons in

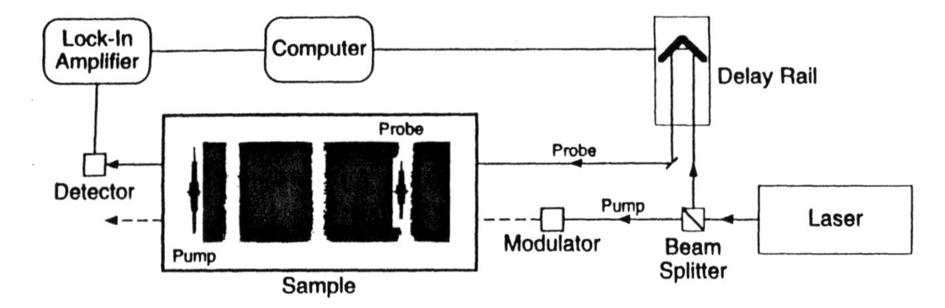

Fig. 15. The geometry of a pump-and-probe transmission experiment (from Merlin [97]).

GaAs. For example von der Linde *et al* [98], working with a picosecond resolution were able to directly measure the lifetime of nonequilibrium LO phonons. In this experiment a first Raman scattering process is made by a pump photon which excites a LO phonon. Then another Raman scattering processes is made by the probe photon, either off the previously excited LO phonon or by exciting a new photon. While the second process is independent of the probe-pump delay time, the first process as a delay-dependent intensity and allows to measure the phonon lifetime. Useless to say that such information has a great relevance in the kinetics of hot electrons and their energy dissipation through LO phonon creation and subsequent decay into acoustic phonons. The energy dissipation cascade of photoexcited hot electrons into LO phonons has been studied as well with the P&P method by Kash *et al.* [99]. This experiment, in which the pump photons excite the carriers and the probe detects LO phonons through a Raman process, required a sub-picosecond resolution and led to determination of an average electron-phonon scattering time of about 165 fs.

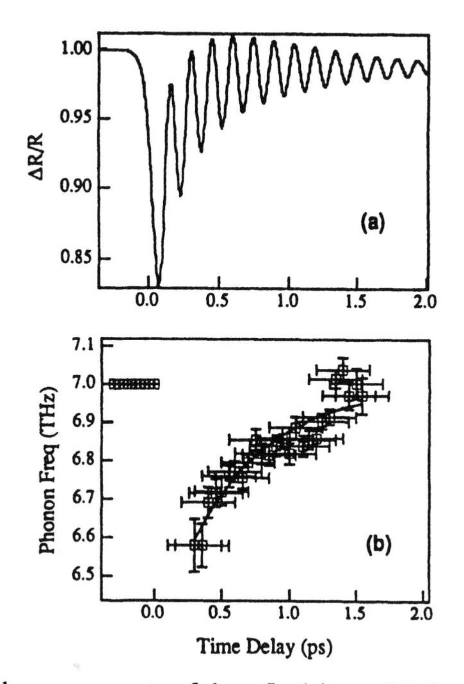

Fig. 16. Pump-and-probe measurements of the reflectivity modulation in Ti_2O_3 produced by the vibrational field of the Raman-active mode of A_{1g} symmetry (a) and the anharmonic change of the vibrational frequency as function of the time-dependent phonon amplitude (b) [from Cheng et al (101)].

These are just two representative early works chosen among the large number of contributions which led to the recent advances in femtosecond laser techniques and the generation of coherent THz phonons in solids. The generation of coherent optical phonons in GaAs has been further demonstrated by Kütt *et at* [100]. The reader is referred to the excellent review by Roberto Merlin [97], where both the experimental and theoretical aspects of the coherent THz phonon generation by laser light pulses are surveyed. Perhaps the most intriguing results are those obtained in the narrow-gap material Ti_2O_3 by Chen *et al.* [101]. The modulation of the reflectivity induced by a phonon displacement field of A_{1g} symmetry has been measured as function of time [Fig. 16(a)].

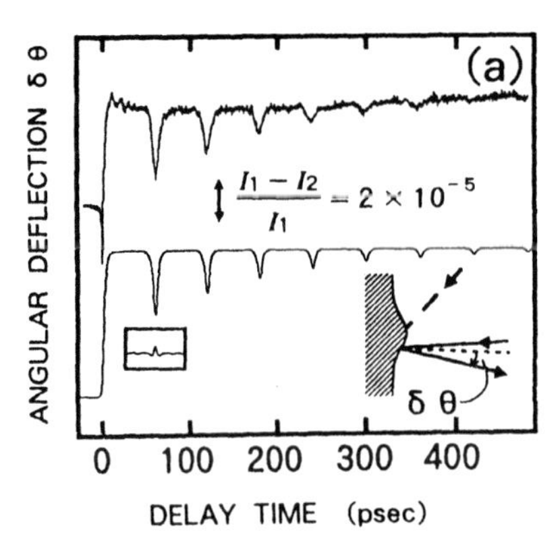

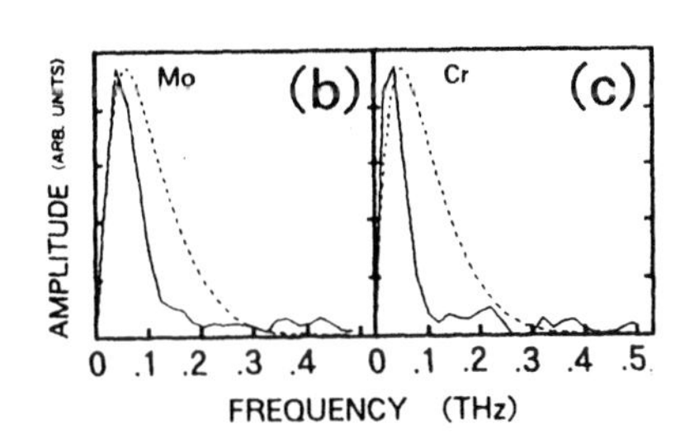

Fig. 17. The angular deflection of the probe beam as a function of the pump-probe delay time (a). The upper curve shows the experimental results, the lower curve a fit from a thermoelastic model. There is in excellent agreement between theory and experiment. The phonon spectra derived from the Fourier transform of the first echo pulse for Mo (b) and Cr (c) favorably compared to the calculated spectra (dashed lines) from the thermoelastic model (from Ref. 102).

The damping of the vibrational amplitude is accompanied by a slight increase of the phonon frequency, from 6.6 THz for the largest amplitude (at short delay times) to the ground-state value of 7.0 THz for small (harmonic) amplitudes at long delay times [Fig. 16(b)]. Since Ti_2O_3 undergoes a transition to a metallic phase with a slight change of the lattice distance, the comparatively large softening of the frequency as function of the amplitude can be interpreted as an effect of enhanced anharmonicity foreshadowing a phase transition.

3.2 Experiments with Surface and Interface Phonons

The above examples, although concerning opaque materials, are all involving bulk optical phonons. However, owing to the limited penetration of photons into opaque materials, the energy transfers from the pump beam as well as the modulation of the probe are concentrated in the surface region, and therefore one expects an enhanced response from surface phonons. Picosecond pump-and-probe experiments revealing surface phonons propagating along the surface of Mo thin films deposited on silica have been carried out by Wright and Kawashima [102]. In these experiments [Fig. 17(a)] the film surface, which is perfectly flat at equilibrium, is perturbed by the propagation of a coherent surface wave which is generated by a 2 ps pulse of the pump and is observed through a time-dependent deflection of the specularly reflected probe beam. Actually a regular sequence of peaks of decreasing intensity is observed as an effect of multiple echo from the buried interface between the thin film and the substrate. The Fourier transform of the angular deflection function provides the spectrum of the surface phonons involved in the process, which is strongly peaked around 80 GHz [Fig. 17(b)]. A similar phonon response spectrum has been obtained for a Cr film [Fig. 17(c)]. Moreover, experiments with a Cr film on an a-Si substrate shows that the phonon reflections from the buried interface carry information on the defect structure of the substrate surface.

Interfaces, like free surfaces, can host localized phonons characterized by a large amplitude and strong interaction with the electron excitations. Lisoli *et al* [103] have been able to study, by means of P&P methods, the ultrafast dynamics of photogenerated plasmons in solid and liquid gallium clusters of nanometric size embedded into a silica matrix. The plasmons are generated by 390 nm 180 fs pump pulses and their decay in time is studied by a 780 nm probe pulse of 150 fs. The plasmon decay time has been measured for three different

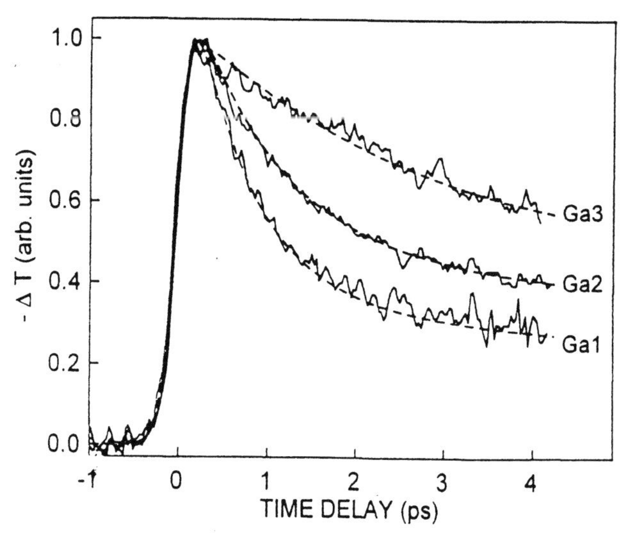

Fig. 18. Plasmon excitation decay as function of pump-probe delay time in Ga nanoparticles of different radii embedded in a silica matrix (from Lisoli *et al* [103]).

cluster sizes as function of their average radius (50, 70 and 90 Å) for both the solid (T = 77 K) and liquid (T = 293 K) phases. The plasmon lifetime is found to increase with the cluster size, indicating a coupling to interface phonons (Fig. 18). The theory of plasmon decay based on the electron-interface phonon coupling mechanism leads to an excellent fit of the measured lifetimes for both phased and to values of the electron-interface phonon coupling constant of 5.9, 4.1 and 3.2 x 10^{16} W m^{-3} K^{-1} for three above radii, respectively.

Nanometric surface defects are responsible for scattering of THz surface phonons and other nonlinear phenomena such as frequency down conversion. Recently Ding *et al* [104] have been able to study this class of phenomena by direct time-of-flight experiments on a high quality (001) surface of silicon, in which a surface phonon generated by a ultra-short laser pulse is detected by a fast aluminum edge bolometer. From studying the propagation and attenuation of high-frequency surface phonons one expects to obtain a fast characterization tool for nano- and mesoscopic surface structures.

3.3 Surfing Carriers

A surface elastic wave produces a periodic acoustic deformation potential which can be able to separate electrons from holes and inhibit recombination of excess carrier pairs. An intriguing demonstration of this effect has been given recently by A. Wixworth [105]. Excess hole-electron pairs are photogenerated by a laser pulse at the surface of a ultrathin InGaAs layer in a region which is crossed by a surface acoustic wave (SAW). The SAW is produced by a radiofrequency generator and travels along the surface carrying a certain amount of excess electrons trapped in its minima and of excess holes trapped at its maxima. The carriers are carried along the surface with the SAW velocity until the SAW energy is absorbed by a transducer and the hole-electron pairs suddendly recombine. Due to radiative recombination photons are re-emitted. Re-emission, however, occurs with a controlled delay from a surface spot away from the absorption site. Many applications in composite devices, combining opto-acoustic with electronic functions, can be envisaged for such a surface-phonon-controlled recombination.

A single-electron transport in a one-dimensional channel by means of high-frequency surface acoustic waves has been recently achieved by V. I. Talyanskii *et al* in a GaAs-AlGaAs heterostructure [106]. In this experiment the well-defined number of electrons trapped in the moving potential well of the SAW is controlled by the electron-electron repulsive potential and may provide, according to the authors, a viable tool for producing a standard of electric current. The influence of surface acoustic wave on the conductance through a ballistic quantum channel has been recently termed *acoustoconductance* [107].

Surface electrons can be trapped by a surface phonon associated with the local lattice deformation that the electron itself produces with its coulomb field (*self-trapping*). This special electron-phonon bound state is known as a *small polaron* and yields a comparatively large electron effective mass. In this case the surface phonon is not injected from outside but is generated by the excess electron itself. The evidence for such a localization of surface excess electrons due to surface phonons has been recently obtained by Ge *et al* [108] in a state-of-the art experiment based on the combination of angle-resolved two-photon photoemission (TPPE) and femtosecond laser techniques.

Ge *et al* experiments have been carried out for a bilayer of *n*-heptane molecules on the (111) surface of silver. One pump photon lifts an electron from silver into a surface image state. This electron is localized in the normal direction at the overlayer-substrate interface but is allowed to freely move parallel to the surface with a parabolic dispersion. A second (probe) photon extracts the electron, after a given pump-probe delay on the *fs* scale, and its kinetic energy at each emission angle. Such a time-resolved, angle-resolved photoemission experiment allows for the determination of the image-state dispersion curve and its evolution with time. The most striking observation concerns the formation of the polaron state (which obviously occurs in the time scale of surface phonon periods) and the consequent self-trapping of the image-state electron. The image-state electron energy as function of parallel momentum is seen to evolve from the parabolic dispersion for free propagation (with an effective mass of only $1.2m_e$) for a 0 *fs* delay, to a dispersionless branch corresponding to a self-trapped state after 1670 *fs*. The authors argue this kind of experiments will provide fundamental information on the carrier dynamics in many low-dimensional systems including organic light-emitting diodes.

From these few examples one expects that the combination of surface phonons with electron dynamics and the photon field can lead to a wide spectrum of new phenomena and to a large variety of new applications and devices. The extension of the surface phonon spectrum to the high-frequency dispersive region will be a natural consequence of the length scale reduction of devices down to the nanometric size.

REFERENCES

[1] Lord Rayleigh, Proc. London Math. Soc., 17 (1887) 4.

[2] G.W. Farnell, in: *Physical Acoustics*, Vol. 6, Eds. W.P. Mason and R.N. Thurston (Academic Press, New York, 1970) p. 109.

[3] A.A. Maradudin, in: *Nonequilibrium Phonon Dynamics*, Ed. by W.E. Bron (Plenum, New York 1985) p. 395.

[4] M. Born and Th. von Kármán, Phys. Zeit. 13 (1912) 297.

[5] M. Born and K. Huang, "*Dynamical Theory of Crystal Lattices*" (Oxford University Press, London, New York 1954).

[6] F.W. de Wette and W. Kress (eds.), *Surface Phonons*, Springer Ser. Surf. Sci., Vol. 21 (Springer, Berlin, Heidelberg 1991).

[7] V. Bortolani, A. Franchini and G. Santoro in: *Electronic Structure, Dynamics and Quantum Structural Properties of Condensed Matter*, Eds. J.T. Devreese and P. van Camp (Plenum Press, New York 1984) p. 401.

[8] For an updated account see the review article: J.P. Toennies and G. Benedek, Surf. Sci. Rep. (1988, in press).
[9] For a compilation of bulk phonon data see H.R. Schober and P.H. Dedrich in Landolt-Börnstein, Neue Serie Bd. 13a (Springer V., Berlin 1981).
[10] I.M. Lifshitz and L.M. Rozenzweig, Zh. Eksp. Teor. Fiz. 18 (1948) 1012.
[11] I.M. Lifshitz: Nuovo Cimento Suppl. 3 (1956) 732.
[12] I.M. Lifshitz and A.M. Kosevich, Rep. Progr. Phys. 29 (I) (1966) 217.
[13] A. A. Maradudin, E.W. Montroll, G.H. Weiss and I.P. Ipatova, *Theory of Lattice Dynamics in the Harmonic Approximation*, Solid State Physics, Suppl. 3 Academic Press, New York 1971).
[14] A.A. Maradudin and J. Melngailis, Phys. Rev. 133 (1964) A1188.
[15] S.W. Musser and K.H. Rieder, Phys. Rev. B2 (1970) 3034.
[16] T.P. Martin, Phys. Rev. B1 (1970) 3480.
[17] G. Benedek, Phys. Status Sol. B58 (1973) 661.
[18] G. Benedek, Surface Sci. 61 (1976) 603.
[19] G. Benedek, M. Miura, W. Kress and H. Bilz, Phys. Rev. Lett. 52 (1984) 1907.
[20] G. Benedek and L. Miglio, in: *Surface Phonons*, Eds. F.W. deWette and W. Kress, Springer Ser. Surf. Sci., Vol. 27 (Springer, Berlin, Heidelberg 1991)
[21] A.A. Maradudin, R.F. Wallis and L. Dobrzynski, *Handbook of Surfaces and Interfaces*, Vol. 3 (Garland, New York 1980).
[22] F. Garcia-Moliner, Ann. Phys. (Paris) 2 (1977) 179.
[23] G. Armand, Phys. Rev. B14 (1976) 2218.
[24] T.E. Feuchtwang, Phys. Rev. 155 (1967) 731.
[25] V. Bortolani, F. Nizzoli and G. Santoro, in: *Lattice Dynamics*, Ed. M. Balkanski (Flammarion, Paris 1978) p. .
[26] J. Szeftel and A. Khater, J. Phys. C20 (1987) 4725; J. Szeftel, A. Khater, F. Mila, S. d'Addato, N. Auby, J. Phys. C21 (1988) 2133; A. Khater, N. Auby and R.F. Wallis, Physica B167 (1991) 273; A. Khater, H. Grimeck, J. Lapujoulade and F. Fabre, Surf. Sci. 251/252 (1991) 291.
[27] S. Trullinger, J. Math. Phys. 17 (1976) 1884.
[28] J.E. Black, in *Dynamical Properties of Solids*, Eds. G.K. Horton and A.A. Maradudin (Elsevier, Amsterdam 1990) Chap. 5.
[29] F.W. de Wette, in: *Surface Phonons*, Eds. F.W. de Wette and W.Kress, Springer Series on Surf. Sci. Vol. 27 (Springer, Berlin, Heidelberg 1991) p.67.
[30] R.E. Allen and F.W. de Wette, Phys. Rev. 179 (1969) 873.
[31] R.E. Allen, F.W. de Wette and A. Rahman, Phys. Rev. 179 (1969) 887.
[32] G.P. Alldredge, R.E. Allen and F.W. de Wette, J. Acoust. Soc. Am. 49 (1971) 1453.
[33] R.E. Allen, G.P. Alldredge and F.W. de Wette, Phys. Rev. Lett. 23 (1969) 1285.
[34] R.E. Allen, G.P. Alldredge and F.W. de Wette, Phys. Rev. B4 (1971) 1648 and 1661.
[35] T.S. Chen, G.P. Alldredge, F.W. de Wette and R.E. Allen, Phys. Rev. Letters 26 (1971) 1543.
[36] T.S. Chen, G.P. Alldredge and F.W. de Wette, Solid State Commun. 10 (1972) 941.
[37] T.S. Chen, G.P. Alldredge and F.W. de Wette, Phys. Lett. 40A (1972) 401.
[38] T.S. Chen, F.W. de Wette and G.P. Alldredge, Phys. Rev. B15 (1977) 1167.
[39] G.P. Alldredge, Phys. Lett. 41A (1972) 291.
[40] V.L. Zoth, G.P. Alldredge and F.W. de Wette, Phys. Lett. 47A (1974) 247.
[41] F.W. de Wette, in: *Dynamical Properties of Solids*, Eds. G.K. Horton and A.A. Maradudin (Elsevier, Amsterdam 1990) Chap. 5.
[42] A. Lock, J.P. Toennies and G. Witte, J. Electr. Spectr. Rel. Phenom. 54/55 (1990) 309.
[43] Ch. Wöll, Thesis (University of Göttingen, 1987); Max-Planck-Institut für Strömungsforschung, Report 18/1987. The calculation of a one layer two dimensional lattice is described in O. Madelung, *Festkörpertheorie II* (Springer, Berlin 1972).

[44] R. Berndt, Diplom thesis (University of Göttingen, 1987).
[45] G. Benedek, J. Ellis, A. Reichmuth, P. Ruggerone, H. Schief and J.P. Toennies, Phys. Rev. Lett. 69 (1992) 2951.
[46] G. Benedek, G.P. Brivio, L. Miglio and V.R. Velasco, Phys. Rev. B26 (1982) 497.
[47] G. Benedek and L. Miglio, in: *Ab-Initio Calculation of Phonon Spectra*, Eds. J.T. Devreese, V.E. van Doren and P.E. Van Camp (Plenum, New York 1983).
[48] G. Brusdeylins, R.B. Doak, and J.P. Toennies, Phys. Rev. B 27 (1983) 3662.
[49] G. Brusdeylins, R. Rechsteiner, J.G. Skofronick, J.P. Toennies, G. Benedek and L. Miglio, Phys. Rev. Lett., 54 (1985) 466.
[50] A.A. Lucas, J. Chem. Phys. 48 (1968) 3156.
[51] L. Miglio and G. Benedek, in: *Structure and Dynamics of Surfaces II*, Eds. W. Duke and P. von Blanckenhagen (Springer, Heidelberg 1987) p. 35.
[52] W.J.L. Buyers, Phys. Rev. 153 (1967) 923
[53] P. Santini, L. Miglio, G. Benedek, U. Harten, P. Ruggerone and J.P.Toennies, Phys. Rev. B42 (1990) 11942.
[54] Y.R. Wang and C.B. Duke, Surf. Sci. 205 (1988) L755.
[55] J.L.T. Waugh and G. Dolling, Phys. Rev. 132 (1963) 2410.
[56] L. Miglio, P. Santini, P. Ruggerone and G. Benedek, Phys. Rev.Lett. 62 (1989) 3070.
[57] M.W. Finnis and V. Heine, J. Phys. F (Metal Phys). 4 (1974) L37.
[58] S. Lehwald, F. Wolf, H. Ibach, B.M. Hall and D.L. Mills, Surf. Sci. 192 (1987) 131.
[59] R.B. Doak, U. Harten, and J.P. Toennies, Phys. Rev.Lett. 51 (1983) 578.
[60] W. Drexel, Z. Phys. 255 (1977) 281.
[61] W. Kohn, Phys. Rev. Lett. 2 (1959) 393.
[62] H.-J. Ernst, E. Hulpke and J.P. Toennies, Europhys. Lett. 10 (1989) 747.
[63] A. Larose and B.N. Brockhouse, Can. J. Phys. 54 (1976) 1819.
[64] E. Hulpke and J. Lüdecke, Phys. Rev. Lett. 68 (1992) 2846.
[65] E. Hulpke and J. Lüdecke, Surface Sci. 287/288 (1993) 837.
[66] J.C. Phillips, Phys. Rev. 166 (1968) 832
[67] S.K. Sinha, Crit. Rev. Solid State Sci. 3 (1973) 273.
[68] C. Falter, Physics Reports 164 (1988) 1.
[69] R.M. Martin, Chem. Phys. Lett. 2 (1968) 268; Phys. Rev. 186 (1969) 871.
[70] F.A. Johnson, Proc. R. Soc. (London) Ser. A 339 (1974) 73.
[71] W. Weber, Phys. Rev. Lett. 33 (1974) 371; Phys. Rev. B15 (1977) 4789.
[72] K.C. Rustagi and W. Weber, Solid State Comm. 18 (1976) 1027.
[73] D. Hanneman, Rep. Prog. Phys. 50 (1987) 1045.
[74] H. Ibach, Phys. Rev. Lett. 27 (1971) 253.
[75] U. Harten, J.P. Toennies, and Ch. Wöll, Phys. Rev. Lett. 57 (1986) 2947.
[76] K.C. Pandey, Phys. Rev. Lett. 47 (1981) 1913; 49 (1982) 223.
[77] E. Northrup and M.L. Cohen, Phys. Rev. Lett. 49 (1982) 1349; J. Vac. Sci. Technol. 21 (1982) 333.
[78] F.J. Himpsel, P.M. Marcus, R. Tromp, I.P. Batra, M.R. Cook, F. Jona, H. Liu, Phys. Rev. B30 (1984) 2257.
[79] T. Takayanagi, Y. Tanshiro, M. Takahashi and S. Takahashi, J. Vac. Sci. Technol. A3 (1985) 1502; Surf. Sci. 164 (1985) 367.
[80] G. Lange, P. Ruggerone, G. Benedek and J. P. Toennies, Europhys. Lett. (1998, to appear)
[81] U. Harten, J. P. Toennies, Ch. Wöll, L. Miglio, P. Ruggerone, L. Colombo and G. Benedek, Phys. Rev. B 38 (1988) 3305.
[82] J. Kim, M. Yeh, F.S. Khan, and J.W. Wilkins, Phys. Rev. B52 (1995) 14709
[83] I. Štich, J. Kohanoff, and K. Terakura, Phys. Rev. B 54, 2642 (1996).
[84] W. Mönch, Semiconductor Surfaces and Interfaces, Springer, Berlin 1993, p. 96

[85] H. Lüth and R. Matz, Phys. Rev. Lett. 46 (1981) 1652
[86] L.H. Dubois and G.P. Schwartz, Phys. Rev. B26 (1982) 794
[87] T.J. Godin, J.P. La Femina and C.B. Duke, J. Vac. Sci. Technol. B 9 (1991) 2282.
[88] J. Fritsch, P. Pavone and U. Schröder, Phys. Rev. Lett. 71 (1993) 4194.
[89] R. de Felice, A. I. Shkrebtii, F. Finocchi, C.M. Bertoni and G. Onida, J. El. Spectr. Rel. Phenom., 64/65 (1995) 697.
[90] W.G. Schmidt, F. Bechstedt and G.P. Srivastava, Phys. Rev. B 52 (1995) 2001.
[91] U. Harten and J.P. Toennies, Europhys. Lett. 4 (1987) 833.
[92] R.B. Doak and D.B. Nguyen, J. El. Spectr. Rel. Phenom., 44 (1987) 205.
[93] L.H. Dubois and G.P. Schwartz, Phys. Rev. B 26 (1982) 794.
[94] M.G. Betti, U. del Pennino and C. Mariani, Phys. Rev. B 39 (1989) 5887.
[95] H. Niehaus and W. Mönch, Phys. Rev. B 50 (1994) 11750.
[96] For an introduction to early P&P experiments see, e.g., S. L. Shapiro, ed., *Topics in Applied Physics* (Springer-Verlag, Heidelberg 1977), Vol. 18.
[97] R. Merlin, in *Highlights In Condensed Matter Physics and Materials Science*, edited by M. Cardona and A. Pinczuk as a special issue of Sol. St. Comm. 102 (1997) 207.
[98] D. von der Linde, J. Kuhl and H. Klingenberg, Phys. Rev. Lett. 44 (1980) 1505.
[99] J. A. Kash, J. C. Tsang and J. M. Hvam, Phys. Rev. Lett. 54 (1985) 2151.
[100] W. Kütt, in *Festkörperprobleme, Advances in Solid State Physics*, Vol. 32, U. Rössler ed. (Vieweg, Braunschweig ,1992); T. Dekorsy, W. Kütt, T. Pfeifer and H. Kurz, Europhys. Lett. 23 (1993) 223.
[101] T. K. Cheng, L. H. Acioli, J. Vidal, H. J. Zeiger, G. Dresselhaus, M. S. Dresselhaus and E. P. Ippen, Appl. Phys. Lett. 62 (1993) 1901.
[102] O. B. Wright and K. Kawashima, Phys. Rev. Lett. 69 (1992) 1668.
[103] M. Nisoli, S. Stagira, S. de Silvestri, A. Stella, P. Tognini, P. Cheyssac and R. Kofman, Phys. Rev. Lett. 78 (1997) 3575
[104] J. Ding, D. M. Photiadis and Z. Barber, Bull. Am. Phys. Soc. 42 (1997) 82.
[105] C. Rocke, S. Zimmermann, A. Wixworth, J. P. Kotthaus, G. Bohm and G. Weinman, Phys. Rev. Letters 78, 4009 (1997).
[106] V. I. Talyanskii, J. M. Shilton, M. Pepper, C. G. Smith, C.J.B. Ford, E.H. Linfield, D.A. Ritchie and G.A.C. Jones, Phys. Rev. B 56, 15180 (1997).
[107] H. Totland, Ø.L. Bø and Y.M. Galperin, Phys. Rev. B 56, 15299 (1997).
[108] N.-H. Ge, C.M. Wong, R.L.Lingle, Jr., J.D. McNeill, K.J. Gaffney and C.B. Harris, Science, 279, 202 (1998). Se also the report by U. Höfer in Science, 279, 190 (1998).

DIELECTRIC DESCRIPTION OF SEMICONDUCTORS:

FROM MAXWELL– TO SEMICONDUCTOR BLOCH–EQUATIONS

Ralph v. Baltz

Institut für Theorie der Kondensierten Materie
Universität Karlsruhe
D–76128 Karlsruhe, Germany

ABSTRACT

A tutorial on the dielectric description of matter is presented with particular attention to semiconductor optics. The first part focuses on general concepts like the construction of macroscopic fields, linear and quadratic response, and simple models how to describe the reaction of matter with respect to the electromagnetic field. Beginning with the two–level approximation of an atom, the second part will lead us to the semiconductor Bloch equations which are the today's standard model of semiconductor optics in the short time regime. Together with the Maxwell equations these form a closed set of dynamical equations for the electromagnetic field, polarization, and electron/hole population of a semiconductor upon optical excitations. Some selected applications and problems are added.

I. INTRODUCTION

For the electrodynamic description of semiconductors near the band edge, matter equations (or constitutive equations) are needed, which relate the charge density and current (or polarization) to the electromagnetic field. The simplest models to describe this coupling are the Lorentz-oscillator and the Drude-free carrier models. For a realistic description, however, the valence - conduction band continuum, excitonic effects, and electron/hole population dynamics must be considered. These phenomena are consistently described by the semiconductor Bloch equations which are a set of nonlinear, coupled differential-integral equations.
The scope of this article is twofold. First, in Chapter II, a survey on the macroscopic electromagnetic description of matter is presented, including some new aspects, discussion of basic models, and fundamentals of linear and nonlinear response.
Second, Chapter III is devoted to the foundation of the microscopic description of the light-matter interaction based on a two-level approach. This will guide us to the

Ultrafast Dynamics of Quantum Systems: Physical Processes and Spectroscopic Techniques, Edited by Di Bartolo and Gambarota, Plenum Press, New York, 1998

Semiconductor Bloch Equations.
To round-off the material, some selected applications and supplements are given in Chapter IV. In particular some properties of the photogalvanic effect are discussed which describes a steady state unidirectional charge transport. Its spectacular properties are the absense of a "driving force" (in the sense of traditional irreversible thermodynamics) and the occurence of photovoltages up to 100 kV in ferroelectrics. Problems (with solutions) are added.

II. MACROSCOPIC ELECTRODYNAMICS

The electromagnetic field (EMF, or just "field") is descibed by the electrical and magnetic fields $\mathcal{E}(\mathbf{r},t), \mathcal{B}(\mathbf{r},t)$ which are coupled to matter through the charge– and current–density fields $\rho(\mathbf{r},t)$ and $\mathbf{j}(\mathbf{r},t)$. To formulate a closed set of equations one needs, besides the Maxwell–equations, either an explicit functional or a set of (differential–) equations for $\rho, \mathbf{j}$ in terms of $\mathcal{E}, \mathcal{B}$[1].
The standard book in the field of macroscopic electrodymamics is Landau and Lifshitz Vol. 8 [1], Wooten [2] gives an excellent introduction to optical properties of solids, and Klingshirn [3] provides a modern introduction to and an overview of semiconductor optics. Linear and nonlinear interactions of electromagnetic waves and matter is covered by several articles of this course [5] and the previous one [4].

II.A. Field Equations

The state of the microscopic electromagnetic field in matter is described by $\mathcal{E}_{\mathrm{m}}, \mathcal{B}_{\mathrm{m}}$ which satisfy the Maxwell–equations:

$$\epsilon_0\mu_0 \frac{\partial \mathcal{E}_{\mathrm{m}}(\mathbf{r},t)}{\partial t} - \operatorname{curl} \mathcal{B}_{\mathrm{m}}(\mathbf{r},t) = -\mu_0 \mathbf{j}_{\mathrm{m}}(\mathbf{r},t), \quad (1)$$

$$\frac{\partial \mathcal{B}_{\mathrm{m}}(\mathbf{r},t)}{\partial t} + \operatorname{curl} \mathcal{E}_{\mathrm{m}}(\mathbf{r},t) = 0, \quad (2)$$

$$\epsilon_0 \operatorname{div} \mathcal{E}_{\mathrm{m}}(\mathbf{r},t) = \rho_{\mathrm{m}}(\mathbf{r},t), \quad (3)$$

$$\operatorname{div} \mathcal{B}_{\mathrm{m}}(\mathbf{r},t) = 0. \quad (4)$$

The first set of Eqs. (1,2) which contain the time–derivatives of the fields are dynamical equations like the Newton–equations for a mechanical system, $\mathbf{j}_{\mathrm{m}}(\mathbf{r},t)$ plays the role of a "driving force". The two other Eqs. (3,4) are of different type and represent "rigid" conditions imposed by $\rho_{\mathrm{m}}(\mathbf{r},t)$ at time t. Together with the Lorentz–force (–density)

$$\mathbf{f}(\mathbf{r},t) = \rho(\mathbf{r},t)\mathcal{E}(\mathbf{r},t) + \mathbf{j}(\mathbf{r},t) \times \mathcal{B}(\mathbf{r},t) \quad (5)$$

these equations define the interacting field–matter system.
$\mathcal{E}_{\mathrm{m}}(\mathbf{r},t), \mathcal{B}_{\mathrm{m}}(\mathbf{r},t)$ contain large, spatially fluctuating contributions on an atomic scale and it is impossible to calculate or measure these fields. (A typical value of such field fluctuations is the field of a nucleus within atomic distances which is of the order of $\mathcal{E}_{\mathrm{m}} \approx 10^9 V/cm$.) To get rid of these fluctuations in a macroscopic description averaging upon so–called physically infinitesimally, small volumes have been known since the Lorentz–era, as a cureable method. If these volumes contain a large number of atoms they can again be treated macroscopically. During the last decade, however, the russian school around Keldysh [6],[7], recognized that this approach is not satisfactory for several reasons, e.g.

[1] vectors and tensors are written in boldface, electric and magnetic fields in caligrafic style

- the wavelength may be considerably reduced by a high refractive index,
- gyrotropy is related to the field gradient on molecular distances,
- the motion of charges is related to the actual field at the position of the particles rather than to the average field.

Therefore, averaging of physically infinitesimally small volumes is abandoned and replaced by the standard method of statistical physics averaging over the Gibbs–ensemble of all possible states of the field and matter, $\mathcal{E} =< \mathcal{E}_{\rm m} >$, $\mathcal{B} =< \mathcal{B}_{\rm m} >$. Owing to the linearity of the Maxwell-equations, this is formally simple and yields equations of the same structure as Eqs. (1-4):

$$\epsilon_0\mu_0\frac{\partial\mathcal{E}(\mathbf{r},t)}{\partial t} - \text{curl }\mathcal{B}(\mathbf{r},t) = -\mu_0\left[\mathbf{j}(\mathbf{r},t) + \mathbf{j}_{\rm ext}(\mathbf{r},t)\right], \tag{6}$$

$$\frac{\partial\mathcal{B}(\mathbf{r},t)}{\partial t} + \text{curl }\mathcal{E}(\mathbf{r},t) = 0, \tag{7}$$

$$\epsilon_0\text{div }\mathcal{E}(\mathbf{r},t) = \rho(\mathbf{r},t) + \rho_{\rm ext}(\mathbf{r},t), \tag{8}$$

$$\text{div }\mathcal{B}(\mathbf{r},t) = 0. \tag{9}$$

For convenience, the external sources $\rho_{\rm ext}(\mathbf{r},t), \mathbf{j}_{\rm ext}(\mathbf{r},t)$ have been separated from the matter fields $\rho(\mathbf{r},t) =< \rho_{\rm m}(\mathbf{r},t) >$, $\mathbf{j}(\mathbf{r},t) =< \mathbf{j}_m(\mathbf{r},t) >$. Here the adjective "external" refers to the control, not to the location of the charges, i.e. we assume that they are not affected by the charges in the medium.

II.B. Matter Equations

In a classical microscopic description the matter fields $\mathbf{j}_{\rm m}(\mathbf{r},t), \rho_{\rm m}(\mathbf{r},t)$ are defined by

$$\rho_{\rm m}(\mathbf{r},t) = \sum_{k=1}^{N} e_k\delta(\mathbf{r}-\mathbf{r}_k(t)), \quad \mathbf{j}_{\rm m}(\mathbf{r},t) = \sum_{k=1}^{N} e_k\mathbf{v}_k\delta(\mathbf{r}-\mathbf{r}_k(t)), \tag{10}$$

where the trajectories $r_i(t)$ of the particles with masses M_k and charges e_k are determined by the Newton-equations:

$$M_k\frac{d^2\mathbf{r}_k(t)}{dt^2} = e_k\mathcal{E}'_{\rm m}(\mathbf{r}_k,t) + e_k\mathbf{v}_k\times\mathcal{B}'_{\rm m}(\mathbf{r}_k,t), \quad \text{k=1...N}. \tag{11}$$

$\mathcal{E}'_{\rm m}, \mathcal{B}'_{\rm m}$ denote the fields without the self-contribution of particle $\#k$. For example, such calculations are presently performed numerically for high–power gyrotrons or particle accelerators by using a "particle in cell code".
For our purposes, a classical description of the EMF is sufficient, however, the matter must be treated quantum mechanically. In this case one has to find the wave–function (or statistical operator) from which the expectation values of the charge– and current–denstiy operators can be calculated. This will be done in Chapter III.
Instead of solving the microscopic equations within some approximation and performing the average afterwards, a much better strategy is to derive and solve manageable equations for $\rho(\mathbf{r},t), \mathbf{j}(\mathbf{r},t)$ in terms $\mathcal{E}(\mathbf{r},t)$, $\mathcal{B}(\mathbf{r},t)$. This is the main issue of this article. A trivial example is the equation of continuity which likewise holds for the microscopic and macroscopic charge- and current–density

$$\frac{\partial\rho(\mathbf{r},t)}{\partial t} + \text{div }\mathbf{j}(\mathbf{r},t) = 0. \tag{12}$$

For stationary fields $\mathcal{E}(\mathbf{r})$ and $\mathcal{B}(\mathbf{r})$ are independent, whereas for time-dependent fields $\mathcal{B}(\mathbf{r},t)$ is fixed by $\mathcal{E}(\mathbf{r},t)$ up to a time-independent field (which will be left–out below). The same holds for $\rho(\mathbf{r},t)$ and $\mathbf{j}(\mathbf{r},t)$:

$$\mathcal{B}(\mathbf{r},t) = -\int_{t_0}^{t} \operatorname{curl}\mathcal{E}(\mathbf{r},t')\,dt', \tag{13}$$

$$\rho(\mathbf{r},t) = -\int_{t_0}^{t} \operatorname{div}\mathbf{j}(\mathbf{r},t')\,dt'. \tag{14}$$

Thus, there is only a single independent matter field, namely $\mathbf{j}(\mathbf{r},t) = \mathbf{j}[\mathcal{E}(\mathbf{r},t)]$ which can be written as a functional solely of $\mathcal{E}(\mathbf{r},t)$.
Instead of using the current density $\mathbf{j}(\mathbf{r},t)$ it is sometimes convenient to work with the polarization $\mathcal{P}(\mathbf{r},t)$ defined by

$$\mathbf{j}(\mathbf{r},t) = \frac{\partial \mathcal{P}(\mathbf{r},t)}{\partial t}, \qquad \rho(\mathbf{r},t) = -\operatorname{div}\mathcal{P}(\mathbf{r},t). \tag{15}$$

This definition includes the continuity equation (12).
Contributions from "free" charges or "magnetic" effects are not simply neglected but they are contained in $\mathcal{P}(\mathbf{r},t)$. Splitting the matter–current into free, bound, and magnitization currents is only useful for quasistationary fields but is neither necessary nor advantageous in solid state optics. At high frequencies the oscillation amplitude of "free" and "bound" charges are of the same order, hence, there is no physical difference. For an experimental investigation of "magnetic" contributions in the IR range see Grosse [8]. The price to pay leaving the magnetization $\mathcal{M}(\mathbf{r},t)$ out of the game is the need of a space–dependent polarization field even when in the convential description $\mathcal{P}$ and $\mathcal{M}$ are (piecewise) constant. But this can be done on equal footing with spatial dispersion (see chapter II.C.).
In the following, we shall preferably work with $\mathbf{j}(\mathbf{r},t)$ to describe "metallic" systems ("free" charges, intraband dynamics) and $\mathcal{P}(\mathbf{r},t)$ for "dielectric" behaviour ("bound" charges, interband dynamics). This is motivated by the fact that for slowly varying fields (with respect to time and space) the following relations hold

$$\mathbf{j}(\mathbf{r},t) = \sigma\mathcal{E}(\mathbf{r},t), \qquad \mathcal{P}(\mathbf{r},t) = \epsilon_0\chi\mathcal{E}(\mathbf{r},t), \tag{16}$$

where constants σ, χ represent the electrical conductivity and susceptibility. These are the simplest form of matter–equations. Systems which contain both types of carriers are conventionally modelled just by adding both contributions.
On a phenomenological level the functional relation between $\mathbf{j}(\mathbf{r},t)$, $\mathcal{P}(\mathbf{r},t)$ and $\mathcal{E}(\mathbf{r},t)$ may be represented by a power-expansion in terms of the field

$$\mathbf{j}(\mathbf{r},t) = \sum_{k=1}^{\infty}\mathbf{j}^{(k)}(\mathbf{r},t), \qquad \mathcal{P}(\mathbf{r},t) = \sum_{k=1}^{\infty}\mathcal{P}^{(k)}(\mathbf{r},t), \qquad \mathbf{j}^{(k)},\ \mathcal{P}^{(k)} \propto \mathcal{E}^k. \tag{17}$$

Expansion (17) is possible if $\mathcal{E}$ is much smaller than typical atomic fields $\mathcal{E}_{at} \approx 10^9$ V/cm. Although such fields cannot be produced in steady state laboratory experiments, ten times larger fields have been recently created in short laser pulses. In addition, fields may be strongly enhanced near resonances as, e.g. in the dynamical Stark–effect.

Within a classical description the dynamics of conduction electrons in a semiconductor or a metal is governed by a hydrodynamic type of equation which is the generalization of the famous Drude–model:

$$\left(\frac{\partial}{\partial t} + \gamma\right)\mathbf{j}(\mathbf{r},t) + \beta\,\mathrm{grad}\,\rho(\mathbf{r},t) = \frac{-e}{m^*}\,\rho(\mathbf{r},t)\mathcal{E}(\mathbf{r},t). \tag{18}$$

m^* is the effective mass, e the charge, γ the relaxation rate of the carriers, and β denotes a dispersion constant which is proportional to the diffusion constant. (For metals $\beta = 3v_F^2/5$, where v_F is the Fermi–velocity.) In addition to the standard Drude model, Eq.(18) includes diffusion. (The coupling to the magnetic field via the Lorentz–force is left–out for simplicity.) For applications to metal–optics near the plasma–edge see Forstmann and Gerhards [9], for plasmons see e.g. v. Baltz [10].
Bound charges like optical phonons can be modelled by an oscillator type of equation which is known as the Lorentz–model [3]:

$$\left(\frac{\partial^2}{\partial t^2} + \gamma\frac{\partial}{\partial t} + \omega_0^2 + \beta\Delta\right)\mathcal{P}(\mathbf{r},t) = \epsilon_0\Omega_p^2\mathcal{E}(\mathbf{r},t), \tag{19}$$

where ω_0 is oscillation frequency, n_0 is the density of oscillators, $\Omega_p^2 = n_0e^2/m^*\epsilon_0$, and $\beta\Delta\mathcal{P}$ accounts for the coupling to neighbouring oscillators [11].
These differential equations are supplemented by boundary conditions like the continuity of the normal component of $\mathbf{j}$ or $\mathcal{D} = \epsilon_0\mathcal{E} + \mathcal{P}$ at surfaces or interfaces. Note, $\mathcal{E}(\mathbf{r},t)$ denotes the total electrical field in matter - rather than the external field.

Problems:
1.) Find the general solutions of Eqs.(18-19) for the homogeneous case. (Neglect nonlinearity in Eq.(18) replacing $\rho(\mathbf{r},t)$ by $-|e|n_0$, where n_0 is the equilibrium electron density.) Use $\mathbf{j}(-\infty) = 0$ and $\mathcal{P}(-\infty) = 0$ as boundary conditions.

2.) Screening of a point–charge by free carriers.
Find the stationary solution for the induced charge–density and potential of a point charge in a metal within the hydrodynamic model as given by Eq. (18). Compare with the bare Coulomb–potential.

II.C. Linear Response

The general form of the linear part of expansion (17) between the current or polarization and the field reads:

$$\mathbf{j}_\alpha^{(1)}(\mathbf{r},t) = \int\int \sigma_{\alpha\beta}^{(1)}(\mathbf{r},\mathbf{r}',t-t')\mathcal{E}_\beta(\mathbf{r}',t')d^3\mathbf{r}'dt', \tag{20}$$

$$\mathcal{P}_\alpha^{(1)}(\mathbf{r},t) = \epsilon_0\int\int \chi_{\alpha\beta}^{(1)}(\mathbf{r},\mathbf{r}',t-t')\mathcal{E}_\beta(\mathbf{r}',t')d^3\mathbf{r}'dt'. \tag{21}$$

For brevity we shall discuss only the $\mathcal{P}$–$\mathcal{E}$ relation (21) in the following, as the current–field relation (20) is analogous.
The "susceptibility–kernel" χ takes into account that the coupling between field and polarization generally is

- nonlocal, i.e. the field at $\mathbf{r}'$ can cause a polarization at another point $\mathbf{r}$,
- has a memory, i.e. $\mathcal{P}$ may exist for some time after the field is switched–off,
- $\mathcal{P}$ may not be parallel to $\mathcal{E}$, i.e. χ is a second rank tensor where α, β denote cartesian components. Summation over repeated indices is implied.

For homogeneous matter susceptibility tensor $\chi_{\alpha,\beta}$ is solely a function of $\mathbf{r} - \mathbf{r}'$ so that the integral relation (21) becomes a convolution:

$$F(t) := [f_1 \otimes f_2](t) = \int_{-\infty}^{\infty} f_1(t-t')f_2(t')dt' \tag{22}$$

which simplifies to a product under Fourier–transformation

$$F(\omega) = f_1(\omega) \cdot f_2(\omega). \tag{23}$$

$f(t)$, $f(\omega)$ denote a Fourier–pair:

$$f(t) = \int_{-\infty}^{\infty} f(\omega) e^{-i\omega t} \frac{d\omega}{2\pi}, \quad f(\omega) = \int_{-\infty}^{\infty} f(t) e^{+i\omega t} dt. \tag{24}$$

The $(\mathbf{r}, t)$–Fourier–transformation is used in the following "plane–wave" form:

$$\chi^{(1)}_{\alpha\beta}(\mathbf{r}, t) = \int\int e^{i(\mathbf{qr}-\omega t)} \chi^{(1)}_{\alpha\beta}(\mathbf{q}, \omega) \frac{d^3\mathbf{q} d\omega}{(2\pi)^4}, \tag{25}$$

$$\chi^{(1)}_{\alpha\beta}(\mathbf{q}, \omega) = \int\int e^{-i(\mathbf{qr}-\omega t)} \chi^{(1)}_{\alpha\beta}(\mathbf{r}, t) d^3\mathbf{r} dt, \tag{26}$$

$$\mathcal{P}^{(1)}_{\alpha}(\mathbf{q}, \omega) = \epsilon_0 \chi^{(1)}_{\alpha\beta}(\mathbf{q}, \omega) \mathcal{E}_{\beta}(\mathbf{q}, \omega). \tag{27}$$

The $\mathbf{q}, \omega$ dependence of the susceptibility (or σ) is termed spatial and temporal dispersion, respectively. $\chi(\mathbf{q}, \omega) = \frac{i}{\epsilon_0 \omega} \sigma(\mathbf{q}, \omega)$.
To keep the presentation simple we omit the tensorial structure of χ, the superscript (1), and spatial dispersion. The most important property of $\chi(t - t')$ is the property of causality: There is no response before the perturbation is turned on. This is one of the fundamental laws of nature.

$$\chi(t - t') \equiv 0, \quad t' > t. \tag{28}$$

Some consequences of causality in the frequency domain will be exploited in the next section.
In contrast to the real response kernel $\chi(t-t')$ in the time–domain its Fourier–transform $\chi(\omega)$ is complex

$$\chi(\omega) = \chi_1(\omega) + i\chi_2(\omega) = \int_{-\infty}^{\infty} \chi(t'') e^{i\omega t''} dt''. \tag{29}$$

The real and imaginary parts of $\chi(\omega)$ are even and odd functions of frequency. (This holds regardless of causality.) Moreover, the real part of $\sigma(\omega)$ and the imaginary part of $\chi(\omega)$ are related to dissipation, whereas the other parts are connected to dispersion. For details see [1], [2] or e.g. an overview given by Di Bartolo in this book [12].
As an illustration of the time and frequency dependence of response functions we state the results for the Drude– and Lorentz–models (omitting spatial dispersion, $\mathbf{q} = 0$. See problems 1,3, and 4), Figs. 1,2.
The Drude conductivity and susceptibility are:

$$\sigma(t - t') = \frac{n_0 e^2}{m^*} e^{-\gamma(t-t')} \theta(t - t'), \tag{30}$$

$$\sigma(\omega) = \frac{n_0 e^2}{m^*} \frac{1}{\gamma - i\omega}, \tag{31}$$

$$\chi(t - t') = \frac{\omega_p^2}{\gamma} \left[1 - e^{-\gamma(t-t')}\right] \theta(t - t'), \tag{32}$$

$$\chi(\omega) = -\frac{\omega_p^2}{\omega(\omega + i\gamma)}, \quad \omega_p^2 = \frac{n_0 e^2}{m^* \epsilon_0}, \tag{33}$$

where n_0 and ω_p denote the density and plasma–frequency, respectively.

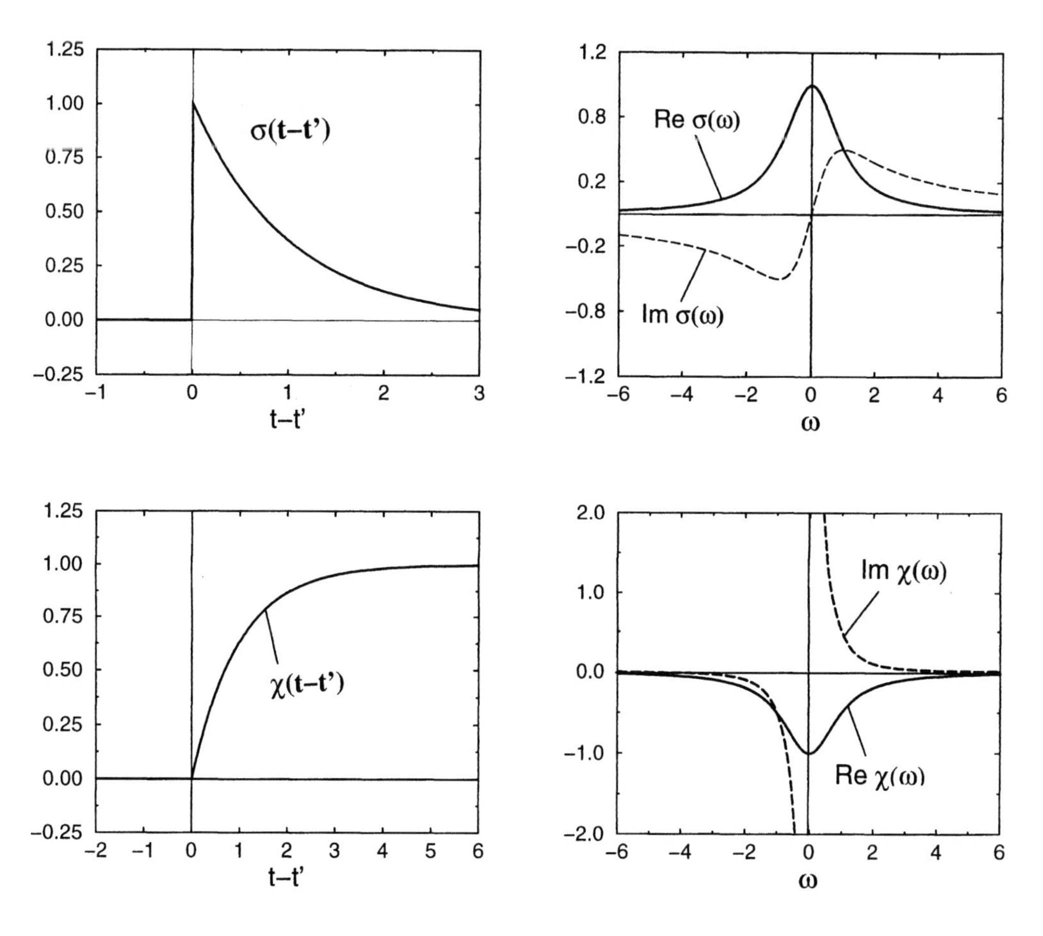

Fig. 1 *Drude–conductivity and susceptibility. (Left) Time domain, (right) frequency domain. (Dimensionless units).*

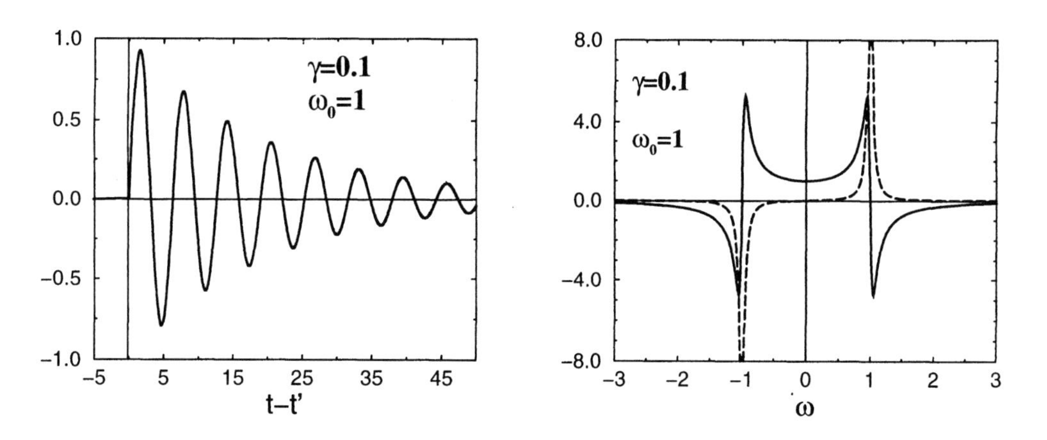

Fig. 2 *Lorentz–susceptibility. (Left) Time domain, (right) frequency domain. (Dimensionless units).*

The Lorentz–susceptibility is given by:

$$\chi(t-t') = \frac{\Omega_p^2}{\Omega_0} e^{-\frac{\gamma}{2}(t-t')} \sin[\Omega_0(t-t')]\,\theta(t-t'), \tag{34}$$

$$\chi(\omega) = \frac{\Omega_p^2}{\omega_0^2-\omega^2-i\omega\gamma}, \quad \Omega_0^2 = \omega_0^2 - (\frac{\gamma}{2})^2. \tag{35}$$

It is convenient to decompose all vector-fields into longitudinal and transverse components with respect to wave vector $\mathbf{q}$, i.e.

$$\mathcal{E}(\mathbf{q},\omega) = \mathcal{E}_\ell(\mathbf{q},\omega) + \mathcal{E}_t(\mathbf{q},\omega) = (\mathcal{E}\cdot\mathbf{n_q})\,\mathbf{n_q} + \mathbf{n_q}\times(\mathcal{E}\times\mathbf{n_q}), \tag{36}$$

with unit vector $\mathbf{n_q} = \mathbf{q}/|\mathbf{q}|$. Even for homogeneous and isotropic systems, like Jellium, the reaction of the charged particles with respect to transverse and longitudinal fields is different, i. e. $\sigma(\mathbf{q},\omega)$ and $\chi(\mathbf{q},\omega)$ are tensors with two different principal components:

$$\mathbf{j}(\mathbf{q},\omega) = \epsilon_0\,\hat{\sigma}(\mathbf{q},\omega)\,\mathcal{E}(\mathbf{q},\omega) = \sigma_\ell(\mathbf{q},\omega)\mathcal{E}_\ell(\mathbf{q},\omega) + \sigma_t(\mathbf{q},\omega)\mathcal{E}_t(\mathbf{q},\omega). \tag{37}$$

Transverse fields, $\mathbf{j}\cdot\mathbf{q}=0$, are source–free, $\mathrm{div}\,\mathbf{j}=0$, and, thus, do not create charge fluctuations: $\rho=0$. On the other hand, longitudinal fields, $\mathbf{j}\parallel\mathbf{q}$, are irrotational and create charge fluctuations: $\rho(\mathbf{q},\omega) = q\,j(\mathbf{q},\omega)/\omega$.

Problems:

3.) Calculate $\sigma(\mathbf{q},\omega)$ and $\epsilon(\mathbf{q},\omega) = 1 + i\sigma(\mathbf{q},\omega)/\omega\epsilon_0$ for longitudinal and transverse fields in a metal by Fourier–transformation of Eq. (18). $\epsilon(\mathbf{q},\omega)=0$ gives the dispersion of longitudinal collective excitations(=plasmons) [10].

4.) Calculate the transverse electrical susceptibility for bound charges by Fourier–transformation of Eq.(19). The poles of $\chi(\mathbf{q},\omega)$ (or $\epsilon = 1+\chi$) gives the dispersion of transverse excitations. Compare with transverse optical phonons [3].

II.D. Nonlinear Response

In intense laser pulses many nonlinear phenomena are observed which are described by the second and higher order terms in the expansion (17). For an introduction and survey see, e.g., Boyd [13].
For simplicity we consider only the second–order current–field relation and omit spatial dispersion. Then, an expansion analogous to Eq.(20) holds:

$$\mathbf{j}_\alpha^{(2)}(\mathbf{r},t) = \int\!\!\int \sigma_{\alpha\beta\gamma}^{(2)}(t-t',t-t'')\mathcal{E}_\beta(\mathbf{r},t')\mathcal{E}_\gamma(\mathbf{r},t'')dt'dt''. \tag{38}$$

As in the linear case, causality restricts the time–integration to $t',t''<t$. In Fourier–space the double time–integral is reduced to a single, convolution–type integral:

$$\mathbf{j}_\alpha^{(2)}(\mathbf{r},\omega) = \int \sigma_{\alpha\beta\gamma}^{(2)}(\omega',\omega-\omega')\mathcal{E}_\beta(\mathbf{r},\omega')\mathcal{E}_\gamma(\mathbf{r},\omega-\omega')\frac{d\omega'}{2\pi}, \tag{39}$$

$$\sigma_{\alpha\beta\gamma}^{(2)}(\omega_1,\omega_2) = \int\!\!\int \sigma_{\alpha\beta\gamma}^{(2)}(t_1,t_2)e^{i(\omega_1t_1+\omega_2t_2)}dt_1dt_2. \tag{40}$$

The quadratic response is described by a third rank tensor which, in contrast to linear response, exists only in noncentrosymmetric crystals like $GaAs$ or $LiNbO_3$. Germanium and Silicon, on the other hand, have a center of inversion and the first nonvanishing nonlinear contribution begins in terms of a cubic fourth rank tensor $\sigma_{\alpha\beta\gamma\delta}^{(3)}$.

As ω_1, ω_2 are dummy variables in Eq.(38), $\chi^{(2)}_{\alpha,\beta\gamma}(\omega_1, \omega_2)$ is symmetric (or can be chosen to be symmetric) with respect to the pair of indices (α, ω_1), (β, ω_2). Further symmetry properties can be found, e.g. in [13].

As an example, we consider the case in which the optical field incident upon a nonlinear optical medium characterized by a quadratic conductivity $\sigma^{(2)}$ consists of two distinct frequency components ω_1, ω_2, $(\omega_2 > \omega_1)$:

$$\mathcal{E}(t) = \mathcal{E}_1 e^{-i\omega_1 t} + \mathcal{E}_2 e^{-i\omega_2 t} + [1 \to -1, 2 \to -2], \quad (41)$$

$$\mathcal{E}(\omega) = 2\pi[\mathcal{E}_1 \delta(\omega - \omega_1) + \mathcal{E}_2 \delta(\omega - \omega_2)] + [1 \to -1, 2 \to -2]. \quad (42)$$

$[1 \to -1]$ means a change of ω_1 by $-\omega_1$ in the preceding expression etc. and $\mathcal{E}_{-1} = \mathcal{E}_1^*$. Then, according to Eq.(38) the second–order current contribution is given by [2]

$$\mathbf{j}^{(2)}(\omega) = 2\pi\Big\{$$

$$+\mathcal{E}_1^2 \sigma^{(2)}(+\omega_1, +\omega_1)\, \delta(\omega - 2\omega_1) + \mathcal{E}_2^2 \sigma^{(2)}(+\omega_2, +\omega_2)\, \delta(\omega - 2\omega_2) \quad (43)$$

$$+2\mathcal{E}_1 \mathcal{E}_2 \sigma^{(2)}(+\omega_1, +\omega_2)\, \delta(\omega - [\omega_1 + \omega_2]) \quad (44)$$

$$+2\mathcal{E}_1^* \mathcal{E}_2 \sigma^{(2)}(-\omega_1, +\omega_2)\, \delta(\omega - [\omega_2 - \omega_1]) \quad (45)$$

$$+[1 \to -1, 2 \to -2],$$

$$+2\left[\mid \mathcal{E}_1 \mid^2 \sigma^{(2)}(-\omega_1, +\omega_1) + \mid \mathcal{E}_2 \mid^2 \sigma^{(2)}(-\omega_2, +\omega_2)\right] \delta(\omega)\Big\}. \quad (46)$$

Via the Maxwell–equations the induced current will in turn be the source of radiation. The various terms describe:

- second harmonic generation (SHG), Eq. (43)
- sum frequency generation (SFG), Eq. (44)
- difference frequency generation (DFG), Eq. (45)
- optical rectification (OR), photogalvalvanic effect (PGE), Eq. (46).

For the description of high–frequency current phenomena SHG, SFG, and DFG the point of view of a "free" charge current $\mathbf{j}(\mathbf{r}, t)$ and a "bound" charge current $\frac{\partial}{\partial t}\mathcal{P}(\mathbf{r}, t)$ are fully equivalent. However, this is different for the $\omega = 0$ component of the current and the polarization. The $\omega = 0$ component of $\mathcal{P}(\mathbf{r}, t)$, known as optical rectification describes an isothermal and isobaric change of the value of the polarization only. It does not give rise to a steady–state current or an electromotive force. In contrast to OR which describes charge separation across a finite distance, the $\omega = 0$ component of $\mathbf{j}(\mathbf{r}, t)$ may be considered as a charge separation across infinite distances. In addition, OR is nondissipative and can occur in the nominally transparent part of the spectrum (where $\text{Im}\, \chi(\omega) = 0$) whereas absorption of light and, hence, dissipation is needed to induce a direct (nonsupra)–current. For instationary excitations OR leads to a transient current whose shape is given by the time derivative of the intensity profile $j \propto \dot{I}(t)$, whereas, the PGE would lead to a current pulse which follows $I(t)$. (As OR is of minor importance in nonlinear optics little attention is usually paid for this subtle distinction and the notation OR is sometimes ambiguous.) The occurence of a direct current upon light absorption in nonlinear crystals is now called photogalvanic effect (or bulk photovoltaic effect) and some of its exciting properties will be discussed in chapter IV.C.

[2] In nonlinear optics a redundant notation with three frequency arguments is used : $\sigma^{(2)}(\omega_3, \omega_2, \omega_1)$. This is technically unnecessary in that ω_3 is always $\omega_1 + \omega_2$. Analogous for the higher order terms.

As in linear response there are Kramers–Kronig relations for second and higher order. However, nonlinear phenomena are almost exclusively studied in the nonabsorptive part of the spectrum so that these relations are of no practical importance.

III. QUANTUM THEORY OF ATOMS AND SEMICONDUCTORS

Our macroscopic description of linear and nonlinear properties of matter has mostly made use of a power series of the current or polarization in terms of the field. Under resonant excitation this approximation fails to provide an adequate description of the response and a description by (nonlinear) equations like Eq.(103) are more appropriate. Under resonant conditions it is usually sufficient to deal only with the two levels which are nearly resonantly excited by the light. Even for semiconductors the optical transitions between valence and conduction band can be visualized as transitions between a collection of two–level systems (TLS).
The discussion of the Semiconductor–Bloch–equations (SBE) will proceed in three steps. First we study the dynamics of atoms near resonance in the two–level approximation and derive the atomic Bloch equations for the polarization. Next, this result is generalized to the case of a semiconductor with noninteracting valence and conduction bands. Eventually, the influence of Coulomb–interaction between the electron–hole excitations is considered. This will lead us to the SBE which presently are the standard model of semiconductor optics in particular to describe nonlinear short pulse phenomena. Presumably Stahl [11] was the first to use such equations in a systematic way to describe the electrodynamics of semiconductors near the band edge.
According to the scope of this article the presentation is kept on an introductory level and "sophisticated" techniques are avoided. A thorough derivation of the SBE is outlined by Haug and Koch [14], which is the standard book in this field, or by Zimmermann [15].

III.A. Dynamics of the Two–Level–System

The optical properies of TLS are presented in many texts, my favorites are the Feynman lectures Vol. 3 in connection with the Ammonia maser [16] and the book by Allen and Eberly [17].
To describe the optical properties of an atom near resonance, we only retain the pair of nearly resonant stationary states $|1>$ and $|2>$ with energies ϵ_1 and ϵ_2, $(\epsilon_2 > \epsilon_1)$, respectively. In particular, we assume that these states have s and p symmetry so that the optical transition is dipole–allowed. In this restricted "base" the state vector of the atom

$$| \psi(t) >= c_1(t)|1> + c_2(t)|2> \tag{47}$$

is represented by the coefficients c_1, c_2 which can be arranged in form of a two–component column vector $\mathbf{c}$. $\mathbf{c}^\dagger = (c_1, c_2)$.
The time–dependence of $c_j(t)$ is governed by the Schrödinger equation:

$$i\hbar\frac{\partial}{\partial t}\begin{pmatrix} c_1 \\ c_2 \end{pmatrix} = \hat{\mathbf{H}}\begin{pmatrix} c_1 \\ c_2 \end{pmatrix}. \tag{48}$$

The first term in the Hamiltonian

$$\hat{\mathbf{H}} = \hat{\mathbf{H}}_0 - \hat{\mathbf{P}}\mathcal{E}(t), \qquad \hat{\mathbf{H}}_0 = \begin{pmatrix} \epsilon_1 & 0 \\ 0 & \epsilon_2 \end{pmatrix}, \qquad \hat{\mathbf{P}} = \begin{pmatrix} 0 & p \\ p^* & 0 \end{pmatrix}. \tag{49}$$

describes the isolated atom, whereas the second term is the interaction with the electrical field of the classical light wave in electrical dipole approximation. p is the dipole matrix–element between $|1>$ and $|2>$.
From the Schrödinger–equation (48) we obtain two coupled first–order differential equations for $c_1(t)$ and $c_2(t)$:

$$\begin{aligned} i\hbar\dot{c}_1(t) &= \epsilon_1 c_1(t) - p\mathcal{E}(t)c_2(t), &(50)\\ i\hbar\dot{c}_2(t) &= \epsilon_2 c_2(t) - p^*\mathcal{E}(t)c_1(t). &(51)\end{aligned}$$

From $c_1(t), c_2(t)$ the energy and the dipole moment of the TLS are fixed by:

$$\begin{aligned} E(t) &= \mathbf{c}^\dagger\hat{\mathbf{H}}_0\mathbf{c} = \epsilon_1|c_1(t)|^2 + \epsilon_2|c_2(t)|^2 = \frac{\epsilon_1+\epsilon_2}{2} + \frac{\epsilon_2-\epsilon_1}{2}I(t), &(52)\\ p(t) &= \mathbf{c}^\dagger\hat{\mathbf{P}}\mathbf{c} = pP(t) + p^*P^*(t), &(53)\\ I(t) &= |c_2(t)|^2 - |c_1(t)|^2, &(54)\\ P(t) &= c_1^*(t)c_2(t), &(55)\end{aligned}$$

where $I(t)$ is the inversion and $P(t)$, the complex dipole moment, which are the basic quantities to describe the physics of the TLS.
For the unperturbed atom ($\mathcal{E} \equiv 0$) the time evolution of $c_j(t)$ is

$$c_j(t) = d_j e^{-i\epsilon_j t/\hbar}, \quad j = 1, 2, \tag{56}$$

with constant prefactors d_1 and d_2. Hence, $I(t)$ =const and $P(t) = d_1^* d_2 \exp(-i\omega_0 t)$, where $\omega_0 = (\epsilon_2 - \epsilon_1)/\hbar$ is the transition frequency between the energy levels ϵ_j.
To solve the coupled system of differential Eqs.(50,51) we first split–off the free time evolution:

$$\begin{aligned} c_j(t) &= d_j(t)e^{-i\epsilon_j t/\hbar}, \qquad j = 1, 2, &(57)\\ \dot{d}_1(t) &= i\frac{p}{\hbar}\mathcal{E}(t)e^{-i\omega_0 t}d_2(t), &(58)\\ \dot{d}_2(t) &= i\frac{p^*}{\hbar}\mathcal{E}(t)e^{+i\omega_0 t}d_1(t). &(59)\end{aligned}$$

These equations are somewhat simpler than Eqs.(50,51), but an analytical solution is still not accessible. Near resonance, however, the product of $\mathcal{E}(t) = \mathcal{E}_0\cos(\omega t)$ and $e^{\pm i\omega_0 t}$ contains a term which is almost constant and another one which oscillates rapidly. This fast oscillating term will be neglected in the following (this is termed "rotating wave approximation", RWA. See the end of this chapter).
Besides the transition frequency ω_0, there are two other characteristic frequencies:

- $\nu = \omega - \omega_0$, which is called "detuning", and
- $\omega_R = p\mathcal{E}_0/\hbar$, the Rabi-frequency.

Within the RWA, the system of differential Eqs.(58,59):

$$\begin{aligned} \dot{d}_1(t) &= i\frac{\omega_R}{2}e^{+i\nu t}d_2(t), &(60)\\ \dot{d}_2(t) &= i\frac{\omega_R}{2}e^{-i\nu t}d_1(t), &(61)\end{aligned}$$

still contains an explicit time dependence. Nevertheless, Eqs.(60,61) transform to the harmonic oscillator when inserting Eq.(60) into Eq.(61). In contrast to the standard oscillator equation, it contains "imaginary" rather than real damping:

$$\left[\frac{d^2}{dt^2} + i\nu\frac{d}{dt} + (\frac{\omega_R}{2})^2\right] d_1(t) = 0. \tag{62}$$

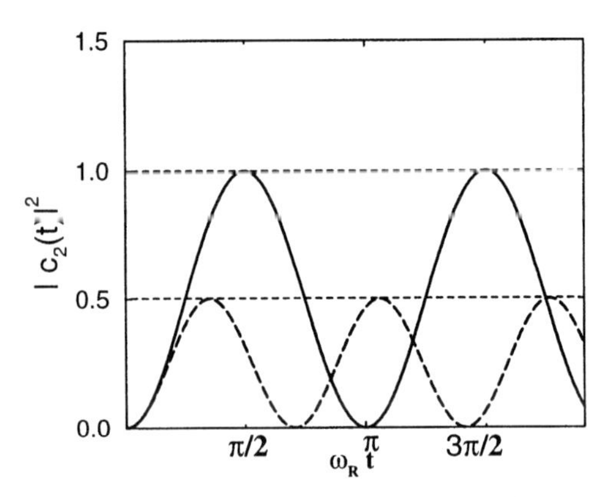

Fig. 3 *Rabi–oscillations of the excited state population. (Full line) Resonant excitation, (dashed line) detuned excitation, $\omega = 2\omega_0$.*

The solution can be found by the standard exponential Ansatz and reads:

$$d_1(t) = \left[a\cos(\frac{\Omega_R}{2}t) + b\sin(\frac{\Omega_R}{2}t)\right]\exp\left(i\frac{\nu}{2}t\right), \tag{63}$$

$$d_2(t) = -i\frac{2}{\omega_R}e^{-i\nu t}\dot{d}_1(t). \tag{64}$$

$\Omega_R = \sqrt{\omega_R^2 + \nu^2}$ is the "detuned" Rabi–frequency.
For example, at resonance and for the initial conditions $c_1(0) = 1$, $c_2(0) = 0$ we have:

$$d_1(t) = \cos(\frac{\omega_R}{2}t), \quad d_2(t) = i\sin(\frac{\omega_R}{2}t). \tag{65}$$

The probability to find the atom in the excited state is given by the absolute square of $d_2(t)$ which oscillates with the Rabi–frequency ω_R. At time $t_1 = \pi/\omega_R$ the atom is in the excited state and at $2\pi/\omega_R$ it is back again in the ground state. For detuned fields, the oscillation period becomes shorter and the amplitude is less than unity, Fig. 3.
For example, for atomic sodium the parameters for the $3s - 3p$ transition are: $p = 2.5a_Be$, $\lambda_0 = 589nm$. For an intensity of 127 Watt cm^{-2} the Rabi–frequency $\omega_R/2\pi = 1$GHz becomes larger than the natural line width [13].
Next, we reformulate the problem and set up an equation for the complex dipole moment and the inversion themselves in terms of the driving field. $\mathcal{P}(t)$ and $I(t)$ likewise fulfill first order differential equations but, in contrast to Eqs. (60,61) these are nonlinear:

$$\left[\frac{d}{dt} + i\omega_0\right]P(t) = -i\frac{p^*}{\hbar}\mathcal{E}(t)I(t) + \dot{P}_{sc}, \tag{66}$$

$$\frac{dI(t)}{dt} = -4\,\text{Im}\left[\frac{p}{\hbar}\mathcal{E}(t)\mathcal{P}^*(t)\right] + \dot{I}_{sc}. \tag{67}$$

The advantage of these equations with respect to Eqs.(50,51) is the possibility to include damping (=collisions, scattering, or incoherent motion). In a simple phenomenological description this is accomplished by:

$$\dot{P}_{sc} = -\frac{P}{T_2}, \quad \dot{I}_{sc} = -\frac{I(t) - I_{eq}}{T_1}. \tag{68}$$

T_1 and T_2 are called "longitudinal" and "transverse" relaxation times. As I is the square of an amplitude we expect $T_2 \approx 2T_1$. I_{eq} is the equilibrium value of the inversion in the absence of the driving field. At zero temperature $I_{eq} = -1$.
Without damping, there is a conserved quantity,

$$4\left|P(t)\right|^2 + I^2(t) = const \tag{69}$$

which may be used to eliminate the inversion from Eq.(66). Its origin becomes obvious from a remarkable analogy between a two-level system and a spin-system in a magnetic field: The level-splitting between the ground state and the excited state of the atom plays the role of a constant magnetic field in z-direction, whereas the light field is equivalent to an oscillatory magnetic field in x-direction. The expectation value of the spin operator, $\mathbf{S} = (S_1, S_2, S_3)$, is closely related to the complex dipole moment and the inversion:

$$S_1 = <\hat{\sigma}_x> = c_1^* c_2 + c_1 c_2^* = 2\,\mathrm{Re}\,P(t), \tag{70}$$

$$S_2 = <\hat{\sigma}_y> = -ic_1^* c_2 + ic_1 c_2^* = 2\,\mathrm{Im}\,P(t), \tag{71}$$

$$S_3 = <\hat{\sigma}_z> = \mid c_1 \mid^2 - \mid c_2 \mid^2 = -I(t), \tag{72}$$

and obeys the atomic Bloch-equations:

$$\frac{d\mathbf{S}(t)}{dt} = \mathbf{\Omega} \times \mathbf{S}(t) + \dot{\mathbf{S}}_{sc}, \quad \dot{\mathbf{S}}_{sc} = \begin{pmatrix} -S_1/T_2 \\ -S_2/T_2 \\ -(S_3 - S_3^{eq})/T_1 \end{pmatrix}, \quad \mathbf{\Omega} = \begin{pmatrix} -\omega_R \cos\omega t \\ \omega_R \sin\omega t \\ -\omega_0 \end{pmatrix}, \tag{73}$$

which describe a rotation of $\mathbf{S}$ around vector $\mathbf{\Omega}$ at each instant of time. The second component of $\mathbf{\Omega}$ is a consequence of the RWA so that the notation eventually becomes obvious.
In the absence of relaxation, the length of the Bloch vector $\mathbf{S}$ is conserved and its motion can be nicely visualized, Figs. 4,5. We consider two limiting cases. Without a time dependent field, $\mathbf{S}$ rotates on a cone around the z-axis which is called Larmor-precession:

$$S_1 = a \sin\omega_0 t, \tag{74}$$

$$S_2 = a \cos\omega_0 t, \tag{75}$$

$$S_3 = \text{const.} \tag{76}$$

If the system is excited at resonance from initial state $\mathbf{S} = (0, 0, 1)$ it performes Rabi–oscillations:

$$S_1 = \sin\omega_R t \sin\omega_0 t, \tag{77}$$

$$S_2 = \sin\omega_R t \cos\omega_0 t, \tag{78}$$

$$S_3 = \cos\omega_R t. \tag{79}$$

Problems:

5.) Calculate $d_j(t)$ according to Eq.(64) for arbitrary detuning and initial conditions $d_1(0) = 1$ and $d_2(0) = 0$.

6.) At resonance there are states of the coupled TLS–electrical field with time-independent probabilities $|d_j(t)|^2 = \text{const}$. Find these states!

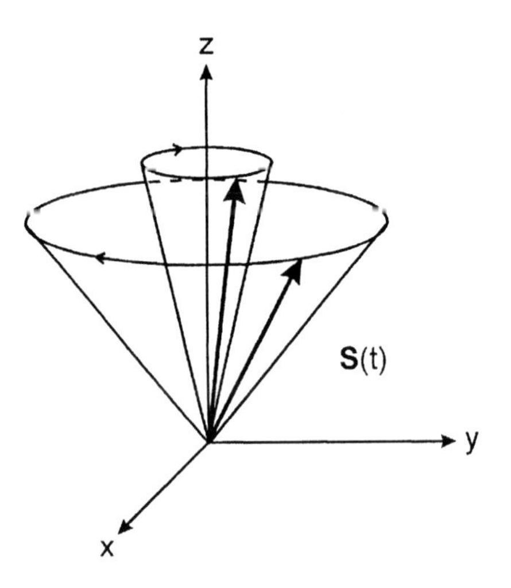

Fig. 4 *Larmor–precession of the Bloch–vector, according to Eqs.(74-76).*

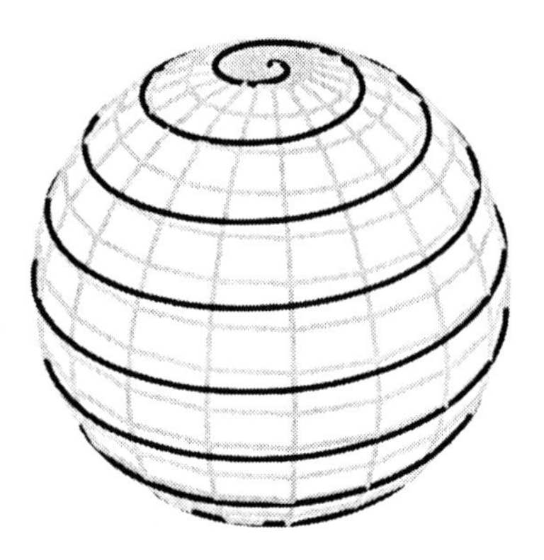

Fig. 5 *Rabi–oscillation: Trace of the Bloch–vector upon resonant excitation (without damping), according to Eqs.(77-79),* $0 \leq \omega_R t \leq \pi$.

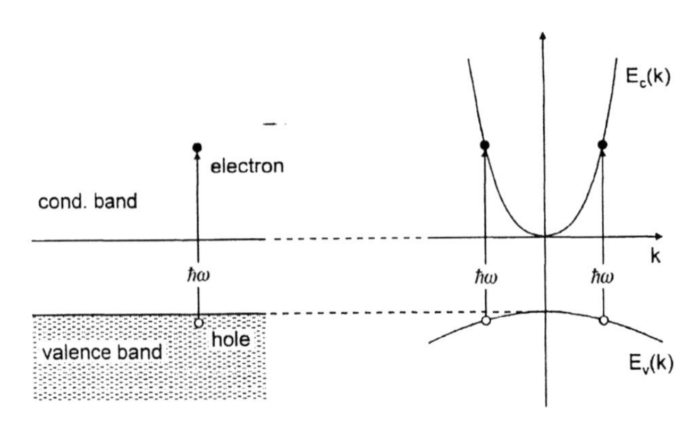

Fig. 6 *Sketch of the band structure and optical transitions in a semiconductor.*

III.B. Semiconductor with Noninteracting Bands

The generalization of the atomic Bloch-equations to the case of a two–band semiconductor is straightforward, Fig 6. In the dipole approximation the optical transitions are vertical in k-space. Without scattering or nonradiative recombination processes, the two band semiconductor is just an assembly of uncoupled TLS and resembles the case of a inhomogeneously broadened line problem in atomic physics:

$$i\hbar\frac{\partial P(\mathbf{k},t)}{\partial t} = \Big[E_c(k) - E_v(k)\Big]P(\mathbf{k},t) + p\mathcal{E}(t)\Big[n_c(\mathbf{k},t) - n_v(\mathbf{k},t)\Big] + i\hbar\dot{P}_{sc}, \quad (80)$$

$$\frac{\partial n_c(\mathbf{k},t)}{\partial t} = -2\,\mathrm{Im}\,\Big[p\mathcal{E}(t)P^*(\mathbf{k},t)\Big] + \dot{n}_c^{sc}, \quad (81)$$

$$\frac{\partial n_v(\mathbf{k},t)}{\partial t} = +2\,\mathrm{Im}\,\Big[p\mathcal{E}(t)P^*(\mathbf{k},t)\Big] + \dot{n}_v^{sc}. \quad (82)$$

Eqs.(80-82) are called optical Bloch equations. In the absence of scattering the equations are uncoupled and $\mathbf{k}$ merely acts as a parameter. From the complex polarization $P(\mathbf{k},t)$ the electronic polarization $\mathcal{P}(\mathbf{r},t)$ of the semiconductor can be obtained from

$$\mathcal{P}(t) = \frac{1}{V}\sum_{k,s}[p_{cv}(k)P(\mathbf{k},t) + cc], \quad (83)$$

where V denotes the crystal volume (normalization volume of the wave–functions) which eventually drops out when performing the sum over wave numbers $\mathbf{k}$ by an integral

$$\frac{1}{V}\sum_{k,s}\ldots = 2\frac{1}{(2\pi)^d}\int\ldots d^dk. \quad (84)$$

$d = 1, 2, 3$ is the spatial dimension and the factor 2 arises from spin. The k-dependence of the dipole matrix element $p_{cv}(\mathbf{k})$ can often be neglected near the band edge.
As an application, we state the linear response result where $n_c = f_c(k)$ and $n_v = f_v(k)$ are the Fermi-functions. As k is merely a parameter, the required solution of Eq.(80) can be found by the Ansatz

$$P(k,t) = Q(k,t)\exp\Big[i(\epsilon_v(k) - \epsilon_c(k)) - \hbar\omega)t\Big] \quad (85)$$

with a simple integration for $Q(k,t)$. When separating the different Fourier–components the susceptibility can be read–off:

$$\chi(\omega) = \frac{1}{V\epsilon_0}\sum_{k,s}|p_{cv}(k)|^2\left\{\frac{f_v(k) - f_c(k)}{E_c(k) - E_v(k) - \hbar(\omega + i\delta)} + \frac{f_v(k) - f_c(k)}{E_c(k) - E_v(k) + \hbar(\omega + i\delta)}\right\}. \quad (86)$$

For parabolic bands and $\mathbf{k}$–independent dipole matrix elements the absorptive part of the susceptibility becomes proportional to the joint density of states which rises as a square–root above the gap $\chi_2(\omega) \propto \sqrt{\hbar\omega - E_g}$.

III.C. Semiconductor Bloch-Equations

We are close to the summit of our tour towards the SBE – which is today's standard model of semiconductor optics. Two features have not yet been taken into account:

- There is a change in Coulomb energy of the interacting many electron ground state when exciting an electron to the conduction band and leaving a hole behind. This (exchange) interaction turns out to be an attractive Coulomb potential.
- With increasing band filling there is a renormalization of the electron/hole band energy by the (repulsive) electron/hole Coulomb interaction:

$$\begin{aligned} i\hbar\frac{\partial P(\mathbf{k},t)}{\partial t} &= \Big[E_g + E_e(k) + E_h(k)\Big]P(\mathbf{k},t) + \\ &\quad \Big[n_e(\mathbf{k},t) + n_h(\mathbf{k},t) - 1\Big]\hbar\Omega_R(\mathbf{k},t) + i\hbar\dot{P}_{sc}, \quad (87) \\ \frac{\partial n_c(\mathbf{k},t)}{\partial t} &= -2\,\mathrm{Im}\left\{\Omega_R P^*(\mathbf{k},t)\right\} + \dot{n}_c^{sc}, \quad (88) \\ \frac{\partial n_h(\mathbf{k},t)}{\partial t} &= -2\,\mathrm{Im}\left\{\Omega_R P^*(\mathbf{k},t)\right\} + \dot{n}_h^{sc}. \quad (89) \end{aligned}$$

For convenience, the change in population of the valence band is formulated within the hole picture as indicated by the index h:

$$n_v(\mathbf{k},t) = 1 - n_h(\mathbf{k},t), \qquad E_v(\mathbf{k},t) = -E_g - E_h(\mathbf{k},t). \quad (90)$$

$E_e(\mathbf{k},t)$, $E_h(\mathbf{k},t)$ are the electron/hole (Hartee–Fock) energies including the interaction with other electrons/holes. For parabolic bands these are:

$$E_j(\mathbf{k},t) = \frac{\hbar^2\mathbf{k}^2}{2m_j} - \frac{1}{V}\sum_q V(\mathbf{k}-\mathbf{q})n_j(\mathbf{q},t), \quad j = e,h. \quad (91)$$

Note, $m_h > 0$. $\Omega_R(\mathbf{k},t)$ denotes the Rabi-frequency function:

$$\hbar\Omega_R(\mathbf{k},t) = p\mathcal{E}(t) + \frac{1}{V}\sum_q V(\mathbf{k}-\mathbf{q})P(\mathbf{q},t). \quad (92)$$

$V(\mathbf{q})$ is the Fourier–transform of the electron–hole Coulomb–potential screened by a "background" dielectric constant $\bar{\epsilon}$:

$$V(\mathbf{q}) = \frac{e^2}{\epsilon_0\bar{\epsilon}q^2}, \qquad V(\mathbf{r}) = \frac{e^2}{4\pi\epsilon_0\bar{\epsilon}r}. \quad (93)$$

In addition, this interaction will be screened by mobile electrons and holes in terms of a dielectric function $\epsilon_\ell(\mathbf{q},\omega)$ as discussed in problem 3.

IV. APPLICATIONS AND SUPPLEMENTS

IV.A. Causality and Kramers–Kronig Relations

There are three equivalent formulations of causality [18]:

1. The original formulation (28) of the polarization or current response just states, that the present value of the polarization does not depend on future fields:

$$\chi(t-t') \equiv 0, \; t' > t. \quad (94)$$

   A function of this type is called a causal function.

2. The Fourier–transform (29) of a causal function

$$\chi(\omega_1 + i\omega_2) = \int_0^\infty \chi(t) e^{i\omega_1 t - \omega_2 t} dt \tag{95}$$

is an analytic function of $\omega = \omega_1 + i\omega_2$ in the upper part of the complex ω–plane, $\omega_2 > 0$. In this half–plane $\chi(\omega)$ has no poles or other singularities. For real ω, $\chi(\omega)$ is the boundary value of this analytic function.

3. Real and imaginary parts of $\chi(\omega)$ are connected by the Kramers–Kronig relations:

$$\chi_1(\omega) = \frac{+1}{\pi} P\!\!\int_{-\infty}^{\infty} \frac{\chi_2(\omega')}{\omega' - \omega} d\omega', \tag{96}$$

$$\chi_2(\omega) = \frac{-1}{\pi} P\!\!\int_{-\infty}^{\infty} \frac{\chi_1(\omega')}{\omega' - \omega} d\omega'. \tag{97}$$

Apart from causality the following assumptions have been made to derive Eqs.(96,97):

- $\chi(\omega)$ has no singularities on the real ω–axis,
- $\chi(\omega)$ tends to zero at large frequencies.

If there are singularities in $\chi(\omega)$ which lie perfectly on the real axis (like the pole of the Drude susceptibility at $\omega = 0$, Eq.(33)) these terms have to be subtracted before applying the Kramers–Kronig relations: $\chi \to \bar{\chi} = \chi(\omega) - \chi_{\rm pole}(\omega)$.
If $\chi(\omega)$ tends to a finite value at $\omega = \infty$, $\chi \to \bar{\chi} = \chi(\omega) - \chi(\infty)$ in Eqs.(96,97). For the magnetic susceptibility such a constant term arises from the diamagnetic contribution. In addition, Im $\chi_{mag} > 0$ is not guaranteed! [20]. However, the electrical susceptibility and the conductivity always fulfill $\chi(\infty) = 0$, $\sigma(\infty) = 0$. To describe the low–frequency properties of semiconductors (e.g. the contribution of optical phonons) it is sometimes convenient to neglect dispersion at high frequencies (e.g. of the electronic interband transitions) by introducing a constant χ_∞. For instance, for $GaAs$, $\epsilon_\infty = 1 + \chi_\infty = 10.6$ which holds up to half of the band edge at $\hbar\omega = 1.4eV$ [3]. The resistivity, on the other hand, has a first order pole at $\omega = \infty$ so that this pole has to be likewise subtracted from $\rho(\omega)$ before applying the Kramers–Kronig relations to $\rho = 1/\sigma$.
In Eqs.(96,97) the "P" denotes "principal value" which is a prescription how to treat the singular integral:

$$g(\omega) = \frac{1}{\pi} P\!\!\int_{-\infty}^{\infty} \frac{f(\omega')}{\omega' - \omega} d\omega' = \lim_{\delta \to 0^+} \frac{1}{\pi} \left(\int_{-\infty}^{\omega - \delta} + \int_{\omega + \delta}^{\infty} \right) \frac{f(\omega')}{\omega' - \omega} d\omega'. \tag{98}$$

In mathematics this relation is termed Hilbert– transformation.
Another way to interpret this limiting process is to replace the singular function $\frac{1}{\omega' - \omega}$ by a "regularized" function $P\frac{1}{\omega' - \omega}$ which are almost identical except near the singularity. There $P\frac{1}{\omega - \omega'}$ becomes zero in an (anti–) symmetrical fashion. For instance, $\frac{1}{\omega' - \omega}$ is put to zero (anti–)symmetrically around the singularity from $\omega - \delta$ to $\omega + \delta$ with $\delta \to 0$. This is just a reformulation of Eq.(98). But any other symmetrical replacement is equivalent and resembles the way the Dirac delta–function is constructed, e.g.

$$P\frac{1}{\omega} \simeq \frac{\omega}{\omega^2 + \delta^2}, \quad \delta(\omega) \simeq \frac{1}{\pi} \frac{\delta}{\omega^2 + \delta^2}, \quad \delta \to 0^+. \tag{99}$$

$P\frac{1}{\omega}$ has a precise meaning only under an intergral and is - like $\delta(\omega)$ - another example of a distribution.

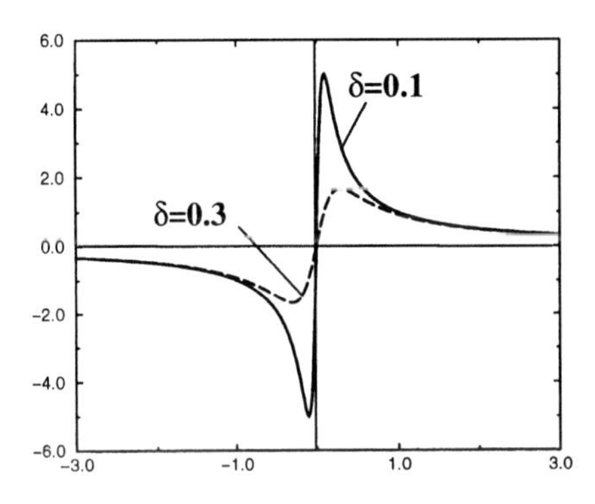

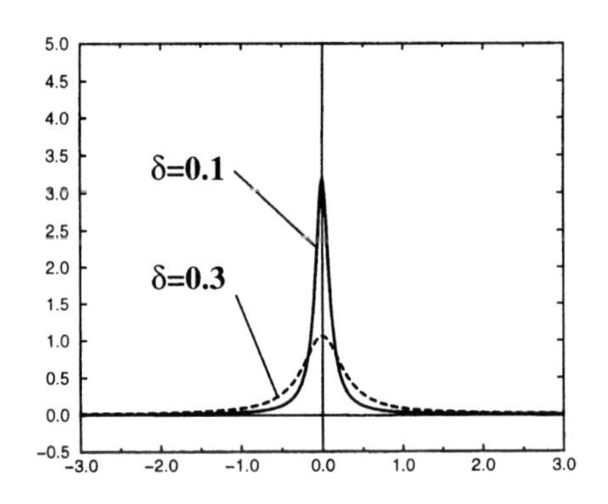

Fig. 7 *Sequence of function which "converge" to the principal value (left) and to the delta function (right).*

Because of the singular structure of the integrand in Eqs. (96-98) these integrals cannot be simply numerically calculated, e.g. by just using a Simpson–routine. For some elementary functions $f(\omega)$, however, the integral and the limiting process in Eq.(98) can be done analytically (see problem 8). In other cases, $f(\omega)$ is too complex or only known numerically. Then, one possibility to perform the Hilbert–transformation numerically is to fit $f(\omega)$ piecewise by a parabolic function (or a cubic spline), do the integral piecewise analytically and eventually sum all contributions numerically. Problem 9 supplies another possibility.
Traditionally, the proof of the Kramers–Kronig relations is by using tools from complex analysis, e.g. [1],[2], or [12]. A much shorter and almost "trivial" proof, however, can be found in Stößel's book on Fourier–Optik [19]:

- A causal function (94) trivially obeys $\chi(t) \equiv \chi(t)\theta(t)$, where $\theta(t)$ is the unit step function. This relation holds if there is no singular part in $\chi(t)$ of the type $\chi_\infty \delta(t)$ which corresponds to a nonvanishing contribution in $\chi(\omega)$ at $\omega = \infty$.

- Using the convolution theorem (22,23), Fourier–transformation yields:

$$\chi(\omega) = \chi(\omega) \otimes \theta(\omega). \tag{100}$$

  As $\theta(t)$ does not converge to zero at $t = \infty$ its Fourier–transform needs, as usual in such cases, an adiabatic switching–off factor $exp(-\delta t)$ to define the integral

$$\theta(\omega) = \int_0^\infty 1 \cdot e^{i\omega t} e^{-\delta t}\, dt = \frac{1}{\delta - i\omega} = P\frac{i}{\omega} + \pi\delta(\omega). \tag{101}$$

- Separation of real and imaginary parts in Eq.(100) immediately leads to the Kramers–Kronig relations (96,97).

For a detailed discussion of Kramers–Kronig relations in connection with sum–rules we refer to the seminal article by Martin [20].

Problems:

7.) Sketch the loci of the singularities of $\chi(\omega)$ and $\sigma(\omega)$ in the complex ω–plane.
(a) For the Drude and (b) the Lorentz–model. (Use results of problems 3,4).

8.) Calculate the Hilbert–transform (98) of the "box–function" $\text{box}(\omega) = 1$ for $|\omega| < 1$, otherwise $\text{box}(\omega) = 0$.

9.) The Hilbert–transformation (98) is again a convolution and, thus, may be transformed to a product in time–domain

$$g(t) = f(t) \cdot s(t), \tag{102}$$

where $s(t)$ is the Fourier–transform of $P\frac{1}{\omega}$.
Show that $s(t) = -\frac{i}{2}\text{sign}(t)$ with $\text{sign}(t) = +1$ for $t > 0$ and $\text{sign}(t) = -1$ for $t < 0$.
Thus the Hilbert–transformation can be performed by two Fourier–transformations: First transform from ω to time–domain, multiply by $s(t)$ and transform back to frequeny domain. Numerically, this can be done efficiently by standard FFT routines [21].
Hint: Prove that the Fourier–transform of $s(t)$ is $P\frac{1}{\omega}$. Use an adiabatic switching–factor $exp(-\delta \mid t \mid)$, $\delta \rightarrow 0^+$.

IV.B. Oscillator with a Quadratic Nonlinearity

We consider the case of charges bound in a noncentrosymmetric crystal which can be modelled by adding a quadratic force term in the Lorentz–model Eq.(19). (For simplicity spatial dispersion and vector properties of $\mathcal{P}$ and $\mathcal{E}$ will be omitted.)

$$\Big[\frac{d^2}{dt^2} + \gamma\frac{d}{dt} + \omega_0^2\Big]\mathcal{P}(t) + \lambda\mathcal{P}^2(t) = \epsilon_0\Omega_p^2\mathcal{E}(t). \tag{103}$$

No analytic solution of Eq.(103) is known, which is not surprising, as this model contains rich physics from periodic to chaotic phenomena. Note that there are two independent parameters λ and Ω_p^2 which can be used to set up perturbation expansions.
If $\mathcal{E}(t)$ is sufficiently weak, the nonlinear term $\lambda\mathcal{P}^2$ will be much smaller than the "restoring-force" $-\omega_0^2\mathcal{P}$ so that a pertubation expansion of $\mathcal{P}(t)$ of the form Eq.(17) may be used

$$\mathcal{P}(t) = \mathcal{P}^{(1)}(t) + \mathcal{P}^{(2)}(t) + \mathcal{P}^{(3)}(t) \ldots. \tag{104}$$

The various orders obey the following chain of differential equations

$$\Big[\frac{d^2}{dt^2} + \gamma\frac{d}{dt} + \omega_0^2\Big]\mathcal{P}^{(1)}(t) = \epsilon_0\Omega_p^2\mathcal{E}(t), \tag{105}$$

$$\Big[\frac{d^2}{dt^2} + \gamma\frac{d}{dt} + \omega_0^2\Big]\mathcal{P}^{(2)}(t) = -\lambda\left[\mathcal{P}^{(1)}(t)\right]^2, \tag{106}$$

$$\Big[\frac{d^2}{dt^2} + \gamma\frac{d}{dt} + \omega_0^2\Big]\mathcal{P}^{(3)}(t) = -2\lambda\Big[\mathcal{P}^{(1)}(t)\mathcal{P}^{(2)}(t)\Big]. \tag{107}$$

The first-order solution is identical with the Lorentz-solution

$$\mathcal{P}^{(1)}(\omega) = \epsilon_0\chi^{(1)}(\omega)\mathcal{E}(\omega), \tag{108}$$

$$\chi^{(1)}(\omega) = \Omega_p^2 G(\omega), \tag{109}$$

$$G(\omega) = \frac{1}{\omega_0^2 - \omega^2 - i\gamma\omega}. \tag{110}$$

In mathematics, $G(\omega)$ is termed (retarded) Green–function of Eq.(105).
The nonlinear susceptibilities are calculated in an analogous manner, where the products on the *rhs* of the equations for $\mathcal{P}^{(k)}(t)$ become convolutions in the frequency

domain, e.g.

$$G^{-1}(\omega)P^{(2)}(\omega) = -\lambda \int \frac{d\omega'}{2\pi}\mathcal{P}^{(1)}(\omega-\omega')\mathcal{P}^{(1)}(\omega'), \tag{111}$$

$$G^{-1}(\omega)\mathcal{P}^{(3)}(\omega) = -2\lambda \int \frac{d\omega'}{2\pi}\mathcal{P}^{(1)}(\omega-\omega')\mathcal{P}^{(2)}(\omega'). \tag{112}$$

Inserting Eq.(109) in Eqs.(111,112) the higher order susceptibilities can be read–off:

$$\chi^{(2)}(\omega_1,\omega_2) = -\lambda\Omega_p^4 G(\omega_1)G(\omega_1+\omega_2)G(\omega_2), \tag{113}$$

$$\begin{aligned}\chi^{(3)}(\omega_1,\omega_2,\omega_3) &= -2\lambda^2\Omega_p^6 G(\omega_1)G(\omega_2)G(\omega_3)G(\omega_1+\omega_2+\omega_3)\\ &\times\frac{1}{3}\Big[G(\omega_1+\omega_2)+G(\omega_1+\omega_3)+G(\omega_2+\omega_3)\Big].\end{aligned} \tag{114}$$

Eq.(114) for the cubic susceptibility has second-order poles in the degenerate case (some of the ω_n are equal). The origin of these singularities is seen from the geometric expansion [7]

$$\frac{1}{x-a} = \frac{1}{x}+\frac{a}{x^2}+\frac{a^2}{x^3}\cdots, \qquad |\,a\,|<|\,x\,|\,. \tag{115}$$

The lhs of Eq.(115) has a first-order pole at $x=a$, but the expansion on the rhs shows poles of all orders at $x=0$, yet the expansion is not valid there. In Eq. (114) the situation is of the same type. The remedy of the "dangerous" terms (poles of second and higher order) in Eq.(114) is a partial resummation of all singular terms in the infinite perturbation series (104). For a monochromatic field with frequency near ω_0 we have

$$P^{(1)}(t)+P^{(3)}(t)+\ldots = \epsilon_0\Big\{1+\frac{\lambda^2}{\omega_0^2-\omega^2-i\gamma\omega}|\chi^{(1)}\mathcal{E}|^2+\ldots\Big\}\chi^{(1)}(\omega)E(\omega). \tag{116}$$

The first terms of this expansion might be thought as the beginning of a geometric series. Summation can be cast in the quasilinear form:

$$P(t) = \epsilon_0\chi(\omega;E)E(\omega), \tag{117}$$

$$\chi(\omega;E) = \frac{\Omega_p^2}{\Omega^2-\omega^2-i\gamma\omega}, \tag{118}$$

$$\Omega^2 = \omega_0^2-\lambda^2\,|P|^2\,, \tag{119}$$

where Ω describes an intensity dependent eigenfrequency. For stationary fields, its quantum analogue is termed Stark–effect.

Problems:

10.) Calculate the amplitude–dependent eigenfrequency $\Omega=\Omega(A_1)$ of the nonlinear undamped oscillator

$$\frac{d^2x(t)}{dt^2}+x+\lambda x^2=0 \tag{120}$$

up to second order in the amplitude A_1 of the fundamental mode in the Fourier expansion

$$x(t) = \sum_{m=-\infty}^{\infty} A_m e^{im\Omega t}, \qquad A_{-m}=A_m^*. \tag{121}$$

Hints: First, derive the set of nonlinear equations for A_m. Then, expand the equations for $m=0,1,2$ to leading order in A_1. Note that A_0, A_2 are proportional to the square of A_1, other coefficients A_m are of higher order.

IV.C. Photogalvanic Effect

According to the standard rules of irreversible thermodynamics a steady-state (non-supra) current in a solid is always driven by the gradient of the electrochemical potential $\eta = \mu + e\phi$ [22],

$$j = -\frac{\sigma}{e}\text{grad } \eta. \tag{122}$$

μ and ϕ denote the chemical and electrical potential of the charge carriers, respectively. Thus, inhomogeneities are necessarily needed which may either reside in the system itself (e.g. a $p - n$ junction) or are imposed by external conditions (e.g. gradients in temperature or electrical potential). In addition, a photo–induced current solely depends on the number of absorbed photons (regardless of their polarization) and the open circuit voltage is limited by the band gap of the semiconductor.
In noncentrosymmetric crystals, however, an additional direct current originates from the quadratic term in the current-field relation (46):

$$j_\alpha = I P_{\alpha\beta\gamma}(\omega) e_\beta e^*_\gamma. \tag{123}$$

In contrast to Eq.(122) this bulk photovoltaic current is intimately connected to the light polarization described by the complex polarization vector $\mathbf{e}$ and intensity I. $P_{\alpha\beta\gamma}(\omega) \sim \sigma^{(2)}_{\alpha\beta\gamma}(\omega, -\omega)$. Such a third rank tensor exists in all noncentrosymmetric crystals. In particular ferroelectrics, like $LiNb0_3$, allow for nonzero tensor elements with equal indices $\beta = \gamma$, and, hence, a photocurrent can occur even for unpolarized light. The phenomenon described by Eq.(123) in now called photogalvanic effect (PGE). For a survey and further references see Ruppel et al. [23] or v. Baltz [24].
For sake of completeness, we note that there is another contribution to the radiation impressed current which, in distinction to Eq.(123), explicitely depends on the direction of light propagation and, hence, is related to the momentum of the absorbed photons. This phenomenon is called photon drag effect but it is mainly important in the IR region.
There were many fingerprints of the PGE before Glass at Bell Laboratories [25] recognized it as a new photovoltaic mechanism whose spectacular property is the occurence of photovoltages larger than 100 kV even under perfect homogeneous conditions. Astonishingly, the main research activities were almost exclusively done later in the former Sovietunion so that this phenomenon is rarely known in the western hemisphere. To illustrate the discovery and some of its unusual properties of the PGE we give some examples.

- Local changes of indices of refraction were observed in ferroelectrics upon illumination. This leads to a (reversible) "damage" of the phase-matching conditions when using these materials in nonlinear optics. This photorefractive effect results from a small imposed current which charges the faces of the crystal. In a high resistive crystal, like $LiNbO_3$ the (intensity dependent) resistivity is in the range of $\rho \approx 10^{15} \ldots 10^{12} \Omega cm$ so that even a tiny current can lead to very large electric fields. According to the Pockels effect, this field causes a change in the refractive index, Fig. 8.

- Austin et al. [27] discovered efficient "optical rectification" in polar crystals due to impurity absorption. The induced current contains two contributions: one from genuine OR following the derivative of the intensity profile of the laser and a photogalvanic current proportional to the light intensity, Fig. 9.

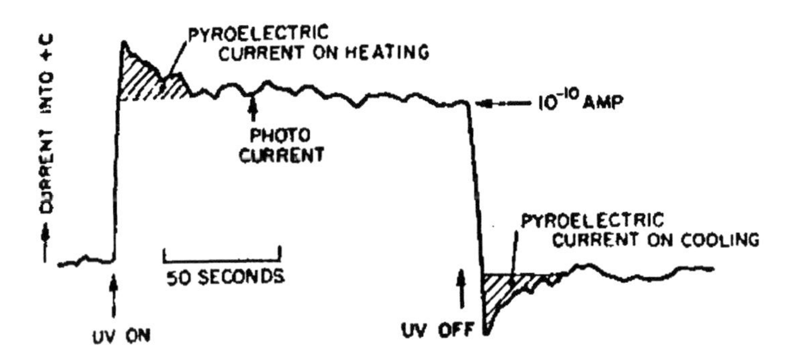

Fig. 8 *Current flow into the c faces of $LiNbO_3$ vs time with uv illumination. From Chen [26].*

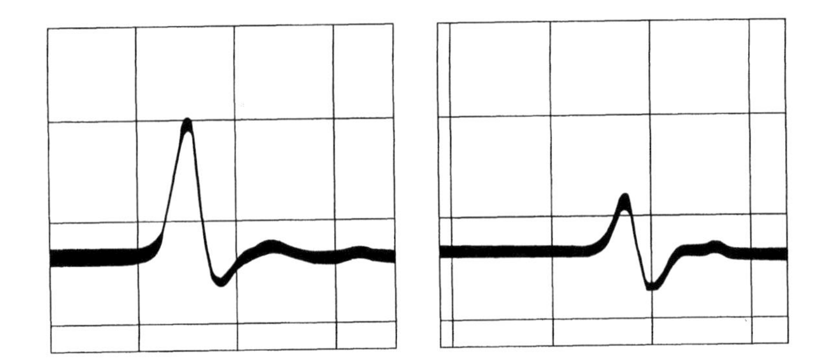

Fig. 9 *Oscilloscope traces of the optical rectification from $LiTaO_3$. (Left) Cu-doped, (right) undoped crystal. Time scale 2 nsec/div. From Auston et al. [27].*

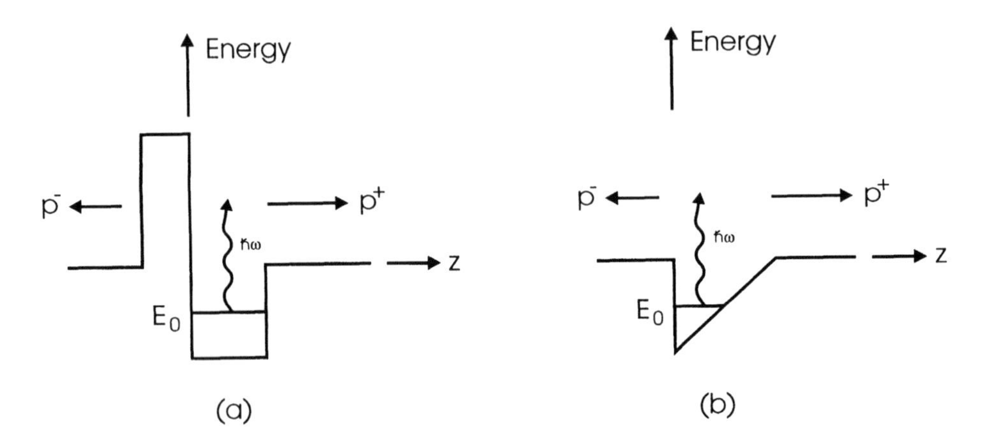

Fig. 10 *Optical transitions from impurities in a ferroelectric crystal resulting in a difference between the generation rates in $\pm z$ direction. Asymmetric potentials, (a) with and (b) without a barrier. From Ruppel et al. [23].*

- Glass [25] recognized the PGE as originating from an asymmetry in the optical transition probabilities, Fig. 10. In a ferroelectric the transition probalitities p_+ and p_- from a bound (Fe, Cu) impurity state to the conduction band can be different in $\pm c$ directions thus leading to a net charge transfer along the c-direction upon optical excitation. This asymmetry is not compensated under recombination because these processes are mostly nonradiative.

  It is convenient to represent the photogalvanic current in the following form:

$$j_{\text{PGE}} = I\,\kappa\,K = e\,\frac{I}{\hbar\omega}\,K\,s, \quad s = (p_+ - p_-)\lambda, \tag{124}$$

  where K is the absorption constant, $\kappa = \frac{e}{\hbar\omega}s$ the Glass–constant, λ the mean free path of the photoexcited carriers, and s the anisotropy distance ("Schublänge"). For $LiNbO_3$ $s \approx 1A$ whereas for $KNbO_3$ $s \approx 18A$. Recent investigations of the photogalvanic tensor components of $LiNbO_3$ were reported by Karabekian and Odulov [28].

- The tensorial dependence of the photogalvanic current in n–doped GaP on the light polarization was studied by Gibson et al. [29] in connection with the investigation of fast responding IR detectors, Fig 11. GaP is a noncentrosymmetric cubic crystal with $\bar{4}3m$ point symmetry. Therefore, the only nonvanishing components of the photogalvanic tensor are $P_{123} = P_{132}$ and cyclic permutations of indices. Linear polarized light propagating along the z–direction induces a photogalvanic current which varies sinusoidally with the polarization of light

$$j_z = I\,P_{123}\,\sin 2\phi. \tag{125}$$

  ϕ is the angle between the polarization vector and the crystal x–axis.

- In GaP the microscopic mechanism of the PGE is different from the "ballistic" mechanism in $LiNbO_3$. The spectral shapes of the photogalvanic current and the interband optical absorption are almost identical which indicates that the PGE is due to transitions near the X point from the conduction band minimum to the next upper band. However, this apparently contradicts the bandstructure theory! According to time–reversal symmetry the band structure obeys $E(-\mathbf{k}) = E(\mathbf{k})$ so that the velocities $\mathbf{v}(\mathbf{k}) = \nabla E(\mathbf{k})$ at $\pm\mathbf{k}$ have opposite sign. As the transition rates are the same at $\pm\mathbf{k}$ there is no net current upon photoexcitation regardless of crystal symmetry. Most remarkably, however, there is a shift in real space of the valence and conduction band wave–packets upon photoexcitation, Fig. 12. In noncentrosymmetric crystals, these shift vectors at $\pm\mathbf{k}_0$ do not compensate each other and lead to a photogalvanic current which can be represented as [30]:

$$\begin{aligned}
\mathbf{j}_{\text{PGE}} &= |e|\frac{I}{\hbar\omega}\frac{e^2}{2\pi^2\epsilon_0 m_0^2 n c\omega^2}\int (f_c - f_v)\,|< c\mathbf{k}\,|\mathbf{ep}|\,v\mathbf{k} >|^2 \times \\
&\quad \mathbf{s}(\mathbf{k})\,\delta(E_c(\mathbf{k}) - E_v(\mathbf{k}) - \hbar\omega)\,d^3k, &(126)\\
\mathbf{s}(\mathbf{k}) &= \mathbf{X}_{cc}(\mathbf{k}) - \mathbf{X}_{vv}(\mathbf{k}) + \nabla_{\mathbf{k}}\Phi_{cv}(\mathbf{k}), &(127)\\
\mathbf{X}_{mn}(\mathbf{k}) &= \int_{unitcell} i\,u^*_{m\mathbf{k}}(\mathbf{r})\nabla_{\mathbf{k}}u_{n\mathbf{k}}(\mathbf{r})d^3r. &(128)
\end{aligned}$$

  $\Phi_{cv}(\mathbf{k})$ denotes the phase of the interband momentum matrix element $< c\mathbf{k}|\mathbf{ep}|v\mathbf{k} >$, n is the refractive index of the material, and $u_{n\mathbf{k}}(\mathbf{r})$ is the periodic part of the electron Bloch function. For the $X_1 \rightarrow X_3$ transition in GaP the shift of wave packets is $s \approx 8A$.

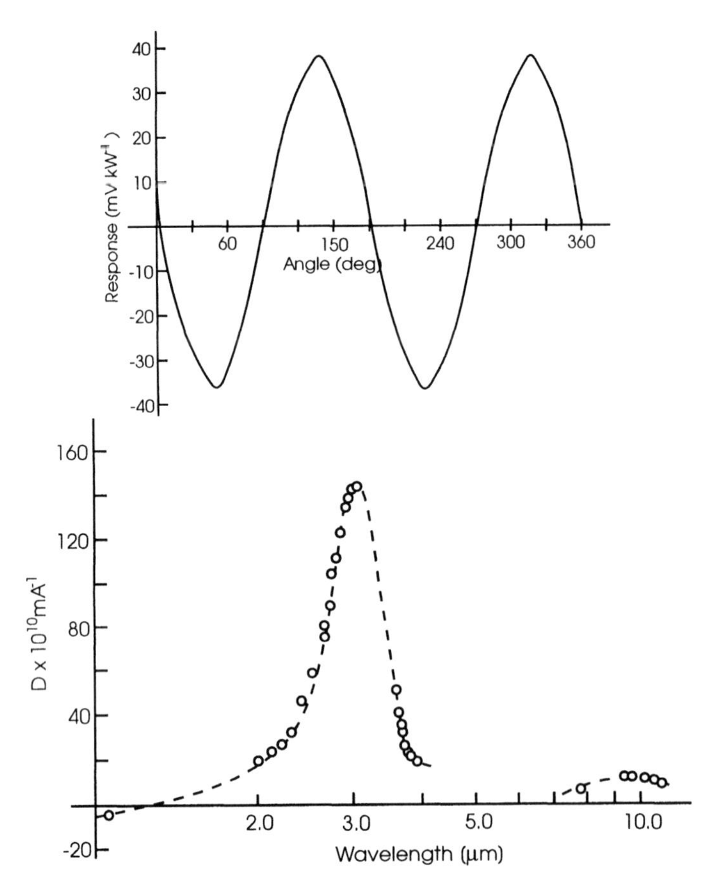

Fig. 11 *Photogalvanic effect in n–doped GaP. (Left) dependence on the light polarization, (right) spectral dependence,* $D = P_{123}/\sigma_{dc}$. *From Gibson et al. [29].*

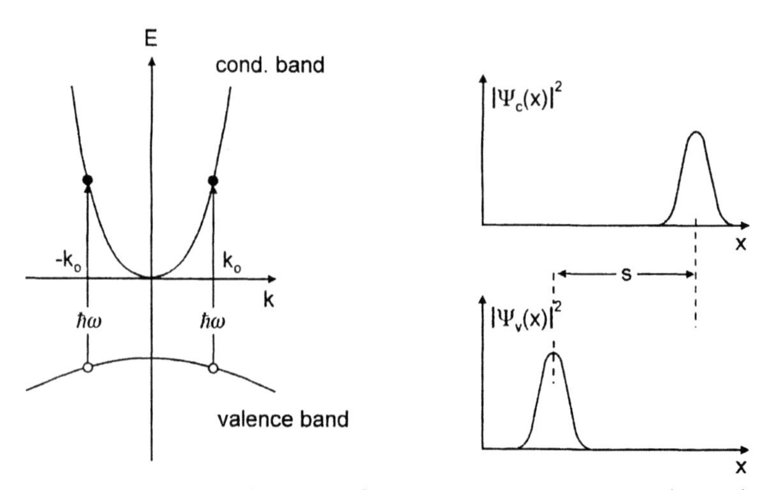

Fig. 12 *Optical transitions of wave packets in a noncentrosymmetric semiconductor. (Left)* **k***–space, (right) real space.*

The shift vectors $\mathbf{X}_{mn}(\mathbf{k})$ are known for almost 40 years from the work of Adams and Blount [31] in connection with the Bloch–representation of the position operator. However, these quantities are rarely explicitely used or even notified. Exceptions are e.g. their connection to Bloch–oscillations, e.g. [32], nonlinear optical susceptibilities [33], or the definition macroscopic polarization in a ferroelectric or piezolectric materials in terms of a Berry–phase [34].
Recently the PGE and related phenomena found new interest. For example Schneider et al. [35] reported on IR photodetectors which were made of asymmetric quantum wells, directed motion of Brownian particles was proposed to occur in "thermal ratchets" (=periodic arrangement of potentials given in Fig. 10b) upon periodic perturbation [36],[37], or small amplitude swimming of a pulsating body [38]. The unifying aspect is the occurence of an unidirectional motion from oscillatory disturbances in noncentrosymmetric structures.

IV.D. Photon–Echo

The analogy of the TLS with the Spin-problem offers the description of an interesting phenomenon which is called photon–echo. Here, we examine the rather marvellous notion that not all decay processes are irreversible. This technique was developped by Hahn [39] for nuclear spin systems and, apart from its beautiful physics, it plays an important role to measure the T_2–time. For a survey and thorough discussion we refer to chapter 9 of Allen and Eberly [17].
In an experiment many TLS are involved and because of different local environments these have individually slightly different transition frequencies (=inhomogeneous line broadening, spectral width is parameterized by $1/T_2^*$). To describe the dynamics of the Bloch–vector it is convenient to transform to a frame rotating with the frequency of the light around the 3–axis:

$$\begin{aligned} R_1(t) &= S_1(t)\cos\omega t - S_2(t)\sin\omega t, & (129)\\ R_2(t) &= S_1(t)\sin\omega t + S_2(t)\cos\omega t, & (130)\\ R_3(t) &= S_3(t). & (131) \end{aligned}$$

(In complex notation the 1, 2 components are summarized by $R = Se^{i\omega t}$.) In this frame the equations of motion become [3]:

$$\begin{aligned} \dot{R}_1(t) &= -\nu R_2(t) - \frac{R_1(t)}{T_2}, & (132)\\ \dot{R}_2(t) &= +\nu R_1(t) + \omega_R R_3(t) - \frac{R_2(t)}{T_2}, & (133)\\ \dot{R}_3(t) &= -\omega_R R_2(t) - \frac{R_3(t) - R_3^{eq}}{T_1}. & (134) \end{aligned}$$

These linear differential equations have constant coefficients so that the solution can be found by an exponential Ansatz. In particular, for $\omega_R = 0$ the Bloch–vector performs a (damped) Larmor–precession around the 3–axis:

$$\begin{aligned} R_1(t) &= [R_1(0)\cos\nu t - R_2(0)\sin\nu t]\, e^{-\frac{t}{T_2}}, & (135)\\ R_2(t) &= [R_1(0)\sin\nu t + R_2(0)\cos\nu t]\, e^{-\frac{t}{T_2}}, & (136)\\ R_3(t) &= R_3(0)\, e^{-\frac{t}{T_1}}. & (137) \end{aligned}$$

[3] Some signs are different from Allen and Eberly [17] who use a different numbering of the ground state and excited state.

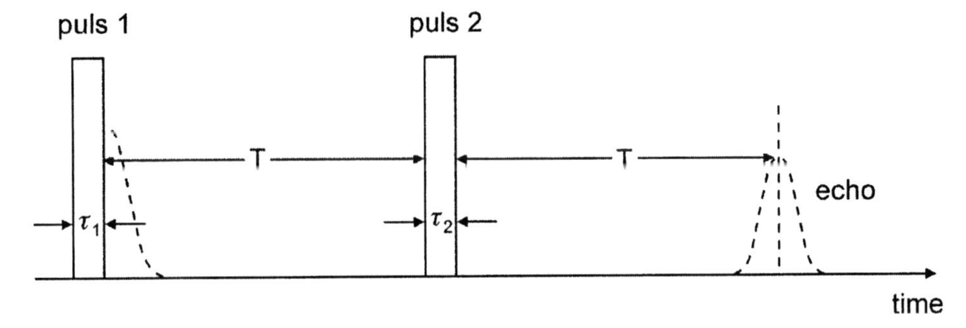

Fig. 13 *Pulse condition for the photon echo experiment. (Full lines) externally applied pulses, (dashed lines) polarization which give rise to the free polarization decay and the echo.*

At resonance, $\nu = 0$, and neglecting damping a light pulse causes a rotation of the Bloch–vector around the 1–axis:

$$R_1(t) = +R_1(0), \tag{138}$$
$$R_2(t) = +R_2(0)\cos\omega_R t + R_3(0)\sin\omega_R t, \tag{139}$$
$$R_3(t) = -R_2(0)\sin\omega_R t + R_3(0)\cos\omega_R t. \tag{140}$$

In the discussion of a photon–echo experiment four periods have to be distinguished, Fig. 13. To simplify matters, we shall assume that the pulse duration is short with respect to T_1, T_2, T_2^* and intense, $\omega_R T_2^* >> 1$, so that the influence of damping and detuning can be safely neglected during the pulses.

1. All TLS start from the same initial state $\mathbf{R}^{(0)} = (0,0,1)$. Then the ensemble of atoms is polarized by a first light pulse (duration τ_1) which leads to a common Bloch vector $\mathbf{R}^{(1)} = (u,v,w)$.

2. After the first light pulse the individual Bloch-vectors $\mathbf{R}^{(2)}$ precess according to Eqs.(135-137). Because of their slightly different frequencies the individual dipolemoments get out of phase and add to zero in a time T_2^* which is much shorter than T_2.

3. After time T a second light pulse is applied (duration τ_2, phase $\phi_2 = \omega_R \tau_2$) which according to Eqs.(138 - 139) "tips" the polarization to $\mathbf{R}^{(3)}$.

4. After the second pulse the Bloch–vectors $\mathbf{R}^{(4)}$ again rotate freely around the 3–axis. As a result, we obtain for the 1–component:

$$R_1^{(4)}(t) = \Big\{u\left[\cos\nu T\cos\nu t - \cos\phi_2\sin\nu T\sin\nu t\right] \tag{141}$$
$$-v\left[\sin\nu T\cos\nu t + \cos\phi_2\cos\nu T\sin\nu t\right]\Big\}\, e^{-\frac{t+T}{T_2}} \tag{142}$$
$$+w\sin\phi_2\sin\nu t\; e^{-\frac{t+T}{T_1}}, \tag{143}$$

where the time t counts from the end of the second pulse. For $\phi_2 = \pi$ the terms in the [..] brackets combine to $\cos\nu(t-T)$ and $\sin\nu(t-T)$ and all individual Bloch–vectors again are in phase at time $2T$ after the first pulse, and add up in phase to a macroscopic polarization. This causes the emission of a light pulse, the photon echo.

The resurrected free polarization signal has the magic quality of something coming from nothing. However, this resurrection is only possible for times which are comparable with the T_2–time. For larger times, the intensity of the photon echo decays as the square of $\exp(-2T/T_2)$ i.e. $\exp(-4T/T_2)$. T_2 is also called dephasing–time.
It is interesting that the existance of the echo is not attributed to the π–character of the second pulse as it is frequently imputed. When decomposing the products of trigonometric functions $\cos\nu T \cos\nu t$ etc., in terms of sum and differences we realize that there is a contribution to the polarization of the form:

$$R_1^{(4)} = \frac{1-\cos\phi_2}{2}\Big\{u\cos\left[\nu(T-t)\right] - v\sin\left[\nu(T-t)\right]\Big\}e^{-\frac{t+T}{T_2}} + \ldots \tag{144}$$

Thus, a second pulse of any duration will induce an echo. However, its intensity is largest for $\phi_2 = \pi$.

Problems:

12.) Find the general solution of the coupled set of differential equations Eqs(132-134) for the Bloch–vector $\mathbf{R}(t)$. Express the integration constants in terms of $\mathbf{R}(0)$.
Hint: The solution can be obtained by an exponential Ansatz $\mathbf{R}(t) = \rho\exp(\lambda t)$, where ρ is a time–independent 3–component vector.

IV.E. Linear Susceptibility: Excitons

To demonstrate the potential and simplicity of the SBE (e.g. compared with an evaluation of the Kubo formula) we derive the linear optical susceptibility of the interacting electron-hole system in the low excitation limit. This will lead us to the exciton and the famous Elliott–formula for the optical absorption.
At zero temperature $n_v = n_h = 0$ and the set of Eqs.(87-89) reduce to:

$$i\hbar\frac{\partial P(\mathbf{k},t)}{\partial t} = \Big[E_g + \frac{\hbar^2\mathbf{k}^2}{2m_r}\Big]P(\mathbf{k},t) - \frac{1}{V}\sum_{\mathbf{q}}V(\mathbf{k}-\mathbf{q})P(\mathbf{q},t) - p\mathcal{E}(t), \tag{145}$$

where m_r denotes the reduced electron-hole mass. In addition to the optical Bloch Eqs.(80-82) there is a interaction part which couples different $\mathbf{k}$'s so that $\mathbf{k}$ is no longer just a parameter. However, this interaction term is of convolution type and the integral equation can be Fourier-transformed to a well known differential equation:

$$i\hbar\frac{\partial P(\mathbf{r},t)}{\partial t} = \Big[E_g + \frac{\hbar^2}{2m_r}\Delta\Big]P(\mathbf{r},t) - V(\mathbf{r})P(\mathbf{r},t) - p\,\mathcal{E}(\mathbf{r},t)\,\delta(\mathbf{r}). \tag{146}$$

This is the (inhomogeneous) Schrödinger equation of the hydrogen atom, $\mathcal{P}(\mathbf{r},t)$ playing the part of the wave–function.
For $\mathcal{E}(\mathbf{r},t) = 0$ the stationary states are well known from standard texts on Quantum Mechanics.

$$P_{\text{stat}}(\mathbf{r},t) = \exp\Big(-i\frac{E_\mu}{\hbar}t\Big)\Psi_\mu(\mathbf{r}), \tag{147}$$

$$E_\mu = -\frac{Ry^*}{n^2}, \tag{148}$$

$$Ry^* = 13.56\frac{m_r/m_0}{\bar{\epsilon}^2}\ \text{eV}, \tag{149}$$

with the quantum numbers $\mu = (n,l,m)$, which have their usual meaning and run over the discrete as well as over the continuum states. In a semiconductor, the (bound) hydrogenic states are called excitons [3]. If the excitonic Rydberg Ry^* is smaller than the LO–phonon energy $\bar{\epsilon}$ is given by the static dielectric constant ϵ_s, otherwise $\bar{\epsilon} = \epsilon_\infty$. For $\mathcal{E}(\mathbf{r}, t) \neq 0$ we seek the solution of $P(\mathbf{r}, t)$ in terms of the complete set of the stationary states (147):

$$P(\mathbf{r}, t) = \sum_\mu Q_\mu(t) e^{-i\frac{E_\mu}{\hbar}t} \Psi_\mu(\mathbf{r}). \tag{150}$$

$Q(\mathbf{r}, t)$ can be found by a simple integration

$$Q_\mu(t) = ip\Psi_\mu(\mathbf{r} = 0) \int_{-\infty}^{t} \mathcal{E}(t') e^{i\frac{E_\mu}{\hbar}t'} e^{\delta t} \, dt', \quad \delta \to 0^+, \tag{151}$$

from which it becomes obvious that only those excitonic states couple to the light which have nonvanishing wave function at the origin. These are the s–states. (This result, however, originates from our assumption $p_{cv}(\mathbf{k}) = const.$ If the dipole element becomes zero at the band edge the coupling is to p–states).

Inserting Eq.(150) in Eq.(84) and using the completeness relation of the stationary states Eq.(147),

$$\sum_\mu \Psi_\mu^*(\mathbf{r}')\Psi_\mu(\mathbf{r}) = \delta(\mathbf{r} - \mathbf{r}'), \tag{152}$$

we finally obtain the electron-hole-pair susceptibility as [14]

$$\chi(\omega) = 2|p_{cv}|^2 \sum_\mu |\Psi_\mu(\mathbf{r} = 0)|^2 \left[\frac{1}{\hbar(\omega + i\delta + E_g + E_\mu)} - \frac{1}{\hbar(\omega + i\delta - E_g - E_\mu)} \right]. \tag{153}$$

Using the Dirac identity (101) the imaginary part is given by:

$$\chi_2(\omega) \propto \sum_{n=1}^{\infty} \frac{4\pi}{n^3} \delta(\Delta + \frac{1}{n^2}) + \theta(\Delta) \frac{\pi e^{\frac{\pi}{\sqrt{\Delta}}}}{\sinh(\frac{\pi}{\sqrt{\Delta}})}. \tag{154}$$

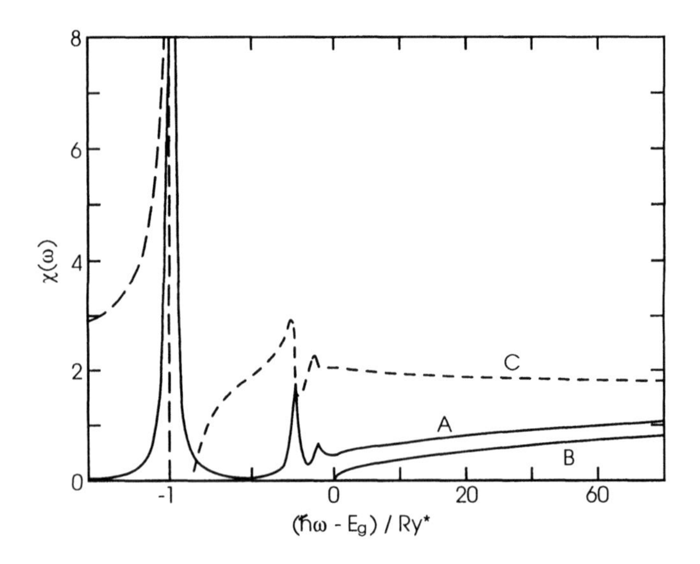

Fig. 14 *Susceptibility of a semiconductor near the gap energy: (A)* Im χ, *(C)* Re χ *of the interacting electron hole system, Eq.(154), (B)* Im χ *of the noninteracting system, Eq.(86). Note the different frequency scale above and below the gap. An appropriate broadening is included. From Stahl and Balslev [11].*

$\Delta = (\hbar\omega - E_g)/Ry^*$ denotes the normalized photon energy. This result is first derived by Elliott [40]. The corresponding real part can be calculated from the Kramers–Kronig relation (96). Recently Tanguy [41] succeeded to express the real part of the susceptibility in terms of known functions.
The optical absorption spectrum of a semiconductor is diplayed in Fig. 14 which gives a lively impression about the importance of the Coulomb interaction and excitonic states near the band gap. The optical absorption spectrum consists of a series with rapidly decreasing oscillator strength $\propto n^{-3}$ and a continuum part. In the very best samples excitonic lines up to $n = 3$ can be resolved, see [3] page 201. Close to the ionization continuum, $\Delta \to 0$ the absorption assumes a constant value, in striking difference with the square–root law for noninteracting bands. Thus, the attractive Coulomb interaction not only creates bound states below the gap but leads to a pronounced enhancement of the absorption above the gap. ("Sommerfeld–enhancement").

V. OUTLOOK

Sofar the damping mechanismn in the SBE was described phenomenologically. In a next step this can be improved by setting up kinetic (Boltzmann) equations for the electron and hole populations with appropriate collision integrals for the carrier–phonon scattering. These equations are coupled nonlinear differential–integral equations which can be solved only by advanced numerical treatments.
The many body effects which are omitted in the SBE as given by Eqs.(87- 89) lead to a further renormalization of the electronic energies, screening of the interactions, and additional collision terms. The success of this theory to descibe numerous linear and nonlinear effects is obvious from Haug and Koch [14] and the contributions presented by Klingshirn [42] and Hvam [43] in this book.
Recent studies by Stahl [44] and his collaborators indicate, however, that the SBE treatment becomes questionable for the tera–Hertz emission in a narrow band superlattice. For ulrashort pulses the formulation of collision integrals in terms of energy conserving processes is no longer possible and quantum kinetic equations are needed.

ACKNOWLEDGEMENTS

I thank Professor B. Di Bartolo for inviting me to Erice which, again, provided a stimulating summer school in a delightful atmosphere.
I also thank Mrs. R. Schrempp, Mrs. M. Bielfeld, Dr. Ch. Fuchs, and Dipl. Phys. A. Mildenberger for their help with the preparation of the article.

REFERENCES

1. L. D. Landau and E. M. Lifschitz, *Electrodynamics of Continuous Media*, Pergamon Press (1984).
2. F. C. Wooten, *Optical Properties of Solids*, Academic Press (1972).
3. C.F. Klingshirn, *Semiconductor Optics*, Springer (1995).
4. *Spectroscopy and Dynamics of Collective Excitations in Solids*, edited by B. Di Bartolo, NATO ASI Series B: Physics Vol. 356, Plenum Press (1997).
5. *Ultrafast Dynamics of Quantum Systems: Physical Processes and Spectroscopic Techniques*, edited by B. Di Bartolo, Plenum Press (This book).

6. L.V. Keldysh, D.A. Kirshnitz, and A.A. Maradudin eds., *The Dielectric Function of Condensed Systems*, North Holland (1989).
7. Yu. Il'inskij and L.V.Keldysh, *Electromagnetic Response of Material Media*, Plenum Press (1994).
8. P. Grosse, *Propagation of Electromagnetic Waves and Magnetooptics*, in *Theoretical Aspects and Developments in Magneto Optics*, edited by J. Devreese, Plenum Press (1980).
9. F. Forstmann and R.R. Gerhards, *Metal Optics Near the Plasma Frequency*, Springer Tracts in Modern Physics, Vol. 109, Springer (1986).
10. R. von Baltz, *Plasmons and Surface Plasmons in Bulk Metals, Metallic Clusters, and Metallic Heterostructures*, in [4].
11. A. Stahl and I. Balslev, *Electrodynamics of the Semiconductor Band Edge*, Springer Tracts in Modern Physics, Vol.110, Springer (1987).
12. B. Di Bartolo, *Propagation of waves in linear and nonlinear media*, in [5].
13. R.W. Boyd, Nonlinear Optics, Academic Press (1992).
14. H. Haug and S.W. Koch, *Quantum Theory of the Optical and Electronic Processes of Semiconductors*, World Scientific (1994).
15. R. Zimmermann, *Theoretical Description of Collective Excitations: Bloch equations and Relaxation Mechanisms*, in [4].
16. R.P. Feynman, R.B. Leighton, and M. Sands, *The Feynman Lectures on Physics*, Addison–Wesley (1966).
17. L. Allen and J.H.Eberly, *Optical Resonance and Two–Level Atoms*, Wiley (1975).
18. H.M. Nussenzweig, *Causality and Dispersion Relations*, Academic Press (1972).
19. W. Stößel, *Fourier Optik. Eine Einführung*, Springer (1993).
20. P.C. Martin, Phys. Rev. **161**, 143 (1967).
21. W.H. Press, B.P. Flannery, S.A. Teukolsky, and W.T. Vetterly, *Numerical Recipes*, Cambridge University Press (1986).
22. S.R. De Groot and P. Mazur, Non–Equilibrium Thermodynamics, North Holland (1962).
23. W. Ruppel, R. von Baltz, and P. Würfel, Ferroelectrics **43**, 109 (1981).
24. R. von Baltz, Ferroelectrics, **35**, 131 (1981).
25. A.M. Glass, D. von der Linde, and T.J. Negran, Appl. Phys. Lett. **25**, 233 (1974).
26. F.S. Chen, Jour. Appl. Phys. **40**, 3389 (1969).
27. D.H.Auston, A.M. Glass, and A.A. Ballman, Phys. Rev. Lett. **28**, 897 (1972).
28. S.I. Karabekian and S.G. Odulov, phys. stat. sol. (b) **169**, 529 (1992).
29. A.F. Gibson, C.B. Hatch, M.F. Kimmitt, S. Kothari, and S. Serafetinidis, J. Phys. **C10**, 905 (1977).
30. D. Hornung, R.von Baltz, and U. Rössler; Sol. State Comm. **48**, 225 (1983).
31. E.I. Blount, in *Solid State Physics*, edited by F. Seitz and D. Turnbull, Academic Press (1962), Vol 13.
32. C.F. Hart and D. Emin, Phys. Rev. **B43**, 4521 (1991) and references cited therein.

33. C. Aversa and J.E. Sipe, Phys. Rev. **B52**, 14 636 (1995).

34. R. Resta, Rev. Mod. Phys. **66**, 899 (1994).

35. H. Schneider et al. Superlattices and Microstructures, **20**, 1 (1996).

36. L.P. Faucheux, L.S. Bourdieu, P.D. Kaplan, and A.J. Libchaber, Phys. Rev. Lett. **74**, 1504 (1995).

37. P. Reimann, M. Grifoni, and P. Hänggi, Phys. Rev. Lett. **79**, 10 (1997).

38. B.U. Felderhof and R.B. Jones Physica, **A202**, 119 (1994).

39. E.L. Hahn, Phys. Rev. **80**, 580 (1950).

40. R.J. Elliott, in *Polarons and Excitons*, edited by C.G. Kuper and G.D. Whitfield, Oliver and Boyd (1969).

41. Ch. Tanguy, Phys. Rev. Lett. **75**, 22 (1995).

42. C. Klingshirn, *Ultrafast Spectroscopy of Semiconductors*, in [5].

43. J. Hvam, *Spectroscopy of Coherent Interactions between Light and Matter*, in [5].

44. V.M. Axt, G. Bartels, and A. Stahl, Phys. Rev. Lett. **76**, 2543 (1996).

SOLUTIONS

1a) The general solution of the first order differential equation for the current Eq.(18) consists of the general solution of the homogeneous equation, which is $C\exp(-\gamma t)$, and a particular solution of the inhomogeneous equation. The boundary condition requires that $C = 0$. The particular solution can be found by "variation of the constant", $C \to C(t)$ and yields:

$$\sigma(t-t') = \frac{ne^2}{m} e^{-\gamma(t-t')}\theta(t-t') \tag{155}$$

1b) The susceptibility in the time domain is the Green–function of Eq.(19), i.e. the particular solution for a delta–pulse $\mathcal{E}(t) = \delta(t-t')$ at fixed time t'. For $t < t'$, $\mathcal{P}(t) \equiv 0$, whereas for $t > t'$ the puls creates a free oscillation which rises continously from zero with slope $\epsilon_0 \Omega_p^2$ so that the second derivative becomes a delta function. As a result, we obtain:

$$\chi(t-t') = \frac{\Omega_p^2}{\Omega_0}\, e^{-\frac{\gamma}{2}(t-t')} \sin\left[\Omega_0(t-t')\right]\theta(t-t'). \tag{156}$$

$\Omega_0^2 = \omega_0^2 - (\gamma/2)^2$. This result holds also for the overdamped and even for the critically damped case, $\omega_0 = \gamma/2$, if the appropriate limit is taken.

2.) In the homogeneous case the negative electron charge density compensates the positive ionic background charge so that $\mathcal{E} = 0$ everywhere. The presence of an external charge Q at $\mathbf{r} = 0$ attracts or repells the mobile electrons and, hence, induces an electrical field $\mathcal{E} = -\text{grad}\,\Phi$. In linear approximation, we obtain from Eq. (18):

$$\beta\, \text{grad}\, \rho(\mathbf{r}) = -\frac{e}{m^*}\, \rho_0\, \text{grad}\, \Phi(\mathbf{r}). \tag{157}$$

The required solution which fulfills the boundary condition $\Phi(\infty) = 0$, $\rho(\infty) = \rho_0$ is

$$\rho(\mathbf{r}) = -\frac{n_0 e}{m^* \beta}\, \Phi(\mathbf{r}) + \rho_0. \tag{158}$$

From the Poisson–equation

$$\Delta\Phi = -\frac{1}{\epsilon_0}\Big[Q\,\delta(\mathbf{r}) + \rho(\mathbf{r}) - \rho_0\Big], \tag{159}$$

we obtain the Thomas–Fermi equation for the potential

$$\Delta\Phi(\mathbf{r}) + \kappa^2\Phi(\mathbf{r}) = -\frac{1}{\epsilon_0}Q\,\delta(\mathbf{r}), \quad \kappa^2 = \frac{n_0 e^2}{\epsilon_0 \beta m^*}. \tag{160}$$

The solution of Eq.(160) reads:

$$\Phi(\mathbf{r}) = \frac{Q}{4\pi\epsilon_0}\frac{1}{r}e^{-\kappa r}, \tag{161}$$

where $1/\kappa$ is the Thomas–Fermi screening length.

3.) Fourier–transformation of Eqs.(18,12) yield (in linear approximation)

$$(-i\omega + \gamma)\,\mathbf{j}(\mathbf{q},\omega) + \beta(-i\mathbf{q})\rho(\mathbf{q},\omega) = \frac{n_0 e^2}{m^*}\mathcal{E}(\mathbf{q},\omega), \tag{162}$$

$$-i\omega\rho(\mathbf{q},\omega) + \mathbf{q}\,\mathbf{j}(\mathbf{q},\omega) = 0. \tag{163}$$

For transverse fields $\mathbf{q}\,\mathbf{j} = 0$ so that $\rho = 0$ whereas in the longitudinal case $\rho = qj/\omega$. As a result we obtain:

$$\sigma_t(\mathbf{q},\omega) = \frac{\frac{e^2 n_0}{m^*}}{\gamma - i\omega}, \quad \sigma_\ell(\mathbf{q},\omega) = \frac{\frac{e^2 n_0}{m^*}}{\gamma - i(\omega - q^2/\omega)}. \tag{164}$$

Equivalently, we may describe $\mathbf{j}(\mathbf{r},t)$ as a displacement current with susceptibility $\chi = i\sigma/\omega\epsilon_0$ and dielectric function $\epsilon = 1 + \chi$ by

$$\epsilon_\ell(\mathbf{q},\omega) = 1 - \frac{\omega_p^2}{\omega(\omega + i\gamma) - \beta q^2}, \qquad \epsilon_t(\mathbf{q},\omega) = 1 - \frac{\omega_p^2}{\omega(\omega + i\gamma)}. \tag{165}$$

Neglecting damping, the dispersion of the longitudinal collective excitations (plasmons) $\omega = \omega(\mathbf{q})$ is determined by $\epsilon(\mathbf{q},\omega) = 0$:

$$\omega(\mathbf{q}) = \sqrt{\omega_p^2 + \beta\mathbf{q}^2} \tag{166}$$

4) Fourier–transformation of Eq.(19) together with $\mathcal{P} = \epsilon_0\chi\mathcal{E}$ yields:

$$\chi_\ell(\mathbf{q},\omega) = \frac{\Omega_p^2}{\omega_0^2 - \omega^2 - i\gamma\omega - \beta q^2}. \tag{167}$$

The dispersion of the transverse collective excitations in given by the poles of $\chi_\ell(\mathbf{q},\omega)$. As a result we have in the limit of small wave–numbers

$$\omega(\mathbf{q}) = -\frac{i\gamma}{2} + \omega_0 + \frac{\beta}{2\omega_0}q^2. \tag{168}$$

This result agrees qualitavely with the dispersion of TO phonons near $\mathbf{q} = 0$ [3].

5.) Off resonance the solution of Eqs.(63-64) reads:

$$d_1(t) = \left[\cos\left(\frac{\Omega_R}{2}t\right) - i\frac{\nu}{\Omega_R}\sin\left(\frac{\Omega_R}{2}t\right)\right]e^{i\frac{\nu}{2}t}, \tag{169}$$

$$d_2(t) = i\frac{\omega_R}{\Omega_R}\sin\left(\frac{\Omega_R}{2}t\right)e^{-i\frac{\nu}{2}t}. \tag{170}$$

The probability finding the atom in the excited state is

$$\mid d_2(t)\mid^2 = \left(\frac{\omega_R}{\Omega_R}\right)^2 \sin^2\left(\frac{\Omega_R}{2}t\right). \tag{171}$$

For $\nu = \omega_R$ the Rabi–amplitude is reduced to $\frac{1}{2}$ and the period is shortened by a factor of $\frac{1}{\sqrt{2}}$ if compared with the resonant case.

6.) At resonance, there are two orthogonal states with $|d_j(t)|^2 = \text{const}$. These are obtained by chosing $a = \pm ib$ in Eqs.(63, 64):

$$d_+(t) = \frac{1}{\sqrt{2}}\begin{pmatrix} +1 \\ -1 \end{pmatrix} e^{-i\frac{\omega_R}{2}t}, \qquad d_-(t) = \frac{1}{\sqrt{2}}\begin{pmatrix} +1 \\ +1 \end{pmatrix} e^{+i\frac{\omega_R}{2}t}. \tag{172}$$

The state of the atom is represented by

$$\mid \Psi_+(t) = \frac{1}{2}\left\{e^{-i(\epsilon_1+\frac{\omega_R}{2})t} \mid 1 > -e^{-i(\epsilon_2+\frac{\omega_R}{2})t} \mid 2 >\right\}, \tag{173}$$

$$\mid \Psi_-(t) = \frac{1}{2}\left\{e^{-i(\epsilon_1-\frac{\omega_R}{2})t} \mid 1 > +e^{-i(\epsilon_2-\frac{\omega_R}{2})t} \mid 2 >\right\}. \tag{174}$$

These time–dependent states are the analoga of the stationary states $\mid 1 >, \mid 2 >$ of the isolated atom where the energies are replaced by $\epsilon_1 \pm \frac{\omega_R}{2}$, $\epsilon_2 \pm \frac{\omega_R}{2}$. Hence, an additional weak perturbing field with variable frequency will induce transitions at three different frequencies ω_0, $\omega_0 + \omega_R$, and $\omega_0 - \omega_R$, where $\omega_0 = (\epsilon_2 - \epsilon_1)/\hbar$. To observe the Rabi–splitting the amplitude of the driving field at frequency ω_0 has to be large enough so that ω_R is larger than the line width.

7.) The singularities of the response functions are located at:
a) Drude–conductivity: $\omega = -i\gamma$.
b) Drude–susceptibility: $\omega = 0$ and $-i\gamma$.
c) Lorentz–susceptibility: $\omega = -i\frac{\gamma}{2} \pm \sqrt{\omega_0^2 - \left(\frac{\gamma}{2}\right)^2}$
In the weak damping case there are two poles which lie closely below the real axis near the frequency of the undamped oscillator. With increasing damping these poles move towards the negative imaginary ω–axis. At critical damping there is a single quadratic pole at $\omega = -i\frac{\gamma}{2}$.

8.) Hilbert–transform of a box–function:
If $\omega \notin [-1, 1]$ the integral is not singular and it can be obtained by elementary means

$$g(\omega) = \frac{1}{\pi}\ln\left|\frac{1-\omega}{1+\omega}\right|. \tag{175}$$

For $\omega \in (-1, 1)$ logarithmic terms $\ln\delta$ appear but, eventually, they drop out and the result given above even holds in this case. Note, the Hilbert–transform of the box function is antisymmetric and has logarithmic singularities at the edges of the box at $\omega = \pm 1$. For $\omega \to \infty$, $g(\omega)$ converges to zero as const/ω.

9.) Kramers–Kronig transformation by double Fourier–transformation.
When using Eq.(99) it is easy to verify that the Fourier–transform of $s(t)$ is $P\frac{1}{\omega}$.

10.) Anharmonic oscillator.

The coupled set of equations for the Fourier–coefficients read:

$$[1 - k^2\Omega^2]A_k + \lambda \sum_{m=-\infty}^{\infty} A_m A_{k-m} = 0. \tag{176}$$

In particular, the equations for $k = 0, 2$ are given by

$$A_0 + \lambda \left[2|A_1|^2 + O(A_1^4)\right] = 0, \tag{177}$$

$$\left[1 - 4\Omega^2\right] A_2 + \lambda \left[A_1^2 + O(A_1^4)\right] = 0. \tag{178}$$

Up to second order in A_1 we obtain:

$$A_0 = -2\lambda|A_1|^2, \quad A_2 = \frac{\lambda}{3}A_1^2. \tag{179}$$

Inserting these results in Eq.(176) with $k = 1$:

$$[1 - \Omega^2]A_1 + \lambda \left[2A_0A_1 + 2A_{-1}A_2 + O(A_1^5)\right] = 0 \tag{180}$$

yields up to second order

$$\Omega^2 - 1 = -\frac{10}{3}\lambda^2|A_1|^2 \tag{181}$$

11.) The general solution of the Bloch Eqs.(132-134) (neglecting damping) are:

$$R_1(t) = R_1(0)\frac{\omega_R^2 + \nu^2 \cos\Omega_R t}{\Omega_R^2} - R_2(0)\frac{\nu}{\Omega_R}\sin\Omega_R t - R_3(0)\frac{\nu\omega_R}{\Omega_R^2}\left[1 - \cos\Omega_R t\right], \tag{182}$$

$$R_2(t) = +R_1(0)\frac{\nu}{\Omega_R}\sin\Omega_R t + R_2(0)\cos\Omega_R t + R_3(0)\frac{\omega_R}{\Omega_R}\sin\Omega_R t, \tag{183}$$

$$R_3(t) = -R_1(0)\frac{\nu\omega_R}{\Omega_R^2}\left[1 - \cos\Omega_R t\right] - R_2(0)\frac{\omega_R}{\Omega_R}\sin\Omega_R t + R_3(0)\frac{\nu^2 + \omega_R^2\cos\Omega_R t}{\Omega_R^2}. \tag{184}$$

In particular, on resonance $\nu = 0$ and $\mathbf{R} = (0, 0, 1)$ we obtain:

$$R_1(t) = 0, \quad R_2(t) = \sin\omega t, \quad R_3(t) = \cos\omega t. \tag{185}$$

Compare with Eqs.(77-79) and Eqs.(129-130).

COHERENT DYNAMICS IN SEMICONDUCTORS

Jørn M. Hvam

Mikroelektronik Centret
The Technical University of Denmark
DK-2800 Lyngby, Denmark

ABSTRACT

Ultrafast nonlinear optical spectroscopy is used to study the coherent dynamics of optically excited electron-hole pairs in semiconductors. Coulomb interaction implies that the optical inter-band transitions are dominated, at least at low temperatures, by excitonic effects. They are further enhanced in quantum confined lower-dimensional systems, where exciton and biexciton effects dominate the spectra even at room temperature. The coherent dynamics of excitons are at modest densities well described by the optical Bloch equations and a number of the dynamical effects known from atomic and molecular systems are found and studied in the exciton-biexciton system of semiconductors. At densities where strong exciton interactions, or many-body effects, become dominant, the semiconductor Bloch equations present a more rigorous treatment of the phenomena. Ultrafast degenerate four-wave mixing is used as a tool to study the coherent exciton dynamics, and the importance of performing transform limited spectroscopy is demonstrated throughout.

I. INTRODUCTION

Coherent interactions between light and matter are basic ingredients in the optical properties of matter, particularly when it comes to nonlinear optical processes.[1,2,3] There are several types of coherence to consider in such processes: i) the coherence of the light field, that is essentially determined by the light source applied, ii) the internal (intra-band) coherence of the quantum system in question, and iii) the coherence of the optical excitations (inter-band polarization) induced by the light.

For non-resonant interaction between light and matter, only the coherence of the light plays any role, provided the coherence times (lifetimes) of the excited states of the system are long enough to allow for well-defined resonances. In this case, the (nonlinear) optical response is instantaneous and follows adiabatically even the fastest laser pulses as long as they can be kept completely non-resonant. This is the preferred regime of nonlinear optical

Ultrafast Dynamics of Quantum Systems: Physical Processes and Spectroscopic Techniques, Edited by Di Bartolo and Gambarota, Plenum Press, New York, 1998

devices for loss-less and fast operation. But then the small non-resonant values of the nonlinear coefficients have to be overcome by long interaction lengths and proper (delicate) phase-matching conditions.

For resonant excitations, the nonlinear coefficients are strongly enhanced and all the above coherences are crucial for the (nonlinear) optical response. This opens up for a large variety of coherent nonlinear optical spectroscopies yielding information about the magnitude and the microscopic origin of the nonlinear optical coefficients and about the coherence and dephasing times of the system. [3, 4]

The simplest systems discussed, are those of non-interacting two-level (or three-level) atoms excited resonantly by an intense laser, describing well the spectroscopy of dilute gases. The dynamics of this system is solved by the optical Bloch equations, illustrating a number of nonlinear optical processes and spectroscopies, such as free polarization decay, transient four-wave mixing, photon echoes, quantum beats and polarization interferences. From these experiments one can determine the dephasing rate, or homogeneous line-width, which is phenomenologically introduced into the optical Bloch equations. In dilute gasses, characteristic dephasing times, T_2, are in the microsecond or even millisecond range, so that nonlinear optical experiments can be performed alternatively in time or in frequency domain. [5]

In solids, e.g. semiconductors, inter-atomic interactions and scattering processes are so fast that it is only in the past decade that ultrafast lasers have provided the opportunity to perform resonant, transient nonlinear experiments in the coherent regime. [4] In particular the excitonic transitions just below the fundamental band edge, have been accessible for transient experiments with picosecond and femtosecond lasers. The discrete nature of the exciton states has made it possible to observe, particularly in low-dimensional semiconductor quantum structures, a number of the above phenomena known from coherent nonlinear spectroscopy of atoms and molecules. [4,6]

From such experiments, exciton-exciton [7], exciton-electron [8], and exciton-phonon [9] interaction parameters have been determined, and exciton localization phenomena due to alloy [10], or heterostructure interface [11], fluctuations have been studied. From quantum-beat experiments in low-dimensional semiconductor nanostructures, biexciton formation has been identified, and binding energies have been determined with high accuracy. [12]

When a high density of excitons, and/or a distribution of continuum states, are excited by a short, intense laser pulse, the true many-body character of the semiconductor is revealed, and the nonlinear signals no longer have the characteristics known from simple two-level systems. In this case, the nonlinear optical response is usually described by the semiconductor Bloch equations, including the Coulomb screening, exchange interactions and local field effects.[13] Still dephasing is introduced "by hand", possibly including excitation-induced dephasing. The most recent development in this field, experimentally as well as theoretically, is to go beyond the Markovian limit of dephasing by an infinite heat bath (T_2), and investigate also truly dynamic dephasing.[14]

In the following section, general principles of coherent interaction between light and simple two-level and three-level systems are being described in terms of the optical Bloch equations. In section III, four-wave mixing (FWM), and in particular transient FWM, is treated in relation to two-beam and three-beam experiments. Section IV briefly discusses some aspects of semiconductors, and in particular low-dimensional semiconductor nanostructures where excitonic features are strongly enhanced. Section V presents some examples of coherent spectroscopy of semiconductor quantum wells, where the importance of spectrally resolving the transient FWM signal becomes obvious. The experimental techniques are briefly mentioned in section VI, and section VII is devoted to a few concluding remarks about temporally and spectrally resolved nonlinear spectroscopy.

II. COHERENT INTERACTION

The interaction between light and a quantum system is described by the Schrödinger equation, e.g. in the density matrix formalism [15]

$$\frac{\partial \hat{\rho}}{\partial t} = \frac{i}{\hbar}\left[\hat{\rho}, \hat{H}\right] - \gamma \hat{\rho} \tag{1}$$

where $\hat{\rho}$ is the density matrix and $\hat{H}$ is the Hamiltonian of the quantum system, interacting with the electromagnetic wave

$$\hat{H} = \hat{H}_0 + \hat{\mu} \cdot \overline{E} \tag{2}$$

$\hat{H}_o$ is the Hamiltonian for the atomic system, $\overline{E}$ is the electric field and $\hat{\mu}$ is the dipole operator. The polarization field of a system with N atoms per unit volume is then described by

$$\overline{P} = N \cdot Tr\left[\hat{\mu}\hat{\rho}\right] \tag{3}$$

where $Tr\left[\hat{A}\right]$ means the trace of the operator $\hat{A}$.

II. A. Two-level System

For a system of non-interacting two-level atoms, equation (1) transforms into the set of optical Bloch equations

$$\frac{\partial \rho_D}{\partial t} = -\frac{2i\mu E}{\hbar}\left(\rho_{10} - \rho_{01}\right) - \frac{\rho_D - \rho_{D0}}{T_1} \tag{4}$$

$$\frac{\partial \rho_{10}}{\partial t} = \frac{\partial \rho_{01}^*}{\partial t} = \frac{i\mu E}{\hbar}\rho_D - \left(\frac{1}{T_2} + i\omega_0\right)\rho_{10} \tag{5}$$

The diagonal element $\rho_D = \rho_{00} - \rho_{11}$ is the population difference between the ground state and the excited state, and the total damping is $\gamma = \frac{1}{T_2} + \frac{1}{2T_1} \approx \frac{1}{T_2}$ since the longitudinal relaxation time (lifetime) T_1 is usually much longer than the transverse relaxation, or dephasing, time T_2. From equation (3) the polarization field is found to be

$$\overline{P} = N\mu\left(\rho_{10} + \rho_{01}\right) = N\mu\left(\rho_{10} + c.c.\right) \tag{6}$$

where $c.c.$ denotes the complex conjugate.

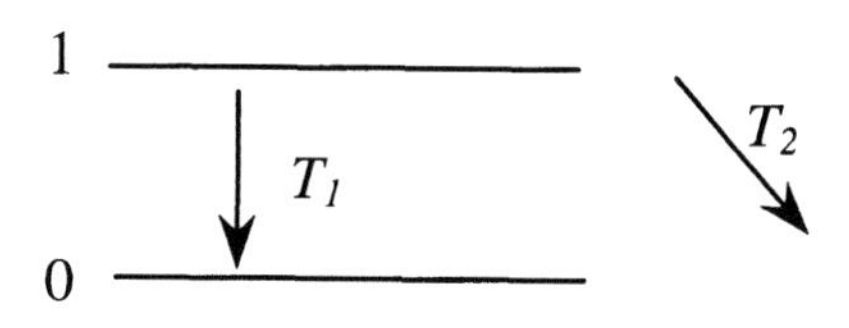

Figure 1. Two-level atom

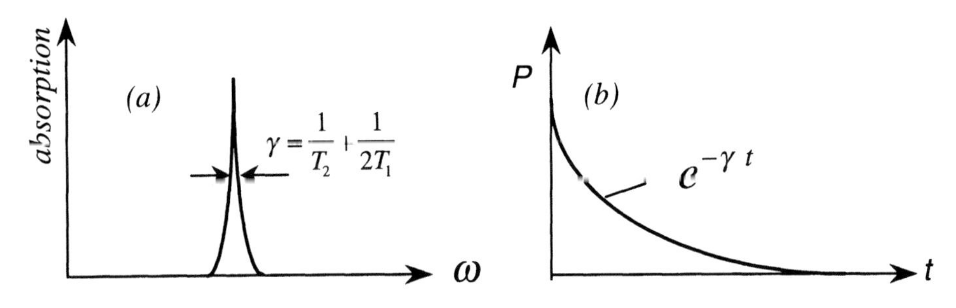

Figure 2. Spectral (a) and temporal (b) determination of the damping (dephasing) in a two-level system.

If the system is left in a coherently excited state at t=0, the polarization field will therefore, according to equation (5), decay as

$$\overline{P}(t) = \overline{P}(0)e^{-\gamma t} \approx \overline{P}(0)e^{-\frac{t}{T_2}} \tag{7}$$

This is the so-called free polarization decay, which is the equivalent of the free induction decay observed in spin systems, being ideal two-level systems.[16]

There are now two equivalent ways of measuring the damping, or dephasing, of such a two-level system. Either by directly measuring the free polarization decay with high temporal resolution after a short-pulse excitation in a resonant fluorescence (luminescence) experiment, or by making a c.w. absorption experiment with high spectral resolution, where the absorption line will have a Lorentzian shape with a width $2\hbar\gamma$ (FWHM). In this simple system any transform-limited combination of spectral and temporal resolution (i.e. limited only by the uncertainty relations, or a derivative of that) will yield the same (maximum) information about the damping of the system.

II. B. Three-level System

If the upper level in Fig. 1 is split into two close-lying levels as in Fig. 3, there are significant changes to be observed in the above experiments. In the spectral domain, two absorption lines will be observed provided the spectral resolution in the experiment is better than the splitting of the two upper levels (Fig. 4a).

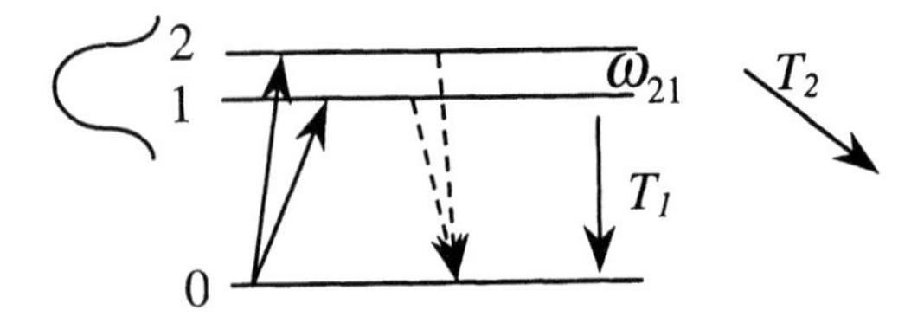

Figure 3. Nearly degenerate three-level system.

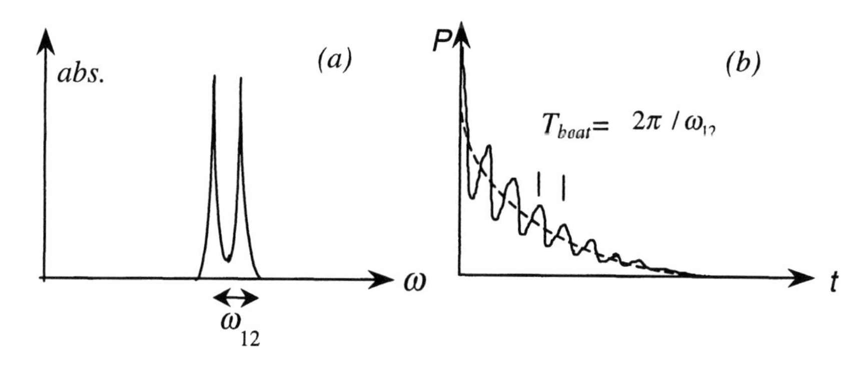

Figure 4. Spectral (a) and temporal (b) determination of the level splitting in a nearly degenerate three-level system.

In a time-resolved experiment, the two upper levels are simultaneously, and coherently, excited by the same laser pulse that is sufficiently short to have a spectral width covering the two close-lying transitions. In this case the free polarization decay from the coherently excited upper levels is superposed an oscillatory component, the so-called quantum beat, which again can be observed in a resonance fluorescence experiment, where the emission intensity after a short-pulse excitation at t=0 will decay as[17]

$$I(t) = \Theta(t) I_0 \left\{ \mu_{10}^4 e^{-2\gamma_{10}t} + \mu_{20}^4 e^{-2\gamma_{20}t} + 2\mu_{10}^2 \mu_{20}^2 e^{-(\gamma_{10}+\gamma_{20})t} \cos \omega_{12} t \right\} \tag{8}$$

$\Theta(t)$ is the Heaviside step function and I_0 is the initial intensity. Note that the overall decay is governed by the upper level with the smallest damping, while the decay of the oscillatory term, the quantum beat, is governed by the level with the largest damping. Hence, in principle the damping of both levels can be determined from this experiment.

It is to some extent a matter of taste whether to perform experiments with spectral or temporal resolution. In atomic and molecular systems, where level splittings can be extremely small, it may be easier to perform time-resolved experiments with characteristic features in the microsecond and millisecond range.[5] In liquids and solids, where dephasing times can be extremely small, i.e. in the picosecond and femtosecond range, it is only recently that ultrafast lasers and detection schemes have made it possible to observe free polarization decays and quantum beats of inter-band transitions.[18] It should be kept in mind, that in a linear optical experiment, simultaneous spectral and temporal resolutions exclude each other beyond the transform limit, set by the fundamental uncertainty principles.

II. C. Inhomogeneous Broadening

In the above, it was assumed that all the atoms had perfectly identical transitions, i.e. that the system is only homogeneously broadened. In many systems, however, this is not the case. There will be randomly fluctuating environmental conditions for the atomic or electronic transitions that will influence the transition energies in a random way. This gives rise to an inhomogeneous broadening of the observed spectral lines that may completely mask the intrinsic homogeneous broadening or dephasing of the system. In that case, the spectral line-width as well as the free polarization decay are governed by the inhomogeneous broadening of the upper atomic level (see Fig. 5).

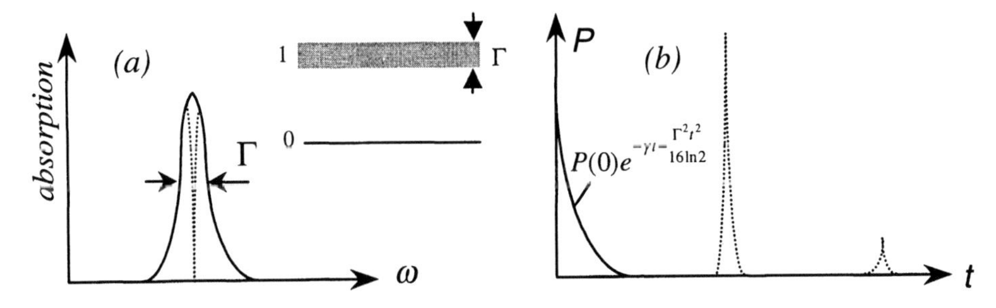

Figure 5. In an inhomogeneously broadened system, the spectral line-width (a) as well as the free polarization decay (b) are governed by the inhomogeneous broadening Γ of the upper atomic level. The dotted curves indicate non-linear spectroscopy in the form of spectral hole burning (a) and photon echo (b).

In linear spectroscopy there is no simple way of getting around this problem. One must therefore resort to non-linear spectroscopies such as spectral hole burning (Fig. 5a) or photon echo (Fig. 5b). [3,19] The latter is the result of transient nonlinear four-wave mixing in an inhomogeneous medium. In the next section, I shall discuss the four-wave mixing process in more detail.

III. FOUR-WAVE MIXING

In classical non-linear optics, the polarization $\overline{P}$ of the medium is usually described as an expansion in powers of the electric field vector $\overline{E}$:[1]

$$\begin{aligned}\overline{P}(\overline{r},t) &= \chi^{(1)}\cdot\overline{E}+\chi^{(2)}\cdot\overline{E}\overline{E}+\chi^{(3)}\cdot\overline{E}\overline{E}\overline{E}+\dots \\ &= \overline{P}^{(1)}(\overline{r},t)+\overline{P}^{(2)}(\overline{r},t)+\overline{P}^{(3)}(\overline{r},t)+\dots\end{aligned} \tag{9}$$

where $\chi^{(n)}$ is the n'th order nonlinear optical susceptibility. Linear optics is governed by the first term. In systems with inversion symmetry the second order term is identically zero, and thus the lowest order non-linearity is given by the third term.

If the total electric field is a superposition of plane waves, the polarization field will also appear as such

$$\overline{E}(\overline{r},t)=\sum_i(\overline{E}_i(\overline{k}_i,\omega_i)e^{i(\overline{k}_i\cdot\overline{r}_i-\omega_i t)}+c.c.) \tag{10}$$

$$\overline{P}(\overline{r},t)=\sum_j(\overline{P}_j(\overline{k}_j,\omega_j)e^{i(\overline{k}_j\cdot\overline{r}_j-\omega_j t)}+c.c.) \tag{11}$$

and the third order polarization can, in the standard non-linear optics notation introduced by Bloembergen.[1], be described as

$$\overline{P}^{(3)}(\overline{k},\omega)=\chi^{(3)}(\omega;\pm\omega_1,\pm\omega_2,\pm\omega_3)\cdot\overline{E}_1(\overline{k}_1,\omega_1)\cdot\overline{E}_2(\overline{k}_2,\omega_2)\cdot\overline{E}_3(\overline{k}_3,\omega_3) \tag{12}$$

In order to have any noticeable energy transfer to the third order polarization wave, energy and wave-vector conservations have to be met in the above equation, i.e.

$$\omega = \sum_{i=1}^{3} \pm \omega_i \quad \wedge \quad \bar{k} = \sum_{i=1}^{3} \pm \bar{k}_i \tag{13}$$

where the plus (minus) sign corresponds to an absorbed (emitted) wave. The equations (13) pose, together with the dispersion relations for all participating waves $\omega_i = \omega_i(\bar{k}_i)$, very severe limitations to efficient generation of non-linear signals. These are the so-called phase-matching conditions. [1,2]

Processes described by equation (12) are generally called four-wave mixing because the three waves on the right-hand side are mixing in the nonlinear medium to form a third order polarization wave, that serves as an antenna for the out-going fourth wave, the signal wave. In degenerate four-wave mixing, the frequencies of all the waves are the same, i.e. $\omega = \omega_1 = \omega_2 = \omega_3 (= 2\omega - \omega)$. If all the wave-vectors are also identical (one beam), phase-matching is automatically fulfilled, and

$$\bar{P}^{(3)}(\bar{k}, \omega) = \chi^{(3)}(\omega; \omega, \omega, -\omega) \left| \bar{E}(\bar{k}, \omega) \right|^2 \bar{E}(\bar{k}, \omega) \tag{14}$$

The non-linear refractive index $n(I) = n_0 + n_2 I$ and the nonlinear absorption coefficient $\alpha(I) = \alpha_0 + \alpha_2 I$ are related to the third order susceptibility in equation (12) by

$$n_2 = \frac{\mathrm{Re}\,\chi^{(3)}}{\varepsilon_0 n_0^2 c} \tag{15}$$

and

$$\alpha_2 = \frac{\omega\,\mathrm{Im}\,\chi^{(3)}}{\varepsilon_0 n_0^2 c^2} \tag{16}$$

These effects are extremely important for the propagation of intense lasers through media with even small or moderate nonlinear coefficients. Self-focusing effects due to a nonlinear refractive index can destroy optical components, but can also be used constructively for the mode-locking of lasers as e.g. in self-mode-locked Ti:sapphire femtosecond lasers.[20] Self-phase modulations in optical fibers and the formation of soliton waves are other examples.[21] Nonlinear absorption (or gain) processes are particularly important in semiconductor lasers and amplifiers for fast optical signal processing in e.g. optical communication systems.

Degenerate four-wave mixing can also occur between different (up to three) beams, and in that case the phase matching conditions become more restrictive. In the case of two beams the situation is sketched in Fig. 6. In this case, exact phase-match is obtained in the directions $\bar{k}_1 = \bar{k}_1 + \bar{k}_2 - \bar{k}_2$ and $\bar{k}_2 = \bar{k}_1 - \bar{k}_1 + \bar{k}_2$. Near phase-match is, on the other hand, obtained in the directions $\bar{k}' = 2\bar{k}_1 - \bar{k}_2 - \Delta\bar{k}$ and $\bar{k} = 2\bar{k}_2 - \bar{k}_1 - \Delta\bar{k}$, where $\Delta\bar{k}$ is a measure of the wave vector mismatch in the forward direction. In the back scattering direction the wave- vector mismatch is measured by $\Delta\bar{k}_b$.

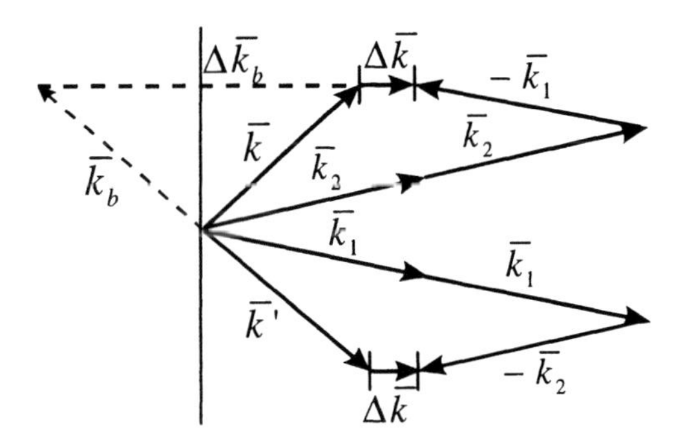

Figure 6. Four-wave mixing between two beams with directions $\bar{k}_1$ and $\bar{k}_2$. Exact phase match in directions $\bar{k}_1$ and $\bar{k}_2$, and near phase match in directions $\bar{k}$ and $\bar{k}'$. The thin vertical line is the sample surface, and the dashed lines indicate back scattering.

The FWM signal propagating in the direction $\bar{k} \approx 2\bar{k}_2 - \bar{k}_1$ can be considered as the diffraction of beam number 2 ($\bar{k}_2$) in the polarization grating formed in the nonlinear medium by interference between the two beams. It has the grating vector $\bar{k}_2 - \bar{k}_1$ with the corresponding line spacing $\Lambda = 2\pi / |\bar{k}_2 - \bar{k}_1|$. If the sample has a slab-geometry with a thickness d, the nonlinear diffracted signal propagating in one of the near phase match directions $\bar{k}$ or $\bar{k}'$ will have the intensity [2]

$$I_s \propto d^2 \frac{\sin^2(\frac{\Delta k \cdot d}{2})}{(\frac{\Delta k \cdot d}{2})^2} \tag{17}$$

For large interaction lengths ($\Delta k \cdot d >> 1$), there will be negligible intensity in the signal beam, but for $\Delta k \cdot d << 1$ (thin sample) the phase mismatch does not lead to severe reduction of the nonlinear signal. The big advantage of the thin-sample geometry for the purpose of nonlinear optical spectroscopy is, that nonlinear signals are propagating as collimated beams in the background free directions $2\bar{k}_1 - \bar{k}_2$ and $2\bar{k}_2 - \bar{k}_1$. By simple spatial filtering one can therefore detect extremely weak nonlinear signals, the detection limit essentially being set by the scattered light intensity. Samples of good optical quality (surfaces) are therefore essential for such experiments.

When the sample is extremely thin or the interaction length for some other reason (linear absorption) is very short, near phase matching can even be achieved in the back scattering direction with intensity comparable to the one in the forward direction although the signal in both directions will diminish as d^2.

III. A. Transient Four-Wave Mixing

In a two-beam transient four-wave mixing experiment, the two incoming beams $(\bar{k}_1, \bar{k}_2)$ each consists of a train of short, e.g. femtosecond, laser pulses from the same coherent laser source. In that case the timing of the two incoming pulse trains is important, since a nonlinear signal in e.g. the direction $2\bar{k}_1 - \bar{k}_2$ is only produced if there is a coherent temporal overlap between the polarization from the first pulse in beam number 1 and the second pulse in beam number two. Hence, information about the coherence and dephasing of optical transitions in a medium can be obtained by recording the nonlinear signal intensity as a function of delay between the incoming pulses. These are the so-called correlation traces. The situation is sketched in Fig. 7. There is an ambiguity with respect to the sign of τ. In the present work, as in most literature on transient FWM, the delay τ is counted positive when the pulse in beam 1 arrives first, and the signal is detected in the direction $2\bar{k}_1 - \bar{k}_2$.

Let us for illustrative purposes look at how an ideal homogeneously broadened two-level system will respond to such a transient FWM experiment. The third-order non-linear signal can be found by solving the optical Bloch equations (3) and (4) in a perturbational way by expanding the density matrix in powers of μE.

$$\rho = \rho^{(0)} + \rho^{(1)} + \rho^{(2)} + \ldots\ldots \tag{18}$$

We then get the successive set of equations

$$\frac{\partial \rho^{(1)}}{\partial t} - \frac{i}{\hbar}\left[\rho^{(1)}, \hat{H}_0\right] + \gamma\rho^{(1)} = \left[\mu E, \rho^{(0)}\right] \tag{19}$$

$$\frac{\partial \rho^{(2)}}{\partial t} - \frac{i}{\hbar}\left[\rho^{(2)}, \hat{H}_0\right] + \gamma\rho^{(2)} = \left[\mu E, \rho^{(1)}\right] \tag{20}$$

etc.

Solving to third order, we obtain the third order nonlinear polarization

$$P^{(3)}(t,\tau) = N\int_0^\infty \mu_{10}\rho_{10}^{(3)}(t,\tau,\omega_{10})d\omega_{10} \tag{21}$$

and its Fourier transform

$$P^{(3)}(\omega,\tau) = \frac{1}{2\pi}\int_0^\infty P^{(3)}(t,\tau)e^{-i\omega t}dt \tag{22}$$

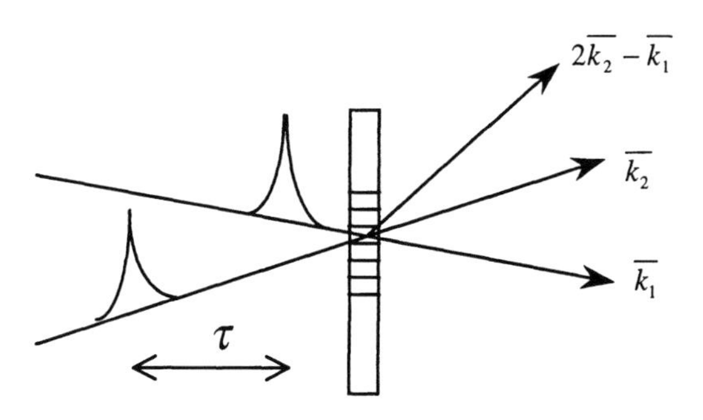

Figure 7. Transient two-beam four-wave mixing

The latter is directly related to what is measured in an experiment where the non-linear signal is spectrally analyzed, integrated by a slow detector and recorded as a function of the delay τ between the incident pulses in the two beams.

For a purely homogeneously broadened two-level system with the complex transition frequency $\Omega_{10} = \omega_{10} - i\gamma_{10}$, the above optical Bloch equations (19)-(21) can be solved analytically for delta-like input pulses to yield [22,23]

$$P^{(3)}(\bar{r}, t, \tau_{12} > 0) \propto \Theta(t)\mu_{10}^4 e^{i(2\bar{k}_2 - \bar{k}_1)\cdot\bar{r} - \Omega_{10}t + \Omega_{10}^* \tau_{12}} \tag{23}$$

This means that the intensity of the transient FWM signal propagating in the $2\bar{k}_2 - \bar{k}_1$ direction is given by

$$I_{TFWM} \propto \left|P^{(3)}(t, \tau > 0)\right|^2 \propto \Theta(t)\mu_{10}^8 e^{-2\gamma_{10}(t+\tau)} \tag{24}$$

i.e. the signal decays in real time as well as a function of the delay with the homogeneous damping rate. In the frequency domain, the signal intensity is expressed by

$$I_{TFWM} \propto \left|P^{(3)}(\omega, \tau > 0)\right|^2 \propto \frac{\mu_{10}^8}{\left|\Omega_{10} - \omega\right|^2} e^{-2\gamma_{10}\tau} \tag{25}$$

from which the homogeneous damping can be determined both from the line-width of the signal and from the decay of the time-integrated signal as a function of the delay between the incident pulses. In the above treatment, the delay has been assumed positive. For negative delay, the pulse in beam 2 has passed when the grating is formed, and its field can therefore not be diffracted into the $2\bar{k}_2 - \bar{k}_1$ direction (but now in the $2\bar{k}_1 - \bar{k}_2$ direction). However, also the pulse in beam 2 leaves a coherent polarization field behind, that can give rise to diffraction into the $2\bar{k}_2 - \bar{k}_1$ direction also for negative delay, provided there is also a polarization interaction in the medium, such as for example exciton-exciton interaction in semiconductors. [24] For negative delay, however, the FWM signal decays twice as fast as for positive delay, because the polarization field enters twice in the nonlinear interaction. The situation is illustrated in Fig 8.

There are other, non-grating, contributions to the transient FWM signal for negative delay, namely such that are associated with two-photon absorption in beam 2 coherently followed by emission stimulated by beam 1. This contribution is strong whenever there are possible two-photon resonances, such as biexcitons. [25]

Transient FWM can also be performed with three incoming beams as in Fig. 9. The first two pulses set up a polarization grating as well as a real density grating (for parallel polarization) and the third pulse is then diffracted in this grating in the directions $\bar{k}_3 \pm (\bar{k}_2 - \bar{k}_1)$. This experimental configuration gives the possibility of separating the truly coherent optical nonlinearities from the incoherent optical nonlinearities arising from an incoherent population of carriers (excitations) in the medium. The coherence and dephasing is monitored by measuring the nonlinear signal e.g. in the direction $\bar{k}_3 + \bar{k}_2 - \bar{k}_1$ as a function of the delay τ_{12} between the two first pulses for a large delay $\tau_{23} >> \tau_{12}$ of the third pulse (see Fig. 10a).

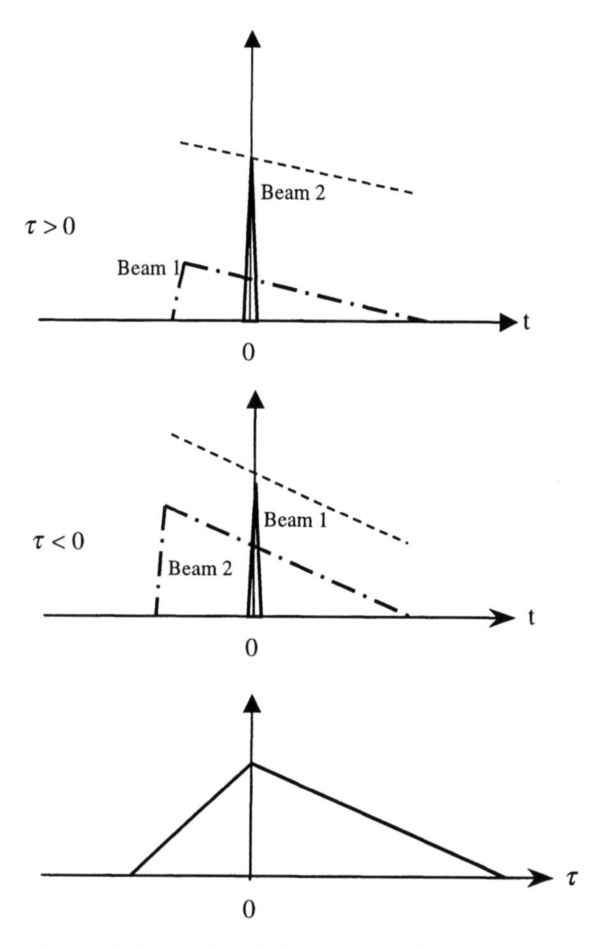

Figure 8. The polarization (dot-dashed) and the field (full) as a function of real time for positive (upper) and for negative (middle) delay. The reason that the signal decays faster for negative delay is that the polarization enters quadratically. For positive delay it enters linearly. The resulting FWM signal as a function of delay is shown at the bottom. The intensity axes are assumed to be logarithmic.

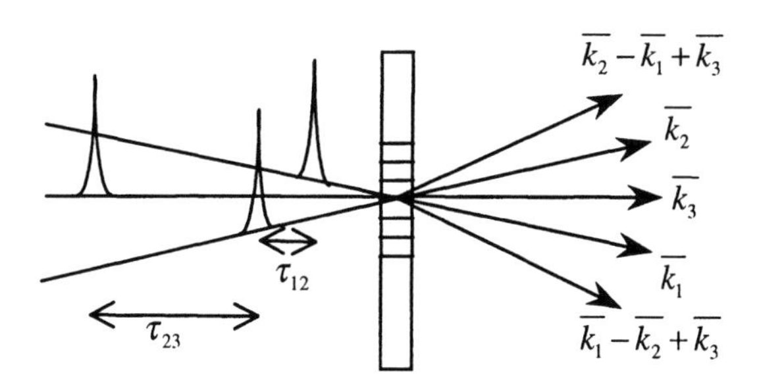

Figure 9. Three-beam transient four-wave mixing, or transient light-induced grating.

To measure the decay of a real density grating the same signal can be measured as a function of τ_{23} for $\tau_{12} = 0$ (see Fig. 10b). These so-called transient grating, or light-induced grating, experiments have been used extensively to measure carrier recombination as well as carrier diffusion in semiconductors. Both phenomena contribute to the decay of the density grating, but they can be separated by performing transient grating experiments with different grating spacing, i.e. for different angles between the two beams setting up the grating. [26]

From such experiments the dynamics of photoexcited carriers in semiconductors can thus be followed from the first scattering events destroying the coherence over carrier relaxation (spectral diffusion) to the final recombination. Since the first experiments of transient FWM by Schultheiss et al. in 1986 [7,8] and until now this has been an extremely powerful way of studying exciton-phonon, exciton-exciton and exciton-electron interaction in semiconductors and semiconductor nanostructures. [27]

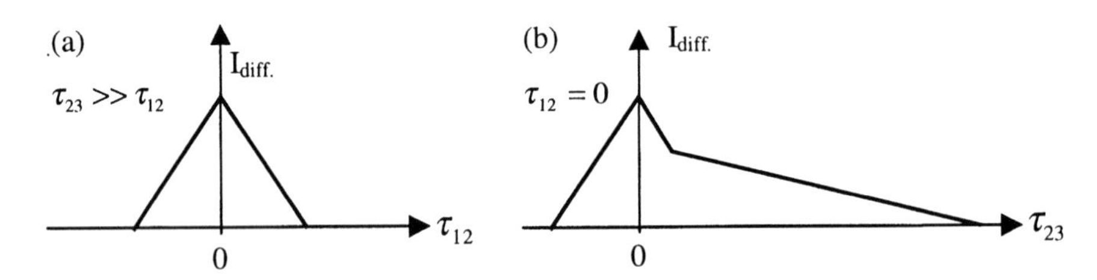

Figure 10. Three-beam FWM or transient grating experiment

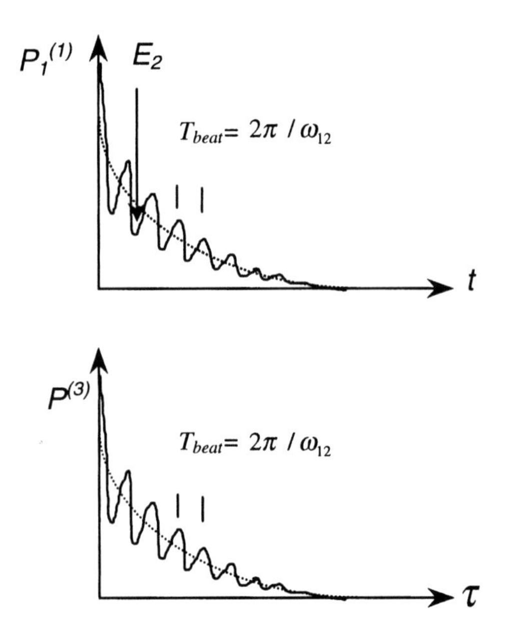

Figure 11. Top curve is the linear quantum beat in the polarization from the first pulse. Bottom curve is the nonlinear quantum beat in the third order polarization as a function of delay, i.e. the time the second pulse arrives.

III. B. Nonlinear Quantum Beats

Interference and beat phenomena, as described in Section II.B, may also occur in nonlinear optical spectroscopy. [6] If the nearly degenerate three-level system of Fig. 3 is exposed to a two-beam transient FWM experiment where the first pulse excites a coherent superposition of the close-lying excited states, quantum beats will appear in the free polarization decay of the first pulse. This will affect the correlation trace of the two-beam FWM that will again show a quantum beat now as a function of the delay, but with the same oscillation period as in the free polarization decay. This is intuitively clear as illustrated in Fig. 11, but it also follows from a solution of the optical Bloch equations for the three-level system described by the complex transition frequencies

$$\begin{aligned}\Omega_{10} &= \omega_1 - \omega_0 - i\gamma_{10} \\ \Omega_{20} &= \omega_2 - \omega_0 - i\gamma_{20} \\ \Omega_{21} &= \omega_2 - \omega_1\end{aligned} \tag{26}$$

The solution to third order with delta-like input pulses, as in section III.A, is for the nonlinear polarization propagating in the $2\bar{k}_2 - \bar{k}_1$ direction [23,28]

$$\begin{aligned}P^{(3)}(t,\tau>0) \propto \Theta(t)\Big[&2\mu_{10}^4 e^{-i(\Omega_{10}t-\Omega_{10}^*\tau)} + 2\mu_{20}^4 e^{-i(\Omega_{20}t-\Omega_{20}^*\tau)} \\ &+ \mu_{10}^2\mu_{20}^2\left(e^{-i(\Omega_{10}t-\Omega_{20}^*\tau)} + e^{-i(\Omega_{20}t-\Omega_{10}^*\tau)}\right)\Big]\end{aligned} \tag{27}$$

where the two first terms express the standard free-polarization decays from the two excited levels and the third term gives rise to the quantum beat terms that are oscillating in real time as well as in delay time.

The true quantum beats from a real three level system, as in Fig. 3, are not always easily distinguished from classical polarization interferences from e.g. a system containing nearly degenerate, but independent two-level atoms. This requires a detailed analysis of the nonlinear signal in equation (27) both as a function of real time and as a function of delay time. Alternatively, one can spectrally analyze the transient FWM signal. Fourier transforming the third order polarization in equation (27) one obtains

$$P^{(3)}(\omega,\tau) \propto \frac{2\mu_{10}^4 e^{i\Omega_{10}^*\tau} + \mu_{10}^2\mu_{20}^2 e^{i\Omega_{20}^*\tau}}{\Omega_{10}-\omega} + \frac{2\mu_{20}^4 e^{i\Omega_{20}^*\tau} + \mu_{10}^2\mu_{20}^2 e^{i\Omega_{10}^*\tau}}{\Omega_{20}-\omega} \tag{28}$$

From this it can be seen that there is a resonant nonlinear contribution for each of the close-lying transitions. For equal dipole matrix elements for the two transitions and assuming distinctly different dampings of the two levels involved in the beating, e.g. $\gamma_{20} >> \gamma_{10}$, a simple expression for the intensity of the transient FWM signal detected near one resonance can be obtained

$$I_{FWM} = \frac{e^{-2\gamma_{10}\tau}\left\{1+e^{-\gamma_{20}\tau}\cos(\omega_{12}\tau)\right\}}{(\omega_{10}-\omega)^2+\gamma_{10}^2} \tag{29}$$

As in the linear case, the average signal decays with the slowest damping rate, whereas the modulation decays with the fastest damping rate.

Note that, in contrast to the linear case, a spectral resolution of the signal from the two resonances does not destroy the quantum beats observed as a function of delay. This reflects the fact that there is no transform relation between the spectrum of the emitted nonlinear signal and the timing of the incident pulses. There is therefore additional real information in the spectral analysis of the transient FWM signal that cannot be found from the delay dependence of the signal alone. The spectral signature of the quantum beat signal in equation (29) is the Lorentzian line shape of the independent transient, which is then modulated as a whole by the quantum interference with the other level. This is in contrast to classical interference between independent transitions where different parts of the spectrum will be modulated with different phases in a similar spectrally resolved transient FWM experiment. [29]

III. C. Photon Echo

If the system of two-level atoms is inhomogeneously broadened, destructive interferences will give rise to a fast decay of the macroscopic polarization in the medium, changing significantly the outcome of a transient FWM experiment. The inhomogeneous broadening can be described by a Gaussian distribution of transition energies around the center frequency ω_0 and with the width Γ_{10} (FWHM)

$$g(\omega_{10}) \propto e^{\frac{-4\ln 2(\omega_{10}-\omega_0)^2}{\Gamma_{10}^2}} \tag{30}$$

The macroscopic inhomogeneous polarization is then found by integration of the homogeneous polarization over the distribution of transition frequencies

$$P_{inh}^{(3)}(t,\tau) \propto \int_0^\infty P^{(3)}(t,\tau)g(\omega_{10})d\omega_{10} \tag{31}$$

For delta-like pulses arriving at $t=-\tau$ and $t=0$, respectively, this problem can again be solved analytically giving rise to a signal intensity with the main features given by [22,28]

$$I_{PE} \propto \left|P_{inh}^{(3)}(t,\tau>0)\right|^2 \propto \mu_{10}^8 e^{-\frac{\Gamma_{10}^2(t-\tau)^2}{16\ln 2}} e^{-2\gamma_{10}(t+\tau)} \tag{32}$$

From equation (32) is seen that the FWM signal in this case appears as a photon echo emitted at the time $t=\tau$. For negative delay there is of course no photon echo. The photon echo signal in the frequency domain can be found from the above by taking the Fourier transform of equation (31). For large delays the result is

$$I_{PE} \propto \left|P_{inh}^{(3)}(\omega,\tau>0)\right|^2 \propto \mu_{10}^8 e^{-8\ln 2\frac{(\omega_0-\omega)^2}{\Gamma_{10}^2}} e^{-4\gamma_{10}\tau} \tag{33}$$

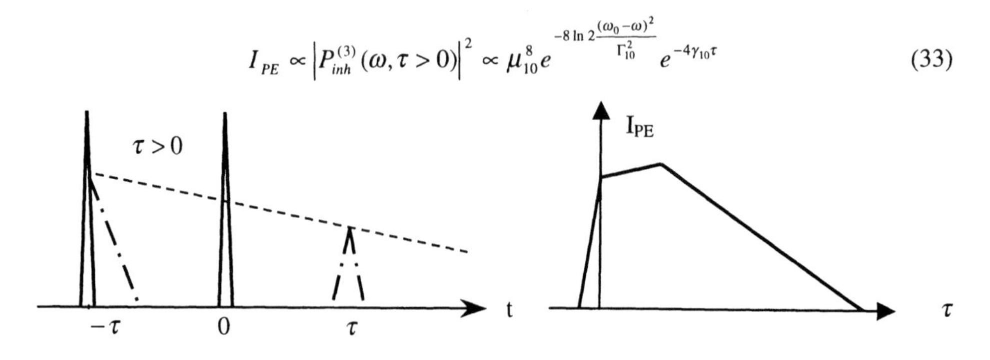

Figure 12. Sketch of photon echo in real time (left) and in delay time (right)

The spectral line width is given by the inhomogeneous broadening Γ_{10}, but the homogeneous dephasing γ_{10} can still be found from the decay of the FWM signal as a function of delay. The reason that this decay is twice as fast as in the homogeneous case is that the signal appears at twice the delay time with respect to the arrival of the first pulse, as sketched in Fig. 12.

IV. SEMICONDUCTORS

One could wonder to what extent the treatment of light-matter interactions given so far has any relevance for semiconductors, i.e. crystalline solids that are usually characterized by continuous energy bands rather than simple two-level and three-level systems. There are, however, a number of effects in semiconductors that modify the optical properties and restore an atom-like behavior in a number of important cases. The most important effect is the Coulomb interaction between photoexcited electrons and holes giving rise to hydrogen-like bound states of electron-hole pairs, or excitons, strongly modifying the optical transitions near the fundamental absorption edge. [30] As sketched in Fig. 13, a number of discrete transition energies appear just below the continuum band edge.

At temperatures that are low compared to the exciton binding energy (exciton Rydberg), excitonic effects dominate the linear optical properties of semiconductors. In the absorption spectrum appears a series of discrete hydrogen-like lines just below the fundamental

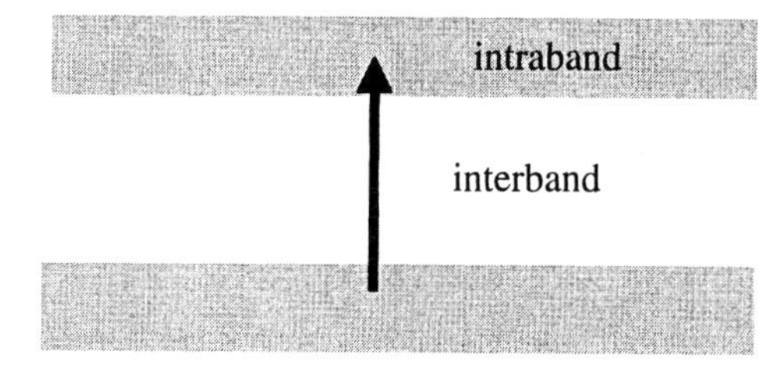

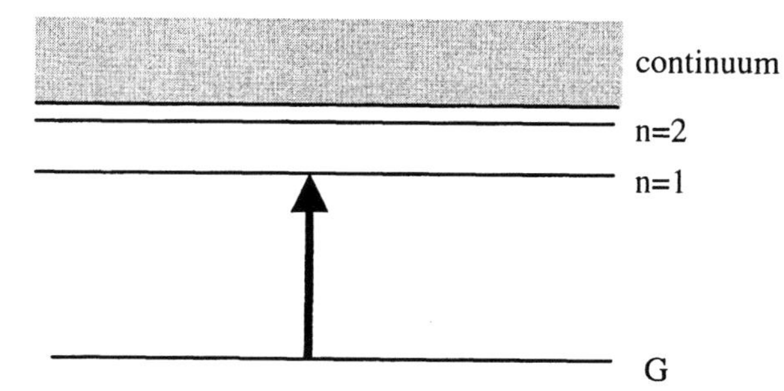

Figure 13. Semiconductor band model. The discrete energy levels and transitions below the band gap are due to the Coulomb interaction between the electron and the hole, forming an exciton.

absorption edge and a strong enhancement of the optical absorption at the edge, the so-called Coulomb enhancement.

Also the nonlinear optical properties are strongly influenced by excitons. One reason is that due to the weak binding of the excitons, they fill a rather large volume and therefore interact strongly even at modest densities. Exciton interactions thus strongly affect the optical nonlinearities in semiconductors around the fundamental absorption edge. Bound states of biexcitons may form and give rise to two-photon resonances just below the lowest exciton energy. Dephasing times of excitons and biexcitons are typically in the picosecond range at low temperature and density. Coherent dynamical studies of excitons can therefore be performed by nonlinear spectroscopies using state-of –the art femtosecond lasers.[31]

IV.A. Low-dimensional Semiconductors

Another way of affecting optical transitions in semiconductors to show a more discrete character is by forming semiconductor nanostructures, where the electronic motion in one ore more dimensions is so strongly confined that the kinetic energy of the carriers is becoming quantized (see Fig. 14). Structures, where one semiconductor with a certain band gap is embedded in another semiconductor with a significantly larger band gap, can now be fabricated by a variety crystal growth techniques, such as molecular beam epitaxy (MBE), metal organic vapor phase epitaxy (MOVPE) etc. [32]

If the carriers are only confined in one direction we speak about quantum wells in which the carriers behave in a two-dimensional (2D) way. Confining the carriers in still more directions, leads to the formation of quantum wires (1D) and quantum dots (0D).

The lowering of the dimensionality of semiconductors has two main effects on the optical properties. Firstly, the density of states is sharpening up, concentrating the optical oscillator strength in a narrower spectral range. Secondly, the excitonic effects are strongly enhanced because the simultaneous confinement of electrons and holes in the same small volume tends to increase the Coulomb interaction (binding) between them. By proper design of the semiconductor quantum structures it is therefore, in principle, possible to fabricate (nonlinear) optical materials in which attractive atom-like optical properties are embedded in a solid host that integrates well with modern electronics.

The inter-band, or joint, density of states governing the absorption in semiconductors is sketched in Fig. 15 without Coulomb interaction. In bulk semiconductors (3D), the excitonic effects give rise to discrete lines below the absorption edge and a significant enhancement of absorption at the edge (Coulomb enhancement). Quantum wells (2D) can be fabricated directly by epitaxial growth of lattice matched heterostructures. Without

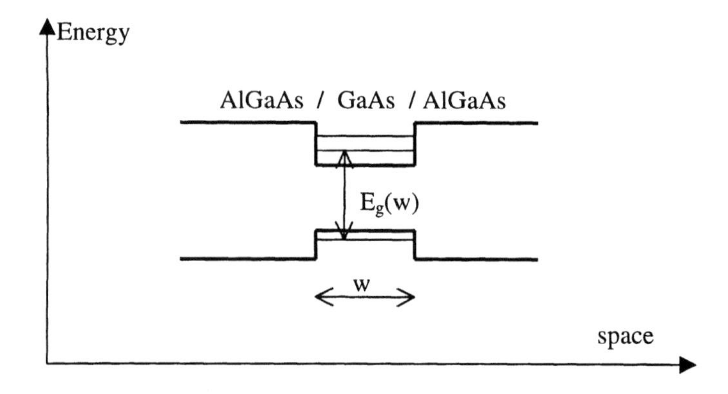

Figure 14. Energy bands/levels for semiconductor nanostructure

Coulomb interaction the absorption is a step function in energy. The 2D exciton problem can also be solved analytically resulting in a four times larger exciton binding energy and a peaking of the absorption at the continuum edge.

Quantum wires (1D) are technologically more difficult to fabricate. There are, however, a number of techniques that are currently being tested, such as etching through a quantum well structure, epitaxial growth on a pre-patterned (V-grooved) substrate and cleaved edge overgrowth. We have had a reasonable success with the latter technique to grow quantum wires with a 1D exciton energy that is about 60 meV lower than the 2D exciton energy in the neighboring quantum wells.[33] The 1D-exciton binding energy is difficult to determine both theoretically and experimentally. The 1D-exciton problem cannot be solved analytically and the optical transitions to the continuum edge disappear because the oscillator strength is in fact transferred to the exciton transition.

Quantum dots (0D) are now most elegantly fabricated by growing a few monolayers of lattice-mismatched material (e.g. InAs) on a substrate (e.g. GaAs). This leads to three-dimensional growth (so-called Stranski-Krastanov) of self-assembled and possibly self-

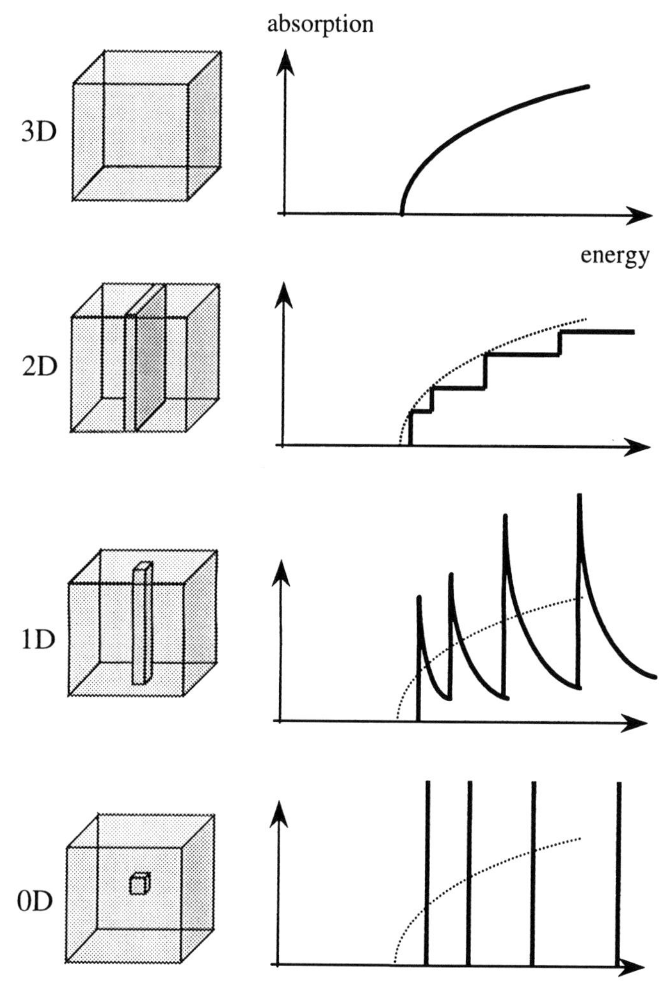

Figure 15. Joint-density of states, or absorption, as a function of energy for bulk semiconductors (3D), quantum wells (2D), quantum wires (1D) and quantum dots (0D).

ordered islands or dots. If the dots are small compared to the 3D-exciton diameter, the confinement energy will be much larger than the exciton binding energy (strong confinement) and exciton effects are small. In the opposite case (weak confinement) only the translational motion of the excitons is restricted, resulting in a quantization of the exciton kinetic energy. The intermediate case of comparable Coulomb and confinement energies is difficult to treat, but in any case the transition energies in a single dot will be discrete as in Fig. 15. Due to size variations of the dots, however, usual quantum dot spectra are strongly inhomogeneously broadened.[34]

V. COHERENT DYNAMICS IN QUANTUM WELLS

In this section the principles of ultrafast coherent spectroscopy on semiconductors will be illustrated by discussing some transient FWM experiments performed on a series of

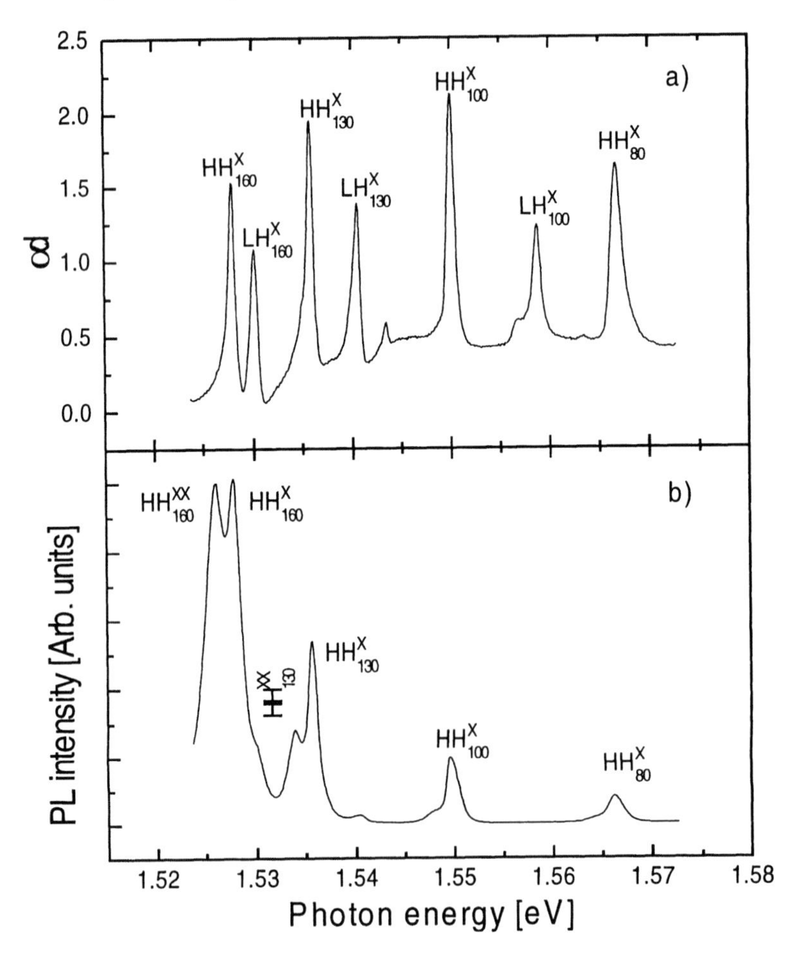

Figure 16. Absorption (a) and photoluminescence spectra (b) of semiconductor sample with quantum wells of different thicknesses. The transitions involve both the heavy-hole (HH) and the light-hole (LH) valence band. The superscript indicates exciton (X) or biexciton (XX) transitions and the subscript indicate the well width in Angstrom.

semiconductor quantum wells with varying well width, and therefore varying degree of confinement in one direction. A sample was grown, in which is embedded a series of multiple quantum wells with the different widths 80, 100, 130 and 160 Å. The linear optical characterization of the sample is shown in Fig. 16

The absorption spectrum (top curve) shows sharp lines from the heavy-hole excitons (HH_w^x) and the light-hole excitons (LH_w^x) for the different wells of width w. For w = 130Å and w = 100Å, also the n=2 HH^x and the continuum edge are observed. In the luminescence spectrum (bottom curve) the light-hole excitons do not appear, because they very quickly relax into the HH_w^x and the HH_w^{xx} states. The latter are the biexciton states discussed above, forming in this case the lowest excited electronic states in these GaAs quantum wells. The narrow absorption lines and the small Stokes shift of the luminescence lines relative to the absorption lines indicate the very good quality of the MBE-grown heterostructure sample.

Transient FWM spectra of the same sample are shown in Fig. 17 together with the absorption spectrum of the sample and the laser spectrum. Heavy-hole and light-hole exciton peaks are seen in the absorption spectrum and are labeled with their corresponding well width w. The maximum of the laser spectrum is tuned in resonance with the HH_{100}^x line. The continuum states of the 130Å and the 160Å wells are excited for this tuning of the laser, whereas the continuum of the 100Å wells is excited marginally. The 80Å wells are excited below resonance. The carrier density in the resonantly excited 100Å wells is estimated to be about $1.5 \cdot 10^9$ cm^{-2}. The time-integrated FWM spectra are recorded as a function of the delay between the incident pulses in a standard two-beam configuration with detection in the $2\bar{k}_2 - \bar{k}_1$ direction.

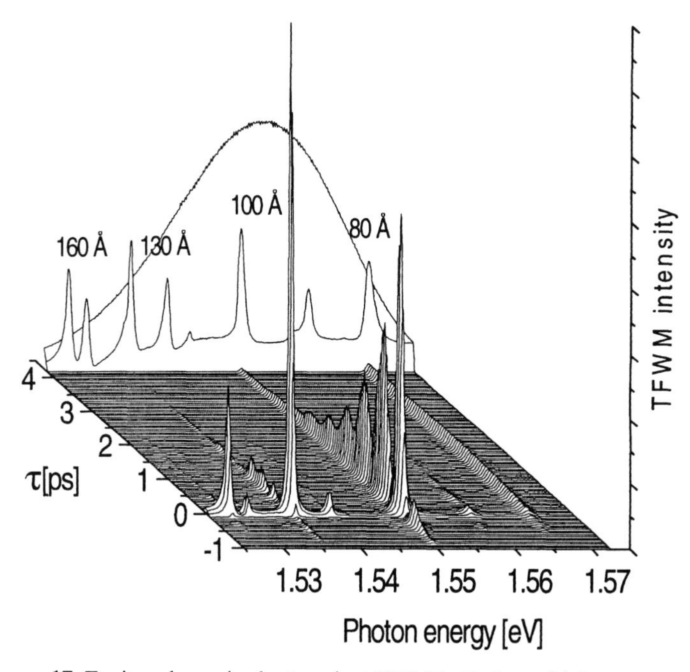

Figure 17. Exciton dynamics by transient FWM in GaAs multiple quantum wells.

The decay of the transient FWM signal in the HH^{x}_{80} line is long and has a delayed maximum, because the laser predominantly excites the localized tail states of the exciton resonance. The transient FWM signal is in this case due to a photon echo involving destructive interference and rephasing of the polarization of the localized states. The FWM signal from the HH^{x}_{100} line shows pronounced HH^{x}_{100}-LH^{x}_{100} beats due to the coherent excitation of both resonances. The transient FWM signals from the HH^{x}_{130} and the HH^{x}_{160} excitons show a dramatic change as compared to those from the 100Å and 80Å wells. In the former cases, where the continuum states are excited coherently with the excitonic states, we observe a continuum contribution to the transient FWM for very short delays.

A beat structure is seen in the FWM signal from the HH^{x}_{130} excitons due to beats with the LH^{x}_{130} excitons and the 2s state of the HH^{x}_{130}, which is well resolved in the absorption spectrum (Fig. 17). However, it is evident from the data that these beats are of much lower intensity than the continuum contribution to the FWM signal. Thus, Fig. 17 shows that the excitonic TFWM signal is dramatically changed depending on whether the continuum states are excited or not. However, it is clearly demonstrated in this experiment that this effect is not related to incoherent scattering processes, that would have lead to spectral broadening and a faster real time decay. We observe no additional broadening of the exciton resonance in the FWM spectra for continuum excitation.

It is obvious that when a broad distribution of continuum state has been excited, the optical Bloch equations treated so far no longer give a valid description of the (nonlinear) optical interaction. Instead one has to resort to solving the semiconductor Bloch equations, that are treated by von Baltz in a separate article in this volume.[35] Here it suffices to say that a numerical solution of the semiconductor Bloch equations could reproduce the above experiments, when including an excitation-induced dephasing of the form

$$\gamma(t) = \gamma_0 + \gamma' \tfrac{1}{2} \sum_k \left[f_{ek}(t) + f_{hk}(t) \right] \tag{34}$$

In equation (34), $\gamma_0 = 1/T_2$, $f_{e,h}(t)$ are the probabilities that the respective electron and hole states are occupied, and γ' is a parameter describing the nonlinear interaction potential. [36]

Looking at the transient FWM spectra in Fig. 17, it is clear that there is a wealth of information in them. It is also clear that in the general case there is very little correlation between the observed decays and the spectral shapes of the nonlinear signals. Thus in all but the simplest case of a homogeneously broadened two-level system it is absolutely vital to spectrally resolve the FWM signal to retrieve the maximum information from the experiment. Alternatively, one has to resolve the real-time behavior the emitted nonlinear signals, e.g. by a new nonlinear up-conversion with a split-off reference pulse from the laser.

VI. EXPERIMENTAL TECHNIQUES

The transient FWM experiments described here are just examples of a more general class of nonlinear optical spectroscopy, the so-called pump-probe, or excite-and-probe, experiments. The general features being that the sample is exposed to a sequence of short optical pulses and the test signal (transmission, reflection, diffraction, luminescence, etc.) is detected and analyzed as a function of the delay between the incident pulses (see Fig. 18).

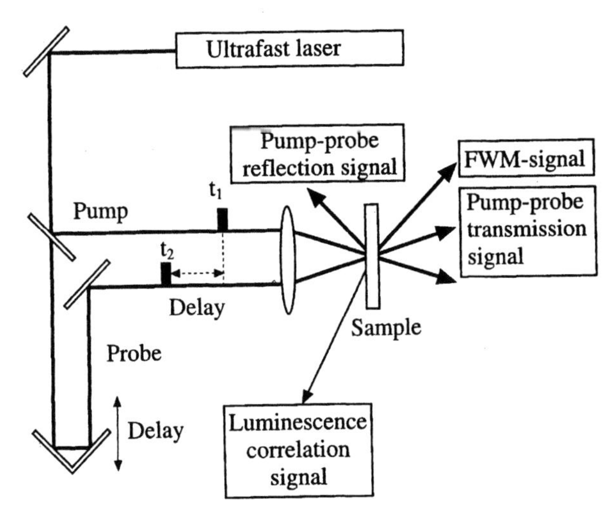

Figure 18. Schematic diagram of pump-probe experiments

In order to retrieve maximum information in a nonlinear pump-probe experiment it is absolutely vital that the incident pulses are well characterized to be either transform limited or e.g. with a known chirp. For transform limited pulses, the choice between temporal and spectral width has to be based on a compromise between the wanted temporal resolution and the distribution of states to be excited in a given situation. It is therefore essential, not only to have access to lasers that can produce the fastest transform limited pulses, but also to be able to perform an adequate and individual pulse-shaping of the incident pulses.

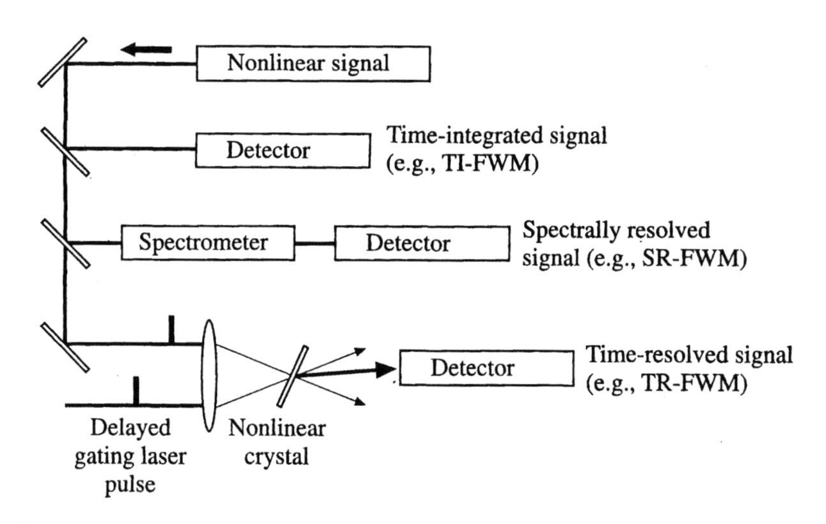

Figure 19. Detection and analysis schemes in nonlinear pump-probe experiments

On the output side it is equally important to perform a transform-limited detection and analysis of the nonlinear signal (see Fig. 19). If the nonlinear signal is detected only by a slow detector (i.e. time-integrated) information about the nonlinear process is lost. Full information about the signal intensity can only be retrieved if the signal is either spectrally, or temporally resolved and analyzed. In Fig. 19, the time resolution of the nonlinear signal is obtained by nonlinear mixing (up-conversion) with a delayed laser pulse in a separate nonlinear crystal. One can even determine both amplitude and phase of the nonlinear signal by further spectrally analyzing the up-converted signal.

VII. CONCLUDING REMARKS

In this article I have tried to illustrate how ultrafast nonlinear optical spectroscopy can be used, and has been used over the last decade or more, to investigate the coherent dynamics of inter-band transitions in semiconductors. The emphasis has been on simple descriptions of theoretical models and experimental techniques, and it has been my intention to point out some important differences between linear and nonlinear optical spectroscopy.

In linear spectroscopy simultaneous temporal and spectral resolution is limited by the uncertainty principles. In nonlinear spectroscopy more than one photon or light-wave is involved and there is no transform relation (or uncertainty relations) between pump and probe photons (or waves). Thus, high temporal resolution of the incident pump pulse does not exclude, or even limit, the high spectral resolution of the outgoing signal pulse. On the contrary, it is necessary to resolve, spectrally or temporally, the outgoing signal in order not to throw away information in the nonlinear pump-probe experiment.

My article is by no means meant as a review article over the field to which many groups from all over the world have contributed through quite a number of years. The examples are to a large extent taken from the work of my own group, even in cases where other groups may have contributed more to the developments of the field, and I apologize for any omissions I have made in giving reference to important contributions to the field. Still, I hope that I have given the reader a flavor of the rich field of nonlinear spectroscopy and the ultrafast coherent dynamics of semiconducting materials and structures.

ACKNOWLEDGEMENTS

I am deeply grateful to the large number of colleagues, students and guests in my group that have worked hard to keep us active in this strongly competitive field. In particular, I would like to thank D. Birkedal, W. Langbein, J. Erland, V.G. Lyssenko, V. Mizeikis, J. Singh, and H.-P. Wagner for their significant contributions to this work, that has been supported by the Danish Ministries of Research and Industry in the framework of CNAST.

REFERENCES

1. N. Bloembergen, *Nonlinear Optics* (Benjamin Inc., New York, Amsterdam, 1965).
2. R.W. Boyd, *Nonlinear Optics* (Academic Press, San Diego, 1992).
3. J.M. Hvam, in *Nonlinear Spectroscopy of Solids: Advances and Applications*, B. Di Bartolo , ed. (Plenum Press, New York, 1994) pp. 91-149.
4. J. Shah, *Ultrafast Spectroscopy of Semiconductors and Semiconductor Nanostructures* (Springer-Verlag, Berlin Heidelberg New York, 1996).
5. S. Haroche, in *High Resolution Laser Spectroscopy,* K. Shimoda ed. (Springer-Verlag, Berlin Heidelberg, 1976) pp. 253-313.
6. K.-H. Pantke and J.M. Hvam, Int. J. Modern Phys. B **8**, 73-120 (1994).
7. L. Schultheis, J. Kuhl, A. Honold, and C.W. Tu, Phys. Rev. Lett. **57**, 1797 (1986).

8. L. Schultheis, J. Kuhl, A. Honold, and C.W. Tu, Phys. Rev. Lett. **57**, 1636 (1986).
9. A. Honold, L. Schultheis, J. Kuhl, and C.W. Tu, in *Ultrafast Phenomena VI*, T. Yajima, K. Yoshihara, C.B. Harris, S. Shionoya, eds. (Springer-Verlag, Berlin Heidelberg, 1989) pp. 307-309.
10. U. Siegner, D. Weber, E.O. Göbel, D. Bennhardt, V. Heuckeroth, R. Saleh, S.D. Barankovskii, P. Thomas, H. Schwab, C. Klingshirn, J.M. Hvam, V.G. Lyssenko, Phys. Rev. B **46**, 4564 (1992).
11. D. Birkedal, V.G. Lyssenko, K.-H. Pantke, J. Erland, and J.M. Hvam, Phys. Rev. B **51**, 7977 (1995).
12. D. Birkedal, J. Singh, V.G. Lyssenko, J. Erland, and J.M. Hvam, Phys. Rev. Lett. **76**, 672 (1996).
13. H. Haug and S.W. Koch, *Quantum Theory of the Optical and Electronic Properties of Semiconductors* (World Scientific, Singapore, 1993).
14. R. Zimmernann, in *Spectroscopy and Dynamics of Collective excitations in Solids*, B. Di Bartolo, ed. (Plenum Press, New York London, 1997) pp. 123-145.
15. See e.g. B. Di Bartolo, in *Nonlinear Spectoscopy of Solids: Advances and Applications,* B. Di Bartolo, ed. (Plenum Press, New York, 1994) pp. 1-74.
16. E.L. Hahn, in *Nonlinear Spectoscopy of Solids: Advances and Applications,* B. Di Bartolo, ed. (Plenum Press, New York, 1994) pp.75-90.
17. P. Meystre and M. Sargent, *Elements of Quantum Optics* (Springer-Verlag, Berlin Heidelberg, 1990)
18. V. Langer, H. Stolz and W. von der Osten, Phys. Rev. Lett. **64**, 854 (1990).
19. R.M. Macfarlane, in *Nonlinear Spectoscopy of Solids: Advances and Applications,* B. Di Bartolo, ed. (Plenum Press, New York, 1994) pp. 151-224.
20. A.I. Ferguson, in *Nonlinear Spectoscopy of Solids: Advances and Applications,* B. Di Bartolo, ed. (Plenum Press, New York, 1994) pp. 225-249.
21. L.F. Mollenauer and J.P. Gordon, in *Nonlinear Spectoscopy of Solids: Advances and Applications,* B. Di Bartolo, ed. (Plenum Press, New York, 1994) pp. 451-480.
22. T. Yajima and Y. Taira, J. Phys. Soc. Japan **47**, 1620 (1979).
23. J. Erland and I. Balslev, Phys. Rev. A **48**, 1765 (1993).
24. K. Leo, M. Wegener, J. Shah, D.S. Chemla, E.O. Göbel, T.C. Damen, s. Schmitt-Rink, W. Schäfer, Phys. Rev. Lett. **65**, 1340 (1990).
25. C. Dörnfeld and J.M. Hvam , IEEE J. Quantum Electron. **QE-25**, 904 (1998).
26. H.J. Eichler, P. Günther, and D.W. Pohl, *Laser Induced Dynamical Gratings* (Springer-Verlag, Berlin, 1986).
27. D. Oberhauser, K.-H. Pantke, J..M. Hvam, G. Weimann, and C. Klingshirn, Phys. Rev. B **47**, 6827 (1993).
28. J. Erland, K.-H. Pantke, V. Mizeikis, V.G. Lyssenko, and J.M. Hvam, Phys. Rev. B **50**, 15047 (1994).
29. V.G. Lyssenko, J. Erland, I. Balslev, K.-H. Pantke, B.S. Razbirin, and J.M. Hvam, Phys. Rev. B **48**, 5720 (1993).
30. N. Peyghambarian, S.W. Koch, and A. Mysyrowicz, *Introduction to Semiconductor Optics* (Prentice Hall, Englewood Cliffs, 1993).
31. A.I. Ferguson, this volume.
32. E.O. Göbel and K. Ploog, Prog. Quant. Electr. **14**, 289 (1990).
33. H. Gislason, W. Langbein, and J.M. Hvam , Appl. Phys. Lett. **69**, 3248 (1996).
34. L. Bányai and S.W. Koch, *Semicondictor Quantum Dots* (World Scientific, Singapore, 1993)
35. R. v. Baltz, this volume
36. D. Birkedal, V.G. Lyssenko, J.M. Hvam, and K. El Sayed, Phys. Rev. B **54**, R14250 (1996).

QUASIPARTICLES AND QUASIMOMENTUM

Ralph v. Baltz[1] and Claus F. Klingshirn[2]

[1]Institut für Theorie der Kondensierten Materie
[2]Institut für Angewandte Physik
Universität Karlsruhe
D–76128 Karlsruhe, Germany

ABSTRACT

Common and different properties of particles and quasiparticles are discussed and, in particular, the difference between real and quasimomentum is clarified.

I. INTRODUCTION

The concept of elementary excitations or quasiparticles has turned out to be an extraordinary powerful tool to describe the low lying excitations of condensed matter: In many cases the excited states can be viewed as a gas of approximately noninteracting particles with energies ϵ_α which contribute to total energy according to

$$E\{n_\alpha\} = \sum_\alpha \epsilon_\alpha n_\alpha. \tag{1}$$

α's label single particle states and $\{n_\alpha\} = \{n_1, n_2, \dots\}$ denote the set of (nonnegative integer) occupation numbers. In many respects, these quasiparticles behave like ordinary particles, e.g. they are bosons or fermions with $n_\mathbf{q} = 0, 1, 2, \dots$ or $n_\mathbf{q} = 0, 1$, respectively.

Some authors reserve the name quasiparticle to cases where α has the property of momentum. Nevertheless, there are differences between particles and quasiparticles, in particular there is a subtle difference between momentum and quasimomentum which is not always fully respected.

The concept of quasiparticles was originally developed by Landau [1] who realized that there is a continuous mapping of the low energy excitation spectrum with interparticle interactions. Amazingly, this description holds even in relatively strong interacting systems like in metals or in superfluid helium.

A well known example of classical "quasiparticles" is the introduction of relative and center of mass coordinates of a two body problem interacting via a central force

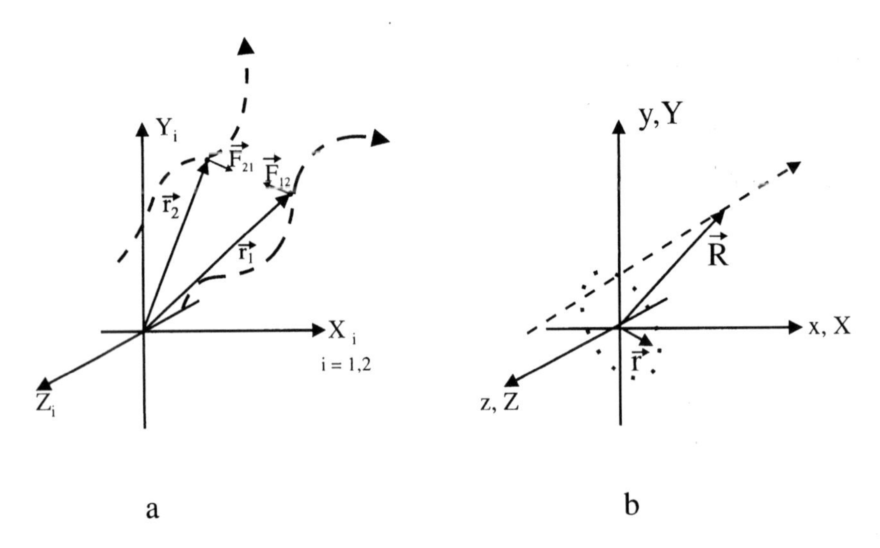

Fig. 1. *(a) Classical two body system interacting via a central force* **F** *and (b) the decoupled motion of the relative– and center of mass coordinates which describe the motion of two noninteracting "quasiparticles". Dashed and dotted lines indicate possible trajectories.*

$\mathbf{F}(\mathbf{r}) = f(|\mathbf{r}|)\,\mathbf{r}$, see Fig. 1. The two particles are located at positions $\mathbf{r}_i(t)$ $(i = 1, 2)$ and the forces between the particles obey Newton's law actio=reactio. We omit for simplicity external forces and obtain:

$$m_1 \frac{d^2\mathbf{r}_1}{dt^2} = f(|\mathbf{r}_1 - \mathbf{r}_2|)\,(\mathbf{r}_1 - \mathbf{r}_2), \tag{2}$$

$$m_2 \frac{d^2\mathbf{r}_2}{dt^2} = f(|\mathbf{r}_2 - \mathbf{r}_1|)\,(\mathbf{r}_2 - \mathbf{r}_1). \tag{3}$$

By introduction of relative $\mathbf{r}$ and center of mass coordinates $\mathbf{R}$

$$\mathbf{r} = \mathbf{r}_1 - \mathbf{r}_2, \quad \mathbf{R} = \frac{m_1\mathbf{r}_1 + m_2\mathbf{r}_2}{m_1 + m_2}, \tag{4}$$

the dynamics of the two interacting particles at positions $\mathbf{r}_1$, $\mathbf{r}_2$ is transformed formally to the dynamics of a system of two noninteracting particles

$$m\frac{d^2\mathbf{r}(t)}{dt^2} = f(|\mathbf{r}|)\,\mathbf{r}, \qquad M\frac{d^2\mathbf{R}(t)}{dt^2} = 0, \tag{5}$$

with reduced mass m and total mass M

$$m = \frac{m_1 m_2}{m_1 + m_2}, \qquad M = m_1 + m_2. \tag{6}$$

We can call the two entities associated with $\mathbf{r}$ and $\mathbf{R}$ quasiparticles. In contrast to the original particles, however, these particles are not material bodies, i.e. there is no matter located at positions $\mathbf{r}$ and $\mathbf{R}$ so that there will be no collision at $\mathbf{R} = \mathbf{r}$. The procedure outlined above for a classical quasiparticle is obviously very similar to the one sketched in [2] when introducing the concept of quasiparticles in solids such as phonons, plasmons, magnons, or excitons.

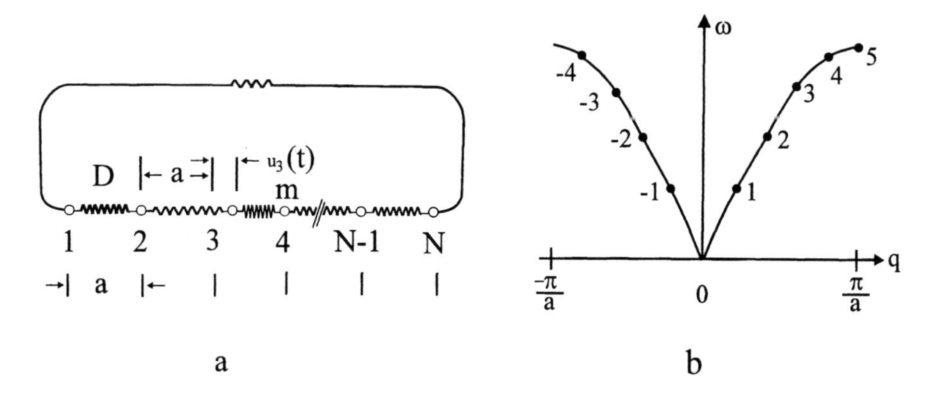

Fig. 2 *(a) Linear monoatomic chain with equal masses m and nearest neighbour springs D and periodic boundary conditions, (b) frequency spectrum for $N = 10$ "atoms".*

The aim of this note is to discuss and to clarify the quasiparticle properties of condensed matter by using the linear chain as a prominent example in section II. Then, the equivalence of bosons and oscillators is sketched in section III. In Section IV we summarize common and different properties of particles and quasiparticles and give two examples.

II. LATTICE VIBRATIONS

As a prominent example for collective excitations in condensed matter we consider the vibrations of a linear chain, Fig. 2a. The longitudinal displacements u_j from the equilibrium positions $x_j^0 = ja, j = 1, 2....N$ where a denotes the lattice constant, obey the Newton-equation of motion:

$$m\frac{d^2u_j(t)}{dt^2} = D\left(u_{j+1} - u_j\right) - D\left(u_j - u_{j-1}\right). \tag{7}$$

Imposing periodic boundary conditions $u_{N+1} = u_1$ we find two different types of solutions:

$$u_j^{cm}(t) = \bar{u} + \bar{v}t, \quad \text{independent of } j \tag{8}$$

$$u_j^{\alpha}(t) = \text{Re } A_\alpha e^{i(q_\alpha ja - \omega_\alpha t)}, \qquad q_\alpha = \frac{2\pi}{Na}\alpha. \tag{9}$$

Eq.(8) describes the rigid motion of the chain, corresponding to the center of mass motion, counting as one degree of freedom whereas Eq.(9) describes $N-1$ vibrations labeled by $\alpha = \pm 1, \pm 2, \pm[\frac{N}{2}]$. (For even N, $-[\frac{N}{2}]$ is excluded.) $\bar{u}$, $\bar{v}$ are two real constants and A_α are complex constants which are fixed by the initial conditions. The frequency spectrum of the chain

$$\omega(q_\alpha) = 2\sqrt{\frac{D}{m}}\left|\sin\frac{q_\alpha a}{2}\right| \tag{10}$$

is depicted in Fig. 2b. Note, $\alpha = 0$ is definitely excluded because the corresponding solution $u_j^\alpha(t) = \text{const}$ is already contained in Eq.(8) as a special case with $\bar{v} = 0$ which describes a static displacement. Nevertheless, this property is often formulated in jargon, even in theoretical texts e.g. Kittel [4] p. 15

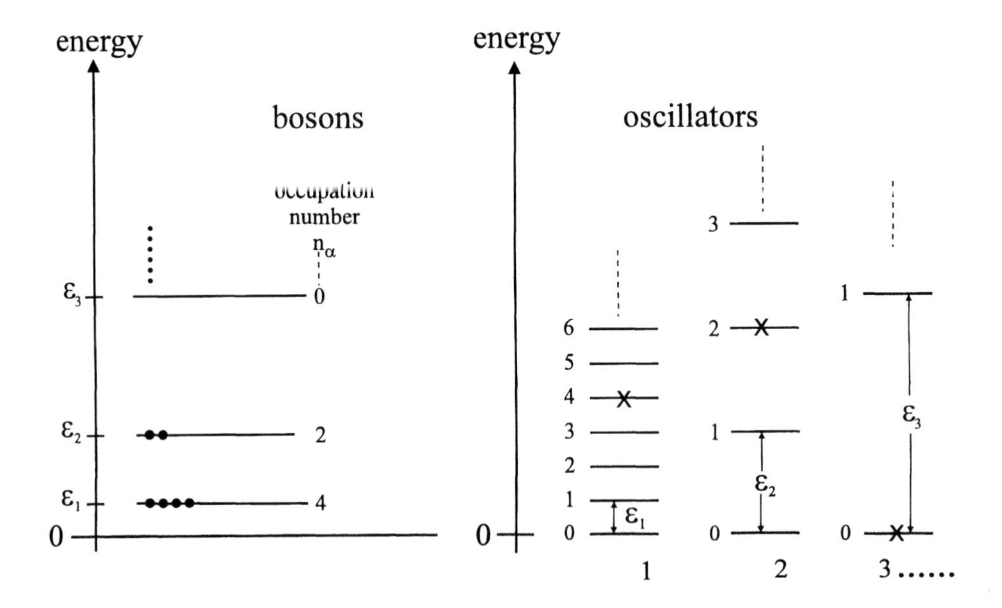

Fig. 3 *Equivalence of a system of N (noninteracting) bosons with single–particle energies ϵ_α and occupation numbers n_α and an infinite (uncoupled) set of harmonic oscillators with frequencies $\omega_\alpha = \epsilon_\alpha/\hbar$. Note that the zero–point energies of the oscillators are omitted. Dots symbolize particles, crosses excited states, respectively.* $N = 6$.

> "The total momentum involves only the $q = 0$ mode which is a uniform translation of the system".

Obviously, this statement is inappropriate as $u_j^\alpha(t)$ describes only the relative motion of the masses so that the total momentum is zero just by construction.
Quantization of the chain is almost trivial as each mode represents a harmonic oscillator. In particular, the total energy of the chain is

$$E\{n_q\} = \frac{p_{cm}^2}{2Nm} + \sum_q \hbar\omega_q \left(n_q + \frac{1}{2}\right), \quad n_q = 0, 1, 2, \ldots, \tag{11}$$

where p_{cm} denotes the total (center of mass) momentum and, as usual, the $N-1$ independent modes are conveniently described by a wave number in the first Brillouin zone, $\mid q \mid \leq \pi/a$

III. BOSONS AND OSCILLATORS

A harmonic oscillator has the unique property that the excitation spectrum is represented in terms of integer multiples of $\hbar\omega$ above the zero point energy $\frac{1}{2}\hbar\omega$. Equivalently, one might view the excited states as realized by adding hypothetical particles with energy $\hbar\omega$ to the "vacuum" state $|0>$ as shown in Fig. 3. For a single oscillator the "life" of these particles is rather "dull" because these paricles have no degrees of freedom.
If we have, however, a set of oscillators (or of quasiparticles) with different quantum numbers like those of the linear chain in Fig. 2b, then a quasi-particle may be forced to "jump" from one (single particle) state α to another one α'. This analogy can be put further and leads to the formulation of "second quantization" (occupation number representation) as sketched in Fig. 3.

Each N–boson state (described by a symmetric wave function) and each operator $\mathcal{O}$ can be translated to the occupation number representation so that the expectation values are the same in both representations.

$$\Psi_{\alpha_1,\alpha_2,\ldots}(\mathbf{r}_1,\mathbf{r}_2,\ldots) \quad \rightarrow \quad |n_{\alpha_1},n_{\alpha_2},\cdots>, \tag{12}$$

$$\mathcal{O}(\mathbf{r}_1,\mathbf{r}_2,\ldots,\mathbf{p}_1,\mathbf{p}_2,\ldots) \quad \rightarrow \quad \hat{\mathcal{O}}(\{a_\alpha,a_\alpha^\dagger\}). \tag{13}$$

Conveniently, operators $\hat{\mathcal{O}}$ are represented in terms of ladder operators $a_\alpha^\dagger, a_\beta$ with $a_\alpha|0>=0$ and commutation relations

$$\left[a_\alpha, a_\beta^\dagger\right] = \delta_{\alpha,\beta}, \quad [a_\alpha, a_\beta] = 0, \quad \left[a_\alpha^\dagger, a_\beta^\dagger\right] = 0. \tag{14}$$

In "particle language", the oscillator quantum numbers are termed occupation numbers, whereas, $a_q^\dagger, q_q$ are called creation and annihilation operators for particles in the (single–particle) states labeled by α.
For example, the operator of total momentum is translated as

$$P = \sum_{j=1}^{N} p_j \quad \rightarrow \quad \hat{P} = \sum_{\alpha,\alpha'} < \alpha'|p|\alpha > a_{\alpha'}^\dagger a_\alpha, \tag{15}$$

where $\mathbf{p}$ is the momentum operator of a single particle, $\mathbf{p} = -i\hbar\nabla$.
However, the equivalence of bosons and oscillators is not one to one! The occupation number representation is more versatile than the traditional representation in terms of wave functions. The particle number is no longer just a parameter but becomes a dynamic variable described by the operator

$$\hat{N} = \sum_\alpha a_\alpha^\dagger a_\alpha. \tag{16}$$

The eigenvalues of $\hat{N}_\alpha = a_\alpha^\dagger a_\alpha$ are just $n_\alpha = 0,1,2,\ldots$. As all $\hat{N}_\alpha$ commute, these operators have simultaneous eigenstates which are labeled by $\{n_\alpha\}$.
Operators which conserve the particle number, e.g. Eqs.(15,16), are composed of products with equal numbers of $a_\alpha^\dagger$ and a_β.
The equivalence theorem gives the possibility to go the other way and associates bosons with a set of oscillators, too. In the case of lattice vibrations, these bosons are called phonons. The displacement–operator of the chain (see [4] p.16), however,

$$u_j^\alpha = \sum_{q'} \sqrt{\frac{\hbar}{2Nm\omega_{q'}}} e^{iq'x_j^0} \left(a_{-q'}^\dagger + a_{q'}\right) \tag{17}$$

changes the phonon number by one and, thus, cannot be handled with wave functions.
Fermionic systems may be mapped to two-level systems, where all commutators Eq.(14) are replaced by anticommutators. For details, we refer to standard textbooks, e.g. Kittel [4].

IV. QUASIPARTICLES

In particular cases, e.g. for vibrations of a crystalline lattice or for electrons in a periodic potential, α displays the characteristic properties of momentum. This is easiest

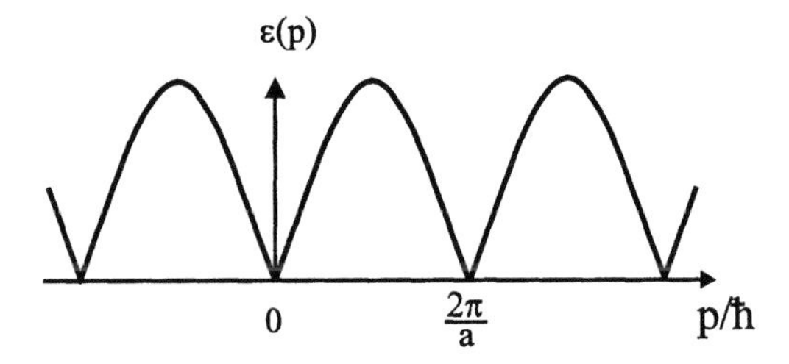

Fig. 4 *The dispersion relation of (longitudinal) acoustic phonons extended over several Brillouin zones.*

recovered by using the extended zone scheme to define the crystal–momentum

$$\mathbf{p} = \hbar(\mathbf{q} + \mathbf{G}), \tag{18}$$

where $\mathbf{q}$ is restricted to the first Brillouin zone and $\mathbf{G}$ denotes an appropriate vector of the reciprocal lattice. Now $\epsilon(\mathbf{p})$ becomes a periodic function of $\mathbf{p}$ which is a general property of excitations in crystals, as shown in Fig. 4 for acoustic phonons. The quantity $\mathbf{p}$ is called quasi–, crystal– or pseudo–momentum and, in the "everyday laboratory jargon" often simply momentum. For most practical purposes a phonon (or other quasiparticles in crystals) acts as if $\mathbf{p}$ is the momentum but there are subtle differences. In the following, we shall present aspects in which quasiparticles behave like real particles and also in which they are different.
As an exotic example, we present the collective excitations in suprafluid and solid helium ^4He which both are of phonon–type at small $\mathbf{q}$. Interestingly, the "roton"–minimum in the liquid phase near $q \approx 2A^{-1}$ seems to be the remniscent of the longitudinal phonons in the crystalline phase near $\mathbf{q} \approx 0$ when shifted to the next Brillouin zone.

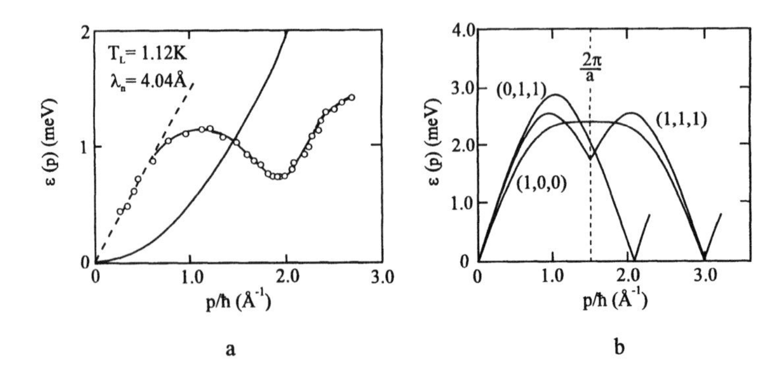

Fig. 5 *Spectrum of (longitudinal) elementary excitations in helium ^4He: (a) superfuid and (b) bcc crystalline phase as determined by inelastic neutron scattering. According to Henshaw et al. [6] and Osgood et al. [7]. The parabola in (a) represents the kinetic energy of free ^4He atoms*

A. Common properties of particles and quasiparticles

1. quasiparticles describe a transport of momentum and energy through condensed matter like ordinary particles do in vacuum. This transport is characterized by an energy - quasimomentum relation also called dispersion relation

$$\epsilon = \epsilon(\mathbf{p}). \tag{19}$$

For excitations in crystals, $\epsilon(\mathbf{p})$ is a periodic function of $\mathbf{p}$ where $\mathbf{p}$ is usually called crystal–momentum. For relativistic particles (in vacuum) there is a universal function

$$\epsilon(\mathbf{p}) = \sqrt{(m_0c^2) + (c\mathbf{p})^2}, \tag{20}$$

where c is the velocity of light in vacuum and m_0 is the rest mass.

2. The transport velocity of energy and momentum is given by

$$\mathbf{v}_T = \frac{\partial\epsilon(\mathbf{p})}{\partial\mathbf{p}}. \tag{21}$$

In the wave–picture this conforms with the group–velocity if ϵ and $\mathbf{p}$ are substituted by $\hbar\omega$ and $\hbar\mathbf{k}$, respectively.

3. Interaction with (temporary and spatially) slowly varying external forces is governed by the equation of motion

$$\frac{d\mathbf{p}(t)}{dt} = \mathbf{F}_{ext}(t). \tag{22}$$

This equation is analogous to the Newton–equation in classical mechanics and it is valid provided the perturbation does not induce transitions between different branches of the quasiparticle spectrum (so-called quasiclassical dynamics).

4. Interaction with other particles (e.g. neutrons) or quasiparticles (e.g. phonons, Bloch electrons etc.) may create, destruct or scatter quasiparticles. These processes are governed by the conservation laws for energy and (quasi)momentum. For example, the creation of a phonon by a neutron scattered from $\mathbf{P}_i$ to $\mathbf{P}_f$ is governed by

$$\mathbf{P}_i - \mathbf{P}_f = \mathbf{p}, \quad E_i - E_f = \epsilon(\mathbf{p}). \tag{23}$$

Note, the participation of reciprocal lattice vectors is already included in Eqs.(23) by using the extended zone scheme.

5. quasiparticles are either bosons or fermions provided the interaction between them is weak i.e., the density of quasiparticles is low. Collective excitations, e.g. phonons, magnons or plasmons are (mostly) bosons and, thus, the change of their number is not restricted. Fermionic quasiparticles, however, can only be created or diminished in pairs as e.g., for electron-hole pair excitations near the Fermi-surface of a metal. In a semiconductor, such electron–hole pairs can form hydrogenic bound states (=excitons) which act as bosons at low density and if the excitation energy is smaller than the binding energy, see e.g., [2].

6. Quasiparticles are described by delocalized states, e.g., by plane waves or Bloch waves.

7. Quasiparticles have a finite life time τ. According to the energy - time uncertainty relation this life time causes a finite width $\Delta\epsilon = \hbar/\tau$ of the dispersion curve $\epsilon(\mathbf{p})$. A consistent quasiparticle description requires $\Delta\epsilon < \epsilon$.

Properties 1-7 strongly support the view that (real) momentum and quasimomentum describe apart from the name the same quantity. However, there are at least three subtle differences.

B. Differences between real momentum and quasimomentum

1. Symmetries and conservation laws [5].
   Associated with every symmetry of a Hamiltonian is a conservation law. This is the famous Noether–theorem: The Hamiltonian of the linear chain (with periodic boundary conditions, see Fig. 2a) is invariant under (arbitrary) translations which are intimately connected to the conservation of total momentum. For example, scattering of a neutron with momentum $\mathbf{P}_i$ to $\mathbf{P}_f$ transfers (real) momentum to the crystal as a whole

$$\mathbf{P}_i - \mathbf{P}_f = \mathbf{p}_{\mathrm{cm}} \tag{24}$$

   which is carried by the center of mass degree of freedom.

   The Hamiltonian of the linear chain is also invariant under the discrete transformation $x_j \to x_j + a$ which, – in contrast to the previously considered continuous transformation – would also be present if the masses would be tied to the "ground" by additional springs. This discrete symmetry (=renumbering of the atoms) leads to the conservation of crystal momentum of the interacting quasiparticles

$$\sum_j \mathbf{p}_j = 0 \pmod{\mathbf{G}}. \tag{25}$$

   Of the two conservation laws Eqs.(24,25), crystal momentum is by far more important in solid state physics than ordinary momentum.

2. Transformation properties [8].
   Energy ϵ and momentum $\mathbf{p}$ of a non-relativistic particle with mass m can be changed by changing the frame of reference $\mathcal{R}$. If $\epsilon, \mathbf{p}$ are defined in $\mathcal{R}$, a Galilei–transformation to $\mathcal{R}'$ (which moves with velocity $-\mathbf{V}$ with respect to $\mathcal{R}$) yields:

$$\epsilon' = \epsilon + \mathbf{p}\mathbf{V} + \frac{m}{2}\mathbf{V}^2, \quad \mathbf{p}' = \mathbf{p} + m\mathbf{V} \tag{26}$$

   so that the relation between energy and momentum is form-invariant in all inertial systems: $\epsilon' = \epsilon(\mathbf{p}')$. This is the principle of classical relativity. Quasiparticles, however, transform differently:

$$\epsilon' = \epsilon + \mathbf{p}\mathbf{V}, \quad \mathbf{p}' = \mathbf{p}. \tag{27}$$

   $\epsilon' - \mathbf{p}'\mathbf{V}$ and $\mathbf{p}'$ are invariant with respect to Galilei-transformations. Eqs.(27) are closely related to the transformation properties of wave phenomena in material media

$$\omega' = \omega + \mathbf{q}\mathbf{V}, \quad \mathbf{q}' = \mathbf{q}. \tag{28}$$

   The second equation of (28) states that (apart from relativistic effects) the wave length $\lambda = 2\pi/|\mathbf{q}|$ remains unchanged whereas the frequency is Doppler-shifted.

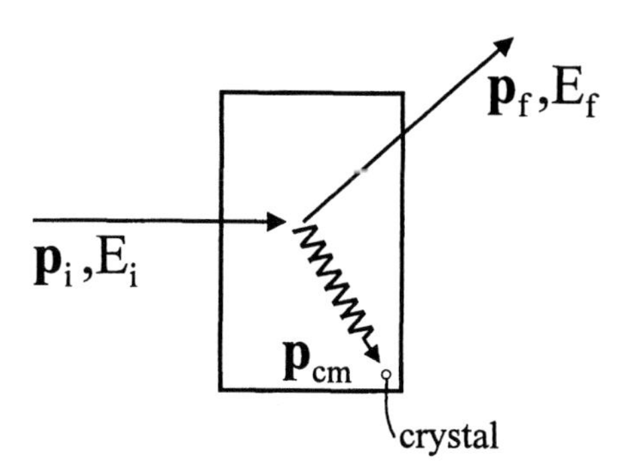

Fig. 6. *Inelastic scattering of a neutron with momentum transfer* $\mathbf{p}_{cm} = \mathbf{P}_i - \mathbf{P}_f$ *to the crystal and the production of a phonon with wave vector* $\mathbf{q} = \mathbf{p}_{cm}$ *mod* $\mathbf{G}$.

3. For phonons, magnons, and other "lattice excitations" the number of modes is determined by the number of different wave vectors within a single Brillouin zone. For real particles, however, e.g. free electrons or photons, momentum is not restricted: There is a minimum wavelength a/π for lattice vibrations but not for free electrons or photons.

C. Example 1: Neutron scattering

As an illustration we first consider the excitation of phonons by neutron scattering as is discussed in Kittel [3], chapter 5. The following conservation laws hold:

$$\begin{aligned} \text{momentum:} \quad \mathbf{P}_i &= \mathbf{P}_f + \mathbf{p}_{cm}, & (29) \\ \text{crystal–momentum:} \quad \mathbf{p} &= \mathbf{p}_{cm}, & (30) \\ \text{energy:} \quad E_i &= E_f + \epsilon(\mathbf{p}) + \frac{\mathbf{p}_{cm}^2}{2Nm}. & (31) \end{aligned}$$

The scattered neutron transfers momentum to the crystal, which is carried by the center of mass degree of freedom. Simultaneously, a phonon with wave number $\mathbf{q} = \mathbf{p}$ mod $\mathbf{G}$ is created provided this process is allowed by the conservation of energy, see Fig. 6. A model calculation is sketched in appendix A whereas Fig. 7 gives some selected experimental results .
Fig. 7a displays the non–primitive cubic *bcc* unit cell, the first Brillouin zone, and the dispersion relations for (metallic) potassium. Since there is only one atom in the primitive cell there are only (three) acoustic branches. Note the different dispersion of the LA and TA branches along various directions in the Brillouin zone. Fig. 7b displays the neutron count rate of a certain (zone–boundary) optical phonon in (insulating) La_2CuO_4 at (reduced) wave number (0.5,0.5,0) when excited in three different Brillouin zones: The maximum energy loss of the neutron is with about $15THz$ (or $65meV$) always at the same frequency (or energy transfer) while the momentum transfer differes by reciprocal lattice vectors $(2,2,0)$, $(2,2,2)$, and $(2,2,4)$ (in units of $\frac{2\pi}{a}$, $\frac{2\pi}{b}$, $\frac{2\pi}{c}$, where a, b, c denote the lattice constants). Therefore, the same phonon may be created in different Brillouin–zones yet with different scattering cross sections. La_2CuO_4 crystallizes in the body centered orthorhombic structure so that the Brillouin zone boundary in $(1,1,0)$ direction is at $(0.5, 0.5, 0)$. The spectral width is due to the finite phonon life time as well as to the instrumental resolution. Note, Sr–doped La_2CuO_4 was the first HTc superconductor which was discovered by Bednorz and Müller [11] in 1986.

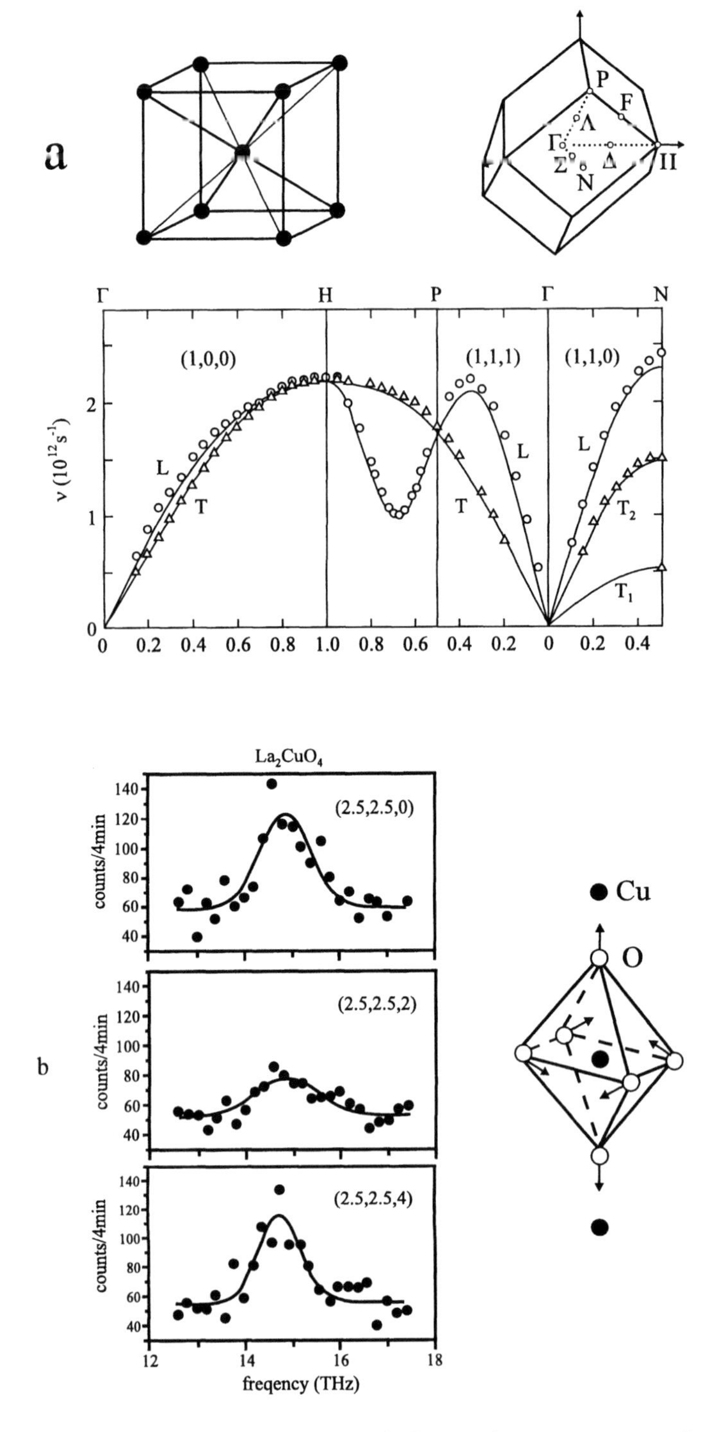

Fig. 7. *(a) bcc structure, Brillouin zone, and phonon dispersion curves for potassium. Along the horizontal axis we plot q, $q/\sqrt{2}$, and $q/\sqrt{3}$ for the $(1,0,0)$, $(1,1,0)$, and $(1,1,1)$ directions, respectively. (b) excitation of a particular optical phonon ("scissor" mode) of La_2CuO_4 when excited with the participation of different reciprocal lattice vectors. According to Cowley et al.[9] and Pintschovius [10].*

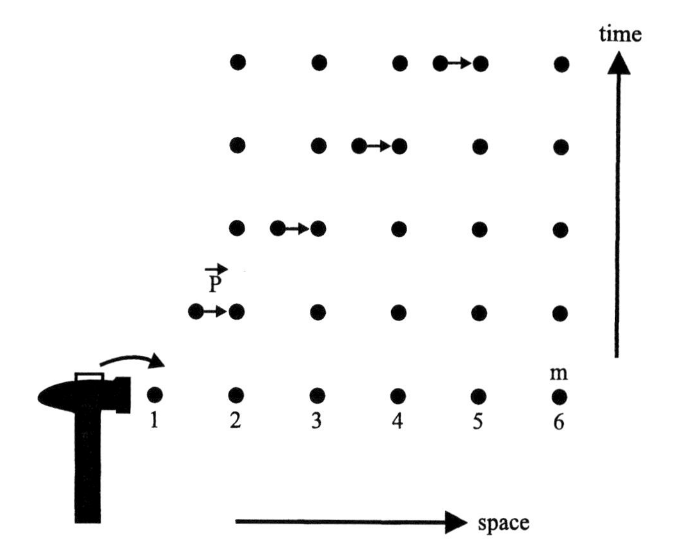

Fig. 8. *Transport of momentum and energy through a linear arrangement of masses which interact only by central collisions.*

D. Example 2: Transport of momentum and energy through a chain [8]

We consider the transport of momentum and energy through a piece of matter which is initially at rest. Next, we transfer a certain amount of momentum to one side e.g., by the hit of a hammer. The question arises how this momentum and the related energy will be passed from one volume element to the neighboring one. As a trivial example we consider the arrangement depicted in Fig. 8. The momentum p, which is transfered to particle # 1, will be transported by successive collisions. Although, this momentum is numerically equal to the total momentum of the chain, there is – in a naive view – no motion of the system as a whole at least for very large systems (see below). For a finite system the center of mass moves to the right however, with a speed proportional to the inverse of the number of masses involved. Yet, there is transport of momentum and energy with velocity $v_T = p/m$ through the system.
In another frame of reference $\mathcal{R}'$ (moving with $-V$ with respect to $\mathcal{R}$) the total momentum and energy are transformed according to (see also Eq.(26)):

$$p'_{\text{tot}} = p + (Nm)V, \quad E'_{\text{tot}} = \frac{p^2}{2m} + pV + \frac{V^2}{2(Nm)}. \tag{32}$$

For fixed p the minimum of the energy can be reached in a frame of reference which moves with velocity V_0.

$$V_0 = -p/Nm, \quad E_0 = \frac{p^2}{2m}(1 - \frac{1}{N}). \tag{33}$$

E_0 is the internal energy. For large systems, E_0 is almost identical with the energy in $\mathcal{R}$ where, by definition, V=0. In all other frames of reference, however, the total energy as well as the total momentum scales to infinity with $N \to \infty$. A meaningful definition of the transported momentum and energy, however, must contain only those parts of Eqs.(32) which are independent of N in the limit of $N \to \infty$. These terms are:

$$\epsilon(p) = E_{\text{tot}} - N\mu = \frac{p^2}{2m} + pV, \quad p' = p. \tag{34}$$

The quantity $\mu = mV^2/2$, which is necessary to add a particle with mass m, is the chemical potential. The system defined by Eqs. (34) is again called a quasiparticle.

V. CONCLUSIONS

Concerning the transport of energy and momentum through matter, quasiparticles behave like ordinary particles and quasimomentum plays the part of ordinary momentum. Nevertheless, quasimomentum is not the generator of translations, and energy and momentum behave differently with respect to Galilei–transformations.

Although the quasiparticle description was originally developed for weakly interacting systems, this concept proved to be fruitful even for moderately strong interactions like real metals and liquid helium ^3He and ^4He at low temperatures. For strong interactions, however, new types of quasiparticles appear. For example the spin and charge degrees of freedom may separate (spinons, holons), charges may become fractional (as in the fractional Quantum Hall effect), or non Bose/Fermi statistics occurs (Anyons in two dimensional systems) [12].

The great advantage and success of the quasiparticle picture lies in the fact that thermodynamic as well as transport properties can be simply described. Examples are the temperature dependence of the specific heat in ordinary and Heavy–Fermion metals [13], de Haas van Alphen effect, cyclotron dynamics in semiconductors, or the Gunn–effect [3].

On the other hand, if a quasiparticle picture is inappropriate, like in amorphous solids or (normal) liquids, even a qualitative description is difficult. Fortunately, phonons still exist in the long wavelength limit where a continuum description always holds. For electrons in amorphous solids, on the other hand, all states become localized above a critical degree of disorder (=Anderson localization) [14]. The situation becomes even worse in (1–dimensional) quasicrystals, i.e. solids with nonperiodic but deterministic structure where the energy spectrum is not even continuous but is a Cantor–set [15].

In appendix B we try to sketch a similar chain of arguments as developed above for the quasimomentum for the angular momentum of quasiparticles.

ACKNOWLEDGMENTS

Stimulating discussions during this school with many colleagues are acknowledged, especially with Prof. Dr. M. Cardona (Stuttgart) on the topics of quasimomentum and avoided level–crossings.

We thank Prof. Dr. H. Rietschel and Dr. L. Pintschovius (INFP, Forschungszentrum Karlsruhe) for Fig. 7b.

REFERENCES

1. L.D. Landau and E.M. Lifshitz, *Course of Theoretical Physics*, Vol. 5, Pergamon Press (1958)
2. C. F. Klingshirn, *Some Selected Aspects of Light–Matter Interaction in Solids.* This volume.
3. Ch. Kittel, *Introduction to Solid State Physics*, John Wiley and Sons, Inc (1967)
4. Ch. Kittel, *Quantum theory of Solids*, John Wiley and Sons, Inc (1963)

5. N.W. Ashcroft and N.D. Mermin, *Solid State Physics*, Holt, Rinehart and Winston (1975)
6. D.G. Henshaw and A.D.B. Woods, Phys. Rev. **121**, 1266 (1961)
7. E.B. Osgood, V.J. Minkiewics, T.A. Kitchens, and G. Shirane Phys. Rev. **A5**, 1537 (1972)
8. G. Falk, *Theoretische Physik, I Punktmechanik*, Heidelberger Taschenbücher Bd. 7, Springer Verlag (1966). (There may be an english edition.)
9. R.A. Cowley, A.D.B. Woods, and G. Dolling, Phys. Rev **150**, 487 (1966)
10. L. Pintschovius, private communication.
11. J.G. Bednorz and K.A. Müller, Z. Physik, **B64**, 189 (1986)
12. E. Fradkin, *Field Theories of Condensed Matter Systems* Addison–Wesley Publishing Company (1991)
13. D.W. Hess, P.S. Riseborough, and J.L. Smith, Encyclopedia of Applied Physics, **7**, 435 (1993)
14. C. di Castro, in *Anderson Localization*, T. Ando and H. Fukuyama eds., Springer Verlag 1988
15. C. Janot, *Quasicrystals*, Clarendon Press, Oxford (1992)
16. C.F. Klingshirn, *Semiconductor Optics*, Springer Verlag, Berlin (1997) and references cited therein.

APPENDIX A

To establish the origin of the conservation laws stated in section IV we present a simple calculation of the phonon excitation rate upon neutron scattering. We consider a three dimensional harmonic crystal with one atom per unit cell which is described by the Hamiltonian

$$\hat{H} = \frac{\hat{p}^2}{2Nm} + \sum_{\mathbf{q}} \hbar\omega_{\mathbf{q}} a^{\dagger}_{\mathbf{q}} a_{\mathbf{q}}. \tag{35}$$

$\hat{\mathbf{p}}$,$\hat{\mathbf{r}}$ are the center of mass momentum and coordinates of the crystal. To lighten the notation we have indicated only one of the three phonon branches $\omega(\mathbf{q})$. The external particle with mass M ("neutron") is described by operators $\hat{\mathbf{P}}$, $\hat{\mathbf{R}}$ and Hamiltonian

$$\hat{H}_{ext} = \frac{\hat{\mathbf{P}}^2}{2M}. \tag{36}$$

Stationary states of the (uncoupled) system "neutron - crystal" are labeled by $\mathbf{K}$ (=neutron momentum), $\mathbf{k}$ (=total momentum of the crystal) with $\hbar\mathbf{k} = \mathbf{p}$, and $\{n_{\mathbf{q}}\}$,

$$|\mathbf{K}, \mathbf{k}, \{n_{\mathbf{q}}\}> \;=\; \frac{e^{i\mathbf{KR}}}{\sqrt{\Omega}} \frac{e^{i\mathbf{kr}}}{\sqrt{\Omega}} |\{n_{\mathbf{q}}\}>, \tag{37}$$

$$E(\mathbf{K}, \mathbf{k}, \{n_{\mathbf{q}}\}) \;=\; \frac{(\hbar\mathbf{K})^2}{2M} + \frac{(\hbar\mathbf{k})^2}{2Nm} + \sum_{\mathbf{q}} \hbar\omega_{\mathbf{q}} \left(n_{\mathbf{q}} + \frac{1}{2}\right). \tag{38}$$

Ω is the volume of the crystal.

The interaction between the neutron and the crystal is:

$$\hat{H}_{int} = \sum_{j-1}^{N} V(\hat{\mathbf{R}} - (\hat{\mathbf{r}} + \mathbf{r}_j^0 + \hat{\mathbf{u}}_j)). \tag{39}$$

$V(\ldots)$ denotes the interaction potential with a single atom of the crystal at position $\mathbf{r} + \mathbf{r}_j^0 + \mathbf{u}_j$, where $\mathbf{r}$ gives the position of the center of mass of the crystal and $\hat{\mathbf{u}}_j$ the displacement of atom $\#j$ from its equilibrium position at $\mathbf{r}_j^0$. According to the Golden Rule, $\hat{H}_{int}$ induces transitions between stationary states Eq.(37) with a rate $\Gamma_{i\to f}$

$$\Gamma_{i\to f} = \frac{2\pi}{\hbar} \left| < i|\hat{H}_{int}|f > \right|^2 \delta(E_i - E_f), \tag{40}$$

where the delta–function displays the conservation of energy.
As an initial state we chose a crystal at rest, i.e. $\mathbf{k} = 0$, $\{n_\mathbf{q} = 0\}$ for all $\mathbf{q}$. Then the transition matrix element becomes:

$$< f|\hat{H}_{int}|i >= \int \frac{d^3\mathbf{R}}{\Omega} \int \frac{d^3\mathbf{r}}{\Omega} e^{i(\mathbf{K}_i - \mathbf{K}_f)\mathbf{R}} e^{-i\mathbf{k}\mathbf{r}} \sum_{j=1}^{N} < \{n_\mathbf{q}\}|V(\mathbf{R} - \mathbf{r} - \mathbf{r}_j^0 - \mathbf{u}_j)|0 > \tag{41}$$

which can be rewritten as:

$$< f|\hat{H}_{int}|i >= \int e^{i(\mathbf{K}_i - \mathbf{K}_f - \mathbf{k})\mathbf{R}} \frac{d^3\mathbf{R}}{\Omega} \int V(\mathbf{r}) e^{-i\mathbf{k}\mathbf{r}} d^3\mathbf{r} \frac{1}{\Omega} \sum_{j=1}^{N} e^{i\mathbf{k}\mathbf{r}_j^0} < \{n_\mathbf{q}\}|e^{i\mathbf{k}\mathbf{u}_j}|0 >). \tag{42}$$

The first integral is nonzero only for

$$\mathbf{K}_i - \mathbf{K}_f = \mathbf{k} \tag{43}$$

which is the conservation of (total) momentum, Eq.(24). The second integral yields the Fourier–transform of the atomic potential, $V(\mathbf{k})$, which is called the structure factor of the atom.
For small displacements, we may expand the exponential operator

$$e^{i\mathbf{k}\mathbf{u}_j} = 1 + i\mathbf{k}\mathbf{u}_j \ldots \tag{44}$$

The matrix element of the first term is nonzero only for $n_\mathbf{q} = 0$ which describes purely elastic (Bragg–) scattering:

$$< f|\hat{H}_{int}|i >= \delta_{\mathbf{K}_i, \mathbf{K}_f + \mathbf{k}} \frac{N}{V} \sum_{\mathbf{G}} \delta_{\mathbf{k},\mathbf{G}} V(\mathbf{k}). \tag{45}$$

The next term in the expansion changes the phonon number by one. To evaluate these matrix elements we use the representation of $\mathbf{u}_j$ in terms of phonon creation and destruction operators $a_\mathbf{q}^\dagger$, $a_\mathbf{q}$, Eq. (17):

$$\mathbf{u}_j = \sum_{\mathbf{q}'} \sqrt{\frac{\hbar}{2Nm\omega_{\mathbf{q}'}}} e^{i\mathbf{q}'\mathbf{r}_j^0} \mathbf{e}_\mathbf{q} (a_{-\mathbf{q}'}^\dagger + a_{\mathbf{q}'}). \tag{46}$$

$\mathbf{e}_q$ denotes the polarization vector (normalized eigenvector of the dynamical matrix).

As a result, we obtain for elastic and single phonon excitation processes:

$$< f|\hat{H}_{int}|i >= \delta_{\mathbf{K}_i,\mathbf{K}_f+\mathbf{k}} \frac{N}{\Omega} \sum_{\mathbf{G}} \left(\delta_{\mathbf{k},\mathbf{G}} + i(\mathbf{k}\mathbf{e}_\mathbf{q}) \sqrt{\frac{\hbar}{2Nm\omega_\mathbf{q}}} \delta_{\mathbf{k}-\mathbf{q},\mathbf{G}} \right) V(\mathbf{k}) \tag{47}$$

The first Kronecker–delta describes again the convervation of total momentum, Eq.(24), whereas the conservation of crystal momentum, Eq. (25), is captured by the Kronecker–deltas in the bracket. Note, that the one–phonon transition rate depends on the total transferred momentum - not on the wave number $\mathbf{q}$ within the first Brillouin zone. For crystals with more than one atom in the primitive cell, $V(\mathbf{k})$ is replaced by the structure factor, see [3]. Higher order terms in the expansion (44) lead to multiphonon processes and to the Deye–Waller factor. Note further, that the transition rate $\Gamma_{i\to f}$ is proportional to N^2 which indicates that for Bragg–scattering as well as for inelastic phonon scattering all atoms of the crystal contribute coherently.

APPENDIX B

In the following, we try to explore if a similar chain of arguments as given above for the quasimomentum of quasiparticles can be found also for the angular momentum. Free space has spherical symmetry i.e. it is invariant against any rotation around any axis. As a consequence (the Noether–theorem applies again) $\mathbf{J}^2$ and its projection onto a quantization axis (usually called z-axis) are conserved commuting quantities, i.e., their eigenvalues are "good" quantum numbers

$$\mathbf{J}^2|j, j_z >= j(j+1)\hbar^2|j, j_z >, \qquad \mathbf{J}_z|j, j_z >= j_z\hbar|j, j_z >, \tag{48}$$

where $j = 0, \frac{1}{2}, 1\ldots$ and $j_z = -j, -j+1, \ldots j$.
The angular momentum $\mathbf{J}$ of a particle can be decomposed into an orbital part $\mathbf{L} = \mathbf{r}\times\mathbf{p}$ and an inner part $\mathbf{S}$ which is in some cases called spin. In a classical picture $\mathbf{L}$ vanishes in a frame of reference in which the trajectory of the center of mass of the particle under consideration passes through the origin.
For the following discussion we consider only the inner part of angular momentum. In contrast to free space (see above), a crystalline solid is invariant only under rotations of $\pm\frac{\pi}{3}$, $\pm\frac{\pi}{2}$, $\pm\frac{2\pi}{3}$, and $\pm\pi$ around selected axes. Strictly speaking, angular momentum is therefore not a "good" quantum number for the classification of quasiparticles. The "good" quantum numbers follow from the irreducible representations of the point group of the solid (see e.g. [16]). However, the inspection of the compatibility tables of the full spherical group and of finite point groups of solids, shows that for crystals with high symmetry and in the vicinity of $\mathbf{k} = 0$ i.e. close to the Γ point and partly also along some directions of high symmetry, the angular momentum is up to $j = \frac{3}{2}$ with $j_z = \pm\frac{1}{2}, \pm\frac{3}{2}$ a "reasonably good" quantum number. Examples are the cubic point groups O_h and T_d. In uniaxial hexagonal systems like C_{6v} similar statements hold at least for the projection of the angular momentum on the hexagonal axis. In an arbitrary direction in $\mathbf{k}$–space or in crystals of low symmetry, the concept of angular momentum cannot be used in solids. Within this limitation we discuss now the absorption of a photon with $\mathbf{q} \approx 0$ incident on a crystal with high symmetry in an analogous way, as we discussed the inelastic scattering of a neutron in chapter IV.C. We assume, that the photon is in a circularly polarized state σ^+ i.e., it carries an angular momentum $\hbar$ in the direction of momentum. When the photon is absorbed in the crystal by creating an optical phonon, an exciton, a plasmon, or an optical magnon, the crystal as a whole

carries the angular momentum $j_z = \hbar$ in analogy to Eq. (29) for the momentum. The rotational energy E_{rot} of the crystals as a whole, however,

$$E_{rot} = \frac{\mathbf{L}^2}{2\Theta} = \frac{\ell(\ell+1)\hbar^2}{2\Theta}, \tag{49}$$

tends to zero for a macroscopic crystal because of the large moment of inertia Θ. Simultaneously, the angular momentum of the absorbed photon appears as the quasi-, pseudo or crystal angular momentum of the created quasiparticle as for the quasimomentum in Eqs.(29), (30).

To elucidate this concept and assuming that the reader has some basic experience with group theory in solids [16], we give the following example. In a crystal with point group T_d i.e., with zincblende type crystal structure, the transitions from the crystal ground state with symmetry Γ_1 are dipole allowed only to excited states with symmetry Γ_5 in one–photon absorption processes. Simultaneously one finds in the above mentioned compatibility tables, that a state in vacuum with $j_z = 1$ is compatible with the irreducible representation Γ_5.

In this sense we may say that a quasiparticle in a crystal carries energy, quasimomentum and even quasi angular momentum. The validity of the latter concept is, however, more restricted than that of quasimomentum. For example, there are $2s+1$ different internal states of a particle with spin s which have the same energy $\epsilon(\mathbf{p})$. Phonon, magnon etc. dispersion curves, however, split with increasing quasimomentum, in particular along low–symmetry directions, see Fig. 7a. Therefore, quasiangular momentum is not an internal degree of freedom of a quasiparticle so that it has to be used with more precautions than the concept of quasimomentum.

ON THE EQUIVALENCE OF COUPLED PENDULA AND QUANTUM MECHANICAL NON-CROSSING RULE

A. Jolk[1], C. F. Klingshirn[1], and R. v. Baltz[2]

[1]Institut für Angewandte Physik
[2]Institut für Theorie der Kondensierten Materie
Universität Karlsruhe
Kaiserstr. 12, 76128 Karlsruhe, Germany

ABSTRACT

The classical system of two coupled ideal harmonic oscillators is solved and its exact solution is compared to the quantum mechanical equivalent of two coupled states. In both cases, the coupling leads to a non-crossing behavior with a "level repulsion" depending on the coupling parameter. A special Hamiltonian is presented that gives identical results in the classical and in the quantum mechanical case.

I. INTRODUCTION

Many physics textbooks explain the quantum mechanical non-crossing rule using the classical analogon of two ideal pendula coupled by a (massless) spring. They usually present the differential equation and its solution, sketch the two proper oscillation modes of the coupled system, often for the case of identical pendula, and proceed with the quantum mechanical case. There are, however, some subtle differences between both systems that are not normally dealt with.

During this summer school at Erice, there have been some exciting discussions on this topic, most of them relating to the question whether these systems are in fact equivalent. This contribution aims to present the complete solution of both the classical as well as the quantum mechanical problem and to compare them. It is found that while there are indeed differences, it is possible to construct a quantum mechanical system that is completely equivalent to the classical one.

II. THE CLASSICAL SYSTEM

We first examine the classical system of two ideal harmonic oscillators coupled by a massless spring. The corresponding mechanical setup is sketched in fig. 1. (Note that this setup represents harmonic oscillators only for small amplitudes.) The equations of

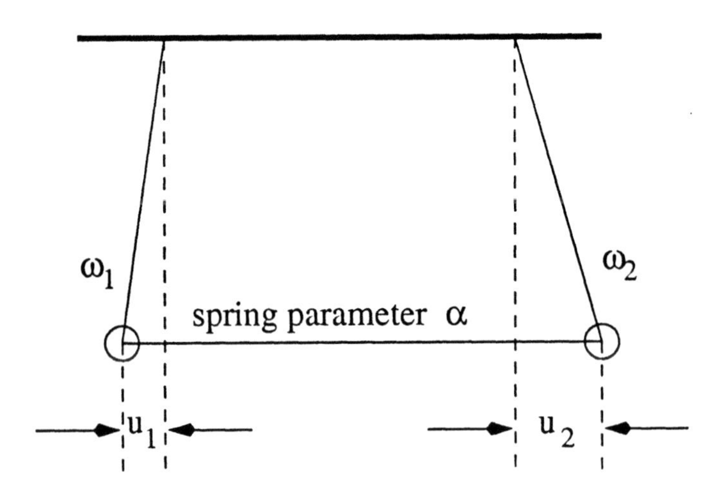

Figure 1: The classical system of two coupled pendula.

motion of the two masses are in the limit of small amplitudes

$$\frac{\partial^2 u_1}{\partial t^2} + \omega_1^2 u_1 + \alpha(u_1 - u_2) = 0 \tag{1}$$

$$\frac{\partial^2 u_2}{\partial t^2} + \omega_2^2 u_2 - \alpha(u_1 - u_2) = 0 \tag{2}$$

where α is positive. The frequencies ω_i depend on the gravitation and the length l_i of the pendula suspension as $\omega_i = (g/l_i)^{1/2}$; in the following, the ω_i are used as variable parameters.

Using the substitution $v_i = \frac{\partial}{\partial t} u_i$, this system of two coupled differential equations of order two transforms into four coupled differential equations of order one:

$$\left(\frac{\partial}{\partial t} + \underbrace{\begin{pmatrix} 0 & 0 & -1 & 0 \\ 0 & 0 & 0 & -1 \\ \omega_1^2 + \alpha & -\alpha & 0 & 0 \\ -\alpha & \omega_2^2 + \alpha & 0 & 0 \end{pmatrix}}_{=:A} \right) \begin{pmatrix} u_1 \\ u_2 \\ v_1 \\ v_2 \end{pmatrix} = 0 \tag{3}$$

The standard procedure for solving such a system of coupled linear differential equations consists of solving the eigenvalue problem of the matrix A. Any solution of the differential equation is a linear combination of functions $\vec{a_i}(t) = \vec{a_i}^0 \exp(-\lambda_i t)$, where $\vec{a_i}^0$ are the eigenvectors and λ_i the respective eigenvalues of A.

In this case, λ_i will be purely imaginary, thus resulting in oscillating solutions. The eigenvalues are calculated from

$$0 = \det(\lambda \cdot I - A) = \begin{vmatrix} \lambda & 0 & 1 & 0 \\ 0 & \lambda & 0 & 1 \\ -c_1 & \alpha & \lambda & 0 \\ \alpha & -c_2 & 0 & \lambda \end{vmatrix} = \lambda^4 + \lambda^2(c_1 + c_2) + c_1 c_2 - \alpha^2 \tag{4}$$

where $c_i := \omega_i^2 + \alpha$. Thus

$$\omega_e^2 := -\lambda^2 = \frac{1}{2}\left(\omega_1^2 + \omega_2^2 + 2\alpha \pm \sqrt{(\omega_1^2 - \omega_2^2)^2 + 4\alpha^2}\right) \tag{5}$$

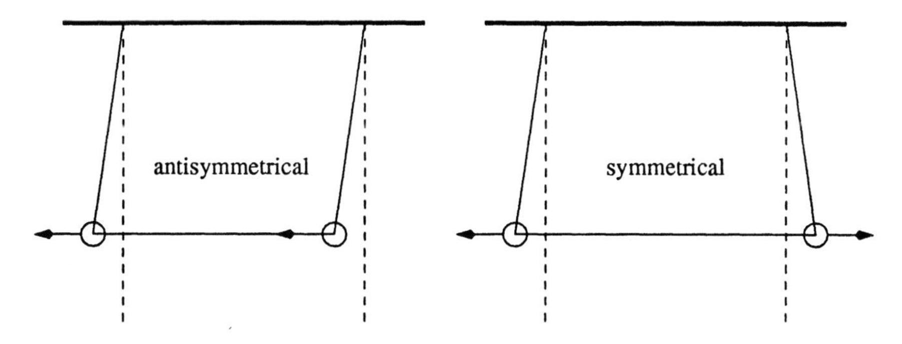

Figure 2: The symmetrical and antisymmetrical oscillation modes for $\omega_1 = \omega_2$.

and ω_e is found to be real for $\alpha \geq 0$ and ω_i real. We thus find two oscillation modes with eigenfrequencies ω_e^+ and ω_e^-. The phase of each eigenmode can be arbitrarily chosen by a linear combination of exponentials $\propto \exp(\pm i\omega_e t)$.

The eigenvectors of A are

$$\vec{a}_i^{\,0} = \begin{pmatrix} \alpha \\ \lambda_i^2 + c_1 \\ -\lambda\alpha \\ -\lambda(\lambda^2 + c_1) \end{pmatrix}. \tag{6}$$

From these eigenvectors, we find the ratio of the amplitudes of the two pendula to be

$$\frac{u_2^0}{u_1^0} = \frac{\lambda_i^2 + c_1}{\alpha} = -\frac{1}{2\alpha}\left(\omega_2^2 - \omega_1^2 \pm \sqrt{(\omega_1^2 - \omega_2^2)^2 + 4\alpha^2}\right). \tag{7}$$

In the special case $\omega_1 = \omega_2$, this reduces to

$$\frac{u_2^0}{u_1^0} = -\frac{1}{2\alpha}(\pm 2\alpha) = \mp 1, \tag{8}$$

with eigenfrequencies

$$\omega_e^2 = \frac{1}{2}\left(2\omega_1^2 + 2\alpha \pm \sqrt{4\alpha^2}\right) = \omega_1^2 + \begin{cases} 2\alpha \\ 0 \end{cases}. \tag{9}$$

The solution $\omega_e^- = \omega_1$ is antisymmetrical with respect to a parity operation, whereas the solution $\omega_e^+ = \sqrt{\omega_1^2 + 2\alpha}$ is symmetrical (fig. 2). It is evident that for the antisymmetrical mode, the oscillation frequency remains unchanged because the spring is not elongated. For the symmetrical mode, the spring leads to an additional force which causes a higher eigenfrequency.

In the limit $\omega_1^2 \gg \omega_2^2, \alpha$,

$$\frac{u_2^0}{u_1^0} \rightarrow \frac{\omega_1^2}{\alpha}. \tag{10}$$

Fig. 3 sketches the ratio u_2/u_1 as a function of ω_1 for $\omega_2 = 1$ and two different values of the coupling parameter α. For moderate coupling, at $\omega_1 \ll \omega_2$ the two eigenmodes are nearly independent oscillations of one of the two pendula. With increasing ω_1, the amplitude of the other pendulum increases. Both amplitudes are equal for $\omega_1 = \omega_2$, and for $\omega_1 \gg \omega_2$, the roles are exchanged.

Fig. 4 shows the eigenfrequencies $\omega_e^{\pm}(\alpha)$ for two different values of ω_1 and constant $\omega_2 = 1$. In the case $\omega_1 = \omega_2$, the lower eigenfrequency remains constant while the

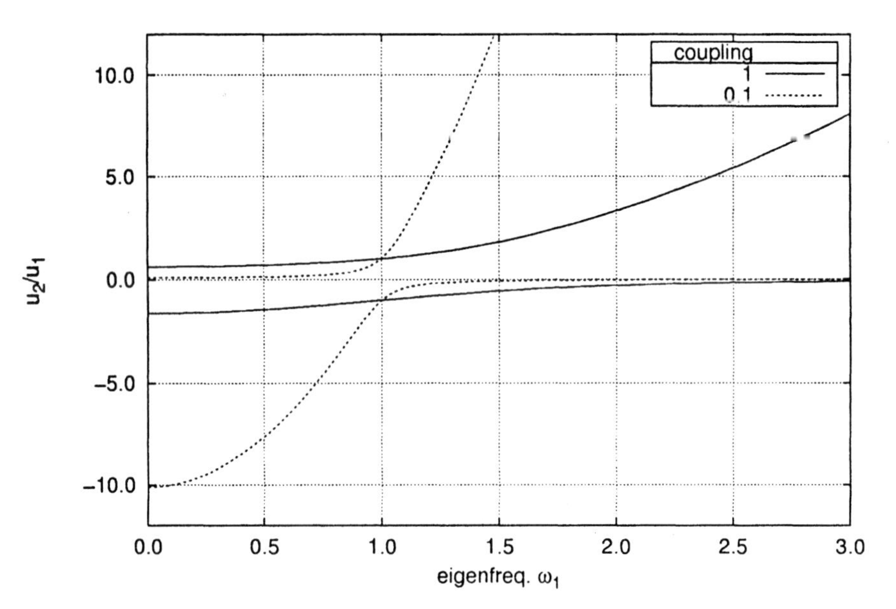

Figure 3: Amplitude ratio u_2/u_1 as a function of the eigenfrequency ω_1 for two different values of α and constant $\omega_2 = 1$.

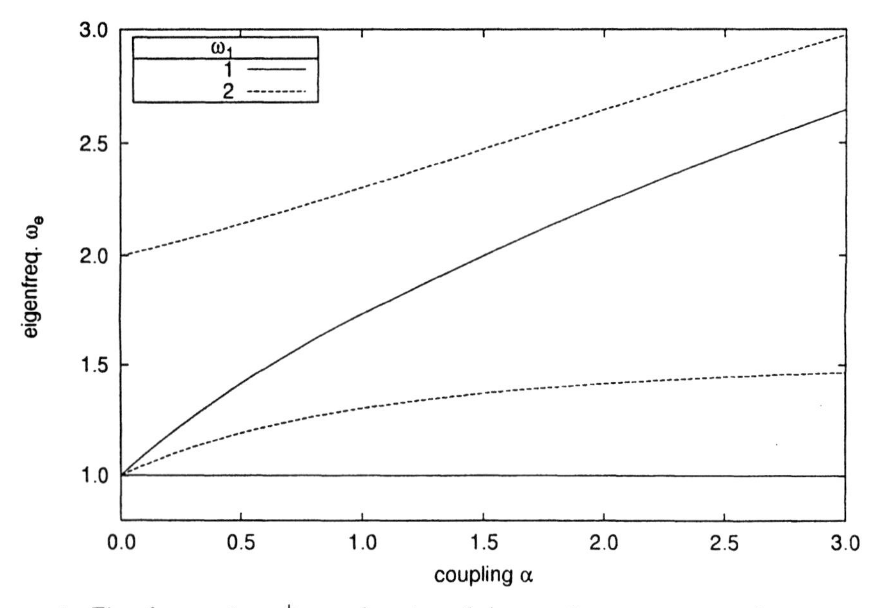

Figure 4: Eigenfrequencies $\omega_e^{\pm}$ as a function of the coupling parameter α for $\omega_2 = 1$ and $\omega_1 = \omega_2$ (solid lines) and $\omega_1 = 2\omega_2$ (dashed lines).

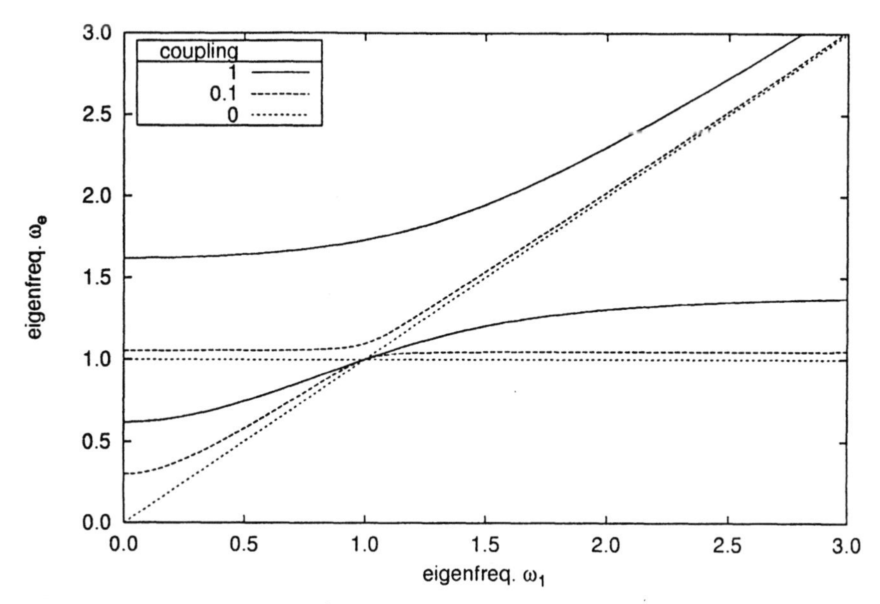

Figure 5: Eigenfrequencies $\omega_e^{\pm}$ as a function of ω_1 at $\omega_2 = 1$ for different values of α.

upper eigenfrequency shifts away with increasing coupling. In the case $\omega_1 \neq \omega_2$, both eigenfrequencies depart from their values at $\alpha = 0$. In this case, there is no mode where the coupling spring does not intervene, as was the case for the antisymmetrical mode in fig. 2 for $\omega_1 = \omega_2$.

Fig. 5 shows the eigenfrequencies $\omega_e^{\pm}(\omega_1)$ for different values of the coupling parameter α. Without any coupling, we observe a clear "level crossing" where the two curves intersect. With increasing coupling, the levels start to repel each other and we get a finite repulsion at $\omega_1 = \omega_2$.

III. THE QUANTUM MECHANICAL SYSTEM

We now analyze the behavior of two coupled states in a quantum mechanical framework. This system is equivalent to the classical system in the sense that it exhibits a level crossing at zero coupling which is avoided as soon as one introduces a finite coupling matrix element.

Let $\binom{1}{0}$ and $\binom{0}{1}$ be the eigenfunctions of the uncoupled oscillators with energies E_1 and E_2, respectively. In this base, the Hamiltonian of the uncoupled system is

$$H^0 = \begin{pmatrix} E_1 & 0 \\ 0 & E_2 \end{pmatrix}. \tag{11}$$

The coupling between the two oscillators is then introduced as a non-vanishing off-diagonal matrix element λ:

$$H = \begin{pmatrix} E_1 & \lambda \\ \lambda & E_2 \end{pmatrix}. \tag{12}$$

The eigenvalues of this operator are found starting from

$$0 = \det(H - \omega \cdot E) = (E_1 - \omega)(E_2 - \omega) - \lambda^2 \tag{13}$$

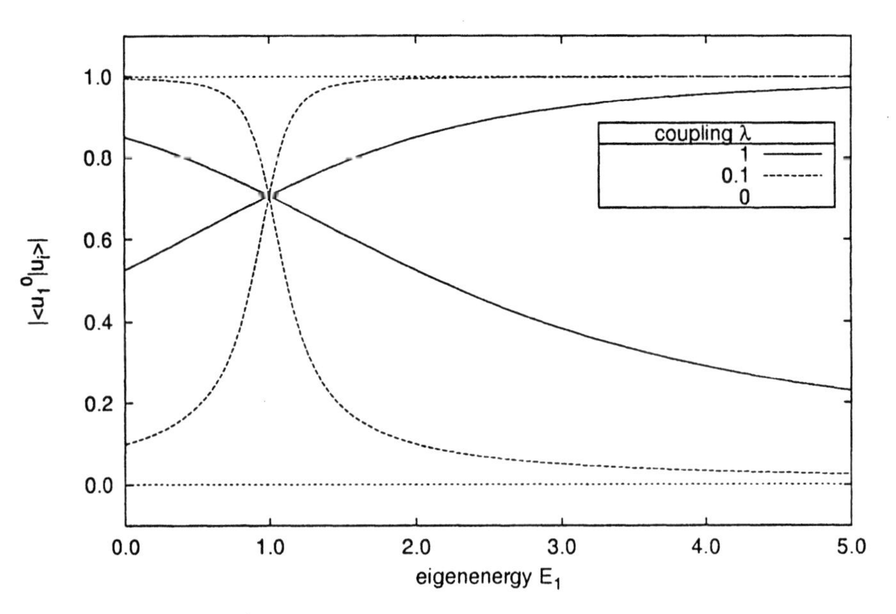

Figure 6: Projection of $\psi^{\pm}$ to the unperturbed eigenfunctions as a function of E_1 for $E_2 = 1$.

to be

$$\omega^{\pm} = \frac{E_1 + E_2}{2} \pm \sqrt{\frac{(E_1 - E_2)^2}{4} + \lambda^2}. \tag{14}$$

The respective eigenfunctions are (not normalized)

$$\psi^{+} = \left(\frac{E_1 - E_2}{2} + \sqrt{\frac{(E_1 - E_2)^4}{4} + \lambda^2} \right) \binom{1}{0} + \lambda \binom{0}{1} \tag{15}$$

and

$$\psi^{-} = -\lambda \binom{1}{0} + \left(\frac{E_1 - E_2}{2} + \sqrt{\frac{(E_1 - E_2)^4}{4} + \lambda^2} \right) \binom{0}{1}. \tag{16}$$

For $\lambda \to 0$, $\psi^{\pm}$ tend to the unperturbed eigenfunctions as expected. Fig. 6 and 7 show the projection of the eigenfunctions $\psi^{\pm}$ to the unperturbed eigenfunction $\binom{1}{0}$ both as a function of the coupling parameter λ and as a function of the unperturbed eigenenergy E_1, for constant $E_2 = 1$. At $E_1 = E_2$, for nonzero coupling λ the projection is $\sqrt{2}/2$ corresponding to the well-known eigenfunctions $\psi^{\pm} = \sqrt{2}/2\binom{1}{\pm 1}$.

Far from the crossover point at $E_1 = E_2$, both solutions tend towards the unperturbed eigenfunctions. The solution ψ^{+} which starts out at $E_1 \ll E_2$ as $\binom{1}{0}$ acquires more and more $\binom{0}{1}$–like character when approaching the crossover point. At $E_1 \gg E_2$, it tends to pure $\binom{0}{1}$ (see fig. 6). With increasing coupling, this mixing process is spread out over increasingly wide energy regions (fig. 7).

Fig. 8 and 9 show the resulting eigenenergies $\omega^{\pm}$ for different values of λ and E_1 and constant $E_2 = 1$. Fig. 8 shows the energy shift of the eigenenergies as a function of the coupling parameter λ. Note the striking similarity of this behavior with the one observed in the classical case, fig. 4, with the only exception that the average energy remains constant in the quantum mechanical case while it moves to higher ω_ϵ in the classical case.

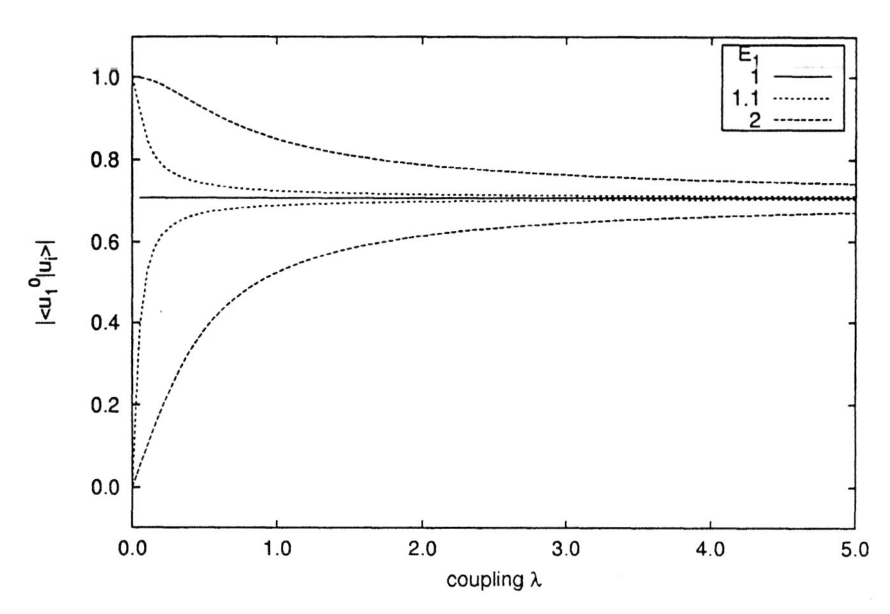

Figure 7: Projection of $\psi^{\pm}$ to the unperturbed eigenfunctions as a function of λ for $E_2 = 1$.

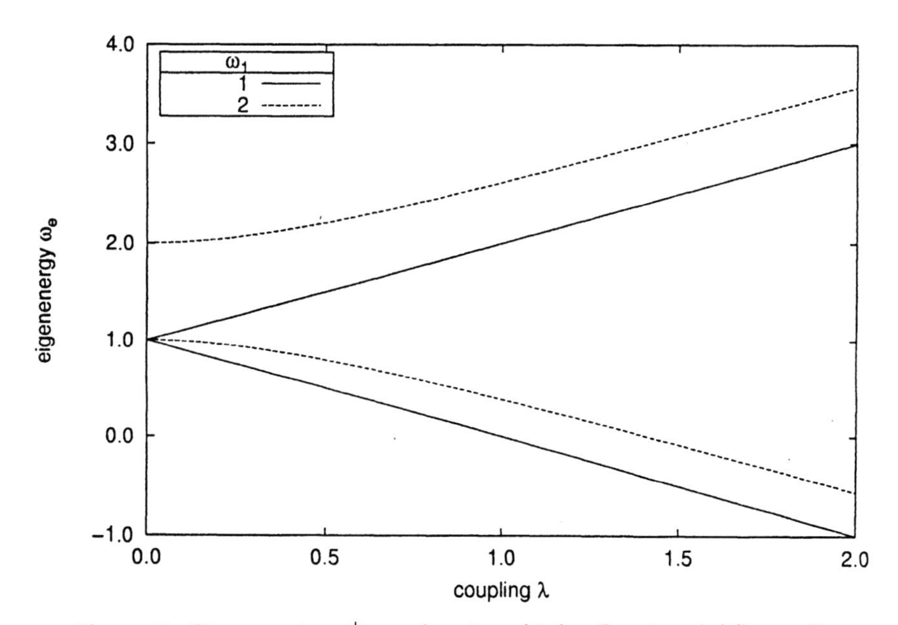

Figure 8: Eigenenergies $\omega^{\pm}$ as a function of λ for $E_2 = 1$ and different E_1.

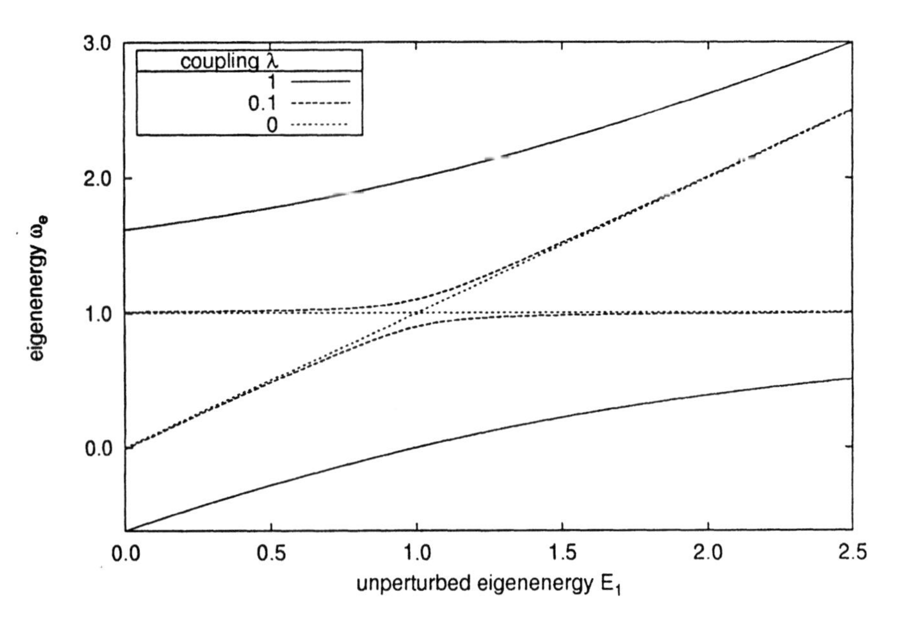

Figure 9: Eigenenergies $\omega^{\pm}$ as a function of the unperturbed E_1 for $E_2 = 1$ and different λ.

The anti-crossing behavior is evident in fig. 9 where we observe an increasing level repulsion with increasing coupling. Contrary to the classical case (fig. 5), both eigenenergies shift symmetrically away from their position at zero coupling.

IV. THE EXACT QUANTUM MECHANICAL EQUIVALENT

From a comparison of the differential equation of the classical system, eq. (3), with the quantum mechanical Hamiltonian eq. (12), we introduce an additional perturbation λ on the diagonal of the Hamiltonian. The resulting system, described by

$$H = \begin{pmatrix} E_1 + \lambda & -\lambda \\ -\lambda & E_2 + \lambda \end{pmatrix}, \tag{17}$$

is completely equivalent to the classical system of two coupled pendula. Its eigenenergies are

$$\omega^{\pm} = \frac{1}{2}\left(E_1 + E_2 + 2\lambda \pm \sqrt{(E_1 - E_2)^2 + 4\lambda^2}\right) \tag{18}$$

while the eigenfunctions $\psi^{\pm}$ remain unchanged. It is interesting to note that these eigenvalues correspond exactly to the classical eigenfrequencies if one replaces λ by α and ω_i^2 by E_i. Especially, in this case the quantum mechanical system is asymmetrically split, as was the classical one.

V. CONCLUSION

We have presented the complete analytical solution to both the classical system of two coupled pendula and the quantum mechanical system of two coupled harmonic oscillators. In a first treatment, we find differences between their behavior that vanish when a Hamiltonian analogous to the classical system is chosen. We thus find that the quantum mechanical system is completely equivalent to the classical one and that

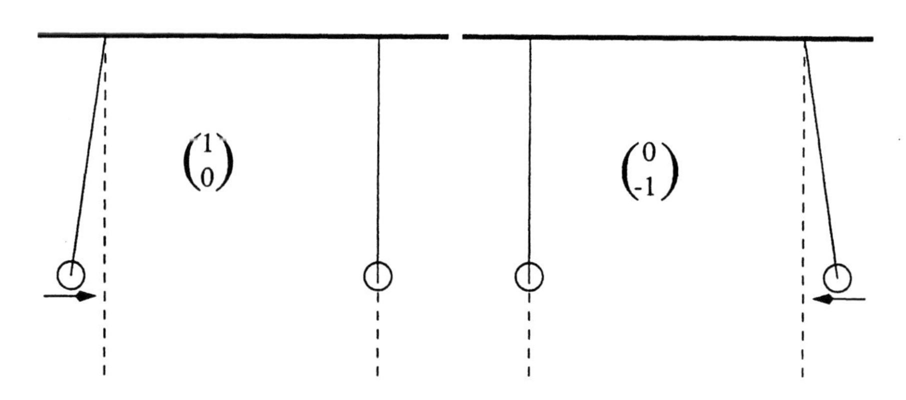

Figure 10: The two base oscillation modes.

quantum mechanical effects such as a non-crossing rule can be observed as well in the classical system.

VI. APPENDIX: SYMMETRY CONSIDERATIONS

The classical system of two identical pendula is symmetrical under a parity operation. In the following, we sketch an argument based on parity considerations that directly results in the proper eigenmodes.

We first write the equations of motion of the coupled pendula, eq. (1) in a matrix notation:

$$\frac{\partial^2}{\partial t^2}\begin{pmatrix} u_1 \\ u_2 \end{pmatrix} + D\begin{pmatrix} u_1 \\ u_2 \end{pmatrix} = 0 \tag{19}$$

where

$$D = \begin{pmatrix} \omega_1^2 + \alpha & -\alpha \\ -\alpha & \omega_2^2 + \alpha \end{pmatrix} \tag{20}$$

is the "spring matrix." The two base functions $\binom{1}{0}$ and $\binom{0}{1}$ correspond to oscillations of only one of the pendula (see fig. 10).

It is evident that these two base functions have no defined parity. Consequently, symmetry or for this example more precisely parity arguments cannot be used to decide whether the off-diagonal coupling terms vanish or not. On the other hand, it is obvious from fig. 10 that a parity operation transforms $\binom{1}{0}$ into $\binom{0}{-1}$ and vice versa. The eigenmodes of the parity operation are thus the sum and the difference of these two base modes,

$$p^+ = \frac{1}{\sqrt{2}}\begin{pmatrix} 1 \\ -1 \end{pmatrix} \tag{21}$$

$$p^- = \frac{1}{\sqrt{2}}\begin{pmatrix} 1 \\ 1 \end{pmatrix}. \tag{22}$$

(23)

The corresponding transformation matrix is

$$\Pi = \frac{1}{\sqrt{2}}\begin{pmatrix} 1 & 1 \\ 1 & -1 \end{pmatrix} \tag{24}$$

and the transformed spring matrix is

$$D' = \Pi' D \Pi = \frac{1}{2} \begin{pmatrix} \omega_1^2 + \omega_2^2 & \omega_1^2 - \omega_2^2 \\ \omega_1^2 - \omega_2^2 & \omega_1^2 + \omega_2^2 + 4\alpha \end{pmatrix} . \tag{25}$$

For identical pendula ($\omega_1 = \omega_2$), this matrix is in fact diagonal. This is consistent with parity considerations. The frequencies of the two proper oscillation modes are those found in eq. (9). For $\omega_1 \neq \omega_2$, the two oscillators are not equivalent and the parity argument does not hold any longer.

ACKNOWLEDGMENTS

Fruitful and stimulating discussions with many participants of the school on this topic, especially with Prof. Dr. M. Cardona (Stuttgart) are gratefully acknowledged.

ULTRAFAST OPTICAL SPECTROSCOPY AND OPTICAL CONTROL

Timothy F. Crimmins, Richard M. Koehl, and Keith A. Nelson

Department of Chemistry
Massachusetts Institute of Technology
Cambridge, MA 02139 USA

ABSTRACT

A pedagogical review of optical spectroscopy and control at the "impulsive" limit, defined as a time scale faster than the fundamental vibrational or molecular motion under examination, is presented. Optical absorption and stimulated Raman scattering in this limit are seen to result in coherent vibrational motion in molecules and materials. A theoretical description of impulsive absorption and stimulated Raman scattering is followed by a survey of illustrative examples in which time-resolved observations of molecular and collective vibrational excitations are recorded. The topic then turns to optical control, beginning with a discussion of femtosecond pulse shaping through which complex ultrafast optical waveforms needed for many control objectives are produced. Theoretical formulations for optical control are reviewed, and a series of examples illustrating optical control over molecules and materials is presented.

I. INTRODUCTION

There has been a surge of activity in ultrafast optical spectroscopy in the last two decades beginning in the early 1980's with the development of the first reliable sub-100-fs light sources and accelerating with the current availability of commercial femtosecond laser systems that include high-power amplification and broadband wavelength tuneability. At the same time, the last decade has seen dramatic extension beyond time-resolved *observation of* material responses into the realm of ultrafast coherent *control over* them. Like ultrafast optical spectroscopy, ultrafast coherent control also has been driven by advances in the underlying optical technology, in particular by the development of practical *femtosecond pulse shaping* methods[1-3] through which ultrafast optical waveforms can be crafted.

Ultrafast Dynamics of Quantum Systems: Physical Processes and Spectroscopic Techniques, Edited by Di Bartolo and Gambarota, Plenum Press, New York, 1998

In this review we describe some of the conceptual and theoretical underpinnings of ultrafast optical spectroscopy, present illustrative experimental examples, and attempt to offer a vision of the prospects for the still developing area of coherent control. A complete elucidation of the relevant theory and a comprehensive review of all or most of the experimental results is impossible even in a lengthy review. Our hope is that the present article will serve as a useful introduction to the field, at roughly the first-year graduate student level, and a guide to further study. We also hope to offer, through the prism of our own research perspectives, a glimpse into what we feel are the tantalizing prospects that are now within reach.

The review begins with a theoretical description of impulsive time-resolved spectroscopy. A fully classical description of impulsive excitation and the responses it induces is given first to illustrate the basic physical principles. A semiclassical treatment (electromagnetic fields treated classically, matter treated quantum mechanically) is developed next, primarily in terms of time-dependent wavepacket [4-7] dynamics. This time-domain formulation, originally developed for treatment of conventional frequency-resolved absorption and light scattering spectroscopies, is particularly well adapted to time-resolved methods. In this picture, excitation light pulses are seen to initiate coherent wavepacket propagation which may be monitored through measurements of various optical observables with variably delayed probe pulses. A series of examples is presented next to illustrate the basic principles. The discussion then turns to the topic of coherent control. Femtosecond pulse shaping methods are described, and examples of their use for control over various molecular and material responses are presented. The review ends with a discussion of some of the prospects and future directions of ultrafast coherent control.

II. INTRODUCTION TO IMPULSIVE TECHNIQUES

The objective of this section is to provide a pedagogical introduction to impulsive stimulated Raman scattering (ISRS) and impulsive absorption. First, a fully classical description of non-resonant ISRS is advanced. For those not wishing to go further into the theoretical development, this classical introduction should provide all that is necessary to understand most of the topics in sections III, IV and V. Next, a semiclassical description of light absorption and stimulated Raman scattering is presented. The well known formulas describing the frequency dependences of the absorption and Raman cross sections are cast into a wavepacket dynamics formalism according to the prescription given by Heller. Finally, ISRS and impulsive absorption are treated in terms of this formulation.

The intent of this review is to provide a comprehensive, pedagogical introduction to impulsive techniques. All results are derived using a perturbative approach and are, consequently, only valid in a weak field limit. Some of the experiments reviewed in section III which discuss experimental attempts to control material responses do not take place in this limit. A non-perturbative treatment of ISRS is straightforward and is given elsewhere [8]. The weak field limit is used in this review because the physical implications of the results are more transparent, befitting the pedagogical motivation of the work.

II. A. Classical Description of Stimulated Raman Scattering

The objective of this section is to derive through classical mechanics an expression for the electric field produced by impulsive stimulated Raman scattering from a collection of polarizable harmonic oscillators. The strategy will be to first derive expressions for continuous-wave (CW) excitation and probing, i.e., ω_0 frequency domain stimulated Raman scattering, following the treatment of Yariv [9], and then to convert these expressions to those for the case of temporally impulsive pumping and probing. This somewhat circuitous route will be taken for two reasons: to draw upon intuition about traditional, frequency-domain Raman scattering and to stress the connection between the time-domain and frequency-domain treatments and information contents.

In a Raman scattering experiment, an electromagnetic wave at frequency ω_1 is scattered by thermally excited vibrations of a sample. The scattered light is frequency-shifted by the natural vibrational frequency , yielding signal that is detected at the Stokes and anti-Stokes scattered frequencies $\omega_2^S = \omega_1 - \omega_0$ and $\omega_2^{AS} = \omega_1 + \omega_0$. See Figure 1. Raman scattering is defined as spontaneous when no light at frequency ω_2 is present initially. The amplitude of the scattered field at frequency ω_2 is linearly proportional to that of the input field at frequency ω_1.

In stimulated Raman scattering, both ω_1 and ω_2 frequency components are initially present in the excitation light field. Usually this is achieved through the use of two distinct laser beams at the two frequencies, as shown in Figure 2. These two components exert a force on Raman-active vibrational modes at the difference frequency $\omega_1 - \omega_2 = \omega$, resulting in coherent vibrational oscillations driven at that frequency. The coherent vibrational response can then scatter a third, probe light beam in a fashion similar to the scattering of light by thermal excitations in spontaneous Raman scattering. However, in this case the vibrations are coherent, with the same spatial and temporal characteristics, so the scattering can be thought of as "diffraction" and the scattered light is a coherent beam that leaves the sample in a well defined direction.

Note that spontaneous and stimulated scattering processes involve change in the light wavevector as well as frequency. The beams at frequency ω_1 and ω_2 are specified also by their corresponding wavevectors $\vec{k}_1$ and $\vec{k}_2$, and the scattering process involves a vibrational excitation at the difference wavevector $\vec{q} = \vec{k}_1 - \vec{k}_2$ as well as the difference frequency ω.

The material will be modeled as a collection of N independent, polarizable harmonic oscillators. This model can be used to accurately represent excitations like phonons and molecular vibrations (and polaritons with several additional considerations). The analysis will be two dimensional, with each oscillator being described by its position x, z and normal vibrational coordinate $Q(t,\vec{r})$. See Figure 3. The equation of motion for a single oscillator is :

$$\frac{dQ}{dt^2} + \gamma \frac{dQ}{dt} + \omega_0^2 Q = \frac{F(t,\vec{r})}{m} \tag{II.1}$$

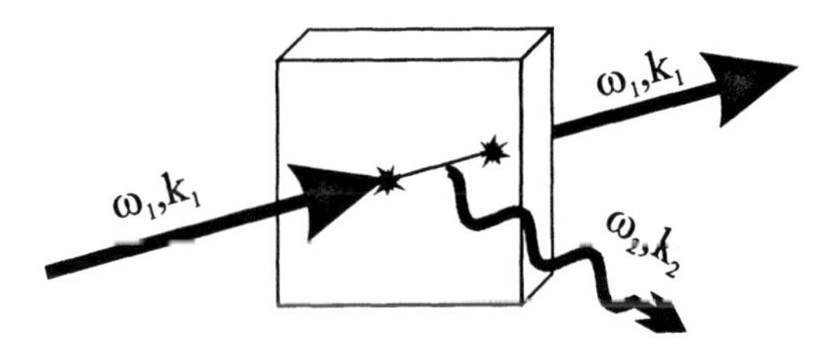

Figure 1. Spontaneous Raman Scattering. Incident light at frequency ω_1 and wavevector $\vec{k}_1$ with is scattered by vibrations at frequency ω_0 and wavevector $\vec{q}_0$ into a new frequency ω_2 and $\vec{k}_1$ where $(\omega_2, \vec{k}_2)=(\omega_1 \pm \omega_0, \vec{k}_1 \pm \vec{q}_0)$.

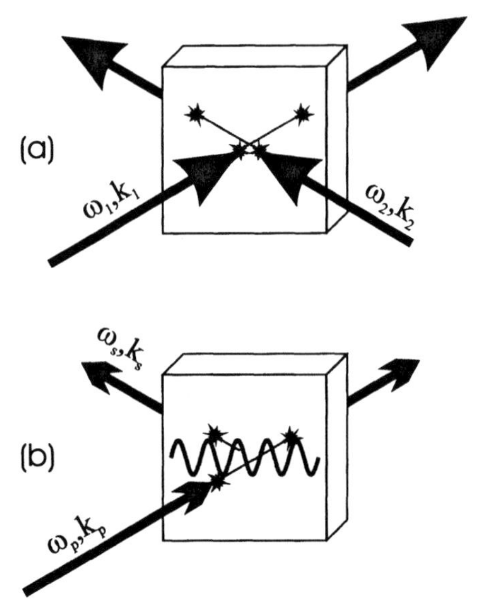

Figure 2. Coherent stimulated Raman scattering with two beams of frequencies and wavevectors $(\omega_1, \vec{k}_1)$ and $(\omega_2, \vec{k}_2)$. The beams drive a material response at the difference frequency and wavevector.

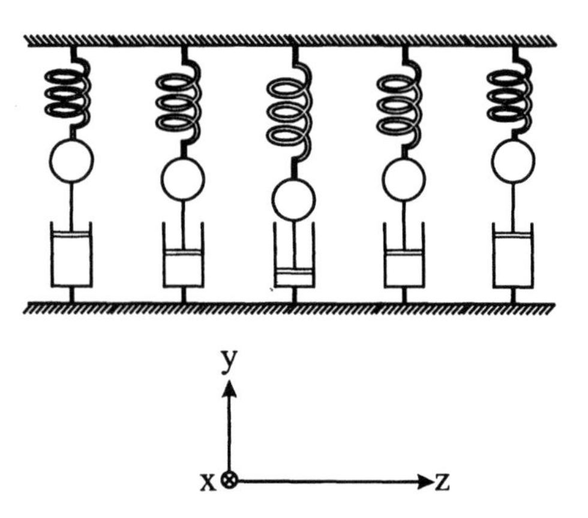

Figure 3. A condensed phase system is modeled as a collection of damped, harmonic oscillators in the derivation of the ISRS response.

Here ω_0 is the natural frequency of the oscillator, γ is the damping constant, $F(t,\bar{r})$ is the driving force and m is the mass or reduced mass of the oscillator.

First, expressions for the response of the oscillators to a CW optical field with two frequency components ω_1 and ω_2 will be found, ignoring all effects of the field except those which will be called the Raman force. The vibrational response to a temporally impulsive optical field will then be derived, and finally the prescription for treating pump pulses with an arbitrary electric field profile (e.g. femtosecond pulse sequences) will be described. Second, the time dependent polarization resulting from the interaction of a CW probe beam with the excited material will be found and used to derive the results for a temporally impulsive probe. Again, the prescription for generalization of these results to arbitrary probe electric field profiles will be laid out. Finally, the optical field produced by this time-dependent polarization, the ISRS signal field, will be found.

The optical driving force produced by the pump beam will be derived by considering the electrostatic stored energy density W in the oscillators:

$$W = \frac{1}{2}\varepsilon(Q)\bar{E}\cdot\bar{E} \tag{II.2}$$

The $Q(t,\bar{r})$ dependence of the dielectric constant ε will be treated by writing the dielectric constant in terms of the polarizibility α which can be expanded around $Q(t,\bar{r}) = 0$, keeping only terms up to first order in $Q(t,\bar{r})$:

$$\varepsilon = \varepsilon_0(1 + N\alpha(Q)) \cong \varepsilon_0\left\{1 + N\left[\alpha_0 + \left(\frac{\partial\alpha}{\partial Q}\right)_0 Q\right]\right\} \tag{II.3}$$

The force is now given by

$$F(t,\bar{r}) = \frac{\partial W}{\partial Q} = \frac{1}{2}\varepsilon_0 N\left(\frac{\partial\alpha}{\partial Q}\right)_0 \bar{E}\cdot\bar{E} \tag{II.4}$$

This force on the oscillators, proportional to the differential polarizability $(\partial\alpha/\partial Q)_0$ (a purely phenomenological quantity in this classical treatment) will be referred to as the Raman force.

In CW stimulated Raman scattering, the initial electromagnetic field contains two optical frequencies ω_1 and ω_2:

$$\bar{E}(t,\bar{r}) = \frac{1}{2}\hat{e}_1 E_e e^{i(\omega_1 t - \bar{k}_1\cdot\bar{r})} + \frac{1}{2}\hat{e}_2 E_e e^{i(\omega_2 t - \bar{k}_2\cdot\bar{r})} + c.c. \tag{II.5}$$

For simplicity, the two components of the electric field will be assumed to be polarized in the x direction (i.e., along the direction of the displacement of the oscillators), reducing the problem to a scalar one. It can be assumed without loss of generality that $\omega_1 > \omega_2$. The driving force is proportional to the square of this field, which contains sum and difference frequency components. The natural frequencies ω_0 of the oscillators, i.e., of the Raman-active molecular or lattice vibrational modes, are much less than the optical frequencies i.e., $\omega_0 << \omega_1, \omega_2$,. Consequently, only the components of the Raman force at the difference frequency $\omega_1 - \omega_2$ may be near resonance with the vibrational frequencies, and only these terms may drive significant vibrational responses. Neglecting other components, the Raman driving force is given by:

$$F(t,\vec{r}) = \frac{N\varepsilon_0}{2}\left(\frac{\partial\alpha}{\partial Q}\right)_0 |E_e|^2 e^{i(\omega_1-\omega_2)t} e^{-i(\vec{k}_1-\vec{k}_2)\cdot\vec{r}} + c.c. \tag{II.6}$$

The problem is simply that of a driven harmonic oscillator, with traveling wave solutions of the form:

$$Q_{CW}(t,\vec{r}) = Q_{CW} e^{i(\omega_1-\omega_2)t} e^{-i(\vec{k}_1-\vec{k}_2)\cdot\vec{r}} + c.c. \tag{II.7}$$

given by insertion of (II.6) into (II.5). The coherent vibrational amplitude is:

$$Q_{CW} = \frac{N\varepsilon_0\left(\frac{\partial\alpha}{\partial Q}\right)_0 |E_e|^2}{2m\left(\omega_0^2 - (\omega_1-\omega_2)^2 + i\gamma(\omega_1-\omega_2)\right)} \tag{II.8}$$

Thus coherent vibrational oscillations at the difference frequency $\omega_1 - \omega_2$ are driven by the two field components through stimulated Raman scattering. The largest vibrational amplitude occurs when the oscillator is driven on resonance, i.e., $\omega_1 - \omega_2 = \omega_0$. The vibrational energy imparted to the oscillator, i.e., the average power P absorbed from the field, is given by the average of the force $F(t,\vec{r})$ times the velocity dQ_{CW}/dt and is described in the usual limit $\gamma << \omega_0$ by a Lorentzian function:

$$\langle P\rangle \propto \frac{1}{\omega_0^2 - \Delta\omega^2 + i\gamma\Delta\omega} \tag{II.9}$$

as shown in Figure 4. In "stimulated Raman gain" measurements, the intensity of light at frequency ω_2 is measured as frequency ω_1, and thus $\Delta\omega$, is scanned, revealing the Lorentzian response with its maximum at the vibrational resonance frequency[10]. Through the stimulated Raman scattering process, vibrational energy is produced as incident light at the higher frequency ω_1 is converted into scattered light at the lower frequency ω_2.

These expressions describe the dynamics of the coherent vibrational response $Q_{CW}(t,\vec{r})$ for the case of CW excitation. The purpose of this review, however, is to introduce impulsive stimulated scattering, in which the excitation field is that of an ultrashort laser pulse. For this purpose it will be useful to describe the response in terms of a vibrational susceptibility χ:

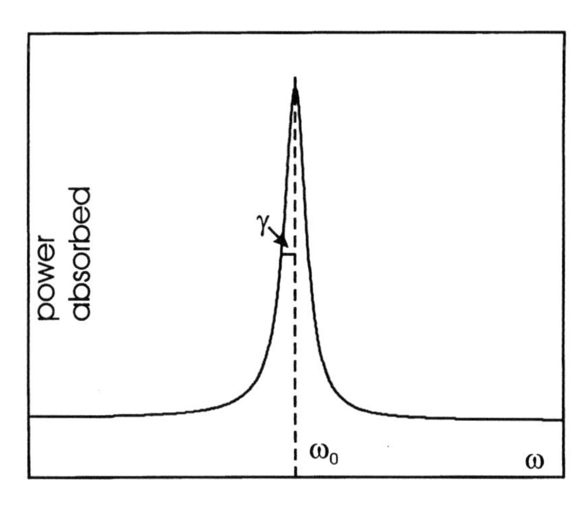

Figure 4. Lorentzian lineshape function of the power absorbed by an oscillator with natural frequency ω_0 and damping rate γ as a function of the frequency ω of the driving force.

$$Q_{CW}(t,\vec{r}) = \frac{N\varepsilon_0\left(\frac{\partial\alpha}{\partial Q}\right)_0}{2}\chi(\omega_1,\omega_2)\left|E_e\right|^2\left[e^{i(\omega_1-\omega_2)t-i\vec{q}\cdot\vec{r}}\right]+c.c. \quad \text{(II.10)}$$

From (II.10), it is clear that:

$$\chi(\omega_1,\omega_2) = \frac{1}{m\left(\omega_0^2-(\omega_1-\omega_2)^2+i\gamma(\omega_1-\omega_2)\right)} \quad \text{(II.11)}$$

The spatial dependence of the response is described by the wavevector of the excitation force $\vec{q} = \vec{k}_1 - \vec{k}_2$. We have assumed the susceptibility itself does not have any explicit spatial dependence in the limit of optical wavelengths. Note that this assumption is valid for molecular vibrations and most optic phonons, but not for acoustic waves or polaritons whose resonance frequencies ω_0 and dephasing or damping rates γ are wavevector-dependent.

Equation (II.8) describes the vibrational amplitude resulting from a driving force that oscillates at any particular frequency $\omega = \omega_1 - \omega_2$. An impulsive (i.e., delta-function) driving force contains equal contributions from all frequencies, i.e., a "white" frequency spectrum. In order to find the response to impulsive excitation, the response to each frequency is found using equations (II.10) and (II.11) and the contributions are summed:

$$Q_\delta(t,\vec{r}) = \frac{N\varepsilon_0\left(\frac{\partial\alpha}{\partial Q}\right)_0}{2}\int_{-\infty}^{\infty}\chi(\omega_1-\omega_2)\left|E_e\right|^2 e^{i(\omega_1-\omega_2)t}e^{-i\left(\vec{k}_1-\vec{k}_2\right)\cdot\vec{r}}\,d(\omega_1-\omega_2)+c.c. \quad \text{(II.12)}$$

Note that a femtosecond pulse is a moving "pancake" of light whose thickness is given by the pulse duration times the speed of light. For example, a 33-fs pulse is just 10 microns thick, even though the spot size (i.e. the transverse dimensions) may be much larger. In the limit of a true delta function excitation pulse, the pulse is infinitely thin! In this limit, two crossed pulses only overlap at any time along a single line, regardless of the spot sizes, as the pulses move forward so does the position of the line of overlap. With a finite pulse duration, the pulses overlap and form an optical interference or "grating" pattern across a region of space with finite width in the transverse dimension, usually still much smaller than the spot sizes. The use of diffractive optics to produce the two pulses and an appropriate imaging system to cross them [11] results in a large region of overlap, essentially equal to the spot sizes, and therefore produces many interference fringes. Here we will assume that an interference pattern is formed in this manner, and we will not consider any effects of finite transverse dimension of the interference region. This arrangement also allows us to assume that the interference fringe spacing, or grating (or stimulated scattering) wavevector, is independent of the frequency of any component of the pulse. This facilitates our use of superposition to calculate the total vibrational response. This approach can be used because the vibrational response, although a nonlinear function of the optical field, is linear in the Raman force. Defining the fourier transform of $\chi(\omega_1 - \omega_2)$ as $G(t)$, equation (II.12) can be written:

$$Q_\delta(t) = G(t)F_\delta(t) \tag{II.13}$$

From this, it is clear that $G(t)$ is the impulse response of the system to the impulsive Raman force. The response function to an impulse at *t=0*, hereafter referred to as the impulse response function, is given by the fourier transform of (II.11):

$$G(t) = \begin{cases} \dfrac{2}{m} e^{-\gamma t/2} \sin(\omega_v t), & t > 0 \\ 0, & t < 0 \end{cases} \tag{II.14}$$

writing the underdamped frequency as $\omega_v = \sqrt{\omega_0^2 - \gamma^2 / 4}$. This gives the response of the material to temporally impulsive excitation.

The material response to a general time-dependent pump electric field profile is given by the convolution of the corresponding time-dependent Raman force $F(t, \vec{r})$ produced by the field profiles with:

$$Q_F(t, \vec{r}) = \int_{-\infty}^{\infty} G(t') \cdot F(t - t', \vec{r}) dt', \tag{II.15}$$

Having derived the vibrational response driven through stimulated Raman scattering, we now turn our attention to coherent scattering of probe light by this response. The polarizability α of the oscillators depends on vibrational displacement $Q(t, \vec{r})$ as expressed in equation (II.3) through the differential polarizability $(\partial\alpha/\partial Q)_0$, and therefore the coherent vibrational oscillations of the sample produce coherent oscillations in the polarizability with the same spatial and temporal dependence as the vibrational displacement. The induced polarization is given by the product of the polarizability and the probe field:

$$P(t,\vec{r}) = \varepsilon_0 N\alpha(Q)E_p(t,\vec{r}) = \varepsilon_0 N\left[\alpha_0 + \left(\frac{\partial\alpha}{\partial Q}\right)_0 Q(t,\vec{r})\right]E_p(t,\vec{r}) \qquad \text{(II.16)}$$

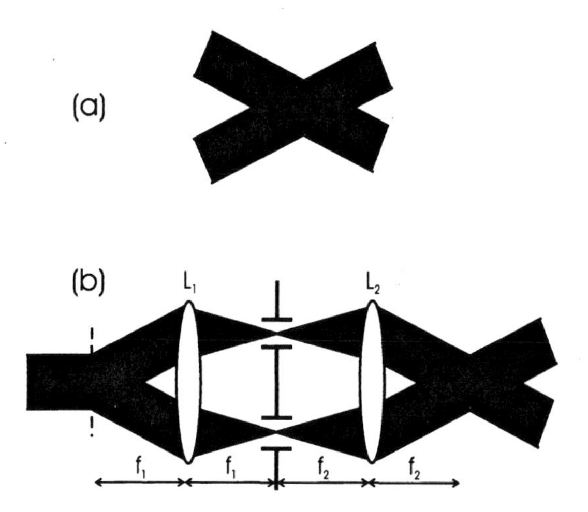

Figure 5(a). Two femtosecond beams produced through partial reflection by a beamsplitter can only overlap over a small region of space, resulting in limited wavevector resolution which diminishes with decreasing pulse duration. (b) Two beams separated with a diffraction grating and combined with a confocal imaging system can be overlapped over the entire spot size, greatly enhancing wavevector resolution and making it independent of pulse duration. (Reprinted from ref [11]. Copyright 1998 Optical Society of America.)

This macroscopic polarization radiates the scattered field. The probe field $E_p(t,\vec{r}) = (1/2)E_p e^{i(\omega_p t - \vec{k}_p\cdot\vec{r})} + c.c.$ is assumed to be weak (i.e., it does not significantly excite any new vibrations). For simplicity, only the polarization induced by the Raman force, or the term in equation (II.16) proportional to the differential polarizability, will be considered. Assuming impulsive pump beams and assuming $\omega_0 >> \gamma$ for simplicity, equations (II.13) and (II.14) can be combined with equation (II.16) to give the following expression for the Raman polarization driven by the Raman force and a CW probe:

$$P_R(t,\vec{r}) = \frac{\varepsilon_0^2 \pi N^2 |E_e|^2 \left(\frac{\partial \alpha}{\partial Q}\right)_0^2}{m\omega_v} e^{-\gamma t/2} \sin(\omega_v t) \cdot \cos(qz) \times \left[E_p e^{i(\omega_p \cdot (t-t_p) - \vec{k}_p \cdot \vec{r})} + c.c. \right]$$

$$= \frac{\varepsilon_0^2 \pi N^2 |E_e|^2 \left(\frac{\partial \alpha}{\partial Q}\right)_0^2}{\omega_v} G(t) |E_e|^2 e^{-i(\vec{k}_1 - \vec{k}_2) \cdot \vec{r}} E_p \times \left[e^{i(\omega_p \cdot (t-t_p) - \vec{k}_p \cdot \vec{r})} + c.c. \right] \tag{II.17}$$

Expanding this result illustrates the new frequency and wavevector components in the Raman polarization:

$$P_R(t,\vec{r}) = \frac{-\varepsilon_0^2 \pi N^2 E_p |E|^2 \left(\frac{\partial \alpha}{\partial Y}\right)_0^2}{4im\omega_v} e^{-\gamma t/2} e^{-i\omega_p t_p}$$

$$\left\{ \exp\left[i(\omega_p + \omega_v)t - i\left(\frac{\omega_v \cos\theta}{c} + k_{px}\right)x - i(k_{pz} + q)z \right] \right.$$

$$+ \exp\left[i(\omega_p + \omega_v)t - i\left(\frac{\omega_v \cos\theta}{c} + k_{px}\right)x - i(k_{pz} - q)z \right] \tag{II.18}$$

$$- \exp\left[i(\omega_p - \omega_v)t - i\left(-\frac{\omega_v \cos\theta}{c} + k_{px}\right)x - i(k_{pz} + q)z \right]$$

$$\left. - \exp\left[i(\omega_p - \omega_v)t - i\left(-\frac{\omega_v \cos\theta}{c} + k_{px}\right)x - i(k_{pz} - q)z \right] + c.c. \right\}$$

It will be shown that this polarization radiates an electromagnetic field which is the ISRS signal. The optical dispersion relation $\omega = |k|c$ requires that the Stokes scattered signal at frequency $\omega_S = \omega_p - \omega_v$ have wavevector $|\vec{k}_p - \vec{q}| = \frac{\omega_S}{c}$. For crossed excitation pulses that form a grating pattern, the probe beam can, at best, be incident at the "Bragg" diffraction angle, i.e., phase matched, such that half of the terms in (II.18) obey this condition. Once a choice is made from which side the probe beam will enter, only two of the forms, e.g. the first and fourth terms in (II.18) meet this condition and radiate a signal field; the other two would give equivalent results with the probe beam incident from the other side. This polarization will radiate a signal at new signal frequencies $\omega_s = \omega_p - \omega_v$ and $\omega_{AS} = \omega_p + \omega_v$ in a direction given by the "diffracted" signal wavevector. The time-dependence of the signal field reveals the material response function $G(t)$ whose elucidation is the usual objective of an ISRS experiment.

In principle, the ISRS measurement might be conducted using ultrashort (impulsive) excitation and a CW probe beam to produce the scattered field whose time dependence would be analyzed by a fast photodetector and digitizing electronics. In practice, this would require femtosecond time resolution in the detection system, and this is not available. Consequently, the experiment is carried out with an ultrashort probe pulse that is delayed by a specified time period t_p following excitation, and the signal generated from this probe pulse is used to measure the material response at just the single time that is probed. Then the excitation-probe sequence is repeated with a different time delay, and then with another, and so on, until the entire time-dependent sample response is determined in a point-by-point fashion along the time axis. What is measured at each time delay is the total amount of coherently scattered light i.e., time-integrated intensity of the signal field is measured by a slow detector. In order to describe this, the polarization produced by a temporally impulsive probe pulse will be found and used to find the polarization response to an arbitrary pulse shape.

The polarization produced by an impulsive probe field E_p^δ is found by integrating over all possible probe frequencies ω_p:

$$P_R^\delta(t-t_p) = G(t)E_p^\delta |E_e|^2 \delta(t-t_p) \tag{II.19}$$

The response of the polarization to an impulsive probe at t_p is an impulse at t_p, with an amplitude proportional to the product of the electric field strengths and the time dependent response evaluated at t_p. The response of the polarization to a general probe temporal profile can be found by convolving the probe field with the response function:

$$P_R^\delta(t,t_p) = \int_{-\infty}^{\infty} G(t')E_p(t')|E_e|^2 \delta(t-t'-t_p)dt' \tag{II.20}$$

Carrying out the integration gives:

$$P_R^\delta(t,t_p) = G(t_p)E_p(t-t_p)|E_e|^2 \tag{II.21}$$

Finally, the ISRS electric field can be found by entering the time dependent Raman polarization into Maxwell's equations as a source term:

$$\frac{\partial^2 E_y(t,\bar{r})}{\partial x^2} + \frac{\partial^2 E_y(t,\bar{r})}{\partial z^2} = \mu\varepsilon \frac{\partial^2 E_y(t,\bar{r})}{\partial t^2} + \mu \frac{\partial^2 (P_R)_y}{\partial t^2} \tag{II.22}$$

Making the usual slowly varying envelope approximation, i.e., $\frac{\partial^2 E_y(t,\bar{r})}{\partial z^2}, \frac{\partial^2 E_y(t,\bar{r})}{\partial x^2} << c^2 \frac{\partial^2 E_y(t,\bar{r})}{\partial t^2}$, the ISRS field is given by:

$$E_y^{ISRS}(t,\bar{r}) = \frac{1}{\varepsilon}(P_R)_y \tag{II.23}$$

As explained above, the response function is typically recovered by repeatedly performing the ISRS experiment as incremental variations in the probe delay t_p are made.

For each t_p value, the intensity of the electric field produced by $P_R(t,\vec{r})$ is integrated by a slow detector:

$$S_{ISRS}(t_p) = \int_{-\infty}^{\infty} \left| G(t_p) E_p(t - t_p) F_\delta(t, r) \right|^2 dt \qquad \text{(II.24)}$$

In other words, the signal recovered in a typical ISRS experiment is the convolution of the system's response following excitation by the Raman force with the probe electric field temporal profile. If the probe pulse as well as the excitation pulses are short enough to be considered "impulsive" i.e., if no significant vibrational motion occurs during the pump and probe pulses, then the result can be simplified to:

$$S_{ISRS}(t_p) = \left| G(t_p) \right|^2 I_p \left| F_\delta(t_p) \right|^2 \qquad \text{(II.25)}$$

In this case the signal directly gives the impulse response function. The signal is proportional to the probe intensity and the square of excitation intensity.

Note that heterodyne methods have been demonstrated recently in which a strong reference beam propagates collinear and in phase with the signal beam. In this case the signal and reference fields add constructively, and the measured intensity is dominated by their produce which is linear in the response function and the excitation intensity, as well as the probe intensity.

Most of the present discussion has emphasized the transient grating geometry in which crossed femtosecond excitation pulses are used and signal is produced by coherent scattering, or diffraction, of probe light. However, in general ISRS excitation occurs even when a single excitation pulse is incident on a Raman-active sample. Stimulated scattering occurs in the forward direction, with higher-frequency components of the incident pulse scattered into lower-frequency components still contained within the pulse bandwidth. The probe pulse is collinear (or in practice, nearly collinear) with the excitation pulse and the signal is still derived from Equation (II.18) with $\theta = 0$ and $k_{pz} = 0$. Whether the coherently scattered probe light is Stokes or anti-Stokes shifted, i.e. which of the terms in Equation (II.18) dominates, depends on the probe delay relative to the excitation pulse. The results can be understood by considering the impulsive force exerted by the probe pulse, which unlike the excitation pulse encounters the sample already undergoing coherent vibrational oscillations. If the probe pulse arrives at the sample after an integral multiple of vibrational periods, then it drives vibrational motion in phase with that already under way, resulting in an increased vibrational amplitude. Therefore the probe pulse imparts vibrational energy to the system and emerges red-shifted. If the probe pulse arrives after, say, one-half the vibrational period, then the force it exerts opposes the motion already under way and the vibrational amplitude is decreased. The probe pulse takes vibrational energy away from the system and emerges blue-shifted. Thus the spectrum of the transmitted probe pulse "wags" back and forth from red to blue at the vibrational frequency. As we shall see, these time-dependent spectral shifts can be detected readily, and represent just one of several observables through which the results of ISRS excitation with a single pulse can be monitored.

II. B. Wavepacket Description of Absorption Spectroscopy and Raman Scattering

The classical description of spontaneous and stimulated Raman scattering, including ISRS, presented in the previous section is appropriate when no electronic resonances are involved. To a limited extent, the same description can still be used even if the excitation or probe wavelength is absorbed significantly. The values of differential polarizabilities are generally wavelength-dependent, and they usually increase substantially near electronic absorption peaks, especially for vibrational modes with substantial vibronic coupling (e.g. molecular vibrational modes for which the equilibrium position is shifted significantly in the electronic excited state). Thus resonances can be accounted for simply by appropriate adjustment of the differential polarizability values. However, a quantum mechanical treatment, in a simple way, is necessary to account for light absorption and generation of electronic excited states, vibrational motion in the electronic excited states, electronic phase coherence between ground and excited electronic states, and resonance effects on the vibrational phase and motion in the ground electronic state. A semiclassical treatment (light fields treated classically, matter treated quantum mechanically) is generally sufficient to describe the important material responses and signals.

The semi-classical wavepacket dynamics formalism is well suited to the problem [4-7]. Its primary advantage is that it provides a particularly clear physical picture that emphasizes the role played by coherent vibrational motion in the electronic excited state. It provides a unified description of light absorption and resonance-enhanced Raman scattering, including resonance-enhanced stimulated Raman scattering and ISRS. The differential polarizability (and for light absorption, the cross section) is not defined phenomenologically, but is the central quantity whose value is derived and is found to depend on the details of time-dependent wavepacket propagation.

1. Absorption Cross Section. First, the frequency dependence of the absorption cross section will be considered. We start from the standard expression for the absorption probability for a transition from initial state $|I\rangle = |i, j\rangle = |i\rangle|j\rangle$ to final state $|F\rangle = |f, n\rangle = |f\rangle|n\rangle$ where the quantum numbers i and f refer to ground and excited electronic states and the quantum numbers j and n refer to vibrational eigenstates of the respective electronic states. The transition probability is proportional to the absorption cross-section σ given by

$$\begin{aligned}\sigma_n(\omega) &= 4\pi^2\alpha a_0\omega\left|\langle n|\langle f|\mu|i\rangle|j\rangle\right|^2 = 4\pi^2\alpha a_0\omega\left|\langle n|\mu_{fi}|j\rangle\right|^2 \\ &\equiv 4\pi^2\alpha a_0\omega\Sigma_n(\omega)\end{aligned} \tag{II.26}$$

where α is the fine structure constant, a_0 is the Bohr radius, ω is the frequency of radiation, and $\mu_{fi} = \langle f|\mu|i\rangle$ is the transition dipole matrix element between the electronic states. This expression can be analyzed by separating the transition dipole matrix element from the vibrational wavefunctions, leaving the Frank-Condon factors (overlap integrals) $\langle n | j \rangle$ whose magnitude determines the strengths of the transitions into different vibrational levels of electronic excited state. Instead, we go to a time-domain picture by retaining the transition dipole operator inside the matrix elements.

A model system will be considered which has two electronic states with vibrational hamiltonians H_g and H_e. $| j \rangle$ is taken to be a vibrational eigenstate of the ground electronic state hamiltonian and $|n\rangle$ is taken to be a vibrational eigenstate of the excited electronic state hamiltonian. For notational convenience, we set $\hbar = 1$. $\Sigma_n(\omega)$ can be rewritten :

$$\Sigma_n(\omega) = \langle \phi | n \rangle \langle n | \phi \rangle \tag{II.27}$$

where

$$|\phi\rangle \equiv \mu_{fi} | j \rangle \tag{II.28}$$

In this picture, $|\phi\rangle$ is the wavepacket produced on the excited electronic state by impulsive excitation of the initial state. Since the nuclei do not have time to move a significant distance during the excitation event, the excited state starts out with the same distribution of vibrational displacements that was present in the ground state just before excitation. In order to find the absorption cross section for a transition from state i to any state with energy E, equation (II.27) must be summed over all degenerate states with energy E:

$$\Sigma(\omega) \equiv \sum_n [\Sigma_n(\omega)] = \langle \phi \left| \left[\sum_n |n(E)\rangle \langle n(E)| \right] \right| \phi \rangle \tag{II.29}$$

The strategy at this point is to rewrite the bracketed term in equation (II.29) in such a way that $\Sigma(\omega)$ can be described by the dynamic propagation of $|\phi\rangle$, so an attempt will be made to convert the bracketed term into some type of time propagation operator. The next several steps will be presented without clear motivation, but the manipulations will produce a clear result. The closure relation can be written in the following fashion:

$$\hat{I} = \int dE' \sum_n |n(E')\rangle \langle n(E')| \tag{II.30}$$

where $\hat{I}$ is the identity operator. A delta function which takes operators as its arguments is introduced and shown to have the following properties:

$$\begin{aligned} \delta(\hat{O}) | \psi \rangle &\equiv \frac{1}{\sqrt{2\pi}} \int_{-\infty}^{\infty} d\tau \cdot e^{i\hat{O}\tau} e^{i\omega\tau} | \psi \rangle \\ &= \frac{1}{\sqrt{2\pi}} \int_{-\infty}^{\infty} d\tau \cdot e^{io\tau} e^{i\omega\tau} | \psi \rangle \\ &= \delta(o) | \psi \rangle \end{aligned} \tag{II.31}$$

where $\hat{O}$ is an operator with eigenvalue o, and $|\psi\rangle$ is an eigenfunction of $\hat{O}$. Consider the properties of the energy selector operator delta function, an operator delta function with $E \cdot \hat{I} - \hat{H} \cdot \hat{I}$ as its argument, where $E = E_i + \hbar\omega$ and E_i is the energy of $|i\rangle$:

$$\delta(E \cdot \hat{I} - \hat{H}_e \cdot \hat{I})|n\rangle = \delta(E \cdot \hat{I} - E_n \cdot \hat{I})|n\rangle = \hat{\delta}_{E,E_n} \tag{II.32}$$

where $\hat{\delta}_{E,E_n}$ is an operator Kronecker delta function with the following definition:

$$\hat{\delta}_{E,E_n} = \hat{I} \quad for\ E = E_n \ = \hat{0} \quad for\ E \neq E_n \tag{II.33}$$

Equation (II.32) can be right multiplied by the energy selector operator delta function:

$$\begin{aligned}\delta(E \cdot \hat{I} - \hat{H}_e \cdot \hat{I}) &= \int dE' \sum_n |\psi_n(E')\rangle\langle\psi_n(E')|\delta(E \cdot \hat{I} - \hat{H}_e \cdot \hat{I}) \\ &= \int dE' \sum_n |\psi_n(E')\rangle\langle\psi_n(E')|\delta(E \cdot \hat{I} - E' \cdot \hat{I}) \\ &= \sum_n |\psi_n(E)\rangle\langle\psi_n(E)|\end{aligned} \tag{II.34}$$

Equation (II.34) can replace the term in brackets in (II.29), yielding:

$$\Sigma(\omega) = \langle\phi|\delta(E \cdot \hat{I} - \hat{H}_e \cdot \hat{I})|\phi\rangle \tag{II.36}$$

While this expression is not yet in the form being sought, it is instructive at this point to recover the connection with standard treatments of absorption. One can expand $|\phi\rangle$ in eigenstates of H_e. The energy selector operator delta function acting on $|\phi\rangle$ will select out only those eigenstates of H_e in the expansion which have energy $E_i + \hbar\omega$, corresponding to states which differ energetically from the initial state by the energy contained in one photon of frequency ω. From energy conservation, one would expect only these states to be populated following absorption of a photon of wavelength ω. This gives

$$\Sigma(\omega) = \langle\phi|c_E|\psi_E\rangle, \tag{II.37}$$

where c_e is the coefficient of eigenstate $|\psi_E\rangle$ with energy E in the expansion of $|\phi\rangle$. Of course $\langle\phi|$ can also be expanded in eigenstates of H_e. Then, by orthogonality,

$$\Sigma(\omega) = |c_E|^2 \tag{II.38}$$

The matrix element in the absorption cross section gives the probability that a vibrational state of the electronic excited state with energy E will be populated following excitation, as one would expect from a standard treatment of absorption.

Returning to the development of the wave packet description, $\Sigma(\omega)$ as written in (II.36) is just the expectation value of $\delta(E \cdot \hat{I} - \hat{H} \cdot \hat{I})$, which can be written as the trace of the product of the delta function operator and the density matrix $\rho_\phi \equiv |\phi\rangle\langle\phi|$:

$$\Sigma(\omega) = Tr\left[\delta(E \cdot \hat{I} - \hat{H}_e \cdot \hat{I})\rho_\phi\right] \tag{II.39}$$

The delta function can be alternatively written as a fourier transform, and this expression can be recast into the form of the fourier transform of the correlation between two wave packets, which is the desired result:

$$\begin{aligned}
\Sigma(\omega) &= \frac{1}{\sqrt{2\pi}} Tr \int_{-\infty}^{\infty} e^{i(E\cdot\hat{I} - H_e\cdot\hat{I})\tau} \rho_\phi d\tau \\
&= \frac{1}{\sqrt{2\pi}} Tr \int_{-\infty}^{\infty} \sum_n \frac{(iE\tau)^n \cdot \hat{I}^n}{n!} e^{-iH_e\cdot\hat{I}\tau} \rho_\phi d\tau \\
&= \frac{1}{\sqrt{2\pi}} Tr \int_{-\infty}^{\infty} \sum_n \frac{(iE\tau)^n}{n!} e^{-iH_e\cdot\hat{I}\tau} \rho_\phi d\tau \\
&= \frac{1}{\sqrt{2\pi}} \int_{-\infty}^{\infty} e^{iE\tau} Tr\left[e^{-iH_e\cdot\hat{I}\tau} \rho_\phi\right] d\tau \\
&\equiv \frac{1}{\sqrt{2\pi}} \int_{-\infty}^{\infty} e^{iE\tau} Tr \rho_\phi(\tau) d\tau \\
&= \frac{1}{\sqrt{2\pi}} \int_{-\infty}^{\infty} e^{iE\tau} \langle\phi|\phi(\tau)\rangle d\tau
\end{aligned} \tag{II.40}$$

where the following definitions are made recalling that $e^{-iH_e t}$ is the time propagation operator:

$$\rho_\phi(\tau) = e^{-iH_e\tau} \rho_\phi \tag{II.41}$$

$$|\phi(\tau)\rangle = e^{-iH_e\tau} |\phi\rangle \tag{II.42}$$

Equation (II.40) is the result which is sought. The absorption cross section is proportional to the fourier transform of the correlation between the static wave packet created from the initial state by impulsive excitation and an identical wave packet propagating on the excited state potential energy surface.

As an illustration, consider the absorption of monochromatic light. See Figure 6. The CW optical field can be broken down into a series of short pulses centered at frequency ω, each of which produces a coherent vibrational wavepacket on the excited state. Call the impulse at time t_1 at a peak in the optical field $\delta(t_1)$. It will create a wavepacket propagating on the excited state with a phase factor $e^{i\omega_D(t-t_1)}$. (Note that ω_D is the energy difference between the ground electronic, ground vibrational state and the excited

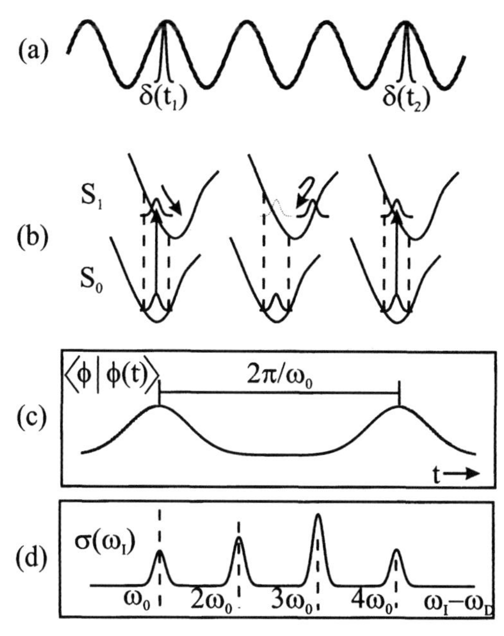

Figure 6(a). An optical field with frequency ω_I can be represented as a sum of delta functions, with $\delta(t_1)$ and $\delta(t_2)$ the delta functions at different peaks in the field. (b). The wavepackets produced at t_1 and t_2 are seen to constructively interfere at t_2 if ω_I matches $\omega_D+\omega_0$, the vibronic frequency of S_1, which intuition predicts will lead to absorption at this frequency. (c). The correlation between the solid and dashed wavepackets on S_1 in (b) $\langle\phi|\phi(t)\rangle$ shows peaks spaced by $2\pi / \omega_0$. (d) The fourier transform of (c), shown in the Heller formalism to be proportional to the absorption cross section $\sigma(\omega)$, has peaks separated by ω_0, i.e., vibrational progressions appear in the absorption spectrum.

electronic, ground vibrational state.). The impulse at a later peak in the optical field at time t_2, call it $\delta(t_2)$, will also produce a wavepacket on the excited state potential energy surface (PES) with phase factor $e^{i\omega_d(t-t_2)}$. The wavepacket produced by $\delta(t_1)$ will interfere with the wavepacket produced by $\delta(t_2)$. If the wavepacket produced by $\delta(t_1)$ has not moved significantly by time t_2, e.g. if time t_2 is the very next peak in the optical field following time t_2, then the interference will be constructive unless the optical frequency is extremely far from a vibronic transition frequency $\omega_D + n\omega_0$ (assuming a harmonic excited-state PES). After the wavepacket moves away from its initial position, say at time t_3, its overlap with a wavepacket created by a third impulse $\delta(t_3)$ is minimal so there is no significant interference, constructive or destructive. At still later times, say t_4, the wavepacket may return to the region of its origin. In this case it may add constructively to the new wavepacket created by $\delta(t_4)$, if the optical frequency is close to a vibronic transition frequency. If the optical frequency is not so close, then the interference may be destructive. If the wavepacket continues to return undiminished to the region of its origin for many vibrational periods, i.e. if the dephasing rate γ is very low, then even a slight mismatch between the optical frequency and a vibronic transition frequency eventually leads to destructive interference. Thus the absorption peaks correspond to vibrational progressions in the excited electronic state, and they are very sharp if the electronic and vibrational dephasing of the excited state is slow. On the other hand, if dephasing is fast then the initial constructive interference that occurs before the wavepacket first leaves its region of origin, and perhaps additional constructive interference over just a few vibrational periods, is never canceled by destructive interference later on and so there is still absorption even thought the optical frequency may not precisely match the vibronic frequency. That is, faster dephasing rates give rise to broader absorption lines. In condensed phases, e.g. molecular liquids, dynamical change of the electronic phase factor ω_D is usually fast enough (femtoseconds) to cause distinct vibrational progressions to broaden and merge into continuous absorption bands.

2. Raman Scattering. It is possible to develop a similar wavepacket formalism for Raman scattering, in particular when the incident light is near an electronic absorption resonance. In this case the influence of the particular electronic excited state into which absorption may occur is predominant. We will see that the Raman polarizability, like the absorption cross-section, depends strongly on the dynamics of vibrational motion in the electronic excited state. As in the case of absorption, we begin with the standard expression for the relevant cross section (or in this case Raman polarizability) and from this we interpret a time-domain formulation, again as laid out by Heller [7].

The Raman scattering process is a second-order process involving two transition dipole operators, one which takes the system from the initial state $|I\rangle = |i, j\rangle = |i\rangle|j\rangle$ to an intermediate electronic state $|k\rangle$ and the other which takes the system from the intermediate state to the final state $|F\rangle = |f, n\rangle = |f\rangle|n\rangle$. We will assume that the initial and final states are two different levels of a vibrational mode in the ground electronic state. The standard expression for the differential cross section for Raman scattering $d\sigma / d\Omega(\omega)$ relates the cross section to the square of the differential polarizability $\alpha(\omega)$[12]:

$$\frac{d\sigma}{d\Omega_2} = \frac{\omega^4}{4\pi\varepsilon_0 hc^2}|\alpha(\omega)|^2 \propto \left|\sum_k \left(\frac{\mu_{0k}\mu_{0k}}{\hbar\omega_{0k} + \hbar\omega} + \frac{\mu_{0k}\mu_{0k}}{\hbar\omega_{0k} - \hbar\omega}\right)\right|^2 \qquad \text{(II.43)}$$

where μ_{0k} is a transition moment function for an electronic transition from $|0\rangle \to |k\rangle$. A wavepacket interpretation can be derived in order to intuitively explain the frequency dependence of $\alpha(\omega)$ for the resonant or near-resonant case, i.e., $\hbar\omega_{ok} \approx \hbar\omega$ in a manner similar to that shown earlier [7]:

$$\begin{aligned} \alpha(\omega) &\propto \left|\int_0^\infty dt e^{iE_i t/\hbar} \left\langle \phi_f \left| e^{-i(H_k - i\gamma_k)t/\hbar} \right| \phi_i \right\rangle\right|^2 \\ &\propto \left|\int_{-\infty}^\infty dt e^{iE_i t/\hbar} \left\langle \phi_f | \phi_i(t) \right\rangle\right|^2 \end{aligned} \qquad \text{(II.44)}$$

A phenomenological electric dephasing rate γ_k has been introduced and where $|\phi_f\rangle$ and $|\phi_i(t)\rangle$ are as given in (II.28) and (II.42). In Equation (II.44), the summation over electronic states k has been omitted because the resonant state has the predominant effect. The probability that a vibrational state $|n\rangle$ will be populated following Raman scattering is given by the square of equation (II.44). The terms in this expression are similar to those for the case of absorption. Again, the matrix elements are proportional to the fourier transform of the correlation function between two wavepackets on the excited state PES. One wave packet $|\phi_i(t)\rangle$ is produced by impulsive promotion of the initial vibrational state $|j\rangle$. This wave packet is propagating on the excited electronic PES. The other wavepacket is produced by impulsive promotion of the final state $|n\rangle$ to the excited state PES where it does not propagate. The frequency dependence of the differential polarizability near resonance can be understood intuitively in terms of wavepacket propagation and overlap.

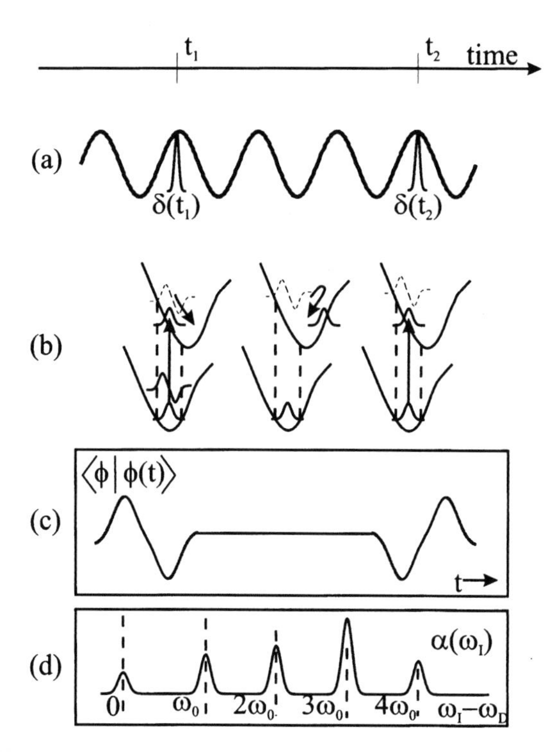

Figure 7(a). An optical field with frequency ω_I can be represented as a sum of delta functions, with $\delta(t_1)$ and $\delta(t_2)$ the delta functions at different peaks in the field. (b). The wavepacket produced at t_1 interferes with the final state (dashed) projected onto S_1 .(c). The correlation between the solid and dashed wavepackets on S_1 $\langle\phi|\phi(t)\rangle$, shows an envelope at ω_0, the natural frequency of S_1, with structure under the envelope related to the overlap between the initial and final states projected onto S_1. (d) The fourier transform of (c), shown in the Heller formalism to be proportional to the Raman scattering differential polarizability $\alpha(\omega_I)$, has a peak at $\omega_0$if the overlap between the initial and final states is strong i.e., a peak occurs in the Raman scattering cross section if the incident field is resonant with a vibronic state in S_I.

Driving frequencies that are on resonance with a level in the excited electronic state vibronic manifold will have the largest Raman cross sections, but only if there is overlap between the final state's projection on the electronic excited state and the propagating projection of the initial state.

Consider again an optical field with frequency ω_I which can be thought of as a series of impulses. See Figure 7(a). The impulse at t_1, $\delta(t_1)$, will promote the initial state from the ground electronic state to an electronically excited PES with phase $e^{i\omega_D t_1}$ where the wavepacket will propagate. If ω_0 is chosen such that this wavepacket is returning to its birthplace at each subsequent peak in the optic field, absorption will of course take place, but also a peak might be seen in the Raman scattering cross section for final state $|j\rangle$ if the static projection of $|j\rangle$ onto this excited state PES overlaps with the propagating initial state. This may be understood as the Stokes frequency ω_2 projecting the propagating wavepacket $|\phi_i(t)\rangle$ down to vibrational states of the ground electronic level with which it overlaps.

II. C. Semi-Classical Treatment of Stimulated Raman Scattering and Impulsive Stimulated Raman Scattering and Impulsive Absorption

Both resonance Raman and absorption have been treated in the impulsive limit by a number of authors [13-17]. This treatment will closely mirror that of Chesnoy and Mokhtari [18]. The Hamiltonian H_0 for a model system consisting of electronic ground and excited states S_0 and S_1 containing harmonic vibronic manifolds is perturbed by the dipolar interaction $H_1(t) = -PE(t)$ with an excitation light pulse $E(t) = E_0(t)e^{-i\omega_0 t + kz} + c.c.$, treated here (neglecting polarization effects) as a scalar. The time evolution of the system can be described by the time evolution of the density matrix which follows the equation of motion:

$$\frac{d\rho}{dt} = \frac{1}{i\hbar}\left[H_0 + H_1(t)\right] + \frac{d}{dt}\rho_{relax} \tag{II.45}$$

with ρ_{relax} describing dephasing and energy exchange between different states.

This equation can be solved pertubatively. At second order, there will be populations in both electronic states and coherences between vibrational levels γ, γ', etc. in S_0 and β, β' in S_1. Defining $R_{\gamma\beta}(t)$ as the response function of the first order density matrix element $\rho^1_{\gamma\beta}$, which is given in the Block approximation as:

$$R_{\gamma\beta}(t) = e^{-(i\omega_{\gamma\beta} + 1/T_{\gamma\beta})t} \quad for \quad t > 0 \tag{II.46}$$

and assuming that the dephasing times $T_{\gamma\beta}$ between the two electronic states are short compared with the optical pulse duration, the density matrix elements to second order in pump optical field describing vibrational coherences in the ground and excited states are:

$$\rho^2_{\gamma\gamma'}(t) = -\frac{1}{\hbar^2}\int_{-\infty}^{\infty} dt' E_0^2(t-t')\sum_{\beta} P_{\gamma\beta}P_{\beta\gamma'}\left[\frac{\rho^0_{\gamma\gamma}R_{\gamma\gamma'}\left(t'-\frac{T^+_{\gamma\beta}}{2}\right)}{i(\omega_0+\omega_{\gamma\beta})+\frac{1}{T_e}}+\frac{\rho^0_{\gamma'\gamma'}R_{\gamma\gamma'}\left(t'-\frac{T^-_{\gamma'\beta}}{2}\right)}{-i(\omega_0+\omega_{\gamma\beta})+\frac{1}{T_e}}\right] \quad \text{(II.47)}$$

$$\rho^2_{\beta\beta'}(t) = -\frac{1}{\hbar^2}\int_{-\infty}^{\infty} dt' E_0^2(t-t')\sum_{\beta} P_{\beta\gamma}P_{\gamma\beta'}\rho^0_{\gamma\gamma}\left[\frac{R_{\beta\beta'}\left(t'-\frac{T^+_{\gamma\beta'}}{2}\right)}{i(\omega_0+\omega_{\gamma\beta})+\frac{1}{T_e}}+\frac{R_{\beta\beta'}\left(t'-\frac{T^-_{\gamma\beta}}{2}\right)}{-i(\omega_0+\omega_{\gamma\beta})+\frac{1}{T_e}}\right] \quad \text{(II.48)}$$

where $T^{\pm}_{\gamma\beta} = 1/\left[1/T_e \pm i(\omega_0 - \omega_{\gamma\beta})\right]$, and $P_{\gamma\beta}$ is the dipolar matrix element. The vibrational coordinate Q_γ is given by:

$$Q_\gamma(t) = \sum_{\gamma,\gamma'} \rho^2_{\gamma\gamma'}q_{\gamma'\gamma} \quad \text{(II.49)}$$

where $q_{\gamma\gamma'}$ are the vibrational matrix elements and an equivalent expression exists for Q_β. In the harmonic potential limit, Raman selection rules dictate that $\Delta\gamma = \pm 1$ and $\Delta\beta = \pm 1$. Assuming oscillators of mass m^γ and m^β with frequencies ω^γ and ω^β, the normal mode amplitudes in the two electronic states following excitation are given by:

$$Q_\gamma(t) = 2\,\mathrm{Re}\int_{-\infty}^{\infty} \frac{-dt'}{\left(2m^\gamma\omega^\gamma\hbar^2\right)^{1/2}} E_0^2(t-t')\sum_{\gamma,\beta}\sqrt{\gamma+1}P_{\gamma\beta}P_{\beta,\gamma+1}$$
$$\times\left[\frac{\rho^0_{\gamma\gamma}R_{\gamma,\gamma+1}\left(t'-T^+_{\gamma\beta}\right)}{i(\omega_0+\omega_{\gamma\beta})+\frac{1}{T_e}}+\frac{\rho^0_{\gamma+1,\gamma+1}R_{\gamma,\gamma+1}\left(t'-T^-_{\gamma+1,\beta}\right)}{-i(\omega_0+\omega_{\gamma+1,\beta})+\frac{1}{T_e}}\right] \quad \text{(II.50)}$$

$$Q_\beta(t) = 2\,\mathrm{Re}\int_{-\infty}^{\infty} \frac{-dt'}{\left(2m^\beta\omega^\beta\hbar^2\right)^{1/2}} E_0^2(t-t')\sum_{\gamma,\beta}\sqrt{\beta+1}P_{\beta\gamma}P_{\gamma,\beta+1}$$
$$\times\left[\frac{\rho^0_{\gamma\gamma}R_{\beta,\beta+1}\left(t'-T^+_{\gamma,\beta+1}\right)}{i(\omega_0+\omega_{\gamma,\beta+1})+\frac{1}{T_e}}+\frac{\rho^0_{\gamma,\gamma}R_{\beta,\beta+1}\left(t'-T^-_{\gamma\beta}\right)}{-i(\omega_0+\omega_{\gamma\beta})+\frac{1}{T_e}}\right] \quad \text{(II.51)}$$

Finally, we consider the interactions of these excitations with a probe beam E_S. Ignoring the coherent artifact, and assuming an electronic dephasing time T_e faster than the pulse duration, the polarization $P^3(t)$ radiated by the probe pulse is given by:

$$P^3(t)=\frac{i}{\hbar^3}\sum_{\substack{\gamma,\beta\\ \gamma',\beta'}} E_S(t-\tau-T^+_{\gamma\beta})e^{-i\omega_0(t-\tau)}\frac{P_{\gamma\beta'}P_{\beta'\gamma'}P_{\gamma'\beta}P_{\beta\gamma}}{i(\omega_0+\omega_{\gamma\beta})+\frac{1}{T_e}}$$

$$\times\int_{-\infty}^{\infty}dt'E_0^2(t-t')\left[\frac{\rho^0_{\gamma\gamma}R_{\gamma\gamma'}\left(t'-\frac{T^+_{\gamma\beta'}}{2}\right)}{i(\omega_0+\omega_{\gamma\beta'})+\frac{1}{T_e}}+\frac{\rho^0_{\gamma'\gamma'}R_{\gamma\gamma'}\left(t'-\frac{T^-_{\gamma'\beta'}}{2}\right)}{-i(\omega_0+\omega_{\gamma'\beta'})+\frac{1}{T_e}}\right]$$

$$\times\left[\frac{\rho^0_{\gamma'\gamma'}R_{\beta\beta'}\left(t'-\frac{T^+_{\gamma'\beta'}}{2}\right)}{i(\omega_0+\omega_{\gamma'\beta'})+\frac{1}{T_e}}+\frac{\rho^0_{\gamma'\gamma'}R_{\beta\beta'}\left(t'-\frac{T^-_{\gamma'\beta}}{2}\right)}{-i(\omega_0+\omega_{\gamma'\beta})+\frac{1}{T_e}}\right]$$

Impulsive Raman force imparts momentum to the oscillators at t=0, after which the vibrational displacement of the form $Q\propto e^{-\gamma t}\sin\omega_v t$ begins. On resonance, impulsive absorption into the excited electronic state produces a wavepacket which is already displaced away from the S_1 minimum, i.e., which already has the maximum vibrational displacement in S_1. In the excited state, the vibrational oscillations therefore take the form $e^{-\gamma t}\cos\omega_D t$, where the frequency ω_D and damping rate γ may be different from those of the ground state S_0. On resonance, ISRS also yields some displacement on the *ground* state PES, in contrast to the nonresonant case. This occurs because the wavepacket may be promoted into a steeply sloped part of the excited state PES, and in this case measurable displacement may occur even within the pulse duration or the electronic dephasing time T_e, during which time the wavepacket must return to the ground electronic state to complete the resonance-enhanced ISRS excitation process. Thus the vibrational phase in the ground state PES are projected back down to the ground state PES. It is this same motion along the excited state PES within the electronic dephasing time that is responsible for the resonance enhancement of conventional Raman scattering and for the appearance of vibrational overtones in the resonance-enhanced Raman spectrum. Interestingly, if the resonant ISRS excitation pulse is extremely short in duration-shorter than the electronic dephasing time-then these effects are reduced, since the excitation process is completed during the pulse duration and the wavepacket is not given time to accelerate fully on the excited-state PES.

III. REVIEW OF IMPULSIVE SPECTROSCOPY EXPERIMENTS

III. A. Introduction

This review will focus primarily on impulsive absorption and ISRS experiments in which coherent vibrational oscillations are observed in condensed phase materials. The impulsive techniques introduced in section II will be shown to be powerful tools for the excitation and detection of a host of different responses in a broad range of materials and phases. The first major subsection will illustrate the diverse applications of nonresonant impulsive stimulated scattering techniques. Results which demonstrate the feasibility of studying coherent optic phonons, acoustic phonons, and molecular in solids and liquids ranging from metal films to protein solutions will be presented. The advantages of time domain techniques over frequency domain techniques in some investigations will be emphasized. The next section will present a similar review of results of resonant impulsive excitation of molecular systems including biological molecules. In these cases, absorption into electronic excited states will play important roles, and in some examples the main information of interest concerning dynamics in the excited electronic state. Finally, impulsive excitation of opaque materials is reviewed. While the last two sections focused on impulsive absorption, the case of collective electronic excitations in opaque materials is sufficiently distinct as to warrant a separate discussion.

III. B. Phonons, vibrations, and rotations across phases and materials

1. Optic Phonons and Collective Structural Rearrangements. Time-resolved observations of quantum beats, i.e. of oscillations in wavefunctions described by coherent a superposition of eigenstates, have been conducted since as early as 1964 [19]. The eigenstates, e.g. $|0\rangle$ and $|1\rangle$, have included magnetic and electronic levels as well as molecular rotations and vibrations and lattice vibrations. Even in the case of vibrations, the focus of this review, the first quantum beat measurements were made on coherent states involving eigenstates of different vibrational modes which had similar frequencies ω_{v1} and ω_{v2}. In this case the time-resolved measurements revealed oscillations at the difference frequency ω_{v1} -ω_{v2}, not at either fundamental frequency. Only impulsive excitation, on a time scale short compared to the vibrational period, produces coherent states involving different eigenstates *of the same vibrational mode.* In this case fundamental vibrational oscillations are observed. Time-resolved measurements are conducted on vibrationally distorted molecules or crystal lattices, a bit like "stop-action" photos recorded at various stages of vibrational distortion.

Early experimental observations of coherent acoustic and optic phonons have been reviewed previously, so a comprehensive review will not be presented here. The first two impulsive stimulated Raman scattering observations of optic phonons were conducted on a nonpolar organic molecular crystal, α-perylene [20], and a polar inorganic ferroelectric crystal, lithium tantalate ($LiTaO_3$) [21-23]. In the former case, a femtosecond pulse was split by a partial reflector and the two resulting excitation pulses were overlapped spatially and temporally inside the sample. The parallel-polarized excitation pulses formed an optical interference or “grating” pattern as discussed earlier and illustrated in Figure 2. Coherent scattering of a variably delayed probe pulse incident at the Bragg angle for diffraction was measured, revealing time-dependent oscillations of two optic phonon modes at 80 cm^{-1} and 104 cm^{-1} energies.

The "transient grating" or four-wave mixing geometry has since been used to record ISRS data from coherent optic phonons in many crystals. Figure 8 shows data recorded

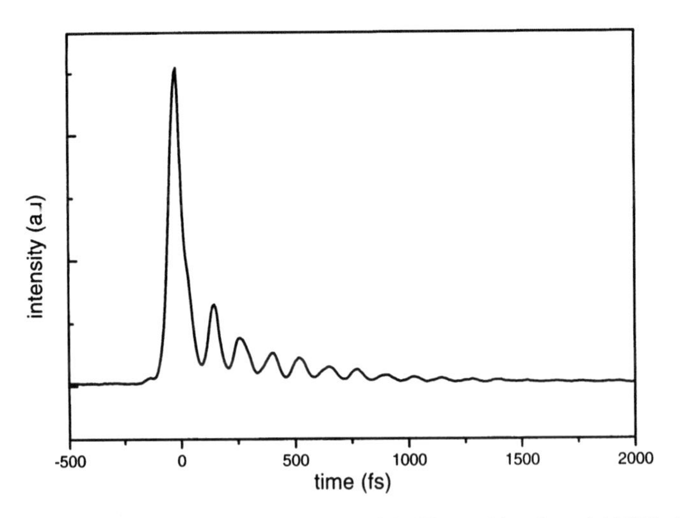

Figure 8. ISRS data from $LiTaO_3$ showing oscillations at 8.8 THz resulting from 4.4 THZ phonon-polariton vibrations in the material. The signal depends on the square of the material response and this gives rise to the doubling of the vibrational frequency.

recently from $LiTaO_3$. Data from this and related crystalline materials will be discussed further below. For now we mention that the vibrational motion involves primarily oscillations of the Ta ion within a surrounding "cage" of oxygen ions. Of some note is the experimental geometry used to record the data shown in Figure 5. Rather than a beamsplitter, a diffractive phase mask pattern (a series of parallel lines etched into a glass substrate) was used to separate the two excitation pulses as shown in Figure 5. The ±1 orders of diffraction were used for the two excitation pulses, which were passed through a simple two-lens imaging system and crossed in the sample. This setup facilitates the generation and overlap of the excitation pulses, and makes it simple to change the scattering wavevector simply by introducing a new mask pattern with a different fringe spacing into the beam path. Of further note is the use of the phase mask for the probe as well as excitation pulses. As shown in the schematic illustration, the probe beam is also passed through the mask pattern to form ±1 orders of diffraction. One of these plays the role of the probe beam which arrives at the sample at the Bragg angle for diffraction and whose coherently scattered (i.e. diffracted) light yields the ISRS signal. If the other beam is fully blocked, then this measurement is as described in section II. If the other beam is only partially blocked, then the part that reaches and passes through the sample is precisely collinear with the ISRS signal, and plays the role of a "reference" beam I_R whose field adds coherently that of the diffracted signal field I_D. In this heterodyne detection scheme, the total light intensity at the sample is $S \propto I_R + \sqrt{I_R I_D} + I_D$. The reference beam amplitude is adjusted to be larger than that of the diffracted signal, and in this case the cross-term $\sqrt{I_S I_D}$ produces the dominant contribution to signal intensity. This can improve the experimental signal and signal/noise levels considerably, and in many cases can simplify the ISRS data analysis since now the signal varies linearly rather than quadratically with the vibrational response as discussed briefly in the theory section. Heterodyned ISRS signal collected from $LiTaO_3$ is shown in Figure 9.

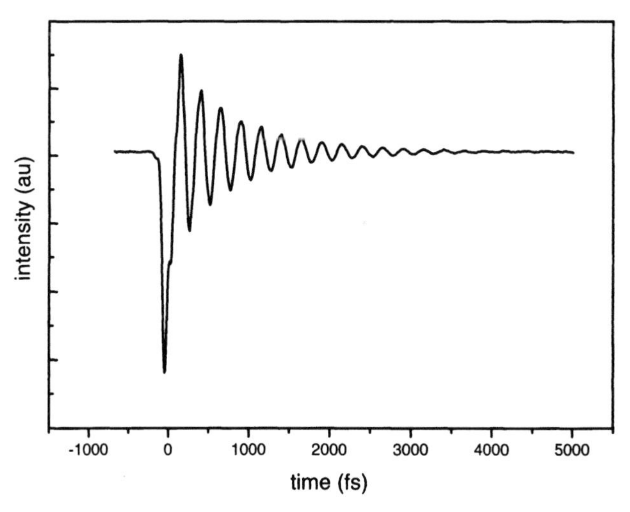

Figure 9. Heterodyne ISRS data taken from $LiTaO_3$ showing oscillations at 4.4 THz resulting from phonon-polariton oscillations in the material. The mixing of signal and reference fields yields a total light intensity which depends linearly on the material response.

The second early ISRS measurement of optic phonons [21-24] was conducted on $LiTaO_3$, but not with a grating excitation geometry. Rather, a single excitation pulse was used. The coherent vibrational response was detected through measurement of time-dependent depolarization of variably delayed probe pulses. Although there are not two crossed excitation pulses to define a scattering wavevector, this measurement is more similar to the transient grating experiment than may be apparent. As in any ISRS excitation process, stimulated Raman scattering takes place through mixing among the continuous spectrum of frequency components contained within the bandwidth of the ultrashort excitation pulse(s). With a single excitation pulse, the stimulated Raman scattering also involves mixing among the continuous distribution of *wavevector* components contained in the excitation beam. For a well collimated beam, all the wavevector components are in the same direction, but there are different wavevector magnitudes $k = \omega n/c$ (with n the refractive index) corresponding to the different frequency components. Stimulated scattering occurs in the forward direction, with higher-frequency components scattered into lower-frequency components (Stokes scattering) while the coherent vibrational response is generated. The process can be envisioned in the time domain. The excitation pulse reaches the front of the sample first, and coherent vibrational oscillations begin there. Later the pulse reaches the middle and then the back of the sample, so vibrational oscillations are initiated later in these regions. The distance that the pulse travels inside the sample during one vibrational oscillation period is the vibrational wavelength $\Lambda = (c/n)(2\pi/\omega_{vib}) = 2\pi/q$, so the vibrational (i.e. scattering) wavevector is in the forward direction and has magnitude is $q = n\omega_{vib}/c$. This can also be written in the form $\omega_{vib}/q = c/n = \omega/k$, relating the vibrational and optical frequency/wavevector ratios. Finally, if the single excitation beam is focused tightly, then the wavevector components are not all in the same direction and the scattering wavevectors include components perpendicular to the forward (light propagation) direction.

The early ISRS experiments on $LiTaO_3$ illustrated an important property of polar lattice vibrations in crystals of this sort, namely their coupling to electromagnetic radiation in the same (Terahertz) frequency and wavevector range to form mixed *phonon-polariton* modes. Since the modes are polar, they are IR as well as Raman-active. For example, when far-IR (THz frequency) radiation enters the crystal it drives the lattice vibrational response, and a mixed electromagnetic - polar vibrational (phonon-polariton) excitation propagates through the crystal. In our case, the mixed mode is not excited by far-IR radiation but by ultrashort optical pulses, through ISRS. The phonon-polariton response propagates away from the excitation region and into neighboring regions of the crystal at the speed of far-IR radiation in the crystal, i.e. a significant fraction of the speed of light in air. In the early experiments, a single excitation pulse was tightly focused into $LiTaO_3$ and the probe pulse was variably delayed in time *and position* away from the arrival time and position of the excitation pulse. As the spatial separation between the excitation and probe pulses was increased, the time of arrival of the phonon-polariton excitation at the probe spot increased. Similar experiments on phonon-polariton propagation had been conducted earlier using stimulated Raman scattering excitation with discrete frequency components ω_1 and ω_2. In this case, individual vibrational oscillations are not observed but phonon-polariton propagation and attenuation can be monitored.

In Figure 10 we show more recent ISRS data from $LiTaO_3$. In this case the probe beam is not tightly focused to a single spot separated by a specified distance from the excitation region, but instead is collimated and passed through a large volume of the crystal that includes the excitation region and all nearby regions into which the phonon-polaritons

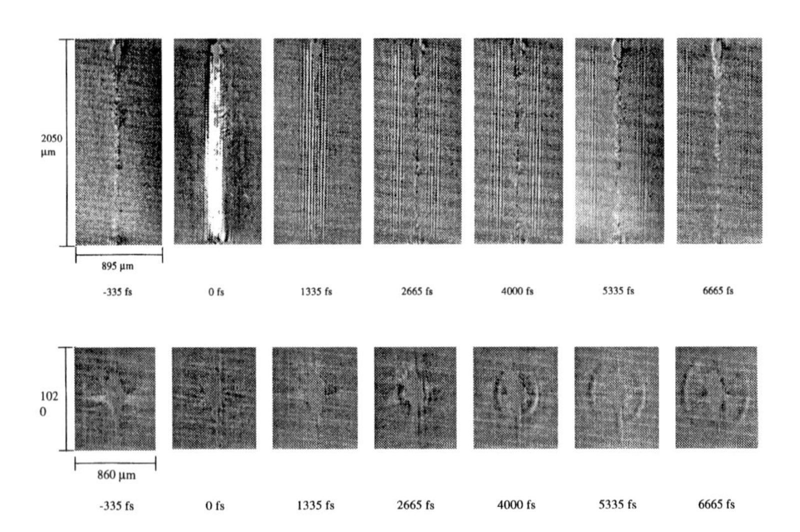

Figure 10. Images of propagating polaritons excited with two crossed pulses (a) and a single excitation pulse (b) through impulsive stimulated Raman scattering. The propagation speed is roughly one-fourth the speed of light in air. (Adapted from ref. [25].)

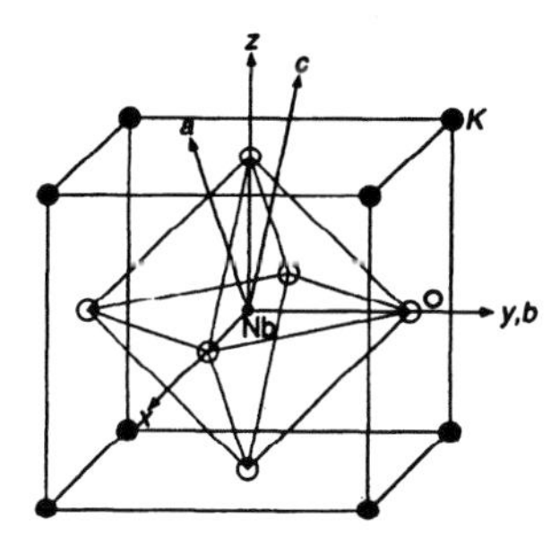

Figure 11. The perovskite structure of paraelectric $KNbO_3$. (Reprinted from ref [26].)

might propagate. The transmitted probe light is then imaged onto a CCD (a multielement photodetector). As is evident from the figure, this permits the entire position-dependent phonon-polariton response to be imaged at each probe delay time. In this manner the time and position-dependent phonon-polariton evolution can be monitored. The images in Figure 10(a) were recorded following excitation with crossed pulses which formed a "grating" pattern. The phonon-polariton response therefore consists of two counterpropagating, spatially periodic "wavepackets". Excitation with a single focused pulse produces a "ripple" response which emanates outward from the excitation spot, as seen in Figure 10(b).

Among the most important applications of ISRS to crystalline solids is in the study of collective structural rearrangements such as structural phase transitions or domain switching. This is possible because in many such rearrangements, one or a small number of lattice vibrational modes plays the role of "collective reaction coordinate" along which molecules or ions in the crystal move from their initial lattice positions toward the positions they occupy in an incipient crystalline phase or domain orientation. In many crystals, motion along such so-called "soft" vibrational modes becomes strongly damped, especially near a phase transition temperature. An excellent example is provided by potassium niobate and related ferroelectric crystals of the perovskite structure illustrated in Figure 11. ISRS data from $KNbO_3$ in its tetragonal ferroelectric phase are shown in Figure 12. Despite the strong damping, the data permit accurate determination of the phonon frequency and dephasing rate. In contrast, the conventional Raman spectrum shows broad Stokes and anti-Stokes features merged with a strong central peak, making reliable characterization of the vibrational mode impossible at many temperatures. Data like that shown in Figure 12 were crucial in showing that motion of the Nb ions among different off-center sites is vibrational in character, as opposed to hopping motion that would prevail if the different local sites were separated by high potential energy barriers. The results supported a partially displacive (as opposed to purely order-disorder) model of the phase transitions in potassium niobate and related materials.

2. Molecular and Intermolecular Vibrations. ISRS excitation of coherent molecular vibrations has also been conducted extensively. The possibilities were first demonstrated on liquid dibromomethane [27, 28]. Coherent vibrational oscillations of the Br-C-Br bending mode of CH_2Br_2 were observed using various excitation geometries and detection

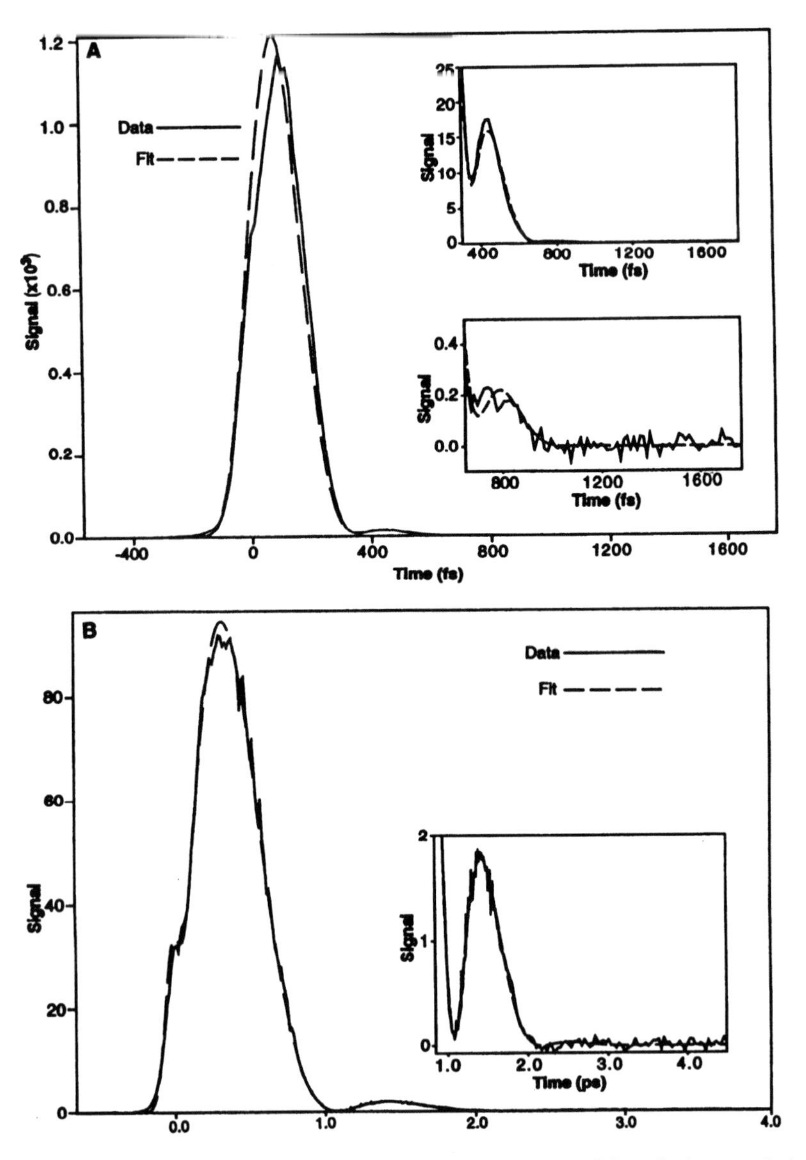

Figure 12(A). B_2 symmetry ISRS data from $KNbO_3$ showing the dynamics of the soft phonon-polariton mode at 442.4 K The response is fit (dashed line) to that of a damped harmonic oscillator with frequency ω_0 =9.00 THz and damping rate γ=6.14 THz. (B) E symmetry phonon-polariton ISRS data, generated by crossing a vertically and a horizontally polarized pump pulse, from $BaTiO_3$ at 321.6 K which is fit (dashed line) as a damped oscillator with ω_0 =2.9 THz and γ=1.8 THz. In both cases oscillatory, not relaxational (i.e., "hopping") dynamics are clearly observed. (Adapted from ref [26].)

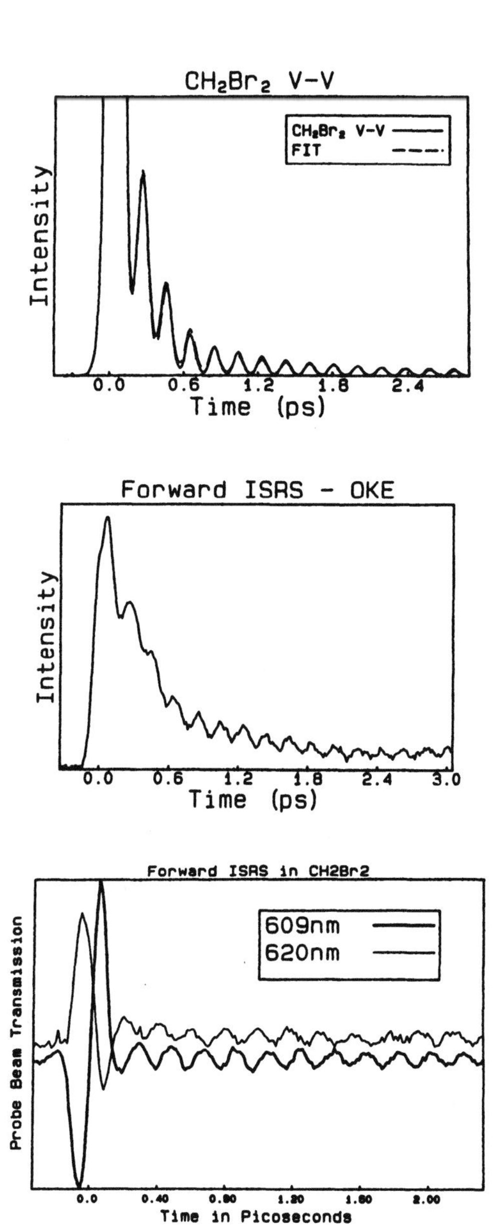

Figure 13(a). ISRS data from CH_2Br_2. The 173 cm^{-1} halogen bending mode was excited with two crossed, vertically polarized excitation pulses and probed, via transient diffraction, with a vertically polarized pulse. (b) ISRS data from CH_2Br_2 excited with a single vertically polarized excitation pulse and probed, via the optical Kerr effect with a 45° polarized pulse. Transmitted signal intensity polarized at -45° was measured. (c) ISRS data from CH_2Br_2 excited with a single vertically polarized pulse and probed via the spectral shifts of a vertically polarized probe pulse. The spectrum of the transmitted probe pulse is alternately red- and blue-shifted at the CH_2Br_2 molecular frequency. (Reprinted from ref. [27]. Copyright 1987 Editions de Physique.)

schemes. The transient grating geometry used for optic phonon ISRS experiments was also used for molecular vibrations, yielding the data shown in Figure 13(a). A single excitation pulse was also used with two different probing methods. In the first, yielding the data shown in Figure 13(b), a nearly collinear probe pulse polarized at 45° relative to the excitation pulse polarization was used, and transmitted probe light whose polarization was rotated to -45° was detected. This measurement of time-dependent depolarization or optical Kerr effect shows vibrational oscillations because the 0°-polarized excitation pulse produces vibrationally excited molecules with a preferred orientation, based on the anisotropy of the Raman-scattering cross-section. The molecular vibrations then induce refractive index changes of different magnitudes for light polarizations at 0° and 90°, giving rise to birefringence for probe light polarized at 45°. The other probing method also used a probe pulse that was nearly collinear with the excitation pulse. In this case time-dependent spectral shifts were measured by passing the transmitted probe pulse through an interference filter centered either on the red side (620 nm) or the blue side (609 nm) of the spectrum of the incident probe pulse. See Figure 13(c). The influence of the coherent vibrations on the probe pulse spectrum can be understood through consideration of the impulsive force exerted by the probe pulse which either amplifies or opposes the coherent vibrational response induced by the excitation pulse depending on the probe pulse delay time, as discussed in section II. Note that in this measurement the largest spectral shifts, and therefore the largest signals, occur when the vibrational velocity is highest and the vibrational displacement is zero, in contrast to the other measurements described in which the largest signals arise when the vibrational displacements are largest. The spectral shifts in this experiment are due to the time-derivative of the refractive index rather than to variations in the refractive index itself.

ISRS measurements with single excitation pulse illustrate the fact that a sufficiently short pulse, when it enters a Raman-active sample of any kind, unavoidably undergoes impulsive stimulated Raman scattering. ISRS is thus a ubiquitous phenomenon, and indeed its effects have now been observed in dozens of materials.

Intermolecular librations and other motions can also be studied in the liquid phase. Carbon disulphide provides a model system in which polarized excitation pulses exert an impulsive torque that drives molecular orientational motion toward molecular alignment with the polarization direction [27, 29-31]. This is impulsive stimulated rotational Raman scattering, which in the gas phase induces free molecular rotation [32] but in the liquid phase induces intermolecular librations. As with molecular vibrations, ISRS data from intermolecular motions has been recorded in transient grating geometries and also through measurement of induced birefringence and other observables. The most important result of note is the clearly vibrational character of the motion, revealed by the inertial lag of signal which continues to increase (showing that molecules continue to move away from their initial, local equilibrium configurations) after the excitation pulses have left the sample. In some liquids, as in CS_2 at low temperatures or elevated pressures, underdamped intermolecular vibration also is apparent at later times, when a weak recurrence of signal is observed. In contrast, if a diffusional model were sufficient to describe molecular rotation in the liquid, then signal would decrease monotonically as soon as the sudden driving force were finished. As in the case of optic phonons in potassium niobate, the conventional Raman spectrum shows a broad central feature which is difficult to analyze uniquely. Subsequent experiments have been conducted on molecular liquids in which multiple excitation pulses are used to manipulate the coherent motion and reveal the origin of the rapid decay of signal [33]. This class of experiment will be discussed briefly in the section on optical control and multiple-pulse femtosecond spectroscopy.

3. Acoustic Phonons. Although the main focus in this article is on femtosecond time-resolved measurements, "impulsive" measurements more generally began with acoustic phonons, which can be observed on slower time scales. A transient grating arrangement is usually used since this permits the acoustic wavevector q to be well specified. The acoustic frequency q is then determined by the sound velocity v, i.e. $\omega/q = v$. The excitation mechanism may be impulsive stimulated Brillouin scattering (ISBS), which is precisely analogous to ISRS with the differential polarizability expressed as a photoelastic constant which gives the change in refractive index as a function of acoustic strain. Through ISBS, an impulsive stress is exerted at the grating maxima. If there is optical absorption of the excitation pulses, this can give rise to sudden heating of the sample in a grating pattern. In this case, thermal expansion launches acoustic waves. This measurement has been called "impulsive stimulated thermal scattering" (ISTS) for notational consistency, but the excitation mechanism is really very different from stimulated light scattering. Through the ISBS mechanism, longitudinal, shear, or mixed-polarization acoustic waves can be generated in transparent samples. If there is even modest light absorption, the ISTS excitation mechanism usually dominates in generation of longitudinal or mixed polarization acoustic waves.

ISBS experiments on shear acoustic waves in crystalline solids and in viscoelastic fluids have been examined in fundamental studies of structural phase transitions [34](in which the structures of the two phases are interchanged through an acoustic shear, i.e. the "collective reaction coordinate" is a soft shear acoustic mode) and liquid-glass transitions [35], in which the onset of solid-like behavior as the temperature is reduced is marked by the onset of a propagating, albeit heavily damped, shear acoustic mode. Experiments of this type of been reviewed previously [13]. In each case, the capability for time-domain observation of low-frequency, heavily damped modes has yielded information previously inaccessible through frequency-domain light scattering spectroscopy which shows Stokes and anti-Stokes features merged around a central peak.

ISBS and ISTS experiments have been conducted on crystals near structural phase transitions and on glass-forming liquids for reasons similar to those described above [37-39]. ISTS produces sufficiently strong signal levels that it is a practical method for examination of acoustic waves in submicron films as well as bulk materials [40]. In this case, acoustic waveguide modes (similar to familiar "drumhead" acoustic modes, or to planar optical waveguide modes) are excited, and the probe light is diffracted in reflection mode by modulation (i.e. "ripple") of the film surface, as it would be from an ordinary grating. ISTS measurements on thin films permit determination of thin film elastic and loss moduli, thermal diffusivities, and thickness, as well as detection of film delamination from substrates. The measurements produce sufficiently high signal/noise levels that a

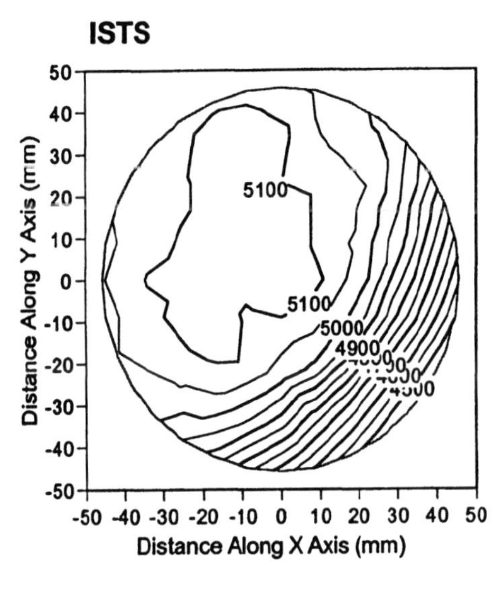

thickness in Angstroms

Figure 14. Contour map showing the thickness variations across a nominal 5000-Å tungsten layer on a silicon wafer as determined by measuring surface acoustic wave frequencies generated and detected via ISTS. (Reprinted from ref [36]. Copyright 1998 American Institute of Physics.)

commercial ISTS instrument is now in use in the microelectronics industry! The primary use is measurement of metal film thickness, which can be deduced from the acoustic frequency with angstrom repeatability in a measurement time of about one second. Figure 14 shows a contour map of a tungsten film on a silicon substrate produced in about one minute with this instrument [36]. So we see that time-domain spectroscopy at the impulsive limit has matured sufficiently to be a practical commercial method!

The transient grating ISTS measurement on thin films can also be conducted with femtosecond or several-picosecond time resolution. In this case the acoustic waveguide modes, which propagate in the plane of the film, are still excited and probed, but these only require subnanosecond time resolution. The additional acoustic response induced through the use of shorter pulses propagates through the film plane [41]. This response is due to sudden thermal expansion from the heated film surface at each grating peak into the film. For submicron film thicknesses, the excitation pulse duration must be substantially shorter than the acoustic propagation time through the film, which for a 100-nm film is on the order of 20 ps. The acoustic responses give rise to surface ripple each time a partial reflection from an interface returns to the surface, and this is observed in the diffracted signal intensity. Figure 15 shows data recorded from a Al/TiW/Si multilayer structure. The partial reflections can be used to determine the film thicknesses and other properties, and the reliability of the values can be ensured through simultaneous measurement of the acoustic waveguide response that propagates in the plane of the film.

Note that the through-plane acoustic response does not have a well defined wavevector in its propagation direction but rather is a wavepacket with many wavevector components. This type of response can be generated with only a single excitation pulse. In this case

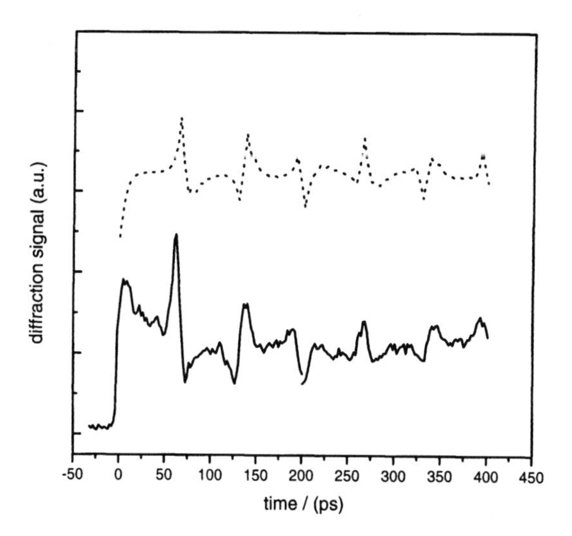

Figure 15. ISTS data (solid) and simulation (dashed) measuring through-plane acoustic pulse dynamics on a picosecond time-scale in an Al/TiW multi-layer film structure on a Si (110) substrate. (Adapted from ref [41].)

there is no diffracted probe beam and no acoustic waveguide response propagating in the film plane, but the through-plane acoustic can be detected upon return of partial acoustic reflections to the film surface through induced changes in probe beam reflectivity or deflection. Measurements of this type have been made on a variety of materials to assess film thicknesses, elastic moduli, and other properties [42, 43]. A commercial system of this type is also available - further proof that time-resolved spectroscopy at the "impulsive" limit is fully ready for real-world use!

III. C. Resonant molecular phenomena

All the results presented so far except for ISTS excitation of acoustic waves have involved nonresonant excitation through impulsive stimulated scattering. Even in ISTS, the only role of absorption is to provide a mechanism for rapid sample heating. More generally, absorption into electronic excited states opens up new mechanisms for impulsive excitation since ISRS may be resonance enhanced (allowing excitation of otherwise Raman-inactive vibrational modes in the ground state and increasing the coherent amplitude of vibration for Raman-active modes) and, in addition, coherent vibrations may be initiated in the electronic excited state. See Figure 6. In general, we will refer to initiation of excited-state molecular vibrational coherences (describable quantum mechanically in terms of a coherent superposition of vibrational levels in the excited electronic state) as "impulsive absorption" (IA) and ground-state vibrational coherences (describable quantum mechanically in terms of a coherent superposition of vibrational levels in the excited ground state), even with resonance enhancement, as ISRS. However, this distinction is less clear for cases (generally in gas-phase molecules) in which the electronic ground and excited states remain phase-coherently coupled well after the excitation process. When electronic phase-coherence persists over a time scale that is significant compared to that of coherent vibrational motion, then the system must be described in terms of a coherent superposition of vibrational and electronic levels; the system is not in either the ground or excited electronic eigenstate. Even in this case, however, the simplification into separate ISRS and IA excitation mechanisms may be useful if the probing process projects out the coherent vibrational motion in just one of the electronic eigenstates. This is common since in gas-phase measurements the probe wavelength is usually absorbed selectively by one of the electronic states.

An illustration of some of these issues is provided by results from organic dyes in liquid solution [44-46]. This example is relatively simple because in liquid solution, electronic dephasing is nearly instantaneous even on the time scale of the femtosecond pulse duration, and so after excitation each molecule can considered to be in either the ground or excited electronic eigenstate. Those remaining in the ground electronic state undergo coherent vibrational oscillations due to (resonance-enhanced) ISRS. Those in the electronic excited state undergo coherent vibrational oscillations due to impulsive absorption. For the former, we have seen that the coherent vibrational amplitude increases linearly with excitation pulse intensity since the ISRS driving force increases similarly. For the latter, the coherent vibrational amplitude in the excited state is independent of excitation intensity, at least for moderate intensities that do not result in excitation into still higher electronic levels. The excited molecule has its equilibrium geometry displaced from that of the ground electronic state along any particular vibrational coordinate by some distance, as shown in Figure 6, and that displacement determines the vibrational amplitude in the excited state. Higher excitation intensities result in more molecules in the electronic excited state, but each electronically excited molecule undergoes coherent vibrational oscillations of the same amplitude that would obtain from lower excitation intensities. A second difference lies in the vibrational phases of coherent oscillations in the ground or excited electronic states. ISRS excitation imparts momentum to molecules in the ground electronic state, leading to sinusoidal vibrational oscillations. Impulsive absorption leaves molecules initially displaced from equilibrium geometries in the electronic excited state, leading to cosinusoidal oscillations.

Many observations of coherent molecular vibrations in electronically excited molecules have been conducted, starting with the initial measurements [44-46]of transient absorption in malachite green and Nile blue dyes in solution. A particularly illustrative example is provided by experiments on malachite green in which resonant (620-nm) excitation and probe wavelengths were used and measurements were made of time-dependent transient birefringence and transient dichroism (polarization dependent absorption) [47]. See Figure 16. The former probes the real part of the time-dependent refractive index and the impulse response function, while the latter probes the imaginary parts. Both measurements show oscillations with the same period of ~150 fsec but with different vibrational phases and dephasing rates. In order to explain these observations, it is useful to consider the spectra of the imaginary part of the refractive index (i.e. the absorption spectrum) and the real part of the refractive index (the dispersion) near an absorption peak, as shown in Figure 17. The coherent vibrational oscillations cause small time-dependent shifts in the absorption and dispersion spectra. If the probe wavelength is centered at the absorption maximum, then its absorption is insensitive to small spectral

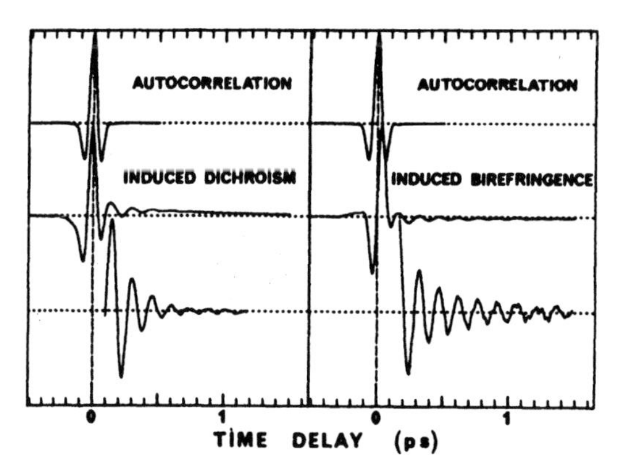

Figure 16. Resonant ISRS signal from malachite green dye in water. The response is probed through the induced dichroism and induced birefringence of the solution. The former monitors molecular vibrations in the excited electronic states and the latter in the ground electronic states. (Reprinted from ref [47]. Copyright 1988 American Physical Society.)

shifts, but the real part of the refractive index changes substantially since it is sharply sloped at this wavelength. Thus vibrational oscillations of molecules in the ground electronic state have a strong influence on the birefringence of a resonant probe, but little influence on the dichroism. For molecules in the electronic excited state, the probe wavelength can induce stimulated emission back to the electronic ground state. However, the stimulated emission spectrum, like the fluorescence spectrum, is red-shifted relative to the absorption spectrum. Thus the probe wavelength that matches the absorption peak lies to the side of the stimulated emission peak, and lies at around a maximum of the dispersion spectrum for excited-state molecules. In this case, vibrational oscillations of the electronically excited molecules have a greater effect on the dichroism than on the birefringence. Therefore the oscillations observed experimentally in the induced dichroism result from vibrations of molecules in the electronic excited state, while the oscillations in induced birefringence reveal the vibrational oscillations of molecules in the ground electronic state. The similar vibrational frequencies show that the malachite green excited state potential energy surface with respect to the vibrational coordinate probed is nearly identical to that of the ground state, although vibrational dephasing is faster in the excited state.

Molecular vibrations have been observed in all phases including crystalline solids and gases. Vibrational motion in iodine molecule has been observed in gas [48], liquid [49, 50], and solid phases [51], and is particularly interesting since photodissociation can occur from the excited state depending on the excitation wavelength and the surrounding environment. A striking example is provided by experiments on I_2 in crystalline argon at low temperatures [51]. Typical data are shown in Fig. 19. In this case, absorption into a dissociative state leads to stretching of the iodine bond, but complete separation of the two atoms is resisted by the surrounding lattice. This produces coherent vibrational oscillations of the molecule and also of the lattice. The experimental results are reproduced well by molecular dynamics simulations.

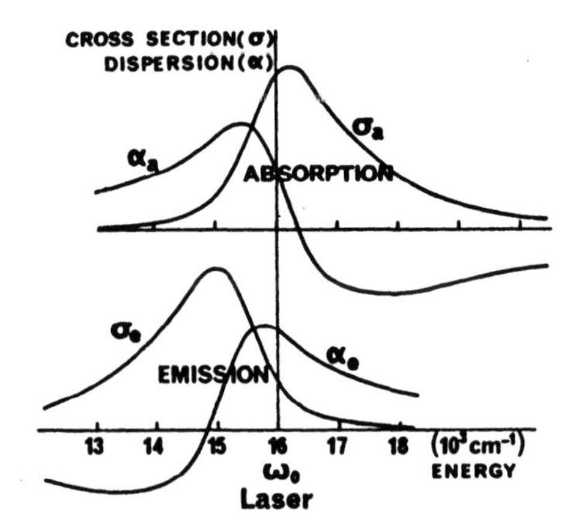

Figure 17. Absorption and emission spectra with corresponding dispersion spectra, of malachite green in water. The laser frequency ω_0 is near the maximum of the absorption cross section where $d\sigma_a / d\omega$ is small, but on a steep section of the dispersion α_a where $d\alpha_a / d\omega$ is large. The opposite situation exists in the emission spectrum. (Reprinted from ref [47]. Copyright 1988 American Physical Society.)

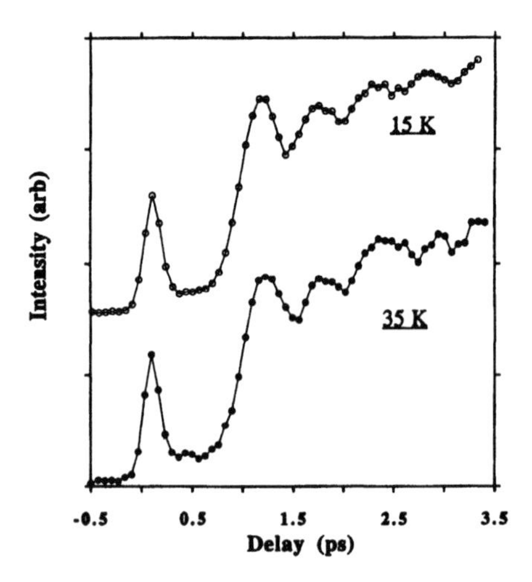

Figure 18. Impulsive absorption data from I_2 molecules in a solid Kr matrix. Following photodissociation which produces the sharp rise in the data, atoms recoil from collisions with the Kr host matrix cage and recombine resulting in coherent vibrations of the nascent molecule. (Reprinted from ref [51]. Copyright 1994 American Institute of Physics.)

1. Impulsive Biology. Coherent vibrational spectroscopy of several photosensitive biological systems, in particular those involved in photosynthesis and vision, has been carried out in the impulsive limit. Coherent motion of molecules may play key roles in these cases. Work on vibrational coherences in photosynthetic reaction centers continues apace following their first observations [52]. Here we review a comprehensive set of experiments on the role of coherent motion in the first step of vision in bacteriorhodopsin [53].

Following photoexcitation, 11-*cis* Rhodopsin isomerizes to an all-trans photoproduct with a high quantum yield of 0.67. Figure 19 illustrates the energetic pathway along which this isomerization is believed to proceed. Transient changes in absorption of probe wavelengths ranging from 450 to 570 nm were observed following impulsive absorption with a 35-fs, 500-nm excitation pulse as seen in Figure 20. The results show the appearance of photoproduct after 200 fs, consistent with previous studies. Following this, coherent oscillations with a ~550 fsec period are observed. The primary mechanism responsible for these oscillations is either oscillations in the ground state of the 11-*cis* Rhodopsin or in the ground state of the photoproduct (Excited state oscillations are unlikely because of the 100 fs lifetime of the S_1 excited state.) The dependence of the phase of the oscillations on the probe wavelength is plotted in Figure 21. A 180° phase shift is seen to occur at 550 nm. This wavelength corresponds to the energy difference between the minimum of the photoproduct ground state energy and the excited state energy. A 180° phase shift is

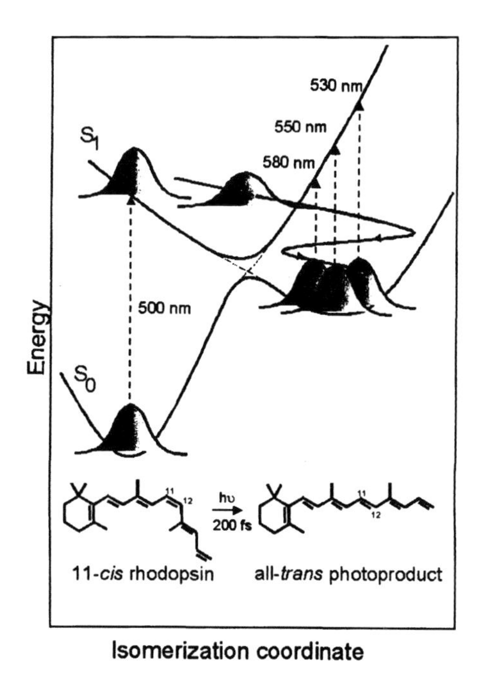

Figure 19. Schematic illustration of the potential energy surfaces for the isomerization of rhodopsin following impulsive excitation from the ground state S_0 to the excited state S_1. The wavepacket motion in the photoproduct potential depicts the effect of coherent vibrational motion on the frequency dependence of the photoproduct absorption. (Adapted from ref [53].)

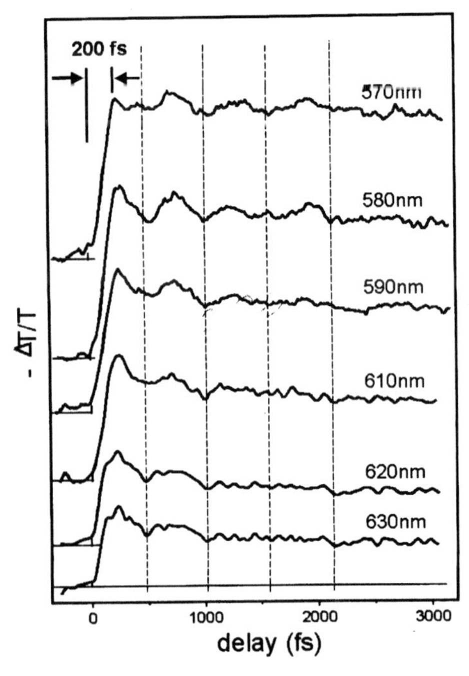

Figure 20. Transient absorption data at several probe frequencies from photo-isomerized rhodopsin following excitation at 500nm. Oscillations in the data are ascribed to coherent vibrational motion of the isomerized photoproduct. (Reprinted from ref [53]. Copyright 1994 American Association for the Advancement of Science.)

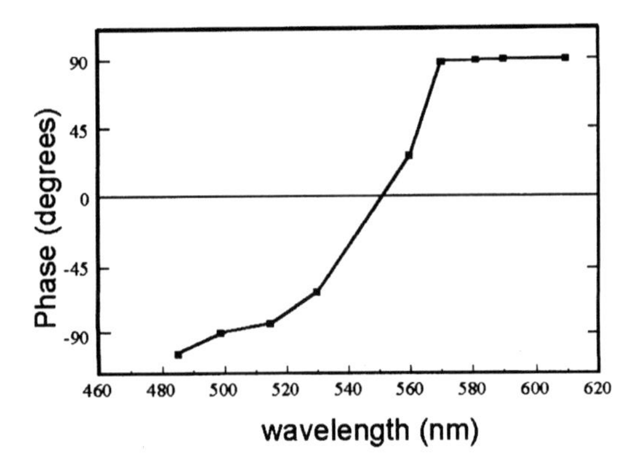

Figure 21. The phases of the oscillations shown in Figure 20 plotted as a function of frequency. (Reprinted from ref [53]. Copyright 1994 American Association for the Advancement of Science.)

expected at this frequency if the coherent motion occurs in the photoproduct as described above. A 180° phase shift at 500 nm would be expected if the oscillations were from coherent motion in the 11-*cis* Rhodopsin. It is thus concluded that photoisomerization, the primary early event of vision, is a coherent process, leaving the product vibrating coherently. This is believed to be partly responsible for the high quantum yield, since coherent motion leaves little chance for relaxation back to the initial state.

Before leaving this brief discussion of photobiology, it is worth mentioning one case in which biological photochemistry was observed to give rise to coherent *acoustic* waves [54]. Photoexcitation of carboxymyoglobin (MbCO) with a transient grating experimental geometry resulted in generation of acoustic waves at the grating wavevector. Unlike the usual case of ISTS or ISBS excitation, in which the acoustic response grows in during the first oscillation period, MbCO excitation appears to lead to an extremely rapid volume change. The results are interesting because light absorption in MbCO leads to photodissociation of CO from iron in a heme group within the protein. This is a highly localized process, yet it apparently leads to shock-like global expansion of the protein on a fast time scale, giving rise to a sudden acoustic response in the surrounding medium.

III. D. Coherent phonon generation in opaque materials

Coherent optic phonons can also be detected in opaque materials, such as metals and semiconductors. The primary excitation mechanism in most cases is impulsive absorption, which has been labeled displacive excitation of coherent phonons (DECP) by some authors [55, 56].

Impulsive absorption into band electronic excited states has some consequences that differ from those of absorption into localized excited states like those of molecules. Phonon excitation can be viewed as resulting from promotion of valence electrons into the conduction band. The removal of valence electrons results in reduced screening of the positively charged nuclei from each other. This gives rise to sudden repulsions that initiate coherent vibrational oscillations, especially in "breathing" modes of the unit cell. In this case, unlike the molecular case, an increase in excitation intensity leads to an increase in the internuclear repulsions and in the coherent vibrational amplitude.

Observations of this type were made through measurements of time-resolved changes in reflectivity in bismuth, antimony [55], tellurium, and titanium oxide [56, 57]. In most cases and at modest excitation intensities, the phonon frequencies matched those observed in Raman spectra. The case of Ti_2O_3 is particularly interesting, however, because of its unusually large signal levels (over 10% change in reflectivity, as opposed to the usual small fraction of a percent) and changes in frequency observed at high excitation intensities. See Figure 22.

Ti_2O_3 undergoes a semiconductor to semimetal phase transition in the T=300-600K temperature range. From x-ray diffraction measurements, it is known that the distance between nonequivalent Ti ions in the lattice changes by three percent upon transformation to the semimetal phase. It was estimated from a comparison of temperature-dependent x-ray scattering and reflectivity measurements that the phonon amplitudes being driven

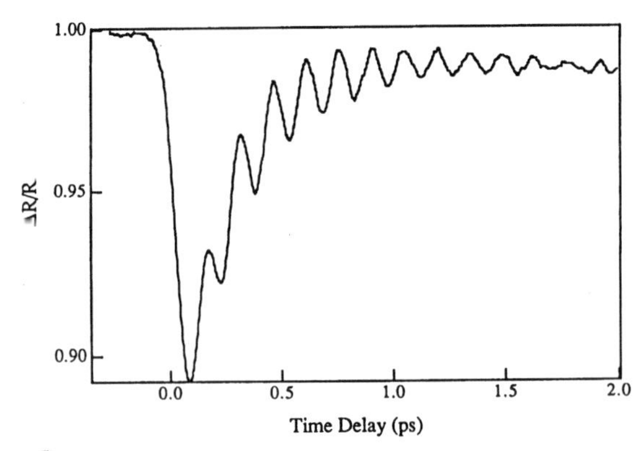

Figure 22. Impulsive absorption data from Ti_2O_3 showing ~7 THz oscillations from an A_{1g} optic phonon. The sample reflectivity prior to excitation has been normalized to unity. The large changes in reflectivity are due to phonon-induced changes in the bandgap. (Reprinted from [57]. Copyright 1993 American Physical Society.)

impulsively in Ti_2O_3 correspond to a change of 2% (0.07 Å) in the nonequivalent Ti ion distance. These results are especially significant for a number of reasons. First, an extensive study of the incompletely understood semiconductor-to-semimetal phase transition was possible because of the sensitivity of the experiments to phonon and electron dephasing times and phonon frequencies[58]. Second, these experiments demonstrate the feasibility of modulation of and control over the phase of matter on a femtosecond time scale. In related experiments, mode softening has been observed in Te[59, 60] following impulsive excitation at high intensities.

Optic phonons induced by impulsive absorption and detected via transient reflectivity measurements were reported in GaAs [61], see Figure 23. These data show an 8.8-THz oscillation superimposed on a relaxational signal. The oscillations are believed to be caused by the coupling between the longitudinal electric field associated with the Γ_{15} LO optic phonon and the index of refraction via the electro-optic effect, while the relaxational signal reveals the thermalization and relaxation of optically excited electronic carriers. In this case, impulsive absorption is believed to give rise to sudden surface charge screening which drives the coherent lattice vibrational response. The excitation mechanism was identified by measuring the change in reflectivity with a single pump beam and with a prepump followed by a pump. It was argued that the prepump, arriving 10 ps before the pump beam, would reduce the surface charge field by the injection of free carriers and would diminish the LO field and the phonon oscillations proportionally, which was the result observed experimentally. The ISRS mechanism was ruled out because of the insensitivity of the results to pump polarization.

Impulsive absorption experiments have been carried out to measure phonon dephasing times in the high-T_C superconductors $YBa_2Cu_3O_{7-x}(x\sim0)$ and $Bi_2Sr_2Ca_2O_{10+\delta}(\delta\sim0)$ [62] [63]. Additionally, coherent phonons have been driven and observed in alkali metal-doped C_{60} [64]. As in the case of ISRS excitation of transparent samples, here we see that impulsive absorption occurs almost always when an ultrashort pulse strikes an absorbing material, resulting in coherent molecular or collective vibrations.

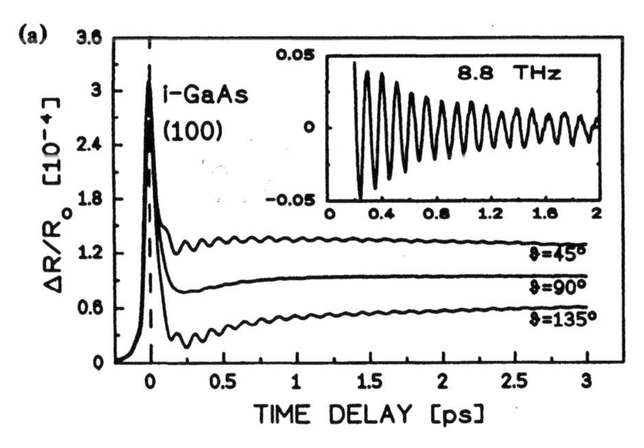

Figure 23. Probe reflectivity changes following impulsive absorption of (100)-oriented GaAs for orthogonally polarized pump and probe beams. The parameter ϑ is the angle between the probe polarization and the [010] crystal axis. The 8.8 THz oscillations correspond to the frequency of the Γ_{15} LO phonon of GaAs. (Reprinted from ref [61]. Copyright 1990 American Physical Society.)

III. E. Summary

Impulsive excitation and time-resolved observation of acoustic phonons, optic phonons, molecular vibrations, and other coherent motions have been made across a cornucopia of materials and phases with a wide range of impulsive excitation and detection schemes. We have illustrated some of the results that have provided information about structural rearrangements and various properties of crystalline solids, liquid-state molecular dynamics, chemical reactivity in all phases of matter, photobiology, and properties of practical interest in advanced materials. Since impulsive stimulated scattering and/or impulsive absorption occur in transparent and absorbing materials, impulsive excitation is a ubiquitous phenomenon which occurs whenever a sufficiently short light pulse interacts with nearly any type of matter. The ultrashort pulse almost always leaves coherent vibrational oscillations in its wake! Impulsive excitation also provides a basis for active optical control over material behavior, as illustrated in the following sections of our review.

IV. PULSE SHAPING

IV. A. Introduction

An active area of optical research in the last decade has been in the field of ultrafast pulse shaping, an intriguing technique in which a single incident femtosecond pulse is converted into a waveform with a desired amplitude and phase profile. The concerted efforts of several groups over a period of years have refined this method to the point that the pulse shaping is almost entirely automated, so that (when everything is set up correctly!) almost arbitrary amplitude and phase profiles may be generated by pushing a few buttons. In this section of the review, we'll summarize the theory of temporal pulse shaping, the device features needed to engineer high-fidelity waveforms, and recent advances in extending this temporal pulse-shaping technique with the addition of an extra spatial dimension. We'll also discuss the limitations of this technique in preparation for the review of control experiments in Section V. The treatment will emphasize pulse shaping in the context that is most familiar to the Nelson group through its own research: the device features of liquid crystal display (LCD) spatial light modulators (SLMs) will be discussed, for example, rather than those of acousto-optic tunable filters (AOTFs) [3, 65]. The mathematical treatment of pulse shaping discussed below will closely follow the analyses of Wefers and Nelson [66, 67]. The Fourier transform convention used here is

$$\bar{F}(k,\omega) = (2\pi)^{-1} \iint f(x,t) \exp(i(kx - \omega t))\, dx\, dt \qquad \text{(IV.1)}$$

$$f(x,t) = (2\pi)^{-1} \iint \bar{F}(k,\omega) \exp(-i(kx - \omega t))\, dx\, dt, \qquad \text{(IV.2)}$$

where a $(2\pi)^{-1/2}$ factor is associated with every transformation. We note that numerous new terms are introduced in the following sections, and some re-use of symbols with new meanings occurs.

IV. B. Temporal Pulse Shaping Theory.

1. The 4-f configuration. Because femtosecond pulses are too short to be shaped in the time domain by acousto-optic or electro-optic devices, they must be shaped in the frequency domain. Early attempts by Heritage, Weiner, Thurston, and Tomlinson using fiber and grating pulse compressors introduced a linear chirp.[68, 69] An important development for pulse shaping as well as for chirped pulse amplification (CPA) was the Martinez antiparallel grating pair,[70] which was used by Weiner, Heritage, and Kirschner to produce the first high-resolution femtosecond shaped waveforms (Figure 24) [2].

Without the lenses, large positive group velocity dispersion is introduced, and with the lenses, no dispersion is introduced or removed. The different frequency components emerging from the diffraction grating on the left are focused by a spherical lens onto a mask, where the pulse is frequency filtered. A second lens recombines the frequency components onto the diffraction grating on the right, from which they emerge temporally shaped. Because the two spherical lenses typically have the same focal length f and the first and second gratings are separated by four times the focal length, this is sometimes referred to as the 4-f configuration.

The mathematical description of the pulse shaper is usually given in terms of Martinez's transfer function description of dispersive elements like gratings or prisms [71]. Suppose the field amplitude just before the first grating is defined by [67].

$$e_1(x,t) = e_{in}(x,t) \exp(i\omega t), \qquad \text{(IV.3)}$$

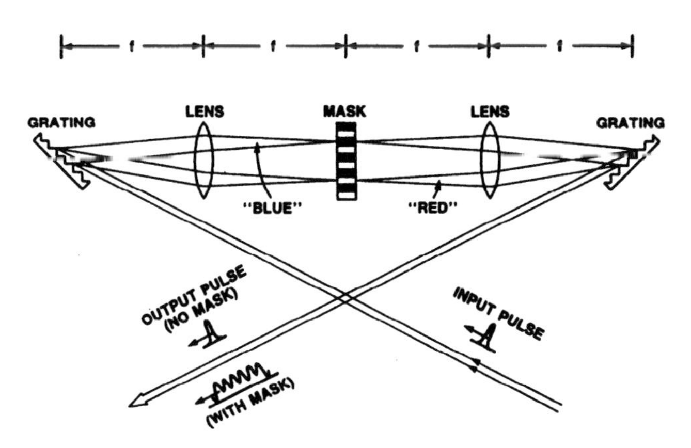

Figure 24. Dispersionless femtosecond pulse shaper design. An input femtosecond pulse is dispersed by the first grating. Its frequency components are collimated by a lens onto a mask, where they may be filtered. A second lens recombines the frequency components onto the second grating, from which the temporally shaped waveform emerges. (Reprinted from ref. [2]. Copyright 1988 Optical Society of America.)

where the optical field's central frequency ω has been factored out from the slowly varying envelope. The diffraction of the first-order beam from the grating is described as

$$\lambda = d(\sin\theta_i + \sin\theta_d). \tag{IV.4}$$

In this familiar expression, λ is the optical wavelength, d is the groove distance of the diffraction grating, θ_i is the incident beam angle, and θ_d is the diffracted beam angle. The method of Martinez is to find a transfer function which reproduces the variation of the emerging angle $\Delta\theta_d = \theta_d - \theta_{d0}$ with the change from the central angle of incidence $\Delta\theta_i = \theta_i - \theta_{i0}$ and central frequency $\Omega = \omega - \omega_0$, subject to Eqn. (IV.4). To first order, this variation is

$$\Delta\theta_d = \beta\Delta\theta_i + \gamma\Omega. \tag{IV.5}$$

The transfer function applied to the frequency Fourier transformed pulse envelope $E_{in}(x,\Omega)$ is

$$E_2(x,\Omega) = b_1 E_{in}(\beta x,\Omega)\exp(i\gamma\Omega x), \tag{IV.6}$$

with $\beta = \cos\theta_{i0} / \cos\theta_{d0}$ and $\gamma = 2\pi / \omega_0 d\cos\theta_{d0}$. b_1 is a coefficient that will disappear when the inverse transfer function is applied at the second grating.[67, 71] Typical values for an 800 nm pulse diffracting from a 1200 lines/mm grating at the Littrow angle (28.69°) are $\beta = 1$, $\gamma = 3.65\times10^{-9}\,\mathrm{s/m}$.

Martinez showed that positive or negative group velocity dispersion may be introduced by placing a telescope between antiparallel gratings. If the gratings are within the two foci of the lenses, positive group velocity dispersion will result; if the gratings are outside the two foci, negative group velocity dispersion will result; if the gratings coincide with the two foci, no group velocity dispersion will result. This scheme also introduces a space-time tilt that Martinez showed may be removed by retroreflecting the beam back in a double-pass geometry. The degree of group velocity dispersion is calculated by introducing a finite beam size and using the Fresnel-Kirchoff integral (which is equivalent to multiplication by a spatial transfer function in Fourier space) to calculate the properties of the beam for a particular lens and grating geometry. We have ignored the finite beam size in this treatment for simplicity, so we are unable to calculate the group velocity dispersion terms—which depend on the finite beam—properly.

Assuming that one focus of each lens coincides with a diffraction grating, it is easy to calculate the field distribution at the mask plane between the lenses. It is the spatial Fourier transform of Eqn. (IV.6) with the substitution $k = 2\pi x / f\lambda$:

$$\begin{aligned} E_3(x,\Omega) &= \frac{1}{\sqrt{2\pi}} \int_{-\infty}^{\infty} E_2(x,\Omega) \exp(ikx)\, dx \\ &= \frac{b_1}{\sqrt{2\pi}} \int_{-\infty}^{\infty} E_{in}(\beta x,\Omega) \exp(i\gamma\Omega x) \exp(ikx)\, dx \\ &= \frac{b_1}{\beta\sqrt{2\pi}} \int_{-\infty}^{\infty} E_{in}(u,\Omega) \exp\left[\frac{i(\gamma\Omega + k)u}{\beta}\right] du \\ &= \frac{b_1}{\beta} \overline{E}_{in}\left(\frac{\gamma\Omega + k}{\beta}, \Omega\right) \\ &= \frac{b_1}{\beta} \overline{E}_{in}\left(\frac{\gamma\Omega + 2\pi x / \lambda f}{\beta}, \Omega\right) \end{aligned} \tag{IV.7}$$

A typical lens focal length is f=20 cm. From the spatial argument we can deduce the spread of frequencies over space.

$$\gamma\Omega + 2\pi x / \lambda f = \text{constant} \tag{IV.8}$$

implies

$$\begin{aligned} \alpha \equiv \frac{\partial x}{\partial \Omega} &= -\gamma\lambda f / 2\pi \\ &= -2\pi c f / \omega^2 d \cos\theta_{d0} \end{aligned} \tag{IV.9}$$

As we will see, α plays a central role in pulse shaping. Without the coupling of frequencies to spatial coordinates, pulse shaping would have to be done with dielectric filters or other photonic band gap materials in the frequency domain. α is -9.29×10^{-17} m s / rad when the parameters have the typical values given earlier. This expression describes the linear dispersion of frequency components in space across a mask; higher order terms have been neglected but may be necessary for very broad bandwidth (sub-20 fs) pulses [66].

Here, at the focal plane of the lens, the mask filters the signal in the frequency domain. Originally either fixed phase or amplitude masks operating in either a binary phase or gray-scale amplitude mode were used by Weiner and co-workers at Bellcore; these were later replaced by liquid crystal masks [1, 72, 73]. Although the liquid crystal masks have discrete pixels, they are programmable with variable phase and amplitude control at each pixel. If an application needs a programmable phase and amplitude mask combination, two 4-*f* configurations can be strung together (with parallel, not antiparallel, gratings at the ends, and without any gratings in the middle) to form the 8-*f* configuration [1]. Aligning and operating this configuration is difficult, so Wefers and Nelson combined programmable phase and amplitude masks into one dual-mask unit that could be used in the 4-*f* configuration [74]. A full study of the mask transfer function $m(x)$ will be given in the next several subsections. For now it is enough to note that the frequency components all traverse the same optical path length from the first grating to the mask. The mask acts as a multiplicative transfer function in the frequency domain.

The frequency components also travel the same optical path length from the mask to the second grating (by way of comparison, imagine a 2-*f* configuration with a grating, a lens, and a reflective mask), so there is no net group velocity dispersion introduced by a properly aligned pulse shaper. Space-time coupling effects, deriving from the space-frequency filtering in Eqn. (IV.9), are also discussed in a later section of this review.

2. Finite beam analysis. Real beams of ultrashort pulses have finite spot sizes. It is very important to consider this fact when constructing a pulse shaper [66, 69, 73]. The spot size at the mask limits the number of effective frequency components, the optimal pixel size if liquid crystal displays are used, and the number of physically realizable distinct temporal features.

We follow the treatment of Wefers and Nelson in starting with the Hermite-Gaussian spatial mode treatment of finite beams developed by Thurston et al [69]. The important assumption in the treatment is that only the lowest-order Gaussian spatial mode of the shaped waveform is kept from after the mask to the output. If the entire frequency spectrum fits on a liquid crystal mask, the mask function m(x) is

$$m(x) = \sum_{n=-N/2}^{N/2} \left\{ B_n \delta(x - nw) \otimes squ(x / wr) + B_g \delta\left(x - \left(n + \frac{1}{2} \right) w \right) \otimes squ(x / w(1-r)) \right\} \quad \text{(IV.10)}$$

where *squ* indicates the rectangular function

$$squ(x) = \begin{cases} 1, & |x| < 1/2 \\ 0, & \text{otherwise} \end{cases} \quad \text{(IV.11)}$$

and $\otimes$ is the usual convolution operator. If only the Gaussian spatial mode is kept, the shaped profile of the output waveform is a function of the Fourier transform of the mask $M(k)$ and a Gaussian temporal distribution $g(t)$ [66].

$$e_{out}(x,y,t) = u_{00}(x,y)\exp(i\Omega t)\left(e_{in}(t) \otimes \left[M(-t/\alpha)g(t)\right]\right) \quad \text{(IV.12)}$$

$$g(t) = \exp\left(-w_0^2 t^2 / 8\alpha^2\right) \quad \text{(IV.13)}$$

with

$$M(k) = \sum_{n=-\infty}^{\infty} [A_{n \bmod N}\, \delta\left(k - n\frac{2\pi}{Nw}\right) \frac{\sqrt{\pi}}{k\sqrt{2}} \sin(krw/2) + B_g \frac{(-1)^n \sqrt{\pi}}{k\sqrt{2}} \sin\left(k(1-r)w/2\right) \delta\left(k - n\frac{2\pi}{Nw}\right)] \quad \text{(IV.14)}$$

where A_n is the discrete Fourier transform element calculated from the series B_n. The fidelity of the generated waveform is clearly better if the gaps are smaller. Even if there are no gaps, a sinc-function modulation of the waveform is introduced by the discrete pixels. The finite spot size introduces a Gaussian envelope to each temporal element. A good choice of spot size is the distance between pixels. This choice minimizes the tradeoff between diffraction into other orders, occurring when the spot size is too large and the Gaussian envelope is narrow, and generation of replica waveforms, which occurs when the spot size is too small. A phase change is also introduced with each temporal feature if the mask is off-center. For high-fidelity waveform generation, it is necessary to compensate for all these effects [66].

3. Space-time coupling effect. An unfortunate pathology of femtosecond pulse shaping is the space-time coupling introduced by the grating's conversion of temporal frequencies into spatial wavevectors. In other words, the space-frequency coupling at the mask leads to space-time coupling effects after the pulse shaper. Martinez considered a space-time tilt in the case of the compressor with finite beams [70]. Wefers and Nelson [66] analyzed this effect in the case of pulse shapers with finite beam sizes. A more transparent interpretation of this effect that does not rely on finite beams was given by Wefers and Nelson [67] and is summarized below.

Returning to the treatment of the 4-*f* configuration presented in Section IV-B-1, consider the electric field profile incident on the mask (Eqn. (IV.7)). This field is filtered by the mask transfer function *m(x)*. (Note that we are no longer working with the Hermite-Gaussian mode model of Thurston et al. used in Section IV-B-2.)

$$E_4(x,\Omega) = E_3(x,\Omega)\,m(x) = \frac{b_1}{\beta}\overline{E}_{in}\left(\frac{\gamma\Omega + 2\pi x/\lambda f}{\beta}, \Omega\right) m(x) \quad \text{(IV.15)}$$

Again, propagation from the object to the image of a lens is accomplished with a Fourier transformation, and propagation from the incident to the diffracted beams of a grating is accomplished with the Martinez formalism.

$$
\begin{aligned}
E_5(x,\Omega) &= \frac{1}{\sqrt{2\pi}}\int_{-\infty}^{\infty} E_4(x,\Omega)\exp(ikx)\,dx \\
&= \frac{b_1}{\sqrt{2\pi}\beta}\int_{-\infty}^{\infty} \overline{E}_{in}\left(\frac{\gamma\Omega + 2\pi x / \lambda f}{\beta},\Omega\right) m(x)\exp(ikx)\,dx \\
&= \frac{b_1}{\sqrt{2\pi}\beta}\left\{\frac{\beta\lambda f}{2\pi}\int_{-\infty}^{\infty}\overline{E}_{in}(u,\Omega)\exp\left(ik\frac{\lambda f}{2\pi}(\beta u - \gamma\Omega)\right)du\right\}\otimes M(2\pi x / \lambda f) \\
&= \frac{b_1}{\sqrt{2\pi}}\left\{\frac{\lambda f}{2\pi}\exp\left(-ik\frac{\lambda f\gamma\Omega}{2\pi}\right)\int_{-\infty}^{\infty}\overline{E}_{in}(u,\Omega)\exp\left(ik\frac{\lambda f\beta}{2\pi}u\right)du\right\} \\
&\quad \otimes M(2\pi x / \lambda f) \\
&= \frac{b_1}{\sqrt{2\pi}}\left\{-\frac{\lambda f}{2\pi}\exp\left(-ik\frac{\lambda f\gamma\Omega}{2\pi}\right)\int_{\infty}^{-\infty}\overline{E}_{in}(-u,\Omega)\exp\left(-ik\frac{\lambda f\beta}{2\pi}u\right)du\right\} \\
&\quad \otimes M(2\pi x / \lambda f) \\
&= \frac{b_1}{2\pi}\left\{\lambda f\exp\left(-ik\frac{\lambda f\gamma\Omega}{2\pi}\right)E_{in}\left(-k\frac{\lambda f\beta}{2\pi},\Omega\right)\right\}\otimes M(2\pi x / \lambda f) \\
&= \frac{b_1}{2\pi}\{\lambda f\exp(-i\gamma\Omega x)E_{in}(-\beta x,\Omega)\}\otimes M(2\pi x / \lambda f)
\end{aligned}
\qquad \text{(IV.16)}
$$

The transfer function for the antiparallel grating would be the inverse of the transfer function defined in Eqn. (IV.6), but the image inversion by the telescope changes the sign of the phase factor.

$$
\begin{aligned}
E_6(x,\Omega) &= b_1^{-1}E_5(x/\beta,\Omega)\exp(i\gamma\Omega x/\beta) \\
&= \left[\frac{1}{2\pi}\{\lambda f\exp(-i\gamma\Omega x/\beta)E_{in}(-x,\Omega)\}\otimes M(2\pi x/\beta\lambda f)\right]\exp(i\gamma\Omega x/\beta). \qquad \text{(IV.17)} \\
&= \frac{\lambda f}{2\pi}\int_{-\infty}^{\infty}E_{in}(-x,\Omega)M(2\pi(x-x')/\beta\lambda f)\exp(i\Omega(\gamma(x-x')/\beta))\,dx'
\end{aligned}
$$

$$
\begin{aligned}
e_6(x,t) &= \frac{\lambda f}{2\pi}\int_{-\infty}^{\infty}E_{in}(-x,\Omega)M(2\pi(x-x')/\beta\lambda f)\exp(i\Omega(t+\gamma(x-x')/\beta))\,dx'\,d\Omega \\
&= \exp(i\omega_0 t)\int_{-\infty}^{\infty}e_{in}(-(x-x'),t)\,g(x',t')\,dx'\,dt'
\end{aligned}
\quad . \text{(IV.18)}
$$

Depending on the dispersion of the gratings, the mask shifts the temporal profile in one transverse spatial dimension as it shapes in time. Thus the early part of a shaped waveform is displaced spatially as well as temporally from the later part of the waveform. Perfectly dispersive gratings would span the whole range of allowable wavevectors with different frequencies and would shape only in time. Mirrors, on the other hand, would send all frequencies through a single wavevector and would shape only in space. Note that no space-time coupling is seen when there is no mask, i.e. $m(x) = 1$.

Martinez noted that double-passing the compressor could be used to alleviate the space-time tilt [70]. However, for a pulse shaper, space-time coupling is also seen in 8-f configurations and doubly-passed 4-f configurations [66, 67]. Despite the presence of space-time coupling, many wonderful things may still be done with shaped femtosecond pulses. In many cases, the spatial displacement between early and late parts of a shaped waveform is small compared to the spot size. Often the impact of the displacement can be minimized by proper focusing of the shaped waveform onto the sample.

4. Programmable phase and amplitude mask design. One student at the Erice school once asked, "I understand how you get a phase mask with liquid crystals, but how do you get an amplitude mask?" It's a good question because it calls up a variety of problems of practical spatial light modulator (SLM) construction, calibration, and use.

The action of a liquid crystal on light is polarization-dependent, so both the characteristics of the liquid crystal and the arrangement of polarizing optics (including gratings) are important. Nematic liquid crystals are usually used in programmable masks [1, 72, 73]. Unlike twisted nematic liquid crystals [75], which couple the amplitude and phase of the outgoing light, nematic liquid crystals are able to impose separable amplitude or phase retardations with applied voltage.

First, consider the phase-only retarder. If incident light traveling in the z direction is polarized along the y, vertical, direction (Figure 25) [73], and the nematic liquid crystals are aligned along the y direction as well, then the light is subject to a refractive index change with applied voltage. The voltage drop swings the nematic liquid crystals from their zero-field net equilibrium alignment along y towards z. This changes the refractive index for y-polarized light, and therefore yields a controllable phase shift. It is preferable to align the zero-field equilibrium direction along y rather than x, despite the preference of diffraction gratings for scattering along x, since electromagnetic edge effects distort the uniformity across pixels in that direction [73]. Polarizers and achromatic zero-order half-waveplates are needed before and after the liquid crystal to convert the polarization from x to y and back to x.

Only these polarizing optics need to be changed to obtain a single mask that has a coupled amplitude-phase profile [1]. If the incident beam is polarized at +45° to the y direction before the mask and passes through a polarizer at -45° after the mask, the SLM operates as a coupled amplitude and phase mask. The phase retardation may be compensated when another phase mask is used in the other part of the 8-*f* configuration [1].

Programmable combined amplitude and phase control was finally obtained by sandwiching two liquid crystal masks next to each other in a 4-*f* configuration (Figure 26).[74] In this design, the liquid crystals are arranged in the first and second masks at ±45° with respect to y, and the outer polarizers are aligned along x. Both amplitude and phase masking of one polarization are possible with this single unit. Two orthogonal polarizations may be filtered by using an 8-f or 4-f configuration with four liquid crystal spatial light modulators and appropriate polarization optics.

The masks must be calibrated for whatever combination is used. Calibration measurements typically involve applying a range of voltages to all the pixels in each mask to see how the output amplitude changes. The results are numerically sorted and combined to find a mapping from voltage to mask filter coefficient B_n.

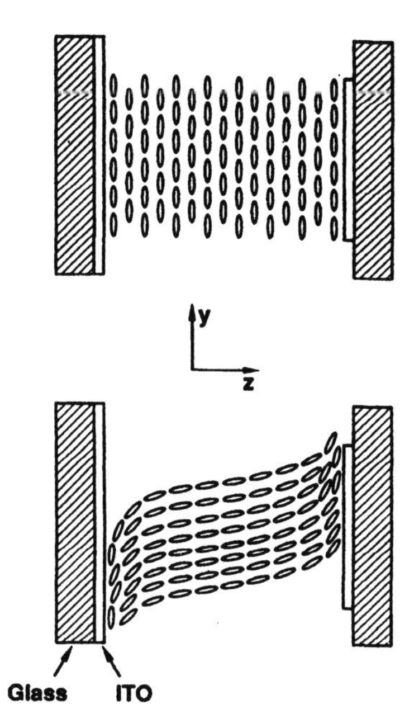

Figure 25. Detail of liquid crystal alignment in programmable pulse shaping mask. The crystals must be aligned appropriately to avoid edge effects. (Reprinted from ref. [73]. Copyright 1992 IEEE.)

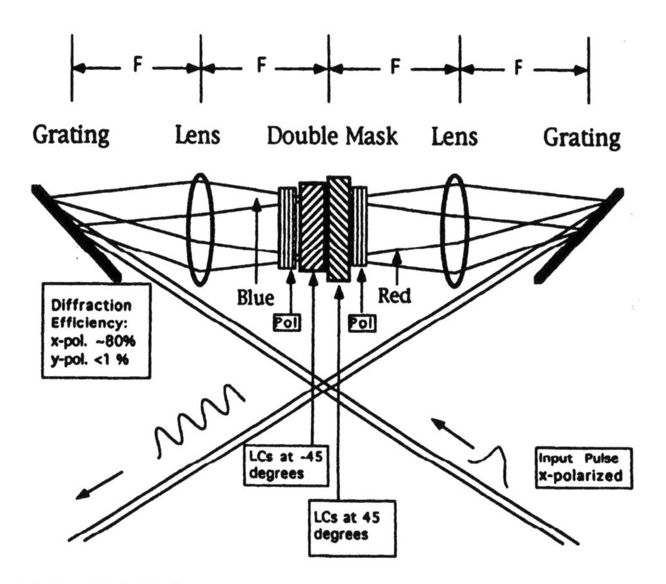

Figure 26. Modern high-fidelity temporal pulse shaper. Two liquid crystal masks, one at +45° and the other at -45°, together shape the amplitudes and phases of the frequency components of a femtosecond pulse, and, therefore, allow the generation of almost arbitrary shaped waveforms. (Reprinted from ref. [74]. Copyright 1995 Optical Society of America.)

5. Spatiotemporal pulse shaping. Recently, examples have been reported in which both spatial and temporal properties of the shaped waveform have been specified [76, 77]. The combination of simultaneous spatial and temporal shaping gives an additional degree of freedom for multiplexing in optical processing applications or for attempting to manipulate propagating excitations in crystals and other hosts. In spatiotemporal shaping (Figure 27), a single incident beam consisting of a single femtosecond pulse can be transformed into many separate outgoing beams, each of which consists of an independently specified pulse sequence or other temporally "shaped" waveform. More broadly, the result should be thought of as a single spatially and temporally coherent output whose position-dependent as well as time-dependent amplitude and phase profiles can be specified. An alternate but equivalent view is in terms of shaping of the wavevector content rather than the spatial profile. As a simple example, the output may consist of two separate beams or, after a focusing element, may consist of an interference pattern resulting from the intersection of those beams in the focal plane. Thus depending on how the output is imaged and where it is viewed, a description in terms of wavevector or position-dependent features may be more convenient.

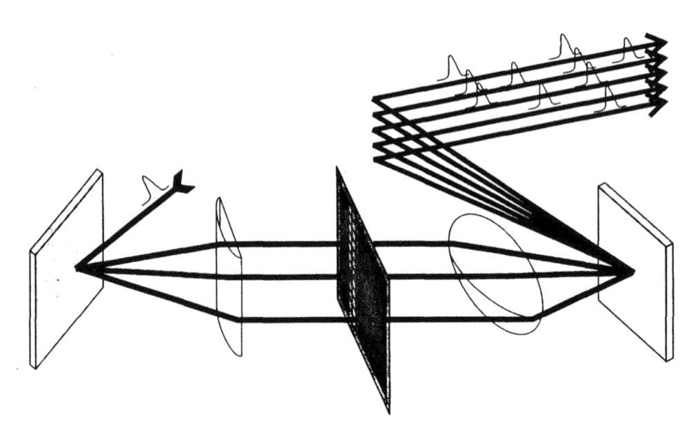

Figure 27. Spatiotemporal pulse shaper. A single incident beam consisting of a single femtosecond pulse can be transformed into many separate outgoing beams, each of which consists of an independently specified pulse sequence or other temporally "shaped" waveform.

Automated spatiotemporal pulse shaping using a liquid crystal mask has been demonstrated.[78] Simulated annealing calculations with the Metropolis algorithm may be used to determine a mask pattern in the frequency domain which minimizes a heuristic potential energy function for the intensity in the time domain. A consequence of the coupled amplitude and phase of this mask's transfer function is that it is easy to form a second pulse on one side of the temporal zero order diffraction peak if the waveform on the other side of the zero order diffraction is allowed to vary freely. In Figure 28a we show the results of calculations made under these restrictions.

These temporal pulse sequences may then be arranged to form a multidimensional waveform shaped in both space and time. Figure 28b shows CCD images of SHG cross-correlation measurements of a multidimensional waveform based on the three temporal pulse sequences shown in Figure 28a. Each of the three temporal pulse sequences has been repeated in three independent strips, using ten rows of pixels per strip, at different vertical positions on the mask. Between pulse sequences, ten rows of pixels have been darkened by setting $u = 1$. The spatial and temporal evolution of the pulse sequence is best visualized with a motion picture.

The temporal evolution of each vertically separated pulse sequence can best be measured by displaying the same mask pattern over all the rows of the mask, so that it only shapes the temporal dimension. Observing the resulting SHG signal with a photodiode and lock-in detection yields experimental SHG intensity profiles of the two-pulse sequences (Figure 28c). Although theoretically the zero order temporal diffraction peak and the delayed pulse are expected to have the same intensity (Figure 28a), that result is not observed, presumably due to calibration errors and variations across the mask.

The development of high-resolution programmable spatiotemporal pulse shapers should allow the manipulation and amplification of propagating excitations, like phonon-polaritons, and it should allow an additional degree of phase control for use in beam-crossing experiments like impulsive stimulated Raman scattering (ISRS).

6. Pulse shaping limitations. It is now possible to generate high-fidelity temporally shaped waveforms, and it will soon be possible to generate high-fidelity spatiotemporally shaped waveforms. What kinds of spectroscopy and control experiments can we do with them? To begin, what kinds of waveforms are possible?

An important experimental condition that is sometimes overlooked is the complexity of a waveform. A bandwidth-limited femtosecond Gaussian pulse, for example, is clearly very simple. Some continuous optical control waveforms, on the other hand, are very complicated, and an optimal path may be surrounded by many suboptimal ones.

It is hard to define a measure of complexity or information content for continuous waveforms like these. How much more or less complex is a square pulse than a triangular pulse, for example? It is easier to quantify waveform complexity in the discrete case. It is also sensible because programmable pulse shapers consist of a finite number of discrete pixels. Consideration of the equivalence of information content in the time and frequency domains or a glance at Eqn. (IV.14) shows that *the maximum number of distinct spatiotemporal features obtainable by a pulse shaper is equal to the number of pixels in the mask.* The nature of each distinct feature depends on the problem. Creating a chirped output pulse (i.e., one whose frequency components are shifted temporally) with a pulse shaper, for example, requires multiple distinct temporal features since the spectral phase of the pulse must shift with time. Temporally displacing a bandwidth-limited pulse, on the other hand, demands only one distinct feature. Temporally displacing a pulse that has already been chirped by a compressor still demands only one distinct feature; we have changed only the temporal basis function.

A large number of variable phase levels in each pixel increases the versatility of the SLM by extending the range of possible sets of distinct features (or, equivalently, the amplitude and phase range of a given distinct feature), but it does not change the maximum number of possible features that can be produced. For spatiotemporal pulse shaping, the maximum number of distinct features in each dimension is set by the number of pixels in

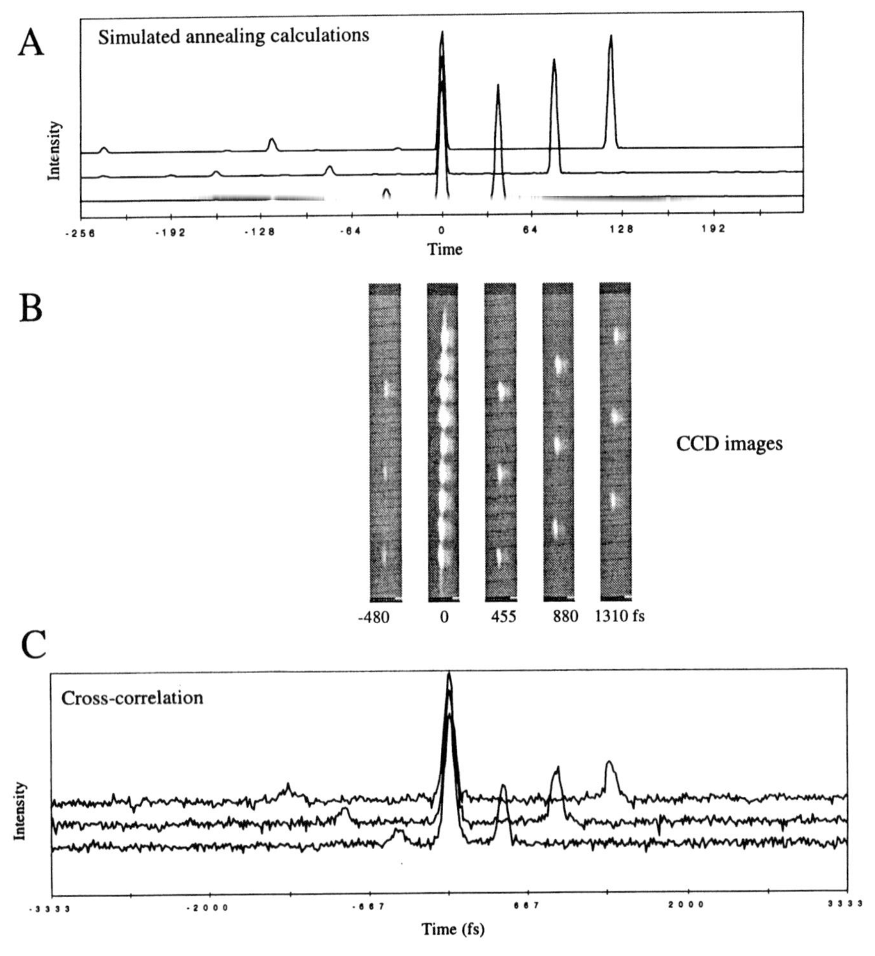

Figure 28. Spatiotemporal pulse shaper output. The output waveforms shown top to bottom are generated simultaneously by a single femtosecond pulse. A liquid crystal mask was used to generate each waveform.

that dimension, and the maximum total number of distinct features is the total number of pixels on the mask. The distance between adjacent spatial features is the same for every temporal feature, and vice versa.

Diffraction effects can also limit the maximum number of features if the spot size is significantly larger than the pixel size for temporal shaping, or if the imaging optics are not sufficiently powerful to image distinct rows of pixels in the spatial domain of spatiotemporal shaping. Pulse shapers based on acousto-optic tunable filters (AOTFs), which are continuous strips without pixels, still obey a limit on the maximum number of distinct features for this reason.

A consequence of this discrete spatiotemporal resolution is that continuous control models are unnecessary, and discrete control models may be used. Current temporal pulse shapers have at most 256 pixels, counting the back-to-back pixels on phase and amplitude SLMs as a single pixel, and spatiotemporal pulse shapers have at most 800x600 (time by space) pixels.

Having surveyed current femtosecond pulse shaping technology, we now review some control strategies and experiments.

V. OPTIMAL OPTICAL CONTROL

V. A. Overview

Many chemists and physicists have long dreamed of being able to manipulate and control matter with light, with varying degrees of success.[79-82] It is certainly possible to direct chemical reactions with light fields, as any photochemist or photophysicist would be happy to demonstrate, but most researchers have a slightly different idea in mind when they speak of control. Rather than see a reaction occur stochastically, as in the classic *cis-trans* photoisomerization of stilbene with ultraviolet light, they would prefer that the reaction be directed or manipulated as a coherent process. In some sense they also would prefer that the path taken from reactant to product be *optimal.*

In engineering, many control problems are not formulated optimally! We must examine what optimal control really is by reviewing the language and formulation of several strains of classical optimal control theory. Because the goal of optimal control experiments is to reach a regime of strong fields or nonlinear motion, we will also be forced to examine some tools used to describe and identify characteristics of nonlinear systems. Luckily, many of these tools will be familiar from nonlinear optical spectroscopy. We will then be able to evaluate the suitability of different theories of quantum optimal control and appreciate the rich variety of possibilities that might be realized with current technologies like multiple pulse ultrafast optical spectroscopy.

V. B. Optimal Control: General Theory

1. First principles. There are several different ways to formulate a control problem, but in the end every technique must be reduced to practice in some way. Every control system must have an *actuator*, a device capable of doing what needs to be done, optimally or not. No magic control box can make pigs fly. In *open loop* control, the actuator is not given any feedback about what the system is doing while it performs. In *closed loop* control, the actuator receives *feedback* about the system from an *estimator*. It is not always possible to separate the estimator, or the feedback loop, from the system to be controlled. In fact, the estimator must usually be included in the description of the complete control problem!

The problem is not just, as might be suspected, that the presence of feedback could lead to a feedback loop. Optical control is not a true example of feedback control since it doesn't happen in real time. Instead, a pulse or sequence of pulses comes along, does something to the sample, the effects are measured, the system reequilibrates, and a new pulse sequence is applied. Using a combination of prior knowledge of the Hamiltonian of the system and the measured responses of the observables, the procedure may be iterated until some goal, such as a reaction, is achieved. True feedback control would require that information about the vibrations of molecules or collective coordinates be fed back into the control system as the vibrations occur in real time. Rather, the response of the estimator must be included in the control system because (like the system itself) the act of measuring has some degree of stochasticity. Thus we need to decide what the best estimator, or statistic, is. Is it unbiased? Does it give the maximum likelihood of yielding the correct parameters?

These seemingly trivial questions will actually become somewhat important when we leave the regime of linear response and move to nonlinear systems because many of the ideas of nonlinear system identification and nonlinear control theory are based on higher-order methods of statistics. One example is the Volterra kernel expansion (equivalent to the Bloembergen formulation of nonlinear optics) that is often used to describe nonlinear systems in quasilinear regimes. Two-photon absorption in the frequency domain, for example, is a nonlinear spectroscopic effect which happens stochastically and can be described in terms of an interaction term between different frequencies, so it can be considered as either a stochastic system or a nonlinear one. These higher-order methods are also conveniently formulated for the application of multiple-pulse spectroscopy. In any case, questions about filtering and prediction of the state of the system are entirely natural. Modern control theory pays significant attention to the problem of finding optimal filters (such as the Wiener or Kalman filters for steady-state or linear systems) for the state variables [83]!

The simplest form of controller used in engineering is the PID (proportional-integral-derivative) controller, which isn't necessarily optimal. In this controller, the state of the actuator *a(t)* is determined by the response *r(t)* using

$$a(t_i + \Delta t) = g_1 r(t_i) + g_2 \int_0^{t_i} r(t)\,dt + g_3 \left(\frac{d}{dt} r(t) \right)_{t=t_i}, \tag{V.1}$$

where the g_i's are the gains for the proportional, integral, and derivative terms, respectively. The integral and derivative terms are used to adjust the selectivity of the gain to low and high frequency responses. Unfortunately, this type of controller doesn't provide much physical insight into the system or the estimator, particularly for nonlinear systems.

The variational formulation of optimal control theory has been popular among scientists because it may be used to generate an optimal path for any kind of system, linear or nonlinear, as long as the parameters of the system are known [83, 84]. The calculus of variations is also very flexible in the kinds of goals and penalties that may be used to decide the optimal path. Usually the dynamics of the system are written in state space. Most studies of nonlinear dynamics also take place in state space, where it is possible to examine phenomena of nonlinear systems like bifurcations and the onset of stochastic motion [85]. We will emphasize the variational aspect of control theory because of the preponderance of calculations and demonstrations of this technique in microscopic systems.

2. The calculus of variations. The calculus of variations relies on a simple quantity, the functional derivative. Given an integral

$$I[f] = \int_a^b f(x, \dot{x}, \ddot{x}, \ldots, t)\, dt, \tag{V.2}$$

its extrema may be found by solving the Euler-Lagrange equations

$$\frac{\delta I[f]}{\delta f} = \frac{\partial}{\partial x} f - \frac{\partial}{\partial t}\frac{\partial}{\partial \dot{x}} f + \frac{\partial}{\partial t^2}\frac{\partial}{\partial \ddot{x}} f - \ldots = 0, \tag{V.3}$$

subject to appropriate boundary and subsidiary conditions. Using the calculus of variations, it is possible to find the invariant quantities of a Hamiltonian system [84-86].

V. C. Optimal Control: Theory for Quantum Systems

Here we discuss several formulations of optimal control, aimed mostly at control over molecular behavior but also applicable to condensed matter. We will not discuss in any detail the difficult issues of vibrational dephasing or intramolecular vibrational energy redistribution, which are predominant limiting factors in any kind of mode-selective chemistry or channel selection [82, 87] and we will omit discussion of how to estimate electronic potential energy surfaces or specify any kind of feedback control.

1. Tannor-Rice formalism. As mentioned earlier, most of the formulations of quantum control rely on variational methods. Tannor and Rice developed a variational formalism for manipulating quantum systems with tailored coherent waveforms [88, 89]. They considered two specific cases: the optimization of two-photon absorption and the moderation of a reaction that can proceed in two channels, ABC -> AB + C vs. ABC -> A + BC. We will consider the two-photon absorption case. The two-photon transition rate is given by

$$\Pr\left(\phi_f \middle| \phi_i\right) = \left|\alpha_{fi}\right|^2 = \left|\int_{-\infty}^{\infty}\int_{-\infty}^{t_2} \left\langle \phi_f \middle| b(t_2) e^{-iH(t_2-t_1)/\hbar} a(t_1) \middle| \phi_i \right\rangle dt_1\, dt_2\right|^2, \tag{V.4}$$

and Tannor and Rice impose the constraint

$$\int_{-\infty}^{\infty} \left|b(t_2)\right|^2 dt_2 = 1, \tag{V.5}$$

where $a(t_1)$ and $b(t_2)$ are the first and second excitation pulses. In other words, they want to maximize the probability of achieving a specified final state given a certain total pulse energy. This is the simplest form of a "pump-dump" scheme in which a desired result is achieved under certain conditions.

Tannor and Rice note that this formulation of the problem relies heavily on the work of E. J. Heller, which describes optical phenomena like absorption or resonant Raman scattering in terms of the motion of wavepackets on Born-Oppenheimer potential energy surfaces, as discussed in Section II of this review. In the cw case for which it was originally applied, the Heller formalism may be viewed as a description of a stochastic process in terms of correlation functions. Many simple cases are covered by two-point correlation functions: absorption is determined by the dipole-dipole correlation function, for example, and Raman activity is determined by the polarizability autocorrelation function. Two photon absorption is a higher-order process [90]. The case of coherent excitation is somewhat different. Rather than having a stochastic process, we have coherent excitation in which all the molecules in the ensemble are following related trajectories.

The Lagrange multiplier formalism may be applied to determine the best form of the phase-sensitive pulse $b(t_2)$ given the constraint in Eqn. (V.5). The quantity to be optimized is [88]

$$F[b(t_2)] = \left| \int_{-\infty}^{\infty} \int_{-\infty}^{t_2} \left\langle \phi_f \left| b(t_2) e^{-iH(t_2-t_1)/\hbar} a(t_1) \right| \phi_i \right\rangle dt_1\, dt_2 \right|^2 + \lambda \left(\int_{-\infty}^{\infty} |b(t_2)|^2\, dt_2 - 1 \right). \quad \text{(V.6)}$$

Taking the functional derivative $\delta F / \delta b$ and resubstituting the earlier form of the constraint (Eqn. (V.5)) yields the optimal waveform, which is phase sensitive:

$$b^*(t) = \frac{f(t)}{\left[\int_{-\infty}^{\infty} |f(t)|^2\, dt \right]}, \quad \text{(V.7)}$$

where

$$f(t) = \int_{-\infty}^{t} \left\langle \phi_f \left| e^{-iH(t-s)/\hbar} a(s) \right| \phi_i \right\rangle dt. \quad \text{(V.8)}$$

Tannor and Rice state that "the optimal stimulating waveform is matched to the convolution of the incident waveform with the dynamics on the excited state potential surface" [88]. For given ground and excited state Born-Oppenheimer potential energy surfaces, the optimal waveform may be calculated using numerical quantum mechanical wavepacket propagation routines [89].

2. Rabitz formalism. Rabitz and collaborators at Princeton have worked to generalize the variational approach to quantum control and refine and extend its application to a wide variety of model systems. Some features common to all control experiments are seen under their analysis. The difference between the weak field (in which the system is still describable by perturbation theory) and the strong field (in which the system has been appreciably changed by the controlling field) becomes evident. As a practical matter, many calculations and experiments are carried out in the weak field limit, but the techniques of optimal control are not limited to weak field responses. Additionally, it is evident that more than one optimal path may exist.

The general theory of quantum mechanical control derived by Rabitz and coworkers is the quantum-mechanical analogue of the classical calculus of variations theory applied to macroscopic dynamical systems. [83] A mathematically rigorous exposition of quantum optimal control, including proofs of the existence of an optimal path and the well-posedness of the problem, was given by Pierce, Dahleh, and Rabitz [91] and applied to a small number of model systems.

A formulation of the general control problem for any observable was derived by Shi and Rabitz in 1989 [92]. Consider the observable $\langle \mathbf{o}(t)\rangle \equiv \langle \psi(t)|\mathbf{o}(t)|\psi(t)\rangle$. At a fixed time t=T, the goal function $\Phi[\langle \mathbf{o}(t=T)\rangle]$ determines whether or not the desired quantum control objective has been achieved. Constraints on the set of observables may be written as a functional $L_1\{[\langle \mathbf{o}(t)\rangle]\} = \int_0^T l_1[\langle \mathbf{o}(t)\rangle]\,dt$, and a cost functional on the controller or the system also may be written as $L_2\{[\langle \mathbf{u}(t)\rangle]\} = \int_0^T l_2[\langle \mathbf{u}(t)\rangle]\,dt$. To minimize the total energy in the laser field with fluence $\varepsilon(t)$, for example, the cost functional would be $L_2 = \frac{1}{2}\int_0^T w_e \varepsilon^2(t)\,dt$.

The standard procedure of variational optimal control theory with a specified final time is to write a new "Hamiltonian" functional, including the observables $\langle \mathbf{o}(t)\rangle$ and the parameters $\langle \mathbf{u}(t)\rangle$, with the equation of motion for the canonical conjugate variables corresponding to the Hamiltonian of the system H (including the interaction terms with laser fields and other parameters $\langle \mathbf{u}(t)\rangle$) included as a constraint that must be satisfied with Lagrange multipliers [83]. In Shi and Rabitz's theory for quantum systems, the functional to be minimized is

$$\begin{aligned}\bar{J}\{\langle \mathbf{o}(t)\rangle,\langle \mathbf{u}(t)\rangle\} &= L_1\{[\langle \mathbf{o}(t)\rangle]\} + L_2\{[\langle \mathbf{u}(t)\rangle]\} + \Phi[\langle \mathbf{o}(T)\rangle] \\ &-\int_0^T \langle \lambda(t)|\dot{\psi}(t)\rangle + \frac{i}{\hbar}\langle \lambda(t)|H|\psi(t)\rangle + \langle \dot{\psi}(t)|\lambda(t)\rangle - \frac{i}{\hbar}\langle \psi(t)|H|\lambda(t)\rangle\,dt\end{aligned} \tag{V.9}$$

Taking the functional derivative with respect to $\langle \mathbf{u}(t)\rangle$ gives a set of coupled differential equations that may be solved numerically

$$\frac{d}{dt}|\psi(t)\rangle = -\frac{i}{\hbar}H|\psi(t)\rangle \tag{V.10}$$

$$|\dot{\lambda}(t)\rangle = -\frac{i}{\hbar}H|\lambda(t)\rangle - \frac{\partial l_1[\langle \mathbf{o}(t)\rangle]}{\partial \langle \mathbf{o}(t)\rangle}\mathbf{o}|\psi(t)\rangle \tag{V.11}$$

$$\frac{\partial l_2[\langle \mathbf{u}(t)\rangle]}{\partial \langle \mathbf{u}(t)\rangle} = -\frac{2}{\hbar}\mathrm{Im}\left\langle \lambda(t)\left|\frac{\partial H}{\partial \mathbf{u}(t)}\right|\psi(t)\right\rangle \tag{V.12}$$

with the boundary condition

$$|\lambda(T)\rangle = \frac{\partial \Phi[\langle \mathbf{o}(T)\rangle]}{\partial \langle \mathbf{o}(T)\rangle}\mathbf{o}|\psi(T)\rangle. \tag{V.13}$$

The numerical calculations involved in approximating a solution to these coupled equations are similar to the wavepacket dynamics methods of Heller and also Tannor and Rice.

It is also possible to calculate the optimal path with an unspecified time or a minimum time (as in the brachistochrone problem or Fermat's principle).[83] In that case, the final time is specified as another control parameter in the auxiliary Hamiltonian $\bar{J}$ [83].

In principle there may be more than one solution that minimizes Eqn. (V.9) when the system is nonlinear. Even when Schrödinger's equation is linear, the interaction terms with the field may be quadratic, and the overall control problem may be cubic [93]. It has been shown that some systems may have an uncountably infinite number of solutions [93, 94].

3. Wilson formalism. A related variational optical control theory has been formulated by Kent Wilson and coworkers at UCSD. In this technique, the integral equation which is solved is based on the Liouville space formalism of quantum dynamics [95]. The Hamiltonian is included implicitly through the target operator rather than explicitly as a term in the functional. This method is claimed to offer improved consideration of the properties of the system, since a density matrix gives information about both the populations and the coherences of the system [96].

The formal solution of the Liouville equation [16, 96]

$$\frac{d\rho(t)}{dt} = -\frac{i}{\hbar}\left[H(t), \rho(t)\right]. \tag{V.14}$$

is the Green's function propagator

$$\begin{aligned} \rho(t) &= G(t,t_0)\rho(t_0) \\ &= \exp\left[-\frac{i}{\hbar}\int_{t_0}^{t} L(t)dt\right] \\ &= \sum_{n=0}^{\infty}\left(-\frac{i}{\hbar}\right)^n \int_{t_0}^{t} dt_n \ldots \int_{t_0}^{t_2} dt_1 L(t_n)\ldots L(t_1) \end{aligned} \tag{V.15}$$

The desired final expectation value is given by evaluating the trace of the operator $\hat{A}$ at time $t=t_f$ [95]

$$A(t_f) = Tr[\hat{A}\rho(t_f)], \tag{V.16}$$

and a penalty function (such as a minimized laser pulse energy, for example) may be represented in the usual manner. The overall functional is then, in this example,

$$J(t_f) = A(t_f) - \frac{1}{2}\int_{t_0}^{t} \lambda(\tau)\varepsilon^2(\tau)d\tau, \tag{V.17}$$

and the variation on the functional is

$$\delta J(t_f) = \delta A(t_f) - \int_{t_0}^{t} \lambda(\tau)\,\varepsilon(\tau)\delta\varepsilon(\tau)\,d\tau, \tag{V.18}$$

By writing the first-order variation on the observable as an integral with kernel K(t,t$_f$),

$$\delta A^{(1)}(t_f) = \int_{t_0}^{t} K(\tau, t_t)\delta\varepsilon(\tau)\,d\tau, \tag{V.19}$$

$$K(\tau, t_t) = -\frac{i}{\hbar} Tr[\hat{A}\,G(t_t, \tau)\,D\,G(\tau, t_0)\rho(t_0)], \tag{V.20}$$

where D is the interaction-picture Liouville space dipole operator, it is possible to find the optimal field in the dipole approximation through numerical calculations.

$$\varepsilon(\tau) = K(\tau, t_f) / \lambda(t). \tag{V.21}$$

It is not necessary to use the full Liouville-space formalism in order to derive this result, which could also be derived using standard (not super-) operators under the Liouville commutator. The advantage of working with density operators is that it is easy to include real-world effects like dephasing and finite temperatures [95]. For example, it is possible to include semiclassically a bath of Brownian oscillators [95]. This calculation has also been performed in a fully quantized fashion [97]. It is also possible to utilize the analogy of Liouville space to the classical Wigner probability space [96] as a basis for statistical analysis.

Like the Rabitz formalism, this analysis has been applied to a number of interesting physical problems. Iodine molecules have been considered: they are particularly suitable experimentally because the absorb near 800 nm. The "cannon" and "reflectron" configurations in which a wavepacket is projected outward or reflected back toward its origin, discussed further below, have been calculated. In these calculations, the position and momentum of outgoing and incoming I_2 wavepackets are focused with chirped optical pulses [98]. Unlike some tailored light fields calculated by optimal control theoretic methods, chirped pulses can be realized in the laboratory with current pulse shaping technology.

4. Genetic Algorithms. Variational control theory requires a great deal of knowledge about the parameters and state of the system to be known beforehand. One long-discussed possible method for coherently controlling the dynamics of a quantum system that does not require this information is the genetic algorithm, a development of artificial intelligence [99, 100].

In a genetic algorithm, a "gene" is encoded with information about the pulse sequence. Many genes are created, and their fitness for survival is evaluated using a merit function. The genes which are most successful at maximizing the merit function are mixed in a form of "mating", with mutations, and the offspring are fit again. This procedure is repeated for many generations until a species has evolved that satisfies the merit function.

The advantage of this approach is that it does not require any information in principle about the Hamiltonian of the system. A Hamiltonian could be included, of course, with parameters that would be determined by the genetic algorithm. In practice, genetic algorithms are most efficient in systems with a relatively small number of degrees of freedom. As the number of degrees of freedom becomes larger, they must stochastically sample a larger volume of parameter space, although it is hoped that a genetic algorithm is an efficient way to perform a large nonlinear optimization. Disadvantages of genetic algorithms include the "black box" nature of the controller, particularly the merit function which must be invented by the user's intuition. It may also take a significant amount of time to evaluate a merit function for each gene in every generation.

5. Summary and discussion. Each of these variants of control theory has been applied, for the most part, to systems with one degree of freedom to be controlled by adjusting a single parameter: a single-input, single-output (SISO) system. Many real-world chemical and physical problems are somewhat more interesting. Nonlinear motion of lattice vibrations in ferroelectric crystals, for example, should be characterized by wavevector overtones: the excitation at a fundamental wavevector in an anharmonic lattice leads to vibrational excitations at integral multiples of the fundamental [101]. Such a system has multiple (coupled) outputs, and, with spatial and/or temporal pulse shapers, could have multiple inputs as well (MIMO). Molecules with many vibrational modes that are anharmonically coupled at large displacements will present comparable challenges and opportunities. We anticipate that the extension of quantum control theories to systems with multiple input and output degrees of freedom, inclusion of estimators for feedback control, and theoretical advances based on statistical methods for nonlinear systems, will follow.

V. D. Control Experiments

Several researchers have demonstrated experimental application of dynamical control theories to quantum systems. We will focus on those experiments which emphasize time-dependent, or variational, control techniques, and refer the reader to reviews elsewhere of quantum interference techniques [102]. Within the realm of time-dependent methods, we will first consider experiments based on two phase-locked pulses and sequences of pulse trains. We will then examine experiments which incorporate the pump-dump or variational control theories to arrive at an appropriately tailored waveform. Finally, we will explore recent applications of genetic algorithms and feedback in quantum coherent control.

1. Phase locked pulses. *Coherent* control requires the ability to manipulate the phase, as well as the amplitude, of a quantum wavepacket. One way to do this is through manipulating the phase of a part of the driving electromagnetic field. Scherer and coworkers generated pairs of phase-locked femtosecond pulses with a Michelson interferometer. One arm of the interferometer is mounted on a piezoelectric stage and may be translated to change the phase of the field envelope in that arm so that the two pulses emerging from the interferometer have well-defined phases relative to each other. Many experiments have been performed with this simple pulse shaper. Scherer and coworkers applied the technique to the study of gas-phase iodine [103-105]. The Nuss group and collaborators used a similar device in controlling terahertz emission from asymmetric double quantum wells.

As Scherer et al. noted, pairs phase-locked pulses are similar to the two colors in the Tannor and Rice pump-dump scheme [105], but the experiment is most easily thought of in terms of quantum mechanical interference [104]. The first pulse creates an electronic coherence between the S0 and S1 states and initiates coherent vibrational motion mainly on the latter. The second pulse creates a second coherence, but there is no interference between the two when the first wavepacket has moved away from the Franck-Condon overlap region, i.e. away from the ground-state iodine bond length. This is because at other geometries the electronic transition frequency is changed, and so the electronic coherence is not at the excitation frequency. But when the first wavepacket returns to near the original geometry, the second coherence interferes constructively or destructively with the first, depending on the relative phases of the two excitation pulses. Constructive interference (no optical phase shift) leads to increased fluorescence while destructive interference (90° phase shift) yields the opposite result. The phase locking is only maintained within a narrow frequency band of the optical spectrum, called the locking frequency, and the spectral characteristics of the interference between the two pulses control the size of the excited state population [104]. Depending on the relative phase of the two pulses, which excite iodine from the ground vibrational level in the X electronic state to a series of vibrational levels in the B excited state, the excited state population may be enhanced, destroyed, or phase shifted. Experimental data, showing the fluorescence as a function of the delay of the second phase-locked pulse relative to the first pulse, are shown in Figure 29 [104]. The fluorescence signal shows several peaks resulting from the coherent addition

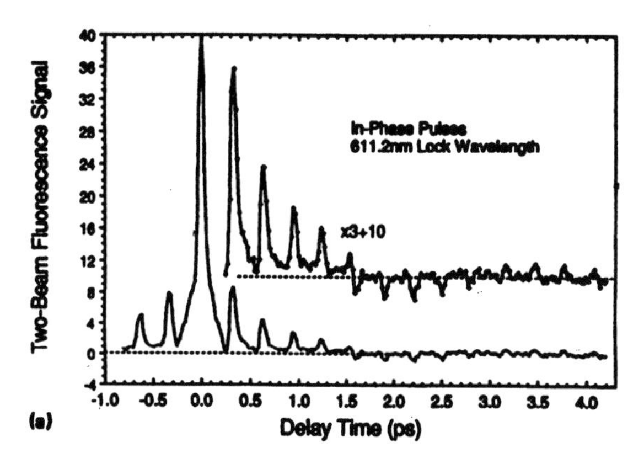

Figure 29. Experimental data showing fluorescence from iodine molecules as a function of the delay of a second phase-locked probe pulse. The recurrences shown in this figure do not appear when phase-locked pulses are not used. The phase of the recurrences depends on the separation between the phase-locked frequency and the resonant absorption frequency. (Reprinted from ref. [104]. Copyright 1991 American Institute of Physics.)

or subtraction of the second pump pulse, and the signal decays rapidly as the delay of the second pump pulse is increased. After about two picoseconds, the sign of the fluorescence signal changes, and Scherer et al. note that this phase reversal is accompanied by a phase shift as the period briefly changes from 300 fs to about 450 fs.

Because this is a molecular system, the iodine fluorescence signal is complicated by the existence of rotational couplings to the vibrations. Scherer and coworkers identify these rotational couplings to be the cause of the rapid decay of the fluorescence signal, but the inverting of the phase after a small number of periods is somewhat harder to describe.

The Nuss group performed a series of experiments to demonstrate the generation of terahertz radiation in quantum wells [106] and its control in asymmetric double quantum wells [107, 108].

The asymmetric double quantum well is shown in Figure 30 [108]. This is a three-level system with allowed dipole transitions between all three levels. The polarization of this system varies with time after an optical pulse, leading to the emission of electromagnetic terahertz radiation. The contributions to the terahertz polarization in a quantum well are from two sources [109]. On the one hand, if there is a dipole moment between the two levels of the excited state, two excitonic levels in the excited state may be produced and interfere, leading to quantum beats. In this case, the source of the polarization is a change in the coherence between the two excited state levels, not a difference in their populations [107, 108]. On the other hand, polarized excitons may be excited into uncorrelated electron-hole pairs, which dephase rapidly [108, 110]. We will focus on the first excitation mechanism in this review.

Using pairs of phase-locked pulses, Planken et al. [107] were able to "enhance, weaken, and induce" substantial phase shifts in the terahertz radiation caused by the

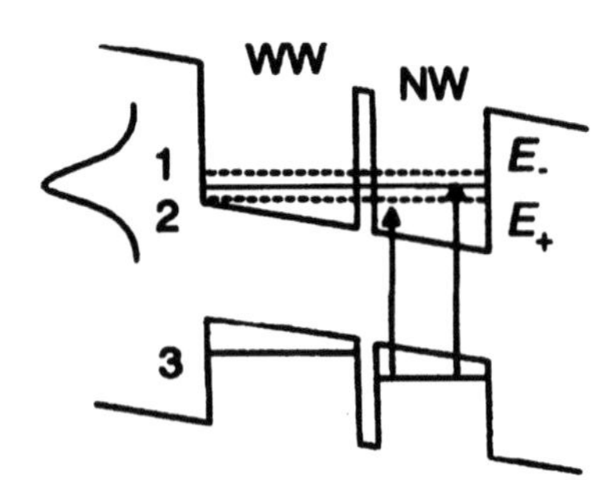

Figure 30. Asymmetric coupled quantum well for the generation of coherent terahertz radiation. Two upper excitonic states lie within the bandwidth of an incident femtosecond laser pulse. The splitting of the two wells, and thus the characteristics of the terahertz radiation, may be adjusted via a bias field. The pulse creates a coherence between the two states, which may also be viewed as a charge oscillation. This coherence/charge oscillation creates a polarization which radiates a terahertz electromagnetic field. (Adapted from ref. [108].)

excitonic charge oscillations (or coherences) by changing the delays and relative phases of the optical pulses. If the second optical pulse arrives at an integral multiple of the period T_{12} of the terahertz oscillation, setting the optical phase to 0 or π will enhance or completely eliminate the terahertz radiation. See Figure 31 [108]. However, if the second optical pulse does not arrive at an integral number of terahertz charge oscillations after the first pulse, the signal depends on the optical phase in an unusual way. If the second pulse arrives at $2.5T_{12}$ after the first pulse with a relative optical phase of 0, then the phase of the terahertz radiation shifts by $-\pi/2$ while the amplitude does not change much. If the relative optical phase is $\pi/2$, the amplitude is diminished but does not die out, and the phase does not change; if the relative optical phase is π, the amplitude is the same as when the phase was 0, but now there is a phase shift of $\pi/2$. See Figure 32 [108]. These experiments are described in terms of a Bloch pseudovector formalism [111]. The coherence evolves according to

$$\rho_{12}(t) = \frac{\mu_{13}\mu_{32}}{\hbar^2} E_0^2 [e^{-i\omega_{12}t}\Theta(t) + e^{-i\omega_{12}(t-T)}\Theta(t-T) + e^{-i\omega_{12}(t-T)}e^{-i\Delta_{13}T}\Theta(t-T) + e^{-i\omega_{12}(t-T)}e^{-i\Delta_{23}T}\Theta(t-T)] \tag{V.22}$$

where $\Theta(T)$ is the Heaviside function, ω_{12} is the quantum beat frequency, and Δ_{13}, Δ_{23} are the detuning frequencies from the central frequency of the exciting optical pulse. The four terms in brackets are labeled 1-4 in the diagram, and evolve as shown in Figure 33.[112]

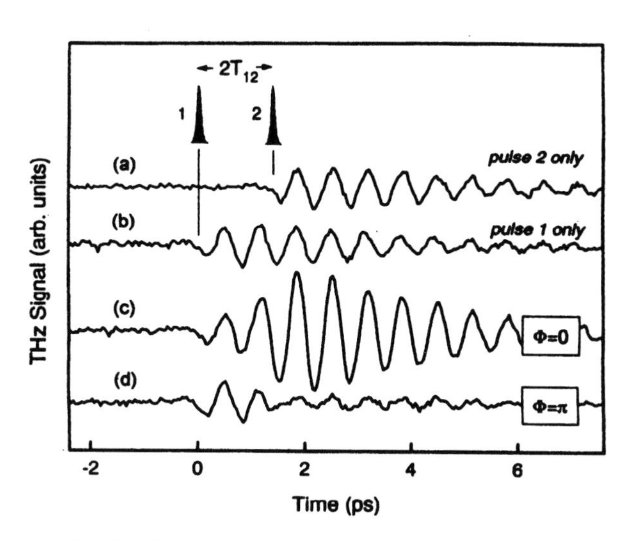

Figure 31. Phase properties of terahertz emission from asymmetric double quantum wells pumped by optically phase-related femtosecond pulses. In this case, excitonic coherences produce the terahertz radiation. Note that the phases shown in the labels are the phases of the optical emission, not the phases of the terahertz emission. When the second optical pulse arrives an *integral* number of terahertz emission periods after the first optical pulse, the results are as shown here. The terahertz emission may be either enhanced or canceled. (Adapted from ref. [108].)

Phase locked pulse pairs could also be used to differentiate between the excitonic coherence and electron-hole pair population contributions to the polarization.[110] The electron-hole pairs dephase much more rapidly than the excitons, and the unusual phase structure seen in the experiment with the coherences is not seen in this population experiment.[110]

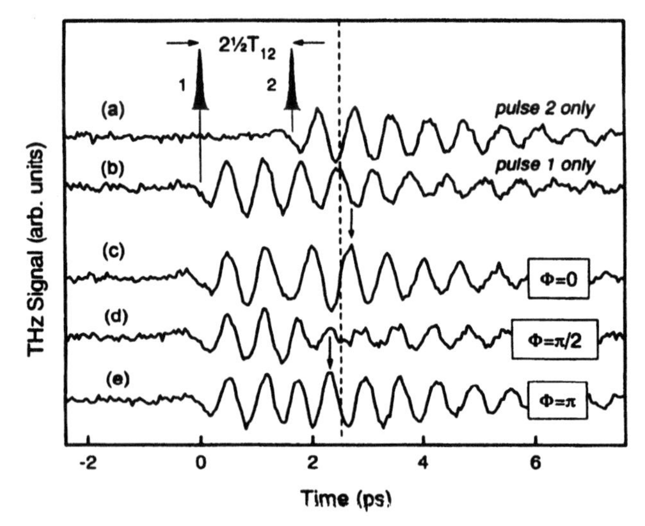

Figure 32. Phase properties of terahertz emission from asymmetric double quantum wells pumped by optically phase-related femtosecond pulses. In this case, excitonic coherences produce the terahertz radiation. Note that the phases shown in the labels are the phases of the optical emission, not the phases of the terahertz emission. When the second optical pulse arrives a *non-integral* number of terahertz emission periods after the first optical pulse, the results are somewhat more complicated. The terahertz emission may be enhanced, but not by so much as in Figure 31. The terahertz emission is not completely canceled when the optical phase is changed, and interesting phase effects appear in the periods of the recurrences. (Adapted from ref. [108].)

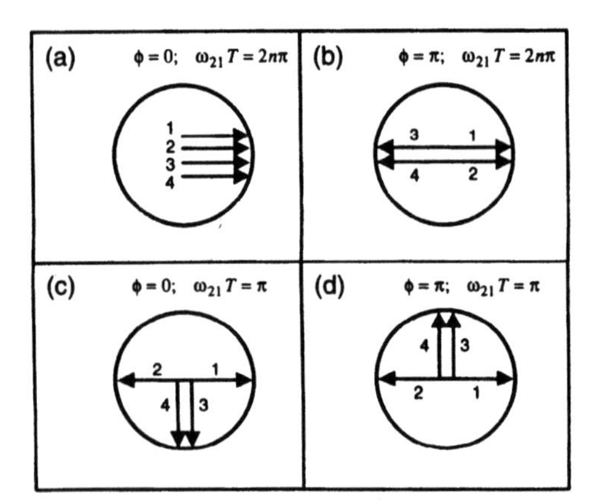

Figure 33. Bloch pseudovector formalism for the analysis of data shown in Figure 31 and Figure 32. The four terms in equation (V.22) are numbered and evolve as shown. In (a) and (b), the two pulses are separted by an integral number of terahertz oscillations, and pseudovectors 1 and 2 are always collinear. In (b) full destructive interference occurs. In (c) and (d), pseudovectors 1 and 2 are antiparallel. Full destructive interference should be possible when $\varphi = \pi/2$ (not shown). (Adapted from ref. [112].)

Although we have focused on phase-sensitive excitation in this section, it is worthwhile to remember that not all possible control experiments are phase sensitive. For one-photon experiments in particular, the time variation or coherence of the optical phase of the driving field clearly does not matter in determining what the most efficient excitation should be; as Brumer and Shapiro noted, whatever result is achieved with a single shaped femtosecond pulse may also be reached with a quasi-cw, or even incoherent, pulse having the same total energy.[113] Results of two-pulse experiments, on the other hand, may clearly depend on the relative phases of the optical driving fields.

2. Pulse shaping experiments. The initial promise of femtosecond pulse shaping has been realized in a number of physical systems in the last several years. Weiner, Leaird, Wiederrecht, and Nelson used a pulse shaper with a fixed phase-only mask to repetitively excite a selected optic phonon in α-perylene [114]. This material has several Raman-active optic phonon modes (Figure 34) [114].

The pulse train produced by the phase-only mask, which has a Gaussian amplitude envelope over a sequence of pulses with varying phases. (In nonresonant impulsive stimulated Raman scattering (see Section II), the optical phase is not important if a single light polarization is used as in this pulse train experiment. When an electronic resonance is present, Kosloff, Hammerich, and Tannor, and, separately, Cina and Smith showed that optical phase may be used to block absorption [115, 116].) As may be seen in Figure 35 and Figure 36 [114], the train of pulses selectively amplifies the 80 cm^{-1} and 104 cm^{-1} vibrational mode.

The main limitation of multiple-pulse spectroscopy is the onset of vibrational dephasing. Additional pulses do not contribute to larger displacements much beyond the vibrational dephasing time. Smith and Cina have suggested that larger amplitudes could be generated in ISRS with pulses that are nearly resonant, but below, an electronic transition.[117] Hiller and Cina have also shown that, for the case of iodine molecules, sequences of pulses with frequency chirp could generate even larger amplitudes.[118]

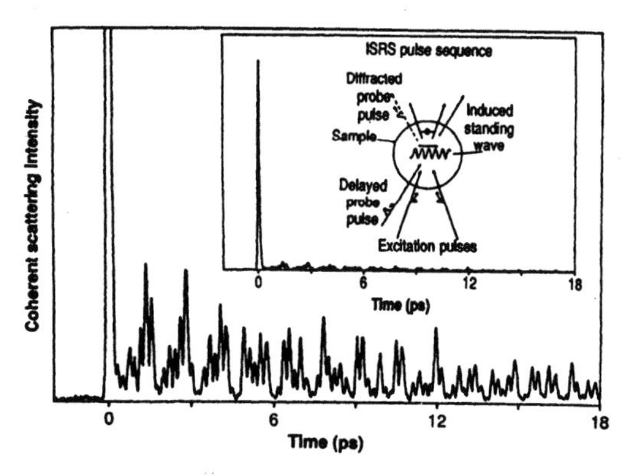

Figure 34. ISRS data from α-perylene excited by a single femtosecond laser pulse pair. Several low-frequency optical phonon modes are contained in this material response. Copied from ref. [114]. Reprinted by permission of the American Association for the Advancement of Science.

For molecular vibrations and solid-state lattice phonons, multiple-pulse spectroscopy does lead to real, larger vibrational amplitudes. A review by Dhar, Rogers, and Nelson discusses the relationship between classical and quantum mechanical interpretations of this experiment.[8]. We also wish to note that multiple-pulse spectroscopy has been used to perform experiments on the nature of the ferroelectric phase transition in lithium tantalate.[119]

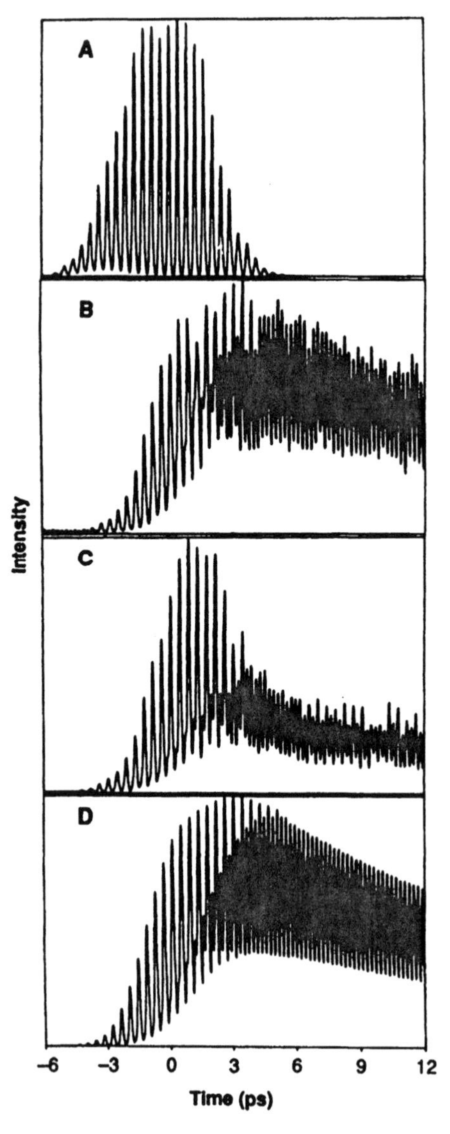

Figure 35. a) Pulse train autocorrelation. b) ISRS data from α-perylene excited by a pair of identical shaped femtosecond laser pulse trains. The frequency of the pulse train was set to match the 80 cm^{-1} mode, which is repetitively excited until the vibrational dephasing barrier appears. c) Similar to b, but the mode is driven slightly off resonance. d) Simulation of data. (Reprinted from ref. [114]. Copyright 1990 American Association for the Advancement of Science.)

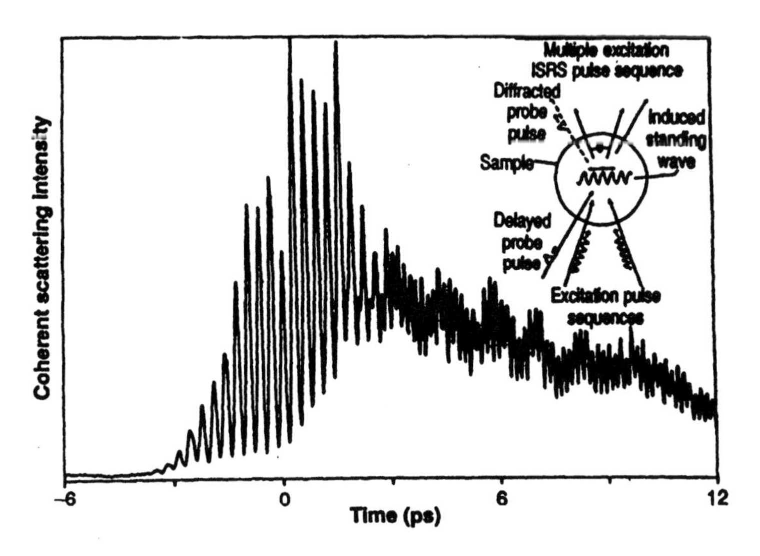

Figure 36. ISRS data from α-perylene excited by a pair of identical shaped femtosecond laser pulse trains. The frequency of the pulse train was set to match the 104 cm^{-1} mode, which is repetitively excited until the vibrational dephasing barrier appears. (Reprinted from ref. [114]. Copyright 1990 American Association for the Advancement of Science.)

Repetitive excitations by trains of pulses using similar phase-only masks matched to the period of excitonic charge oscillation quantum beats have also been performed by the Bell Labs group on single and double quantum wells [120]. An interesting feature of the response, when the phase mask is horizontally centered in the pulse shaper, is that it does not appear until half the pulse train has already passed. When the mask is displaced to change the optical phase pattern of the pulse train, unusual terahertz phase effects occur. See Figure 37 [108].

In many cases, the full potential of simultaneous phase and amplitude shaping of femtosecond pulses in optimal control problems has yet to be realized. Although it is now possible to create high-fidelity tailored waveforms, with specified amplitudes and phases at a large number of discrete spatiotemporal points, most control experiments to date have avoided loss issues by working with phase-only spectral masks or very simple amplitude masks. One exception has been the experiment of Wefers, Kawashima, and Nelson [121], in which two- and three-pulse sequences with specified optical phase relationships were used to prepare specified coherent states. Potassium vapor has an optical transition that is split by spin-orbit coupling into two lines, one at 769 nm and one at 766.5 nm. Again, depending on the delay, the optical phase relationships of the pulse pairs could be used to determine the occupation numbers of the excited states and control the fluorescence intensity. See Figure 38 and Figure 39. The extension to a three-pulse phase-locked sequence is shown in Figure 40 [121].

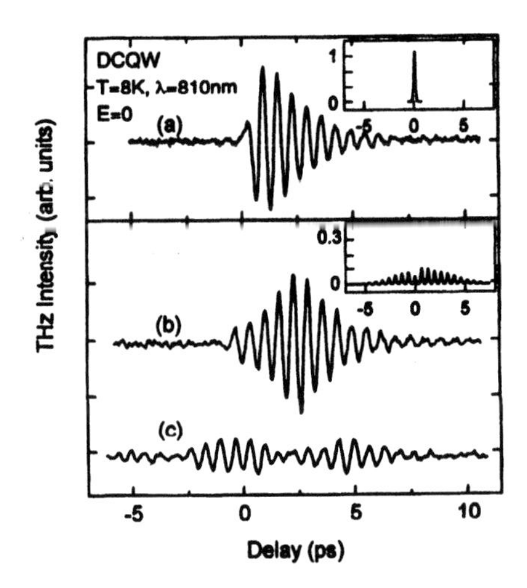

Figure 37. Terahertz emission from an asymmetric double quantum well after excitation by a pulse sequence similar to that shown in Figure 35a. Note the placement of zero time in this figure, which corresponds to the middle of the pulse train. Horizontal displacement in the pulse shaper of the phase mask used to form this pulse train changes the overall optical phase of the excitation, leading to unusual phase effects in the response. Adapted from ref. [108].

3. Variational methods. Recently, several groups have demonstrated the application of the variational control techniques discussed in Section IV to the design and generation of tailored optical waveforms. An important example of this technique has been the wavepacket focusing experiment performed by Kohler et al. with gas phase iodine [122]. This experiment was based on theoretical work published by the Wilson group in 1993 [98]. In this "reflectron" experiment a minimum uncertainty wavepacket is chosen as the target state on the B excited state surface, and the optimal field needed to reach it by focusing the wavepacket is chosen using optimal control theory. The ability of the pump pulse to achieve this goal is measured by laser induced fluorescence (LIF) from a higher lying electronic state that is reached via a "windowing" probe pulse [122].

The potential energy surface of molecular iodine is very anharmonic [122], (Figure 41) [122] so it is reasonable to expect that the optimal field which reaches the target—a focused wavepacket—is one which is strongly negatively chirped. The higher frequency components of the wavepacket are excited first, but they will take longer to travel to reach the target focused wavepacket than the lower frequency components, which are excited later but have less distance to go [122]. Theory predicted that the optimal field should have linear chirp and quadratic chirp components [98], but the optimal field achieved experimentally had only linear chirp [122], as measured with frequency-resolved optical gating [123]. The second peak of the LIF signal, corresponding to the return trip of the excited state wavepacket after it has been reflected, is enhanced by negative chirp and diminished by positive chirp, as confirmed quantitatively by experiment [122] (Figure 42) [122].

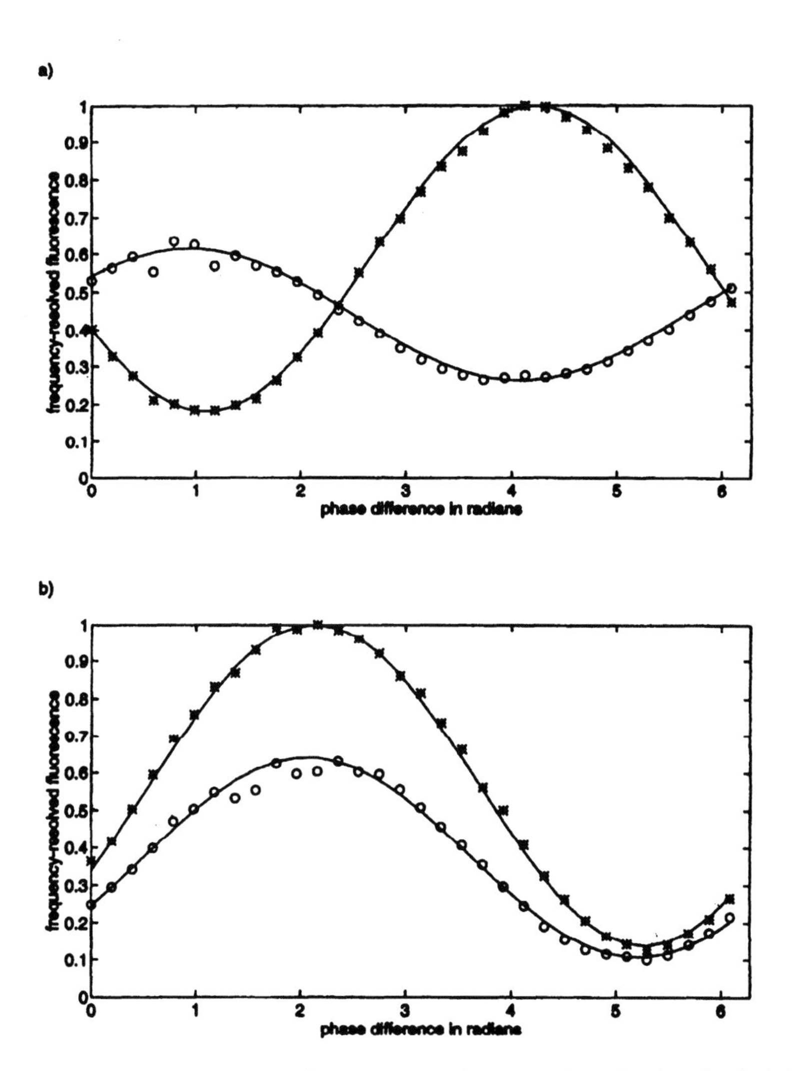

Figure 38. Frequency-resolved potassium fluorescence intensity measured as a function of optical phase difference between two pulses separated by a) 288 fs and b) 576 fs. Atomic potassium has an optical transition that is split by spin-orbit coupling into two lines at 766.5 nm and 769 nm wavelengths. The stars (circles) correspond to fluorescence measured with the monochromator set at 766.5 nm (769 nm). The solid curves correspond to the best sinusoidal fits through the data. The results show constructive or destructive interference between the electronic coherences induced by the two pulses, depending on the relative optical phase between them. (Reprinted from ref. [121]. Copyright 1995 American Institute of Physics.)

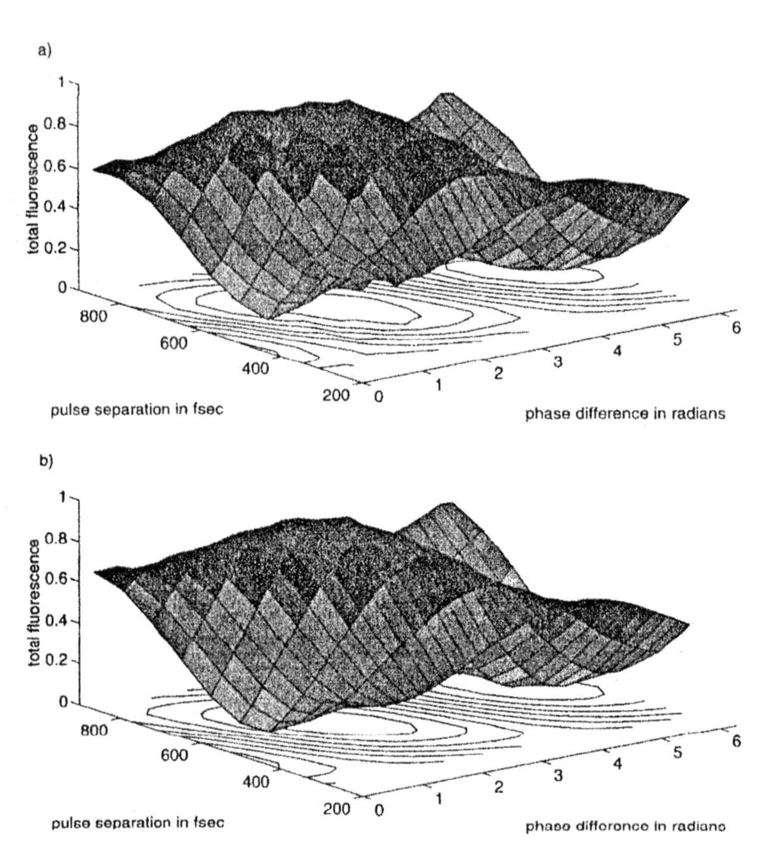

Figure 39. Total potassium fluorescence intensity a) measured experimentally and b) predicted theoretically as a function of the relative delay (288-864 fs) and optical phase (0-2π radians) of a two-pulse sequence. Experiment and theory agree with a standard deviation of 0.6% across the two-dimensional manifold. (Reprinted from ref. [121]. Copyright 1995 American Institute of Physics.)

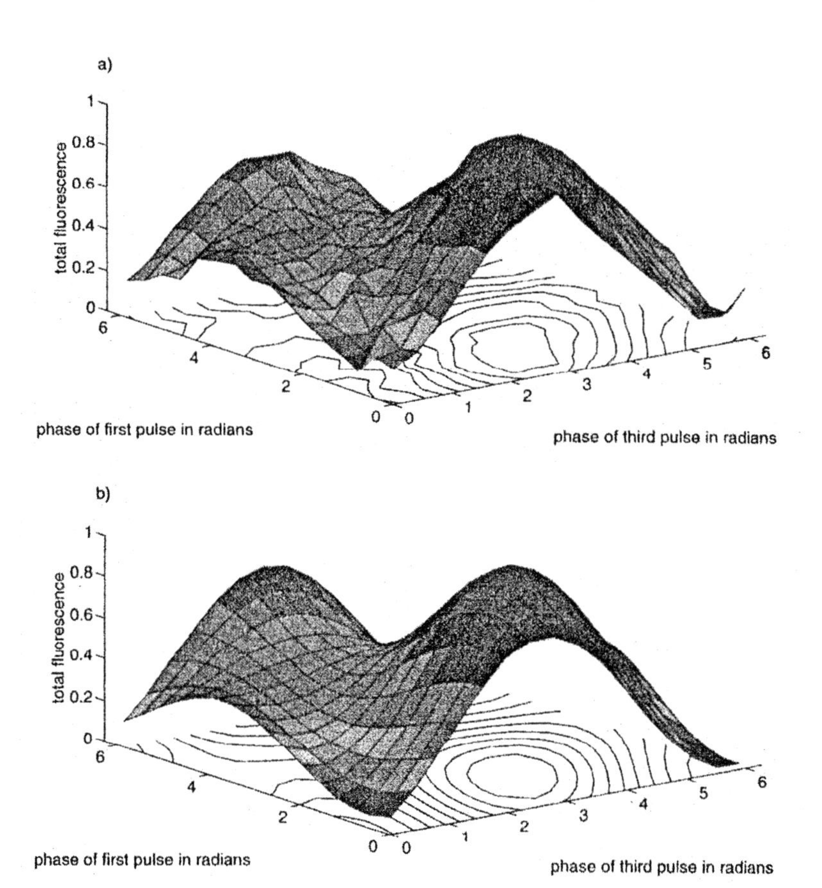

Figure 40. Total potassium fluorescence intensity a) measured experimentally and b) predicted theoretically as a function of the optical phase difference (0-2π radians) between the first and second, and second and third, pulses of a two-pulse sequence. The arrival times of the three pulses are -450, 0, and 300 fs. Experiment and theory agree with a standard deviation of 1.9% across the two-dimensional manifold. (Reprinted from ref. [121]. Copyright 1995 American Institute of Physics.)

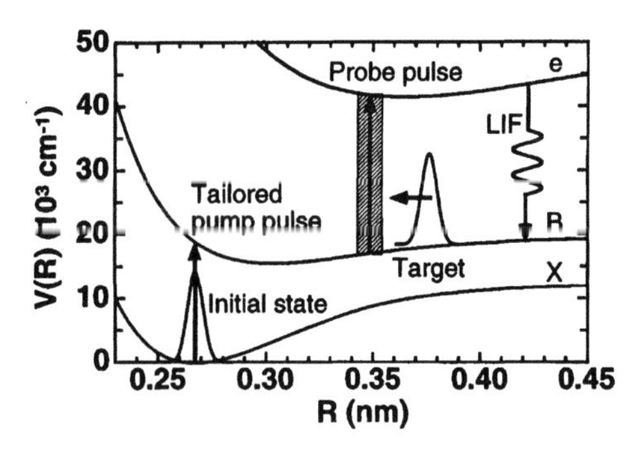

Figure 41. Anharmonic potential energy surface of molecular iodine for the X->B transition, showing the design of the reflectron optimal control experiment. An incident femtosecond laser pulse creates a vibrational wavepacket on the B excited state surface. The negatively chirped pulse first excites higher frequency components in the wavepacket, which have a longer round-trip time due to the anharmonicity. After reflection from the right hand side of the potential energy surface, the different frequency components arrive at the shaded area at approximately the same time to produce a focused minimum-uncertainty wavepacket. The size of the wavepacket is monitored by laser-induced fluorescence at a site near, but not on, the region of the potential energy surface where the minimum-uncertainty wavepacket is to be created. (Reprinted from ref. [122]. Copyright 1995 American Physical Society.)

Subsequently, Bardeen and coworkers have shown the ability to control the product branching ratio in the photodissociation of NaI by means of linearly chirped pulses, in agreement with the predictions of optimal control theory [124]. A variety of experiments using shaped pulses to control wavepacket motion or linearly chirped pulses to control multiphoton ionization have been reported [125-127]. In an unrelated experiment, Bardeen, Wang, and Shank [127] have performed a resonant impulsive stimulated Raman experiment on a dye molecule in liquid solution. The excited electronic state is displaced relative to the ground state, so the electronic resonance frequency changes as a function of

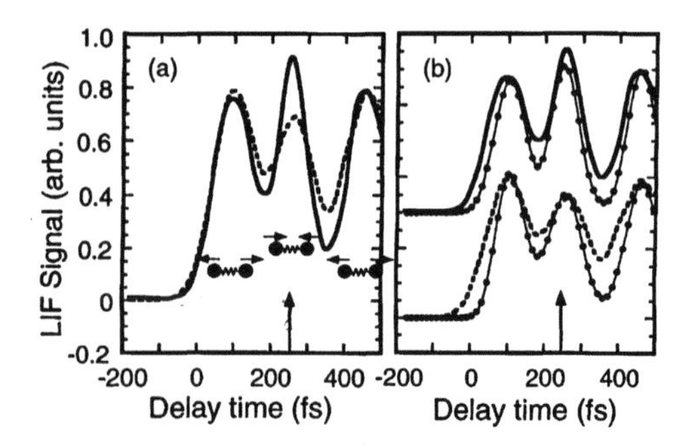

Figure 42. a) Experimental results and b) simulations of data for the reflectron experiment. The fields needed to achieve the minimum-uncertainty wavepacket were calculated by the use of optimal control theory and experimentally achieved by varying the chirp of the femtosecond excitation pulse. The size of the second peak shows enhancement with negative chirp (solid line) and diminishment with positive chirp (dashed line) of the incident laser pulse, indicating that the frequency components of the wavepacket were more tightly focused after reflecting from the right hand side of the potential energy surface when negative linear chirp was used. (Reprinted from ref. [122]. Copyright 1995 American Physical Society.)

the nuclear coordinate (Figure 43) [127]. A negatively chirped pulse enhances the creation of a vibrational wavepacket on the ground electronic state, and a positively chirped pulse retards it. Chirped pulses have also been used to enhance multiphoton absorption in molecular iodine [128].

4. Feedback control. As mentioned in Section IV-C, feedback control using genetic algorithms has been demonstrated in a real system by Bardeen, Yakovlev, Wilson, Carpenter, Weber, and Warren [129]. A genetic algorithm is a method for searching for the solution to a problem in a nonlinear parameter space. In brief, the genetic algorithm encodes some set of parameters of an output waveform into a genome. The suitabilities of randomly generated sets of parameters for achieving some specified goal are measured using a heuristic fitness function, and the best genomes are mated with each other (with the possibility of mutations) to create a new generation of genes. This cycle of testing and mating leads to the survival of the fittest parameters for solving the problem. Although it is best suited to problems with only a small number of parameters, genetic algorithms are "black boxes" that require the experimenter to set up only(!) the organization of the parameter space and the heuristic fitness function.

Bardeen et al. [129] used an acousto-optic modulator (AOM) pulse shaper to optimize the excited state population in the dye molecule IR125. The output waveform, a single pulse, is parametrized in terms of five parameters, including the amplitude, central frequency, pulse width, and linear and quadratic chirp. Each parameter may take one of 256 possible values, and the fitness of the genome is determined by measuring the fluorescence intensity for each pulse. The solution is reached after about ten generations with thirty gene sequences in each generation. Actually, Bardeen and coworkers performed two experiments: one in which the genetic algorithm optimizes the *effectiveness* of excited state population transfer, and another one in which the algorithm optimizes the *efficiency* of the excited state population transfer. See Figure 44 and Figure 45 [129]. Surprisingly, the solutions to these two problems are different. The most *efficient* pulse, that is, the one leading to the greatest fluorescence for a given pulse energy, is a somewhat narrowband

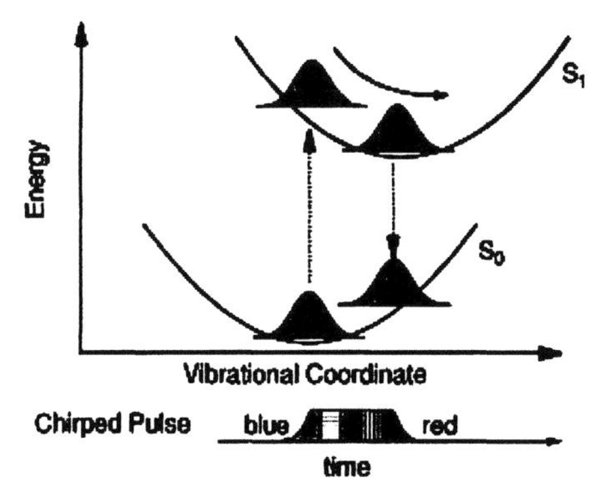

Figure 43. Chirped pulse excitation of a dye molecule in solution. A negatively chirped pulse enhances the creation of a vibrational wavepacket on the ground electronic state, and a positively chirped pulse retards it. (Reprinted from ref. [127]. Copyright 1995 American Physical Society.)

pulse centered on the absorption line of the dye. Bardeen and coworkers note that it is perhaps not surprising that the chirp parameters of the pulse do not matter much in this case in light of Brumer and Shapiro's "Emperor's New Clothes" analysis, which states that one-photon transitions are phase-insensitive [113]. The most *effective* pulse is a broadband

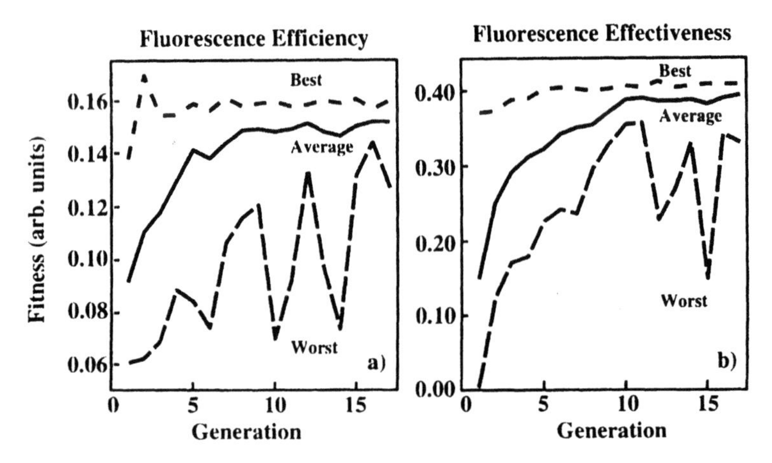

Figure 44. Genetic algorithm feedback control achievement for the cases of most efficient and most effective fluorescence of a dye molecule in solution. The heuristic achievement function measures how closely the members of a population come to reaching a specified goal. Successful members of each generation are chosen to spawn the next generation. Convergence occurs when the achievement function is stable. (Reprinted from ref. [129]. Copyright 1997 Elsevier Science B. V.)

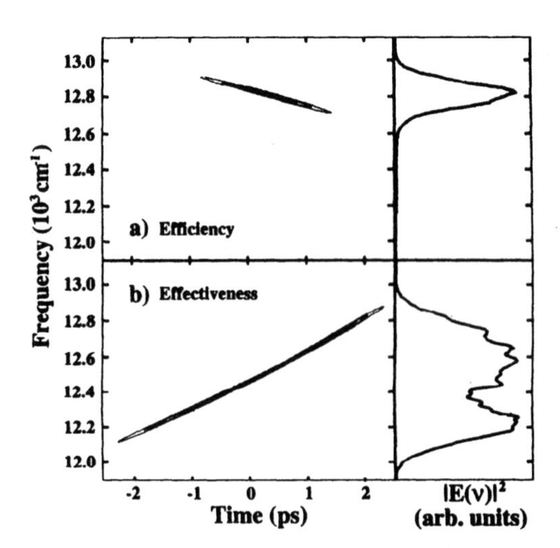

Figure 45. Typical solutions to the control problems of most efficient and most effective generation of fluorescence. The most efficient generator of fluorescence is a pulse, which need not be chirped, centered on the fluorescence absorption line. The most effective pulse is a broadband pulse with substantial linear chirp. Note that results would be different if a different control objective were specified. For example, if it were desired to create a minimum-uncertainty wavepacket on the excited state surface of the dye molecule, the most efficient pulse would probably not be the one that corresponds to maximum absorption. (Reprinted from ref. [129]. Copyright 1997 Elsevier Science B.V.)

pulse with substantial spectral and temporal structure and, generally, substantial positive linear chirp. In this case, positive linear chirp reduces stimulated emission from the excited state back to the ground state as the nuclear wavepacket moves on the excited state potential energy surface. The most effective pulse is not limited to the absorption line of the dye because the genetic algorithm controller is free to use additional energy from off resonance to drive the transition.

Krause, Reitze, Sanders, Kuznetsov, and Stanton [130] have simulated the control of wavepacket motion and terahertz emission from an asymmetric double quantum well structure using both variational optimal control theory and genetic algorithms.

Although the genetic algorithm has so far only been applied to simple systems with a single excitation pulse, in the future it could easily be applied to an experiment with a small number of control parameters.

VI. SUMMARY

Ultrafast spectroscopy at the impulsive limit has opened a new and important window to the world of fast chemical and physical events. The ability to observe molecules and materials as they undergo coherent vibrational motion, essentially recording time-resolved "stop-action" observations at various stages of vibrational distortion, has provided new information about molecular and collective behavior. The extension of this capability to observation of elementary motions that are not vibrational (e.g. during chemical reactions or structural rearrangements) has provided and will continue provide a wealth of information about irreversible changes which are monitored as they occur.

Femtosecond pulse shaping and multiple-pulse femtosecond spectroscopy now offer a set of extraordinary new capabilities for manipulation of elementary molecular and collective behavior. Most experimental demonstrations to date have been on relatively simple model systems, but rapid advances in the underlying technologies, the theoretical underpinnings of the field, and in our appreciation of what kinds of objectives are now coming into view, foreshadow a period of profound change in our ability to manipulate matter with light. We look forward to this period with great anticipation.

REFERENCES

1. M. M. Wefers, K. A. Nelson, *Opt. Lett.* 18, 2032 (1993).
2. A. M. Weiner, J. P. Heritage, E. M. Kirschner, *J. Opt. Soc. Am. B* 5, 1563 (1988).
3. C. W. Hillegas, J. X. Tull, D. Goswami, D. Strickland, W. S. Warren, *Opt. Lett.* 19, 737 (1994).
4. E. J. Heller, *J. Chem. Phys.* 68, 2066-2076 (1978).
5. E. J. Heller, R. L. Sundberg, D. Tannor, *J. Phys. Chem.* 86, 1822-1833 (1982).
6. E. J. Heller, *Acc. Chem. Res.* 14, 368-375 (1981).
7. S.-Y. Lee, E. J. Heller, *J. Chem. Phys.* 71, 4777-4788 (1979).
8. L. Dhar, J. A. Rogers, K. A. Nelson, *Chem. Rev.* 94, 157 (1994).
9. A. Yariv, *Quantum Electronics* (John Wiley & Sons, New York, 1989).

10. Y. R. Shen, *Principles of Nonlinear Optics* (Wiley, 1984).

11. A. A. Maznev, T. F. Crimmins, K. A. Nelson, accepted, *Opt. Lett.* (1998).

12. W. Struve, *Fundamentals of Molecular Spectroscopy* (John Wiley & Sons, New York, 1989).

13. L. Dhar, J. A. Rogers, K. A. Nelson, *Chem. Rev.* 94, 157-193 (1994).

14. T. J. Smith, L. W. Ungar, J. A. Cina, *J. Luminescence* 58, 66-73 (1994).

15. V. Romero-Rochin, J. Cina, *Phys. Rev. A* 50, 763-778 (1994).

16. S. Mukamel, *Principles of Nonlinear Optical Spectroscopy* (The Clarendon Press, Oxford, 1995).

17. W. T. Pollard, S.-Y. Lee, R. A. Mathies, *J. Chem. Phys.* 92, 4012-4029 (1990).

18. J. Chesnoy, A. Mokhtari, *Rev. Phys. Appl.* 22, 1743-171747 (1987).

19. R. Y. Chiao, C. H. Townes, B. P. Stoicheff, *Phys. Rev. Lett.* 12, 592 (1964).

20. S. D. Silvestri, J. G. Fujimoto, E. P. Ippen, E. B. Gamble, L. R. Williams, K. A. Nelson, *Chem. Phys. Lett.* 116, 146-152 (1985).

21. D. H. Auston, *Appl. Phys. Lett.* 43, 713-715 (1983).

22. D. H. Auston, K. P. Cheung, J. A. Valdmanis, D. A. Kleinman, *Phys. Rev. Lett.* 53, 1555-1558 (1984).

23. D. H. Auston, M. C. Nuss, *IEEE J. Quantum Electron.* 24, 184-197 (1988).

24. D. A. Kleinman, D. H. Auston, *IEEE J. Quantum Electron.* QE-20, 964-970 (1984).

25. R. M. Koehl, S. Adachi, K. A. Nelson, submitted, *J. Chem. Phys.* (1998).

26. T. P. Dougherty, G. P. Wiederrecht, K. A. Nelson, M. H. Garrett, H. P. Jensen, C. Warde, *Science* 258, 770 (1992).

27. S. Ruhman, A. G. Joly, B. Kohler, L. R. Williams, K. A. Nelson, *Rev. Phys. Appl.* 22, 1717-1734 (1987).

28. S. Ruhman, A. G. Joly, K. A. Nelson, *J. Chem. Phys.* 86, 6563 (1987).

29. S. Ruhman, B. Kohler, A. G. Joly, K. A. Nelson, *IEEE J. Quantum Electronics* 24, 470-481 (1988).

30. S. Ruhman, L. R. Williams, A. G. Joly, B. Kohler, K. A. Nelson, *J. Phys. Chem.* 91, 2237-2240 (1987).

31. S. Ruhman, B. Kohler, A. G. Joly, K. A. Nelson, *Chem. Phys. Lett.* 141, 16-24 (1987).

32. C. H. Lin, J. P. Heritage, T. K. Gustafson, R. Y. Chiao, J. P. McTague, *Phys. Rev. A* 13, 813 (1976).

33. N. F. Scherer, D. F. Jonas, G. R. Fleming, *J. Chem. Phys.* 99, 153 (1993).

34. L.-T. Cheng, K. A. Nelson, *Phys. Rev. B* 37, 3603-3610 (1988).

35. S. M. Silence, S. R. Goates, K. A. Nelson, *Chem. Phys.* 149, 233-259 (1990).

36. M. J. Banet, M. Fuchs, J. A. Rogers, R. Logan, A. A. Maznev, K. A. Nelson, *Appl. Phys. Lett.* (1998).

37. Y. Yang, K. A. Nelson, *J. Chem. Phys.* 103, 7732-7739 (1995).

38. Y. Yang, K. A. Nelson, *J. Chem. Phys.* 103, 7722-7731 (1995).

39. Y. Yang, K. A. Nelson, *J. Chem. Phys.* 104, 1-8 (1996).

40. A. R. Duggal, J. A. Rogers, K. A. Nelson, M. Rothschild, *Appl. Phys. Lett.* 60, 692-694 (1992).

41. T. F. Crimmins, A. A. Maznev, K. A. Nelson, submitted, *Appl. Phys. Lett.* (1998).
42. C. Thomsen, H. T. Grahn, H. J. Maris, J. Tauc, *Phys. Rev. B* 34, 4129-4138 (1986).
43. H. T. Grahn, H. J. Maris, J. Tauc, *IEEE J. Quantum Electron.* 25, 2562-2568 (1989).
44. M. J. Rosker, F. W. Wise, C. L. Tang, *Phys. Rev. Lett.* 57, 321-324 (1986).
45. M. J. Rosker, M. Dantus, A. H. Zewail, *J. Chem. Phys.* 89, 6113-6127 (1988).
46. K. A. Nelson, L. R. William, *Phys. Rev. Lett.* 58, 745 (1987).
47. J. Chesnoy, A. Mokhtari, *Phys. Rev. A* 38, 3566-3576 (1988).
48. M. Dantus, R. M. Bowman, A. H. Zewail, *Nature* 343, 737-739 (1990).
49. M. Cho, J.-Y. Yu, T. Joo, Y. Nagasawa, S. Passino, G. R. Fleming, *J. Phys. Chem.* 100, 11944 (1995).
50. T. Joo, W. Jia, G. R. Fleming, *J. Chem. Phys.* 102, 4065 (1995).
51. R. Zadoyan, Z. Li, C. C. Martens, V. A. Apkarian, *J. Chem. Phys.* 101, 6648-6657 (1994).
52. M. H. Vos, F. Rappaport, J. C. Lambry, J. Breton, J.-L. Martin, *Nature* 363, 320-325 (1993).
53. Q. Wang, R. W. Schoenlein, L. A. Peteanu, R. A. Mathies, C. V. Shank, *Science* 266, 422-424 (1994).
54. L. Genberg, L. Richard, G. McLendon, R. J. D. Miller, *Science* 251, 1051-1054 (1991).
55. T. K. Cheng, S. D. Brorson, A. S. Kazeroonian, J. S. Moodera, G. Dresselhaus, M. S. Dresselhaus, E. P. Ippen, *Appl. Phys. Lett.* 57, 1004-1006 (1990).
56. T. K. Cheng, J. Vidal, H. J. Zeiger, G. Dresselhaus, M. S. Cresselhaus, E. P. Ippen, *Appl. Phys. Lett.* 59, 1923-1925 (1991).
57. T. K. Cheng, L. H. Acioli, J. Vidal, H. J. Zeiger, G. Dresselhaus, M. S. Dresselhaus, E. P. Ippen, *Appl. Phys. Lett.* 62, 1901-1903 (1993).
58. H. J. Zeiger, T. K. Cheng, E. P. Ippen, J. Vidal, G. Dresselhaus, M. S. Dresselhaus, *Phys. Rev. B* 54, 105-123 (1996).
59. S. Hunsche, K. Wienecke, T. Dekorsy, H. Kurz, *Phys. Rev. Lett.* 75, 1815-1818 (1995).
60. S. Hunsche, H. Kurz, *Appl. Phys. A* 65, 221-229 (1997).
61. G. C. Cho, W. Kütt, H. Kurz, *Phys. Rev. Lett.* 65, 764-766 (1990).
62. J. M. Chwalek, C. Uher, J. F. Whitaker, G. A. Mourou, J. Agostinelli, M. Lelental, *Appl. Phys. Lett.* 57, 1696-1698 (1990).
63. W. Albrecht, T. Kruse, H. Kurz, *Phys. Rev. Lett.* 69, 1451-1454 (1992).
64. S. B. Fleischer, B. Pevzner, D. J. Dougherty, H. J. Zeiger, G. Dresselhaus, M. S. Dresselhaus, E. P. Ippen, A. F. Hebard, *Appl. Phys. Lett.* 71, 2734-2736 (1997).
65. M. Haner, W. S. Warren, *Appl. Phys. Lett.* 52, 1458 (1988).
66. M. M. Wefers, K. A. Nelson, *J. Opt. Soc. Am. B* 12, 1343-1362 (1995).
67. M. M. Wefers, K. A. Nelson, *IEEE J. Quantum Electron.* 32, 161-172 (1996).
68. J. P. Heritage, A. M. Weiner, R. N. Thurston, *Opt. Lett.* 10, 609 (1985).
69. R. N. Thurston, J. P. Heritage, A. M. Weiner, W. J. Tomlinson, *IEEE J. Quantum Electron.* QE-22, 682-696 (1986).
70. O. E. Martinez, *IEEE J. Quantum Electron.* QE-23, 59 (1987).
71. O. E. Martinez, *J. Opt. Soc. Am. B* 3, 929-934 (1986).

72. A. M. Weiner, D. E. Leaird, J. S. Patel, J. R. Wullert, *Opt. Lett.* 15, 326 (1990).
73. A. M. Weiner, D. E. Leaird, J. S. Patel, J. R. Wullert, *IEEE J. Quantum Electron.* 28, 908-920 (1992).
74. M. M. Wefers, K. A. Nelson, *Opt. Lett.* 20, 1047 (1995).
75. C. H. Gooch, H. A. Tarry, *J. Phys. D* 8, 1575 (1975).
76. M. M. Wefers, K. A. Nelson, A. M. Weiner, *Opt. Lett.* 21, 746 (1996).
77. M. C. Nuss, R. L. Morrison, *Opt. Lett.* 20, 740-742 (1995).
78. R. M. Koehl, T. Hattori, K. A. Nelson, to be published.
79. W. S. Warren, H. Rabitz, M. Dahleh, *Science* 259, 1581 (1993).
80. H. Kawashima, M. M. Wefers, K. A. Nelson, *Ann. Rev. Phys. Chem.* 46, 627 (1995).
81. B. Kohler, J. L. Krause, F. Raksi, K. R. Wilson, V. V. Yakovlev, R. M. Whitnell, Y. Yan, *Acc. Chem. Res.* 28, 133-140 (1995).
82. N. Bloembergen, A. H. Zewail, *J. Phys. Chem.* 88, 5459 (1984).
83. A. E. Bryson Jr., Y.-C. Ho, *Applied Optimal Control* (Hemisphere Publishing Corp., Washington, 1975).
84. I. M. Gelfand, S. V. Fomin, *Calculus of Variations* (Prentice-Hall, Inc., Englewood Cliffs, N. J., 1963).
85. A. J. Lichtenberg, M. A. Lieberman, *Regular and Chaotic Dynamics* (Springer-Verlag, New York, 1992).
86. H. Goldstein, *Classical Mechanics* (Addison-Wesley Publishing Co., Reading, MA, 1980).
87. D. J. Nesbitt, R. W. Field, *J. Phys. Chem.* 100, 12735 (1996).
88. D. J. Tannor, S. A. Rice, *J. Chem. Phys.* 83, 5013 (1985).
89. D. J. Tannor, R. Kosloff, S. A. Rice, *J. Chem. Phys.* 85, 8505-8520 (1986).
90. E. J. Heller, *J. Chem. Phys.* 68, 2066 (1978).
91. A. P. Pierce, M. A. Dahleh, H. Rabitz, *Phys. Rev. A* 37, 4950 (1988).
92. S. Shi, H. Rabitz, *J. Chem. Phys.* 92, 364-376 (1990).
93. M. Demiralp, H. Rabitz, *Phys. Rev. A* 47, 809-816 (1993).
94. M. Demiralp, H. Rabitz, *Phys. Rev. A* 47, 831-837 (1993).
95. Y. Yan, R. E. Gillilan, R. M. Whitnell, K. R. Wilson, S. Mukamel, *J. Phys. Chem.* 97, 2320 (1993).
96. U. Fano, *Rev. Mod. Phys.* 29, 74-93 (1957).
97. J. Cao, M. Messina, K. R. Wilson, *J. Chem. Phys.* 106, 5239-5248 (1997).
98. J. L. Krause, R. M. Whitnell, K. R. Wilson, Y. Yan, S. Mukamel, *J. Chem. Phys.* 99, 6562 (1993).
99. K. F. Man, *Genetic Algorithms for Control and Signal Processing* (Springer-Verlag, Berlin, 1997).
100. R. L. Haupt, S. E. Haupt, *Practical Genetic Algorithms* (Wiley, New York, 1998).
101. C. Brennan, K. A. Nelson, *J. Chem. Phys.* 107, 9691-9694 (1997).
102. P. Brumer, M. Shapiro, *Scientific American* 272, 56 (1995).
103. N. F. Scherer, A. J. Ruggiero, M. Du, G. R. Fleming, *J. Chem. Phys.* 93, 856 (1990).

104. N. F. Scherer, R. J. Carlson, A. Matro, M. Du, A. J. Ruggiero, V. Romero-Rochin, J. A. Cina, G. R. Fleming, S. A. Rice, *J. Chem. Phys.* 95, 1487 (1991).
105. N. F. Scherer, A. Matro, M. Du, R. J. Carlson, J. A. Cina, G. R. Fleming, *J. Chem. Phys.* 96, 4180 (1992).
106. P. C. M. Planken, M. C. Nuss, I. Brener, K. W. Goossen, M. S. C. Luo, L. Chuang, L. Pfeiffer, *Phys. Rev. Lett.* 69, 3800 (1992).
107. P. C. M. Planken, I. Brener, M. C. Nuss, M. S. C. Luo, S. L. Chuang, *Phys. Rev. B* 48, 4903 (1993).
108. I. Brener, P. C. M. Planken, M. C. Nuss, M. S. C. Luo, S. L. Chuang, L. Pfeiffer, D. E. Leaird, A. M. Weiner, *J. Opt. Soc. Am. B* 11, 2457-2469 (1994).
109. M. S. C. Luo, S. L. Chuang, P. C. M. Planken, I. Brener, H. G. Roskos, M. C. Nuss, *IEEE J. Quantum Electron.* 30, 1478 (1994).
110. P. C. M. Planken, I. Brener, M. C. Nuss, M. S. C. Luo, S. L. Chuang, L. N. Pfeiffer, *Phys. Rev. B* 49, 4668 (1994).
111. M. C. Nuss, P. C. M. Planken, I. Brener, H. G. Roskos, M. S. C. Luo, S. L. Chuang, *Appl. Phys. B* 58, 249-259 (1994).
112. M. S. C. Luo, S. L. Chuang, P. C. M. Planken, I. Brener, M. C. Nus, *Phys. Rev. B* 48, 11043 (1993).
113. P. Brumer, M. Shapiro, *Chem. Phys.* 139, 221 (1989).
114. A. M. Weiner, D. E. Leaird, G. P. Wiederrecht, K. A. Nelson, *Science* 247, 1317 (1990).
115. R. Kosloff, A. D. Hammerich, D. Tannor, *Phys. Rev. Lett.* 69, 2172 (1992).
116. J. A. Cina, T. J. Smith, *J. Chem. Phys.* 98, 9211 (1993).
117. T. J. Smith, J. A. Cina, *J. Chem. Phys.* 104, 1272 (1996).
118. E. Hiller, J. A. Cina, *J. Chem. Phys.* 105, 3419 (1996).
119. G. P. Wiederrecht, T. P. Dougherty, L. Dhar, K. A. Nelson, D. E. Leaird, A. M. Weiner, *Phys. Rev. B* 51, 916 (1995).
120. I. Brener, P. C. M. Planken, M. C. Nuss, L. Pfeiffer, D. E. Leaird, A. M. Weiner, *Appl. Phys. Lett.* 63, 2213 (1993).
121. M. M. Wefers, H. Kawashima, K. A. Nelson, *J. Chem. Phys.* 102, 9133-9136 (1995).
122. B. Kohler, V. V. Yakovlev, J. Che, J. L. Krause, M. Messina, K. R. Wilson, N. Schwentner, R. M. Whitnell, Y. Yan, *Phys. Rev. Lett.* 74, 3360 (1995).
123. R. Trebino, D. J. Kane, *J. Opt. Soc. Am. A* 10, 1101 (1993).
124. C. J. Bardeen, J. Che, K. R. Wilson, V. V. Yakovlev, P. Cong, B. Kohler, J. L. Krause, M. Messina, *J. Phys. Chem. A* 101, 3815 (1997).
125. A. Assion, T. Baumert, J. Helbing, V. Seyfried, G. Gerber, *Chem. Phys. Lett.* 259, 488 (1996).
126. J. M. Pananikolas, R. M. Williams, S. R. Leone, *J. Chem. Phys.* 107, 4172 (1997).
127. C. J. Bardeen, Q. Wang, C. V. Shank, *Phys. Rev. Lett.* 75, 3410 (1995).
128. Y. V. Yakovlev, C. J. Bardeen, J. Che, J. Cao, K. R. Wilson, *J. Chem. Phys.* 108, 2309-2313 (1998).

129. C. J. Bardeen, V. V. Yakovlev, K. R. Wilson, S. D. Carpenter, P. M. Weber, W. S. Warren, *Chem. Phys. Lett.* 280, 151-158 (1997).
130. J. L. Krause, D. H. Reitze, G. D. Sanders, A. V. Kuznetsov, C. J. Stanton, *Phys. Rev. B* 57, 9024 (1998).

LONG SEMINARS

SOLITONS IN OPTICAL FIBERS AND APPLICATION TO HIGH BIT RATE TRANSMISSION* (Abstract only)

Linn F. Mollenauer

Bell Labs
Lucent Technologies
Holmdel, NJ, USA

In optical fibers, the soliton is a completely nondispersive pulse made possible by the small but significant χ_3 in silica glass. As such, solitons enable one of the most technologically important applications of nonlinear optics, viz., high bit rate, ultra long distance data transmission. The lectures introduce the elegant, fundamental physics of solitons, followed by a brief sketch of the corresponding transmission technology.

I. The Non-linear Schrodinger Equation and Solitons:
Phase and Group Velocities; Some pertinent facts about Fourier transforms; Derivation of the NLS equation; Solitons units; Effect of the dispersive term; Effect of the non-linear term; Origin of the Soliton; "Path-average" solitons; The split-step Fourier transform method for numerical solution of the NLS equation.

II. Amplifier Spontaneous Emission and other Rate Limiting Effects:
Growth of ASE noise in a broad-band transmission line; Amplitude noise; Jitter in pulse arrival times (The Gordon Haus effect); Calculation of BER rates; Amelioration of rate limiting effects with "guiding" optical filters.

III. Wavelength Division Multiplexing in Soliton Transmission:
Soliton-soliton collision and their effects (cross-phase modulation, four wave mixing); Control of collision effects with guiding filters; Experimental results: Nx10 Gbit/s WDM.

----

* For a complete text corresponding to these lectures, see L. F. Mollenauer, J. P. Gordon, and P. V. Mamyshev, in *Optical Fiber Telecommunications III*, I.P. Kaminov and T.L. Koch, eds.(Academic Press, San Diego, Calif., 1997), Chapt.12 (pp. 373-460).

HOT LUMINESCENCE AND FAST RELAXATION OF OPTICAL CENTERS IN SOLIDS

Giuseppe Baldacchini

ENEA
Divisione Fisica Applicata
Centro Ricerche di Frascati
C.P. 65, 00044 Frascati (Roma), Italy

ABSTRACT

Optical centers in solids have been in the past and are still nowadays at the center of attention by the solid state community for their basic properties and noteworthy applications mainly in optoelectronics. Indeed after excitation they usually display a complex optical cycle, whose end product can be an efficient emission. However several processes like hot luminescence, crossover, nonradiative relaxation, energy transfer, etc. may occur during the cycle, and decrease the emission quantum yield, sometimes down to a complete quenching. On this regard one fundamental question is whether all these processes, with the exception of the hot luminescence, occur before or after the configurational relaxation of the excited state. This question has been studied extensively in color centers in alkali halides, mainly in the F center, which is a model case for most of the optical centers in other materials. The results obtained up to now show in the majority of the cases that the processes mentioned above occur after the electronic excitation reaches the relaxed excited state. However there are a few experimental evidences with doubtful interpretation, which require further investigations by using appropriate techniques, for instance ultrafast spectroscopy.

1. INTRODUCTION

Vibronic systems are today a very important class of optoelectronic materials mainly for their optical cycle which can be described by a configurational coordinate diagram showing directly the implications of the electron-phonon interaction. Figure 1 displays the case when only two states are taken into consideration, the ground and excited states, together with the conduction band which corresponds to the electron of the

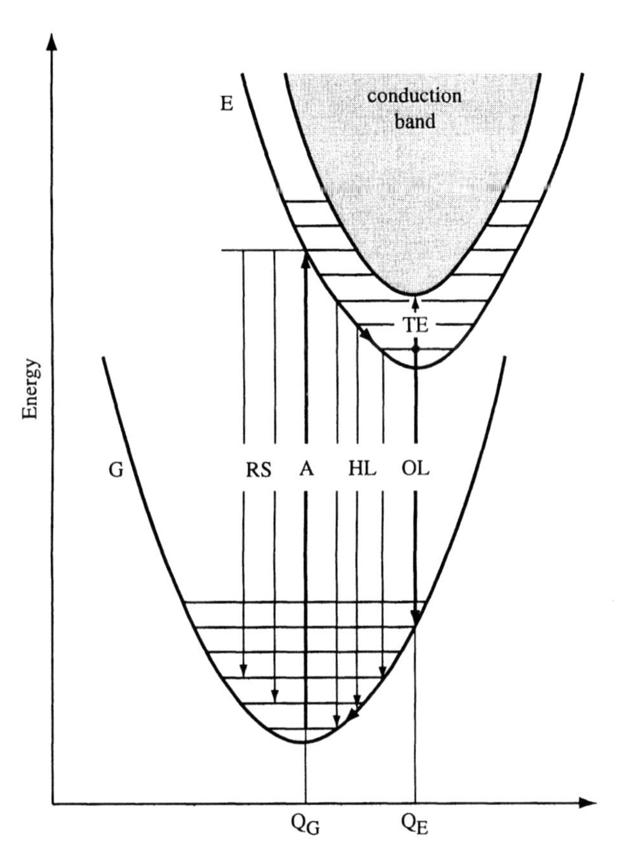

Figure 1. Configuration coordinate diagram illustrating the radiation and radiationless transitions in a simplified optical cycle of a typical vibronic material.

point defect (impurity ions, color center, etc.) completely delocalized in the crystal. The two parabolas represent the ground and excited states of the vibronic system, while the normal mode coordinate Q represents the lattice displacement around the point defect. The difference Q_E-Q_G is a measurement of the electron-lattice coupling, which is usually very strong for color centers and transition metal ions, and very weak for rare earth ions. It is immediately understood why the former have bigger Stokes shifts and wider emission bands than the latter ones. Anyway for both of them optical excitation produces a vertical transition from the ground state, GS, to the excited state in a highly excited vibrational level, URES, i.e. unrelaxed excited state. The vibronic system relaxes in a matter of a few ps to the minimum of the excited parabola which is known as relaxed excited state, RES. The excitation can remain in this peculiar state a relatively long time, from ms to ns, before returning to the unrelaxed ground state, URGS, with a vertical transition accompanied very often by light emission. Soon afterwards a lattice relaxation completes the optical cycle to the GS. However the emission, which is also called ordinary luminescence, OL, is not the only radiative process which takes place after excitation. Indeed, little beyond the obvious Rayleigh line, a structured emission reveals the presence of Raman scattering, RS, which contains detailed information on the lattice modes coupled to the electronic transition. Still beyond that and up to OL there is a very weak emission tail known as hot luminescence, HL, which originates from decaying processes during the relaxation. In the case the temperature is high enough, $kT \geq \Delta E$, where ΔE is the energy distance of the RES form the conduction band, a thermal excitation process proportional to $\nu \exp(-\Delta E/kT)$ can subtract a fraction of excitations from the active optical cycle. Moreover several radiationless processes like crossover, nonradiative relaxation, energy transfer, etc.... may occur during the cycle, which decrease the emission quantum yield, sometimes down to a complete quenching.

One important problem regarding the radiation and radiationless processes just described, is whether they occur before or after the configurational relaxation of the excited state. Apart the RS, strongly related to the absorption process at the intense excitation wavelength, and the HL, by definition emitted during relaxation, both of which are several order of magnitude less intense with respect to the OL when it is not quenched, all the other processes in principle can happen both during and after the relaxation, when the vibronic system has reached the relatively long living RES. This question is not a trivial one, or rather it is of paramount importance not only for basic studies but also for optoelectronic applications. Indeed the final quantum yield of the emission is strongly related to the processes happening during the optical cycle, and the only hope to increase it in old and new active materials is to know in detail when and where these processes actually occur.

This problem has been discussed at length in color centers in alkali halides, mainly F centers, which are well known model cases for all kind of vibronic materials. As such they are well described by the energy diagram of Fig. 1, where the lifetime of the RES is never longer than a few μs [1]. A beautiful experimental verification of the processes described by this configurational coordinate diagram, with the exception of TE which is not observed because of the low temperature used through the work, is reported for the F center in KCl in Fig. 2 [2]. It is composed of the RS which is expanded in the inset, a long tail extending for thousands of wavenumbers and decreasing in intensity toward lower energy, the HL, and eventually the well known Stokes shifted luminescence at ~ 1 μm. The extremely weak luminescence between the RS and the OL has been discovered during measurements of RS [3], and its intensity relative to the OL should be roughly given by the ratio of the relaxation time to the RES lifetime, i.e. $\sim 10^{-12}/10^{-6} \sim 10^{-6}$ [4], which is approximately the value shown in Fig. 2.

Although the hot luminescence phenomenon is not completely understood from the theoretical point of view up to now, it is evident at a first analysis that the hot luminescence is competing with the fast relaxation process just during its development. So, it seems logical to extend also to other phenomena the same approach, i.e. the occurrence during relaxation of any other radiation or radiationless process, allowed by the quantum mechanic rules, is regulated by its probability compared with that of the relaxation. Because the relaxation is one of the fastest process known up to now, it is concluded that significant effects cannot occur during lattice relaxation but only after the excitation reaches the RES. However this topic has been always the subject of a lively debate which dates back since the Dexter-Klick-Russell (DKR) criterion for the occurrence of luminescence [5]. Anyway the very basic idea of the crossover, which is a change of quantum state when two energy levels cross each other, was and is still so radicated in the scientific culture that it is still discussed commonly nowadays, while it should be used with extreme caution.

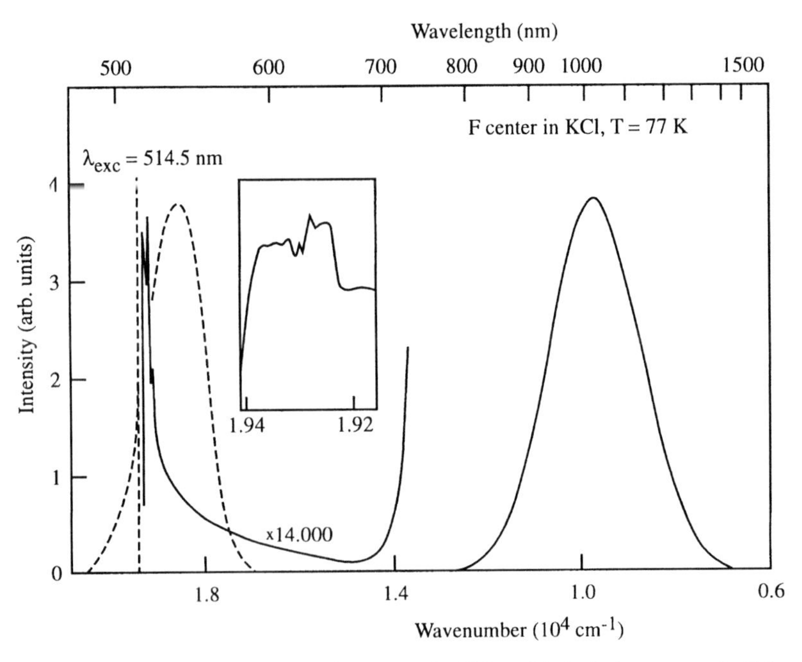

Figure 2. Resonant secondary radiation of F centers in KCl at 77 K. It is composed of the Raman scattering which is expanded in the inset, the hot luminescence and the ordinary F luminescence. The absorption of the F center and the position of the excitation radiation are also shown by dashed lines. The relaxation proceeds from left to right in a time scale of the order of ps [2].

In this paper I will describe three examples, taken from color centers in alkali halides, which have been carefully studied in the last decades with classical tools and lately by using ultrafast spectroscopic techniques. Although the former experiments did show consistently that the excitations reached in the majority of cases the RES first, and only afterward they could be engaged by other processes, only the latter experiments gave the final proof of that, and opened the new and rich field of investigating the details of lattice relaxation. The three examples presented in the next Sections are regarding the crossing of the 2s- and 2p-like excited states of the F center, the intrinsic quenching of the F-center luminescence, and the fast exchange of energy between F center and molecular ions like CN^- and OH^-. Eventually Section 5 deals with the lesson learned by the results described in the previous Sections.

2. CROSSING OF THE 2s AND 2p EXCITED STATES

All optical excitations in the various absorption bands, which correspond to the atomic-like levels 2s, 2p, 3d, 3p,...., of F centers in alkali halides, produce only one emission band which proceeds from a well defined state, the RES, where the electrons rest for a relatively long time. However, when people started to investigate the lifetime of this state, they got the first surprise. In KCl at low temperatures it was found $\tau = 580$ ns against the expected value of $\tau = 10$ ns from the known atomic theory [6]. In order to explain such unexpected long value for the lifetime, which is common to all F centers in the various alkali halides, essentially two models for the RES have been taken in consideration: the large-orbit excited-state wavefunction [7], and the crossing of the 2s and 2p energy levels [8].

Without going in too much details, it is clear that the overlapping between the RES and the URGS wavefunctions, which is inversely proportional to the lifetime in question, decreases significantly in the case the RES wavefunction is more diffuse than the URGS one. Precise calculations have shown that the experimental values of τ can be easily justified in this manner [7], and optical-ENDOR experiments have demonstrated that the RES wave-function is still significant at five cation-anion distances from the vacancy [9], while the GS one is vaning just at one cation-anion distance [10].

On the other hand, it has been found by Stark effect measurements that the 2s level lies above the 2p level in absorption [11] and, by taking into account the structure of the ions together with the dielectric polarization effects, that the 2s and 2p levels cross each other during relaxation, so that when the equilibrium is reached in the RES the 2s state lies below the 2p state by a few tens of meV [8] ,as shown in Fig. 3. In this case the RES is not anymore a pure 2p state but rather a 2s-like state with some admixture through lattice vibrations with the higher 2p state. Because only the latter gives a non zero dipole matrix element for emission, the small amount of it in the resulting RES can justify the long value of the lifetime.

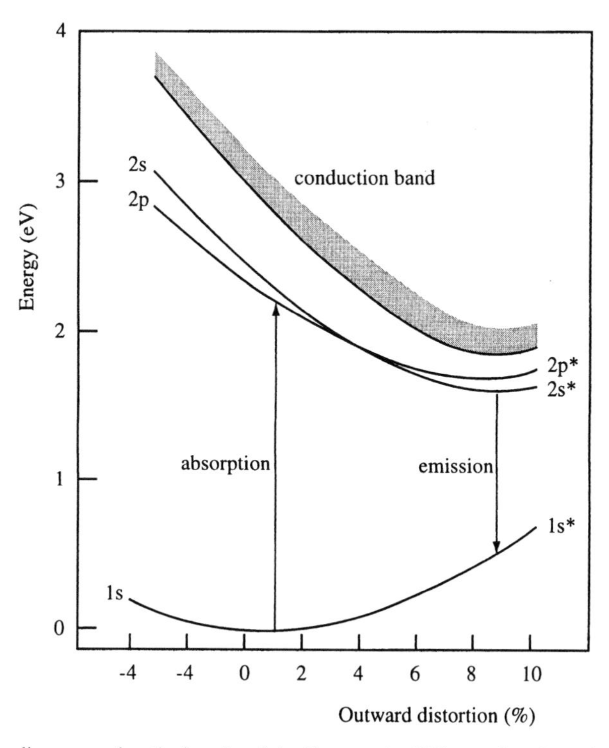

Figure 3. Energy diagram and optical cycle of the F center in KCl as a function of the position of the neighbouring positive ions, which shows the crossing between the 2s and 2p energy levels [8].

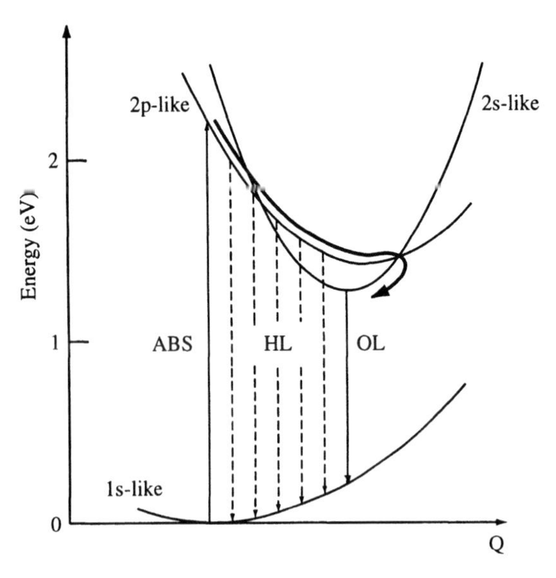

Figure 4. Optical cycle of the F center in KCl showing the crossing of the 2s and 2p states and the relaxation process without crossover [12].

Anyway the two previous models are not excluding each other, and most probably both diffuse wavefunction and 2s-2p crossing coexist and account together for the long value of the lifetime. In addition, it is also commonly believed that the excitation follows the 2p potential curve at the beginning and the 2s one after the crossover point during the relaxation, see Fig. 3. Because this process did not disturb any well defined theoretical and experimental tenet, it was always accepted without any critical discussion. However recent measurements on hot luminescence have produced results which do not agree with such point of view [12].

It was known since its discovery [3] that the HL was retaining a partial polarization, if the excitation was polarized, for thousands of wavenumbers. Recently [12] this still unexplained phenomenon was investigated again by using time-resolved fast spectroscopy in order to better disentangle HL from OL. By using excitation pulses 8 ps long, it has been possible to confirm the previous observations obtained by using cw excitations, and to obtain new results in the region of the OL. More exactly in KCl at 77K the polarization of the HL remains practically constant at a value of about 40% down to ~ 13,000 cm^{-1} at the onset of OL, while at the same time its intensity undergoes a decreasing of more than an order of magnitude. By decreasing further the wavenumber, the polarization of the HL decreases down to 0%, which is reached at 11,000 cm^{-1}, well inside the OL band, see Fig. 2, while its intensity remains more or less constant. Moreover the pulse shapes of the HL and excitation laser pulses are the same down to the onset of OL, and the whole relaxation process lastes less than 15 ps, which is the time resolution of the experimental apparatus.

These new and outstanding results are not compatible with the excitation crossing during the relaxation as implied by Fig. 3, but rather with an excitation relaxing unimpeded along the same potential energy 2p-like state down to its minimun value, from where a transition to the lower 2s state occurs afterward. Figure 4 shows clearly this new proposed mechanism, where the crossing is avoided and the relaxing excitation packet is so fast that the presence of the potential energy of the 2s-like state is practically not felt during the relaxation.

3. QUENCHING OF THE F-CENTER LUMINESCENCE

In F-center systems, having cared for the purity of the starting materials in the preparation and growth of crystals, avoided the formation of aggregated centers during coloration techniques, chosen low temperatures in order to prevent the thermal excitation in the conduction band, and ignored the extremely weak RS and HL, only absorption, relaxation (twice), and luminescence processes are expected, as described in Fig. 1. However this is not the most common case nowadays, and it was not at the time when emissions in F centers were discovered, and the first criterion for the lacking of luminescence in potentially emitting materials was laid down [5]. Indeed the processes in the optical cycle of F centers and vibronic materials as well, are better described by the configurational coordinate diagram of Fig. 5. It shows in particular the case when the GS and ES energy potentials cross each other, and the position of the crossing point X is responsible for the occurrence or quenching of the emission process. Indeed it was postulated that emission will always occur if $E_X > E_R$, while for $E_X < E_R$ the system can cross over between the two states during the oscillation at one of the vibrational level nearby point X. In this way the RES in C is never reached and the luminescence transition C→D is killed. Indeed the excitation from X relaxes directly to A following the

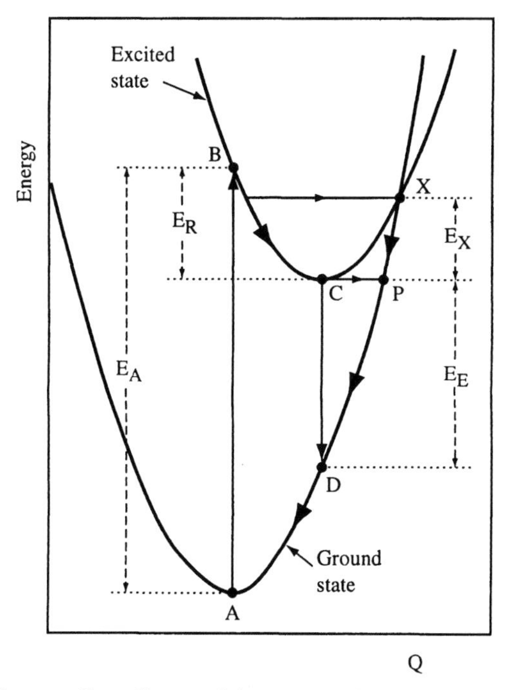

Figure 5. Configuration coordinate diagram of the ground and excited states of an F center with the appropriate transitions for the occurrence or non occurrence of luminescence.

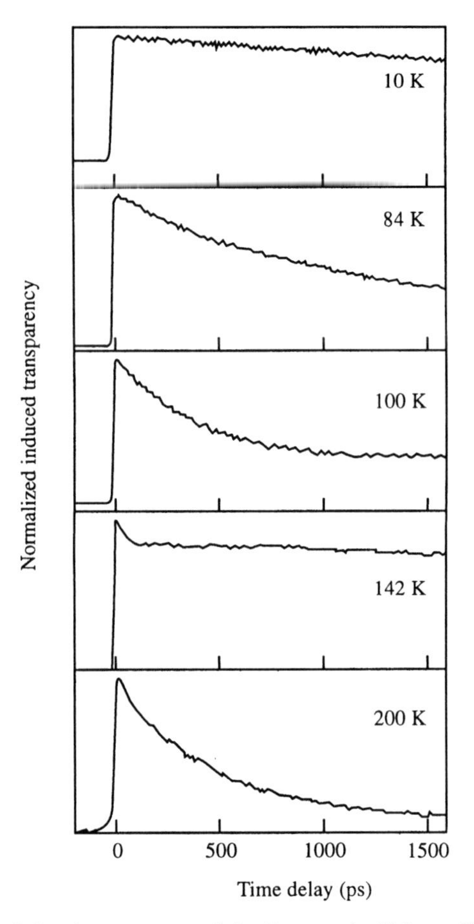

Figure 6. Decay of the induced transparency of the F center in NaBr at different temperatures. The wavelength of the excitation pulses is 580 nm and the time resolution 7 ps [22].

GS parabola. The DKR criterion was later on revisited [13] when more data were available, and its was found that for $E_R/E_A < 1/4$ there was luminescence, while for $E_R/E_A > 1/4$ there was a complete quenching of it. This sharp model was subsequently smoothed, after the discovery of weak but well defined emissions in NaI and NaBr [14] where luminescence was not expected, by introducing the concept of the branching ratio [15]. The latter is given by the probability that the excitation will remain in the excited state at the crossing point X with respect to the probability that the excitation will follow the ground state potential. In this way the theory could account for the just discovered weak luminescences more than satisfactory [4,16].

However, the quenching of luminescence caused by the crossover phenomenon requires that in NaI and NaBr just a few electrons should reach the RES. But the same experiments which have shown the weak luminescence, have also indicated that all the electrons reach the RES [14], from where internal processes like thermal ionization and $F \rightarrow F^-$ formation occur with full efficiency [4]. This point of view was further supported by proper magneto-optical experiments [17], which also led to the discovery of a long

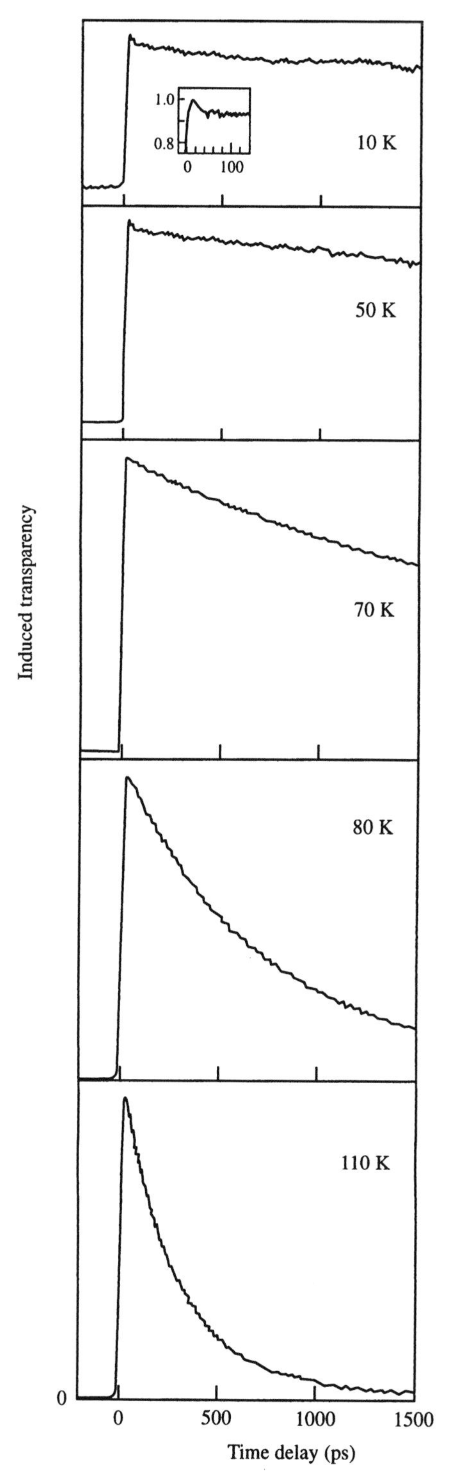

Figure 7. Decay of the induced transparency of the F center in NaI at different temperatures. The wavelength of the excitation pulses is 600 nm and the time resolution 7 ps. The inset at 10 K shows the structure of the decay curve for short time delay [23]

living F^- triplet state [18] and a well defined diamagnetic effect in emission [19]. So, as far as the experimental results in NaI and NaBr are concerned, they support a nonradiative de-excitation from the RES and not at the crossover point X during relaxation. The nonradiative de-excitation from the RES could be a tunneling process from C to P [20,21], which, depending on height and width of the barrier as the crossover process depends on E_X, gives similar result as the latter [4,16].

Although several experimental findings were converging in disproving the crossover model, all the same it continued to be widely used in the scientific literature, most probably because it was easily grasped for its intrinsic simplicity. So, only a very direct measurement of the radiative and nonradiative processes could solve the question in a definitive way. Unfortunately the weak emissions are in the near infrared region of the spectrum (2.0 μm in NaBr and 2.25 μm in NaI), where there are not detectors sensitive and fast enough to be used in spectroscopic measurements resolved in time from μs to ps. However absorption measurements in the visible satisfy these requirements, and so the technique of pump and probe of the ground state can be used to follow in time the recovery of the GS population after a sudden saturating pulse. In this way it is so possible to reconstruct the radiative and nonradiative processes occurring during the optical cycle. Figure 6 shows the time recovery of the GS of the F center in NaBr with a tine resolution of 7 ps after a pumping pulse. Two decay components have been observed, and the fastest one at low temperatures, 6 ns, has been interpreted as the lifetime of the RES [22]. Figure 7 shows the same in NaI, where a time constant of tens of ns has been measured [23]. The short delay time structure shown in the inset has been studied carefully with a 200 fs time resolution, and a 9 ps fast component has been observed. Leaving aside this last result which refers to a small fraction of the excitations and whose attribution is not yet completely clear, the bulk of the time dependent measurements both in NaBr and NaI indicate unambiguously for the kinetic processes values of time of the order of ns. By concluding, if the majority of the electrons would have crossed over at point X they should have reached the GS in a matter of ps, while the relatively long decay times measured in NaBr and NaI suggest the signature of the RES, with the consequence that the majority of electrons reach this peculiar excited state after relaxation, and do not cross over during relaxation.

4. ENERGY EXCHANGE BETWEEN F CENTERS AND MOLECULAR IONS

The optical cycle of the F center can be influenced not only by intrinsic processes like those described in Sections 2 and 3, but also by extrinsic processes [24]. Among the latter ones, molecular ions like CN^- and OH^- have been at the center of the attention of color center experts lately for their unusual properties. These ions can occupy a lattice position in alkali halides replacing a negative ion. They can also attach themselves to F centers leading to the formation of $F_H(CN^-)$ and $F_H(OH^-)$ centers. These new centers can be formed by optical aggregation, i.e. pumping with F-band light, at a temperature slightly below O °C in order to avoid the formation of F_2 and other aggregated centers. The presence of the ions and in particular their aggregation to F centers leads to wastly different optical properties with respect to those of the original F centers. $F_H(CN^-)$ centers in KCl have slightly broadened absorption and emission electronic bands, and a new luminescence centered at ~ 4.8 μm corresponding to the stretching frequency of the

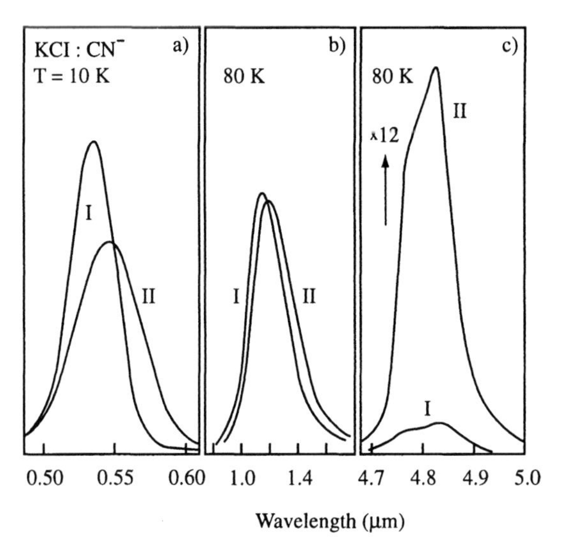

Figure 8. Electronic absorption at T = 10 K (a), emission at T = 80 K (b), and vibrational emission at T = 80 K (c) of additively colored KCl:CN^- crystals ($8\cdot10^{-3}$ CN^-). I, quenched crystal; II, after optical aggregation of F centers to CN^- defects [25].

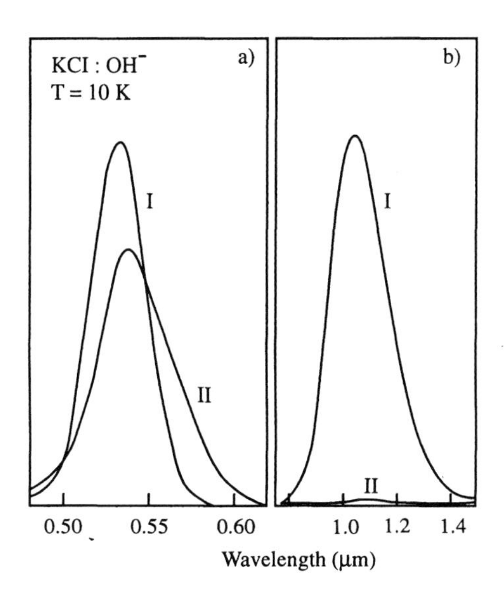

Figure 9. Electronic absorption (a) and emission (b) of additively colored KCl:OH^- crystals ($2.8\cdot10^{-3}OH^-$) at T = 10 K. I, quenched crystal; II, after optical aggregation of F centers to molecular ions [26].

CN^- ion and having a quantum yield of a few percent [25], see Fig. 8. On the contrary $F_H(OH^-)$ centers in KCl not only do not display any vibrational luminescence at the stretching frequency of the OH^- ion at ~ 2.7 μm, but also the electronic emission is almost completely quenched, while the absorption band changes as in the case of $F_H(CN^-)$[26],

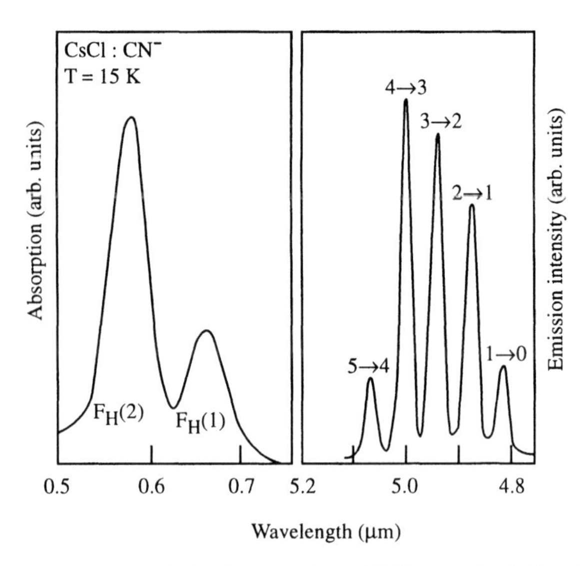

Figure 10. Absorption and emission bands of the $F_H(CN^-)$ center in CsCl at T = 15 K [30].

see Fig. 9. The same properties hold for the other F_H centers in alkali halides with NaCl structure [27,28]. On the contrary, in crystals with Cs halide structure F_H centers display a notable splitting of the absorption band far both type of ions, the electronic emission is almost quenched for CN^- and observed with various efficiencies for OH^-, and the vibrational emission has a high efficiency for CN^-, while is completely quenched, with the exception of a tiny emission in CsI, for OH^- [27,29]. Figure 10 shows the absorption and emission bands of the $F_H(CN^-)$ center in CsCl, where the vibrational emission is so efficient [30] that a laser has been realized at the peak wavelengths of the five observed vibrational bands [31].

Leaving aside most of the properties described above which would require a lengthy discussion outside the limited purpose of this work, it is evident that the electronic excitation is transferred with various degree of efficiency to the vibrational stretching levels of the molecular ions. In the case of CN^- ions this energy-transfer process is very weak in crystals with NaCl structure and strong in crystals with Cs halide structure, while it is very effective in both types of crystals for OH^- ions. In conclusion, when CN^- and OH^- ions are aggregated to F centers the electronic excitation is transferred partially or totally to the vibrational levels of the ions, and as a consequence an efficient vibrational luminescence can appear, as for CN^-, or the electronic luminescence can be completely quenched, as for OH^-. This state of matter is reported in Fig. 11 which refers to the F center in KCl doped with CN^- or OH^-, and where the Electron (E)-Vibrational (V) energy transfer is displayed together with the appropriate transitions. Although the E-V transfer process has been studied thoroughly experimentally [27] and theoretically [32,33,34], the detailed mechanisms and the inner nature of it remain obscure. In particular it was not known until recently, and in some respects it is still so, whether the energy exchange takes place during the relaxation or after the excitation reaches the RES, see Fig 11. Also in this case, in order to explain the high efficiency of the E-V energy transfer and the occurrence of preferential energy paths during relaxation [27], it was

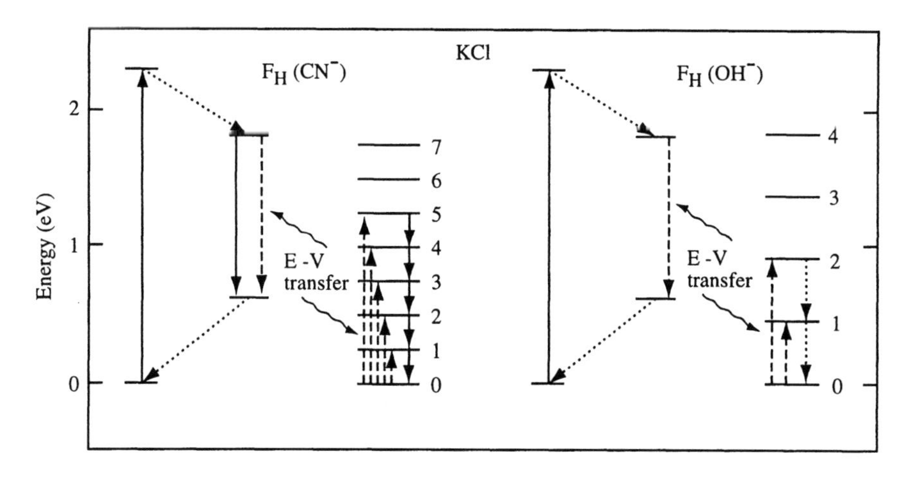

Figure 11. Optical cycle of the F center in KCl perturbed by CN^- and OH^-. The electronic and vibrational transitions are radiative (————), nonradiative (---------), and relaxation (················) processes.

suggested that in several systems the transfer happens during the relaxation processes, and the RES ceases to have a physical meaning [27].

Among several techniques which have been used during the past years to broaden our knowledge about F_H centers, also magnetic circular dichroism (MCD) measurements have been performed in $F_H(OH^-)$ in Cs- [35] and K-halides [36], and in $F_H(CN^-)$ in KCl,CsCl, and CsBr [36]. The spin-orbit parameters of the F_H centers confirm the uniaxial symmetry and the one-electron perturbed F-center character of this defect, and their values are only slightly different from those of the unperturbed F centers. On the contrary, the spin-lattice relaxation time, T_1, and the spin-mixing parameter, ε, have shown relevant changes before and after aggregation [36]. In particular the value of ε, which is defined as the fraction of electrons that flip their spins during an optical cycle, increases drastically for $F_H(CN^-)$ while decreases slightly for $F_H(OH^-)$ defects with respect to that of F centers. Again CN^- ions behave differently from OH^- ions!

These experimental findings have been explained with the very simplified and well tested assumption that the contributions to ε come only from the absorption, ε_1, as it is the case in F centers [37], and from the nonradiative (nr) de-excitation process due to the E-V energy transfer, ε_3. Figure 12 shows three possible optical cycles of the F_H center, where the molecular ion energy ladder, connected to the nonradiative de-excitation transition, τ_{nr}, has been omitted for simplicity. However of the three optical cycles, only the scheme (a) seems plausible from the physical point of view [37]. Indeed in (b) there is not the RES, while in (c) τ_{nr} has to compete with τ_{rel}, and an attempt to solve the rate equations in this last case has not given reasonable results. So, by following the scheme (a), Fig. 13 shows the results of the calculated values of ε as a function of τ_r/τ_{nr} [38]. The experimental values of ε are 0.5 for CsCl, CsBr(CN^-) and 0.016 for KCl(CN^-), with the consequence that $\tau_r/\tau_{nr} \cong 10^2$ and $3 \cdot 10^{-2}$, respectively. These figures agree very well with the measured degree of quenching of the electronic luminescence, which is quite complete in CsCl and CsBr and only partial in KCl. Moreover the radiative lifetimes are 2.9 and 1.5 μs in CsCl and CsBr, so that nonradiative de-excitation times of 29 and 15 ns are

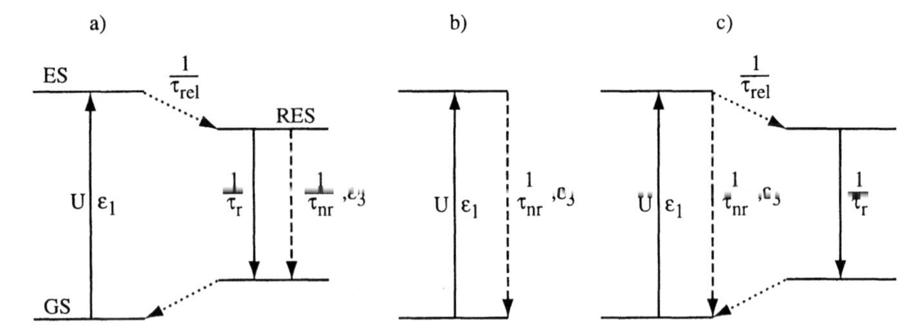

Figure 12. Optical cycle of the F_H center with radiative (————),nonradiative (---------), and relaxation (···············) processes for three different de-excitation pathways. U is the optical pumping rate out of the ground state, ε_1 and ε_3 are the spin mixing parameters during the respective transitions.

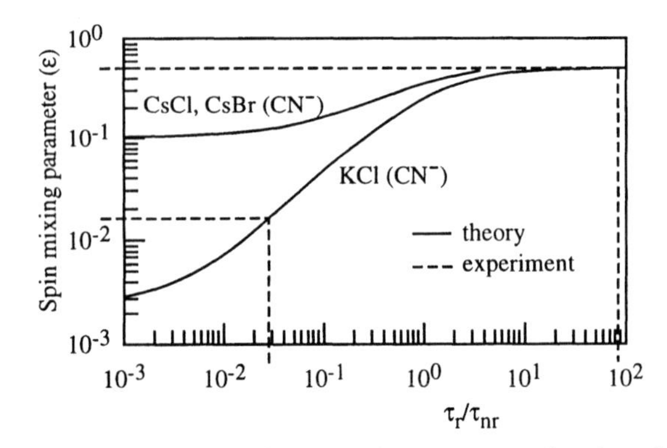

Figure 13. Total spin-mixing parameter for the $F_H(CN^-)$ center as a function of the ratio of radiative and nonradiative lifetimes during the optical cycle of Fig. 12a.

expected for $F_H(CN^-)$ centers in these two crystals, respectively. A completely different approach must be used for $F_H(OH^-)$ centers. Indeed, the fact that ε decreases with respect to indisturbed F centers can be explained only by supposing that the non-radiative process is so fast that the spin has not the time to flip during the transition [37]. Following the well known relationships of the magnetic resonances, it is easily obtained $\tau_{nr} < 10^{-9}$s, which explains why $F_H(OH^-)$ defects are a class apart with respect to the $F_H(CN^-)$ ones.

Although the previous results rested on sound physical mechanisms and gave a clear picture of the processes involved, they still are an indirect way of answering to the central question of whether the excitation reaches the RES or not before the energy transfer to the molecular ion. The only way to address properly this question is given by time resolved measurements which have been attempted earlier with limited results [27] and recently with much better ones. Recently it has been possible to study the dynamics of the very weak electronic and much stronger vibrational luminescences of $F_H(CN^-)$ in CsCl [39], which are reported in Fig. 14. The two curves show a strong correlation, and

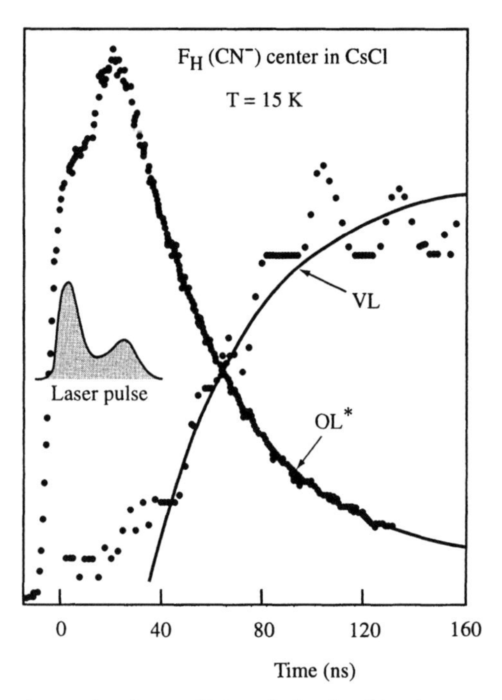

Figure 14. Time dependence of ordinary, OL*, and vibrational luminescence, VL, of $F_H(CN^-)$ centers in CsCl at 15 K under a laser-pulse excitation. The OL* is a new weak emission appearing after the aggregation process [39].

decay and building-up times of ~40 ns, which indicate clearly that the E-V energy transfer process occurs after relaxation in the RES in competition with the electronic luminescence. It is worthwhile to observe that the decay time measured here is very similar to the one obtained with the spin-mixing parameter. The same conclusions have been reached more recently by using a pump and probe technique, where the time of the energy transfer takes 70 ns and little less in CsBr and CsCl, respectively, with the process taking place from the RES [40].

At odd with the previous results are those found with a similar technique in $CsBr(CN^-)$ where a recovery time of ~0.5 ps at 10 K has been measured [41] and the crossover model, this time between the electronic excited state and the vibrational modes, has been resumed. The same authors have measured a recovery time of ~3 ps in $KCl(OH^-,OD^-)$ and $RbCl(OH^-)$, and although this value is very short, it is comparable with those determined more recently in $KBr(OH^-)$ and $RbCl(OH^-)$ at 5 K, ~3 ps and ~2 ps, respectively [42]. However these figures are two order of magnitude shorter than the value of 310 ps measured a few years ago with the same technique in $KBr(OH^-)$ at 77 K [42]. As a consequence some doubts remains on these last measurements, which possess various degrees of complexity, although it seems well established that the energy transfer process for the $F_H(OH^-)$ centers occurs in the ps regime, much shorter than for the $F_H(CN^-)$ centers in the ns regime. Anyway, it is clear that the majority of the experimental data in $F_H(CN^-)$ indicate that the electronic excitation reaches the RES, and only afterwards there is an energy transfer to the vibrational states of the molecular ions. The state of the matter is more confused, instead, for the $F_H(OH^-)$ centers, where the energy transfer seems to be a very fast phenomenon. However, contrary to the easy explanation of a crossever model, this energy transfer can occur after the relaxation which is phenomenon lasting even less than a few ps [44].

5. CONCLUSIONS

The three examples discussed in the previous Sections have shown that crossover models during relaxation, however appealing for their simplicity and understanding, do not work properly. On the contrary all the processes, like activation, tunneling and energy transfer, accur only after the excited system has reached the RES after a fast relaxation. It is true that there are still some doubts in the case of $F_H(OH^-)$ centers, due to a real complex situation, but the overwhelming majority of available and sound experimental data indicate that the RES is reached in the first instance by the electronic excitations, with the exception of the very weak hot luninescence discussed at the beginning. As a consequence it seems natural to extend this model not only to all the other color centers not yet studied in detail, but also to the vibronic systems in general, although some caution should be used.

Anyway in all the examples discussed previously, the fast optical techniques, used to test the dynamical properties of the physical systems, have been often determinant in solving doubtful cases. So they represent a powerful tool for studying the vibronic systems and they will be even more important when femtosecond resolution is used. Indeed very recent ultrafast measurements have shown the details of hidden processes during relaxation. By using 10 fs resolution, a pump and probe experiment on F centers in KBr has revealed oscillations of the electron wave packet during the relaxation along the energy potential surface [45], as expected by recent theoretical models [46].The same technique applied to an impurity laser ion, Cr^{4+} in Mg_2SiO_4 (forsterite), has revealed the nonradiative relaxation pathways to local modes firstly, ~ 3 ps, and to lattice phonon modes in second place, ~ 5 ps [47].

It is evident that this high temporal resolution will both help in clarifying old doubtful cases, as we have seen still exist, and be a magnifying glass on still unknown physical processes which will lead to a new golden age of discoveries.

6. ACKNOWLEDGMENTS

I am indebted with L. Crescentini, A. Pace, C. Piergentili, and P. Riske for their skillful assistance during the preparation of the manuscript.

REFERENCES

1. Physics of Color Centers, W.B. Fowler, ed, Academic Press, New York, 1968
2. Y. Mori and H. Ohkura, J. Phys. Chem. Solids **51**, 663 (1990)
3. C.J. Buchenauer, Ph. D. thesis, Cornell University (1971)
4. F. Lüty, Semicond. Insul. **5**, 249 (1983)

5. D.L. Dexter, C.C. Klick, and G.A. Russell, Phys. Rev. **100**, 603 (1955)
6. R.K. Swank and F.C. Brown, Phys. Rev. **130**, 34 (1963)
7. W.B. Fowler, Phys. Rev. **135**, A 1725 (1964)
8. R.F. Wood and U. Opik, Phys. Rev. **179**, 783 (1969)
9. L.F. Mollenauer and G. Baldacchini, Phys. Rev. Lett. **29**, 465 (1972)
10. H. Seidel and H.C. Wolf, in Ref 1. Chap. 8.
11. G. Chiarotti, U.M. Grassano, and R. Rosei, Phys. Rev. Lett. **17**, 1043 (1966)
12. N. Akiyama, F. Nakahara, and H. Ohkura, Radiat. Eff. Defects Solids **134**, 345 (1995)
13. R.H. Bartram and A.M. Stoneham, Solid State Commun. **17**, 1593 (1975)
14. G. Baldacchini, D.S. Pan, and F. Lüty, Phys. Rev. **B24**, 2174 (1981)
15. R.H. Bartram and A.M. Stoneham, Semicond. Insul. **5**, 297 (1983)
16. G. Baldacchini, in Luminescence: Phenomena, Materials, and Devices, R.P. Rao, ed., Nova Science Publ. Inc., Commack, 1992, p. 23
17. G. Baldacchini, G.P. Gallerano, U.M. Grassano, and F. Lüty, Phys. Rev. **B27**, 5039 (1983)
18. G. Baldacchini, G.P. Gallerano, U.M. Grassano, and F. Lüty, Lett. Nuovo Cimento **36**, 495 (1983); same autors, Radiat. Eff. **72**, 153 (1983)
19. G. Baldacchini, U.M. Grassano, A. Scacco, and K. Somaiah, Nuovo Cimento **12D**, 117 (1990)
20. P.F. Moulton, in Laser Handbook, Vol. 5, M. Bass and M.L. Stitch, eds., North Holland, Amsterdam, 1985, p. 203, and references cited therein
21. L. Gomes and S.P. Morato, J. Appl. Phys. **66**, 2754 (1989)
22. F. De Matteis, M. Leblans, W. Joosen, and D. Schoemaker, Phys. Rev. **B45**, 10377 (1992)
23. F. De Matteis, M. Leblans, W. Slootmans, and D. Schoemaker, Phys. Rev. **B50**, 13186 (1994)
24. G. Baldacchini, in Advance in Nonradiative Processes in Solids, B. Di Bartolo, ed., Plenum Press, New York, (1991), p. 219
25. Y. Yang and F. Lüty, Phys. Rev. Lett. **51**, 419 (1983)
26. L. Gomes and F. Lüty, Phys. Rev. **B30** , 7194 (1984)
27. F. Lüty and V. Dierolf, in Defects in Insulating Materials, Vol. 1, O. Kanert and J.-M. Spaeth, eds, World Scientific, Singapore, 1993, p. 17
28. L. Gomes and F. Lüty, Phys. Rev. **B10**, 7094 (1995)
29. M. Krantz and F. Lüty, Phys. Rev. **B37**, 8412 (1988)
30. Y. Yang, W. von der Osten, and F. Lüty, Phys. Rev. **B32**, 2724 (1985)
31. W. Gellermann, Y. Yang, and F. Lüty, Opt. Commun. **57**, 196 (1986)
32. G. Halama, S.H. Lin, K.T. Tsen, F. Lüty, and J.B. Page, Phys. Rev. **B41**, 3136 (1990)
33. S. Pizer and W.B. Fowler, Materials Science Forum, Vols. **239-241**, G.E. Matthews and R.T. Williams, eds, Trans Tech Publ., Switzerland, 1997, p. 473
34. G. Halama, K.T. Tsen, S.H. Lin, F. Lüty, and J.B. Page, Phys. Rev. **B39**, 13457 (1989)
35. V. Dierolf, H. Paus, and F. Lüty, Phys. Rev. **B43**, 9879 (1991)
36. G. Baldacchini, S. Botti, U.M. Grassano, and F. Lüty, Il Nuovo Cimento **15D**, 207 (1993)
37. G. Baldacchini, U.M. Grassano, and F. Lüty, J. Lumin. **53**, 439 (1992)

38. G. Baldacchini, U.M. Grassano, and F. Lüty, in Defects in Insulating Materials, Vol. 1, O. Kanert and J.-M. Spaeth, eds, World Scientific, Singapore, 1993, pag. 452
39. F. Lüty and V. Dierolf, Abstracts 7th Europhysical Conf. On Defects in Insulating Materials, EURODIM 94, Lyon, France, 5-8 July 1994, p. 243
40. D. Samiec, H. Stolz, and W. von der Osten, Phys. Rev. **B53**, R8822 (1996)
41. Du-Jeon Jang and Jaeho Lee, Sol. State Commun. **94**, 539 (1995)
42. E. Gustin, M. Leblans, A. Bouwen, and D. Schoemaker, Material Science Forum, Vols. 239-241, G.E. Matthews and R.T. Williams, eds, Trans Tech Publ., Switzerland, 1997, p. 465
43. M. Casalboni, P. Prosposito, and U.M. Grassano, Solid State Commun. **87**, 305 (1993)
44. M. Dominoni e N. Terzi, Europhys. Lett. **15**, 515 (1991)
45. N. Nisoli, S. De Silvestri, O. Svelto, R. Scholz, R. Fanciulli, V. Pellegrini, F. Baltram, and F. Bassani, Phys. Rev. Lett. **77**, 3463 (1996)
46. M. Dominoni and N. Terzi, Europhys. Lett. 23, 39 (1993); J. Lumin. **58**, 11 (1994)
47. D.M. Calistru, S.G. Demos, and R.R. Alfano, Phys. Rev. Lett. **78**, 374 (1997)

JAHN-TELLER DRIVEN MOVEMENTS OF EXCITED IONS IN SOLIDS

C.R. Ronda

Philips GmbH Forschungslaboratorien Aachen
P.O. Box 500145
D-52085 Aachen, Germany
e-mail: ronda@pfa.research.philips.com

ABSTRACT

This contribution begins with a tutorial on the nomenclature of spectroscopic terms and the splitting of such terms by low symmetry distortions. The core of this paper concentrates on the Jahn-Teller (JT) effect, with emphasis on its experimental observation in optical spectra of the transition metal ion Mn^{2+} and the s^2 ions Sb^{3+} and Bi^{3+}. It will be shown how the JT effect operating in either the ground- or the excited state causes ionic movement relative to its environment as a consequence of the optical transition. This not only has consequences for the line shape of absorption- and emission bands, the JT effect may also contribute to the Stokes Shift between absorption and emission, relaxing the contradiction between a large Stokes Shift and a high quantum efficiency.

I. A TUTORIAL ON THE NOMENCLATURE OF SPECTROSCOPIC TERMS AND THE SPLITTING OF SPECTROSCOPIC TERMS BY LOW SYMMETRY DISTORTIONS

In this section, we will shortly discuss the way how to derive spectroscopic terms for light and heavy elements. Subsequently, we will discuss the splitting of these terms by distortions of lower symmetry.

A rather simple case to start with is an ion with a d^1 electronic configuration. This is the situation encountered for e.g. Ti^{3+}. The number of electronic states for this configuration is given by $\binom{10}{1}$. Consequently the number of these states is 10.

First we calculate the terms originating from this configuration. The number of states then simply results from the degeneracy of the terms derived and adding these up. The generalised way to derive the terms is sketched in table I. In this table, the columns represent the possible m_l values for d-electrons (number $2l + 1$). The electron itself is symbolised by an arrow.

Table I. Electronic state originating from a d^1 configuration

m_l	2	1	0	-1	-2
	↑				

The terms are found by looking for independent states. By coupling the individual orbital and spin momenta $\mathbf{l}_i$ and $\mathbf{s}_i$ to total **L** and **S** values the so called Russell-Saunders (RS) terms are obtained. The Russell-Saunders terms calculated this way represent the situation without spin orbit coupling. For our d^1 configuration, the highest value of m_l is $m_l = 2$. Consequently the term with the highest orbital momentum L is a term with L = 2. The spin is 1/2, as we have only one electron in this configuration. Therefore, the spin multiplicity of this term is (2S+1) = 2. The general shape of the terms is ^{2S+1}L. The resulting term is therefore a ^2D term (For L = 0, 1, 2, 3, 4, ..., the terms are S, P, D, F, G, ...). The degeneracy of this term is (2L + 1).(2S + 1) = 10. We observe that the d^1 electronic configuration generates only one term. Apart from the state given in table I, the ^2D states generates 9 other states. As can be deduced from table I, these states are the ones with lower m_l values and states with opposite spin.

Next we discuss an ion with d^2 electronic configuration, like the V^{3+} ion. We proceed in the same way as above. The number of electronic states for this configuration is given by $\binom{10}{2}$. Consequently the number of these states is 45.

Table II. Electronic states generated by a d^2 configuration. The total number of states per term (degeneracy) is given in the last column.

m_l	2	1	0	-1	-2	L	Term	#
	↑↓					4	^1G	9
	↑	↑				3	^3F	21
		↑↓				2	^1D	5
		↑	↑			1	^3P	9
			↑↓			0	^1S	1

It might not be obvious to the reader how the independent configurations can be found. The reader might for instance ask why the configuration assigned to the ^1S term is not generated by the ^1G state. The ^1G state indeed generates a term with $m_l = 0$. The state with $m_l = 0$ which can be generated by the ^1G state only is the state with one electron with $m_l = 2$ and the other electron with $m_l = -2$. Similar arguments hold for the other states. By writing down all 45 configurations, generated by the d^2 configuration, the interested reader can verify table II.

In case of heavier elements (like Pb or Tl), this procedure is not valid anymore, due to the strength of spin-orbit interaction. In such cases, the terms are characterised by J values, being calculated from J = L + S, L + S -1, ..., | L - S |. These terms are easily found: their general shape is $^{2S+1}L_J$. Consider for example a ^3P term (L = 1 and S = 1). In the case without spin orbit coupling only one term is generated. In case of spin orbit coupling three terms result from these term: $^3P_2$, $^3P_1$ and $^3P_0$.

From table II, we observe that some of the terms have a very high orbital degeneration. These terms will split when the ion possessing this electronic configuration is incorporated in a crystal. The nature of the splitting will, in general, depend on the symmetry properties of the lattice site on which the ion is incorporated and can be calculated using group theory. This, however, is beyond the scope of this contribution. In view of its importance, we will discuss the nature of the splitting of a D term in an environment with octahedral symmetry. In doing so, we actually treat the crystal field splitting of d orbitals in an octahedral environment. In the treatment we will use symmetry arguments only, no group theory is needed in this case.

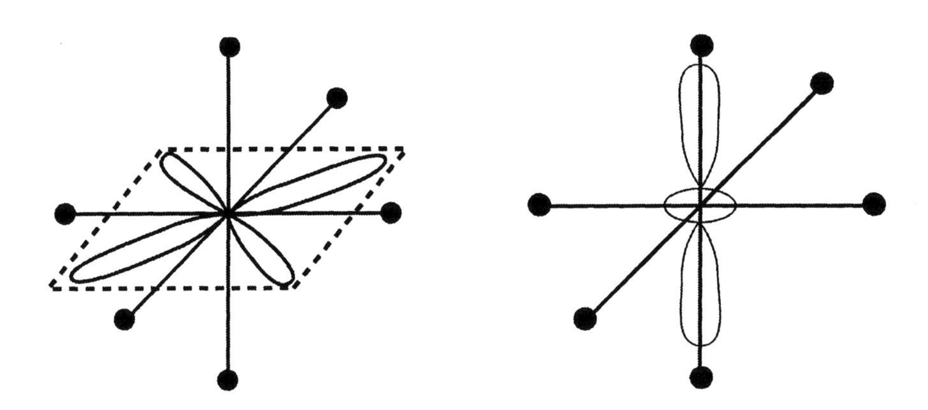

Figure 1. Graphical representation of d_{xy} and d_{z^2} orbitals

The general shape of the d orbitals is given in fig. 1. Two types of orbitals are encountered. The d_{xy}, d_{xz} and d_{yz} orbitals are directed in between the bonding axes of the metal ion. The $d_{x^2-y^2}$ and d_{z^2} orbitals are located on the bonding axes of the metal ion. Therefore, these latter orbitals are pointing to the negatively charged ligands. As a result, the energy of these latter two levels is higher than the energy of the former three levels. By inspection, we deduce that the $d_{x^2-y^2}$ and d_{z^2} orbitals are connected by symmetry and that the same applies to the d_{xy}, d_{xz}

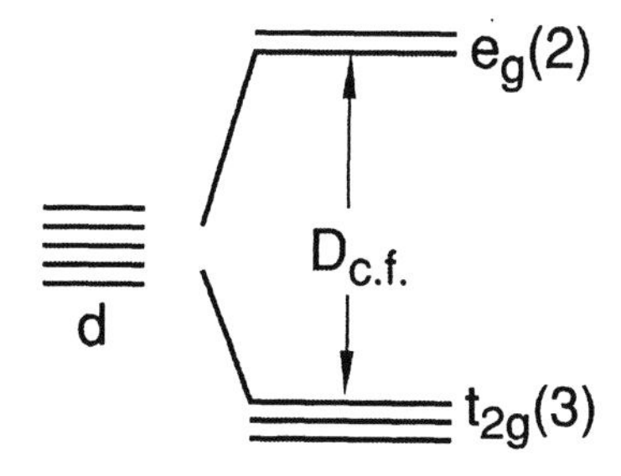

Figure 2. Splitting of d orbitals in field of octahedral symmetry

and d_{yz} orbitals. Based on symmetry arguments only, we now understand that the d levels, which are degenerate in case of a free ion split into two groups when the ion is placed on a site with octahedral symmetry, see fig. 2. As this is a symmetry argument only, the same must also hold for the D term.

The crystal field splitting always occurs, independent of the occupation of the different orbitals. In some cases, crystal field splitting stabilises the system, in other cases there is no benefit (when all orbitals are occupied with the one or two electrons). The crystal field splitting generates two types of degenerate orbitals in octahedral symmetry. We now investigate what happens if at least one type of the degenerate orbitals are not half or completely filled, i.e. if the system has orbital degeneracy. In our treatment, we take the d_{xy}, d_{xz} and d_{yz} orbitals. In octahedral symmetry, the coordination is the one given in fig. 3.

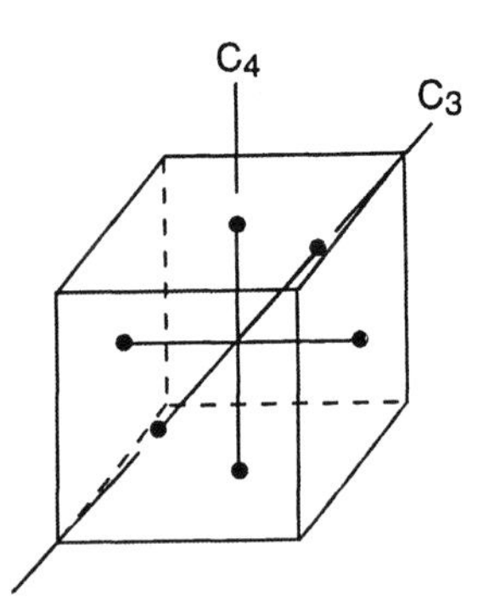

Figure 3. Octahedral arrangement of ligands, together with three- and fourfold rotational axes

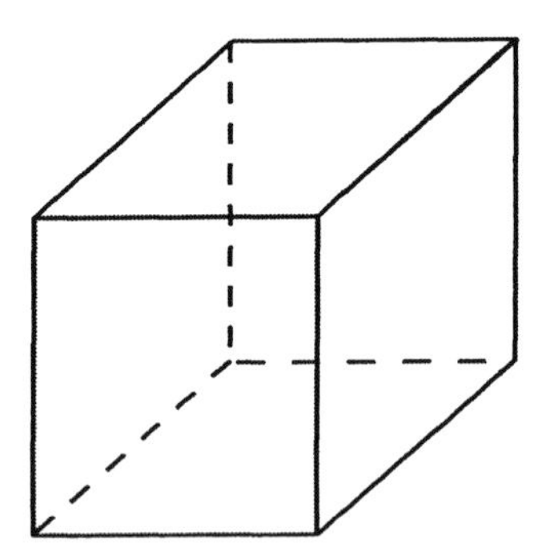

Figure 4. Tetragonally distorted system

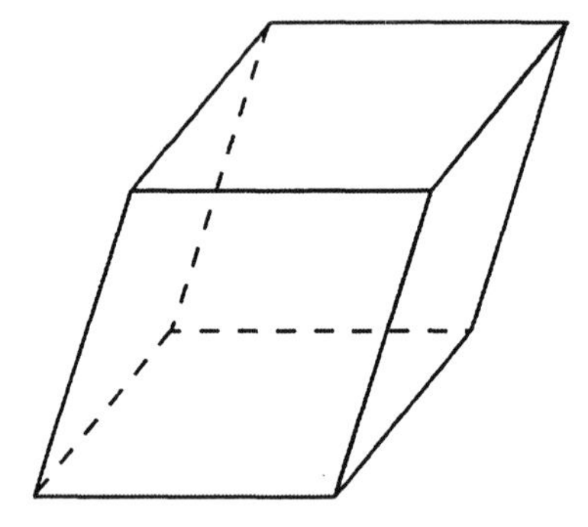

Figure 5. Trigonally distorted system

In a symmetry lower than cubic symmetry, the degeneracy of these orbitals will be lifted. As a consequence, at least one of the orbitals will then have a higher energy than in the undistorted case but at least one of the other orbitals will decrease in energy. If the orbital(s) which increase(s) in energy is not populated, the system as a whole will decrease in energy by such a low symmetry distortion. This can be done by e.g. a distortion along the four- or threefold axis (in octahedral symmetry) of the ion, see fig. 4 and fig. 5. A distortion along the threefold axis is a trigonal distortion, a distortion along the fourfold axis is a tetragonal distortion. In both distorted cases, the x, y and z-axes are not equivalent by symmetry anymore, leading to a partial lifting of degeneracy, see e.g. fig. 6. The phenomenon described here is the Jahn-Teller effect: the stabilisation of an orbital degenerate system by low symmetry distortions. In the following parts we first give a more physical description of the Jahn-Teller effect, followed by some experimental results.

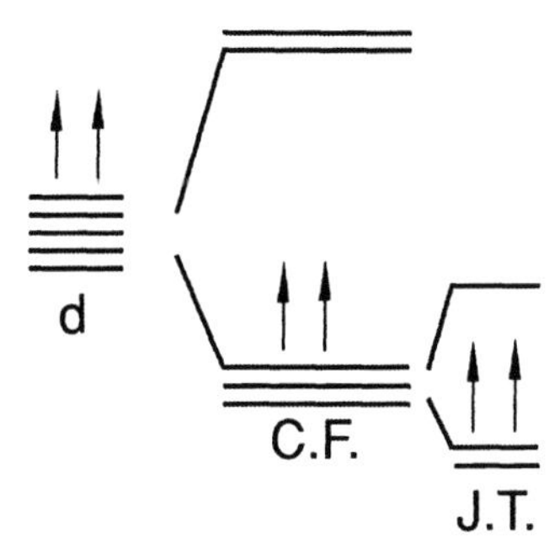

Figure 6. Splitting of degenerate levels in octahedral symmetry by a Jahn-Teller distortion

II. The Jahn-Teller effect

In this section, the physical origin of the JT effect will be discussed briefly.

IIA. The Jahn-Teller effect: an introduction

In many cases, point defects in solids are subject to lattice distortions. By this, it is meant that ions in solids are displaced with respect to their environment, compared to their perfect crystal position. This, in general, leads to a lowering of the point group symmetry of the point defect. The lowering of the symmetry in ground- or excited state is observable in e.g. optical spectroscopy and in many cases (for ground state splitting) even in X-ray diffraction.

A very well known factor, leading to such a lattice distortion is the existence of electronically degenerate states (JT effect). In such a case, the system can lower its energy by lifting the degeneracy, i.e. by lowering its symmetry. Due to the fact that the energetic position of centroid remains fixed, there is at least one sub level which is lowered in energy by lifting the degeneracy. In general, there is more than one equivalent stable new atomic configuration. These configurations are equivalent by symmetry in the new point group of the point defect.

When the defect is frozen in one of its stable positions, the JT effect is called static. One speaks of the Dynamic Jahn-Teller effect (DJT), when the defect reorientates between its different equilibrium positions. In such a case, no static distortion will be observed. As a consequence, one and the same defect can apparently be distorted in some experiments and undistorted in other experiments at the same temperature. In addition, the JT effect may be static at low temperature and dynamic at higher temperature. Even in cases where the DJT effect does not produce a static distortion, the occurrence of the JT effect is accessible from optical spectroscopy: Van Vleck has shown already in 1932 that removing the orbital degeneracy is equivalent to quenching the orbital momentum[1]. The JT effect in general removes at least partly the orbital degeneracy. In cases where the orbital degeneracy is removed only partly, the orbital momentum as well is only partly quenched, resulting in a reduction of spin-orbit interaction (which relies a.o. on the orbital momentum). This is the so-called Ham effect[2]. In the literature, beautiful examples of optical spectra with a strongly reduced energy separation between spin-orbit components as compared to situations where the JT effect is not occurring have been published[3], see e.g. fig. 7 for the $^4A_2 \rightarrow {}^4T_2$ optical transition on V^{2+} in $KMgF_3$. For the parameters used in the calculation, the reader is referred to the original paper.

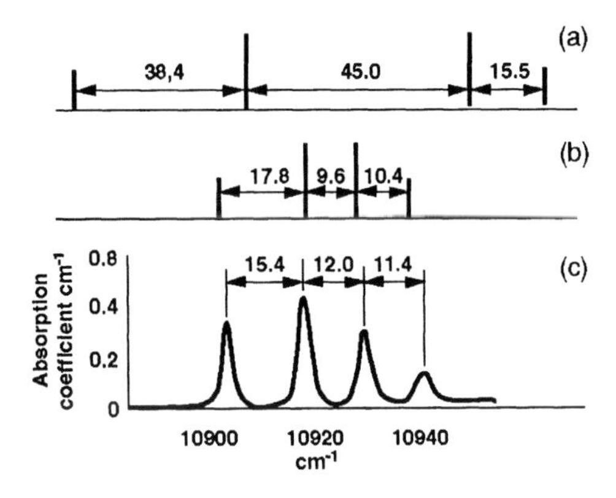

Figure 7. Absorption spectrum of $KMgF_3:V^{2+}$, showing Jahn-Teller induced reduction of spin-orbit splitting: (a) splittings in cm^{-1}, calculated using static crystal field theory, (b) calculated intensity and calculated splitting reduced by the Ham effect and (c) observed spectrum at 2K

In what follows first a formal description of the Jahn-Teller effect will be given. Then, the observation of the JT effect in optical spectroscopy will be discussed on a few selected examples. (The interested reader is referred to more extended reviews on the influence of the Jahn-Teller effect on optical spectra[4].) Here we will see that the absorption of light may induce movements of optically excited ions in crystals, due to the JT effect. Finally, some consequences of the JT effect will be discussed.

IIB. Formal treatment of the Jahn-Teller effect.

The complete Hamiltonian of the system of interest is a function of the complete set of electron co-ordinates **r** and nuclear co-ordinates **R.** In general, the electronic motion and the motion of the nuclei cannot be described separately. An approximate separation can be obtained, using the fact that the electron mass m is much smaller than the nuclear mass M. In that case the Schrödinger equation for the system can be written as:

$$H_e(\mathbf{r},\mathbf{R})\,\psi_e(\mathbf{r},\mathbf{R}) = E_e(\mathbf{R})\,\psi_e(\mathbf{r},\mathbf{R}) \qquad (1)$$

In this equation, the total energy of the nuclei has been neglected. Therefore, this equation describes the motions of the electrons for nuclei fixed at given positions **R**.
For $\mathbf{R} = \mathbf{R}_o$ (the point defect being located at the position with perfect symmetry of the crystal), the solution of (1) can be written as:

$$H_e(\mathbf{r},\mathbf{R}_o)\,\psi_e(\mathbf{r},\mathbf{R}_o) = E_e(\mathbf{R}_o)\,\psi_e(\mathbf{r},\mathbf{R}_o) \qquad (2)$$

Usually, the point defect has a high symmetry, consequently there is degeneration. Important is, that the cases we describe here are states within the energy gap, i.e. localised states. We now treat such a case, degenerate by the symmetry of the point defect in the crystal for $\mathbf{R} = \mathbf{R}_o$, and try to describe the possible splitting for $\mathbf{R} \neq \mathbf{R}_o$. To this end we expand (1) to second order (i.e. the nuclear configuration **R** is still close to $\mathbf{R}_o$):

$$H_e(\mathbf{r},\mathbf{R}) = H_e(\mathbf{r},\mathbf{R}_o) + \sum_s (\delta H_e(\mathbf{r},\mathbf{R})/\delta Q_s))_o\, Q_s + 1/2 \sum_{s,s'} (\delta^2 H_e(\mathbf{r},\mathbf{R})/\delta Q_s \delta Q_{s'}))_o\, Q_s Q_{s'} \qquad (3)$$

In this equation Q_s and $Q_{s'}$ are the normal displacements: suitable linear combinations of the Cartesian co-ordinates of the ionic displacement vectors which describe the ionic displacements from the high symmetry position in the crystal. In (3), the last two terms define a perturbation V. Solutions of (3) are found in terms of ψ_{e} ($\mathbf{r}$,$\mathbf{R}_o$), applying first-order perturbation theory to a degenerate state. The resulting splitting of the degenerate level in first order in V is obtained by diagonalising the perturbation matrix. Its general matrix element is given by:

$$V_{\alpha\beta} = \sum_{s} Q_s < \psi_{e_\alpha}(\mathbf{r},\mathbf{R}_o) \mid (\delta H_e / \delta Q_s)_o \mid \psi_{e_\beta}(\mathbf{r},\mathbf{R}_o) > +$$

$$+ 1/2 \sum_{s,s'} Q_s Q_{s'} < \psi_{e_\alpha}(\mathbf{r},\mathbf{R}_o) \mid (\delta^2 H_e / \delta Q_s \, \delta Q_{s'})_o \mid \psi_{e_\beta}(\mathbf{r},\mathbf{R}_o) > \quad (4)$$

The wavefunctions $\psi_{e_\alpha}(\mathbf{r},\mathbf{R}_o)$ and $\psi_{e_\beta}(\mathbf{r},\mathbf{R}_o)$ belong to the subspace of the degenerate level. Generally, assumptions in order to simplify this expressions are made with respect to the second term, resulting in the fact that the Q_s can be looked upon as the normal modes of the system containing the defect. In that case (4) can be written as:

$$V_{\alpha\beta} = (1/2 \sum_{s} k_s Q_s^2) \mathbf{I}_{\alpha\beta} + \sum_{s} < \psi_{e_\alpha}(\mathbf{r},\mathbf{R}_o) \mid (\delta H_e / \delta Q_s)_o \mid \psi_{e_\beta}(\mathbf{r},\mathbf{R}_o) > Q_s \quad (5)$$

In this expression, k_s are the force constants and **I** is the unit matrix. The Eigen-values of the system (E_i (Q_s)) have to be minimised with respect to Q_s. This procedure leads to the new stable positions of the system.

This treatment describes the physical origin of the JT effect: the splitting of degenerate states under low symmetry distortions. The same kind of mechanism is also present in metals, e.g. in case of Peierl's distortions which may lead to charge density waves like in VSe_2[5]. In what follows, we will discuss the influence of JT distortions on optical spectra.

III. The Jahn-Teller effect in action I : absorption- and emission spectra of $MnCl_2$, $MnBr_2$ and MnI_2

In this section, we will discuss absorption- and emission spectra of MnX_2 (X= Cl, Br, I). We will see that the optical spectra are strongly influenced by the JT effect, operating in the excited state.

The compounds MnX_2 crystallise in two different layer type structures. $MnCl_2$ has the $CdCl_2$ structure with spacegroup P-3m1. $MnBr_2$ and MnI_2 crystallise in the $Cd(OH)_2$ structure with spacegroup R-3m. In both structure types, the Mn^{2+} ions occupy slightly trigonally distorted octahedral sites. In the $Cd(OH)_2$ structure, the anions are hexagonally closely packed, in the $CdCl_2$ structure, the anions form a cubic closed packed stacking.

The Mn^{2+} ions have 5 d electrons. In the halides discussed here, the crystal field is weak, consequently the d electron configuration is $t_{2g})^3 e_g)^2$ in the ground state with an electron spin of S=2.5. The corresponding state is $^6A_{1g}(S)$. The emitting excited state has d-electron configuration $t_{2g})^4 e_g)^1$, with electron spin S= 1.5 ($^4T_{1g}(G)$) state. As the ground state is orbitally non degenerate, the ground state cannot be stabilised by a JT effect. The excited state discussed here is triply degenerate, consequently, a JT interaction in the excited state might be expected.

In the literature, absorption- and emission spectra of MnX_2 in the paramagnetic temperature region have been reported[6,7], see figures 8 and figures 9.

Both the $^6A_{1g}(S) \rightarrow {}^4T_{1g}(G)$ absorption band and the $^4T_{1g}(G) \rightarrow {}^6A_{1g}(S)$ emission band have a Gaussian shape. In the emission spectrum of $MnBr_2$ a shoulder is observed at higher temperatures at low energy. The physical origin of this shoulder is not understood. In

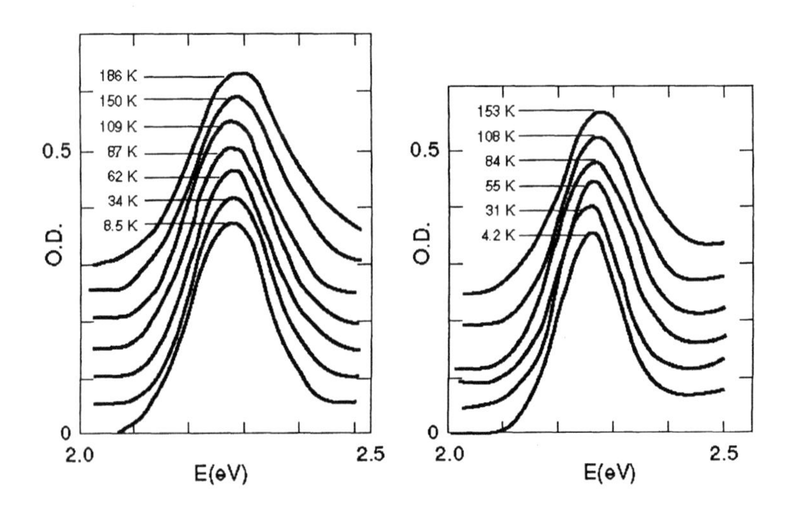

MnCl$_2$ MnBr$_2$

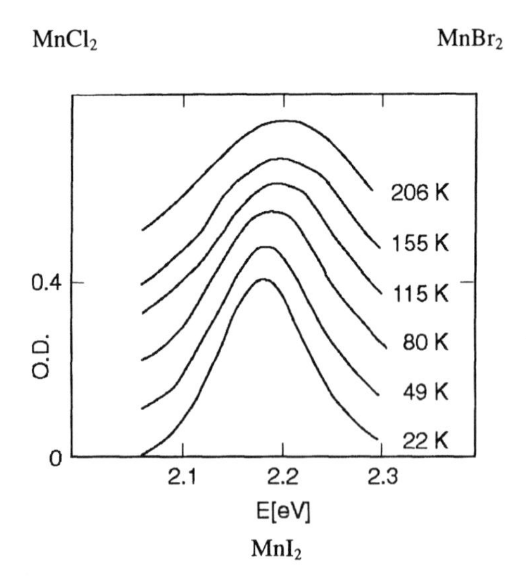

MnI$_2$

Figures 8. Absorption spectra of the $^6A_{1g} \rightarrow {}^4T_{1g}$ optical transition in single crystalline MnX_2
Crystal thicknesses: $MnCl_2$: 315 μm, $MnBr_2$: 390 μm; MnI_2: 180 μm

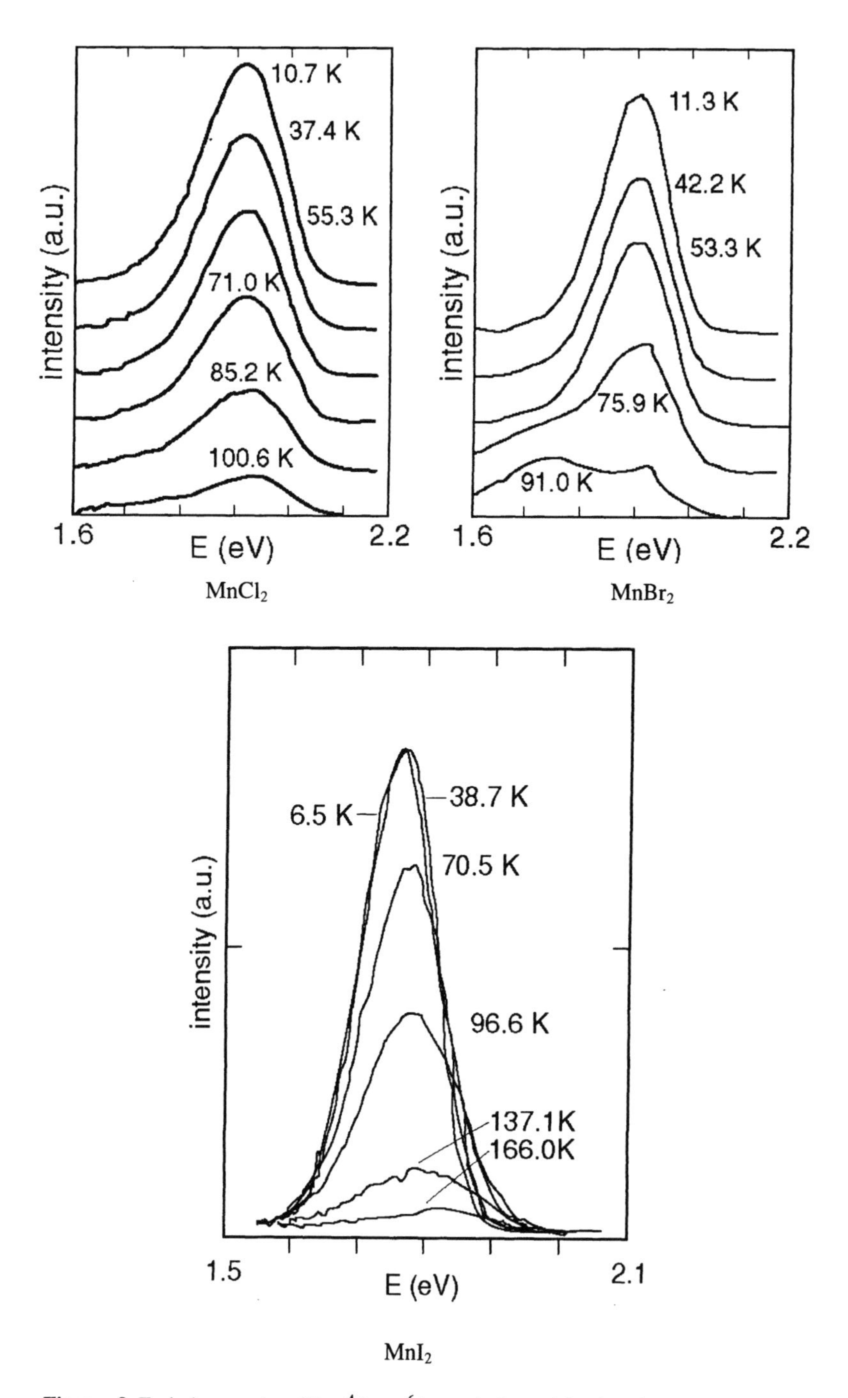

Figures 9. Emission spectra of the ${}^4T_{1g} \rightarrow {}^6A_{1g}$ optical transition in polycrystalline MnX_2

particular the observation of a Gaussian absorption band is surprising as one would actually expect the $^4T_{1g}(G)$ excited state to be split in first order in four spin-orbit components. As the calculated spin-orbit splitting is quite large, at least some structure, due to spin-orbit splitting, would be expected to remain, even in the case of strong exciton-phonon coupling. For the emission a Gaussian band would result if only one spin orbit component of the excited state is involved in the optical transition.

It has been shown in the literature that the $^6A_{1g}$ (S) $\rightarrow$ $^4T_{1g}$ (G) absorption band does not split and has a Gaussian shape when the DJT effect quenches the orbital angular momentum by coupling to phonons with E_g symmetry; coupling to T_{2g} vibrations leads to a three-fold split band[8,9]. In cases where the JT effect quenches the orbital momentum, no spin-orbit splitting will be observed. However, the inverse is also true: a strong spin-orbit coupling in a system stabilises the system against a JT splitting[1].

It has been shown by Ham, that even in cases where the JT effect does not lead to a static distortion of the system, the orbital momentum is quenched partly[2]. Now we will discuss the so-called Ham effect for the JT interaction of a T_{1g} state, interacting with a phonon with E_g symmetry. We assume that the E_g vibrational mode in an octahedron couples stronger to the lattice than phonons with T_{2g} symmetry, as the E_g mode involves radial motion of the ligands and interacts with σ-orbitals, whereas the T_{2g} modes involve tangential motion of the ligands and interact only with π-bonding orbitals[3], see fig. 10.

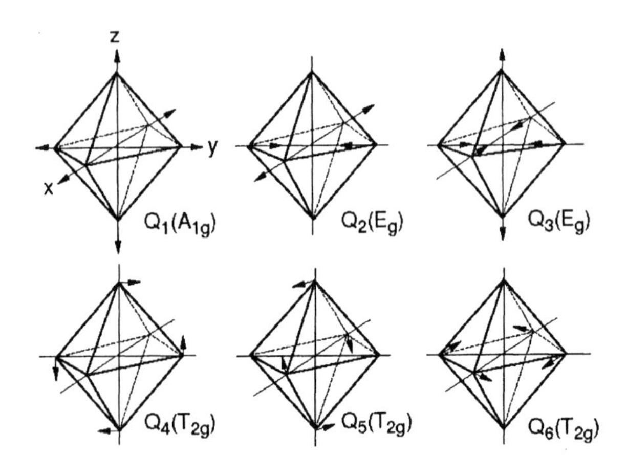

Figure 10. Graphical representation of A_{1g}, E_g and T_{2g} phonon modes in octahedral symmetry

The experimental observation of a Gaussian line shape also indicates that the dominant interaction is with E_g vibrational modes: strong coupling to T_{2g} phonons would lead to a JT splitting of the absorption band[8,9].

For the $^4T_{1g}(G)$ state, a basis set of real electronic functions φ_ξ, φ_η, φ_ζ, transforming as x, y and z under rotations of the octahedral group is adopted. Neglecting all effects associated to the spin, the vibronic Hamiltonian H is given by[2]:

$$H = E_o\mathbf{I} + (1/2\mu)\,[P_\theta^2 + P_\varepsilon^2 + \mu^2\omega^2\,(Q_\theta^2 + Q_\varepsilon^2)]\mathbf{I} + V[Q_\theta\varepsilon_\theta + Q_\varepsilon\varepsilon_\varepsilon] \qquad (6)$$

where E_o again is the energy of the degenerate state in the symmetrical configuration, **I** is the unit matrix, μ is the effective mass of the vibrational E_g mode and ω its angular momentum, P_θ and P_ε are the nuclear energy momenta conjugate to Q_θ and Q_ε and V is the JT coupling coefficient.

The vibrational modes Q_θ and Q_ε transform as $z^2 - 1/2(x^2 + y^2)$ and $(x^2 - y^2)$, respectively. ε_θ and ε_ε are the matrices describing the coupling of the E_g vibrations (Q_θ and Q_ε) to the T_{1g} state:

$$\varepsilon_\theta = \begin{pmatrix} 1/2 & 0 & 0 \\ 0 & 1/2 & 0 \\ 0 & 0 & -1 \end{pmatrix} \qquad \varepsilon_\varepsilon = \begin{pmatrix} -\sqrt{3}/2 & 0 & 0 \\ 0 & \sqrt{3}/2 & 0 \\ 0 & 0 & 0 \end{pmatrix} \tag{7}$$

The vibronic Eigen-functions $\psi_{in}(Q, \tau)$ ($i = \xi, \eta, \zeta$, the components of the T_{1g} state) are harmonic oscillator functions for a displaced two dimensional oscillator. The equilibrium positions are given by:

$$Q_{\theta i} = - V e_{i_\theta} / \mu\omega^2, \; Q_{\varepsilon i} = - V e_{i_\varepsilon} / \mu\omega^2 \tag{8}$$

where e_{i_θ} and e_{i_ε} are the appropriate diagonal components of the matrices ε_θ and ε_ε. The potential energy of the oscillator at these positions is lowered by the JT energy $E_{JT} = V^2 / 2\mu\omega^2$. The vibronic Eigen-functions are:

$$\psi_{in}(Q, \tau) = \psi_i(\tau)\, F_{n_\theta}(Q_\theta + V e_{i_\theta} / \mu\omega^2)\, F_{n_\varepsilon}(Q_\varepsilon + V e_{i_\varepsilon} / \mu\omega^2) \tag{9}$$

where $F_n(y)$ is a standard harmonic oscillator function. The corresponding energies are:

$$E_{in} = E_o - V^2 / 2\mu\omega^2 + (n_\theta + n_\varepsilon + 1)\, \hbar\omega \tag{10}$$

$n_\theta, n_\varepsilon = 0, 1, 2, 3, \ldots$

The vibronic spectrum in this case is the same as in the absence of the JT interaction, except for the displacement $V^2 / 2\mu\omega^2$ for all states.

However, the equilibrium position for the displaced oscillators (8) is different for the three electronic functions $\psi_i(\tau)$. The separation is proportional to $V / \mu\omega^2$ and therefore, the region of overlap between corresponding oscillator states associated with different electronic functions is diminished. This means that matrix elements between these states become smaller. Consider for example an electronic operator O, independent of Q_θ and Q_ε. A vibronic matrix element of O is given by:

$$\langle \psi_{in_\theta n_\varepsilon} | O | \psi_{jn_\theta' n_\varepsilon'} \rangle = \langle \psi_i | O | \psi_j \rangle \langle in_\theta | jn_{\theta'} \rangle_\theta \langle in_\varepsilon | jn_{\varepsilon'} \rangle_\varepsilon \tag{11}$$

where

$$\langle in_\theta | jn_{\theta'} \rangle_\theta = \int_{-\infty}^{\infty} dQ_\theta F_{n_\theta}(Q_\theta + V e_{i_\theta} / \mu\omega^2)\, F_{n_\theta'}(Q_\theta + V e_{j_\theta} / \mu\omega^2) \tag{12}$$

and similarly for $\langle in_\varepsilon | jn_{\varepsilon'} \rangle_\varepsilon$

Evaluation of these integrals (which also appear in the theory of the optical line shape) for the vibronic ground state leads to the following expression for the off-diagonal elements for the operator O:

$$\langle \psi_{i00} | O | \psi_{j00} \rangle = \langle \psi_i | O | \psi_j \rangle \exp[-3E_{JT} / 2\hbar\omega] \tag{13}$$

The exponential factor in (13) is the orbital reduction factor; it is the consequence of the JT quenching of the operator O, the quenching is almost complete when E_{JT} is much larger than $\hbar\omega$. In many cases it is necessary to take second order effects of coupling of the triplet to excited vibronic states into account. The values of these second order reduction factors have been calculated by Ham[2] and MacFarlane[10,11]. The calculations show a strong quenching, even for relatively small values of $E_{JT}/\hbar\omega$.

We remark that in principle odd fermion systems have to be treated differently from even fermion systems. In even fermion systems, the JT active mode has to be contained in the symmetric product of the state of interest. In odd fermion systems, the JT active mode has to be contained in the antisymmetric product of the state of interest. If, however, the spin-orbit coupling is weak, one can restrict oneself to the orbital part of the wavefunction alone, which is always described by a single-valued representation[12].

The considerations given above show for the $^4T_{1g}(G)$ electronic state, coupled to an E_g phonon that in the presence of a strong JT effect, the vibronic state without spin-orbit splitting remains triply degenerate. The (Dynamic) JT effect results in a reduction of the spin-orbit splitting.

The major line broadening mechanism in case of MnX_2 is the coupling of the exciton to the lattice. This implies that the lineshape can be calculated using the well known expressions available from the theory of the configuration co-ordinates[4,8,9,12]. An important contribution to the line width of the $^6A_{1g}$ (S) $\rightarrow$ $^4T_{1g}$ (G) absorption band and the $^4T_{1g}$ (G)$\rightarrow$ $^6A_{1g}$ (S) emission band comes from the very strong dependence of the energy of the $^4T_{1g}$ (G) state on Dq (see fig. 11).

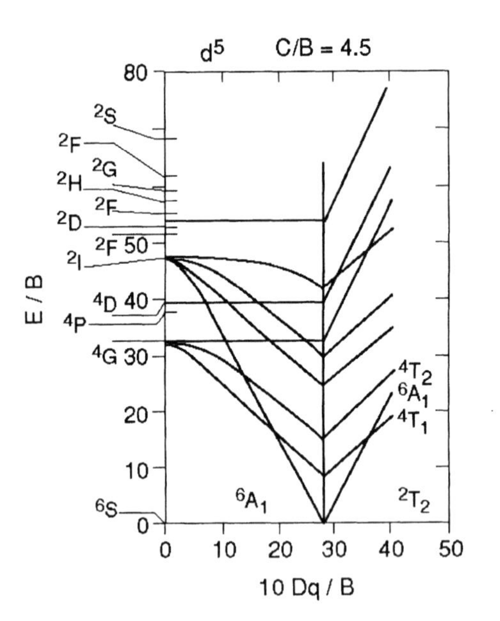

Figure 11. Tanabe-Sugano diagram for a d^5 ion

If the coupling with the E_g and A_{1g} vibrational modes to the lattice is much stronger than the coupling of the T_{2g} vibrational mode to the lattice, the absorption- or emission spectrum is determined by the convolution of the contributions of the E_g and A_{1g} vibrational modes[8]. In the strong coupling case, the Full Width at Half Maximum (FWHM) is given by[4] :

$$\text{FWHM} = \sqrt{8\ln 2}\,[S_1\,(\hbar\omega_1)^2 \coth(\hbar\omega_1/2kT) + S_2\,(\hbar\omega_2)^2 \coth(\hbar\omega_2/2kT)]^{1/2} \qquad (14)$$

where we take $\hbar\omega_1$ for the frequency of the A_{1g} vibrational mode and $\hbar\omega_2$ for the frequency of the E_g vibrational mode. S_1 and S_2 are the average number of A_{1g} and E_g phonons, involved in the optical transition, respectively. This equation is valid only in the harmonic approximation. As the FWHM in the absorption- and emission bands at the same temperature are not equal to each other, the potential energy functions are not the same in ground- and excited state. The FWHM in the absorption band is smaller than the FWHM of the emission band (at the same temperature), indicating that the potential energy of the ions in the excited state is less strongly dependent on the metal ligand distance than in the ground state.
From the FWHM of the emission bands of MnX_2 at high and low temperature, we calculate the number of A_{1g} and E_g phonons involved in the emission process. The phonon frequencies in the ground state are known from literature[13,14], see table III.

Table III. Phonon frequencies of MnX_2 in ground- and excited state. The values for the phonon frequencies in the excited state have been estimated, see text.

Material	$\hbar\omega_g(A_{1g})$ (eV)	$\hbar\omega_g(E_g)$ (eV)	$\hbar\omega_e(A_{1g})$ (eV)	$\hbar\omega_e(E_g)$ (eV)
$MnCl_2$	0.0291	0.0179	0.0275	0.0169
$MnBr_2$	0.0187	0.0112	0.0180	0.0108
MnI_2	0.0139	0.0105	0.0125	0.0095

From these values we can calculate the relaxation energies in the ground state SG, being equal to:

$$SG = S(A_{1g})_g\,\hbar\omega_{1g} + S(E_g)_g\,\hbar\omega_{2g} \tag{15}$$

where the subscript g denotes that the parameters are given for the ground state.

From the difference in the maxima of absorption and emission bands, we also calculate the relaxation energy in the excited state (SE). The smaller relaxation energy in the excited state again indicates a more shallow potential well in the excited state (see table IV).

Table IV. Spectral characteristics of MnX_2, deduced from the emission spectra

Material	$FWHM_{em}$ (eV)	Huang Rhys	$E_{relax,\ ground\ state}$ (eV)	$E_{relax,\ excited\ state}$ (eV)
$MnCl_2$	0.179 (10.7 K) 0.191 (85.2 K)	$S(A_{1g})_g = 2.4$ $S(E_g)_g = 11.7$	0.28 eV	0.09
$MnBr_2$	0.145 (11.3 K) 0.162 (75.9 K)	$S(A_{1g})_g = 6.5$ $S(E_g)_g = 12.1$	0.26 eV	0.10
MnI_2	0.133 (6.5K) 0.198 (150 K)	$S(A_{1g\,g} = 11.3$ $S(E_g)_g = 9.0$	0.25 eV	0.15

We can also calculate SE from the width of the absorption band at high temperature. For such a case, equation (14) can be written as:

$$FWHM = \sqrt{8\ln 2}\,\sqrt{2kT}\,(S(A_{1g})_e\,\hbar\omega_{1e} + S(E_g)_e\,\hbar\omega_{2e})^{1/2} \tag{16}$$

where the subscript 'e' denotes that the parameters are given for the excited state.
The results are given in table V.

Table V. Spectral characteristics of MnX_2, deduced from the absorption spectra

Material	$FWHM_{abs}$ (eV)	Huang Rhys	$E_{relax, excited state}$ (eV)	E_{JT} (eV)	ORF
$MnCl_2$	0.168 (8.5 K) 0.210 (186 K)	$S(A_{1g})_e = 5.4$ $S(E_g)_e = 3.5$	0.21	0.059	5.10^{-3}
$MnBr_2$	0.140 (4.2 K) 0.185 (153 K)	$S(A_{1g})_e = 10.2$ $S(E_g)_e = 2.0$	0.21	0.022	5.10^{-2}
MnI_2	0.110 (20 K) 0.173 (150 K)	$S(A_{1g})_e = 10.3$ $S(E_g)_e = 6.5$	0.19	0.061	6.10^{-5}

We observe that the relaxation energies SE calculated using equation (16) differ from the ones calculated using the Stokes Shift in the ground state. The difference must be due to the presence of anharmonic contributions to the width of at least one of the bands. As a consequence, we can calculate the JT stabilisation energy in an indirect way only. To this end we assume that the phonon frequencies in the excited state are reduced by some (to a certain extent arbitrary) parameter (Table III). From the FWHM of the absorption band at low and high temperature we calculate $S(A_{1g})_e$ and $S(E_g)_e$. From these values we can calculate the JT stabilisation energy and the orbital reduction factor (table V). The calculated orbital reduction factors indicate that the spin-orbit splitting of the absorption band is expected to be very small, as observed. Due to the strong exciton-phonon phonon interaction the separate spin-orbit components cannot be resolved. The variation of the JT stabilisation energy in the series $MnCl_2$, $MnBr_2$, MnI_2 is not understood. Many factors can play a role: spin-orbit interaction increases on going from $MnCl_2$ to MnI_2. The same applies to the polarisability of the lattice. Crystal field strength is decreasing in this sequence. The packing of the halide ions differs in $MnCl_2$ on the one hand and $MnBr_2$ and MnI_2 on the other hand although the local symmetry of the Mn^{2+} ions remains the same. The contribution of anharmonicities is larger in case of MnI_2 than in case of $MnCl_2$ and $MnBr_2$, as judged from the difference in FWHM in the absorption and emission spectra.

The same sequence as for the JT stabilisation energy has been found for the activation energy leading to thermal quenching of the luminescence in this series[7], indicating that a strong JT interaction shifts thermal quenching to higher temperature. This indicates that the JT effect causes a change in the dependence of the potential energy curve in the excited state, see e.g. fig. 12.

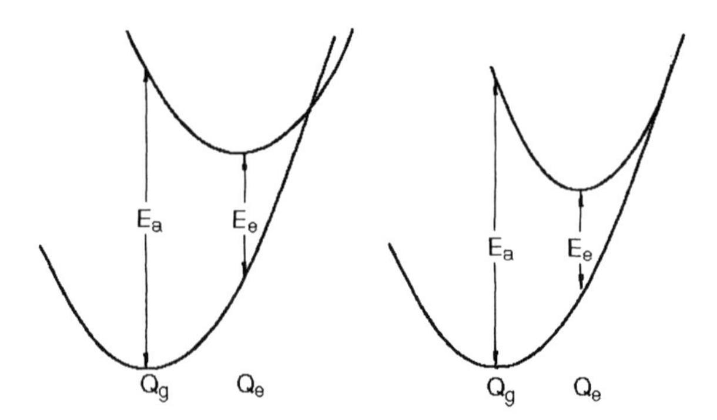

Figure 12. Configuration coordinate diagram, showing change in potential energy curve in the excited state due to the JT effect (left part: no JT effect; right part: JT stabilised)

In[6] a very small net MCD signal is reported for the $^4T_{1g}(G)$ state. This indicates that the spin-orbit interaction is not completely quenched. For completeness, we remark that in

MnX_2 absorption spectra involving excited states that are not expected to show a strong JT interaction indeed show a clear spin-orbit splitting[6].

IV. The Jahn-Teller effect in action II: Sb^{3+} in $Ca_5(PO_4)_3F$

The optical properties of ions with s^2 configuration have been studied for many years, as the optical properties of these ions are very interesting, both from a scientific and an applied point of view. Well known luminescent s^2 ions are Tl^+, Pb^{2+}, Bi^{3+}, Sn^{2+}, Sb^{3+}, and even Te^{4+}. Many of these ions have found their way in industrial products. To mention only a few examples: $Ca_5(PO_4)_3(F,Cl)$:Sb,Mn is a luminescent material, emitting white light, applied in fluorescent lamps, $BaSi_2O_5$:Pb is a luminescent material, emitting in the UV, used in suntanning lamps and CsI:Tl is applied in X-ray detectors.

The energy level scheme of and the optical transitions on an s^2 ion are given in fig 13.

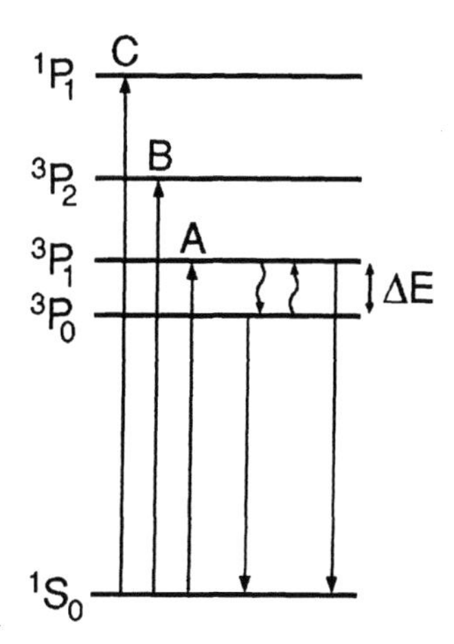

Figure 13. Energy level scheme of and optical transitions on an s^2 ion

The s^2 ground state has ^1S_o configuration (orbitally non degenerate), the excited sp state gives rise to three triplet states in sequence of increasing energy ^3P_o, $^3P_1$, $^3P_2$ and a singlet state $^1P_1$. In octahedral symmetry, these terms are classified according to irreducible representations of the O_h group as follows:

$^1S_o \rightarrow {}^1A_{1g}$

$^3P_o \rightarrow {}^3A_{1u}$; $^3P_1 \rightarrow {}^3T_{1u}$; $^3P_2 \rightarrow {}^3E_u + {}^3T_{2u}$

$^1P_1 \rightarrow {}^1T_{1u}$

We will stay with the atomic nomenclature, however, as this is most frequently used in the literature.

The transition $^1S_o \rightarrow {}^3P_o$ is parity forbidden, the transition $^1S_o \rightarrow {}^3P_1$ is partially allowed by spin-orbit interaction, the transition $^1S_o \rightarrow {}^3P_2$ again is parity forbidden, but gains intensity by coupling to phonons with proper symmetry. The only transition which is electric dipole allowed is the $^1S_o \rightarrow {}^1P_1$ transition.

Emission can be observed from the ^3P_o and $^3P_1$ levels. At low temperature, usually the excitation relaxes non-radiatively to the ^3P_o level which emits to the ^1S_o ground state with a long decay time in view of the forbidden character of the transition. At higher temperatures, emission mainly originates from the $^3P_1 \rightarrow ^1S_o$ transition with a shorter decay time as this transition is parity allowed.

In the literature, the optical absorptions between the s^2 ground state and the sp excited state are often denoted as A, B, and C transitions (see fig. 13). In general, four absorption bands are observed: the fourth state, the so-called D state involves the host lattice as well, see [15] for a recent review. The A band shows a doublet splitting, the C band a triplet splitting. The splitting in the absorption bands is due to coupling of the excited states of the s^2 ions with the E_g and T_{2g} vibrational modes in a regular octahedron (fig. 10), another example of Jahn-Teller interaction in absorption spectra.

Apart from the absorption spectra, the emission spectra of quite a few s^2 compounds also show features due to the JT effect. In some cases, two emission bands are observed on excitation in the A band (see fig. 14).

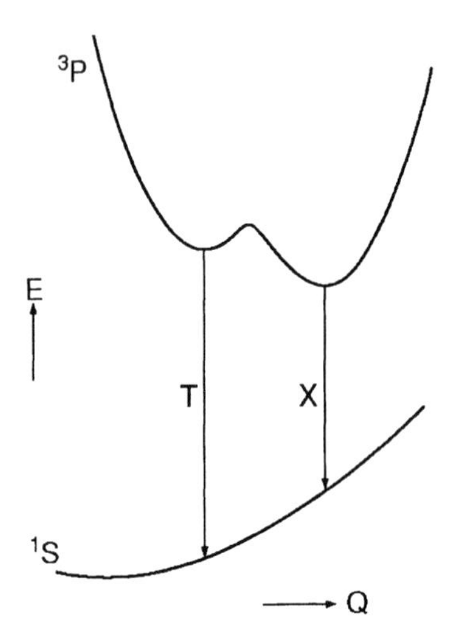

Figure 14. Emission of an s^2 ion showing JT interaction in the excited state

There are several possibilities to describe the occurrence of two bands in the emission spectra.

- The splitting of this band can be due to interaction of the orbital triplet with again E_g vibriational modes, leading to a tetragonal distortion and T_{2g} vibrational modes leading to a trigonal distortion of a regular octahedron. In such a case, either there should be an anharmonic contribution to the potential energy or there is a quadratic JT effect (in that case there are non-zero contributions quadratic in the configuration co-ordinate).
- The splitting of this band can also be due to coupling to E_g (tetragonal) phonons only when there is strong spin-orbit mixing between $^3T_{1u}$ and $^1T_{1u}$ states.
- If the totally symmetric co-ordinate Q_1 varies strongly from one kind of distortion to the other, both can be minima.

A complete description would go beyond the scope of this contribution. A more general outline is given in[16]. For the interested reader: a vast amount of information on the experimental and theoretical investigations, concerning optical spectra of s^2 ions is available in the literature[16,17], to mention a few references only.

From fig. 14, it follows that one expects the intensity of the two emission bands to be temperature dependent on excitation in the A band. At low temperature, emission is observed from the T minimum (with tetragonal symmetry) only, at somewhat higher temperature, the second X band (with trigonal symmetry) shows up, gaining intensity at the cost of the T emission band. At even higher temperatures, emission is expected originating from both bands.

The optical properties of $Ca_5(PO_4)_3F:Sb$ have been studied by a.o. Oomen et al[18]. The emission- and excitation spectra of $Ca_5(PO_4)_3F:Sb$ at 4.2K and room temperature are given in figs. 15.

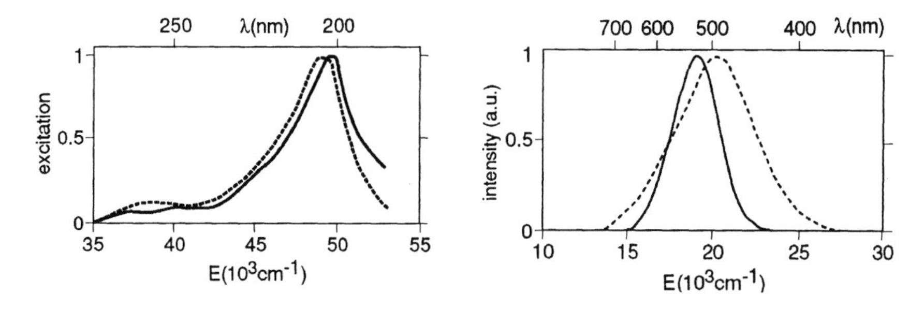

Figures 15. Excitation- and emission spectra of $Ca_5(PO_4)_3F:Sb$ at 4.2K (continuous line) and room temperature (dashed line)

Remarkable are the blue shift of the emission spectrum with increasing temperature and the very large value of the Stokes Shift (19000 cm^{-1}), the latter because of the very high quantum efficiency of the luminescence of this material at room temperature (> 80%). In general, thermal quenching sets in at lower temperature for increasing values of the Stokes Shift, as is easily understood from the configuration co-ordinate model. In addition, the decay times of the emissions, obtained by parameter fitting as only one emission band is observed, are unusually long ($^3P_o \rightarrow {}^1S_o$: τ = 990 ± 20 μs ; $^3P_1 \rightarrow {}^1S_o$: τ = 4.9 ± 0.8 μs). For comparison, see table VI[19] .

Table VI. Luminescence characteristics of Sb^{3+} in a number of host lattices. T_q is the temperature at which the luminescence quantum efficiency is equal to ½ of its maximum value found at low temperature.

Compound	Stokes Shift (cm^{-1})	T_q (K)	τ_o ($^3P_o \rightarrow {}^1S_o$) μs	τ_1 ($^3P_1 \rightarrow {}^1S_o$) μs
$LaPO_4$:Sb	14000	240	185 ± 2	0.18 ± 0.06
$LiLaP_4O_{12}$:Sb	16000	250	720 ± 40	-
LaP_3O_9:Sb	19500	240	690 ± 10	1.6 ± 0.3
$Ca_5(PO_4)_3F$:Sb	19000	> 300	990 ± 20	4.9 ± 0.8

These phenomena are explained by the presence of a JT effect. In order to obtain high quenching temperatures in case of a large Stokes Shift, the curvature of the so-called adiabatic potential energy surface (APES) in the excited state has to be changed. In $Ca_5(PO_4)_3F:Sb$, the curvature is changed by the JT effect.

The blue shift of the emission spectrum with increasing temperature is also explained by the presence of a JT effect in the excited state. Even when there is only one minimum in the APES, nevertheless a blue shift can be expected as is easily recognised from fig. 16.

Finally, the unusually long decay times of emission of Sb^{3+} in this lattice is ascribed to the JT effect too, as the JT effect produces a reduction of the orbital momentum and

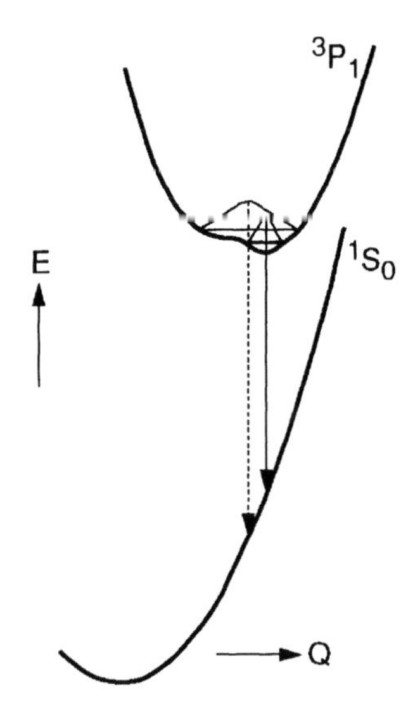

Figure 16. Blue shift of emission in case of JT stabilised excited state

consequently reduces spin-orbit coupling which enables the spin forbidden transitions, leading to emission.

V. The Jahn-Teller effect in action III: Sb^{3+} and Bi^{3+} chloride complexes in solution

The reader might think that the JT effect is encountered in the solid state only. This, however, is not true. As an example, we now treat the optical properties of Sb^{3+} and Bi^{3+} chloride complexes in solution published in the literature a few years ago[19].
The absorption and emission spectra of $[SbCl_4]^-$ and $[BiCl_4]^-$ in acetonitrile are given in figs. 17 and 18.

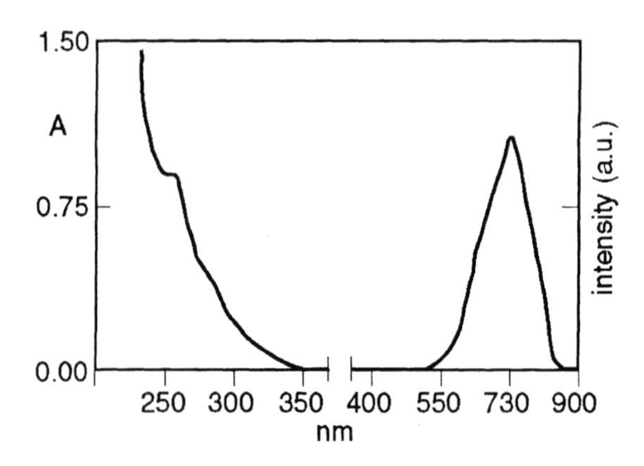

Figure 17. Absorption- and emission spectra of $[SbCl_4]^-$in acetonitrile at room temperature (1 cm cell, $2.54*10^{-4}$ M)

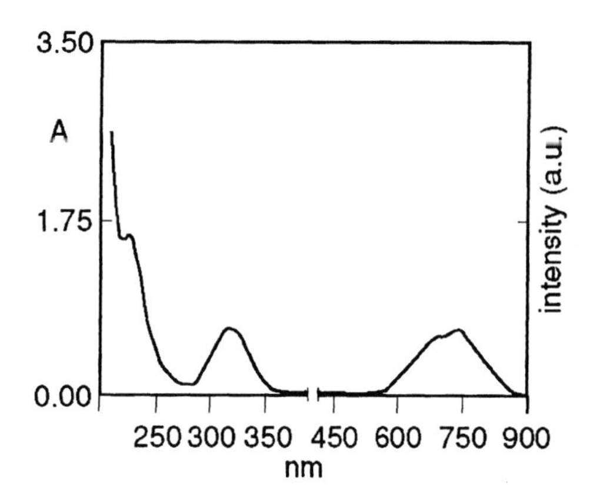

Figure 18. Absorption- and emission spectra of $[BiCl_4]^-$ in acetonitrile at room temperature (1 cm cell, $1.08*10^{-4}$ M)

Table VII. Optical data of $[SbCl_4]^-$ and $[BiCl_4]^-$ complexes in acetonitrile

	$[SbCl_4]^-$	$[BiCl_4]^-$
Absorption λ_{max} (nm)		
A band	283	319
B band	255	
C band	235	227
Emission λ_{max} (nm)	740	720
Stokes Shift (cm^{-1})	21800	17500

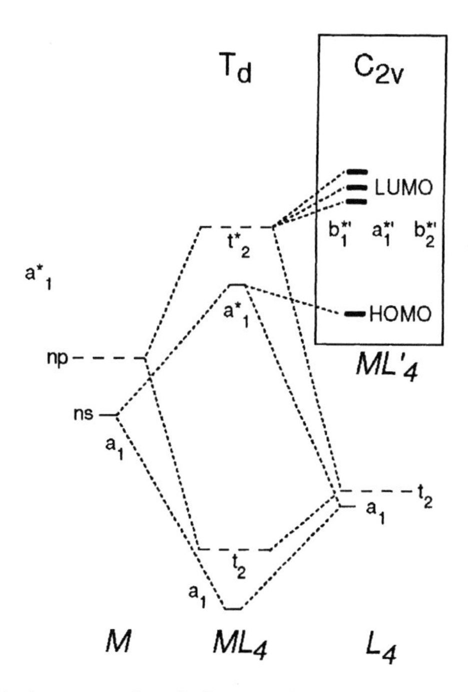

Figure 19. Group theoretical representation of relevant molecular orbitals of the complexes $[SbCl_4]^-$ and $[BiCl_4]^-$ in T_d and C_{2v} symmetry

Here we again encounter three absorption bands. In case of $[BiCl_4]^-$, the B band is not clearly resolved. Only one emission band is observed. The results found are summarized in table VII.

Here again, very large values for the Stokes Shifts are found, yet the complexes show luminescence at room temperature even in a liquid. The large Stokes Shift, accompanied by emission at room temperature again indicates a large structural change, associated with the electronic transition.

The results found can be explained very elegantly with the Valence Shell Electron Pair Repulsion model (VSEPR)[20]. The structures of s^2 complexes with coordination numbers smaller than 6 (here we have coordination number 4) are in agreement with this model. The symmetry of the complex deviates from the highest symmetry possible, due to the presence of a lone pair. As a consequence, the complexes considered here are not tetrahedral but are distorted in the ground state, most likely to C_{2v} symmetry (butterfly structure).

A qualitative molecular orbital scheme is given in fig. 19. The Highest Occupied Molecular Orbital (HOMO) is stabilised by sp(a'_1) orbital mixing (second-order JT effect)[20]. Optical absorption is due to the following transitions (in C_{2v} symmetry): $a_1^* \rightarrow b_1^{*\prime}$, $a_1^{*\prime}$, and $b_2^{*\prime}$.

The emission originates from the transition $b_1^{*\prime} \rightarrow a_1^*$. The authors suggest that the ground state distortion of the complexes $[SbCl_4]^-$ and $[BiCl_4]^-$ is eliminated in the excited state. The excited state rearranges to the symmetrical tetrahedral geometry, driven by the optical excitation. In this way, also the huge Stokes Shift is explained easily.
For completeness: similar arguments have also been developed for s^2 ions in solids[21].

VI. Conclusions

Illustrated by a few examples, the influence of the JT effect on the absorption- and emission spectra of defects in a matrix has been demonstrated. The JT effect produces a change in the optical properties e.g. by reduction of the magnitude of spin-orbit coupling, resulting in a change of the line shape of the spectra and an increase of the emission decay time. In addition large values of the Stokes Shift accompanied to a high quantum efficiencies have been found, which cannot be understood on the basis of a simple configuration co-ordinate diagram and are also due to the JT effect.

VII. References

1. J.H. van Vleck, Electric and Magnetic Susceptibilities, Univ. Press, Oxford (1932)
2. F.S. Ham, Phys. Rev. 138(6A), 1727 (1965)
3. M. D. Sturge, Phys. Rev. B1(3), 1005 (1970)
4. M. Lannoo and J. Bourgoin, Point Defects in Semiconductors, Springer Series in Solid-State Sciences 35, Springer Verlag Berlin, Heidelberg, New York (1981)
5. C.R. Ronda, G.W. Bus, C.F. van Bruggen and G.A. Wiegers, Studies in Solid State Chemistry Vol. 3, Edited by R. Metselaar, H.J.M. Heijligers and J. Schoonman, Elsevier Scientific Publishing Company, 717 (1983)
6. H.J.W.M. Hoekstra, P.R. Boudewijn, H. Groenier and C. Haas, Physica 121B, 62 (1983)
7. C. R. Ronda, H.H. Siekman and C. Haas, Physica 144B, 331 (1987)
8. Y. Toyozawa and M. Inoue, J. Phys. Soc. Japan 20, 1289 (1965)
9. Y. Toyozawa and M. Inoue, J. Phys. Soc. Japan 21, 1663 (1966)
10. R.M. McFarlane, J.Y. Wong and M.D. Sturge, Bull. Am. Phys. Soc. 12, 709 (1967)
11. R.M. McFarlane, J.Y. Wong and M.D. Sturge, Phys. Rev. 166, 259 (1968)
12. R. Englman, The Jahn-Teller Effect in Molecules and Crystals, John Wiley and Sons Ltd., New York (1972)
13. L. Patrinieri, L. Piseri and I. Pollini, Proc. 8th Conference on Raman Spectroscopy, 479 (1982)
14. L. Piseri and I. Pollini, J. Phys. C 17, (1984), 4519

15. H. v.d. Brand, Dissertation, State University of Utrecht, the Netherlands (1996)
16. A. Ranfagni, D. Mugnai, M. Bacci, G. Viliani and M.P. Fontana, Advances in Physics 32, 823 (1983)
17. G. Blasse, Topics in Current Chemistry 171, Springer-Verlag Berlin, Heidelberg 2, (1994)
18. E.W.J.L. Oomen, W.M.A. Smit and G. Blasse, Mat. Chem. and Phys. 19, 357 (1988)
19. H. Nikol and A. Vogler, J. Am. Chem. Soc. 113(23), 8988 (1991)
20. A.T. Albright, J.K. Burdett, M.H. Whangbo, Orbital Interactions in Chemistry, Wiley, New York (1985)
21. G. Blasse, Prog. Solid State Chem. 18, 79 (1988)

RECENT RESULTS IN MULTIPHONON NON-RADIATIVE PROCESSES AND IN NOVEL MATERIALS FOR LASERS AND AMPLIFIERS

F.Auzel

France Telecom, PAB/BAG,
BP. 107, F-92225 Bagneux, France.

ABSTRACT

In the first part we shall present some new results in the field of multiphonon non-radiative transitions. They are known to rule the quantum efficiency of rare earth doped solid state materials used for lasers and optical amplifiers. We show that in certain cases such transitions can be saturated so giving higher quantum efficiencies under high pumping conditions. In the second part we shall present vitreous materials with reduced in homogeneous line width which though obtained by glass technics present crystal-like spectra. Studied materials have the further advantage that they can be design in such a way that they may have weaker non-radiative transition than can be estimated from the highest phonon energy of the glass former.

Such results give expectations for new approaches for solid state lasers and amplifiers; the cases of powder lasers and waveguide amplifiers shall be briefly mentioned in due places as examples.

I. INTRODUCTION

This lecture describes two new aspects of the physics of rare-earth doped systems which could lead to improved laser and amplifier actions.

In the first part, we start from a curious result: the fact that it has been possible to obtain the laser effect of an hydrated powder of $NdCl_3{:}6H_2O$ [1,2], that is with 100% of active ion (Nd) concentration (N_0), in a medium for which the weak excitation quantum efficiency is as low as $\eta=10^{-4}$ [2] due to the multiphonon decay introduced by the high energy O-H vibration at 3600 cm-1. Knowing that the laser threshold is inversely proportional to the product ηN_0 [3], which is here 4×10^{-17}cm^{-3}, to be compared to 1.9 $\times10^{-21}$ cm^{-3} for a x=75% doped $Nd_xLa_{1-x}\ P_5O_{14}$ powdered crystal, the energy threshold should have been about 2×10^4 times larger than for the pentaphosphate powder, whereas we found about the same threshold of 30 mJ under same excitation conditions [2]. This induces us to question the usually assumes constancy of the multiphonon non-radiative decay at high excitation levels needed to reach the threshold of powdered materials.

As well known for many years, the multiphonon non-radiative decay of Transition Metals (TM) and Rare-Earth ions (RE) can be described by two types of approaches. Either through an Nth-order development of the dynamic crystal field at the ion site [4-7], which would apply to quasi zero electron phonon Huang Rhys coupling parameter (S_0), or through the use of the non-adiabatic hamiltonian, in a Condon [7-10] or non-Condon approach [11], which rather corresponds to non-zero S_0. Both methods have been compared [7]. The second type of approach shows that a non-radiative transition can be described besides an electronic matrix element, by the product of two terms : a "promoting mode" term usually considered as a scaling factor experimentally determined and an "accepting mode" term [7,10,11]. It was shown that the accepting mode term can be obtained in a statistical way whatever the details of the physical approach [7]. Recently a new theory attempted to reconcile both methods [12,13], in fact expressing the promoting term through the dynamical crystal field including non-linear terms.

All above theories have in common that they consider each ion as individually coupled to an effective mode of the phonon distribution density of states. This, even though non-linear coupling terms have been introduced [13], provides non-radiative decay probabilities which are excitation density independent. Our present purpose is to go beyond such a limitation.

In a second part, we shall consider a new kind of laser material which lies in between the microcrystalline powder form and the glass structure, namely vitroceramics. In such a composite material, the active rare-earth ion is in still only in the microcrystalline phase but the microcrystal is inside a glassy phase of about same index of refraction which provides an overall transparent material. Such material though obtained by glass tecnology provides the reduced inhomogeneous linewidth of single crystals.

II. RECALL OF THE MULTIPHONON APPROACHES FOR NON-RADIATIVE TRANSITIONS IN RARE-EARTH (RE) SYSTEMS.

II.A-A simple statistical derivation of the Huang-Rhys function for the accepting mode term.

The non-radiative decay probability, as given by the Fermi Golden Rule, can be written [7] :

$$W_{nr} = \frac{2\pi}{\hbar} \left|\langle i|H_{\text{int}}|j\rangle\right|^2 R_N \, \delta\left(E_j - E_i\right) \tag{1}$$

where $R_N \, \delta\left(E_j - E_i\right)$, given by the accepting modes term, represents the final density of states; R_N is the function first derived by Huang-Rhys [14] :

$$R_N = \exp\left[-(2\bar{n}+1)S_0\right] \left(\frac{\bar{n}+1}{\bar{n}}\right)^{N/2} I_N\left(2S_0\sqrt{\bar{n}(\bar{n}+1)}\right) \tag{2}$$

where I_N is the Bessel function of order N with imaginary argument, $\bar{n}$, the phonon occupation number. Including the promoting term, it has been shown [7,8,11], that in the RE weak coupling case ($N>>S_0$), equ. (2) has to be replaced by :

$$R' = e^{-S_0(2\bar{n}+1)} \frac{S_0^N}{N!}\left(\frac{N_p}{S_0}\right)^2 (\bar{n}+1)^N \tag{3}$$

where N_p stands for the promoting modes.

Except for the promoting factor, $(N_p / S_0)^2$, the other factor (the accepting mode part) of this formula can be also derived by combinatorial analysis [7], showing that it only has a statistical content.

II.B-The exponential energy gap law with the promoting term.

Using Stirling's approximation, the *N!* produces the well known exponential gap behaviour :

$$W_{nr} = W_0 \exp[-\alpha(\Delta E - 2.6\hbar\omega_M)] \qquad (4)$$

where ΔE is the energy gap to the next lower level, and α is given by:

$$\alpha = \hbar\omega_M^{-1}\left[\log\left(\frac{N}{S_0(\bar{n}+1)}\right)-1\right] \qquad (5)$$

where N is the multiphonon order, S_0 the Huang-Rhys coupling parameter at 0K, and $\bar{n}$ the phonon occupation number at temperature T. Such exponential behaviour with an energy scale displaced by a 2.6 $\hbar\omega$ phonon energy is well observed experimentally [15], the displacement coming from the promoting term [7].

The following meaning can be given to the accepting factor in (3) [7]: it represents the probability to fill simultaneously N of the possible accepting modes of the host from the electronic energy of one ion.

III EXPERIMENTAL RESULTS FOR DECAYS AT VARIUOUS EXCITATION LEVELS IN RE DOPED GLASSES AT WEAK CONCENTRATION [16].

III.A-Opposite behaviour for experimental decays under various excitation levels for RE ligthy doped glasses.

We present here the results for Er^{3+}, Tm^{3+} and Nd^{3+} excited states in a germanate glass. Concentration of RE ions has been kept to a rather low level (0.2% in weight i.e. 2.5×10^{19} cm^{-3}) in order to avoid any important up-conversion effect by energy transfer [17] which could reduce the lifetimes measured at high excitation intensity. The determination of the non-radiative probability is based on a comparison between radiative decays and experimental lifetimes [6]. As in [6] pulsed emission, here 7 ns duration, from a Nd-YAG laser with frequency doubling is used as well as pulsed Ti-Sapphire laser emissions in order to reach, by focusing, a maximum of 15 J/cm^2 of incident excitation. The glass host has the same composition in weight percentage: GeO_2 66% ; K_2O 17% ; BaO 17% as in [18].The highest phonon frequency ($\hbar\omega_M$) for such a glass is at about 900 cm^{-1}[18]. The experimental non-radiative decay probabilities (W_{nr}) are determined by the usual method [6], that is by measuring experimental decays (τ) from directly excited levels of the considered RE, and calculating the radiative ones (τ_0) from absorption measurements and application of Judd's theory [19]. The non-radiative decay probability is given by :

$$W_{nr} = \tau^{-1} - \tau_0^{-1} \qquad (8)$$

As an example, the general behaviour observed for non-radiative decay rates versus excitation density for ΔE > $3\hbar\omega_M$, (ΔE given in term of the multiphonon order N), is presented on Fig. 1. The excited ions density has been calculated from the incident excitation density, the concentration of the Rare Earth and the absorption cross-section of the

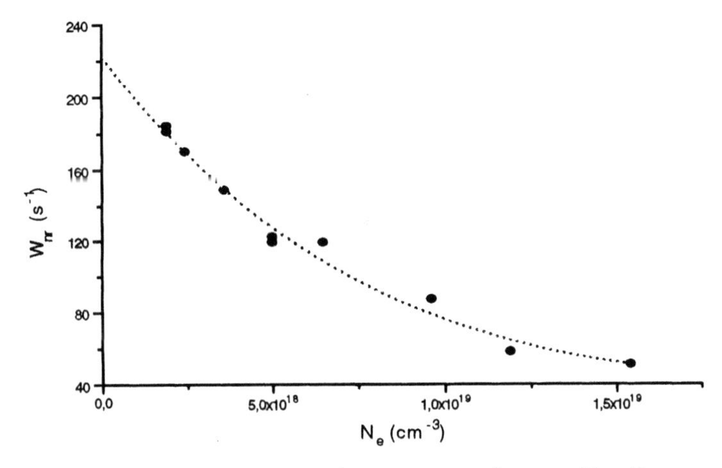

Fig.1 Non-radiative decay probability (W_{nr}) of the $^4F_{3/2}$ level of Nd^{3+}(2.5 10^{19}cm^{-3}) in a germanate glass at room temperature versus excited state density (N_e). After [16].

transitions. A decrease by a factor five in the non-radiative probability from level $^4F_{3/2}$ of Nd^{3+} is clearly observed, with a saturation in the decrease reflecting the absorption saturation at 1.25 x 10^{19}cm^{-3}. The quantum efficiency for this state is found to increase from 92% to 98% within our excitation range.

Interestingly, the non-radiative decay is independent of excitation for gaps corresponding to $3.2\hbar\omega_M$; this value, is about the predicted $2.6\hbar\omega_M$ one for the promoting mode contribution [7] found in Eq.(4); below this value, the non-radiative decay rate, on the contrary, increases with excitation state density, as is clearly demonstrated by the comparison of Fig 2a and 2b, for the $^3H_4$ (Tm^{3+}) and $^4S_{3/2}$ (Er^{3+}) transitions with respective gaps of about 4.25 and 2.1 phonon energy [20].

III. B-Description of behaviour as a variation of the exponential parameter.

The results obtained for Er^{3+}, Tm^{3+} and Nd^{3+} are synthetized on Fig.3, which shows that the exponential gap law is followed with an α parameter (here, the slope on the semi-log plot), increasing with the absorbed excitation intensity i.e. with the excited ions density (N_e).

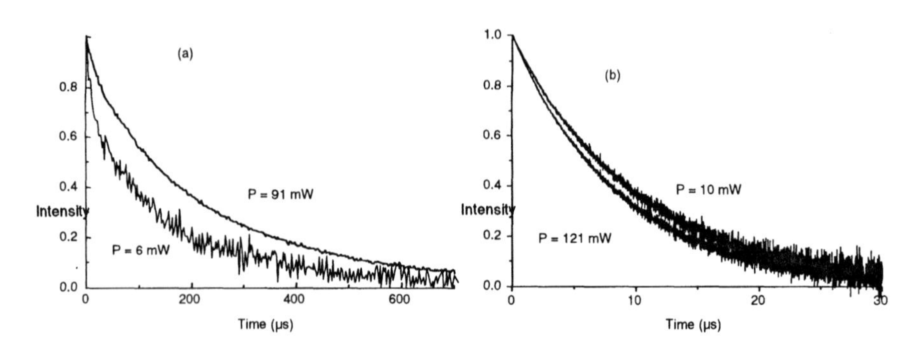

Fig. 2 Experimental decay for $^3H_4$ (Tm^{3+})(a) and for $^4S_{3/2}$ (Er^{3+}) (b) at two different average power of pulsed excitation [20]. Note opposite behaviour due to respective gap of 4.25 and 3.1 phonon energy as explained in text.

On the other hand, it can be noted that the scattering of the experimental points on Fig.3, because of the excitation normalised plot, is much reduced in comparison with unnormalised ones [6]. A rotation of the line around the point $3.2\hbar\omega$ which is then fixed is observed as well as an increase for the α parameter. This obervation for the studied germanate glass at room temperature is a general trend since a systematic study has shown it for glasses with different compositions and maximum phonon frequencies. This demonstrates that contrary to what has been believed up to now, one should specify excitation levels in α parameter determination, specially when high peak excitation powers are used. Besides this, our observed decrease in non-radiative relaxation rate with increasing pump intensity could explain why the threshold can be reached in the laser effect observation in hydrated neodymium chloride powder as mentioned in the introduction.

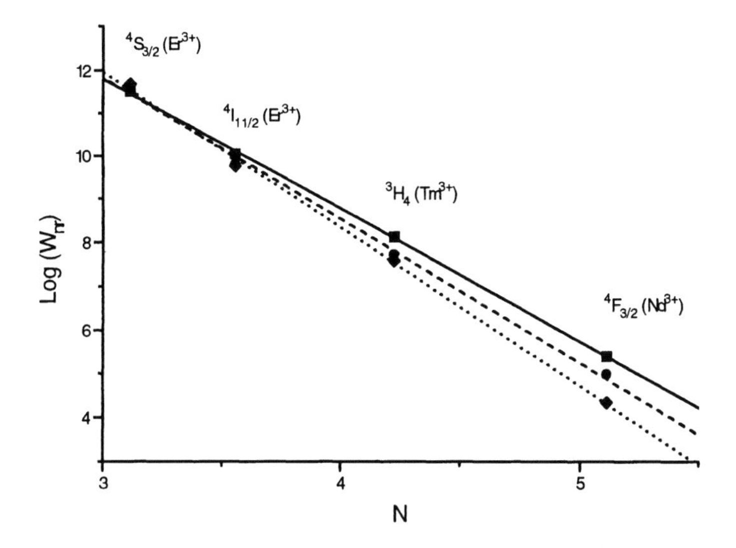

Fig.3 Variation of the exponential gap law versus the multiphonon order N, for three excitation densities: (■) $N_e=10^{17}cm^{-3}$ (l_c=54 Å) ; (●) $N_e=4x10^{18}cm^{-3}$ (l_c=22 Å) ; (◆) $N_e=10^{19}cm^{-3}$ (l_c=20 Å), continuous lines are the fits obtained from Eq.(7); N is phonon order. After [16].

IV. A PROPOSED MODIFICATION FOR THEORY AND DESCRIPTION OF EXPERIMENTAL RESULTS.

IV A.-A model based on the existence of a common diffusion volume for accepting modes.

We propose to now consider the following process: once an ion has given up its energy to an effective accepting mode, another excited ion nearby finds an already partially filled accepting term which leads to a reduction of the non-radiative decay rate for this ion. Because the number of nearby excited ions depends on the excitation density, the non-radiative decay rate shall in turn be reduced at higher excitation level. Since promoting modes play a role only to initiate the non-radiative transition at the expense of the electronic energy, we consider that they are not modified by the process under study. This is justified by what was observed in the experiments described in III.A.

Now, if instead of one ion, an average of $\overline{x}$ other ions simultaneously contribute to the filling of the common accepting mode set shared with the first considered ion, the filling of the accepting modes shall proceed by $(1+\overline{x})N$ at a time instead of N. $\overline{x}$ is related to the excited state density, N_e, by [16]:

$$\overline{x} = N_e\, v_l \left(1 - e^{-N_e v_l}\right) \tag{6}$$

where $v_l = 4\pi\, l_c^3 / 3$ is the volume for phonon diffusion, common at least to two nearby excited ions and l_c is the phonon diffusion length. $\overline{x}$ represents the probability that a given excited ion can share with another excited ion a common sphere of radius l_c. The effect can be described by simply substituting N by $(1+\overline{x})N$ in (3) only for the accepting term and not for the promoting one, assuming them to stay localised. One obtains :

$$W_{nr} = W_0 e^{-S_0(2\overline{n}+1)} (\overline{n}+1)^{(1+\overline{x})N} \frac{S_0}{\left[(1+\overline{x})N\right]!} \left(\frac{N_p}{S_0}\right)^2 \tag{7}$$

for the excitation density dependent nonradiative decay probability.

IV.B-Comparison of theory with experimental results.before and after the exponential gap law approximation

To our knowledge, there has been no experimental investigation of the influence of excitation density on multiphonon non-radiative decay [10]. The existing theoretical consideration of an intensity dependence would predict, on the opposite, an increase of the non-radiative decay rate through a proposed "super non-radiant" effect [21].

In order to explain our experimental observations, we shall consider our microscopic process developed above, which involves only the accepting term readily described by the energetical behaviour of experiments. This approach is justified by the observed "rotation" of the experimental curves of Fig.3 around the point $N = 3.2$. It shows that the promoting term, as given here by the $N = 3.2$ ordinate is not modified by the excitation intensity, each excited ion then individually contributes to the promoting term.

The least squares fit of multiphonon emission rates $W_{nr}^{(N_{exc})}$ measured at fixed excited state density were performed using equ. (7), with l_c as the only free parameter. The theoretical fits are plotted as the continuous lines on Fig.3, in rather good agreement with the experimental points. This shows that (7) is still equivalent to the exponential gap law with an α parameter increasing with excitation density as derived from (5) and (6):

$$\alpha(N_e) = \frac{(1+\overline{x})}{\hbar\omega} \left(\log \frac{(1+\overline{x})N}{S_0(\overline{n}+1)} - 1 \right) \tag{9}$$

The result is given on Fig.4, showing that a 30% variation in α can be found between low and high excitation density (10^{20} cm^{-3}); the continuous line corresponds to the theoretical fit of $\alpha = f(N_e)$ using Equ.(9) with a phonon diffusion length l_c equal to 29Å.

On the other hand, the l_c values, as obtained from the fits of $W_{nr}^{(N_{exc})}$, (for N_e ranging from 10^{17} to 10^{19}) without the exponential approximation , i.e. directly through equ.(7), are plotted on Fig.5. The phonon diffusion length is found to decrease with excitation density down to a saturation value of 20Å, not far from the 29Å already obtained within the exponential approximation.

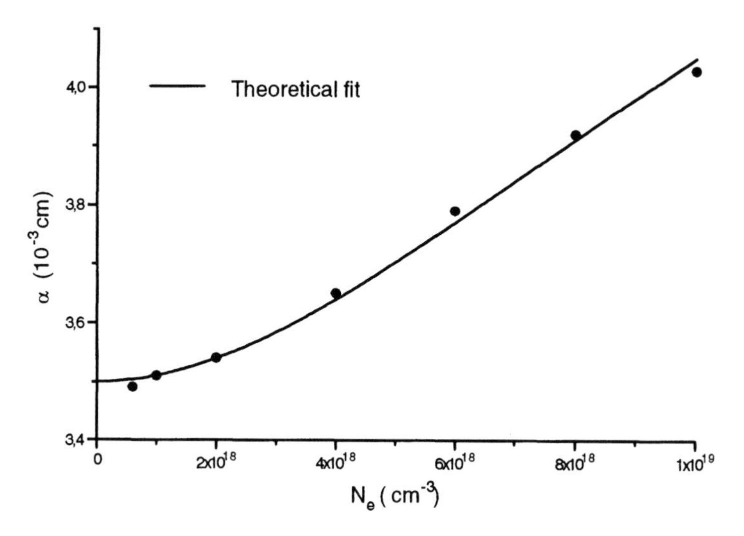

Fig. 4 Exponential parameter α versus excitation density N_e in a germanate glass. Fit is obtained from Equ.(9) with l_c=29Å. After [16].

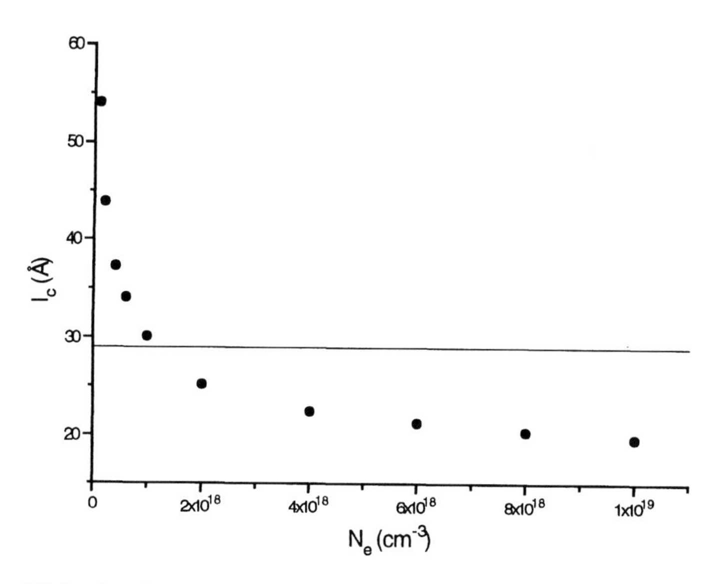

Fig. 5 Phonon diffusion length (l_c) versus excitation density (N_e) derived from non-radiative fits; the value, l_c=29Å, independently derived from sound velocity and thermal properties, is given by the horizontal line. After [16].

IV.C-Comparison of phonon diffusion lengths obtained through the excitation density dependant theory and direct experimental determination based on sound velocity measurements.

We also found that the above diffusion length value corresponds to a low frequency acoustic phonon diffusion length of 29Å which has been determined from a specific heat estimation of 0.445 cal/cm^3 deg , a heat capacity of 1.64x10^{-3} cal/cm deg s [22] and an acoustic velocity of 3.85x10^5 cm/s for a germanate glass [23], l_c being derived from the classical formula for the low frequency phonon diffusion length [24]:

$$l_c = \frac{3K_c}{V_s c_p} \tag{10}$$

where K_c is the heat capacity, c_p the specific heat, and V_s the sound velocity.

This shows that the crude effective mode model for non-radiative relaxation rates describes rather well the whole process including the energy dissipation from the 900 cm^{-1} effective mode towards lower energy modes. This can be understood in the following way : since the lower energy modes are connected with the larger diffusion lengths, the bottleneck we observe contains the slowest step in the energy dissipation. It may take place between the high energy modes ruling the multiphonon process and the lower energy ones providing the final heating process. This last type of bottleneck is well documented [25] in connection with spin-lattice relaxation and is included here through our effective l_c.

IV. D.-Role of Concentration: the Yb^{3+} Case [26].

As an example of the role of RE ions concentration, we shall present the results for Yb^{3+} excited states in a borate glass where the Yb^{3+} ions concentration ranges from a rather low .2% in weight (8 10^{19} cm^{-3}) to 8% in weight (3.2 10^{21}cm^{-3}). The choice of Yb^{3+} was in order to avoid any important up-conversion effect by energy transfer [17] which could reduce the life-times measured at high excitation densities. The choice for a borate glass was to have a large enough phonon energy in order to have a non-negligible non-radiative decay with respect to the large electronic energy gap for Yb^{3+}. The glass host has the following composition in molecular percentage: B_2O_3 65%; Na_2O 10%; BaO 25% .Such a glass has its highest phonon frequency ($\hbar\omega_M$)at about 1400 cm^{-1}[6].The experimental non-radiative rate probabilities (W_{nr}) have been determined by equ.(8) as above,

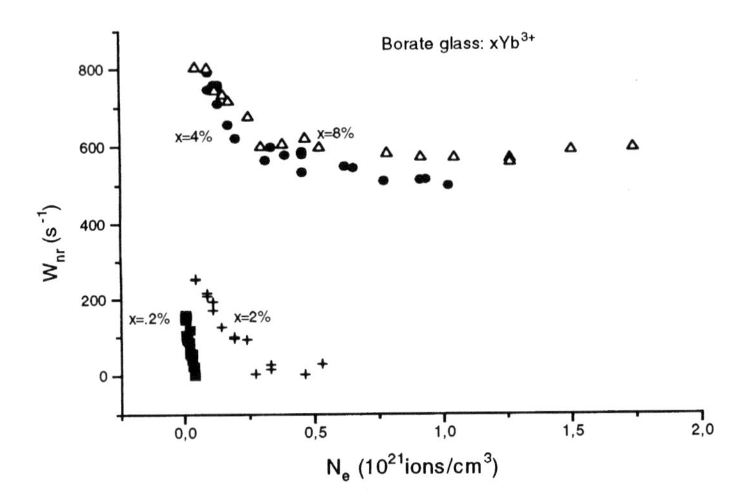

Fig.6. Measured points at room temperature for $^2F_{7/2}$ non-radiative decay probability W_{nr} versus excitation states density, N_e, for different Yb^{3+} concentrations in a borate glass: (■) .2% (8 10^{19} cm^{-3}); (✦) 2% (8 10^{20} cm^{-3}); (●) 4% (1.6 10^{21} cm^{-3}); (Δ) 8% (3.2 10^{21}cm^{-3}). After [26].

that is here by measuring experimentally decays (τ) from the directly excited level ($^2F_{5/2}$) of Yb^{3+}, and calculating the radiative ones (τ_0) from absorption measurement. Care has been taken in order to avoid any photon trapping effect which is known to exist for Yb^{3+} [27]. As an example, the general behaviour observed for non-radiative decay rate versus excitation density for a fixed $\Delta E > 3.2\hbar\omega_M$, here at ~ 10^4 cm^{-1}, that is for $\Delta E = 4.4\hbar\omega_M$, is presented on Fig.6. for four Yb^{3+} concentrations. It is clearly observed that the non-radiative decay probability decreases for excited states densities which of course grow with doping densities.

For the two lower ion concentrations (0.2 and 2 %), the mathematical fit of W_{nr} dependence on N_e, the excited ions density, with l_c as a parameter, gives the continuous curves on Fig. 7, showing that our experiments are well described for phonon diffusion length of respectively 23Å and 9.3 Å at Yb^{3+}concentrations of 0.2% and 2%. The observed reduction in diffusion length can be understood as an increase in phonon scattering with the incorporation of the heavy mass Yb^{3+} ion. The obtained diffusion lengths are in agreement

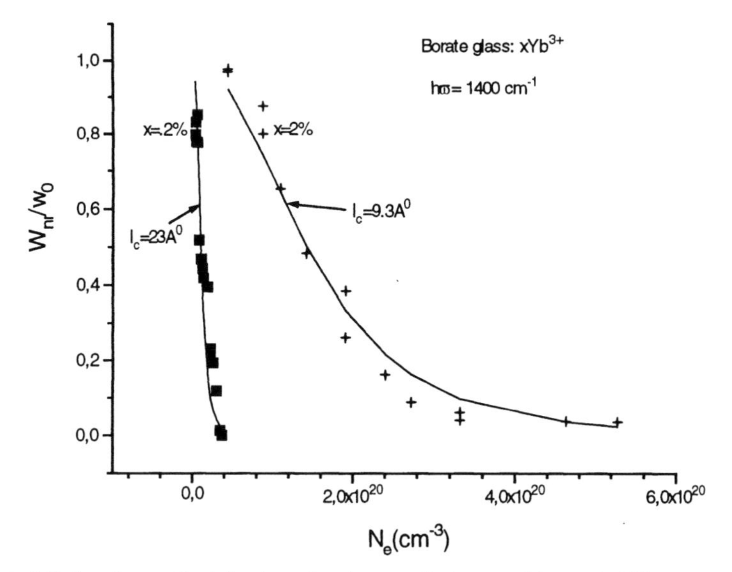

Fig.7. Theoretical fit (continuous line) for the reduced non-radiative transition probability W_{nr}/w_0 of a borate glass doped with 8 10^{19}cm^{-3}(.2%) and 8 10^{20}cm^{-3}(2%) Yb^{3+}, versus N_e the excitation states density. The common experimental non -radiative decay probabilty for weak excitation (w_0) is 220 s^{-1}. After [26].

with the preceeding values for other types of glasses. In the case of the heavily doped samples (4 and 8 %), higher excitation densities cannot reduce to zero the non-radiative relaxation rate showing that asymptotes, respectively at 500 s^{-1} and 575 s^{-1}, are about to be reached for excitation densities around 8.10^{20} cm^{-3}. If we note W_{nr} the non saturated part of non radiative rate, this prompts us to write :

$$W_{nr}(N_e) = W_{nr} - W_{nr}(N_0) \quad (14)$$

where $W_{nr}(N_0)$ is the low excitation density multiphonon non-radiative rate assumed to be a function of the chemical concentration (N_0).

Fitting $W_{nr}(N_e)$ by eq.(14), with l_c as a free parameter, gives the results of Fig.8 showing that here, as for 0.2 and 2% doped samples, eq.(13) is rather well followed.

It reveals that for higher Yb^{3+}concentrations, i.e. 4 and 8%, the phonon diffusion length reaches now a limiting value of about 7.9 Å. This corresponds to the typical value of 8Å given for example for a silica glass [24]. As for $W_{nr}(0)$, it can equivalently be viewed as resulting from an increase in multiphonon rate linked with an increase in the electron-phonon coupling strength becoming large enough not to be saturable by the excitation bottleneck. From eq.12, $W_{nr}(0)$ can be estimated to correspond to an increase in S_0 given by:

$$log\left(\frac{S(N_0)}{S_0}\right)=\frac{\hbar\omega_M}{(\Delta E-2.6\hbar\omega_M)}log\left(\frac{W_{nr}(N_0)}{W_0}\right) \tag{15}$$

From Fig. 6, taking W_0, the low excitation value on the 2% curve, as $220s^{-1}$ and $W_{nr}(N_0)$ as $800s^{-1}$ from the low excitation value on the 8% curve, one get $S(8\%) = 1.34\ S(2\%)$, that is a 33% increase in coupling strength. The asymptotical value $W_{nr}(0)$, 575 s^{-1}, is about given by the difference between the low excitation values for weak and heavy concentration: 580 s^{-1}, confirming a common origin.

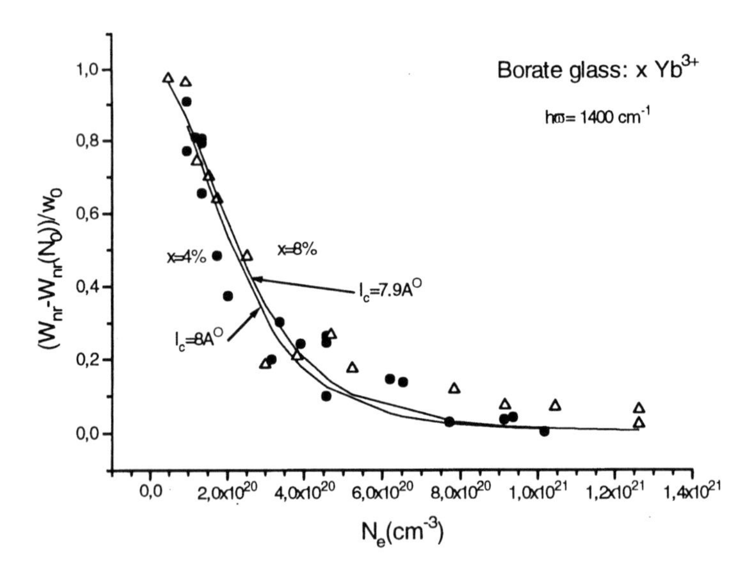

Fig. 8. Theoretical fit (continuous line) for the reduced non-radiative transition probability $(W_{nr}-W_{nr}(N_0))/w_0$ for higher concentrations: (●) for 1.6 $10^{21}cm^{-1}$ (4%) and $W_{nr}(N_0)$=$500s^{-1}$; (Δ) for 3.2 $10^{21}cm^{-1}$ (8%) and $W_{nr}(N_0)$=575s-1. After [26].

V NEW PROPOSED MATERIALS FOR LASER AND AMPLIFIERS.

Though the present size and specification of usual Erbium Doped Fiber Amplifiers (EDFA) are well suited for long distance systems, they are still too bulky and too costly to be usable in the distribution network where there is a need to compensate the inherent losses of multiways splitters. Clearly optical amplifiers need size reduction and technology simplification in order to be generalised in the optical distribution network. One way to look at this problem is to look towards Erbium doped short waveguide amplifiers. In this part, we shall consider the basic limitations impose by physics in reducing the amplifiers active length and look at likely solutions which may be taken into account even before any practical realisation of the waveguides themselves.

V. A.-Typical Specifications of Present Optical Fiber Amplifiers.

1- Comparison and specifications for semiconductor and EDFA optical amplifiers. The competition between semiconductor and EDFA optical amplifiers, was won by these last not because of the gain values themselves which are about equivalent (30-40dB), but essentially because of two specific characters of EDFA: namely the gain saturation power, which reaches +11dBm in EDFA to be compared to -3 to +8dBm in semiconductors and the polarisation effect defined as the ratio of the gain for two orthogonal polarizations which is 0db for fiber and 1to 4 dB for semiconductor amplifiers.

2- Typical lengths and doping concentrations for fiber amplifiers It is interesting to define an upper limit to the gain (G_{lim}) of an Erbium doped optical amplifier, whatever the technology and the pumping efficiency, by simply considering the single path gain for total inversion, which of course can never be reached: if such value is found lower than the researched gain then it shall be hopeless whatever the pumping. It is given by:

$$G_{\lim} = 10 Log_{10}(\exp(\sigma NL)) \qquad (1)$$

where σ is the induced emission cross-section, N the active ion concentration, L the length of the optical amplifier. For typical values of Erbium in a fiber, with σ=5 $10^{-21}cm^2$, N=1.4 $10^{18}cm^{-3}$ ($\cong$200ppm), L=10m, one obtains:

$$G_{\lim} = 10 Log_{10}(\exp 7) = 30.4 dB \qquad (2)$$

Such value, with above length and concentration, being a limit, shall of course never be reached whatever the pumping if a higher gain is looked for. Defining such a limit usefully helps defining the minimum requirements for a given gain value. Let us look at the analogue for the compact waveguide amplifier we are interesting in.

V.B. The Fundamental Parameter for RE Optical Amplifiers.

1- The cross-section-concentration product necessary for a compact waveguide amplifier. Assuming that a compact waveguide amplifier should have a length of about 1cm, that is a length 1/1000 the typical one of an EDFA, and in order to keep a gain of 30.4dB, one need from (2) to still have: $\sigma NL = 7$, that is an increase in σN by a factor 1000 or:

$$\sigma N = 7 cm^{-1} \qquad (3)$$

This is to be compared to 7 10^{-3} cm^{-1}, the usual value for fiber

2- The basic limitations for cross-section and concentration increase in glasses. Because Rare-Earth (RE) 4f-4f transitions are forbidden at first order, the integrated cross-sections over frequencies, that is the oscillator strength are about constant $\cong 10^{-6}$ except for the so-called hypersensitive transitions characterized by $\Delta J \leq 2$. The case of interest to us is the Er^{3+} transition $^4I_{13/2} \rightarrow {}^4I_{15/2}$ at 1.54µm. It could be an hypersensitive one, however because ΔJ=1 is also the selection rule for magnetic dipole transition which are allowed, a large part of the oscillator strength for the above transition is of magnetic dipole origin and so independent of the host. It can be concluded that they are no much chance to find a material with an oscillator strength much larger than the ones we already know. Consequently high cross-section can only be found in materials showing narrow effective line width Δ_{eff} with:

$$\Delta_{eff} = \frac{1}{\sigma_{max}} \int \sigma(\nu) d\nu$$

In glasses the narrower effective linewidth shall be obtained for narrower inhomogeneous width.

As for the concentration increase, the limitations are first at the chemical level; either because one-constituent glass leaves too little space to the large RE ions (1A°), which is rejected by the glass; this is the case for pure silica glass, where concentration is chemically limited to about 100ppm($0.7\ 10^{18}cm^{-3}$) and this directs us towards multicomponent glasses; or because the RE elements behaving usually as glass modifiers and not as glass formers tend to devitrify glasses when they are incorporated at too high concentrations. The highest chemical concentration we know for a phosphate glass reaches $3\ 10^5$ ppm ($2.2\ 10^{21}cm^{-3}$).

The concentration limitation takes place also at the physical level usually before the maximum chemical concentration is reached. This is due to the ion-ion interactions giving rise to "concentration quenching" which is excitation independent. There is also clustering of ions revealed by energy transfers with one of both considered ions being in an excited state; the role of this effect is excitation dependant. Energy transfers by dipole-dipole interactions can take place at distance reaching 20A°, whereas the cluster effect, by definition, exists for nearby ions separated by distances less than 6A° and for such reason is very effective. The clustering problem shall not be addressed here.

V.C. A Proposed Hint for a Partial Solution:.The Inhomogeneous Width Reduction in Glassy Materials.

1- A New Likely Amplifier Material: The RE Doped Vitroceramics. In order to increase the useful cross-section, as observed above, one can decrease the inhomogeneous

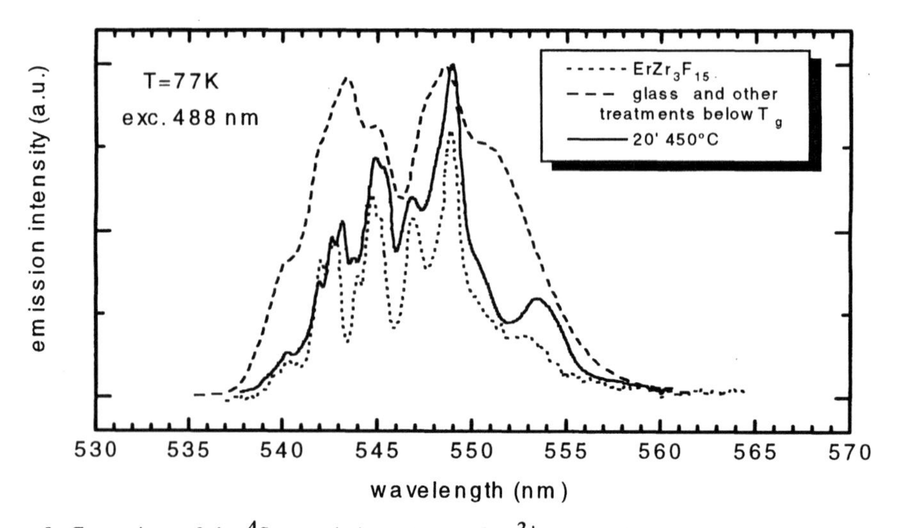

Figure 9. Comparison of the $^4S_{3/2}$ emission spectra of Er^{3+} at 77K in the original glass before heat treatment (----); the vitroceramic as obtained after heat treatment (____); a reference sample of crystalline powder of $ErZr_3F_{15}$ (.....). From [32].

linewidth. We had proposed a long time ago to consider oxyfluoride vitroceramics (GeO_2-PbF_2-REF_3; SiO_2-PbF_2-REF_3), where the doping RE ions was only be in a fluoride microcrystalline phase surrounded by an oxide glassy phase[28]. Because application was for visualisation, the crystallites were larger than the wavelength and these vitroceramics were not transparent. However fluorescent properties of such materials were excellent because cross-sections were the one of crystals and efficiencies were high because the non-radiative transitions of the RE ions were reduced, being ruled by the lower energy phonons of the fluoride surrounding and not by the oxygenic glassy host. For such reason we had then proposed [28] such vitroceramics could be usefully used for lasers, provided the size of the cristallites could be reduced below the considered wavelength.

Since then, these vitroceramics have been modified in order to become transparent with the idea to obtain bulk up-conversion lasers[29,30,31]. Because we are here interested in short path amplifiers, we want to increase the crystalline phase with respect to the glassy one. This leads us to consider fluoride vitroceramics (ZrF_4-LaF_3-AlF_3) doped with Er^{3+}[32]. Being obtained by spinodal decomposition, such vitroceramics show large size crystallites which, being constituted of the phase $Zr(La,Er)_3F_{15}$, shows spectra reflecting the reduced inhomogeneous linewidth of the crystal and not the inhomogeneous linewidth of the initial glass before the heat treatment which provokes the demixtion, see Fig. 9.

Because the masses of the constituents of the glass phase are very near the ones of the crystallites the obtained vitroceramic is still transparent though it has some added scattering losses. Such vitroceramic being cristallised at about 70% of its volume, shows a large active path length. Besides, as shown on Fig. 10, the cross-section is increased, both because the spectra is somewhat narrower than for the initial glass by reduction of the inhomogeneous width, but also because the integrated cross-section is increased by the change in the RE ion environment. Cross-section at 1.5µm can be doubled in such matrix and reach $10^{-20}cm^2$.

2- Present best results in waveguide amplifiers and what could be gain with proposed hints. In silica glass waveguide amplifiers, gain of 9.4 dB have been obtained with L=23cm in S shape for N=3800ppm[33], this corresponds to $\sigma N=0.13cm^{-1}$ and to G_{lim}=13db. It shows that the obtained gain is not far from the limit and improvement without increasing σN is not likely.

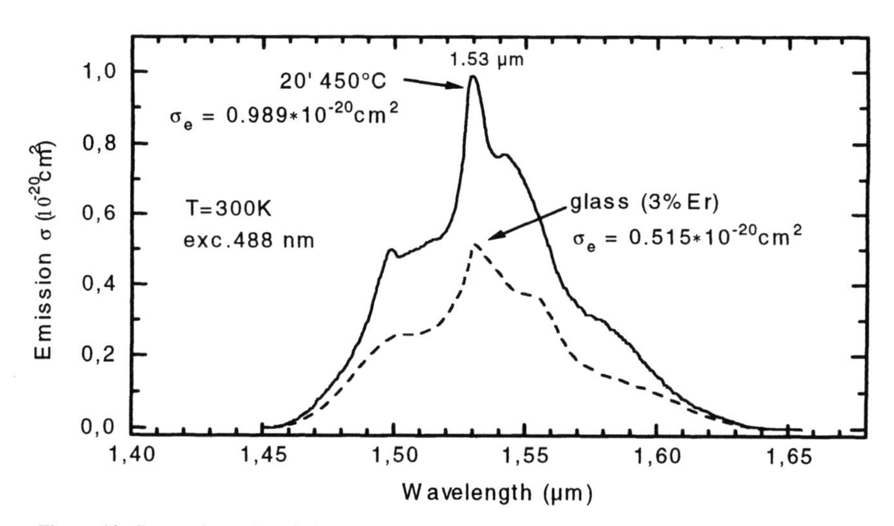

Figure 10. Comparison of emission spectra and cross-sections for the initial glass (---) and the vitroceramic (—). After [32].

In phosphate glass a gain of G=10db has been obtained [34] for N=4 10^4ppm and L=4cm. This corresponds to $\sigma N=1.4cm^{-1}$ and to G_{lim}=24db. It shows that the obtained result is still far from the gain limit and some engineering improvements are still possible.

From the results of 4.1 and 4.3 one can expect a factor 2 increase from cross-sections and a factor 7 for concentration with respect to the values considered for the phosphate amplifier wave guide of ref.[34].
The simultaneous increase in cross-section and in concentration would be a problem within the microcrystals. This is not really link to the clustering process because in the crystallites, contrary to the glass case, the distances are well defined through the crystal structure itself.

VI CONCLUSION.

We have shown that multiphonon non-radiative transitions may be reduced at high excitation density, this being due to a saturation of the accepting modes. Because of the generality of the process, it could be applied to all kind of multiphonon processes involving accepting modes and for such reason it allows us to predict, for example, the eventual following effects: multiphonon side-bands bleaching, quenching of multiphonon assisted energy transfers under high sensitizer excitation. Besides this, a concentration effect on the saturable non-radiative transitions has been observed which behaves as an increase in the electron-coupling strength with RE concentration.

Though the study has, up to now, been limited to glasses, it is being extended to crystals. There, because of the largest phonon diffusion length than in glasses, the saturation effect is expected to be larger.

Such would be the case in particular for the vitoceramics with spinodal demixtion, the new kind of amplifying materials we have described in the second part of the lecture. In these materials the RE ions stay only in the crystallites. This allows to take profit of eventual lower energy phonons and of longer phonon diffusion length with respect to glasses, that is higher quantum efficiency. Also because of the reduced inhomogeneous linewidth, higher cross-sections can be expected from the vitroceremics. However, the scattering losses have still to be reduced and it is the purpose of our present research effort on this subject.

REFERENCES.

[1] C.Gouedard, D.Husson, C.Sauteret, F.Auzel, and A.Migus, IAEA Conf. on Drivers for Inertial Confinement Fusion, Osaka, Japan, 15-19 April 1991, paper E04.

[2] C. Gouedard, D. Husson, C. Sauteret, F. Auzel, A. Migus, J. Opt. Soc. Am. B 10, 2358 (1993).

[3] F.Auzel, "Materials for Ionic Solid State Lasers", in "*Spectroscopy of Solid State Laser-Type Materials*", ed. B.DiBartolo (Plenium Press, New York) (1987) P.293.

[4] A. Kiel, in "*Quantum Electronics*", Eds. P.Grivet and N.Bloembergen, (Dunod,Paris) (1964), p.765.

[5] W.E. Hagston and J.E. Lowther, Physica 70, 40 (1973).

[6] C.B. Layne, W.H. Lowdermilk, and M.J. Weber, Phys. Rev. B16, 10 (1976).

[7] F. Auzel, "*Multiphonon interaction of excited luminescent centers in the weak coupling limit: non-radiative decay and multiphonon side-bands*", in "*Luminescence of Inorganic Solids*", Ed. B.DiBartolo, (Plenum, New-York), (1978), p.67.

[8] T. Miyakawa and D.L. Dexter, Phys. Rev. B1, 2961 (1970).

[9] D.J. Diestler, in "*Radiationless Processes in Molecules and Condensed Phases*", Ed. F.K. Fong, (Springer-Verlag, Berlin), (1976).

[10] R. Englman, "*Non-Radiative Decay of Ions and Molecules in Solids*", (North-Holland, Amsterdam), (1979).
[11] E. Gutsche, Phys. Stat. Sol. (b)109, 583 (1982).
[12] Y.V. Orlovskii, R.J. Reeves, R.C. Powell, T.T. Basiev, and K.K. Pukhov, Phys. Rev. B49, 3821(1994).
[13] Y.V. Orlovskii, K.K. Pukhov, T.T. Basiev, T. Tsuboi, Optical Materials 4, 583 (1995).
[14] K. Huang and A. Rhys, Proc., R. Soc., A204, 406 (1950).
[15] J.M.F. Van Dijk and M.F.H. Schuurmans, J. Chem. Phys.78, 5317 (1984).
[16] F.Auzel and F.Pellé, C.R.A.S. (Paris) 322, IIb, 835 (1996), and Phys.Rev. B55, 11006, (1997).
[17] F. Auzel, in "*Radiationless Processes*", Ed. B.Di Bartolo, (Plenum, New York),(1980), p.213.
[18] R. Reisfeld, L. Boehm, Y. Eckstein, and N. Lieblich, J. Lumin. 10, 193 (1975).
[19] B.R. Judd, Phys. Rev. 127, 750 (1962).
[20] F.Pellé and F.Auzel, to be published.
[21] M.M. Miller, F.K. Fong, J. Chem. Phys. 59, 1528 (1973).
[22] E.H. Ratcliffe, Glass technol.,4, 113 (1963).
[23] M.J. Weber, "*Optical properties of glasses*", in "*Materials Science and Technology*" Eds. R.W. Cahn, P. Haasen, E.J. Kramer,Vol. 9, VCH Publishers, New York,(1991), p.659.
[24] C. Kittel, "*Introduction to Solid State Physics*", Ed. John Wiley (NewYork),4th edition, (1971) p.232.
[25] A.M. Stoneham, Proc. Phys. Soc. 86, 1163 (1965).
[26] F.Auzel and F.Pellé, J.Lumin.,69, 249, (1996).
[27] F.Auzel, Ann. Telecom. 24 (1969) 363.
[28].F.Auzel ,D.Pecile, and D.Morin, J. Electrochem. Soc. 122, 101, (1975).
[29].Yuhu Wang and Junici Ohwaki, Appl. Phys. Lett.63, 3268, (1993).
[30] Kazuyuki Hirao, Katsuhisa Tanaka, Masaya Makita, and Naohiro Soga, J. Appl. Phys. 78, 3445, (1995).
[31] P.A.Tick, N.F.Borelli, L.K.Cornelius, and M.A.Newhouse, J.Appl.Phys., 78, 6367, (1995).
[32] F.Auzel, K.E.Lipinska, and P.Santa-Cruz, Optical Materials, 5, 75, (1996).
[33] K.Hattori, T.Kitagawa, M.Oguma, M.Wada, J.Temmyo, and M.Horiguchi, Electron. Lett., 29, 357, (1993).
[34] D.Barbier, "*Compact 1.54μm optical planar amplifiers*", leaflet issued by GEEO, F-38031 Grenoble cedex France, (1996).

NASA'S DIAL AND LIDAR LASER TECHNOLOGY DEVELOPMENT PROGRAM

James C. Barnes

NASA Langley Research Center
Remote Sensing Technology Branch
Mail Stop 474
Hampton, VA 23607

INTRODUCTION

Extremely stringent NASA DIAL and lidar laser requirements for high energy, high efficiency, specific wavelengths, various pulsewidths, narrow linewidths, and long operational reliability, all in a compact system, is the impetus for NASA's advanced solid-state laser technology development program. The Remote Sensing Technology Branch (RSTB) at NASA Langley Research Center (LaRC) leads the development effort to advance new solid state laser and nonlinear optics technologies to meet the performance, operation, and systems requirements for advanced aircraft and spaceborne remote sensing applications. These development efforts seek to provide significant improvements in efficiency and reduction in the size and weight of laser-based sensors used to measure ozone, water vapor, aerosols and clouds, and wind velocity with high vertical and horizontal resolution. These atmospheric constituents are very important with respect to global climate change: Measurements of water vapor is recognized as a key strategic step towards meeting the NASA Mission To Planet Earth (MTPE) objective of making "observations to monitor, describe, and understand seasonal-to-interannual climate variability, with the aim of improving skill in long-range weather forecasting and seasonal climate predictions." Water vapor is not as well characterized at the higher altitudes, hence, the gap in our knowledge base is considerable (Ismail, 1989). Aerosols, especially those arising from combustion of fossil fuels and biomass burning, reduce the Earth's surface temperature by reflecting a portion of the incident solar radiation back to space, countering some of the effects of increased greenhouse gas concentrations. Clouds also reflect solar radiation, and are considered the major uncertainty in understanding global climate change. Ozone has received increasingly more attention in recent years as a result of its reported depletion in the stratosphere and its importance to absorption of harmful ultraviolet solar radiation. Instruments that can measure these four atmospheric constituents with high spatial resolution and accuracy will make important contributions

Ultrafast Dynamics of Quantum Systems: Physical Processes and Spectroscopic Techniques, Edited by Di Bartolo and Gambarota, Plenum Press, New York, 1998

to understanding the mechanism of global climate change (Browell, 1994; Grant, 1989). Global climate change scientists, atmospheric dynamics researchers, and numerous agencies such as NASA, NOAA, DOD, and DOE greatly desire continuous, global measurements of tropospheric wind with good horizontal and vertical coverage and resolution, i.e. a horizontal vector accuracy of about 1 m/s (Kavaya, 1996). Descriptions of the development of advanced solid state laser light sources for NASA DIAL and lidar measurements of ozone, water vapor, aerosols, clouds and wind velocity are discussed in more detail below.

Laser Materials Modeling and Spectroscopy

A quantum mechanical laser materials model is used by NASA RSTB researchers to guide the search for new laser materials that can meet the stringent and unique space-based DIAL and lidar laser requirements; these include very specific wavelengths, narrow linewidths, high energy output with high efficiency, high repetition rates, and long operational lifetimes with high reliability. Purely experimental approaches for the development of new laser materials result in impractical developmental cost and in time consuming, high risk research and development efforts. Consequently, a quantum mechanical modeling and spectroscopy approach is used by RSTB. The quantum mechanical model is used to calculate specific wavelengths, transition rates, and thermal occupation factors of new laser materials. Researchers use published spectroscopic data for known laser materials that having similar material structure to the new materials being investigated to help validate the quantum mechanical model. Validation of the model through the existing data on similar materials lends credence to the model's extrapolation in predicting the optical and laser characteristics of new materials. Next the researchers perform spectroscopic evaluation of the new laser materials using spectroscopic samples which are relatively inexpensive compared to laser grade samples. The evaluation tests the predictions of the quantum mechanical model with regards to material parameters such as emission wavelength, fluorescence lifetime, and transition strength. The evaluation also provides spectroscopic data with which to fine tune the results, for example changing the ratios of the material composition, if necessary. The results of spectroscopic evaluations are then used in RSTB laser oscillator and amplifier models to aid in the design of specific laser systems. Finally, based on model results, a final laser material composition is chosen to demonstrate laser performance (Jani et al., 1994; Hart et al., 1996). If warranted, the material is further developed to laser grade quality to meet the particular DIAL or lidar system's requirements.

2-μm, Ho:Tm:YLF Laser System for Wind Velocity Measurements

For coherent wind lidar measurements from ground, air, and space-borne platforms, an energetic, low divergence, narrow linewidth all solid-state laser system in the eye-safe region is required as a transmitter. RSTB researchers have led the development of laser diode array pumped, 2-μm, Ho:Tm:YLF laser system for this application, as it emits radiation in an eye-safe region and can achieve efficient diode pumped operation using commercially available laser diodes. An extensive theoretical model (Barnes et al., 1996; Barnes et al., 1996) has been established at NASA LaRC and was used to identify the laser material for optimum gain at 2-μm. The Ho:Tm:YLF crystal, operating at 2.05 μm, was selected as the laser gain medium based on the model. This group previously used a flashlamp pumped oscillator and five diode pumped amplifiers to generate 700 mJ at 1 Hz (Williams-Byrd et al., 1997). Since then, an injection-seeded diode pumped oscillator has been developed and used in the laser system. The pulse repetition rate has been increased to 10 Hz. Amplifier gain at 10 Hz is measured to be comparable to that at 1 Hz. Performance of the system has been fully characterized. A schematic of the injection-seeded Ho:Tm:YLF laser system is shown in Fig. 1. It consists of a CW master oscillator, power oscillator and amplifiers. The diode pumped Q-switched power oscillator utilizes a three meter long, figure-eight ring resonator configuration and is injection-seeded by a CW microchip master oscillator. This combination provides a single-frequency and high beam quality extraction beam for introduction to the amplifiers. The amplifier chain comprises four diode pumped amplifiers which produce a gain of about 9 in the single pass configuration used for this work. The He-Ne laser shown in the schematic is used to align the 2-μm, Ho:Tm:YLF laser system.

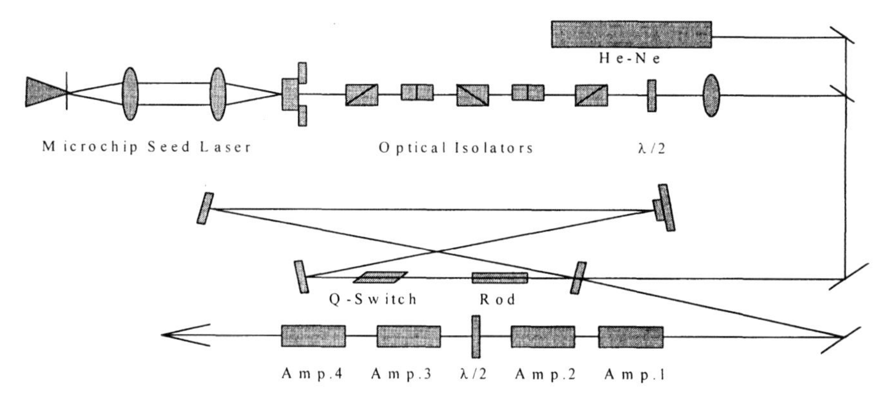

Fig 1. Schematic of injection-seeded Ho:Tm:YLF laser system.

Ti:Sapphire Laser Transmitter for Airborne Water Vapor DIAL Instrument

NASA Langley researchers and several industry partners pioneered Ti:Al_2O_3 (titanium-doped sapphire) laser material development in the mid- to late 1980's, as well as laser pumping and injection seeding techniques (Brockman et al., 1986; Barnes et al., 1988; Rines et al., 1989). The material quality was improved by an order of magnitude, allowing the researchers to demonstrate the superior performance of tunable (from 0.66 μm to 1.1 μm) Ti:Al_2O_3 lasers over conventional tunable dye lasers. With these developments, Ti:Al_2O_3 became a viable laser device with a multitude of potential uses including medical, industrial, communication, military, and scientific applications.

With the advantage of having performed the earlier work, NASA laser engineers designed and built a fully autonomous tunable laser system (TLS), based on Ti:Al_2O_3 laser technology, for NASA's Lidar Atmospheric Sensing Experiment (LASE). Since 1994, the LASE Instrument has flown on over 26 successful DIAL water vapor measurement missions on NASA's high altitude ER-2 aircraft without a single failure of the Ti:Al_2O_3 laser (Browell, 1995). The TLS (a schematic is shown in figure 2 below) was designed to operate in a double-pulse mode at 5Hz, with energy outputs up to 150mJ per pulse. The TLS power oscillator is continuously wavelength tunable from 813 to 819nm, limited by the wavelength reflectivity of the graded reflectivity mirror (GRM). To achieve high fidelity in the DIAL water vapor measurement, the TLS operates with

99% of the output energy within a spectral interval of 1.06 pm. The Ti:Sapphire power oscillator uses a Nd:YAG MOPA laser, frequency-doubled by CD*A, as the pump source and a single mode diode laser as a injection seeder for the Ti:Al_2O_3 laser (J. Barnes et al., 1993; N. Barnes et al., 1993; J. Barnes et al., 1993). The pump laser system has about 600 mJ of 532nm output which is split into equal energy to pump the two Ti:Al_2O_3 laser bricks of the power oscillator. The diode seed laser is autonomously tuned and maintained at a desired water vapor feature through temperature and current controls. A 4-plate bi-refringent filter (BRF) is used to tune the power oscillator to overlap the diode seed wavelength. The LASE instrument was recently reconfigured to fly on a NASA P-3 Aircraft in order to participate in the Southern Great Plains Hydrology Experiment (SGP97) in the summer of 1997. The SGP97 was a joint NASA, DoE, NOAA, NSF, USDA, NRC of Canada, and Oklahoma Mesonet interdisciplinary investigation designed to establish that retrieval algorithms for surface soil moisture developed at high spatial resolutions for ground and airborne sensors can be extended to the coarser resolutions expected from satellites (Jackson et al., 1995).

Cr:LiSAF and Nd:Garnet Laser Development for Space-Based Water Vapor DIAL

NASA researchers and contractors are currently developing a water vapor DIAL laser transmitter for unmanned aerial vehicle deployments as a precursor to future spaceborne DIAL instrument deployments. The current laser transmitter is based on the promising new, laser material, Cr:$LiSrAlF_6$, which has a broad tuning range and is directly laser diode pumpable. Cr:$LiCaLiF_6$ and Cr:$LiSrGaF_6$, two similar materials with slightly different wavelength and thermal properties are also being investigated for future applications. These materials are also attractive because of they have broad absorption (~550nm - 750nm) and emission (~750nm - 1000nm) bands and because they can be heavily doped with the active, Cr^{3+}, ion. The broad tuning range resulting from the broad emission allows for multiple NASA DIAL applications. Tropospheric water vapor measurements can be addressed at wavelengths around 815nm and 940nm, frequency tripling of 900nm yields wavelengths that address atmospheric ozone and the residual 450nm and 900nm wavelengths would provide aerosol and cloud measurements. The broad pump absorption bands of the materials peak in the 670 - 690nm region but extend into the 750nm region where well developed AlGaAs laser diode arrays (LDA's) can be used to pump heavily doped materials (Payne et al., 1992). NASA has supported the development of advanced visible 675nm InGaP laser diode array technology to optimize

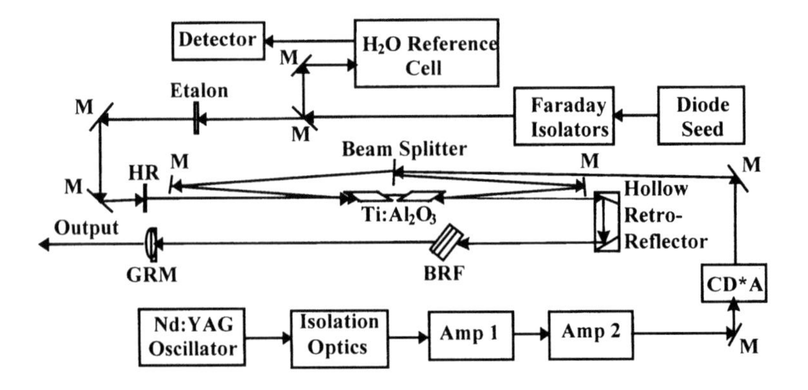

Fig. 2. Schematic representation of the LASE TLS optical layout.

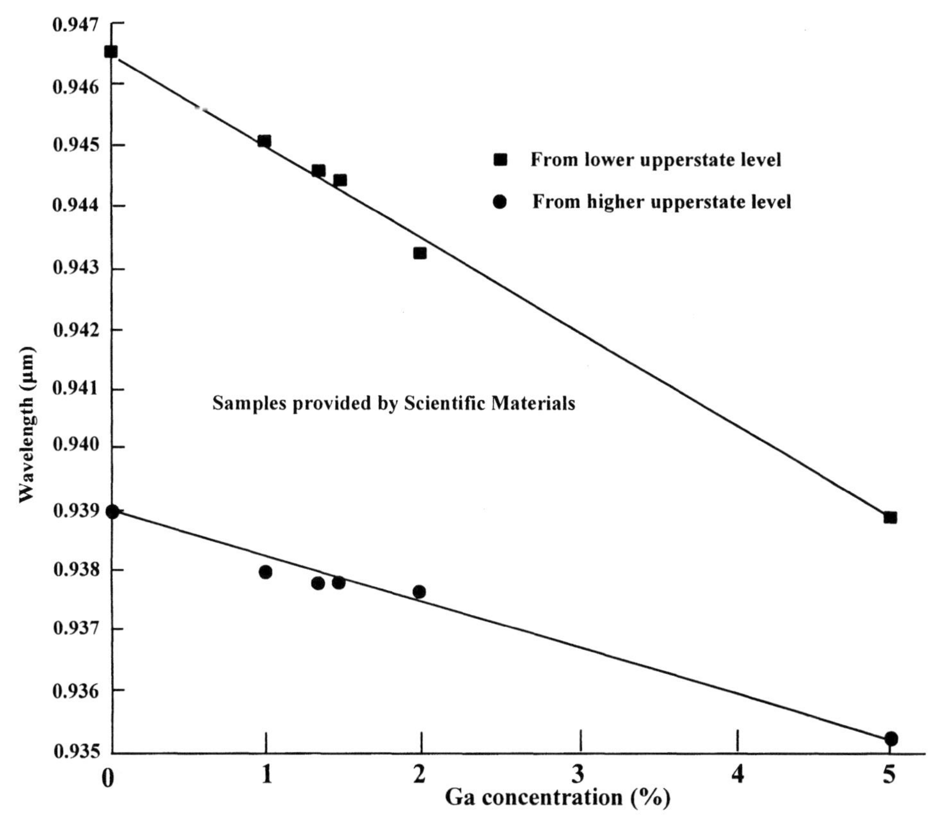

Fig. 3. Peak emission wavelength as a function of Ga concentration in a garnet.

efficiency in pumping the Cr^{3+} lasers. LDA pumping increases the wall-plug efficiency and operational lifetime over current flashlamp pumping, both of which are critical parameters for NASA space missions. The researchers used 3 each six bar stacks of the new 675nm LDA's that operate with 360W output in 200µs pulses to demonstrate the first ever 675nm, LDA side-pumped $Cr:LiSrAlF_6$ laser oscillator (Johnson et al., 1996). The oscillator produces 33mJ of energy in the normal mode and has 3mJ of q-switched energy. The oscillator has been used to measure hard targets and water vapor in a joint NASA/Los Alamos National Laboratory demonstration (Early et al, 1996). Because of the relatively short upper state lifetime of these materials (approximately 65 to 70µs), they require high peak power pumping for high efficiency. Investigations are underway to identify host with longer lifetimes and to improve both the efficiency and the reliability of the 675nm LDA pumps when operating at short pulsewidth and high power.

NASA researchers have demonstrated compositional tuning of Nd:Garnet laser emission to match strong water vapor absorption features in the 940 nm wavelength region for water vapor DIAL applications. Garnet laser materials are pumpable by well developed, less costly, AlGaAs laser diode array technology and promise to provide the efficiency, reliability, and affordability required for future space-based DIAL missions. In the demonstration, the emission peak of the garnet was tuned by adjusting the fractional composition of Ga in Nd:YGAG, a garnet laser material. Tuning of the fluorescence emission peak of the garnet, Nd:YGAG, laser transition was demonstrated from 935nm to 945nm. In this wavelength region, the water absorption lines are 20 times stronger than the lines currently used at ~ 815nm, hence, these wavelengths are attractive for measuring low water vapor concentrations in the stratosphere and upper troposphere accessible from space. Figure 3 shows peak emission from a garnet laser material as a function of Ga concentration.

UV Transmitter for Ozone DIAL

NASA RSTB researchers have begun the development of ultraviolet laser transmitter technology for a space-based ozone DIAL. Atmospheric models predict the transmitter energy output must be 500mJ at wavelengths of 305nm and 315nm and must operate at a minimum 5 Hertz, double-pulsed rate. The current NASA design for demonstration of the UV laser transmitter technology is depicted in figure 3. The frequency doubled output of a Nd:YLF laser is split into three 524nm beam lines. One beam line is used to pump an optical parametric oscillator (OPO). The OPO output is tunable between 740nm and 820nm and its output is amplified in a titanium-doped sapphire laser amplifier being pumped by the second of the 524nm beams. The output of the titanium-doped sapphire laser amplifier and the third 524nm beam will be simultaneously transmitted into a BBO sum frequency mixer to generate the tunable ultraviolet output. Initial demonstrations will be in the 100mJ - 200mJ energy range but with designs to show scalability to 500mJ or more. The technology tall poles for this development include eliminating optical damage to the ultraviolet optics and mixing crystal, providing the precise timing required for an efficient mixing process, and obtaining good beam quality in the OPO and amplifier output.

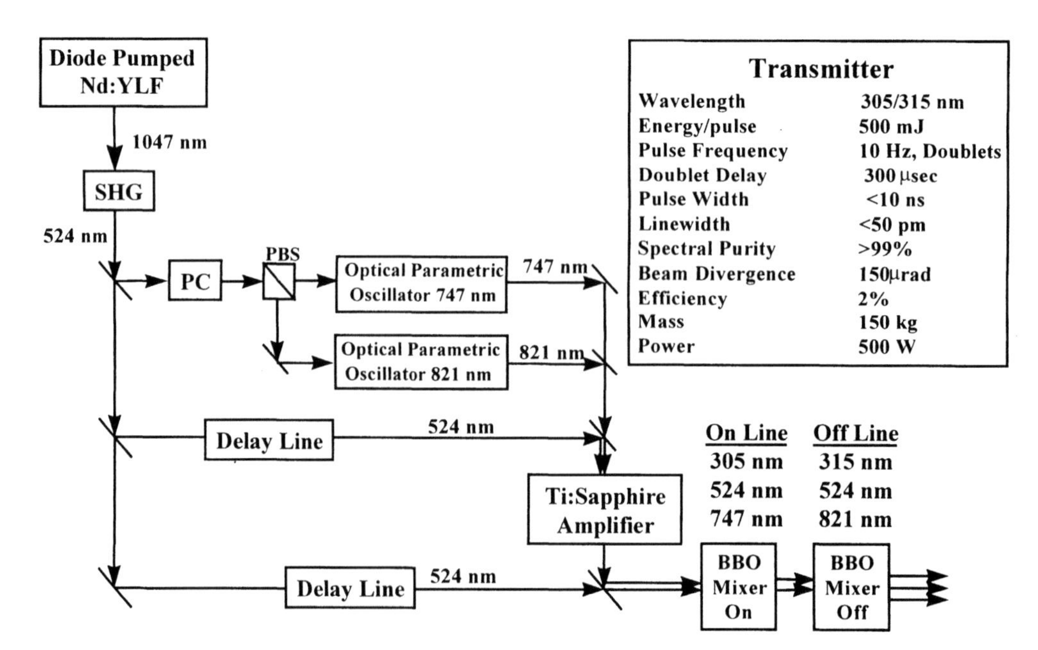

FIGURE 3. Schematic of UV system being developed for spaceborne DIAL ozone measurements.

Acknowledgments

The laser and DIAL research are supported by NASA Headquarters' Codes X, S and Y. The author recognizes the contributions to this paper by Norman Barnes, William Edwards, Larry Petway, Christyl Johnson, Julie Williams-Byrd, Upendra Singh, Waverly Marsh, Jirong Yu, Mark Storm, Brian Walsh, Elka Ertur, and Richard Campbell.

References

Barnes, J. C., N. P. Barnes, and G. E. Miller, 1988, Master oscillator power amplifier performance of Ti:Al_2O_3, *IEEE J. Quantum Electron.,* 24, (6):1029-1038.

Barnes, J. C., W. C. Edwards, L. Petway, and L. G. Wang, 1993, NASA Lidar Atmospheric Sensing Experiment's (LASE) titanium-doped sapphire tunable laser system, *OSA Optical Remote Sensing of the Atmosphere Technical Digest,* 5:459-565.

Barnes, J. C., N. P. Barnes, L. G. Wang, and W. C. Edwards, 1993, Injection seeding II: Ti:Al_2O_3 experiments, *IEEE Journal of Quantum Electronics,* 29, (10):2683-2692.

Barnes, N. P., and J. C. Barnes, 1993, Injection seeding I: theory, *IEEE Journal of Quantum Electronics*, 29 (10):2670-2683.

Barnes, N. P., Filer, E. D., Morrison, C. D., and Lee, C. J., 1996, Ho:Tm lasers I: theoretical, *IEEE J. Quant. Elect.* QE-32, 92-103.

Barnes N. P., Rodriguez, W. J., and Walsh, B. M., 1996, Ho:Tm:YLF laser amplifiers, *J. Opt. Soc. Am. B*, Vol.13, No. 12, 2872-2882.

Brockman, P., C. H. Bair, J. C. Barnes, R. V. Hess, and E. V. Browell, 1986, Pulsed injection control of titanium-doped sapphire laser, *Opt. Lett.,* 11 (11):712-714.

Browell, E. V., 1994, Remote sensing of trace gases from satellites and aircraft, in *Chemistry of the Atmosphere: The Impact on Global Change,* edited by J. Calvert, Blackwell Scientific Publications, 121-134.

Browell, E. V. and S. Ismail, 1995, First lidar measurements of water vapor and aerosols from a high-altitude aircraft, *Proceeds of the OSA Optical Remote Sensing of the Atmosphere Conference,* 2:212-214.

Early, J. W., C. S. Lester, N. J., Cockroft, C. C. Johnson, D. J. Reichle, and D. W. Mordaunt, 1996, Dual-rod Cr:LiSAF oscillator-amplifier for remote sensing applications, *Trends in Optics and Photonics on Advanced Solid State Lasers,* 1:126 -129.

Grant, W. B., ed. , 1989, *Ozone Measuring Instruments for the Stratosphere, Optical Soc. Am.*, 438.

Grant, W. B., et al., 1994, Aerosol-associated changes in tropical stratospheric ozone following the eruption of Mount Pinatubo, *J. Geophys. Res.*, 99:8197- 8211.

Hart, D. W., M. G. Jani, and N. P. Barnes, 1996, Room-temperature lasing of end-pumped Ho:$Lu_3Al_5O_{12}$, *Opt. Lett.,* 21 (10):728-730.

Ismail, S. and E. V. Browell, 1989, Airborne and spaceborne lidar measurements of water vapor profiles: A sensitivity analysis, *Appl. Opt.,* 28:3603-3615.

Jani, M. G., N. P. Barnes, K. E. Murray, and R. L. Hutcherson, 1994, Diode-pumped Ho:Tm:$Lu_3Al_5O_{12}$ room temperature laser, *OSA Proceedings on Advanced Solid State Lasers,* 20:109-111.

Jackson, T. J., LeVine, D. M., Swift, C. T., Schmugge, T. J., and Schiebe, F. R., 1995, Large area mapping of soil moisture using the ESTAR passive microwave radiometer in Washita'92, *Remote Sensing Environment*, 53:27-37

Johnson, C. C., D. J. Reichle, N. P. Barnes, G. J. Quarles, J. W. Early, and N. J. Cockroft, 1996, High energy diode side pumped Cr:LiSAF laser, *Trends in Optics and Photonics on Advanced Solid State Lasers,* 1:120 - 125.

Kavaya, M. J., 1996, Novel technology for satellite based wind sensing, *AIAA Space Programs and Technologies Conference,* AIAA 96-4276:1-3.

Payne, S. A., W. F. Krupke, et al., 1992, 752nm Wing-pumped Cr:LiSAF laser, *IEEE Journal of Quantum Electronics*, 28 (4):1188-1196.

Rines, G. A., P. F. Moulton, and J. Harrison, 1989, Narrowband high energy Ti:Al_2O_3 lidar transmitter for spacecraft sensing, *OSA Proceedings on Tunable Solid State Lasers*, 5:2-8.

Williams-Byrd, J. A., Singh, U. N., Barnes, N. P., Lockard, G. E., Modlin, E. A. and Yu, J., 1997, Room-temperature, diode-pumped Ho:Tm:YLF laser amplifiers generating 700 mJ at 2 μm, *OSA Trends in Optics and Photonics Series,* Advance Solid State Lasers, 10:199-201.

APPLICATIONS OF ULTRAFAST SPECTROSCOPY TO SOLID STATE LASER MATERIALS

John M. Collins

Department of Physics and Astronomy
Wheaton College
Norton, MA 02766 USA

ABSTRACT

Workers in the field of luminescence spectroscopy have, for the most part, been constrained to the role of observer: following excitation with light, the crystal is allowed to tell its own story while we merely record its unfolding. With the advent of femtosecond lasers and new pulse-shaping techniques, we now have the possibility of controlling the outcome of the story. This article will focus on some possible uses of ultrafast spectroscopy for uncovering the physics of rare earth- and transition metal-doped ionic insulating crystals, or solid state laser materials. Specifically, we will show how coherent wavepacket excitation and mode selective excitation can be used to investigate and to control the nonradiative processes, such as multiphonon decay, vibronic transitions, and phonon-assisted energy transfer.

I. SUMMARY OF NONRADIATIVE PROCESSES

Solid state laser materials, in the sense we mean it here, include ionic insulators doped with either transition metal ions or rare earth ions. Although the radiative properties are directly responsible for luminescence, the nonradiative processes, which compete with the radiative processes, often have a profound influence on the emission characteristics of these systems. The reason for this influence is twofold. First, in some systems the ion-phonon interaction is strong, so that small distortions in the local environment can cause a change in the electronic state, a change that is usually followed by a relaxation process, i.e. the emission of phonons. Second, even in those systems in which the ion-phonon interaction is small, the phonon density is so large that the

Ultrafast Dynamics of Quantum Systems: Physical Processes and Spectroscopic Techniques, Edited by Di Bartolo and Gambarota, Plenum Press, New York, 1998

probability of such processes remains large, and they are still able to have a significant affect on the luminescent properties of the system. Thus, paramount to understanding these systems is a clear grasp of the effect of phonons on the ion, and in particular how each phonon mode couples to the ion in its various electronic states. Our goal here is to examine the applicability of ultrafast lasers to probing the physics of nonradiative processes in this important class of materials.

For the purposes of this article, the term nonradiative process refers to any process involving the absorption or emission of one or more phonons, including multiphonon decay between electronic states, multiphonon decay within an electronic state, phonon-assisted energy transfer, and vibronic transitions. We begin with a brief summary of the main features of each of these processes.

I. A. Multiphonon Decay between Electronic States

Transitions between electronic states of the same ion can be bridged by the emission of one or more phonons (see Fig. 1); such a process is called multiphonon decay if it involves more than one phonon. The decay is a result of the interaction between the atomic system and the phonon bath. Generally speaking, the more phonons needed to bridge the the energy gap between the two levels the less probable is the process, so that if level 1 in Fig.1 is close in energy to level 2, level 2 will prefer to decay nonradiatively. If level 1 is separated from level 2 by a large energy (corresponding to many times the Debye energy of the solid), then level 2 may prefer to decay radiatively. The process is, however, a complicated one; the interaction strengths of the various phonon modes with the ion depend strongly on the details of the nuclear and electronic wavefunctions, which are not well known, and so calculations of the multiphonon decay rates are not possible. To date, the most successful comparison of the theory with experimental data has been on the dependence of the multiphonon decay rate versus energy gap for f-levels in rare-earth-doped solids. The decay rate is found to obey a phenomenological energy gap law:

$$W_{mp}(\Delta E) = W_{mp}(0) \exp(-\alpha \ \Delta E). \tag{1}$$

In this formula, $W_{mp}(\Delta E)$ is the multiphonon decay rate, and ΔE is the energy difference to the next lowest level, and must be larger than the highest phonon energy of the material [1]. $W_{mp}(\Delta E)$ is typically determined experimentally by doing kinetic studies. In the simplest cases it is given by

$$W_{mp}(\Delta E) = 1/\tau - W_r \tag{2}$$

where τ is the measured lifetime and W_r is the radiative decay rate of the level in question. W_r may be estimated from absorption measurements, or in those cases in which the energy gap is large enough so that multiphonon transitions are negligible, a lifetime measurement can give directly the radiative rate.

In many instances, the nonradiative decay via phonon emission is so fast that traditional kinetic studies do not reveal anything about the rates from the levels in question. It is in such systems that ultrafast lasers may prove to be important: Section II.A. details an experimental methods by which the decay rates can be measured; Section II.B. discusses the possibility of using ultrafast techniques to control which lower state a level will decay into; Section II.D explores how ultrafast lasers can be exploited to investigate which phonons are most responsible for the

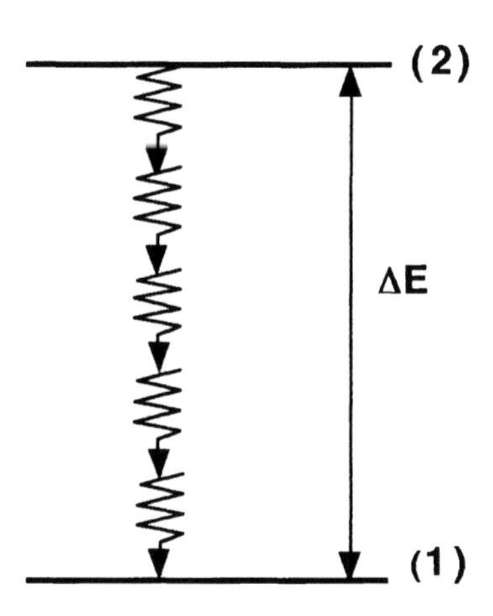

Fig. 1. A schematic diagram of multiphonon decay. An ion in energy level (2) decays via the emission of many phonons to energy level (1).

multiphonon decay process, and discusses the possibility of controlling the decay rates.

I. B. Multiphonon Decay within an Electronic State

In addition to nonradiative decay among electronic states, emission and absorption of phonons by an ion are also possible without the ion changing electronic levels; these processes are responsible for the thermalization of the ions among the vibrational levels of the electronic state. This thermalization is very rapid, and in transition metal-doped systems takes places in times on the order of tens picoseconds or less. (Work done by Alfano et al. [2,3,4] on Cr^{3+}-doped emerald, alexandrite, and Al_2O_3 demonstrate that the thermalization of the $^4T_2$ levels occurs in times of 23 ps, 17 ps, and < 5 ps, respectively.) Although knowing the decay rates is important, that knowledge does not contribute much to a basic understanding of the decay process. In ref. 2, it was speculated that the thermalization rate increases as the crystal field parameter, *Dq*, increases, but a more fundamental understanding will require more experimental data, as well as on a detailed theoretical model. Ultimately, the thermalization of ions among vibrational levels depends on two factors: (1) the ability of the ion to give up vibrational energy to the lattice, and (2) the efficiency with which energy in a certain vibrational mode is distributed among the other normal modes of the solid. Femtosecond lasers again may prove useful in measuring these very fast decays: Section II.C. and II.D. dicusses methods for determining how quickly the phonons reach equilibrium, and which phonon modes participate in the decay, respectively.

I. C. Phonon-Assisted Energy Transfer

Nonradiative energy transfer involves the transfer of energy from an excited sensitizer ion to an activator ion. Although the activator may also be in an excited state, for simplicity we will assume that is it initially in its ground state. According to the treatment of Forster [5] and Dexter [6], the probability of such a process is proportional to the overlap of the absorption band of the activator and the emission band of the sensitizer.

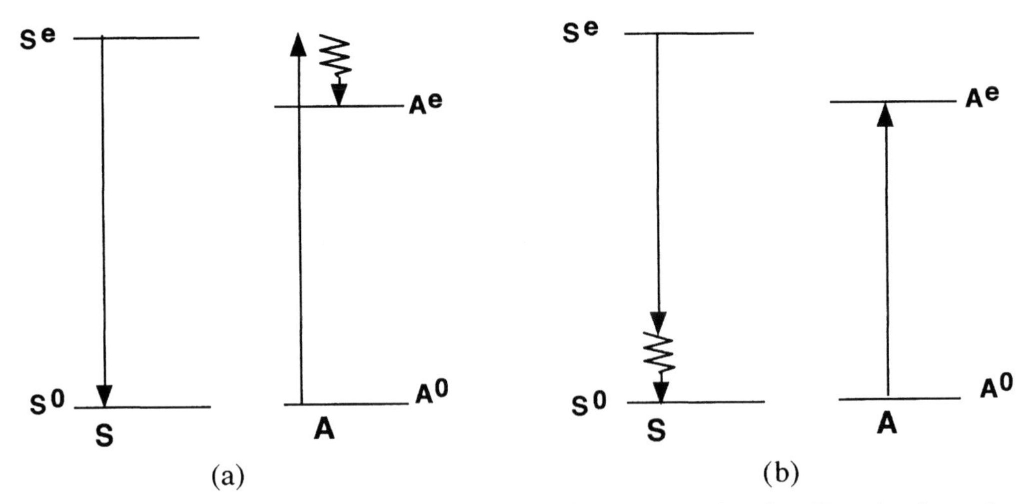

Fig. 2. Phonon-assisted nonradiative energy transfer between sensitizer ion (S) and activator ion (A). The phonon can be created either at (a) the site of the sensitizer or (b) the site of the activator.

When there is an energy mismatch between the pairs of energy levels, energy transfer may still occur if the energy difference is made up for by the emission or absorption of a phonon. Such a process is called phonon-assisted energy transfer, and is shown schematically in Fig. 2. Considering the case of emission of a single phonon accompanying the energy transfer, the probability of transfer was considered by Orbach, and is given by [7]:

$$P_{S\rightarrow A} = \frac{12\pi^2\omega_p}{h^3\rho v^5}\left|\left\langle S^e A^0 \,|\, H_1 \,|\, S^0 A^e \right\rangle\right|^2 Q(V_1)\,(n_p + 1) \tag{3}$$

The process is driven by the interaction (H_1) of the electrons on the sensitizer with those on the activator, usually dipole-dipole, and thus the transfer rate is proportional to square of the matrix element of H_1 between the initial state (S^eA^0 - sensitizer in an excited state and activator in the ground state) and final state (S^eA^0 - sensitizer in an ground state and activator in the excited state). The reader is referred to ref. 7 or to the treatment by Di Bartolo [8] for the further details of the formula. For the present purposes, the main point is the linear dependence on the phonon density, n_p. (For cases involving N phonons, a perturbational approach leads to a dependence that goes according to $(n_p)^N$.) Because the process is phonon-assisted, it can be enhanced or quenched by varying the temperature [9]; the transfer rate usually increases with temperature, but in some cases where only one phonon is involved, the transfer has been observed to decrease [10].

At first glance, if a particular phonon mode is responsible for the transfer, then the generation of phonons with the correct frequency (which can be accomplished using ultrafast lasers - see Section II.C.) could increase the transfer rate. The rate of such processes, however, are typically on the order of 10^3 - 10^6 sec^{-1}, considerably slower than the rate at which phonons in a solid reach equilibrium (lifetime of nonequilibrium phonons are in the nanosecond to picosecond regime). Thus, any nonequilibrium phonons generated will likely decay into other modes before participating in the nonradiative energy transfer. Because of this vast difference in time scales, it is unlikely that ultrafast lasers will be a useful tool for uncovering the physics of this important process, and it will not be considered any further in this article.

I.D. Vibronic Transitions

Vibronic transitions are a class of transitions in which the ion changes both its electronic and vibrational state following the absorption of emission of a photon. These are very common among transition metal-doped solids where the charge distribution of the optically active electrons are exposed to the ligands. In such systems, when the ions absorb a photon, its charge distribution is altered considerably, and the ion then vibrates about a new equilibrium position with a different frequency and amplitude than prior to the absorption. This absorption is said to be vibronic, because the ion has changed not only its electronic state but also its vibrational state. A typical absorption (or emission) spectrum of transition metal-doped systems show most of the absorption on the high energy side of the zero-phonon line, indicating that the absorption is dominated by vibronic transitions.

In rare earth-doped systems, absorption (and emission) of radiation is almost completely concentrated in the zero-phonon line, meaning that vibronic transitions are very unlikely. This behavior is a result of the weak coupling between the optically active electrons and the lattice. Vibronic transitions do, however, occur and can be of the Stokes or anti-Stokes variety, the latter of which are proportional to the number of available phonons. We shall see in Section II. C. that the anti-Stokes vibronic absorption process, which can be enhanced with nonequilibrium phonons generated using ultrafast lasers, can play an important role in monitoring the phonon dynamics in these systems.

II. APPLICATIONS OF ULTRAFAST TECHNIQUES

II. A. Monitoring Decay Rates of Nonluminescent States

One class of experiments in which ultrafast lasers could be an important tool is, rather obviously, in measuring the decay rates of short-lived electronic states. Such short-lived states (lifetime < 1 ps) are found in a number of different solid state systems: transition metal-doped insulators, F-centers, charge transfer states in rare earth-doped systems, and closely spaced electronic states in rare earth-doped systems. The lifetime of these levels, which are on the order of 10^{-11} to 10^{-14} seconds, can be determined indirectly [11], but usually remain undetermined with some upper limit being set which is related to the time resolution of the experimental apparatus.

The method for directly measuring the lifetime of a short-lived state is the pump-probe technique, shown in Fig. 3. Suppose level 3 is the short-lived electronic state whose lifetime is to be determined. In the experiment, level 3 is pumped with an ultrafast pulse, after which the ions decay rapidly back to level 2. A second laser, the probe, is then used to monitor the population of level 2; as level 3 decays the population to level 2 increases. By tuning the probe to the transition from level 2 to 4, absorption of the probe gives information on the population of level 2, and hence the decay of level 3. Experiments of this type were done by Gayen et al. [2,3,4] on the dynamics of the decay of the $^4T_2$ to the ^2E levels of Cr^{3+} in alexandrite, emerald and ruby. In alexandrite and in emerald, the $^4T_2$ level was determined to have lifetimes of about 27 and 20 ps, respectively,

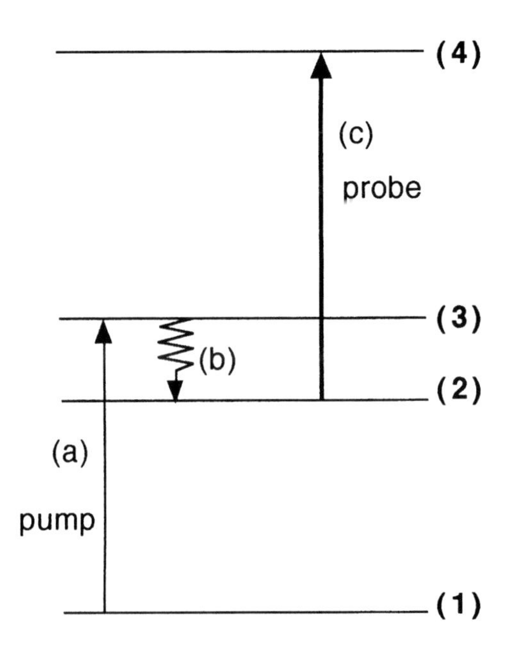

Fig. 3. The pump-probe technique for monitoring the decay from a short lived state 3 to a metastable level: (a) absorption of a photon into level 3; (b) fast nonradiative decay to metastable level 2; and (c) a probe pulse with a variable delay is tuned to absorption from level 2 to level 4. The population of level 2 as a result of feeding from level 3 is monitored by the absorption of the delayed probe pulse.

while in ruby only an upper limit to the lifetime was determined due to the fact that the lifetime is less than the pulsewidth (7 ps) of the pump pulse. It is for these very short-lived states that femtosecond lasers may prove useful.

It is noted here that this decay monitoring experiment is a rather limited application of the ultrashort pulse; whereas these pulses carry a broad spectrum and a strong coherence among the different modes in addition to its small temporal width, this application capitalizes only on the small temporal width. In the application discussed in the following section, where we propose the possibility of controlling the feeding from one electronic state to lower lying states, all of the properties of the ultrashort pulses will be utilized.

II. B. Controlling Nonradiative Feeding Fractions to Lower Levels

The field of optical luminescence has been driven in a large part by two applications: phosphors (for lamp and CRT devices) and lasers. In both cases, work has been concerned with obtaining emission at as many wavelengths as possible and with maximizing the efficiency of the emission. Strategies toward these goals include changing the temperature of the sample, varying the concentration and type of the optically active ion, selecting different pump wavelengths, and fabricating of new materials. What is proposed here is a method of changing the emission characteristics of an ion by pumping with an ultrashort pulse. That is, it will be shown that pumping into a certain electronic state with an ultrashort pulse may result in a different emission spectrum than pumping that same electronic state using the more standard pumping techniques (e.g. pumping continuously or with nanosecond pulses).

In order to see how this could happen, consider the configurational coordinate diagram shown in Fig. 4. The figure shows three electronic states, each with a parabolic potential and each having vibrational levels. There is a crossover between the two upper states, which in the single configurational coordinate model is where the transition from one state to another is most probable. The reader is referred to the configurational coordinate model of Struck and Fonger [11,12] for the details, but in such a model, the probability of a nonradiative transition from one electronic state to another is given by:

$$W_{n \to m} = N \sum_{m=0}^{\infty} (1 - r_v)\, r_v^{\,m} \langle u_n | v_m \rangle \tag{4}$$

where r_v is the Boltzmann factor for the initial excited electronic state, and v_m (u_n) represents the mth (n^{th}) vibrational wavefunction of the electronic state v (u). The meaning of this formula is clear: the transition rate is given by the overlap integral of the vibrational wavefunctions u_n and v_m weighted by the Boltzmann distribution. Because the overlap integrals reach a maximum near the crossover point, it is via the crossover that nonradiative transitions occur, but only if the appropriate vibrational levels are occupied. Thus, in order to maximize this transition, one would simply pump the crystal in such a way so that the ion were most likely to reach the crossover point.

The obvious question then is: what techniques are available that could direct the ion to the crossover point following excitation? Direct pumping from the the ground electronic state (w) to the crossover point cannot be accomplished efficiently, since it is forbidden by the Franck-Condon principle. Continuous pumping (or pumping with a long pulse) also cannot offer any control over the placement of the ion on the diagram, since these methods result exclusively in a Boltzmann distribution of the ions among the vibrational levels of the electronic state v. This distribution leads to what we might call the typical emission spectrum, and is independent of the precise pumping wavelength as long as it is within the w -> v absorption band.

In order to steer the ion to the crossover point, it is necessary to use coherent wavepacket excitation. As discussed earlier in this volume, pumping with an ultrashort pulse places the ion in a superposition of vibrational states, where the probability of being in a certain vibrational state is described by a distribution which is specifically non-Boltzmannian – for parabolic potentials, the distribution is Poissonian. This pumping into a coherent superposition of vibrational states results in the formation of a localized wavepacket which is immediately launched toward the opposite side of the parabola [13,14]. As the wavepacket moves, the probability of a transition to the lower electronic state (which is proportional to the overlap integrals in eq. (4)) is constantly changing, but it is not significant until the wavepacket arrives at the crossover point, at which point the transition probability increases dramatically. In order for the maximum amount of decay to occur at the crossover, the superposition state that makes up the wavepacket must have a distribution such that the most populated vibrational level is at the same energy as the crossover point. Fig. 4. illustrates what in being described here. In Fig. 4a the system is pumped with an ultrafast laser; the resulting distribution (the population of the vibrational levels is represented by the horizontal lines) is Poissonian, and a wavepacket is formed. The wavepacket moves toward the opposite side of the parabola, and if the Poissonian distribution is such that the vibrational level with the highest probability for occupation is at the crossover energy, then the nonradiative decay to state u via the crossover will be maximized. Since u -> w nonradiative decay is unlikely (the u-w crossover is too high), the electronic energy is ultimately released by radiative decay to w. Also, notice that tuning the ultrafast laser up and down the parabola of electronic state v produces wavepackets that shift relative to the crossover energy, so the rate of nonradiative decay should change as the laser is tuned.

In Fig. 4b, the pumping is done with a continuous (narrow bandwidth) laser. In this situation, there is no wavepacket formed; following excitation the ions simply decay by giving up their energy to the lattice until thermal equilibrium (Boltzmann distribution) is reached. From the lowest vibrational levels (where most of the population resides) nonradiative decay is not probable since those levels are far below the crossover point, and the system will prefer to decay radiatively to u or w. Thus, narrow band pumping results primarily in emission from level v, whereas coherent wavepacket excitation leads to emission mainly from level u.

As an illustration of how the transition probability changes as the wavepacket traverses the parabola, we have calculated overlap integrals for two hypothetical electronic states using the following parameters: vibrational quanta energy = 400 cm^{-1}; spring constant = 200 for both parabolas; wavepacket width (Δx) = width of m=0 gaussian wavefunction; parabola offset = 3 Δx; and the crossover is at m=3 and n = 13. The vibrational wavefunctions n and m used to calculate the overlap integrals are the standard harmonic oscillator wavefunctions. The nonradiative transition probability from v to u is the sum of the overlap integrals in eq. (4) weighted according to the Poissonian distribution rather than the Boltzmann distribution. Table 1 shows the result of such calculations with the nonradiative rates measured relative to their value at the initial position of the wavepacket. The rate, it should be noted, is roughly four orders of magnitude larger at the crossover point than at the initial point of excitation. The main point is that the nonradiative decay is considerably more probable at the crossover, so that as long as the wavepacket remains intact there is a nonequilibrium distribution of ions in the vibrational levels of the emitting state, and the location of the crossover relative to the distribution will affect the probability of decay.

Before concluding this section, the following thoughts are offered regarding the limitations and assumptions involved in the above analysis.

(1) A significant alteration of the emission spectrum by coherent wavepacket excitation should not be expected in all systems, since not all systems have the parabolas arranged so conveniently as in Fig. 4.

(2) The method relies on the wavepacket maintaining its integrity as it traverses the parabolic potential well. In fact, there are many factors that can cause the wavepacket to lose its coherence, including the specific shape of the potential well and phonon dephasing mechanisms. This dephasing will be briefly considered in the next section, though it is not expected to be significant on the time scale of a single vibration.

(3) The whole idea proposed here is based on the single configurational coordinate model, which is a great oversimplification of the system since there are about 3N normal coordinates, N being the number of ions in the crystal. Thus, it is possible that the wavepacket may be launched into a more complicated phase space trajectory. The single coordinate model has,

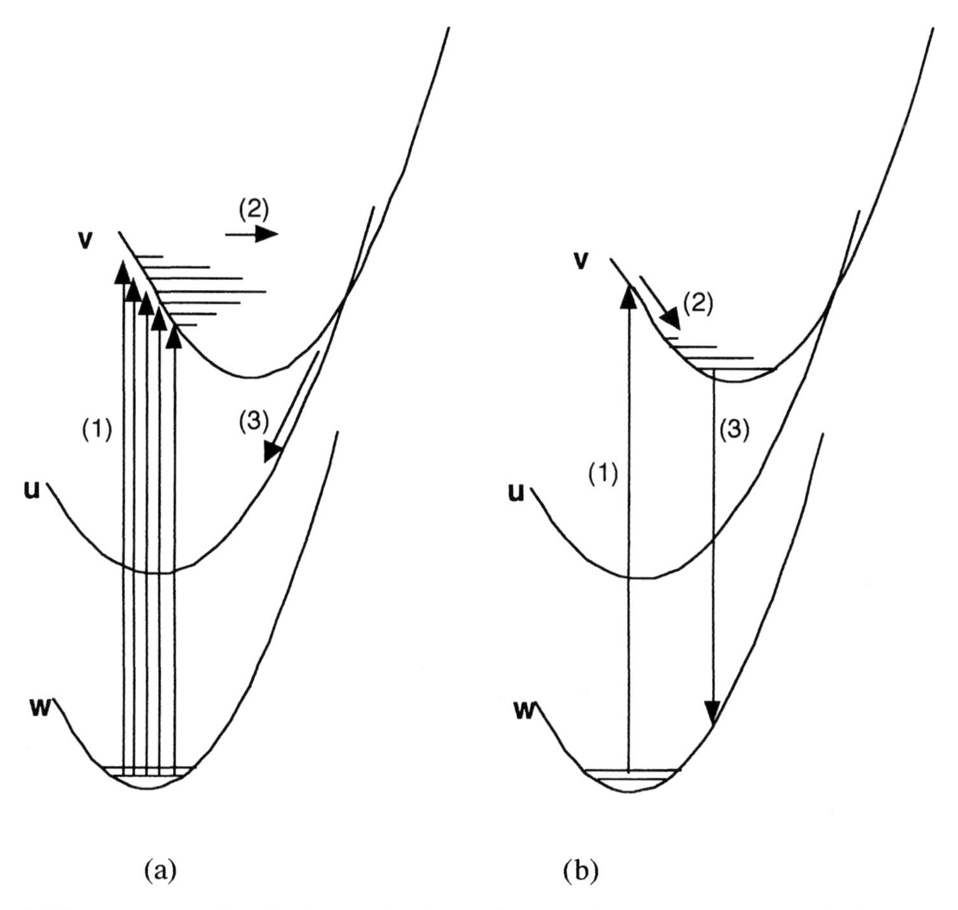

Fig. 4. Diagram comparing the decay of an ion under (a) coherent wavepacket excitation and (b) narrow band laser pumping. In (a): (1) the ion is excited with an ultrafast pulse; (2) the wavepacket is launched toward the crossover point; and (3) at the crossover the ion decays nonradiatively to level u. In (b): (1) narrow band excitation to level v; (2) nonradiative decay to a Boltzmann distribution in the vibrational levels of v; and (3) level v decays radiatively.

TABLE 1

POSITION (Δx)	0	1	2	3	4	5	6	7
$W_{u\text{->}v}$	1	1.4	6.1	290	330	2500	11000	16000

Table 1. Calculations of the u to v transition rate ($W_{u\text{->}v}$) given by the overlap integrals weighted according to a Poissonian distribution of ions in the vibrational levels of the initial electronic state, v. The position is in units of Δx, the width of the gaussian wavefunction in the m=0 vibrational state. The units of the $W_{u\text{->}v}$ are arbitrary.

however, successfully modeled absorption, emission, and nonradiative decay in the past, so there is some hope that it will remain a viable model for the experiment proposed here. For those systems where we may expect significant alteration of the spectrum under coherent wavepacket excitation, the experiments may even be regarded as a test of the single configurational coordinate model.

II. C. Measuring the Decay Rates of Nonequilibrium Phonons

In order to better understand the limits of the kind of control discussed in the last section, it is important to understand the dynamics of the phonons in the solid. Phonons, because of various mechanisms such as scattering from impurities and defects, and the anharmoticity in the crystal, decay in time by dispersing into other modes. It is of interest to know the lifetime of all phonon modes and the responsible decay mechanisms.

Work on this problem in solid state laser materials has been limited primarily due to the difficulty in creating the necessary nonequilibrium phonon distributions at arbitrary phonon energies. A substantial amount of work was done on ruby, where the ^2E level is split by an energy of 29 cm^{-1}. Pumping the upper level then leads to a nonequilibrium phonon distribution of 29 cm^{-1} phonons, and the lifetime of the phonons can be determined. An excellent summary of this work can be found in ref. 15.

An important advancement in this class of investigations was made by Yen and coworkers [16,17,18,19] in which they were able to create nonequilibrium phonon distributions over a continuous range of energies, and monitor the temporal and spectral characteristics of the phonon population via a process known as Anti-Stokes absorption vibronic sideband spectroscopy. The process is shown in Fig. 5. Creation of the nonequilibrium phonon distribution was accomplished by hitting a $Pr:LaF_3$ crystal with a far infrared (FIR) laser pulse having a pulsewidth of 40 ns. Because the Pr ions destroy translational symmetry in the crystal, conservation of momentum is relaxed and the FIR radiation can be converted into vibrational energy; the result is a nonequilibrium distribution of phonons at the same frequency of the FIR pulse. A second laser is then used to pump the Pr from the ground $^3H_4$ level to the excited $^3P_0$ level. This second laser pump is delayed from the FIR laser, and is of slightly lower in energy than the energy difference between levels so that direct absorption from level 1 to level 3 cannot occur. If a phonon participates in the process, however, the photon can be absorbed; this in an example of the vibronic transitions discussed in Section I.D. Because it involves only one phonon, the probability of such a transition depends linearly on the density of the phonons having the correct energy, and so the number of phonons is simply proportional to the intensity of the luminescence from the level (3). By varying the delay between the FIR pulse and the second laser pulse, the decay of the nonequilibrium phonons can be monitored in time. The method is quite general, but is limited by the FIR pulse; it pulsewidth is 40 ns, and it is able to produce phonons only up to about 100 cm^{-1}, the regime of the acoustic phonons.

Often it is the optical phonons which play a dominant role in the luminescent properties of the system, and so it is crucial to have the ability to generate the nonequilibrium phonon distribution at higher energies, and it is in this role where femtosecond lasers may prove to be a valuable tool. It has been shown by Wefers and Nelson [20] that current pulse shaping techniques allow for the generation of pulse trains with equally-spaced pulses. Such a pulse train acts as a series of

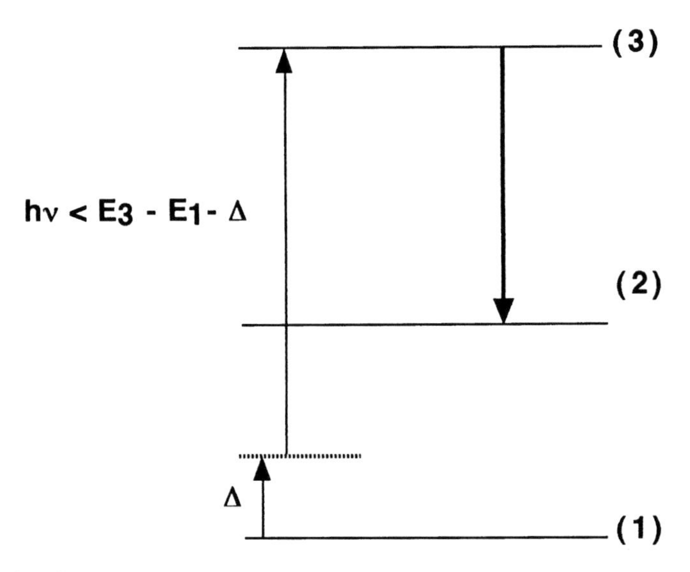

Fig. 5. Schematic diagram of the Anti-Stokes absorption vibronic sideband spectroscopy experiment used to monitor the nonequilibrium phonon population in Pr-doped crystals. The vibronic transition includes a photon ($h\nu < E_3 - E_1$) and a phonon of energy Δ.

impulses which individually excite many phonon modes in the crystal, but collectively select phonons in only a very narrow frequency range, and a standing wave is established in the crystal. This standing wave is the manifestation of a non-equilibrium population of phonons. The frequency of the central phonon mode, ω_0, excited by the pulse train is determined by the spacing between consecutive pulses, Δt:

$$\omega_0 = 2\pi/\Delta t. \tag{5}$$

A typical frequency of an optical phonon in solid state laser materials is a few hundred cm^{-1}, corresponding to a Δt of 100 to 200 fs. With pulses of 20 fs quite common and sub 5 fs pulses possible, such pulse trains should be attainable.

Once the standing wave is established in the crystal, the crystal behaves like a grating and a probe laser pulse passing through the crystal is diffracted. Monitoring the intensity of the diffracted pulse as a function of the delay time between the pulse train and the probe, the decay time of the nonequilibrium phonon population can be measured. This method of exciting and probing the phonon populations in called Multiple Pulse Impulsive Stimulated Raman Scattering [14].

Of course, one could use ultrashort pulses to generate the phonons, and detect their presence using the previously described method of Anti-Stokes absorption vibronic sideband spectroscopy, but care must be taken. Phonons decay via a variety of processes, but the dominant process at high frequencies is anharmonic decay which has a rate that goes as $(1/\nu)^5$, where ν is the phonon frequency [16]. Extrapolating from the data in ref. 16, phonons with energies of 200 to 300 cm^{-1} are expected to decay on the order of tens picoseconds or less, and so the probe laser must be at least that fast. At first glance one would think that this would not be a problem, since femtosecond laser are available. One should be cautious, however, because a femtosecond laser in the optical region has a bandwidth that extends over several hundred cm^{-1}, which is so broad that it will be problematic to detune the probe laser from the absorption band, and still tune it so that vibronic transitions can occur. The probe laser must be chosen so that is short enough to measure the the decay process, but long enough so that the spectral bandwidth is limited to well below the frequency of the phonons generated.

II. D. Measuring and Controlling Decay Rates of Nonradiative Transitions

The final application of ultrafast techniques to solid state laser materials to be considered here is the control the nonradiative decay from higher to lower electronic states. In spite of the fact that such transitions occur in most systems and are taken for granted by those working in the field of luminescence, current understanding of these processes is limited; the models that can be fit to data are phenomenological [1,9], while formulas derived from first principles are very complicated and have many as yet unknowable quantities (such as the precise electronic wavefunctions and the phonon density of states), and so are not easily compared with experiments. Clearly, more experimental data would be helpful if they could illuminate some of basic interactions that are responsible for these nonradiative transitions.

To proceed logically, the first point of discussion should be to determine what kind of information would be helpful, and so a brief discussion of the theory is in order. Treatments of ions in solids often begin with the adiabatic approximation: this involves forming the eigenfunctions of the system (ion + lattice) as Born-Oppenheimer products of the electronic and vibrational wavefunctions, and often also entails neglecting terms in the Hamiltonian that describe how the electronic wavefunctions depend on the position of the nearby nuclei. Thus, in this form the adiabatic approximation forbids nonradiative transitions from one electronic state to another; the system has been separated into "fast" (electronic) and "slow" (vibrational) subsystems; the electronic states depend only parametrically on the position of the ions in the host, so as those ions vibrate the electronic wavefunction follows them adiabatically. Explaining the existence of nonradiative transitions between electronic states demands reintroducing the so-called nonadiabatic terms in the hamiltonian, which leads to a nonradiative decay given by [21]:

$$W_{v \to u} = \frac{h^3}{4\pi^2} \sum_{m,n} (1 - r_v)(r_v)^m \sum_s \left| \left\langle u \left| \frac{\partial}{\partial Q_s} \right| v \right\rangle \right|^2 \left| \left\langle u_{n_s} \left| \frac{\partial}{\partial Q_s} \right| v_{m_s} \right\rangle \right|^2 \times \prod_{j \neq s} \left| \left\langle u_{n_j} | v_{m_j} \right\rangle \right|^2 \delta(E_{u_n} - E_{v_m}) \qquad (6)$$

where u and v are the electronic wavefunctions of the ion, and u_n and v_m are the vibrational wavefunctions of corresponding states. Under certain assumptions, this equation reduces to eq. (4), but here are two main differences: (1) the role of the nonadiabatic operator ($\partial/\partial Q_s$) is explicit in eq. (6), and (2) the sum over vibrational modes has been split into a sum over s and a product state over $j \neq s$. The modes represented by s are those which drive the transition from one electronic state to another, and are sometimes referred to as "promoting modes". All other modes ($j \neq s$) are referred to as "accepting modes."

Because promoting modes connect different electronic states, they are subject to selection rules which are determined by the local symmetry of the ion. Evidently, the transition rates will be driven mainly by those modes for which the initial electronic wavefunction depends strongly on the normal coordinates; these are the high frequency local modes which cause maximum distortion of the local symmetry. The matrix element involving the accepting modes are simply given by the overlap integrals of all other modes in the solid, and it is into these modes that the electronic energy is ultimately lost. The important point for this article is that different phonon modes play distinct roles in the decay process, and ultrafast techniques may make it possible to determine the roles of the various modes in the multiphonon decay process.

What we seek, then, is basic information on the role of the different phonon modes: which are accepting modes?, which are promoting modes?, and which modes do not participate at all? Since we can use ultrafast lasers to generate nonequilibrium phonon distributions (see the previous section), it should be possible to drive the multiphonon decay process; if the frequency of the nonequilibrium phonons match the frequency of the promoting modes, then the nonradiative decay can be enhanced. Physically, creating the nonequilibrium phonons causes the local distortions to become larger, and the effect of the nonadiabatic operator increases accordingly. Of course, eq.(6) does not strictly hold under these conditions since the system is no longer in thermal equilibrium – the ions are no longer in a Boltzmann distribution – but it is still the nonadiabatic operator that drives the transition. In effect, creating a nonequilibrium phonon distribution at the correct frequency moves the ions out of the Boltzmann distribution and places them in vibrational states from which the transition to the final electronic state can occur.

Alternatively, it is possible to affect the decay rate by creating phonons at frequencies of the accepting modes. In the case of multiphonon decay discussed in Section I.A., the process is generally treated as a Nth order process, N being the number of phonons required to bridge the energy gap. Since lower ordered processes are generally preferred, the decay is dominated by phonons of the highest frequency, ν, such that $N = \Delta E/h\nu$ must be as small an integer as possible [9]. In the standard perturbation treatment, the decay rate can be either stimulated or spontaneous, and so will be proportional to $(n_p + 1)^N$; thus the presence of phonons at the correct frequency should enhance (by stimulated emission of phonons) the multiphonon decay rate.

This enhancement depends, however, on whether indeed the multiphonon decay is via just a few phonon modes. Work done by Demos et. al [22,23] of hot luminescence and Raman studies of Cr^{4+}-doped forsterite shows the presence of some favored modes during the multiphonon decay process, and identify those modes as accepting modes. Other work referred to earlier [1,9] lend some experimental evidence to the assumption that multiphonon decay occurs via emission of phonons at a single frequency (one that minimizes the order of the process). The models that have been developed thus far to fit the evidence (such as the energy gap law in eq. (1)) are, however, strictly phenomenological; there is a little direct evidence showing which phonon modes are preferred in multiphonon decay. And with the exception of the work referred to above, there is no evidence to this author's knowledge of work that clearly identifies certain modes act as accepting or promoting modes. Thus, there seems to be a whole body of investigations into the basic physics of the decay process waiting to be carried out, and ultrafast lasers in combination with new pulse shaping techniques may be the appropriate tool; by creating nonequilibrium phonons at a variety of frequencies, and using the pump-probe or Raman techniques, it should be possible to determine the role of the various phonon modes in the multiphonon relaxation process.

III. CONCLUSIONS

In this article we have considered some possible applications of ultrafast spectroscopy to the study of nonradiative processes in solids. Because of the importance of nonradiative decay between electronic states to the luminescent properties of many systems, much of the focus here has been on the how ultrafast techniques could be used to study the basic physics of multiphonon decay mechanisms, and also to monitor and control the decay process. Ultrafast lasers offer a unique and versatile tool – they are much more than just fast pulses. Their broadband nature allows for the possibility of coherent wavepacket excitation, a pump mechanism which we have shown can be used to control, at least in some systems, the pathways of decay from an excited state to the ground state, and may even be used to change the intensity of the various emission lines. Also, these lasers in conjunction with new pulse-shaping technology can allow for the creation of nonequilibrium phonon distributions in solids, which can in turn be used to: (1) study the dynamics of the phonons in the solid as they decay to equilibrium, (2) enhance the nonradiative decay if the nonequilibrium phonons are at the correct frequency, and (3) be used to investigate basic physical phenomena, such as the role of the various phonon modes (whether they act as promoting or accepting modes) in nonradiative decay between electronic states. All of these ultrafast techniques show tremendous promise for workers to gain basic information on these very important decay processes, information that has thus far been elusive.

REFERENCES

1. M.J. Weber, Phys. Rev. 138, 54 (1973)

2. S.K. Gayen, W.B. Wang, V. Pertricevic, S.G. Demos, and R.R. Alfano, *J. Luminescence* 47, 181 (1991)

3. S.K. Gayen, W.B. Wang, V. Pertricevic, and R.R. Alfano, *Appl. Phys. Lett*. 49, 437 (1986)

4. S.K. Gayen, W.B. Wang, V. Pertricevic, K.M. Yoo, and R.R. Alfano, *J. Luminescence* 50, 1494 (1986)

5. T. Forster, Ann. Phys. 2, 55 (1948)

6. D.L. Dexter, J. Chem. Phys. 21, 836 (1953)

7. R. Orbach, in: *Optical Properties of Ions in Crystals*, eds., H.M. Crosswhite and H.W. Moos, Interscience, Ney York, (1967)

8. *Optical Interactions in Solids*, B. Di Bartolo, John Wiley & Sons, New York (1968)

9. N. Yamada, S. Shionoya, and T. Kushida, *J. Phys. Soc. Japan* 32, 1577 (1972)

10. J.M. Collins and B. Di Bartolo, *Phys. Stat. Sol (b)* 189, 441 (1995)

11. C.W. Struck and W.H. Fonger, *J. Luminescence* 1,2, 456 (1970)

12. C.W. Struck and W.H. Fonger, *J. Luminescence* 10, 1 (1975)

13. B.M. Garraway and K.A. Suominen, *Rep. Prog. Phys.* 58, 365 (1995)

14. L. Dhar, J.A. Rogers, and K.A. Nelson, *Chem. Rev.* 94, 157 (1994)

15. K.F. Renk in: *Nonequilibrium Phonons in Nonmetallic Crystals*, W. Eisenmenger and A.A. Kaplyanskii, eds., North Holland, Amsterdam (1986)

16. W.A. Tolbert, W.M. Dennis, and W.M. Yen, , *Phys. Rev. Lett.* 65, 607 (1990)

17. X. Wang, J. Ganem, W.M. Dennis, and W.M. Yen, *Phys. Rev. B* 44, 900 (1991)

18. X. Wang, W.M. Dennis, and W.M. Yen, *Phys. Rev. Lett.* 67, 2807 (1991)

19. W.A. Tolbert, W.M. Dennis, R.S. Meltzer, J.E. Rives, and W.M. Yen, *J. Luminescence* 48 & 49, 209 (1991)

20. M. M. Wefers and K.A. Nelson, *Opt. Lett.* 20, 1047, (1995)

21. F. Auzel in: *Luminescence of Inorganic Solids*, B. Di Bartolo, ed., Plenum Press, New York, (1977)

22. S.G. Demos, Y. Takiguchi, and R.R. Alfano, *Opt. Lett.* 18, 522 (1993)

23. S.G. Demos, J.M. Buchert, and R.R, Alfano, *Appl. Phys. Lett.* 61, 660 (1992)

ULTRAFAST ELECTRON-LATTICE RELAXATION OF OPTICALLY EXCITED CENTERS IN CRYSTALS

Nice Terzi and Matteo Dominoni

Dipartimento di Scienza dei Materiali
and Istituto Nazionale per la Fisica della Materia
Universita' degli Studi di Milano
via Cozzi 53
20126 Milano, ITALIA

ABSTRACT

The excited-state dynamics of a localized center (defect, molecule, cluster) in a nonconducting crystal is discussed, and the Pulse Model is introduced to study its tempora behavior on a scale from few femtoseconds to tens of picoseconds after ultrafast electroni excitation. The simple Pulse Model and a modified version, both elaborated few years ago b the authors, allow a numerical selfconsistent evaluation of the electron-phonon couple dynamics during the relaxation, i.e. the values of position and momentum of the ions as well a of the wavefunction and energy of the electron as function of time. In particular the process o relaxation of the ions surrounding the center is related to the motion of the multimod wavepacket of lattice phonons generated by the ultrafast optical excitation of the center Numerical data for defects in alkali halides in the limit of strong electron-phonon coupling ar presented in two cases: that of a non degenerate electronic excited state and that of a doubl degenerate excited state (*Exe* Jahn-Teller case). In the former case, we show that the decay tim to the relaxed configuration is related to the autocorrelation time of the multimode phono wavepacket coupled to the excited electron. In the second case, the Jahn-Teller case, w observe a process of dynamical trapping and hopping between distorted lattice configurations with features depending on the symmetry, electron-phonon coupling strength and temperature.

1. INTRODUCTION

In recent years, due to the increasing availability of fs pulsed light sources, grea attention has been devoted to the transient processes in condensed matter in the time range fror few fs to several ps. The growing interest to this topic is just shown by the works of the present Erice school.

One of the most interesting feature is the persistent coherent dynamics induced by ultrafast excitation of electrons bound to centers or molecules in crystals, which lasts for many ps before thermalization becomes appreciable [1,2,3]. Experimentally, this coherence is seen as an oscillating time behavior of the differential absorption coefficient in pump and probe measurements. However, few cases of fast relaxation have also been found, with relaxation times smaller than 100 fs (See Par. 3E).

From the theoretical side, studies on the relaxation after fs excitation of electrons in molecules and solids have produced a vast literature [4]. The modality of the excited state relaxation seem now clear for small molecules with few (two, at the most) vibrational modes [5,6]. It is commonly agreed that in the time interval from few fs up to some tens of ps and when long lived electronic excitations are involved, the response functions give information about the motion of the phonon wavepacket generated by the excitation on the excited adiabatic surface(s).

Much less is known on bound electronic excited states in solids, because of the increasing complexity of the relaxation modalities with the number of the involved vibrational modes and electronic levels. Theories accounting for the coupling with a continuum of frequencies have been suggested which give the right order of magnitude of the relaxation time [7,8] and of the oscillating behavior.

In the present lecture we report on the Pulse Model (PM), proposed by the present authors few years ago [8]. First we show how the PM works and then how it can be extended (Modified Pulse Model, MPM) to describe the relaxation when a bound electron is excited to a many level state. This is a necessary extension for bound electrons in high symmetry, because of the intrinsic degeneracy of several of its electronic excited states. Furthermore, an electron with at least two excited interacting levels is essential to describe the phenomenology of the level crossing, in solids as well in molecules.

By using the PM and the MPM the coupled dynamics of the lattice and of the excited state electron can be evaluated from few fs up to tens of ps after excitation. The important quantities one can deduce are the electronic energies and states, with the relative occupation probability, as well as the positions and momenta of the lattice ions.

The system hereafter considered is a center in a ionic crystal, i. e. a molecule or a cluster or a substitutional impurity strongly interacting with the ions of the host harmonic lattice. The center is assumed to give rise to bound long lived electronic states in the forbidden gap. The excitation is Condon-like, very fast on the scale of the ionic motion. During the transient following the excitation, the electron is assumed to interact strongly with the surrounding harmonic crystal via the electron-phonon (EP) interaction suddenly switched by the excitation. By strong coupling one usually means that the pertinent absorption band has an half width of few meV, corresponding to a Huang-Rhys factor of some units.

2. THE SEMICLASSICAL APPROACH

We use the classical mechanics to describe both the lattice dynamics of a harmonic crystal at thermal equilibrium and the propagation inside the same crystal of the phonon wavepacket generated by the fs excitation. A classical description of the lattice dynamics is an accepted procedure in the study near to the thermal equilibrium of the electron-phonon (EP) coupled systems, where only the electron dynamics is treated quantum mechanically, as for instance in the ab-initio calculations. Here we bring reasons to justify this approach also far from the equilibrium, as in the special case of the propagation at ultrashort times (in a range from few fs to about ten ps) of intense phonon pulses generated by an ultrafast excitation: the classical picture comes out as a correct approximation of the whole quantum picture.

The core argument resides on the particular nature of the phonon pulse initially prepared by the ultrashort light pulse. We remember that the excitation process is very fast on the time scale of the ionic motion, so fast, indeed, that the uncertainty principle must be carefully taken into account. Because a light pulse as short as few tens of fs has an energy uncertainty of the order of the *meV,* the whole vibronic excited state it generates must have an equivalent energy uncertainty. In other words, the absorption of a fs light pulse cannot excite a selected stationary state inside the vibrational manifold relative to the electronic excited state. On the contrary, a whole phonon wavepacket is generated whose composition depends jointly on the excitation modalities and on the electronic excited state.

In the special case when the light uncertainty is as large as the whole absorption band, this phonon wavepacket is actually a *phonon coherent state* [10,11]. In two previous editions of Di Bartolo's Schools (1983 and 1991) one can find discussed in some details the properties of the phonon coherent wavepackets and the dynamics of classical phonon pulses in a harmonic crystal. Here, when necessary, we refer directly to the two lectures and to the equations there contained. The former lecture [11] is about the properties of Phonon Coherent Wavepackets and the second lecture [12] is about the Classical Dynamics of Phonon Pulses.

For the present purposes, the most interesting property of the phonon coherent states is that the expectation value of the position $\hat{q}$ and momentum $\hat{p}$ quantum variables over a coherent state behaves like the corresponding classical variables:

$$q^{Cl}(t) = \langle\psi(t)|\hat{q}|\psi(t)\rangle \qquad (1.a)$$

$$p^{Cl}(t) = \langle\psi(t)|\hat{p}|\psi(t)\rangle, \qquad (1.b)$$

where $|\psi(t)\rangle$ is the appropriate coherent state of the phonon field. This is the reason why the coherent states are also called semiclassical states.

A classical description of the phonon pulse propagation can therefore be considered as a full quantum approach in which the phonon field is always in a coherent state and only the expectation values of the quantum variables are of interest. One can then more appropriately speak of a propagation of an elastic pulse from a source (the excited center), leaving the expression "phonon wavepacket" to a second quantization treatment. Such an assumption gives probably a good account of the transient dynamics for the excitation conditions here considered.

In [11] the expectation values of ionic position and momenta over coherent states are discussed for 1-dim and 3-dim crystals. In [12] the motion of a classical pulse in 3-dim is studied and its propagation modalities are related to the possible shapes of the phonon spectral densities. For centers in crystals the pertinent spectra are the projected densities of states, related to the motion of the local symmetry coordinates. They are hereafter called SDOS.

3. THE PULSE MODEL (PM)

3.A. Introduction

The *configurational coordinate* (c.c.) model is the most widely accepted picture used to describe the complete optical cycle of bound defects in crystals such as the color centers [13].

We remember that he optical cycle consists of the ordered sequence of the following four processes: light absorption, followed by lattice relaxation, then light emission and finally the recovering of the lattice to the initial configuration. The second and fourth steps are non radiative, in the sense that phonons and not photons are emitted during the processes. The c.c. model is actually a *pseudomolecular model*, the pseudomolecule being the cluster composed by the defect and its few neighbors (often only the n.n.), which are assumed to interact weakly with the surrounding host crystal, acting then as the thermal bath.

After the optic ultrafast excitation (the first step of the optical cycle), the energy excess locally stored by the excitation relaxes via the EP coupling. The relaxation is the second step of the optical cycle and very often only one mode is considered significant. In the c.c. model and using a quantum description of the process, it is usually said that the relaxation occurs via a phonon cascade of the phonons which have been excited over the few local normal modes of the pseudomolecule. The host crystal is there assumed to play an indirect role: it is capable of absorbing in a due time the local energy excess, which is eventually dissipated, but without specifying the processes leading to it. The c.c. model copies the picture used to describe the relaxation after excitation of a molecule in a gas, with two important changes: the cluster is bound to a lattice site (no Doppler effect) and the collisions are absent. Therefore it is not clear what is the mechanism through which the phonon energy excess is supplied to the thermal bath.

In the PM, on the contrary, neither local modes nor a weak interaction with a thermal bath are assumed, the excited state relaxation being driven essentially by the dynamics of the whole perturbed crystal, as felt by the ions of the local cluster. The price to pay is the use of the classical lattice dynamics, which can however be justified for ultrashort times, as discussed in the previous section.

In the following we show some consequences of the classical approach. The most significant is that the local relaxation probed by pump and probe experiments reflects a deterministic propagation inside the lattice of the initial phonon pulse: the surrounding crystal is treated as an elastic medium coherently interacting with the pulse, and not as a thermal bath. This is the reason why we have called the model the *pulse model*.

In the PM the lattice phonon densities of states (DOS), suitably symmetrized (SDOS) when necessary, characterize the propagation modalities of the phonon pulse into the crystal. The initial composition of the wavepacket is determined by the EP coupling, between the bound excited electron and the surrounding ions, switched on by the electronic transition. Notice that the same SDOS and EP coefficients are the physical quantities characterizing the quantum study of the phonon properties of imperfect crystals, such as the Raman scattering, the absorption bandshape, etc.. Now we see them entering in a natural way even when a part of the system (the lattice) is treated classically.

3.B. The Hamiltonian

In this section we report the results obtained by introducing the minimal number o approximations that are still significant for the problem. In particular we assume that: i) th lattice dynamics is harmonic; ii) the excited electronic state is not degenerate; iii) the EI interaction acts on the n.n. of the center. Point i) has already been discussed previously. W comment it again afterwards in the present section. In the next section we modify point ii) t exemplify how to deal with electronic degeneracy. Point iii) is actually the least cogent and ha been introduced to simplify mathematics and to reduce the parameters (coupling coefficients) o the problem; EP interaction with a larger, more realistic number of neighbor shells can be easil dealt with. However it does not introduce new didactic concepts and it is not discussed here.

The harmonic hamiltonians of the perturbed crystal, before and after a Condon excitation, are respectively::

$$H_0 = \sum_{l\alpha} \frac{p_\alpha^2(l)}{2m_l} + \frac{1}{2} \sum_{l\alpha, l'\beta} \phi_{\alpha\beta}(l,l') q_\alpha(l) q_\beta(l') \tag{2}$$

$$H = H_0 + \sum_{l\alpha} f_\alpha(l) q_\alpha(l), \tag{3}$$

where l labels the atoms in the lattice and α the cartesian components; $q_\alpha(l)$ and $p_\alpha(l)$ are the α component of the displacement and momentum, respectively, of the l -th ion; $\phi_{\alpha\beta}(l,l')$ are the lattice force constants; $f_\alpha(l)$ are the forces entering EP coupling. In the quantum approach $q_\alpha(l)$ and $p_\alpha(l)$ in eqs. 2 and 3 are actually operators. The linear term in eq. 3, when acting on a phonon state describing the crystal at the thermal equilibrium (i.e. the lattice before excitation), generates a phonon coherent state $|\psi(t)\rangle$ of the phonon field [11]. As already discussed in Par. 2, the relaxation after excitation of the lattice can be described classically via the expectation values $q^{Cl}(t)$ and $p^{Cl}(t)$ over $|\psi(t)\rangle$ (Eq. 1).

We believe that this is a reliable approximation in the intermediate and strong EP coupling limit, i.e. when the displacements of the relaxing ions is larger than their quantum fluctuations. The expectation value of the higher moments of q and p gives the measure of the quantum fluctuations of the system, hereafter neglected. The time dependence of $q^{Cl}(t)$ can be obtained either by evaluating of the phonon field $|\psi(t)\rangle$ and then $\langle\psi(t)|\hat{q}|\psi(t)\rangle = q^{Cl}(t)$, or directly by solving eqs. (2,3) as classical equations [12]. In both cases one obtains:

$$q_{\Gamma\gamma}^{Cl}(t) = \sum_{\gamma\gamma'} f_{\Gamma\gamma} \int_0^{\omega_{Max}} \rho_{\gamma\gamma'}^{\Gamma}(\omega) \frac{\cos(\omega t)}{\omega^2} dt, \tag{4}$$

$$q_{\Gamma\gamma} = \sum_{l\alpha} c(\Gamma\gamma, l\alpha) q_\alpha(l), \qquad f_{\Gamma\gamma} = \sum_{l\alpha} c(\Gamma\gamma, l\alpha) f_\alpha(l). \tag{5}$$

Here $q_{\Gamma\gamma}(t)$ are the so-called symmetry coordinates, i.e. the $\Gamma\gamma$-symmetry combination of the n.n. displacements, Γ being the suitable irr. rep. of the point group and γ a degeneration index (how many times Γ is contained in the symmetry decomposition of the n.n. displacements). $\rho_{\gamma\gamma'}^{\Gamma}(\omega)$ indicates the SDOS, i.e. the phonon DOS projected on the symmetry coordinates of the center. In Ref. [12] one can find in some details how SDOS are related to the normal modes (perturbed and unperturbed) of the whole crystal and to the symmetry of the local center. Eq. 4 is actually a Fourier transform, which relates the time-dependent displacements to the appropriate SDOS, suitably weighted by the coupling coefficients. Notice that experimentally the procedure is the opposite: by Fourier transforming a signal, obtained in the ultrafast pump and probe experiments, one gets $\rho_{\gamma\gamma'}^{\Gamma}(\omega)$ and the EP coefficients, both giving information on local properties.

3.C. Temperature.

A serious point is how to model the lattice temperature while working in a classical frame. We have introduced the temperature into the dynamics as a thermal noise always present in the ionic motion before and after the excitation. Before the excitation, since the lattice is at thermal equilibrium, ion positions and momenta fluctuate in time around average values dictated by the temperature alone, but obviously independent from the EP coupling, still not acting. After the excitation, the thermal fluctuations still evolve freely because in the equations of motion of the lattice the homogeneous part cannot be modified by a linear coupling such as the EP coupling. Then the noise evolves unperturbed and independent from the electron dynamics, and at each time we can identify and separate the thermal ($q^{Th}(t)$ and $p^{Th}(t)$) and the classical ($q^{Cl}(t)$ and $p^{Cl}(t)$) contributions to the total displacements and momenta $q_{\Gamma\gamma}(t)$ and $p_{\Gamma\gamma}(t)$

$$q_{\Gamma\gamma}(t) = q_{\Gamma\gamma}{}^{Cl}(t) + q_{\Gamma\gamma}{}^{Th}(t), \qquad (6a)$$

$$p_{\Gamma\gamma}(t) = p_{\Gamma\gamma}{}^{Cl}(t) + p_{\Gamma\gamma}{}^{Th}(t). \qquad (6b)$$

The thermal noise has been formed in two steps. First we have built at a given time t_0 the Γ-symmetry displacements $q_{\Gamma\gamma}{}^{Th}(t_0)$ and momenta $p_{\Gamma\gamma}{}^{Th}(t_0)$ as linear combinations of the lattice normal modes . We have weighted each mode with random deviates belonging to a normal distribution, whose variance is related to the quantum statistics of the harmonic lattice, so to have the corrected mean square values of positions and momenta at the temperature considered. Then we have left them to evolve harmonically.

3.D. General features of the relaxation.

In what follows, unless expressly stated, we always report on the behavior of the classical component only, as it results after cleaning away the thermal noise from $q_{\Gamma\gamma}(t)$ in eqs.7.

We have numerically evaluated the relaxation process for different centers and parameters (EP coupling, initial conditions, temperature). The process always shows a structure in two phases. In the first phase, usually very short (10 - 100 fs), the n.n. displace of a sizable quantity towards the new equilibrium configuration pertaining to the electronic excited state. During the second phase the ions, adjusted to the new displaced positions, perform small oscillations around them.

During the first phase the energy stored by the excitation (via the EP coupling) on the n.n. moves into the surrounding crystal in a very fast way. This large and fast initial displacement is due to the ballistic propagation into the crystal of the phonon wave packet, initially located inside the cluster local volume. The escape time from the local volume is ruled jointly by the EP couplings and by the dynamical characteristics of the host lattice, i.e. the harmonic force constants (including the forces to which the cluster is linked to the crystal) and the masses (here the n.n. masses). The stronger the forces and the smaller the masses, the quicker the pulse propagates into the crystal from the excitation volume. The time spent in the first phase by the atoms to reach the new equilibrium sites gives a measure of the relaxation time.

3.E. Comparison with the experiments: the F_2^+ color center in LiF.

The center we discuss here in connection with the Pulse Model is a color center of the ionic crystal LiF (symbolized as $LiF : F_2^+$), which consists of an electron linked to a couple of anionic vacancies in the [110] crystallographic direction. Its symmetry group is D_{2h}. The interest for this color center is due to the fact that its relaxation time has been one of the first to

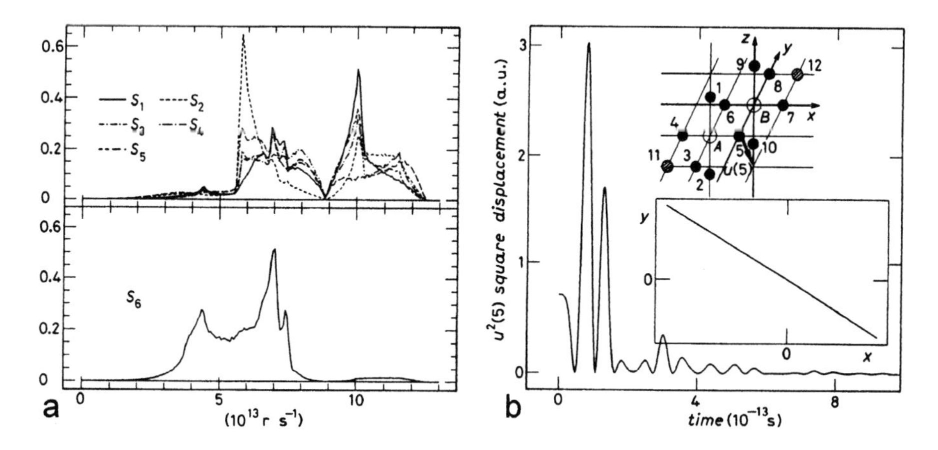

Figure 1. (a). The SDOS $\rho_{N1,\gamma}$ (ω) (γ=1,...,6) of symmetry N_1 of the D_{2h} group, pertinent to the color center $LiF : F_2^+$ and relative to the Li^+ motion (see the upper insert in Fig.1b). **(b)** Square displacement of the ion 5 belonging to the cluster of neighbors of the F_2^{++} color center, i.e. the couple of anionic vacancies (white open dots) outlined in the upper part. Black dots indicate Li^+ ions, gray dots F^- ions. In the insert the trajectory of the ion 5 in the plane (x,y) is reported

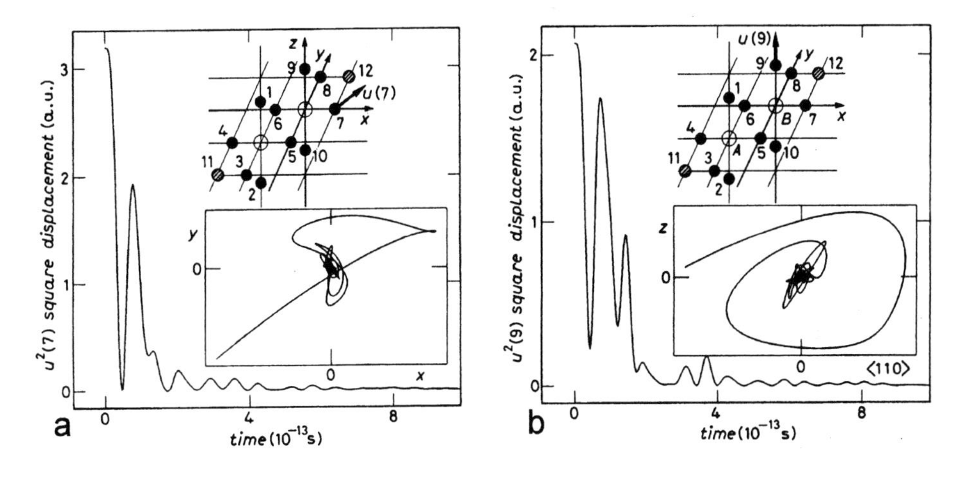

Figure 2 . (a) Square displacement of the ion 7. Symbols are the same as in Fig.1b. In the insert it is shown the trajectory of the ion 7 in plane xy, as indicated. **(b)** Square displacement of the ion 9. In the insert it is shown the trajectory of the ion 9 in plane <110 >z.

be measured using ultrafast pulse laser and pump and probe measure technique. A relaxation time of about 75 fs has been found [14]. This time is very short, of the order of the period of the optical modes of the host crystal, whose frequency is usually taken as a sound value of the "local" mode frequency, when working within the c.c. model. In the c.c. model such a very short relaxation time cannot be explained as a decay process from an excited vibrational level of the local mode, because the time the oscillator takes to reach the bottom of the parabola via non linear processes must be much longer than the vibrational period.

In the PM this short phenomenon can be easily explained because it is, as said, the time taken by the pulse to leave the cluster, owing to the action of the (perturbed) force constants linking the cluster to the host crystal ions. Notice that the n.n. of the F_2^+ center are the light and then very in very good agreement with the experimental data, given the assumptions made.

Notice that this result is somewhat independent from the EP coupling coefficients. In fact the local site symmetry is so low that only the breathing displacements, transforming as the irr. rep. N_1 of the point group, is probably significant.. It then follows that the relaxation process (but not the value of the total displacement) is ruled by the SDOS of N_1 symmetry, where the Li^+ ion mass and the lattice force constants between Li^+ and F^- are the determining quantities. Figs.1-2 show the geometry of the cluster formed by F_2^+ and its n.n., the pertinent SDOS of N_1 symmetry and the squared displacements of few significant ions Li^+ n.n. of the F_2^+ center. One can see that the frequency spectra are distributed among all the allowed frequencies. In particular, owing to the ω^{-2} factor, the acoustic branch gives a considerable contribution at small times. We have taken as an estimate of the relaxation time τ the time necessary to Li^+ ions to loose about 90 % of the energy. A value of about 105 fs is found, corresponding to the first 6 oscillations of the Li^+ ions.

4. THE MODIFIED PULSE MODEL (MPM)

In the present section we show how to modify the PM in order to take into account the possible degeneration of the excited state of the center, all the other approximations remaining the same as in the previous PM. The simplest case is that of a twofold degenerate state, typically a E-symmetry state, of a center with the same cubic symmetry of the host crystal. The lattice symmetry displacements with which such a state interacts must be of e-symmetry, as group theory states. This case is usually called in the literature the *many mode Exe Jahn-Teller (JT) case.*

4.A. The many mode Jahn-Teller system.

The Exe JT case with a single frequency is the simplest of all the JT cases and as such has been deeply studied in the past [15,16]. However, we think that there are still unexpected features of the JT case that deserve attention, the main among them being the consequence of having several modes involved into the dynamics. This is the case of some substitutional impurities in ionic crystals, as well as of molecules and clusters, strongly interacting with the lattice ions when in their degenerate excited states. In these strong coupled systems the ions interacting with the degenerate electron oscillate and move according to a large and structured frequency spectrum, the SDOS. In such cases the dynamics of the electron, entangled to that of the lattice phonons, cannot be reduced to that of an isolated JT center with a single suitable frequency, however cleverly averaged over the whole spectrum and combined with a convenient damping factor.

By summarizing, a *many-mode JT system* is characterized by a phonon SDOS which cannot be reduced to a delta like function centered at one single average frequency, mainly when studied in conditions far from the equilibrium, as during its transient relaxation after excitation. The following dynamical and thermodynamical considerations further support the need of an extended pulse model. First we notice that the frequency spectrum gives the measure of the coupling of the impurity with the host crystal ions: a strong interaction gives rise (via the dynamical force constants) to a broad frequency spectrum, as for impurities in crystals. A very significant consequence is that any excitation localized near the JT center propagates into the crystal and viceversa, or, in other words, that energy and angular momentum of the JT center are no more conserved as, on the contrary, it occurs in the one mode JT centers.

Secondly, we point out the thermodynamical implications of a continuous spectrum. For JT centers whose interaction with the surrounding crystal ions ranges from medium to strong values, it is impossible to define a thermodynamical boundary enclosing a dynamical system (such as a suitable cluster of ions around the JT center) weakly interacting with a thermal reservoir, the surrounding crystal. Therefore a thermodynamical approach to the response functions in the usual way is impossible and one has to resort to a statistical approach: first solve the deterministic motion of the excitation switched at $t=0$ on the JT center and then average over all the possible initial electron and lattice configurations compatible with the parameters (polarization, coupling parameters, symmetries, etc.) and the temperature of the whole system.

To take into account both dynamical and thermodynamical effects on the relaxation modality is not straightforward and can be done only computationally. Here we discuss first in par. 4B the equations of motion of the excited electron and of its n.n. ions in presence of the thermal ionic fluctuations (par. 4C). Then we solve computationally and selfconsistently the system of equations so derived. We select few realizations of the relaxation of a positive substitutional impurity in the ionic KI crystal, excited in a E-symmetry state and interacting with the e-symmetry components of the JT n.n displacements. We discuss them in par. 4C.

We leave the statistical treatment and discussion of the ensemble of the realizations to a future lecture.

4.B. The Equations of Motion and their Numerical Solution.

We use an iterative selfconsistent scheme to solve the equation of motion of the lattice (classical slow subsystem) coupled to the electron (quantum fast subsystem) at finite temperature [9]. The electronic excitation is assumed to occur instantaneously at $t = 0$. The EP interaction then acts at $t > 0$ via the hamiltonian $\hat{H}^{EP}$, which at each time t is assumed linear in the symmetry coordinates $q_{\Gamma\gamma}(t)$. The scheme is summarized as follows.

The whole time interval (from 0.5 to 33.5 ps) has been divided in N tiny intervals Δ . Δ are taken so small that during Δ the lattice acts (via $\hat{H}^{EP}$) as a static deformation field on the electron motion. We take care to be always in the limit for which the relation $\Delta << \hbar/\Delta E << \hbar/\omega_M$ is satisfied, where ω_M is the maximum frequency of the host crystal lattice and ΔE is the lattice induced energy splitting of the twofold electronic level (of symmetry E). In turn, the electron density matrix $\rho(t)$ evaluated during Δ determines the potential felt by the ions during the next interval, via the expectation value of H^{EP}.

$$\ddot{q}_{\Gamma\gamma}(t) = -\nabla U_{eff}(q, t_0) \qquad t_0 < t < t_0 + \Delta \qquad (8a)$$

$$\downarrow$$

$$q_{\Gamma\gamma}(t_0 + \Delta), \dot{q}_{\Gamma\gamma}(t_0 + \Delta) \qquad (8b)$$

$$\downarrow$$

$$i\hbar \frac{\partial}{\partial t}\rho(t) = \left[H^{EP}(t_0 + \Delta), \rho(t)\right], \qquad t_0 + \Delta < t < t_0 + 2\Delta \qquad (8c)$$

where:

$$H^{EP}(t)=\sum_{\Gamma\gamma} V^{\Gamma\gamma} q_{\Gamma\gamma}(t), \qquad (9)$$

$$U_{eff}(q,t)=U_{harmonic}+tr\left(H^{EP}(t)\rho(t)\right). \qquad (10)$$

The density matrix $\rho(t)$ is relative to the degenerate electronic subspace. $V^{\Gamma\gamma}$ are the forces (actually matrices in the degenerate electronic subspace, see Par. 4C below), corresponding to the scalar forces $f_\alpha(l)$ in Eq.(6).

Eq. (8a) and (10) show the two causes according to which the ions move: the interaction with the lattice ions (via the harmonic potential $U_{harmonic}$) and the interaction with the excited electron (via the JT interaction $H^{EP}(t)$).

By Eq. (8c) the electron evolves during each time interval according to the value that the JT interaction $H^{EP}(t)$ takes at the end of the previous interval.

The coupled dynamics of eqs. (4) are non linear, and as such depend critically on all the parameters, in particular on the *initial conditions* (i.e. electronic wavefunction and lattice deformation at $t=0$) and on the temperature.

As said in Par. 3.C., we assume that there is no feedback on the crystal temperature coming from the electronic dynamics in the time interval considered, because the JT interaction is linear in the ionic displacements. Then the parting of the total displacement (see eqs.7) into a thermal and a classical components is still meaningful. In the following we discuss on the relaxation as it results without the thermal noise.

Before the excitation, the crystal ions are assumed to be oscillating at thermal equilibrium around a cubic symmetry configuration. Fig.3b shows an example of the time-dependent displacements due to the thermal fluctuations alone. We chose at random the initial time $t=0$ at which the instantaneous excitation occurs. At $t=0$ positions and momenta take the values $q_{\Gamma\gamma}(0)=q_{\Gamma\gamma}{}^{Th}(0)$ and $p_{\Gamma\gamma}(0)=p_{\Gamma\gamma}{}^{Th}(0)$, so that the electronic levels, degenerate when $q_{\Gamma\gamma}(0)=0$,

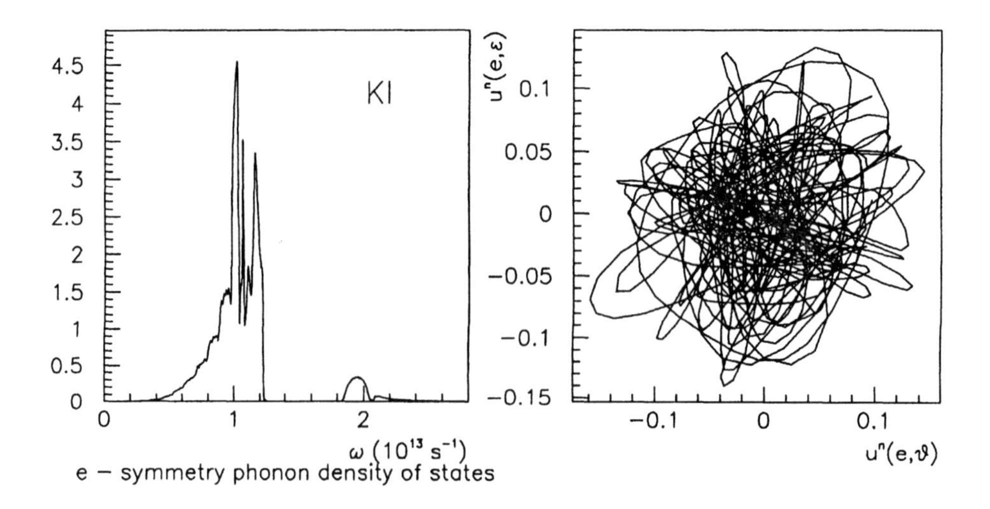

Figure 3. (a) . The e-symmetry DOS relative to the I^- ions motion. **(b)** Trajectory formed at T= 5 in the e-symmetry plane by the symmetrized thermal displacements (i.e. $q_\varepsilon{}^{Th}(t)$ vs $q_\theta{}^{Th}(t)$).

are slightly split by the thermal field. At further times the energy splitting evolves in time as given in the Eq. 8c. and the n.n. iodine ions move according to Eq. 8a.

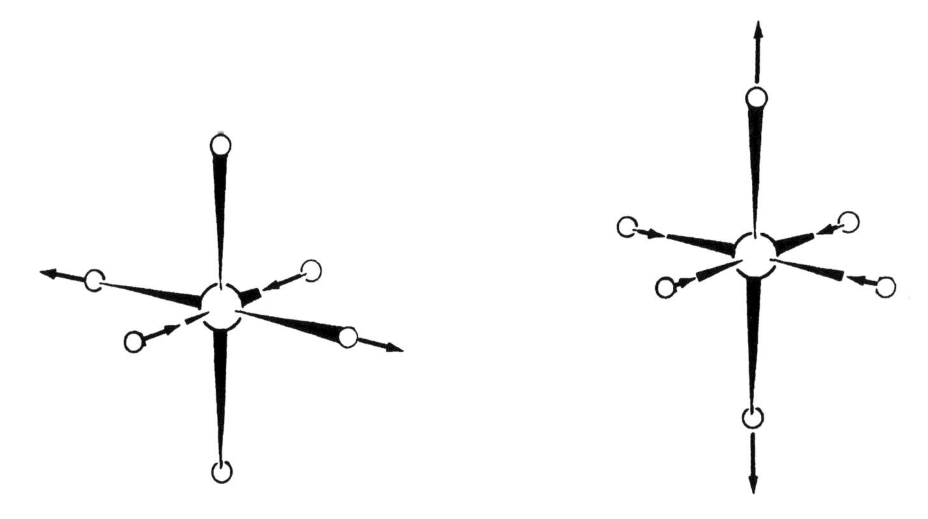

Figure 4. The two combinations of e-symmetry of the displacements of the ions I^-, n.n. of a positive substitutional impurity in *KI*. At left: tetragonal displacement $q_\theta(t)$. At right: rhombic displacement $q_\varepsilon(t)$.

4.C. Numerical Results and Discussion.

The (Exe) case here studied is that of a positive substitutional impurity in a *KI* crystal at T=5, excited in a *E*-symmetry electronic state which interacts with the first shell of the impurity neighbors (iodine ions). Such could be the case of $KI:Tl^+$, where the perturbed lattice dynamics is well known [17] and the excited states have been since long well identified and studied, also in relation with the JT effect [16,17,9]. In this case the electron states transform like the axes of the twofold irreducible representation *E*, i.e. $|E,1\rangle = |(2z^2 - x^2 - y^2)\rangle$ (tetragonal symmetry θ) and $||E,2\rangle = |(x^2 - y^2)\rangle$ (rhombic symmetry ε) and the electron dynamical matrix elements are $\rho_{ij}(t) = |E,i\rangle\langle E,j|$ (*i,j*=1,2). According to the group theory, Γ=e in Eq. 9: the symmetrized displacements belong to the twofold degenerate irreducible representation e of the cubic point group. They are hereafter indicated by $q_{e\gamma}(t)$ (γ =θ,ε), transforming respectively as $(2z^2 - x^2 - y^2)$ and $(x^2 - y^2)$ and corresponding to the deformations shown in Fig.4. The JT forces $V^{\Gamma\gamma}$ in Eq. 9 are then the following 2x2 matrices:

$$\hat{V}^\theta = b\hat{\sigma}_z = b\begin{pmatrix} 1 & 0 \\ 0 & -1 \end{pmatrix}, \qquad \hat{V}^\varepsilon = b\hat{\sigma}_x = b\begin{pmatrix} 0 & 1 \\ 1 & 0 \end{pmatrix}. \qquad (11)$$

In what follows we focus our discussion mainly on the features shown by the classical contributions $q^{Cl}{}_{e\varepsilon}(t)$ and $q^{Cl}{}_{e\theta}(t)$ to the two e-symmetry displacements, obtained by subtracting at each instant the thermal contribution from the total displacements, as given in eqs. 7.

We draw the trajectories in the plane $q^{Cl}{}_{e\varepsilon}(t)$ vs. $q^{Cl}{}_{e\theta}(t)$, in a Lissajous-like plot. In Fig.5 we report a single result, pertinent to a strong coupling (S = 6) at the temperature T = 5. However, the other data here not reported, relative to the strong coupling limit, support the

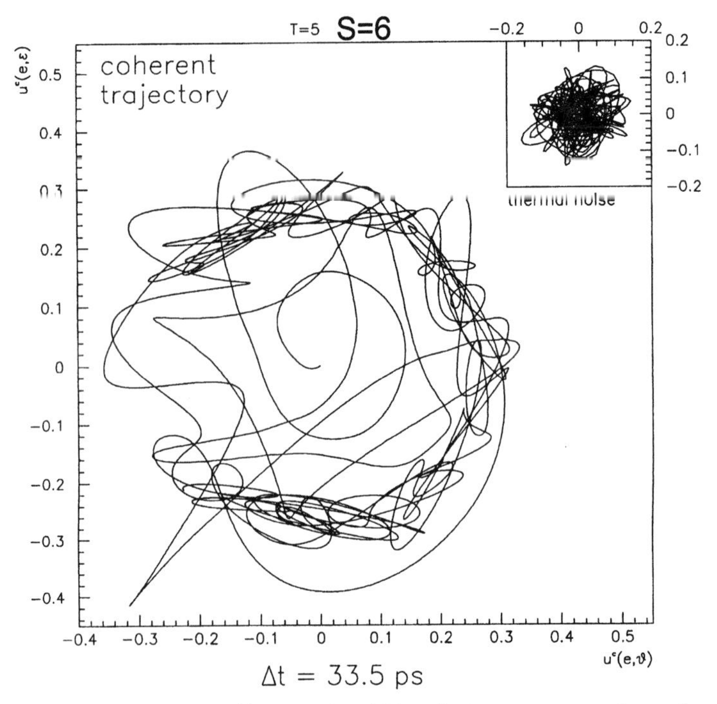

Figure 5. Lissajous trajectory traced between 0 and 33 ps in the e-symmetry plane, when the rhombic displacement $q_\varepsilon^{Cl}(t)$ is reported vs the tetragonal displacement $q_\theta^{Cl}(t)$ (see Fig.4). The JT coupling corresponds to a Huang-Rhys factor S= 6 , a value which is of the order of the JT coupling in the A excited state of Thallium in *KI*. (see ref. 17).

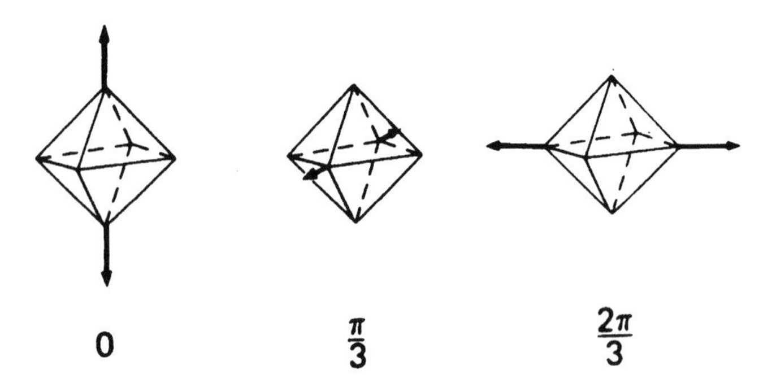

Figure 6. Elongations of the octahedron formed by the ions I^- n.n. of a positive substitutional impurity in *KI*, corresponding in the e-symmetry plane to the three angles of Fig. 5 where the trajectory becomes denser.

following general statements. In all the cases the first of the two phases of the relaxation, found in the non JT cases and previously described in Par.3.A, is still present: the point, soon after the excitation, moves very fast from the origin towards a new dynamical configuration. The second phase, however, is now different and shows a new characteristics typical of the JT effect: the point circles around the origin, the stronger the coupling the plainer the feature.

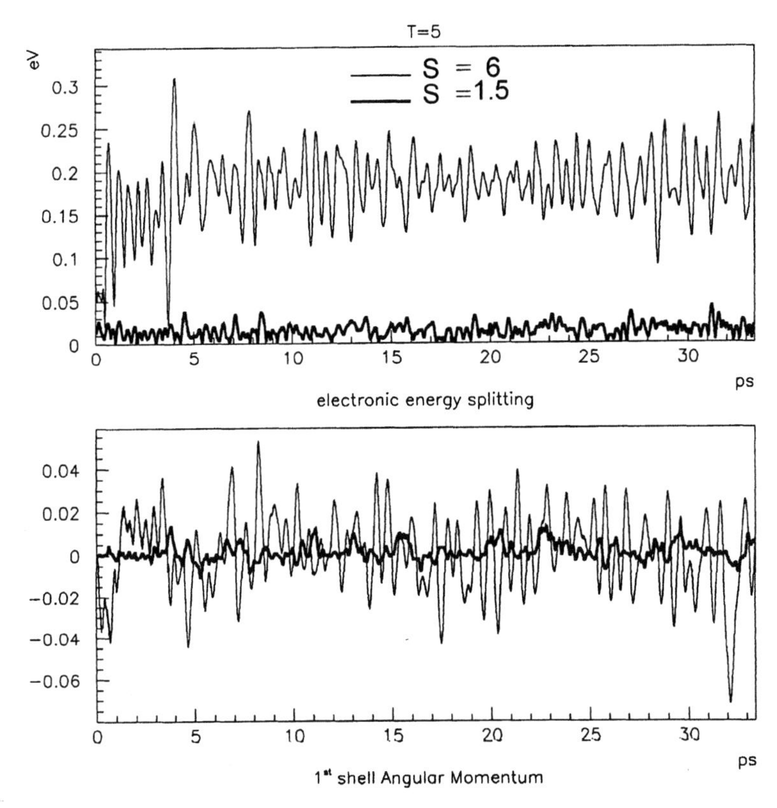

Figure 7. a) Time dependent electronic energy splitting for two values of the JT coupling: S=1.5 (intermediate) and S=6 (strong). **(b)** JT pseudo-angular momentum of e-symmetry ($p^{Cl}_{\varepsilon}\, q^{Cl}_{\theta} - p^{Cl}_{\theta}\, q^{cl}_{\varepsilon}$) relative to the motion of the ions I^- n.n. of a positive substitutional impurity.

With respect to Fig.5 one can summarize this behavior in three points: i) the shell of the n.n. relaxes in less than 1 ps towards the new dynamical configuration. ii) The point in the Lissajous plot circles and oscillates in an irregular way around the origin, as it were confined to move around the bottom of the brim of a Mexican hat. This is an excellent result, because this form of the potential surface is the typical features of the one mode JT Exe case. iii) The potential surface faintly shows three minima, at the angles of 0, 1/3 π , 2/3 π, because the trajectory is denser there than in the other regions of the brim. Notice that these three regions of the symmetry plane correspond in a representation of the n.n. displacements to have the n.n. octahedron elongated along the x,y,z, directions, as shown in Fig.6.

It is worth mentioning that point iii) is usually obtained within the "one-frequency JT scheme", as a static effect due to non linear EP interactions [15,16]. Here it appears as a dynamical effect, caused by the linear (but many mode!) JT interaction, whic therefore contains in itself an intrinsic non linearity. Further explicit non linearities, here not shown, can make deeper the three dips. By increasing the temperature the motion in the symmetry plane becomes much more regular, the oscillations less important and the presence of the minima uninfluential. It is worthwhile to notice that in the JT systems only a part of the total energy supplied by the EP interaction is dissipated in time into the host crystal, mainly during the first phase. In fact a consistent part of it, that corresponding to the off diagonal JT coupling and responsible of the transversal motion inside the brim of the Mexican hat, remains trapped inside the local volume. In subsequent times it could serve as a source of energy at disposal for other electron-phonon processes, as those related to suitable electronic transitions to non degenerate states. Finally, because the JT system is not isolated the angular momentum $(p_\varepsilon q_\theta - p_\theta q_\varepsilon)$ is not strictly conserved, as it does in the one mode (E x e) JT case. We compare the cases of intermediate (S=1.5) and strong (S=6) couplings in Fig7. We see that conservation holds at lower coupling. The regular ionic motion still found in the strong coupling limit is actually due to a persistent hidden coherence (tunneling between adiacent minima in the potential surfaces) in the motion of the electron, which determines the JT forces acting on the ions.

This is however a rather long story, which we postpone to a forthcoming paper and, hopefully, to a future lecture in Erice.

REFERENCES

1. See the proceedings of the recent Conference on *Ultrafast Processes in Spectroscopy*, eds. O.Svelto, S. De Silvestri and G. Denardo, (Plenum, 1996)
2. See for instance W.T.Pollard et al , J. Chem. Phys. 96, 6147 (1992); Q. Wang et al , Science **266**, 422 (1994), and references therein. See also M. Nisoli, S. De Silvestri, O. Svelto, R. Scholz, R. Fanciulli, V. Pellegrini, F. Beltram, and F. Bassani, *Phys. Rev. Lett.* **77**, 3463-3466 (1996).
3. See the issue of Journal of Luminescence, vol. **58,** 1-428 (1994) on *Dynamical Processes in Excited States of Solids.*
4. See for instance for polyatomic molecules: W.T. Pollard, H.T. Fragito, J.-Y. Bigot, C.V. Shank and R.A. Mathies, *Chem. Phys Lett.* **168**, 239-245 (1990), and for defects in solids: R. Sholz, M. Schreiber, F. Bassani, M. Nisoli, S. De Silvestri and O. Svelto, *Phys. Rev.* **B 56**, 1179-1195 (1997).
5. M. Gruebele and A.H. Zewail, Phys. Today 43, 24 (1990).
6. K.-A. Suominen, B.M. Garraway and S. Stenholm, *Phys. Rev.* **A 45**, 3060 (1993); B.M. Garraway, S. Stenholm and K.-A. Suominen, *Physics World,* 46, (1993).
7. G.S. Zavt, V.G. Plekhanov, V.V. Hizhnyakov and V.V. Shepelev, *J. Phys.* **C 17**, 2839 (1984).
8. M. Dominoni and N. Terzi, *Europhys. Lett.* **15**, 515-520 (1991).
9. M. Dominoni and N. Terzi, *Europhys. Lett.* **23**, 39-44 (1993); *J. Lumin.* **58**, 11-14 (1994); *Zeitsch. Phys. Chem.* **200**, 245 (1997)
10. A. Giorgetti and N. Terzi, *Solid St. Comm.* **46**, 635 (1981).
11. N. Terzi, in *Collective Excitations in Solids*, ed. by B. di Bartolo (Plenum, 1983) 149-182.
12. M. Dominoni and N. Terzi, in *Non radiative Processes in Solids*, ed. B. Di Bartolo (Plenum, 91) 443-482.
13. See for instance the introductory lectures in *Collective Excitations in Solids*, ed. by B. di Bartolo (Plenum, 1983).
14. W.A. Knox, L.F. Mollenauer and R.L. Fork, *Chem. Phys. Series* (Springer, Berlin, 1986) **46**, 287.
15. R. Englman, *The Jahn-Teller Effect in Molecules and Crystals* (Wiley, London, 1972).
16. Yu.E. Perlin and M. Wagner eds., *The Dynamical Jahn-Teller Effect in Localized Systems* (North-Holland, Amsterdam, 1984).
17. G. Benedek and N.Terzi, *Phys. Rev.* **B 8**,1746 (1973).

LASER CRYSTAL EMITTING FROM ULTRAVIOLET TO INFRARED AND THE IMPACT OF NONLINEAR OPTICS

Georges Boulon

Physico-Chimie des Matériaux Luminescents
Université Claude Bernard Lyon I
Unité Mixte de recherche CNRS n° 5620
43, boulevard du 11 novembre 1918
69622 Villeurbanne cedex, France

ABSTRACT

There is a renewal in the search of new laser materials from ultraviolet to infrared and the area of nonlinear crystals is very useful to cover new capabilities of laser wavelengths. One tendency is to find efficient and compact all-solid-state laser sources in the modular approach with a diode-laser, a laser crystal, a nonlinear crystal and even a saturable absorber. The objective of this article is to report the main trends both for laser crystals and nonlinear crystals.

I. INTRODUCTION

In the past decade, the renewal in the search of new laser materials covering the widest spectrum from ultraviolet to infrared has been very strong in both government and high technology private company communities. Today great efforts are devoted to the development of all-solid-state laser sources that offer advantages of compactness, long life, reliability, low noise, safety and easier maintenance with respect of gas laser systems.

Ultrafast Dynamics of Quantum Systems: Physical Processes and Spectroscopic Techniques, Edited by Di Bartolo and Gambarota, Plenum Press, New York, 1998

Among the multitude of applications in laser systems, we can mention scientific instruments, optical communications, environmental monitoring by differential absorption laser radar (lidar), medical diagnosis and treatment, materials processing, optical signal processing, data storage, color printing, displays for the future generation of TV, remote sensing, underwater communication, chemical and biological species detection and more.

The two main trends are the development of both doped-oxyde or doped-fluoride inorganic solid-state lasers and nonlinear optical materials converting an available laser wavelength which is needed for a particular application. It is interesting to see that in the approach of laser crystals, the pumping energy is converted to lower values, excepted with very new up-conversion systems, whereas on the contrary, nonlinear crystals in general convert the laser beam energy to higher ones reducing thermal loading process and then, extending the range of laser sources from the mid-ultraviolet to the far-infrared.

II. LASER DOPED-CRYSTALS

Since the first laser demonstration in 1960 by T. Maiman with $Cr^{3+}:Al_2O_3$ ruby crystals [1], hundreds of doped-crystals or doped-glasses have been demonstrated for lasing. The dopants are either transition metal ions having a 3d configuration exposed to the crystal field of the surrounding ions in the host :

$Ti^{3+}(3d^1)$, $Cr^{4+}(3d^2)$, $Cr^{3+}(3d^3)$, $Cr^{2+}(3d^4)$, $Co^{2+}(3d^7)$ or rare-earth metal ions having a $4f^n$ configuration shielded from crystal field effects by outer configurations :

Ce^{3+}, Pr^{3+}, Nd^{3+}, Dy^{3+}, Ho^{3+}, Er^{3+}, Tm^{3+}, Yb^{3+}.

Obviously the easiest substitutions are met between activators and host cations of both similar ionic radii and oxydation states. A general schematic energy diagram can be seen in Fig. 1. The best insulator hosts are characterized by a large gap or a ground state of the active center distant enough from the conduction band in order to eliminate any losses by excited-state absorption. The ^2Eg unique excited state of $Ti^{3+}(3d^1)$ as a dopant in Al_2O_3 is a fine example of this model in view of the great success of this tunable solid-state laser. A similar approach can be observed for Yb^{3+} -doped crystals with only one $^2F_{5/2}$ excited-state.

Chemistry and Physics of this area can be found in review articles and books [2-3-4-5-6-7]. Fig. 2 gives the main domain of laser wavelengths of commercial crystals and a few

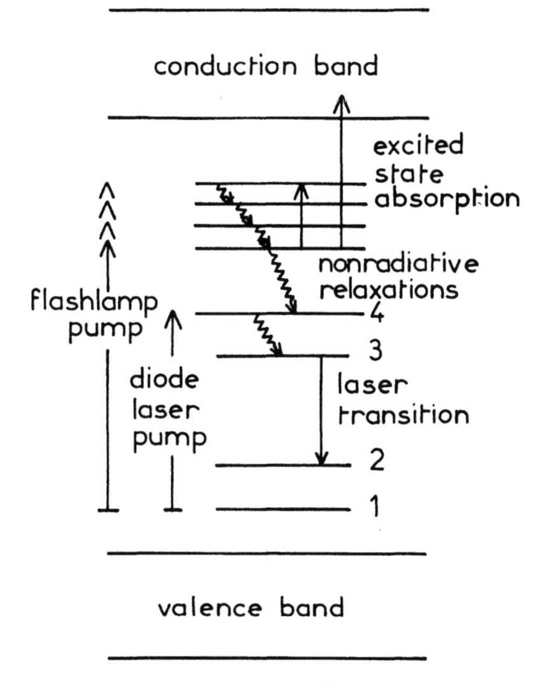

Figure 1. Model of the laser center energy diagram in the band gap of the host

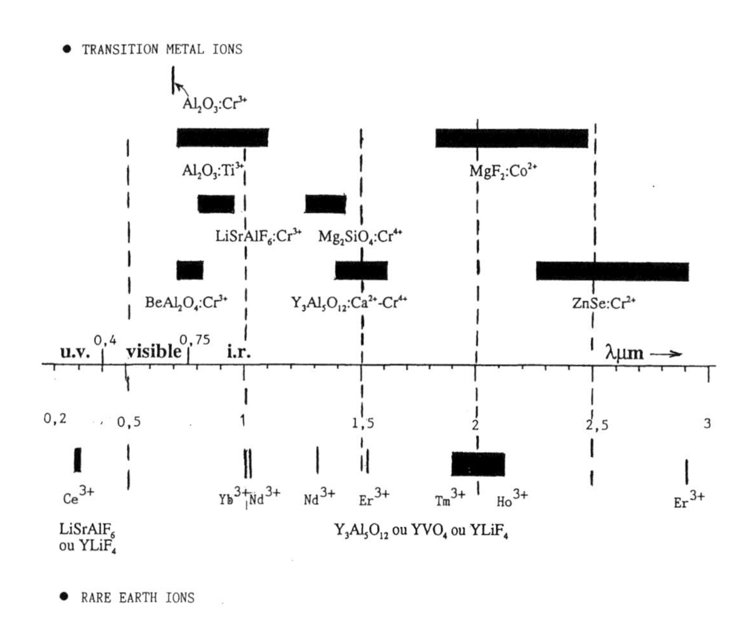

Figure2. Laser emission spectral ranges of transition metal ions and rare-earth ions

TABLE 1. Optical and thermal characterizations of the main lasers crystals

	λ absorption (nm)	λ emission (nm)	cross-section (10^{-20} cm^2)	lifetime μs	thermal conductibility W/m.K
Nd^{3+}:YAG	807.5 diode	1064.2	30	244	10,3
Nd^{3+}:$LiYF_4$		1047 1053	37 26	480	
Nd^{3+}:phosphate glass		1054	4	290-330	
Nd^{3+}:YVO_4	808.5 diode	1064.3	15.6	100	
Nd^{3+}:$Sr_5(VO_4)_3F$	809.6 diode	1065	5		
Yb^{3+}:YAG	943 diode	1030	2	950	13
Yb^{3+}:$Sr_5(PO_4)_3F$	900 diode	1047	7.3	1100	2
Ce^{3+}:$LiSrAlF_6$	260 (YAG)	285-315	950	0.028	3.1
Ce^{3+}:$LiYF_4$	KrF excimer 248	309-325			
Cr^{3+}:$LiSrAlF_6$	670 diode	780-990	3.2	67	3.1
Cr^{4+}:Mg_2SiO_4	1064 (YAG)	1130-1367	20	3	5
Cr^{2+}:ZnSe	1800 diode	2200-3000	80	8	18
Ti^{3+}:Al_2O_3	Ar^+ ion 488,514.5	660-1180	30	3,2	

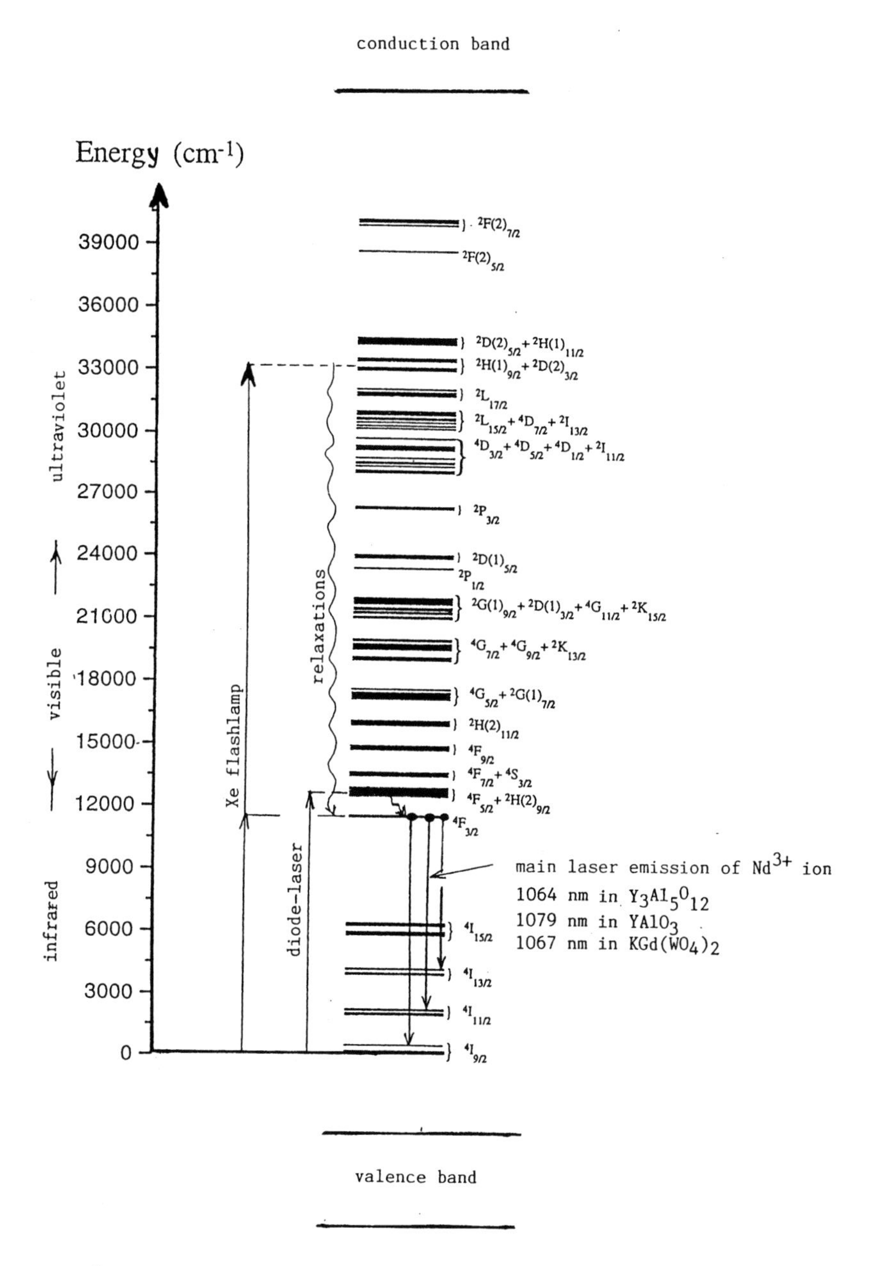

Figure 3. Nd^{3+} -ion energy diagram. The main laser wavelengths of the ${}^4F_{3/2}\rightarrow{}^4I_{11/2}$ transition are given in three commercialized crystals

values of absorption, emission, stimulated emission cross-section, lifetime of the excited state and thermal conductibility have been reported in Table 1. The number is severely restricted depending not only on spectroscopic properties but also on good mechanical and thermal properties.

II. A. Rare-Earth Ions

The dominant laser crystal is, without any doubt, Nd^{3+} -doped $Y_3Al_5O_{12}$ (YAG) a four-level scheme working at room temperature efficiently pumped either by broadband flash-lamps in the visible range or now by narrowband diode lasers around 807.5 nm (Fig.3). 1064 nm is a very well known emission from the $^4F_{3/2} \rightarrow {}^4I_{11/2}$ transition, with the highest intensity but two other emissions can be used at 1320 nm ($^4F_{3/2} \rightarrow {}^4I_{13/2}$) and an even nearly resonant one at 945 nm ($^4F_{3/2} \rightarrow {}^4I_{9/2}$) (Fig. 3-4-5-6-7). The emission cross-section at 1064 nm is high ($\approx 30.10^{-20}$ cm^2) and the lifetime of the $^4F_{3/2}$ metastable excited state is long enough for the energy storage needed to produce Q- switching short pulses. In addition, the emission bandwidth is quite wide for a modelocking operation. However the Nd^{3+} -doped YAG shows thermal lensing which is not accepted for specific applications of high power lasers.

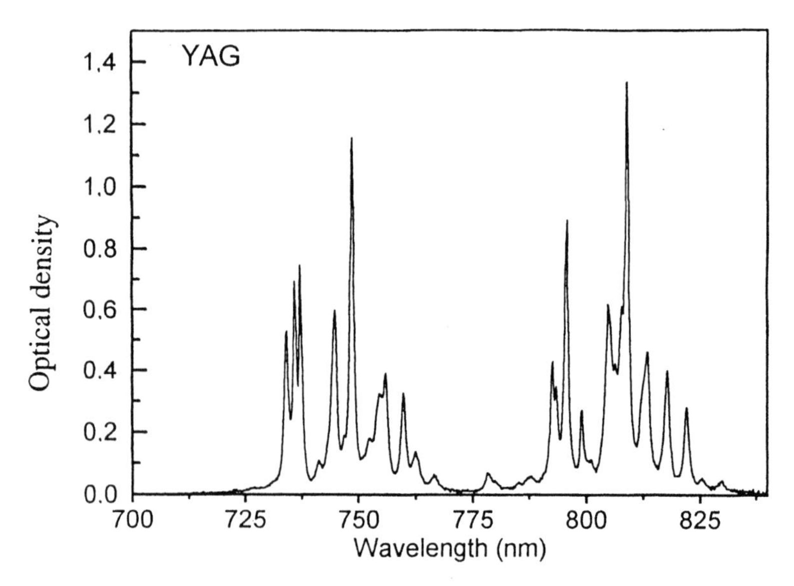

Figure 4. Absorption spectrum of Nd:YAG at room temperature in the region of the diode pumping

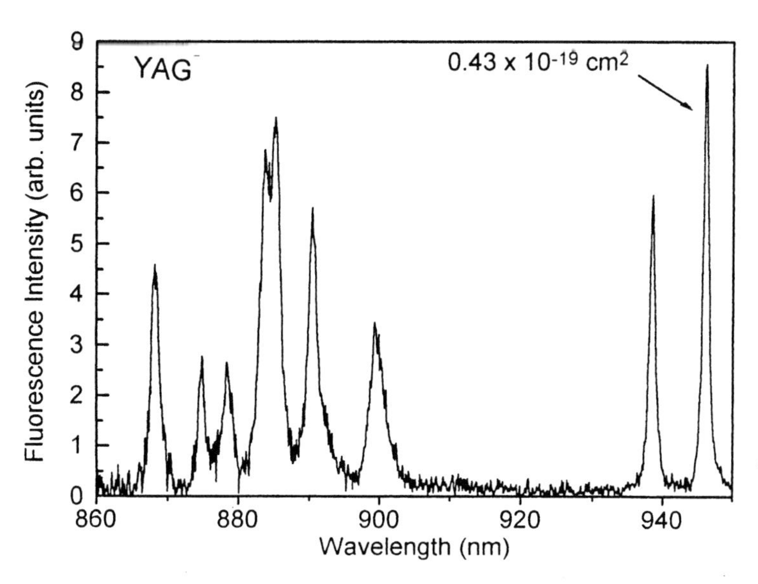

Figure 5. Emission spectrum of Nd:YAG at room temperature of the $^4F_{3/2}\rightarrow{}^4I_{9/2}$ transition region

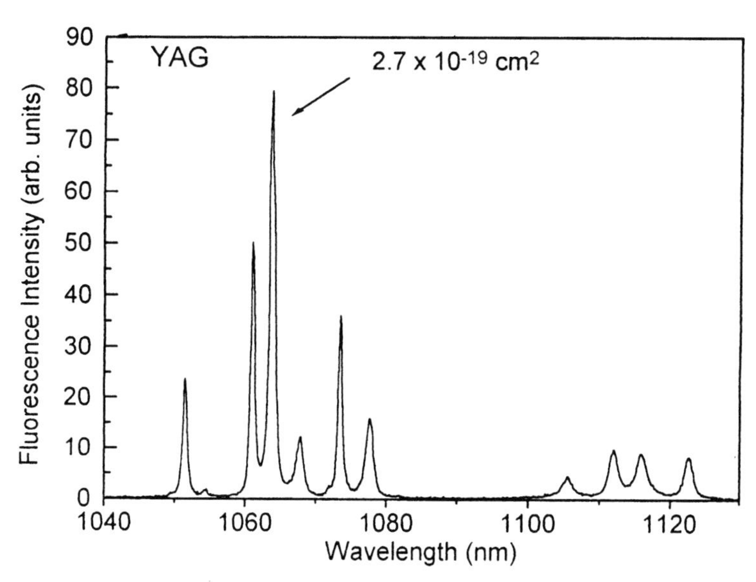

Figure 6. Emission spectrum of Nd:YAG at room temperature of the $^4F_{3/2}\rightarrow{}^4I_{11/2}$ transition region

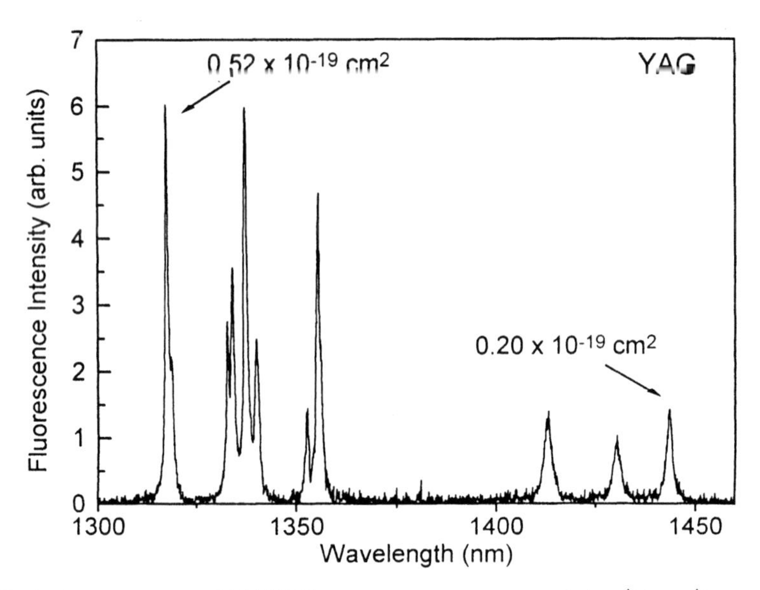

Figure 7. Emission spectrum of Nd:YAG at room temperature of the $^4F_{3/2} \rightarrow ^4I_{13/2}$ transition region

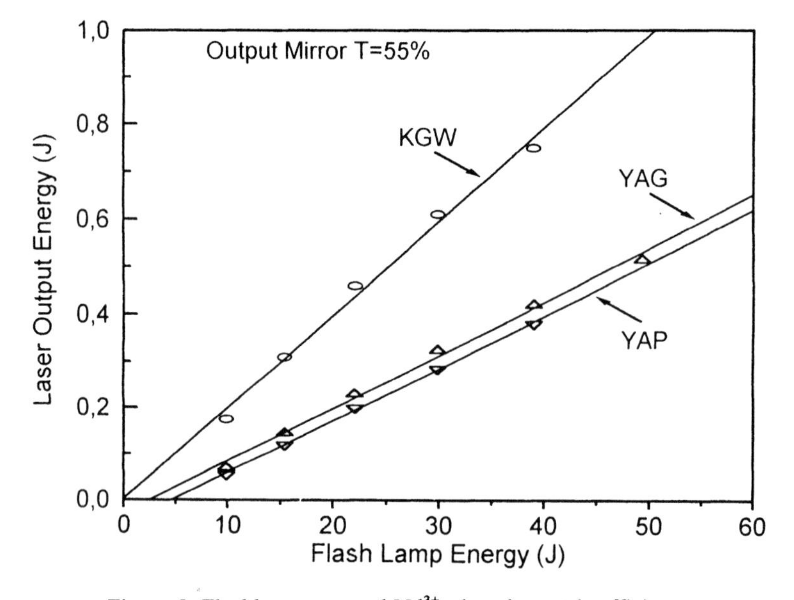

Figure 8. Flashlamp-pumped Nd^{3+} -doped crystals efficiency

Such behavior explains the development of fluoride crystals such as Nd^{3+} -doped $LiYF_4$(YLF) but unfortunately with lower thermal and mechanical properties. Other matrices, Nd^{3+} -doped YVO_4 or $GdVO_4$ have been grown due to their low threshold, high gain cross-section, broad homogeneous absorption bandwidth which is well adapted to diode pumping and therefore the efficiency compares favourably with the Nd^{3+} -doped YAG only for small sizes due to big difficulties in growing it with high homogeneity (Fig.9).

So, it is difficult to compete with the Nd^{3+} -doped YAG and to find better sources emitting around 1 μm and 1.32 μm. As this laser crystal remains an excellent reference, we have shown in Fig. 5, 6, 7 and 8 absorption and emission spectra which are used by mentioning stimulated emission cross-sections. We think that such results are important in

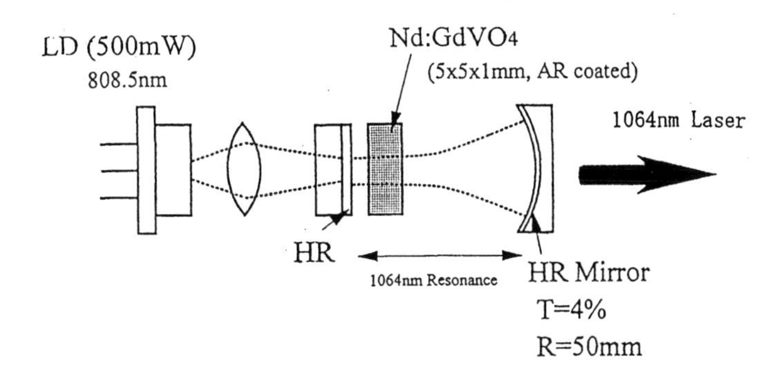

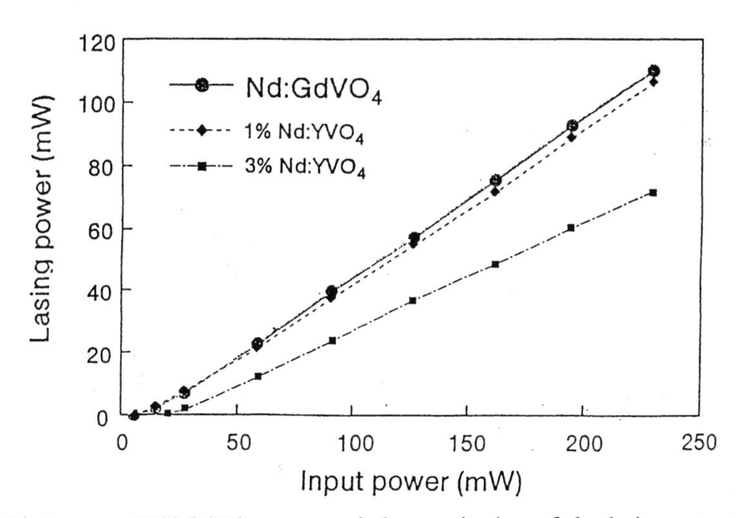

Figure 9. Diode-pumped $Nd:GdVO_4$ set up and characterization of the lasing output power as a function of the input power

TABLE 2 : Main optical properties of Nd:YAG, Nd:YAP and Nd:KGW. K_{Nd}:Nd^{3+} segregation coefficient, τ_R and τ_F Radiative and Fluorescence Lifetimes, σ_{em} at λ_L) Stimulated emission cross section at laser wavelength, α (at λ_p) Absorption coefficient at pump wavelength, $\Delta\lambda_p$ and $\Delta\lambda_L$ FWHM of pump and laser transition lines.

Material	K_{Nd} and Max. Nd^{3+} concentration	Refractive index n and dn/dT (10^{-6}/°C)	α_p (at λ_p) and $\Delta\lambda_p$	τ_R and τ_F	σ_{em} (at λ_L) and $\Delta\lambda_L$
Nd:YAG $Y_3Al_5O_{12}$	0.18 and 1.25at%Nd^{3+} (1.725×10^{20} cm^{-3})	1.82 (at 1.06 μm) and 9.9	9 cm^{-1} (at 808 nm) and 0.9 nm for 1.1at%Nd^{3+}	255 μs and 220 μs at 1.1%atNd^{3+}	2.7×10^{-19} cm^2 (at 1.064 μm) and 0.6 nm
Nd:YAP $YAlO_3$	0.8-0.95 and 1.8at%Nd^{3+} (3.546×10^{20} cm^{-3})	1.93 (at 1.06 μm) and 9.7	11.5 cm^{-1} (at 813 nm, E//c) and 1.6 nm for 0.77at%Nd^{3+}	180 μs at 0.77%atNd^{3+}	2.4×10^{-19} cm^2 (at 1.0795 μm, E//c) and 1.6 nm
Nd:KGW $KGd(WO_4)_2$	- 10at%Nd^{3+} (6.4×10^{20} cm^{-3})	2 (at 1.06 μm) and 0.4	8 cm^{-1} (at 810nm, E//n_m) and 2 nm for 3 at%Nd^{3+}	120 μs at 3%atNd^{3+}	3.4×10^{-19} cm^2 (at 1.067μm, E//n_m) and 1.2 nm

the search of other Nd^{3+} -doped crystals which can play an important role. In this sense, two hosts are also developed : $YAlO_3$ (YAP) and KGd $(WO_4)_2$ (KGW). Table 2 gives their main optical properties and a discussion of the comparison of each type of crystal can be found in reference [8]. The laser performances shown in Fig. 9 which have recently been obtained under the same non-optimized flash-lamp pumping conditions are in favor of the Nd^{3+} -doped KGW only by considering optical properties but unfortunately thermal conductivity (3.8 W/m.K) is three times lower than with the Nd^{3+} -doped YAG (13 W/m.K).

The future of solid-state lasers is strongly connected with diode-laser or diode-laser array pumping. Besides Nd^{3+}, few rare-earth ions can be diode pumped: Er^{3+} is efficiently pumped under three spectral ranges to get a 1540 nm eye-safe laser emission from $^4I_{13/2} \rightarrow ^4I_{15/2}$, transition as can be seen in Fig. 10. Morever, Tm^{3+} and Ho^{3+} emissions around 2010, 2089 and 2127 nm are also diode pumped either for an eye-safe laser or for medical applications due to strong absorption of biological tissue within this domain.

The Yb^{3+} ion is another rare-earth ion emitting in the same region as Nd^{3+} which is diode pumped with advantages such as a low quantum defect (absorption around 920-980 nm, emission around 1030 nm), a long lifetime (≈ 1ms), no excited-state absorption (only

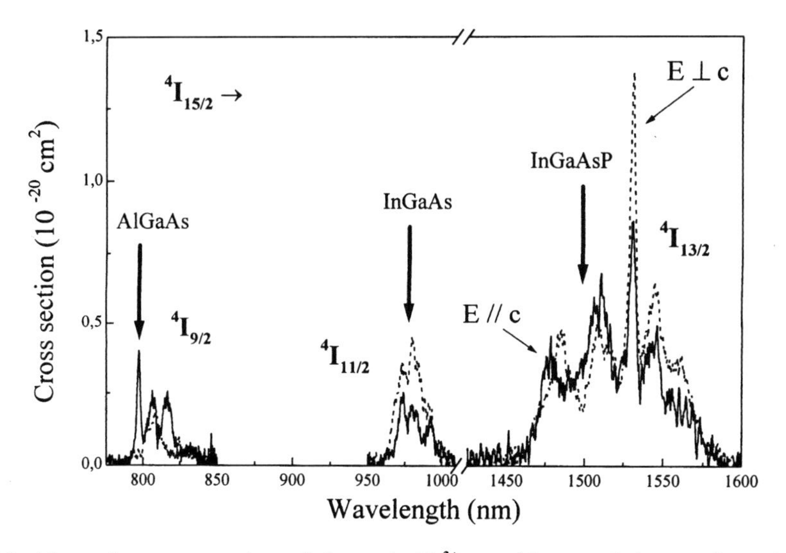

Figure 10. Absorption cross-section of the main Er^{3+} transitions and the overlap with diode laser wavelengths

one $^2F_{5/2}$ excited level), weak concentration quenching. The disadvantage is due to a quasi-four level scheme with Stark levels of the excited and ground states leading to thermal quenching processes. Yb^{3+} is really one of the most promising laser ions for the future under diode pumping in the approach to finding a better system than the Nd^{3+} active center. Another challenge with both Yb^{3+} and Nd^{3+} ions emitting near 1030-1070 nm is the alternative to CW-ion lasers, providing blue-green light, by Second-Harmonic Generation (SHG). Intracavity-doubled, diode-array-pumped Nd^{3+} -doped crystals (YAG, YVO_4, YLF) are now generating greater than 6W output at 532 nm with 20W diode pump power. This represents an efficiency of 30% with the great advantages of much better amplitude stability, typically ten times lower than that of CW-Argon laser, and also lower noise levels, five times those of a multiple-watt output level. This intracavity doubling takes advantage of the far greater fundamental intensity near the beam waist. Third-harmonic (355 nm) at powers higher than 1W and even fourth-harmonic (266 nm) at around 100 mW of 1064 nm emission have also been demonstrated with a Nd:YVO_4 [9-10]. However, despite these advantages, the prices of diode-pumped CW green sources are still very expensive due to the use of sophisticated optical, thermal and mechanical technics in order to keep single-longitudinal mode operations with right polarization (linearly or circularly polarized), a very stable temperature in the absence of mechanical fluctuation.

Today affordable diode-pumped green lasers with powers less than 10W are commercialized with similar cost to that of Argon ion gas laser but the cost of a diode-pumped solid-state laser is still prohibitive for many industrial and defense applications like laser range finders, laser designators, laser radars, even in the absence of doubling operation. Although the cost between a diode-pumped laser system (≈\$1000/W) and a flashlamp pumped system (\$100/W) is decreasing, it will take many years before diode-pumped solid state lasers replace flashlamp-pumped lasers. This means that research of the spectroscopic properties of doped-crystals absorbing most of the broadband power output of flashlamps have to continue.

Recently, the Ce^{3+} rare-earth ion has even been reported for tunable ultraviolet lasers in $LiSrAlF_6$ (LiSAF) between 285 and 315 nm under the 266 nm Nd^{3+}:YAG pump with the highest cross-section of inorganic crystals (950.10^{-20} cm^2) [11]. In the physical point of view, it is interesting to note that usual solarisation processes can be avoided by adding another harmonic of the Nd^{3+}:YAG, at 532 nm, whose effect is to eliminate color centers. The total efficiency of this system is 47%.

II. B. Transition Metal Ions

For many years, atomic physics laboratories have used the tunability of dye-lasers to pump solid, liquid or gas phases. Now, tunable solid-state laser sources are removing dye-lasers. Transition-metal ions are mainly used in their configuration. The levels of the d configuration in transition metal ions are exposed to the crystal field of the surrounding ions in the host. Thus the interaction between the electronic energy levels and lattice vibrations induces both broad band absorption and fluorescence, and so tunability is possible.

The Ti : sapphire is the most famous one in the 650-1100 nm spectral region but its cost remains high due to the use of the CW-Ar^+ ion laser pump. Among new tunable sources, Cr^{3+} -doped colquirite fluoride such as $LiSrAlF_6$ (LiSAF) can be flash-lamp pumped in the visible range or diode pumped at 670 nm [12]. The $^4T_2 \rightarrow ^4A_2$ broad emission between 780 and 990 nm leads to the demonstration of an all-solid-state femtosecond source. Cr^{4+} emission has been discovered by analysies of Cr^{3+} -doped Mg_2SiO_4 (forsterite). The presence of Cr^{4+} was clear on the tetrahedral Si^{4+} in addition to the Cr^{3+} substitution with the Mg^{2+} octahedral site [13]. Similar results were obtained with the two tetrahedral and octahedral sites of garnets [14]. In forsterite, the tunability varies from 1130 to 1370 nm whereas in YAG the domain is 1370-1580 nm. The recent prospect in tunable mid-infrared has also been enriched by Cr^{2+} -doped ZnSe with broad band emission between 2150 and 2800 nm [15]. These materials are just at the start of the development, and high quality crystals are needed in the future.

III. NONLINEAR CRYSTALS

III. A. Nonlinear Optical Effects

Nonlinear optical effects have been discovered a long time ago by Kerr (1875), Pockells (1893) and Raman (1930) but the real potentialities of these phenomena appeared with the advent of laser [1]. The first paper of Franken was dealing with the generation of optical harmonics [16]. It was not surprising to wait for laser sources because nonlinear optics needs very strong and monochromatic electric fields. In usual linear optics, the oscillating electric field E of a light wave propagating through a transparent dielectric medium induces an oscillating polarization field P in the collection of atomic dipoles responding at the same frequency as the original light, which can be written as follows :

$$P_i = \varepsilon_\circ \sum_j \chi_{ij} \mathrm{E}_j$$

$\varepsilon_\circ$: permittivity of free space

χ_{ij} : dimensionless second-rank tensor is called linear susceptibility coefficient

With the strong electric field of lasers, the dipoles of a dielectric medium no longer respond linearly and the induced dipole moment P_i can be expressed as a power series expansion as follows :

$$P_i = \varepsilon_\circ \sum_j \chi_{ij} \mathrm{E}_j + \varepsilon_\circ \mathop{\mathrm{E}}_{jk} \chi_{ijk} \mathrm{E}_j \mathrm{E}_k + \varepsilon_\circ \mathop{\mathrm{E}}_{jkl} \chi_{jk\ell} \mathrm{E}_j \mathrm{E}_k \mathrm{E}_\ell + \ldots$$

linear optics nonlinear optics

The classical schematic way to express P is :

$$P = \varepsilon_\circ \left(\chi^{(1)} \mathrm{E} + \chi^{(2)} \mathrm{E}^2 + \chi^{(3)} \mathrm{E}^3 + \ldots \right)$$

III. B. Birefringence Phase Matching

The second term corresponds to nonlinear phenomena as Second-Harmonic Generation (SHG), Pockells effect, sum and difference frequency mixing, optical parametric amplification (OPA) and optical parametric oscillation (OPO). The third term corresponds to Third Harmonic Generation (THG), Kerr effect, stimulated Raman scattering, four-wave mixing, phase conjugaison, self-focusing, two-photon absorption. Although such optical phenomena can be observed in gases and liquids, the nonlinear crystals are usually used in optical systems. Among the 32 point groups of the 7 crystallographic structures, only 21 are not centrosymmetric and can play a role for SHG or, in general, with the non-vanishing second term of $P(2\omega)$. The 11 other point groups cannot be active for second order linear action.

The expression of $P(2\omega)$ power output of the laser beam at the frequency 2ω can be seen in Fig. 11. The demonstration of this formula belongs to optics education and several books can be mentioned in this field [17-18]. It is clear that the energy output is maximum when

$$n(\omega) = n(2\omega)$$

ans so $P(2\omega)$ is quadratically dependent on the ℓ crystal length. Such a condition is available with birefringent materials. For example, in the very well known $LiNbO_3$ crystal, Fig. 12 shows the angle θm giving rise to expected result by taking account of ordinary (o) and extraordinary (e) beam refractive index variations.

Second Harmonic Generation

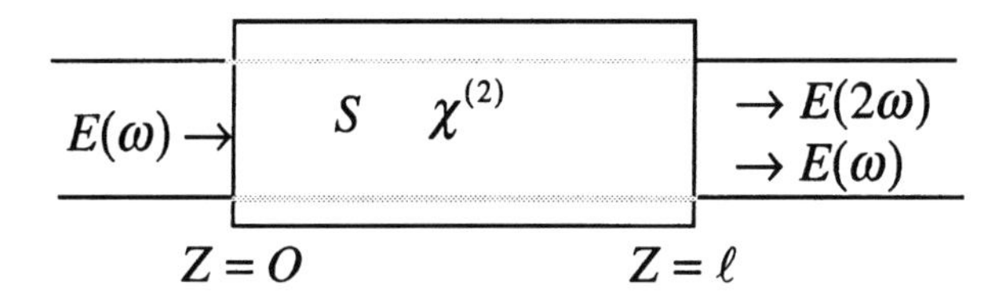

Electric Field Amplitude at 2ω :

$$E(2\omega) = \left(\frac{2\omega}{c}\right)^2 \frac{\chi^{(2)} E^2(\omega)}{2k_{2\omega}(2k_\omega - k_{2\omega})} \exp[i(2k_\omega - k_{2\omega})\ell - 1]$$

$$k_\omega = n(\omega)\frac{\omega}{c} \qquad k_{2\omega} = n(2\omega)\frac{2\omega}{c}$$

Power output of the beam at 2ω :

$$P(2\omega) = \frac{4\left[\chi^{(2)}\right]^2 \omega^5}{\varepsilon_o c^6 S (2k_\omega)^2 k_{2\omega}} \cdot \frac{\sin^2 (2k_\omega - k_{2\omega})\frac{\ell}{2}}{\left[(2k_\omega - k_{2\omega})\frac{\ell}{2}\right]^2} P^2_{(\omega)} . \ell^2$$

Figure 11. $\chi^{(2)}$ nonlinear crystal used for Second Harmonic Generation : expression of the electric field amplitude and the power output of the beam at 2 ω

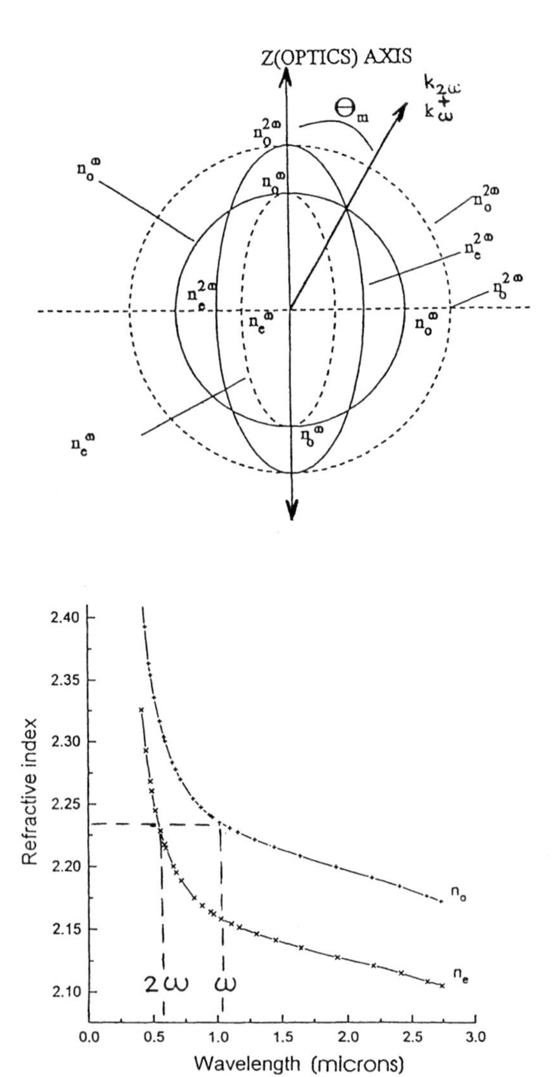

Figure 12. Birefringence phase matching in $LiNbO_3$ crystal

If such a condition is not respected, *P(2ω)* oscillates with the spatial period z_o :

$$z_o = \frac{\lambda}{2(n_\omega - n_{2\omega})}$$

the maximum being reached for the coherence length ℓ_c :

$$\ell_c = \frac{z_o}{2} = \frac{\lambda}{4(n_\omega - n_{2\omega})}$$

Therefore, the birefringence phase-matching is the usual way to get Second Harmonic Generation with bulky crystals which have to fulfill requirements such as convenient birefringence in the used acceptance bandwidths and also adequate nonlinear susceptibility, the lowest losses by scattering and absorption, high damage resistance, good thermal properties and, finally, they have to be growable with high chemical stability, without aging. For the laser wavelength of interest, the physical conditions are governed by both energy conservation and momentum conservation. $\chi^{(2)}$ nonlinear susceptibility is proportional to d_{31} coefficient and another feature is that waves are orthogonally polarized. Most set-ups use ordinary waves at 1064 nm and extraordinary waves at 532 nm with the Nd:YAG.

III. C. Birefringence Phase Matching in bulky $\chi^{(2)}$ - Type Nonlinear Crystals

After 30 years of research, there are still only a few crystals that are used in commercial products : $LiNbO_3$, KDP(KD_2PO_4), KTP($KTiOPO_4$), KTA($KTiOAsO_4$), (KN) $KNbO_3$ for visible, BBO(β-BaB_2O_4) and LBO(LiB_3O_5) more adapted for ultraviolet and $AgGaSe_2$ for infrared range. Each of them exhibits of course advantages and drawbacks. Articles in reference [7] are especially relevant for the state of art.

In ultraviolet region, there is a need for new materials. KDP is widely used but its nonlinear susceptibility and birefringence are small and it is high temperature sensitive. KTP is another popular material with a spectral transmission range of O.35 to 4.5 μm, high SHG efficiency, a weak walk-off angle of 1 mrad and an angular banwidth of 15-68 mrad-cm [19]. Important limitation of KTP is the occurrence of gray-track damage, optical damage in which the crystal darkens and absorbs the optical waves. Research is active to detect any responsability of impurety traces and to understand the damage processes in order to avoid permanent problems. The best performances are observed with KTA, a new nonlinear crystal of the same orthorhombic point group as KTP still under development in laboratories : a wider infrared transmission spectrum as can be seen in Fig. 13 and a higher nonlinear susceptibility coefficient that is 1.6 times that of KTP [20]. KTA is thus a good candidate for the substitution of KTP.

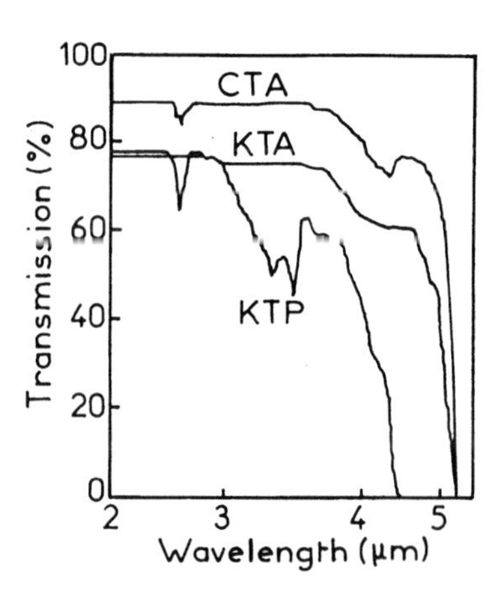

Figure 13. Infrared spectral range absorption of three isomorphic nonlinear crystals CTA:$CsTiOPO_4$ - KTA:$KTiOAsO_4$ - KTP:$KTiOPO_4$

BBO and LBO also have wide transmission ranges especially used in ultraviolet for SHG and in infrared for OPO and OPA, but they are difficult to grow.

$KNbO_3$ orthorhombic structure has the highest d_{31} nonlinear susceptibility of used inorganic crystals for SHG (Table 3). Noncritical SHG produces blue between 430 and 490 nm with Ti:Sapphire tunable laser and THG produces 440 nm from 1340 nm and 670 nm with high damage threshold. The greatest disadvantage is depoling by inversion of the ferroelectric domains. The consequence is the use of a temperature controller and the application of a continuous electric field applied during growth processes. The difficulty to obtain high crystalline quality is associated with structural phase transitions in perovskites as it is illustrated in Fig. 14 for $KNbO_3$. Fig. 15 shows, as an example, $KNbO_3$ recent potentiality of intracavity, frequency-doubled miniaturized blue laser at 465 nm by using 930 nm laser emission of $^4F_{3/2} \rightarrow {}^4I_{9/2}$ Nd^{3+} ion in $YAlO_3$ perovskite. We can say it is not only all-solid-state laser source but almost all-perovskite laser source [21].

New ultraviolet nonlinear crystals are emerging : CLBO, GdCOB and YCOB. $CsLiB_6O_{10}$ (CLBO) transparent till 180 nm in ultraviolet but only 2750 nm in infrared can melt congruently at 848°C and can be grown in a large size (14x11x11 cm^3 in 21 days) by the top seeded solution growth method. This can generate 4th (266 nm) and even 5th (213 nm) harmonics of the Nd:YAG [22].

$YCa_4O(BO_3)_3$ (YCOB) [23] and $GdCa_4O(BO_3)_3$ (GdCOB) [24] monoclinic biaxial crystals, recently discovered are excellent candidates for SHG of the Nd:YAG laser output. The main reasons are that they exhibit high chemical stability, good mechanical properties and, in addition, they are grown by usual Czochralski method. The Sellmeier equations

TABLE 3. Nonlinear Optical Susceptibilities d_{ij} (10^{-12} m/V)

Crystal	d_{31}	d_{33}	d_{32}
$LiNbO_3$	5.8	28-41	
$LiTaO_3$	1.6	34	
$KNbO_3$ (KN)	11.3	23.4	12.8
$KTiOPO_4$ (KTP)	6.5	14-18	5
$BaTiO_3$	18	6,7	
$Ba_2NaNb_5O_{15}$(BNN)	14.6	20	
$K_3Li_{2-x}Nb_{5+x}O_{15+2x}$(KLN)	14		

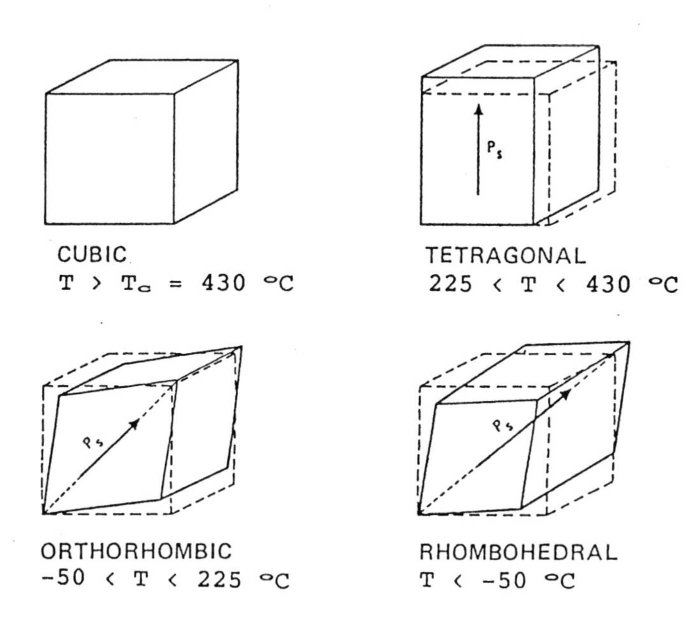

Figure 14. Schematic representation of phase transitions in $KNbO_3$ as a function of temperature

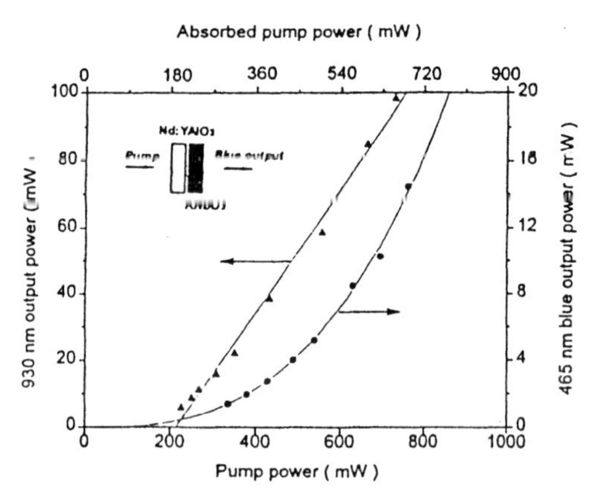

Figure 15 Laser performance of the intracavity frequency-doubled miniaturized $Nd:YAlO_3$ (1.2 mm)/$KNbO_3$ (1.3 mm) blue emission at 465 nm

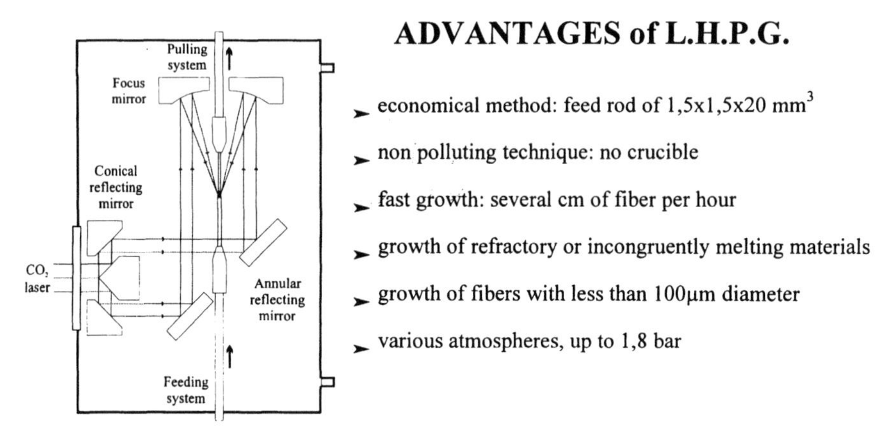

Figure 16 The schematic set-up of LHPG technique used to grow fiber crystals for nonlinear optical crystal and laser crystal applications

however are in favor of YCOB predicting the limits of the phase-matching wavelength to be 840 nm for GdCOB and 720 nm in YCOB respectively. Moreover, the Third Harmonic Generation (THG) of the Nd:YAG laser is possible with YCOB whereas not in GdCOB due to the larger birefringence of YCOB (Δn = O.O41) in comparison with that of GdCOB (Δn = 0.033) at 1064 nm.

The example of our scientific activity in the $\chi^{(2)}$ nonlinear crystal family is described in reference [25]. We are involved in the search of new systems based upon undoped and doped niobate bulky crystal family grown either by Czochralski technique ($LiNbO_3$, $LiTaO_3$) [26] or by the Laser Heated Pedestal Growth (LHPG) technique ($LiNbO_3$, $Ba_2NaNb_5O_{15}$, $K_3Li_{2-x}Nb_{5+x}O_{15+2x}$) [27] which is very useful for any scientific program on optical materials as it is explained in Fig. 16.

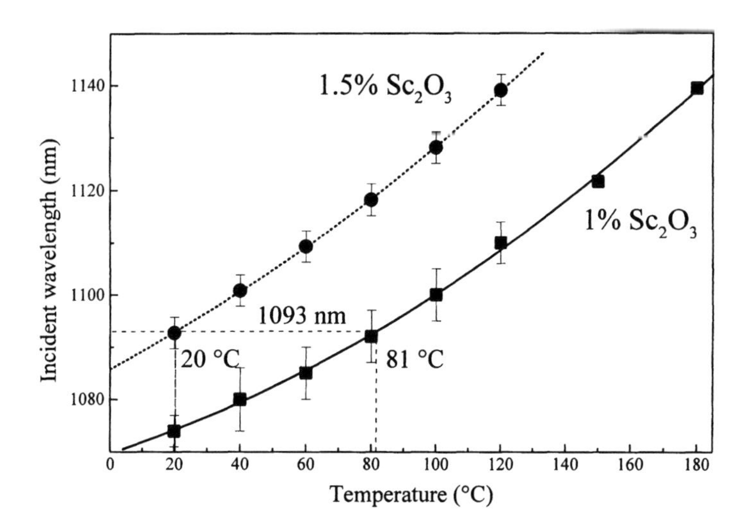

Figure 17. Evolution of phase-matching optimal wavelengths with temperature for 1%Sc_2O_3 and 1.5%Sc_2O_3:1%Nd:$LiNbO_3$ crystal fibers

We have completed previous results and also given new data relevant for laser applications of Nd^{3+} -doped $LiNbO_3$, MgO:$LiNbO_3$, Sc_2O_3:$LiNbO_3$ and $LiTaO_3$ crystal hosts such as polarized ground state and excited-state absorption spectra, polarized fluorescence spectra, in both infrared and self-frequency doubling domains [26]. For self-frequency doubling of the 1093 nm, σ-polarized laser radiation of Nd^{3+} in a 1.5% Sc_2O_3:1% Nd_2O_3:$LiNbO_3$, it is interesting to mention in Fig. 17 the possibility to work at room temperature avoiding the usual high temperature domain with $LiNbO_3$ crystals [28].

Among promising $\chi^{(2)}$ -type nonlinear crystals, we have choosen to grow by LHPG the stable tungsten bronze-type structure such as $Ba_2NaNb_5O_{15}$(BNN), to get green emission, and $K_3Li_{2-x}Nb_{5+x}O_{15+2x}$(KLN,0.15<x<0.5), to get blue emission, which are difficult to grow by Czochralski technique without cracks caused by a strong lattice change occuring during cooling of the crystals. In addition, microwins in BNN crystals and compositional inhomogeneities in KLN ones are often encountered. The obtention of good quality and crackless undoped and new uniaxial tetragonal symmetry of BNN found above 3at% Nd^{3+} (Fig. 18) is highly promising to renew this old nonlinear material whose properties are much improved through the orthorhombic-tetragonal structural transition, avoiding the microwinning, a and b cell parameters being similar in the tetragonal phase.

III. D. Quasi-Phase Matching in $\chi^{(2)}$ -Type Nonlinear Crystals

In contrast to birefringence phase matching, which is dependent on the dispersion of the crystal being used, quasi-phase matching (QPM) uses a periodic grating of ferroelectric

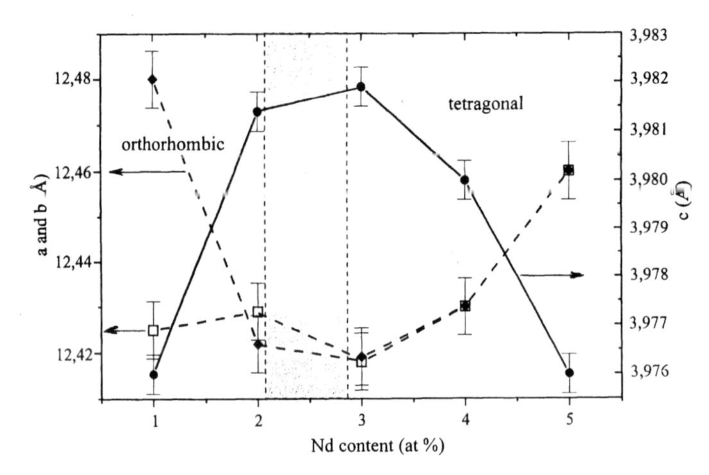

Figure 18. Evolution of a, b and c cell parameters of $Ba_2NaNb_5O_{15}$ single-crystal fibers as a function of Nd^{3+} rare-earth concentration

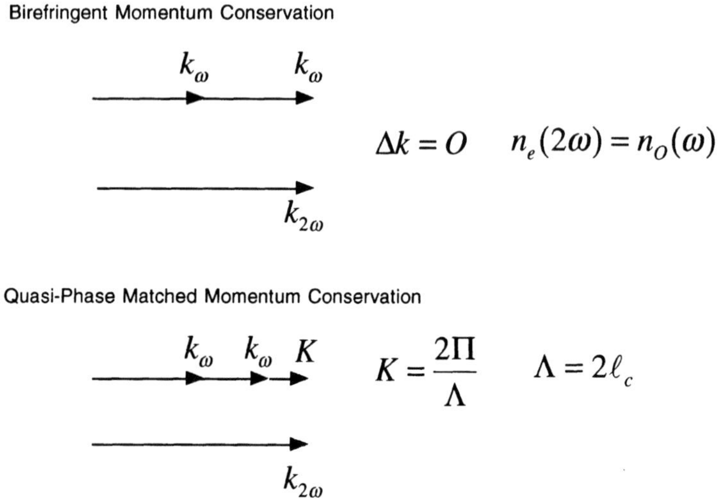

Birefringent Momentum Conservation

k_ω k_ω

$$\Delta k = O \qquad n_e(2\omega) = n_O(\omega)$$

$k_{2\omega}$

Quasi-Phase Matched Momentum Conservation

k_ω k_ω K

$$K = \frac{2\Pi}{\Lambda} \qquad \Lambda = 2\ell_c$$

$k_{2\omega}$

Figure 19. Illustration of both birefringence phase patching and quasi-phase matching from the simple model of the momentum conservation

domains to achieve phase-matching [29]. The approaches are different in the sense that the first one is only depending on the crystallographic properties of the material whereas the second one is depending on the engineering of the material. A simple illustration of this comparison is seen in Fig. 19 by the momentum conservation. Quasi-phase matching is a function of the momentum $K = 2\Pi/\Lambda$ that is to say of the spatial frequency $\Lambda = 2\ell_c$ shown in Fig. 20 whose the calculation yields a few μm. The advantage of this system over birefringence phase matching, in addition of the choice of Λ to match the momentum

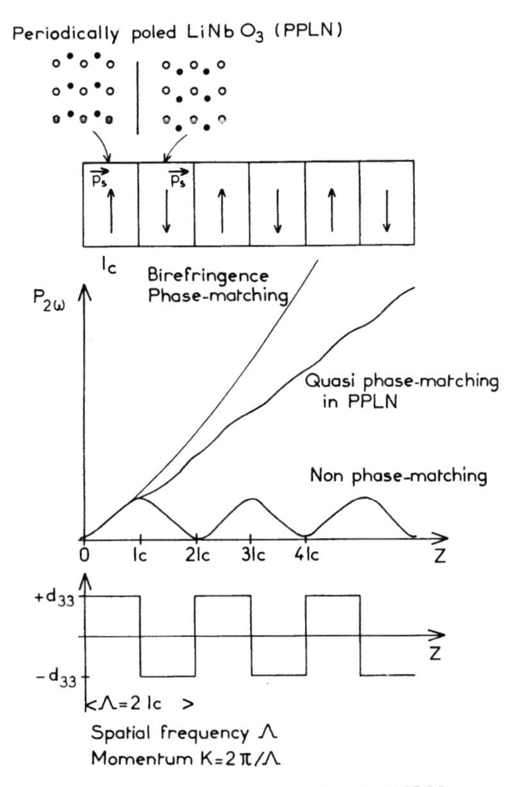

Figure 20. Quasi-phase matching in PPLN

conservation, is that d_{33} dielectric susceptibility parameter is occuring instead of d_{31}, increasing the effective gain of roughly 20 (see Table 3). Finally, in contrast of birefringence phase matching, noncritical phase matching is always possible with quasi-phase matching for which there is no birefringence-induced walk-off of beams, both input and output rays being polarized in the extraordinary direction. The challenge is thus to make a periodic grating of d_{33} parameters in the scale of a few μm.

Recently, an electric-field poling technique has been used to reproducibly manufacture periodically poled $LiNbO_3$(PPLN) as represented in Fig. 20. OPO and SHG operations have been reported [30-31] and one of the most significant result is that more than 2.7 W of green output was generated at a conversion efficiency exceeding 42% when the output of a diode-pumped Nd:YAG laser was focused into a 52 mm long PPLN [32]. OPO has worked between 1.5 and 4.5 μm. Tunability is achieved either by producing multiple gratings with different Λ or by changing temperature of PPLN. The best systems are produced by subjecting a lithographically patterned wafer of optical quality $LiNbO_3$ to a high-voltage pulse. We also have described attempts to obtain periodic inversion of ferroelectric domains in $LiNbO_3$:1%MgO single crystal fibers during growth by LHPG technique [33]. In those first trials, the obtained domains (ℓ_c = 3.4 μm) can be clearly seen but do not lead to a single quasi-phase matching condition for SHG around 1060 nm.

III. E. $\chi^{(3)}$ -Type Nonlinear Crystals

This is a very interesting area since the first demonstration of a $Ba(NO_3)_2$ crystal Raman shifter in the first Stokes line pumped by 1320 nm Nd:YAG laser [34-35]. An intracavity Raman shifter produces 1535-1556 nm laser output in the eye-safe spectral range. In these systems, the stimulated Raman scattering is used with high third order dielectric susceptibilities [18]. The output intensity $I(\ell)$ is a function of the input intensity I(o) as follows :

$$I(\ell) = I_{(o)}\exp[G.I(\ell).\ell]$$

ℓ : length of the crystal

G : gain varies as the inverse of the half-bandwidth $\Delta\omega_R$ of the Raman line at the frequency ω_R

$$\omega_{output} = \omega_{input} - \omega_R$$

In $Ba(NO_3)_2$:

$\omega_R = 1047.3\ cm^{-1}$

$\Delta\omega_R = 0.6\ cm^{-1}$

G = 11 cm/GW

ω_R corresponds to A_g mode of the breathing oscillation of the 3 Oxigen anions in $(NO_3)^{-1}$ [36]. Up to now, a few crystals have been studied to be applied. Among them, double tungstate $KY(WO_4)_2$ and $KGd(WO_4)_2$ and double molybdate $KGd(MoO_4)_2$ have been selected. Their Raman active modes are 901 (Fig. 21) and 960 cm^{-1} respectively, they have good chemical and mechanical properties but crystal growth is difficult. In addition, they can be Nd^{3+} -doped as explained previously in this article, giving the potentiality to build up self-Raman shifters [2]. Further investigations on such crystals are in progress to know their capabilities to fulfill the requirements of new laser sources [37-38-39].

CONCLUSION

Research in the field of laser crystals is very active due to the increasing of applications in many areas. Promising new systems are under development depending both of the nature of the optical pumping given by flashlamps, lasers or diode lasers and also of the performance specifications like ultrashort pulses or continuous waves from ultraviolet to

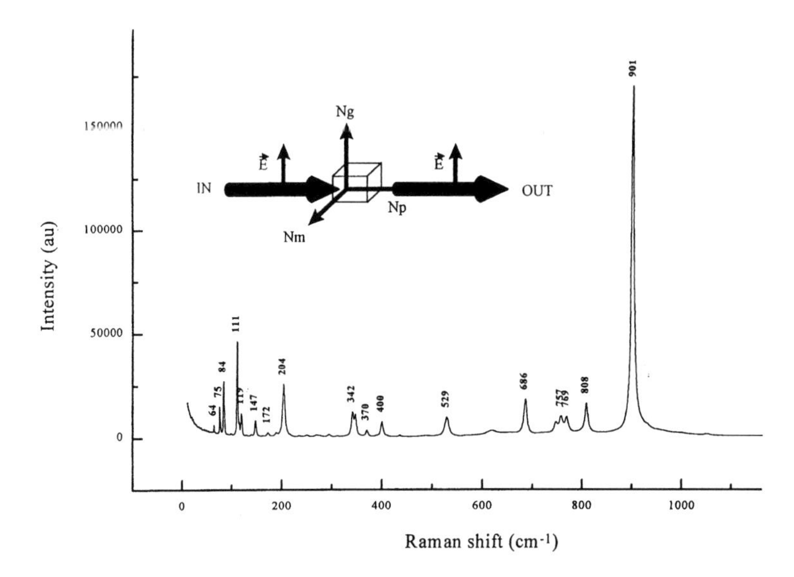

Figure 21. Raman spectrum of $KGd(WO_4)_2$ crystal at room temperature under Ar^+ -ion laser excitation at 514.5 nm. The highest intensity line is occuring at 901 cm^{-1} for the configuration shown by the scheme. Under other polarization states, several high intensity Raman lines are observed

infrared with a good efficiency. The renewal is spectacular with diode-pumped rare-earth lasers in the approach of all-solid-state laser sources. Especially upconversion schemes are very attractive candidates for the future of these laser crystals.

The understanding of nonlinear optical crystals is not so advanced as laser crystals and the requirements of $\chi^{(2)}$ and $\chi^{(3)}$ susceptibility materials are much more difficult to fulfill. Such materials contribute to all-solid state laser sources not only with usual bulky shapes but also with the optical engineering approach like in periodically poled nonlinear crystals and, finally, with multifunctional crystals as, for example, self-frequency doubling crystals or self-stimulated Raman-scattering crystalline lasers.

The general advance of both laser crystals and nonlinear crystals justifies complementary scientific collaborations and illustrates the necessity to continue and to improve crystal growth techniques, crystal qualities, spectroscopic properties, nonlinear optical properties and lasing experiments.

References

[1] T.H. Maiman, Nature 187, 493 (1960)

[2] A. Kaminskii, Crystalline Lasers : physical processes and operating schemes. CRC press (Laser and Optical Sciences and Techology Series), Edition Marvin Weber (1996)

[3] Fuxi Gang, Laser Materials, World Scientific Publishing Co, London (1995)

[4] S.A. Payne and W.F. Krupke, A glimpse into the laser crystal ball, Optics and Photonics News (OSA) 31 (August 1996).

[5] G. Boulon and R. Reisfeld, French-Israeli Workshop on Apatites and Lasers (Jerusalem - November 25-26, 1996). A special review in Optical Materials 8, n°1/2, July 1997

[6] G. Boulon « Matériaux pour lasers à solide » (in French), Les lasers et leurs applications scientifiques et médicales, Les Editions de Physique. Editeurs : C. Fabre, J.C. Pocholle, p. 259-285 (1996)

[7] Proceedings of International Symposium on Laser and Nonlinear Opticla Materials (Singapore, November 3-5, 1997). Edited by Takatomo Sasaki (Dept. of Electrical Engineering, Osaka University, Japan)

[8] R. Moncorgé et al. see ref [5], Optical Materials 8, 109 (1997)

[9] M. Oka and S. Kubota CLEO 1995, Washington D.C., Technical Digest. CTuMI, OSA

[10] Y. Kaneda et al. CLEO 1995, Washington D.C., Technical Digest. CPD22, OSA

[11] C. Marshall, S. Payne, J. Speth, J. Tassano, W. Krupke, G. Quarles, V. Castillo, B. Chai. SPIE Visible and UV lasers 2115, 7 (1994)

[12] S.A. Payne et al. IEEE J. Quant. Electron. 24, 2243 (1988)

[13] V. Petricevic, S. Gayen and R. Alfano, OSA Proc. on Tunable Solid State Lasers 5, 77 (1989)

[14] R. Moncorgé, H. Manaa and G. Boulon, Optical Materials 4, 139 (1994)

[15] L. Deloach, R. Page, G. Wilke, S. Payne and W. Krupke, OSA Proc. On Advanced Solid State Lasers 24, 127-131 (1995)

[16] P.A. Franken et al. Phys. Rev. Lett. 7, 118 (1961)

[17] A. Yariv, Optical waves in crystals, John Wiley and Sons (1984)

[18] J.Y. Courtois, Optique non linéaire (in French), see ref [6], p. 87-172 (1996)

[19] J. Bierlein and V. Vanherzeele, J. Opt. Soc. Am. B6 (4), 622 (1989)

[20] J. Bierlein et al., Appl. Phys. Lett. 54 (9), 783 (1989)

[21] J.H. Zarrabi et al., Appl.Phys. Lett. 67, 2439 (1995)

[22] T. Sasaki I., Kuroda S., Proc. of Advanced Solid-State Lasers 24, 91 (1995) OSA

[23] M. Iwai, T. Kobayashi, H. Furuya, Y. Mori and T. Sasaki, Jpn. J. Appl. Phys. 36, 276 (1997)

[24] G. Aka, A. Kahn-Harari, D. Vivien, J.M. Benitez, F. Salin and J. Godard, Eur. J. of Solid State Inorg. Chem. 33, 727 (1996)

[25] G. Foulon, M. Ferriol, A. Brenier, M.T. Cohen-Adad, M. Boudeulle, G. Boulon, Optical Materials 8, 65-74 (1997)

[26] R. Burlot, R. Moncorgé, H. Manaa, G. Boulon, Y. Guyot, J. Garcia-Solé, D. Cochet-Muchy, Optical Materials 6, 313-330 (1996)

[27] G. Foulon, M. Ferriol, M.T. Cohen-Adad, G. Boulon, Chem. Phys. Lett. 245, 55-58 (1995)

[28] R. Burlot, R.Moncorgé and G. Boulon, J. of Luminescence 72-74, 812-815 (1997)

[29] J. Armstrong, N. Bloemberger, J. Ducuing and P. Pershan, Phys. Rev. 127, 1918 (1962)

[30] M. Yamada, N. Nada, M. Saitoh and K. Watanabe, Appl. Phys. Lett. 62, 435 (1993)

[31] L. Myers et al, J. Opt. Soc. Am. B12, 2101 (1995)

[32] G. Miller et al, OSA Advanced Solid State Laser (1997)

[33] A. Brenier, G. Foulon, M. Ferriol and G. Boulon, J. Phys. D : Appl. Phys. 30, L37 (1997)

[34] P. Zverev, J. Murray, R. Powell, R. Reeves, R. Basiev, Opt. Commun. 97, 59 (1993)

[35] J. Murray, R. Powell, N. Peyghambarian, D. Smith, W. Austin and R. Stozenberger, Opt. Lett. 20, 1017 (1995)

[36] J. Basiev, P. Zverev, Sov. J. Quant. Electr. 17, 1560 (1987)

[37] A. Métrat, N. Muhlstein, A. Brenier and G. Boulon, Optical Materials 8, 75-82 (1997)

[38] F. Bourgeois, short seminar in this School.

[39] Y. Terada, Kiyoshi Shimamura, T. Fukuda, Y. Urata, H. Kan, A. Brenier and G. Boulon, OSA Advanced Solid State Laser 10, 458-461 (1997)

NON - PERTURBATIVE UP - CONVERSION TECHNIQUES: ULTRAFAST MEETS X-RAYS

Chris B. Schaffer

Harvard University
Department of Physics
Cambridge, MA 02138
schaffer@physics.harvard.edu

ABSTRACT

Femtosecond laser pulses are now available from the ultraviolet (200 nm) all the way to the mid-infrared (12 μm) in compact (often commercially available) systems. This tunability is achieved through "perturbative" nonlinear optical wavelength conversion techniques in crystals. To reach further into the ultraviolet and X–ray regions of the spectrum, a new set of techniques becomes necessary. In this paper, we will review some of these "non–perturbative" nonlinear optical methods. Specifically, we will consider the high harmonic generation process in detail, and go through the essentials of the semi-classical theory. Next, we will review a new technique, based on Thomson scattering, which has produced 0.4-Angstrom, 300-fs radiation. Finally we will consider means of measuring femtosecond pulses in this short wavelength regime.

INTRODUCTION

Recent progress in ultrafast technology has led to a wide range of available wavelengths with sub–50-fs laser pulses. The Titanium: Sapphire laser oscillator provides a stable, efficient source for generating 800-nm radiation which is used as a seed pulse for amplification and wavelength conversion. [1] Chirped pulse amplification techniques allow amplification up to 1 mJ per pulse at 1 kHz with approximately 25-fs pulse widths. [2] Optical parametric amplifiers allow wavelength conversion of the amplifier output into the near IR, with subsequent nonlinear conversion (either sum or difference frequency generation) allowing access to the near ultraviolet, the entire visible, and the near to mid infrared regions of the spectrum (200 nm – 12 μm). [3] All these frequency conversion techniques are based upon "perturbative" nonlinear optical techniques in nonlinear crystals. These techniques have been established well enough that they have been commercialized. "Turn -

key" systems capable of accessing the entire visible and near infrared regions of the spectrum are now available. [4] Given crystals with appropriate transparency regions, the ability to phase match, and high enough damage thresholds (so higher intensities can be realized in the crystal) one could go further into the ultraviolet and infrared with these techniques. However, crystals transparent to wavelengths shorter than a couple hundred nanometers or longer than a few tens of microns are difficult to find. On the infrared side, radiation out to beyond 10 μm has been generated in $AgGaSe_2$. [5] Fourth harmonic generation is essentially the limit of crystal based nonlinear ultraviolet generation techniques. [6] For the shortest pulses, it is problematic using crystals even for third harmonic generation due to finite phase matching bandwidths and temporal walkoffs associated with dispersive crystals. [7]

Conspicuously lacking is any convenient source in the extreme ultraviolet or X–ray regions of the spectrum. A short pulse X–ray source would be of great interest because it would allow for time-resolved diffraction experiments. Recent experiments on GaAs indicate that for 100 fs excitation pulses with fluences near the damage threshold, there is a rapid (500 fs) semiconductor to metal phase transition. Excitation of a large percentage (10%) of the valence electrons to the conduction band leads to a destabilization of the bonding in the lattice and finally a bandgap collapse and thus metallic properties. [8] These conclusions were drawn from measurements of the dielectric function from 1.5 to 3 eV by broadband pump probe techniques over a wide range of fluences. A sub–Angstrom, femtosecond X–ray source would make a definitive proof of the result possible by allowing one to measure the rapid loss of long range order due to the lattice destabilization by watching diffraction peaks disappear as a function of time. An extreme ultraviolet light source would also find numerous applications. For X–ray biological imaging, the spectral region between 23 and 44 Angstroms is very interesting. This is the so-called "water window," where absorption by water (oxygen) is small compared to that by carbon, allowing for significant transmission of the radiation into tissues, and differentiation of tissue types by their carbon absorption signatures. [9] – [11]

This paper will discuss a few techniques for converting ultrafast laser pulses into the extreme ultraviolet and X–ray regions of the spectrum. First we will consider in detail the high harmonic generation process, and discuss the semi-classical model developed by Corkum. [12] Next, a technique developed recently based on 90 degree Thomson scattering of a femtosecond laser pulse off relativistic electrons will be discussed. [13] Finally, we will discuss a technique for measuring pulse widths in this spectral range based on the absorption or emission of infrared photons by X–ray photoionized electrons. [14]

HIGH HARMONIC GENERATION

I. Experimental Overview

High harmonic generation is observed when high intensity laser light is focused into a gas jet as shown in Fig. 1a. [15] Most state-of-the-art setups for high harmonic generation experiments are based on Titanium: Sapphire laser systems. A self mode locked Titanium: Sapphire laser oscillator produces pulses which are temporally stretched in a grating stretcher to avoid optical damage and undesirable nonlinear effects (self focusing, self phase modulation, etc.) in the amplifier chain. The pulses are subsequently amplified in a Titanium: Sapphire regenerative or linear amplifier. [8] The output is then temporally compressed with a grating compressor. Typical laser intensities are on the order of 10^{14} – 10^{16} W/cm^2. [16] The beam is linearly polarized and is slowly focused into the gas jet target, contained in a high vacuum chamber. [17] The gas is typically He or Ne, sometimes Ar or Xe. [15] The harmonics are generated in the forward direction. A spectral filter blocks the

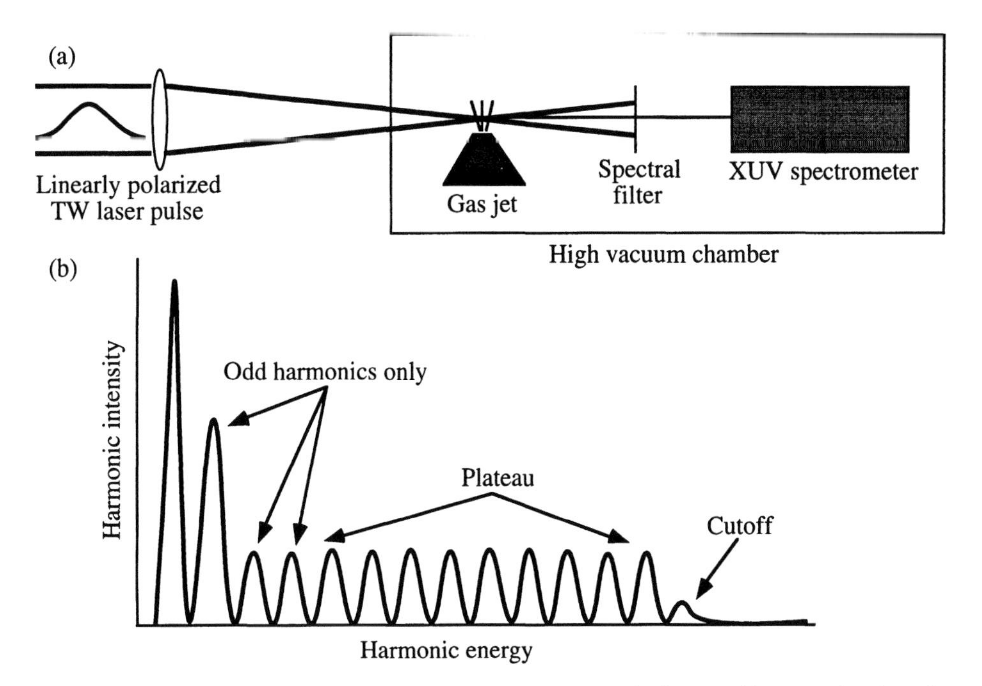

Fig. 1. (a) High harmonic experimental setup. A Terawatt (TW) laser pulse is focused into a gas jet where the harmonics are produced. A filter blocks the pump laser and the harmonic spectrum is analyzed in an extreme ultraviolet spectrometer. (b) Typical high harmonic spectrum. Only odd harmonics are produced. The intensity of the harmonics falls off over the first few orders, then remains constant through the plateau region out to the cutoff frequency.

unconverted laser light and transmits the harmonics which are analyzed with an extreme ultraviolet spectrometer. Spectra are recorded as a function of pulse energy, gas pressure, gas species, pulse width, etc.

Original experiments were preformed with ultraviolet laser pulses (248 nm) with typical durations of 1ps. [15] With these pulse widths, harmonics were observed out to about the 17th order ($\omega_{harmonic} = 17\omega_{laser}$), as well as fluorescence from excited electronic states of the gas atoms and their ions. [15] It was observed that above about the 7th harmonic, all harmonics were produced with approximately the same efficiency out to some maximum, cutoff harmonic. [18] This is in stark contrast to standard non-linear optics where higher order processes are less important than lower order ones — this is what is meant by the "perturbative" regime. The process cannot be a standard sum frequency process, because from about 7th order to 17th there is no change in the efficiency. In recent experiments, with 25-fs or shorter, 800-nm laser pulses from Ti:Sapphire lasers, harmonics are produced up to the 155th order in Ne and 221st order in He with near equal efficiency to the first few. [10], [11], [17] Clearly we are far outside the regime of perturbation theory, where each additional process we include makes an ever smaller contribution to the overall picture. [19] Figure 1b shows a schematic of a typical high harmonic spectrum illustrating some of their universal characteristics. There is a sharp drop-off in the intensity of the harmonics over the first few orders followed by a "plateau" where all harmonics are of approximately the same intensity out to some "cutoff" harmonic where the intensity once again falls rapidly. [15] – [18] Only odd harmonics are produced. [15] The energy of the highest harmonic produced is empirically found to obey the scaling relation:

$$E_{max} = I_p + 3U \quad (1)$$

where I_p is the ionization potential of the atom and

$$U = (e^2 A^2) / (4m\omega^2) \quad (2)$$

is the pondermotive or quiver energy of an electron of mass m and charge ($-e$) in a field of amplitude A and frequency ω. [16] The pondermotive energy is the average kinetic energy of a charge oscillating with an electromagnetic field.

II. Semiclassical Theory

The features of high harmonic spectra can be understood in terms of a semi-classical model of the single atom response to intense radiation. This model was first developed by Paul Corkum in the early 1990's. [12] Consider an isolated hydrogenic atom (Fig. 2a). Now apply a strong electric field and consider the potential the electron feels. The electric field "tilts" the atomic potential allowing for unbound states at the same energy as the bound states (Fig. 2b). In the bound state the electron is localized near the nucleus, while in the unbound state the electron would accelerate away from the nucleus along the electric field. There is a barrier between the bound and unbound states. The electron is initially in one of the bound states of the atomic potential, but there is a finite probability that it will tunnel through the barrier to an unbound state on the other side (Fig. 2c). This tunneling probability was first calculated by Keldysh in 1964. [20]

For oscillatory electric fields (such as our laser field) with a frequency smaller than the electron orbital frequency, we can have tunneling ionization just as in the static case. High harmonic generation results from the stimulated recombination of electrons that have undergone tunneling ionization in the laser field. The process is illustrated schematically in Fig. 3. An electron which is ionized by the laser field is accelerated away from its parent ion by the field. A short time later, the field reverses and the electron is accelerated in the other direction. If the laser is linearly polarized, then the electron can return to the atom it was

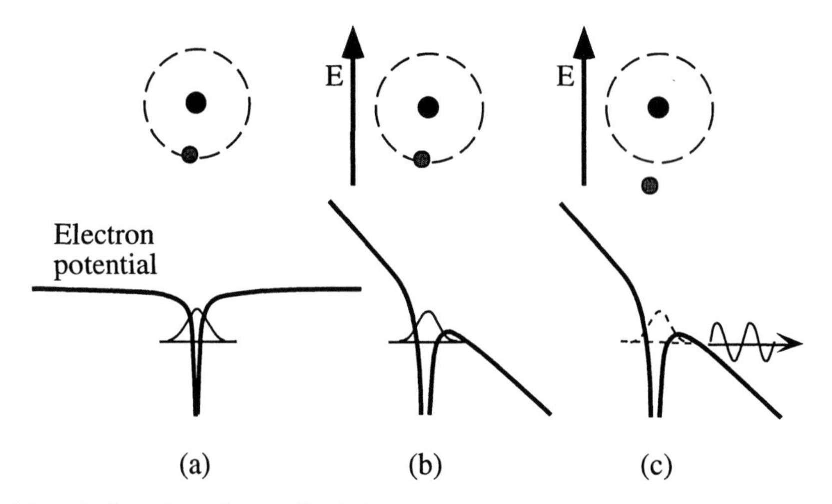

Fig. 2. Schematic illustration of the tunnelling ionization process. (a) An atom has an electron in a bound state. (b) A strong electric field (E) is applied which "tilts" the potential. (c) The electron can then tunnel to the unbound states on the other side of the barrier formed by the atomic and electric field potentials.

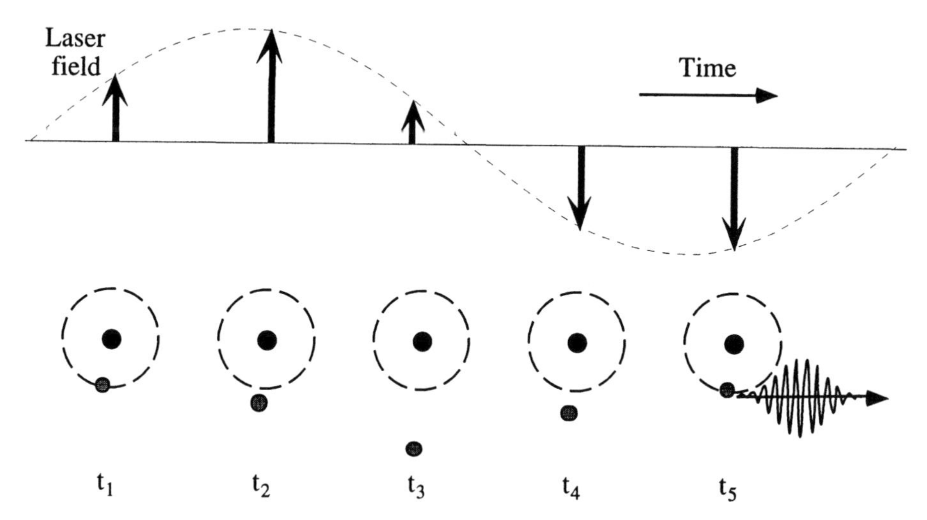

Fig. 3. Schematic of the quasi-classical picture of high harmonic generation. An atom is exposed to a strong oscillatory laser field (t_1). The electron is ionized by tunneling ionization (t_2) and is accelerated away from its parent ion by the laser field (t_3). Later the laser field reverses direction and accelerates the electron back toward its parent ion (t_4). There is a probability that the electron will recombine with the ion and release radiation with energy equal to the ionization potential plus the kinetic energy the electron gained from the laser field (t_5).

ionized from. There is a (small) probability that the electron will then radiatively recombine with its parent ion. [12], [21] The energy of the photons emitted in the radiative recombination will be equal to the ionization potential of the atom plus the kinetic energy gained by the electron in the laser field. [12]

It should be noted that this is not simply ionization and recombination of the electron. This process is entirely laser driven — the laser ionizes the atom, then brings the electron back for recombination within the same cycle of the laser field, it is a coherent process. We can now explain the qualitative features of high harmonic spectra; namely the odd harmonic only emission, the plateau, and the cutoff. Because the atom ionization and the electron recombination are driven by the laser and occur at the laser frequency, the emission upon recombination must be at a harmonic of the laser frequency. Emission occurs every half-cycle of the laser pulse, and any radiation that is not at a harmonic of the laser frequency will be out of phase with the emission from other half-cycles of the pulse and will destructively interfere. Furthermore, the even harmonics will also be out of phase from one half-cycle to the next (and since the gas has inversion symmetry they will also be of the same amplitude from one half-cycle to the next) and will also cancel. [15] We are left with only odd harmonics of the laser frequency in the emission — these interfere constructively.

The electrons can tunnel from the bound state to the unbound state at different times during the driving laser's oscillation period. It is this tunneling phase that determines the kinetic energy that the electron returns to its parent ion with. Figure 4b shows the electron distance from the nucleus as a function of time for four different tunneling phases. The slope of the distance vs. time curve at the electron's reencounter with the ion determines the kinetic energy. The largest kinetic energy an electron can gain determines the cutoff harmonic. [22] The plateau results from having a similar number of electrons tunnel at different phases of the drive laser, giving similar numbers of electrons recombining with different kinetic energies. To understand these last two properties properly requires some calculation.

The calculation we present is similar to, but simpler than, the calculation by Corkum

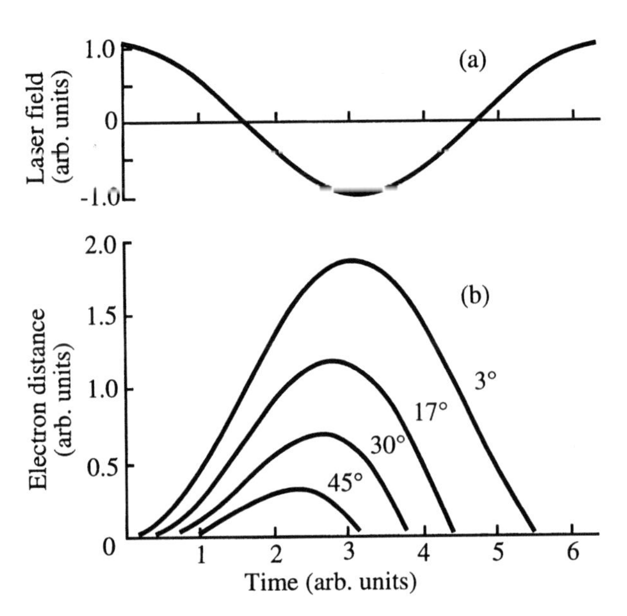

Fig. 4. (a) One period of the drive laser field. (b) Plots showing the distance the electron is from the nucleus as a function of time for different tunneling phases. The slope of the curve when it comes back to zero (back to its parent ion) is directly related to the radiated harmonic energy. The slope is steepest (highest energy harmonic) for a tunneling phase of 17°.

which first explained high harmonic generation. Consider a hydrogen atom in a low frequency laser field. We calculate the tunneling rate of electrons from the Coulombic well to the free states outside in the static limit. This rate will then be used to calculate the probability of tunneling as a function of time during a half-cycle of our laser field. We then treat the electron in the field classically, and giving the time of tunneling, compute the kinetic energy the electron acquires on its path away from, then back to its parent ion. We then know the probability of ionization vs. the kinetic energy gained by the electron. This is a long way from the induced dipole as a function of energy, but this simple calculation seems to capture the essentials.

We use the WKB approximation to calculate the tunneling rate. The WKB wavefunctions are given by

$$\Psi = \Psi_0 \frac{1}{\sqrt{p(x)}} \exp\left[\pm \frac{i}{\hbar} \int_{x_0}^{x} p(x)\, dx\right], \tag{3}$$

where $p(x)$ is the momentum of the particle. [21] Now, consider a hydrogen atom initially in the ground state in the presence of an electric field along the z-axis. The resulting potential is shown in Fig. 2c. We wish to calculate the tunneling probability, P, which is the square of the ratio of the wavefunctions just outside and just inside the potential barrier. We use WKB wavefunctions everywhere, and evaluate their amplitudes at the classical turning points for motion of an electron in the potential. The potential for our problem is:

$$V(r) = -\frac{e^2}{r} - eAr, \tag{4}$$

where A is the electric field amplitude. The energy of the ground state of hydrogen in the absence of the applied field is given by

$$E = \frac{-me^4}{2\hbar^2}. \quad (5)$$

The inner classical turning point is simply taken to be the Bohr radius, a_0, here we neglect the potential of the electric field since r is small. For the other turning point, b, we neglect the Coulomb potential (here r is large) to get

$$b = \frac{me^3}{2A\hbar^2}. \quad (6)$$

Recalling the expression for the momentum of a particle,

$$p = \sqrt{2m(E-V)}, \quad (7)$$

we can write the tunneling probability as

$$P = \left|\frac{\Psi(b)}{\Psi(a_0)}\right|^2 = \left[\frac{m^2e^5}{4A\hbar^4}+\frac{1}{2}\right]\exp\left[\frac{-2\sqrt{2m}}{\hbar}\int_{a_0}^{b}\left[\frac{e^2}{r}+eAr+\frac{-me^4}{2\hbar^2}\right]^{1/2}dr\right]. \quad (8)$$

Now the electron (classically) "collides" with the barrier with a frequency given by

$$f = \frac{v}{2a_0} = \frac{me^4}{\hbar^3}. \quad (9)$$

The tunneling rate is just the product of the tunneling probability and the collision frequency. After evaluating the integral and simplifying we find (the negligible 1/2 in the prefactor was dropped)

$$W = \frac{fC}{4A}\exp\left[-\frac{2C}{3A}\right], \quad (10)$$

where

$$C = (m^2e^5)/\hbar^4 \quad (11)$$

is the strength of the Coulombic electric field evaluated at the Bohr radius. We take the applied field, A, to be oscillatory, giving the time dependent ionization probability due to the laser pulse.

Now consider the motion of the ionized electron in the linearly polarized electric field. For a field

$$\boldsymbol{A}(t) = A_0\cos(\omega t)\hat{z} \quad (12)$$

the velocity and position of the electron (treated classically) are

$$\boldsymbol{v} = v_0[\sin(\omega t) - \sin(\omega t')]\hat{z} \quad (13)$$

$$\boldsymbol{x} = \{x_0[\cos(\omega t') - \cos(\omega t)] + v_0\sin(\omega t')[t'-t]\}\hat{z}, \quad (14)$$

where

$$v_0 = -eA_0/m\omega \quad (15)$$

$$x_0 = -eA_0/m\omega^2 \quad (16)$$

and t' is the time the electron was "born" into the field, i.e. the tunneling time.

For harmonic generation, the electron must accelerate away from the ion until the field reverses direction, then accelerate back. The energy which goes into the harmonic radiation will be the kinetic energy of the electron at the time that it returns to the nucleus ($x = 0$). This gives us a relationship between the kinetic energy and the phase at which the electron tunnels, $\omega t'$. In Fig. 5a, we show a plot of the energy of the returning electron as a function of the tunneling phase, Fig. 5b shows the tunneling probability as a function of the tunneling phase (static limit). Note that the tunneling probability is nearly constant over a large range of tunneling phases, and therefore over a large range of return kinetic energies of the electron. Putting these two pieces of information together, we can more fully understand the plateau and cutoff phenomena. Over a large range of return kinetic energies the ionization probability is approximately constant. This means there is a nearly uniform distribution of electrons vs. return kinetic energy up to the cutoff energy. Hence, all harmonics emitted are of similar intensity, giving a plateau in the harmonic spectrum. The maximum possible return kinetic energy is 3.17 U, where U is the pondermotive potential introduced before (this maximum occurs at a tunneling phase of 17°). With this energy, the radiated harmonic

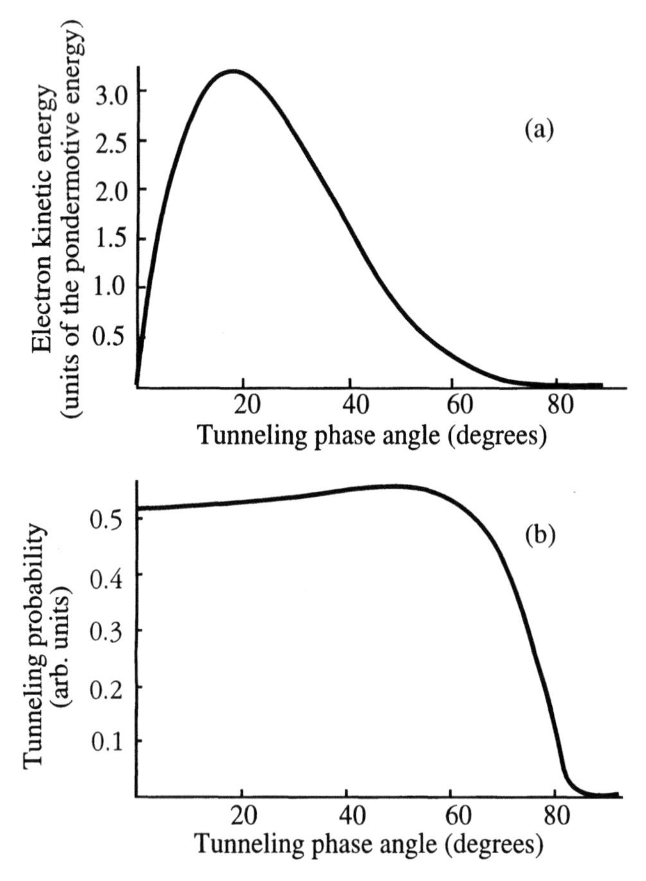

Fig. 5. (a) Kinetic energy of electron when it returns to its parent ion as a function of its tunneling phase. (b) Ionization probability as function of tunneling phase.

would have an energy of 3.17 U plus the ionization potential of the atom (the electron has to hop back into the atomic potential). Thus we have arrived at an understanding of the cutoff law for harmonic generation,

$$E_{max} = I_p + 3.17U. \tag{17}$$

To calculate the emission explicitly, we would need the dipole matrix element between the unbound state and the atomic state the electron recombines into. We have found the number of electrons with a given kinetic energy (and thus leading to a given harmonic) returning to the nucleus, but have not dealt with the recombination probability. The main problem is that there is an energy dependence to the probability of recombination. The electron wavepacket spreads in the directions perpendicular to its trajectory thus decreasing the probability for recombination. The degree of this spreading will depend on how long the electron is in the field before it returns to its parent ion and therefore on the energy. Corkum's theory explicitly takes this into account, with the transverse spreading of the wavefunction taken as a parameter. [12]

Propagation effects are less significant in high harmonic generation than in usual nonlinear optics. [24] The primary reason is that the gas densities usually used are on the order of 10^{18} atoms/cm^3. At these low densities there is effectively no dispersion across a large frequency range. The only phase matching considerations come from a geometrical phase introduced by focusing. [24] As long as the focusing is slow, this too can be neglected, and the process is phase matched for most of the harmonics. [17] Recent experiments indicate that dispersion and geometrical phase errors can become important for the shortest wavelengths (less than 10 nm). [10], [11]

For the case of very short (< 25 fs) driving pulses, some new effects appear. First, the harmonic spectrum extends to beyond the cutoff energy described above. [10], [11], [25] This is because the rise time of the pulse is so short that the electron ends up in a stronger field when it finishes tunneling than the field which induced the tunneling, and the electron can gain a little more energy from the field as a result. The odd harmonic only structure of the spectrum is also distorted, and the harmonics merge into a continuum. [25] This is due mainly to the fact that for short pulses, most of the emission comes from only a few cycles of the driving laser pulse. The emission of only harmonics of the driving field and the cancelation of even harmonic radiation, both of which are due to multiple interactions of atoms with the field, disappear. Observation of continuum radiation has been made only very recently. [10] Calculations suggest that the emission is limited to a single cycle (or a few cycles) because the dipole moment no longer exactly tracks the field for a pulse with a fast rise time. [26] The dipole gets out of phase with the field and emission is suppressed. For a 5-fs pulse, calculations suggest that different harmonic frequency bands will be emitted during different half-cycles of the pulse. [26] Perhaps most interestingly, the temporal coherence properties of the harmonics is vastly improved for short driving pulses. Since the harmonics in a given frequency range are all emitted during a single half-cycle of the pulse, they all have the same phase characteristics. There is no dispersion in this system, so what one has is very broadband, coherent, and phase locked radiation in the ultraviolet. The bandwidth of the shortest wavelength spectral region which is emitted together and the phase properties of the radiation may support pulses as short a 100 attoseconds (0.1 fs). [25] Only very recently has high harmonic radiation been produced with such short driving pulses, and no measurements have been made of the pulsewidth of the shortest wavelength radiation. [9]

90 DEGREE RELATIVISTIC THOMSON SCATTERING

Here we discuss a technique for generating short pulsed, short wavelength radiation which has been demonstrated very recently. [13], [27] The technique is based on Thomson scattering of near–IR photons off relativistic electrons. A schematic of the setup is shown in Fig. 6. A Terawatt laser pulse from a Titanium: Sapphire chirped pulse amplifier is intersected at right angles with a 50-MeV electron beam. The polarization of the laser field is perpendicular to the propagation direction of the electron beam. Photons are scattered in the forward direction at an energy of 30-keV. The duration of the X–ray pulse is determined by the transit time of the 100-fs laser pulse across the 90-μm diameter of the electron beam, giving a 300-fs X–ray pulse width. [13] The 30-keV photons have a wavelength of only 0.4 Angstroms, making them ideal for time resolved diffraction studies. In addition, the X–rays are absolutely synchronized to the pumping laser, thus optical pump, X–ray probe measurements should be possible with temporal resolution limited by only the X–ray pulse width.

The wavelength conversion process can be understood as follows. The electric field of the laser pulse forces the electrons in the electron beam to oscillate back and forth and thus to radiate. In non-relativistic Thomson scattering, the electron radiates at the same frequency it is driven at (we neglect inelastic effects, i.e. Compton recoil). To understand the drastic frequency conversion, one must consider the relativistic Doppler shifts of the driving and radiated fields. The frequency of light observed in a moving frame is given by:

$$\omega' = \gamma\omega\left(1 - \frac{v}{c}\cos\theta\right), \tag{18}$$

where

$$\gamma = 1/\left(\sqrt{1 - v^2/c^2}\right) \tag{19}$$

is the standard factor from special relativity, and θ is the angle of the k–vector of the radiation with respect to the velocity, v, of the moving frame. [28] In the laboratory frame, the infrared pulse has a wavelength λ. In the rest frame of the electron, which is traveling at a velocity, v, perpendicular to the propagation direction of the light, the laser field has a wavelength that is shorter by a factor of γ. The electrons oscillate and radiate at this Doppler

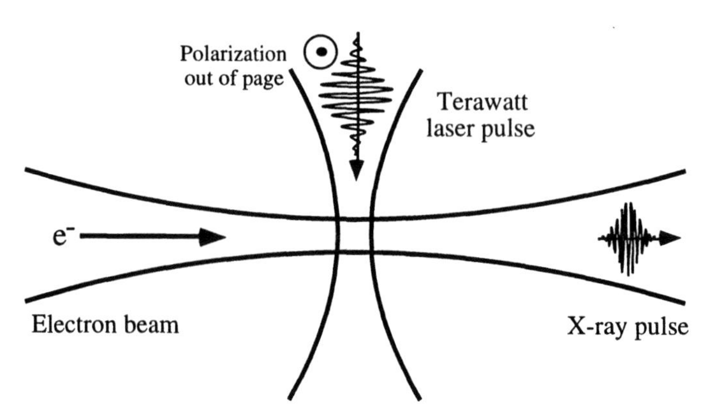

Fig. 6. Schematic of the Thomson scattering setup for generating femtosecond X-ray pulses. Infrared femtosecond laser pulses are scattered at right angles with a 50-MeV electron beam up to X-ray energies. The duration of the scattered X-ray radiation is determined by the transit time of the 100-fs pulse across the 90-μm electron beam focus.

shifted wavelength in the electron rest frame. In the laboratory frame, this radiation is peaked in the direction the electrons are propagating due to relativistic effects (as in a synchrotron). Boosting by velocity, v, in the opposite direction that the radiation is peaked brings us back to the lab frame. This boost Doppler shifts the scattered radiation propagating directly along the electron beam to a wavelength shorter by a factor of 2γ (where we take $v/c \approx 1$ for the 50-MeV electrons). So, the on-axis radiation in the laboratory frame is shorter by a factor of 2γ than the electron rest frame radiation which is in turn shorter than the drive radiation by a factor of γ. Radiation that does not lie exactly along the electron propagation direction ends up at longer wavelengths (because off the electron axis, θ in Eq. 18 is not exactly 180^0). The highest energy scattered radiation produced has a wavelength given by:

$$\lambda_{scatt} = \frac{\lambda}{2\gamma^2} = \frac{800nm}{2\,(98)^2} = 0.4\,Angstroms \qquad (20)$$

where we have used the fact the $\gamma = 98$ for the 50-MeV electron beam.

This incredible source will soon be used for just the kind of time-resolved diffraction experiments mentioned above. [13] Right now the X–ray pulse is not very bright, only about 5×10^4 photons per pulse. [9] There is no reason why the whole device cannot be scaled up to achieve a greater luminosity, however. Improvements will include shorter electron bunches, higher electron currents, and higher power laser fields. [13]

PULSE WIDTH MEASUREMENT TECHNIQUES

No electronics are fast enough to capture the evolution of femtosecond laser pulses. The pulse width can only be measured using cross-correlation techniques — the laser pulse must be used to measure itself. First the laser pulse is split into two pulses and one is delayed with respect to the other by an optical delay line. The two laser pulses are overlapped in space in an interaction region. Next, we find a system to put in the interaction region which has a noticeably different signature when both laser pulses arrive at the same time rather than one after the other, and which has a sufficiently short-lived interaction. These requirements for the interaction of the laser pulse with the system comes down to needing an interaction which depends on the instantaneous laser intensity — a nonlinear interaction. Varying the optical delay line through zero delay and measuring the magnitude of the difference of interaction gives a measure of the pulse width. With infrared and visible laser pulses, one often uses one of the standard perturbative nonlinear optical interactions such as second harmonic generation or sum-frequency generation. These interactions for pulsewidth measurement have all the same difficulties for extreme ultraviolet and X–ray pulses as generating these wavelength with the standard techniques has. There are no crystals transparent to the X–ray radiation that can be used for the nonlinear interaction medium.

To measure the pulse width of extreme ultraviolet and X–ray femtosecond pulses, we need an interaction that can be used at these short wavelengths. There are several techniques most of which involve photoionizing an electron from a noble gas with the extreme ultraviolet pulse and looking at sidebands introduced by infrared pulses. [14], [29], [30] The process could be described as two-color, two-photon absorption in noble gas atoms. A schematic is shown in Fig. 7. The X–ray radiation is incident on a jet of noble gas atoms, and the photoelectrons which are produced are analyzed with a time of flight electron spectrometer. One sees a peak in the electron energy spectrum corresponding to the energy of the X–ray radiation. A remainder of the infrared laser pulse that was used to produce the short wavelength radiation is routed through an optical delay line and overlapped with the X–ray radiation in the gas jet. When the X–ray and infrared pulses arrive at the same time

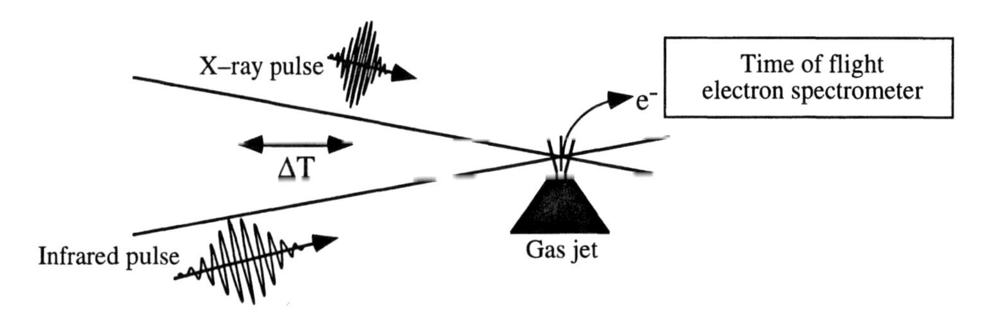

Fig. 7. Schematic of the experimental setup for cross-correlating X–ray and infrared laser pulses. An optical delay, ΔT, is introduced between the X–ray and infrared pulses. A time of flight electron spectrometer measures the energy of the electrons photoionized from the atoms in the gas jet by the X–ray pulse. When the X–ray and infrared pulses overlap in space and time in the gas, the infrared pulses induce sidebands in the photoelectron spectrum whose magnitude can be used as a cross-correlation signal.

in the gas jet, the electron photoionized by the X–ray pulse can absorb an infrared photon or undergo stimulated emission of an infrared photon. This gives sidebands in the photoelectron spectrum, the size of which depends on the overlap in time of the X–ray and infrared pulses. The magnitude of these sidebands can be used as a cross-correlation signal for X–ray pulsewidth measurement if the infrared pulsewidth is known. [14]

CONCLUSIONS

We have discussed techniques for frequency conversion of ultrafast laser pulses to the extreme ultraviolet and X–ray regions of the spectrum. High harmonic generation has provided radiation all the way down to ~ 3 nm, shortening the wavelengths available with a (big) femtosecond system by almost two orders of magnitude. In addition, high harmonics produced with very short (5 fs) drive pulses may possess the coherence properties necessary to support attosecond laser pulses, thus opening the window on a new realm of dynamics. Very recently, short pulse X–rays of a wavelength appropriate for X–ray diffraction experiments have been produced by scattering infrared photons off relativistic electrons. Measurement of ultrafast X–ray pulsewidths can be accomplished by monitoring sidebands introduced by infrared pulses in X–ray photoionized electron spectra.

REFERENCES

1. J. Zhou, G. Taft, C. Huang, M. M. Murnane, and H. C. Kapteyn, *Optics Letters*, **19**, 1149 (1994).
2. J. Zhou, C. Huang, M. M. Murnane, and H. C. Kapteyn, *Optics Letters*, **20**, 64 (1995).
3. C. J. Bardeen, K. R. Wilson, and V. V. Yakovlev, *CLEO '96 Technical Digest*, **9**, 544 (1996).
4. Both Coherent and Spectra-Physics offer such systems for a few hundred thousand dollars!
5. H. M. van Driel, A. Hache, and G. Mak, *Proc. SPIE*, **2041**, 50 (1993).
6. F. Seifert, J. Ringling, F. Noack, V. Petrov, and O. Kittelmann, *Optics Letters*, **19**, 1538 (1994).

7. S. Backus, J. Peatross, E. Zeek, K. Reed, H. C. Kapteyn, and M. M. Murnane, *CLEO '96 Technical Digest*, **9**, 544 (1996).
8. L. Huang, J. P. Callan, E. N. Glezer, and E. Mazur, *Phys. Rev. Lett.*, accepted for publication; E. N. Glezer, Y. Siegal, L. Huang, and E. Mazur, *Phys. Rev. B*, **51**, 6959 (1995).
9. P. Eisenberger, S. Suckewer, *Science*, **274**, 201 (1996).
10. Ch. Spielmann, N. N. Burnett, S. Sartania, R. Koppitsch, M. Schnurer, C. Kan, M. Lenzner, P. Wobrauschek, and F. Krausz, *Science*, **278**, 661 (1997).
11. Z. Chang, A. Rundquist, H. Wang, M. M. Murnane, and H. C. Kapteyn, *Phys. Rev. Lett.*, **79**, 2967 (1997).
12. P. B. Corkum, *Phys. Rev. Lett.*, **71**, 1994 (1993).
13. R. W. Schoenlein, W. P. Leemans, A. H. Chin, P. Volfbeyn, T. E. Glover, P. Balling, M. Zolotorev, K. -J. Kim, S. Chattopadhyay, C. V. Shank, *Science*, **274**, 236 (1996).
14. T. E. Glover, R. W. Schoenlein, A. H. Chin, and C. V. Shank, *Phys. Rev. Lett.*, **76**, 2468 (1996).
15. A. McPherson, G. Gibson, H. Jara, U. Johann, T. S. Luk, I. A. McIntyre, K. Boyer, and C. K. Rhodes, *J. Opt. Soc. Am B*, **4**, 595 (1987).
16. J. L. Krause, K. J. Schafer, and K. C. Kulander, *Phys. Rev. Lett.*, **68**, 3535 (1992).
17. J. Zhou, J. Peatross, M. M. Murnane, and H. C. Kapteyn, *Phys. Rev. Lett.*, **76**, 752 (1996).
18. A. L' Huillier, Ph. Balcou, *Phys. Rev. Lett.*, **70**, 774 (1993).
19. J. J. Macklin, J. D. Kemetec, and C. L. Gordon III, *Phys. Rev. Lett.*, **70**, 766 (1993).
20. L. V. Keldysh, *Sh. Eksp. Teor. Fiz.*, **47**, 1945 (1964) [*Sov. Phys. JEPT*, **20**, 1307 (1965)].
21. M. Lewenstein, Ph. Balcou, M. Yu. Ivanov, Anne L'Huillier, and P. B. Corkum, *Phys. Rev. A*, **49**, 2117 (1994).
22. P. B. Corkum, N. H. Burnett, and F. Brunel, *Phys. Rev. Lett.*, **62**, 1259 (1989).
23. E. Merzbacher, *Quantum Mechanics*, 2nd ed., John Wiley (1970).
24. A. L'Huillier, K. J. Schafer, and K. C. Kulander, *Phys. Rev. Lett.*, **66**, 2200 (1991).
25. I. P. Christov, J. Zhou, J. Peatross, A. Rundquist, M. M. Murnane, and H. C. Kapteyn, *Phys. Rev. Lett.*, **77**, 1743 (1996).
26. I. P. Christov, M. M. Murnane, and H. C. Kapteyn, *Phys. Rev. Lett.*, **78**, 1251 (1997).
27. M. M. Murnane, H. C. Kapteyn, and R. W. Falcone, *IEEE Qant. Elec.*, **25**, 2417 (1989).
28. J. D. Jackson, *Classical Electrodynamics*, Sect. 11.3(d), John Wiley (1975).
29. A. Bouhal, R. Evans, G. Grillon, A. Mysyrowicz, P. Breger, P. Agostini, R. C. Constantinescu, H. G. Muller, and D. von der Linde, *J. Opt. Soc. Am. B.*, **14**, 950 (1997).
30. J. M. Schins, P. Breger, P. Agostini, R. C. Constantinescu, H. G. Muller, A. Bouhal, G. Grillon, A. Antonetti, A. Mysyrowicz, *J. Opt. Soc. Am. B.*, **13**, 197 (1996).

NONLINEAR INTERACTIONS IN INSULATING LASER CRYSTALS UNDER FEMTO- AND PICOSECOND EXCITATION

Alexander A. Kaminskii*

Institute of Crystallography
Russian Academy of Sciences
117333 Moscow
RUSSIA

ABSTRACT

New manifestations of optical nonlinear interactions have been discovered when femto- and picosecond light pulses are propagated through different types of insulating laser host crystals having considerable high $\chi^{(2)}$, as well as simultaneously $\chi^{(2)}$ and $\chi^{(3)}$ nonlinearities. Among them are: efficient multiple anti-Stokes and Stokes generation by collinear stimulated Raman scattering (SRS) and coaxial ring emission by Raman induced four-wave mixing (effect of laser "Raman rainbow"); femtosecond continuously tunable second harmonic generation (SHG) over the entire-visible spectral range; phenomenon of SHG due to "Cerenkov"-type phase matching (effect of laser "Cerenkov rainbow"); as well as new CW self-frequency doubling and self-SRS crystalline lasers. In this seminar I shall discuss several new possible applications of these newly discovered nonlinear optical potentialities for the creation of a new generation of crystalline lasers.

I. INTRODUCTION

The structural diversity of known laser crystals provides a wide variety of physical properties. The resultant inexhaustible spectroscopic potential constitutes a good mine of opportunities to develop and create new operating schemes and principles for excitation of stimulated emission (SE) under different experimental conditions. Among the approximately three hundred laser crystalline hosts ever known, acentric compounds with high quadratic nonlinear susceptibility $\chi^{(2)}$ are distinguished by a unique set of properties [1]. Thus, they have already provided the elegant advancement in quantum electronics of self-frequency doubling in bulk generating crystals [2-4] and in crystalline waveguide structures [5]. A no less interesting nonlinear interaction, the SRS, is also exhibited by laser insulating crystals both with self-frequency conversion in generating and pumping channels [6-8]. Recently, we showed that several laser host-crystals offer both relatively high $\chi^{(2)}$ and $\chi^{(3)}$

---

* Results were obtained within Joint Open Laboratory for Laser Crystals and Precise Laser Systems at Institute of Crystallography and Institute of Laser Physics of the Russian Academy of Sciences with Professors H. J. Eichler, K. Ueda, S. N. Bagayev, A. Z. Grasyuk and Doctors D. Grebe, R. Macdonald, A. A. Pavlyuk, H. Nishioka, A. V. Butashin et al. (see also Reference list).

Ultrafast Dynamics of Quantum Systems: Physical Processes and Spectroscopic Techniques, Edited by Di Bartolo and Gambarota, Plenum Press, New York, 1998

nonlinearities, which we can use for the creation of second harmonic generators, crystalline laser Raman shifters, or both devices simultaneously.

In this seminar, I will comprehensively report several new manifestations of high $\chi^{(2)}$ and $\chi^{(3)}$ susceptibilities in a wide variety of laser host crystals of different atomic structure under femto- and picosecond excitations: highly efficient (more than 50%) multiple anti-Stokes and Stokes generation by collinear SRS and coaxial ring emission by Raman induced four-wave mixing (RFWM); femtosecond continuously tunable SHG in the $\beta' - Gd_2(MoO_4)_3$ crystal; the phenomenon of SHG due to "Cerenkov"-type phase matching in the trigonal $LaBGeO_5$ crystal; and the new CW self-frequency doubling $\beta' - Gd_2(MoO_4)_3$ and self-SRS ($KGd(WO_4)_2$": Pr^{3+})crystalline lasers. The experimental techniques used and several new possible applications of these newly discovered nonlinear optical potentialities were also discussed.

II. SRS AND RFWM PROCESSES IN LASER HOST-CRYSTALS

In general, two kinds of nonlinear optical mechanisms can give rise to the second- and higher-order collinear and coaxial ring Stokes generation in Raman-active crystalline media. The first one is the SRS generation using threshold cascade pumping by the first Stokes laser beam in the same propagation direction and stimulating the gain properties of the crystal (Fig. 1a). The intensity of the second Stokes collinear generation (for the steady-state mode, when pump pulse duration τ Ø $T_2=1/\pi \, \Delta\nu_R$, where T_2 and $\Delta\nu_R$ are the dephasing time and linewidth of SRS-active crystal vibration ω_R , respectively) will grow according to the well-known equation:

$$I_{St2}(L) = I_{St2}(0)\exp(g\, I_{St1}L) = I_{St2}(0)\exp(g\, I_f L). \qquad (1)$$

Here I_{st2} (0) is the intensity of the spontaneous Raman emission, g is the Raman gain coefficient depending on the corresponding coefficient of the $\chi^{(3)}$ tensor of the crystal, I_{st1} is the intensity of the first Stokes Raman generation, and L is the length of the gain sample. To reach the SRS generation threshold from spontaneous Raman-emission level, a g $I_f L$ value of about 25 is required [11]. When the second Stokes generation is strong enough, it generates the third Stokes emission by the next cascade SRS process. The phase array of the atomic vibrations (i.e. polarization gratings of atomic vibrations) is excited between the fundamental (pump) and the Stokes waves. To preserve momentum conservation, k_Q must obey

$$k_Q = k_f - k_{St1} \qquad (2)$$

where g k_f and k_{St1} are the wave vectors corresponding to the pump and the first Stokes components, respectively. This phase matching condition is automatically satisfied between any pump and first Stokes waves so that the Stokes wave can be amplified in any direction. In many cases, above the threshold, the Stokes emission can be observed as a conical emission.

The second possible mechanism for the second and higher-order Stokes generation is the threshold-less RFWM with the pump and first Stokes beams driving the nonlinear polarization in the SRS-active crystals. The second Stokes wave is produced through the interaction of the first Stokes wave and the phase array k_Q which is generated between the pump wave and the self-seeded first Stokes emission, as shown in Fig. 1b. Of course, the latter process needs the phase-matching condition

$$2k_{St1} - k_f - k_{St2} = 0 \qquad (3)$$

and generates a coaxial ring emission. In this equation k_{St2} is the wave vector of the second Stokes wave. The phase-matching angle is given by the dispersion of the refractive indices of the Raman crystal. In this case, the second Stokes component grows according to

$$dI_{St2}/dz = \varepsilon g_{RFWM}(\omega_{St2}/\omega_f)' \; I_{St2}(I_f I_{St2})', \qquad (4)$$

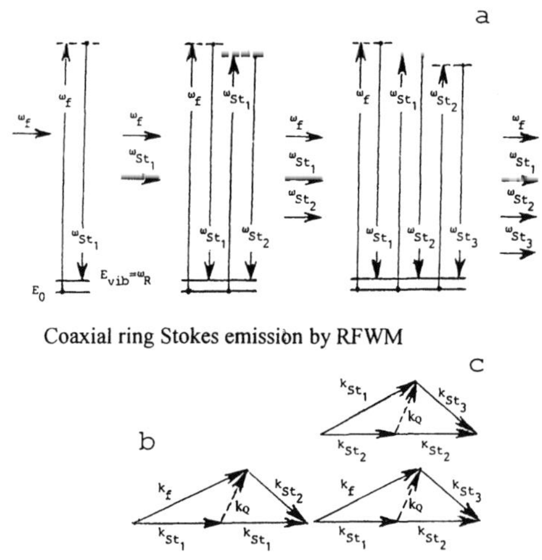

Figure 1. The energy and momentum conservation in stimulated Raman scattering: (a) the energy conservation; E_o and $E_{vib} = \omega_R$ are the ground and vibrational levels of the Raman-active medium; ω_f, ω_{St1}, ω_{St2} and ω_{St3} are the optical frequencies of the pump, first, second, and third SRS emission: (b) and (c) show the momentum conservation and phase-matching condition in the RFWM for the coaxial ring-type second Stokes generation and two possible conditions for the coaxial ring-type third Stokes generation, respectively. The wave vectors k_f, k_{St1}, k_{St2}, and k_{St3} are for the pump, first, second and third Stokes components [13].

where I_f, I_{St1}, and I_{St2} are the fundamental, the first, and the second Stokes emission intensity, g_{RFWM} is the Raman gain coefficient of the RFWM process. The factor ε is the coefficient of gain reduction due to the phase mismatching Δk,

$$\Delta k = (2k_{St1} - k_f - k_{St2})z \tag{5}$$

and

$$\varepsilon = \mathrm{Re}\,\exp(i\Delta kz), \tag{6}$$

where z is the unit vector parallel to the wave vector k_{St2}.

For the third Stokes RFWM case, two different phase-matching conditions (Fig. 1c) with the pump, and the co-propagating first and second Stokes components are possible for the wave vector of the third Stokes wave k_{St3}, such that

$$2\,k_{St2} - k_{St1} - k_{St3} = 0 \tag{7}$$

and

$$k_{St1} + k_{St2} - k_f - k_{St3} = 0 \tag{8}$$

The latter case is dominant when the pump depletion is small, because the amplitude of the exciting phase array between the fundamental and first Stokes emission significantly exceeds that between the first and second Stokes pulse generations.

IIA. High Efficient Multiple Anti-Stokes And Stokes Srs In Ky(Wo4)2, Kgd(Wo4)2, Ky(Moo4)2, Kla(Moo4)2, And Linbgeo5 Laser Host-Crystals

For single-pass SRS excitation in monoclinic $KY(WO_4)_2$, $KGd(WO_4)_2$, and $KLa(MoO_4)_2$ and orthorhombic $KY(MoO_4)_2$ and $LiNbGeO_5$ single crystals, we used a high-power picosecond laser designed at the Optical Institute of the Technical University Berlin [12]. The experimental set-up is shown in Fig. 2. An active-passively modelocked

$Nd^{3+}:Y_3Al_5O_{12}$ master oscillator produces pulse trains with energies of about 5 mJ at 1.06415 μm wavelength ($^4F_{3/2} \rightarrow {^4I_{11/2}}$ lasing channel of Nd^{3+} activators). An external Pockels cell pulse slicer transmits only the central ≈120 ps pulse, with an energy of about 200 μJ. The following two-stage $Nd^{3+}:Y_3Al_5O_{12}$ amplifier can increase the energy of this one-micron single pulse up to ≈10 mJ. A KD*P doubler converts the beam emission to 0.53207 μm wavelength with a pulse duration of about 80 ps and a Gaussian profile. The oriented (along crystallographic axes) crystalline samples of centimetre size are placed in the focus of a lens with f=500 mm leading to a beam waist diameter of about 75 mm. The spectral composition of the Stokes and anti-Stokes generation is measured with a Si-CCD Spectrometric Multichannel Analyzer (CSMA), consisting of a modified McPherson Model 218 scanning grating monochrometer without exit slit and a Princeton Instruments Model ST-130 Analyzer. The wavelength dependence of the sensitivity for analyzing system is shown in Fig. 3.

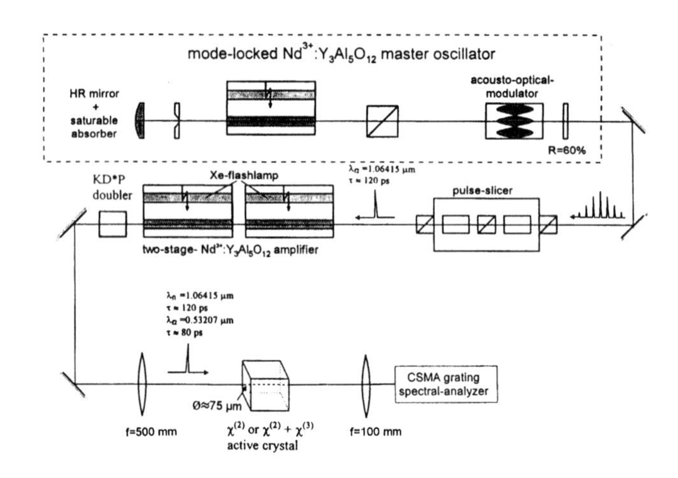

Figure 2. Schematic experimental set-up for SRS excitation [9]. For explanation see text.

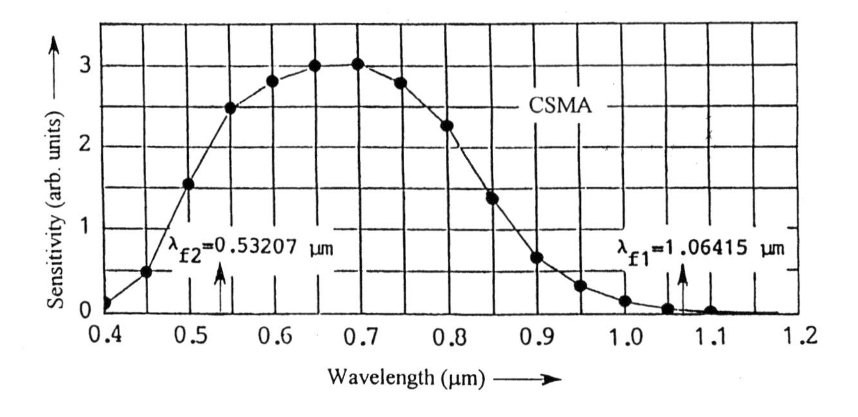

Figure 3. Wavelength dependence of a sensitivity for the spectral analyzing CSMA system.

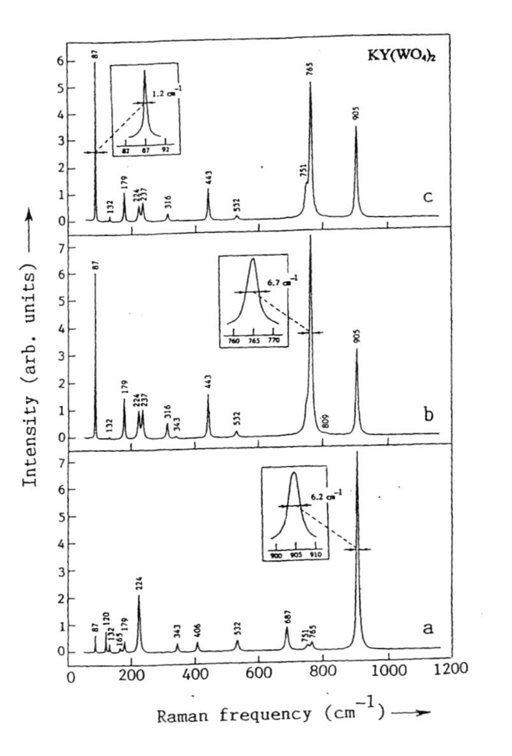

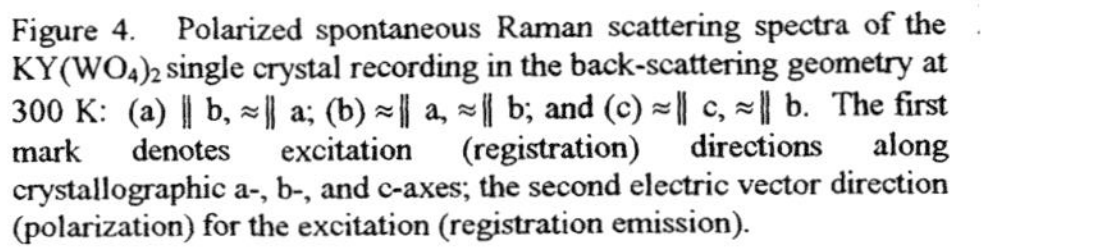

Figure 4. Polarized spontaneous Raman scattering spectra of the $KY(WO_4)_2$ single crystal recording in the back-scattering geometry at 300 K: (a) ‖ b, ≈‖ a; (b) ≈‖ a, ≈‖ b; and (c) ≈‖ c, ≈‖ b. The first mark denotes excitation (registration) directions along crystallographic a-, b-, and c-axes; the second electric vector direction (polarization) for the excitation (registration emission).

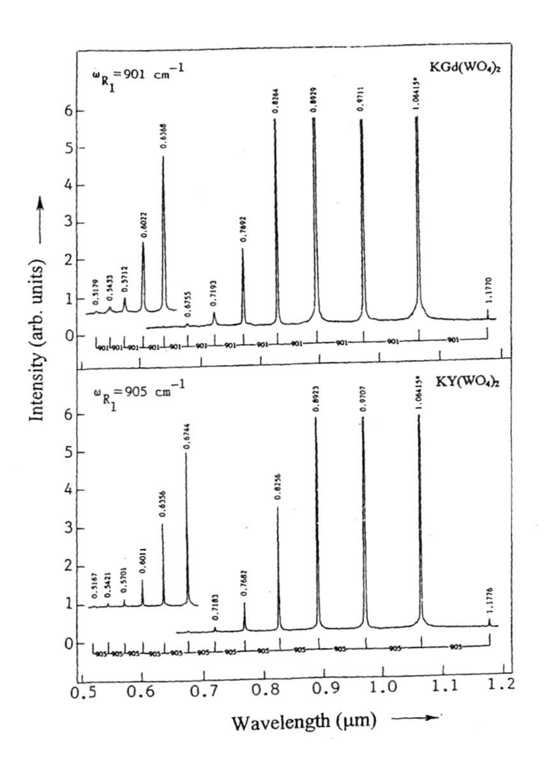

Figure 5. Orientational SRS spectra of the monoclinic $KY(WO_4)_2$ and $KGd(WO_4)_2$ single crystals recording in the collinear single-pass geometry at ‖ b, ≈‖ a and 300 K. Intensities of the SRS components and excitation lines are shown without a correction on spectral sensitivity of analyzing system. The pump laser lines are marked by asterisks. The connection of the Stokes and anti-Stokes generation components with the SRS-active vibration modes of crystal with the frequencies of 905 cm^{-1} for $KY(WO_4)_2$ and 901 cm^{-1} for $KGd(WO_4)_2$ are indicated by asterisks. Results on SRS and RFWM-line attribution are listed in Table 1.

For obtaining some information about the frequencies of atomic vibrations of investigated crystals, before SRS experiments we have measured the spontaneous Raman scattering spectra at 300 K by the standard method. As an example, Fig. 4 shows three such polarized spectra for monoclinic $KY(WO_4)_2$ crystal, where the linewidth of their SRS-active Raman lines are also presented.

Besides the multiple ordinary cascading SRS generation, we have observed coaxial ring emission (laser "Raman rainbow") of the Stokes wavelength in the visible by a RFWM parametric process. The main results of our SRS experiments are presented in Figs. 5-16. Data of identifications of SRS and RFWM lines for two related monoclinic $KY(WO_4)_2$ and $KGd(WO_4)_2$ are listed in Table 1. Our measurements showed also that the total nonlinear conversion efficiency of the fundamental (pump) pulsed emission to the Stokes and anti-Stokes components exceeds 50%. For $KY(WO_4)_2$, $KGd(WO_4)_2$, and β'-$Gd_2(MoO_4)_3$ the conversion efficiencies are more than 70%.

Table 1

Spectral composition of the Stokes and anti-Stokes room-temperature generation of Raman lasers on the base of monoclinic $KY(WO_4)_2$ and $KGd(WO_4)_2$ crystals under picosecond $Nd^{3+}:Y_3Al_5O_{12}$ laser excitation at 0.53207 (SHG) and 1.06415 μm wavelengths

Pumping condition		Stokes and anti-Stokes emission		Stokes and SRS- and RFWM-line attribution	Frequency of SRS-active crystal vibration mode (cm^{-1})		
λ_f (μm)	Geometry of excitation*	Wavelength** (μm)	Line***		ω_{R_1}	ω_{R_2}	ω_{R_3}
			$KY(WO_4)_2$ crystal				
1.06415 (Fig. 5)	‖b, ≈‖a	0.5167	ASt_{11-1}	$\omega_f+11\omega_{R_1}$	905		
		0.5421	ASt_{10-1}	$\omega_f+10\omega_{R_1}$	905		
		0.5701	ASt_{9-1}	$\omega_f+9\omega_{R_1}$	905		
		0.6011	ASt_{8-1}	$\omega_f+8\omega_{R_1}$	905		
		0.6356	ASt_{7-1}	$\omega_f+7\omega_{R_1}$	905		
		0.6744	ASt_{6-1}	$\omega_f+6\omega_{R_1}$	905		
		0.7183	ASt_{5-1}	$\omega_f+5\omega_{R_1}$	905		
		0.7682	ASt_{4-1}	$\omega_f+4\omega_{R_1}$	905		
		0.8256	ASt_{3-1}	$\omega_f+3\omega_{R_1}$	905		
		0.8923	ASt_{2-1}	$\omega_f+2\omega_{R_1}$	905		
		0.9707	ASt_{1-1}	$\omega_f+\omega_{R_1}$	905		
		1.1776	St_{1-1}	$\omega_f-\omega_{R_1}$	905		
1.06415 (Fig. 6)	≈‖a, ‖b	0.6779	ASt_{7-2}	$\omega_f+7\omega_{R_2}$		765	
		0.7149	ASt_{6-2}	$\omega_f+6\omega_{R_2}$		765	
		0.7563	ASt_{5-2}	$\omega_f+5\omega_{R_2}$		765	
		0.8028	ASt_{4-2}	$\omega_f+4\omega_{R_2}$		765	
		0.8553	ASt_{3-2}	$\omega_f+3\omega_{R_2}$		765	
		0.9152	ASt_{2-2}	$\omega_f+2\omega_{R_2}$		765	
		0.9841	ASt_{1-2}	$\omega_f+\omega_{R_2}$		765	
		1.1585	St_{1-2}	$\omega_f-\omega_{R_2}$		765	
1.06415 (Fig. 8a)	≈‖c, ≈‖b	0.6779	ASt_{7-2}	$\omega_f+7\omega_{R_2}$		765	
		0.7149	ASt_{6-2}	$\omega_f+6\omega_{R_2}$		765	
		0.7563	ASt_{5-2}	$\omega_f+5\omega_{R_2}$		765	
		0.7682	ASt_{4-1}	$\omega_f+4\omega_{R_1}$	905		
		0.8028	ASt_{4-2}	$\omega_f+4\omega_{R_2}$		765	
		0.8256	ASt_{3-1}	$\omega_f+3\omega_{R_1}$	905		
		0.8553	ASt_{3-2}	$\omega_f+3\omega_{R_2}$		765	
		0.8617	$St_{1-3}ASt_{3-2}$	$\omega_f+3\omega_{R_2}-\omega_{R_3}$		765	87
		0.8923	ASt_{2-1}	$\omega_f+2\omega_{R_1}$	905		
		0.9036	$ASt_{1-2}ASt_{1-1}$	$\omega_f+\omega_{R_2}+\omega_{R_1}$	905	765	
		0.9080	$ASt_{1-3}ASt_{2-2}$	$\omega_f+2\omega_{R_2}+\omega_{R_3}$		765	87
		0.9152	ASt_{2-2}	$\omega_f+2\omega_{R_2}$		765	
		0.9226	$St_{1-3}ASt_{2-2}$	$\omega_f+2\omega_{R_2}-\omega_{R_3}$		765	87
		0.9301	$St_{2-3}ASt_{2-2}$	$\omega_f+2\omega_{R_2}-2\omega_{R_3}$		765	87
		0.9707	ASt_{1-1}	$\omega_f+\omega_{R_1}$	905		
		0.9757	$ASt_{1-3}ASt_{1-2}$	$\omega_f+\omega_{R_2}+\omega_{R_3}$		765	87
		0.9841	ASt_{1-2}	$\omega_f+\omega_{R_2}$		765	

	0.9926	$St_{1-3}ASt_{1-2}$	$\omega_f+\omega_{R_2}-\omega_{R_3}$		765	87
	1.0012	$St_{2-3}ASt_{1-2}$	$\omega_f+\omega_{R_2}-2\omega_{R_3}$		765	87
	1.0544	ASt_{1-3}	$\omega_f+\omega_{R_3}$			87
	1.0741	St_{1-3}	$\omega_f-\omega_{R_3}$			87
	1.0842	St_{2-3}	$\omega_f-2\omega_{R_3}$			87
	1.1585	St_{1-2}	$\omega_f-\omega_{R_2}$		765	
	1.1776	St_{1-1}	$\omega_f-\omega_{R_1}$	905		
1.06415 ≈‖a, ≈‖b (Fig. 8b)	0.6142	ASt_{9-2}	$\omega_f+9\omega_{R_2}$		765	
	0.6444	ASt_{8-2}	$\omega_f+8\omega_{R_2}$		765	
	0.6779	ASt_{7-2}	$\omega_f+7\omega_{R_2}$		765	
	0.7149	ASt_{6-2}	$\omega_f+6\omega_{R_2}$		765	
	0.7563	ASt_{5-2}	$\omega_f+5\omega_{R_2}$		765	
	0.8028	ASt_{4-2}	$\omega_f+4\omega_{R_2}$		765	
	0.8085	$St_{1-3}ASt_{4-2}$	$\omega_f+4\omega_{R_2}-\omega_{R_3}$		765	87
	0.8142	$St_{2-3}ASt_{4-2}$	$\omega_f+4\omega_{R_2}-2\omega_{R_3}$		765	87
	0.8490	$ASt_{1-3}ASt_{3-2}$	$\omega_f+3\omega_{R_2}+\omega_{R_3}$		765	87
	0.8553	ASt_{3-2}	$\omega_f+3\omega_{R_2}$		765	
	0.8617	$St_{1-3}ASt_{3-2}$	$\omega_f+3\omega_{R_2}-\omega_{R_3}$		765	87
	0.8682	$St_{2-3}ASt_{3-2}$	$\omega_f+3\omega_{R_2}-2\omega_{R_3}$		765	87
	0.8748	$St_{3-3}ASt_{3-2}$	$\omega_f+3\omega_{R_2}-3\omega_{R_3}$		765	87
	0.9009	$ASt_{2-3}ASt_{2-2}$	$\omega_f+2\omega_{R_2}+2\omega_{R_3}$		765	87
	0.9080	$ASt_{1-3}ASt_{2-2}$	$\omega_f+2\omega_{R_2}+\omega_{R_3}$		765	87
	0.9152	ASt_{2-2}	$\omega_f+2\omega_{R_2}$		765	
	0.9226	$St_{1-3}ASt_{2-2}$	$\omega_f+2\omega_{R_2}-\omega_{R_3}$		765	87
	0.9301	$St_{2-3}ASt_{2-2}$	$\omega_f+2\omega_{R_2}-2\omega_{R_3}$		765	87
	0.9376	$St_{3-3}ASt_{2-2}$	$\omega_f+2\omega_{R_2}-3\omega_{R_3}$		765	87
	0.9454	$St_{4-3}ASt_{2-2}$	$\omega_f+2\omega_{R_2}-4\omega_{R_3}$		765	87
	0.9675	$ASt_{2-3}ASt_{1-2}$	$\omega_f+\omega_{R_2}+2\omega_{R_3}$		765	87
	0.9757	$ASt_{1-3}ASt_{1-2}$	$\omega_f+\omega_{R_2}+\omega_{R_3}$		765	87
	0.9841	ASt_{1-2}	$\omega_f+\omega_{R_2}$		765	
	0.9926	$St_{1-3}ASt_{1-2}$	$\omega_f+\omega_{R_2}-\omega_{R_3}$		765	87
	1.0012	$St_{2-3}ASt_{1-2}$	$\omega_f+\omega_{R_2}-2\omega_{R_3}$		765	87
	1.0100	$St_{3-3}ASt_{1-2}$	$\omega_f+\omega_{R_2}-3\omega_{R_3}$		765	87
	1.0189	$St_{4-3}ASt_{1-2}$	$\omega_f+\omega_{R_2}-4\omega_{R_3}$		765	87
	1.0281	$St_{5-3}ASt_{1-2}$	$\omega_f+\omega_{R_2}-5\omega_{R_3}$		765	87
	1.0448	ASt_{2-3}	$\omega_f+2\omega_{R_3}$			87
	1.0544	ASt_{1-3}	$\omega_f+\omega_{R_3}$			87
	1.0741	St_{1-3}	$\omega_f-\omega_{R_3}$			87
	1.0842	St_{2-3}	$\omega_f-2\omega_{R_3}$			87
	1.0946	St_{3-3}	$\omega_f-3\omega_{R_3}$			87
	1.1051	St_{4-3}	$\omega_f-4\omega_{R_3}$			87
	1.1582	St_{1-2}	$\omega_f-\omega_{R_2}$		765	
0.53207 ‖b, ≈‖a (Fig. 9a)	0.4853	ASt_{2-1}	$\omega_f+2\omega_{R_1}$	905		
	0.5076	ASt_{1-1}	$\omega_f+\omega_{R_1}$	905		
	0.5590	St_{1-1}	$\omega_f-\omega_{R_1}$	905		
	0.5888	St_{2-1}	$\omega_f-2\omega_{R_1}$	905		
	0.6219	St_{3-1}	$\omega_f-3\omega_{R_1}$	905		
	0.6590	St_{4-1}	$\omega_f-4\omega_{R_1}$	905		
	0.7008	St_{5-1}	$\omega_f-5\omega_{R_1}$	905		
	0.7482	St_{6-1}	$\omega_f-6\omega_{R_1}$	905		
0.53207 ≈‖a, ‖b (Fig. 9b)	0.4920	ASt_{2-2}	$\omega_f+2\omega_{R_2}$		765	
	0.5113	ASt_{1-2}	$\omega_f+\omega_{R_2}$		765	
	0.5547	St_{1-2}	$\omega_f-\omega_{R_2}$		765	
	0.5792	St_{2-2}	$\omega_f-2\omega_{R_2}$		765	
	0.6061	St_{3-2}	$\omega_f-3\omega_{R_2}$		765	
	0.6356	St_{4-2}	$\omega_f-4\omega_{R_2}$		765	
	0.6680	St_{5-2}	$\omega_f-5\omega_{R_2}$		765	
0.53207 ≈‖a, ≈‖b (Fig. 9c)	0.5346	St_{1-3}	$\omega_f-\omega_{R_3}$			87
	0.5371	St_{2-3}	$\omega_f-2\omega_{R_3}$			87
	0.5396	St_{3-3}	$\omega_f-3\omega_{R_3}$			87
	0.5520	$ASt_{1-3}St_{1-2}$	$\omega_f-\omega_{R_2}+\omega_{R_3}$		765	87
	0.5547	St_{1-2}	$\omega_f-\omega_{R_2}$		765	
	0.5574	$St_{1-3}St_{1-2}$	$\omega_f-\omega_{R_2}-\omega_{R_3}$		765	87
	0.5601	$St_{2-3}St_{1-2}$	$\omega_f-\omega_{R_2}-2\omega_{R_3}$		765	87
	0.5628	$St_{3-3}St_{1-2}$	$\omega_f-\omega_{R_2}-3\omega_{R_3}$		765	87
	0.5792	St_{2-2}	$\omega_f-2\omega_{R_2}$		765	

		0.5821	$St_{1-3}St_{2-2}$	$\omega_f-2\omega_{R_2}-\omega_{R_3}$	765	87	
		0.5851	$St_{2-3}St_{2-2}$	$\omega_f-2\omega_{R_2}-2\omega_{R_3}$	765	87	
		0.6061	St_{3-2}	$\omega_f-3\omega_{R_2}$	765		
		0.6356	St_{4-2}	$\omega_f-4\omega_{R_2}$	765		
			$KGd(WO_4)_2$				
1.06415 (Fig. 5)	$\Vert b, \approx\Vert a$	0.5179	ASt_{11-1}	$\omega_f+11\omega_{R_1}$	901		
		0.5433	ASt_{10-1}	$\omega_f+10\omega_{R_1}$	901		
		0.5712	ASt_{9-1}	$\omega_f+9\omega_{R_1}$	901		
		0.6022	ASt_{8-1}	$\omega_f+8\omega_{R_1}$	901		
		0.6368	ASt_{7-1}	$\omega_f+7\omega_{R_1}$	901		
		0.6755	ASt_{6-1}	$\omega_f+6\omega_{R_1}$	901		
		0.7193	ASt_{5-1}	$\omega_f+5\omega_{R_1}$	901		
		0.7692	ASt_{4-1}	$\omega_f+4\omega_{R_1}$	901		
		0.8264	ASt_{3-1}	$\omega_f+3\omega_{R_1}$	901		
		0.8929	ASt_{2-1}	$\omega_f+2\omega_{R_1}$	901		
		0.9711	ASt_{1-1}	$\omega_f+\omega_{R_1}$	901		
		1.1770	St_{1-1}	$\omega_f-\omega_{R_1}$	901		
1.06415 (Fig. 6)	$\approx\Vert a, \Vert b$	0.5856	ASt_{10-2}	$\omega_f+10\omega_{R_2}$		768	
		0.6132	ASt_{9-2}	$\omega_f+9\omega_{R_2}$		768	
		0.6435	ASt_{8-2}	$\omega_f+8\omega_{R_2}$		768	
		0.6769	ASt_{7-2}	$\omega_f+7\omega_{R_2}$		768	
		0.7140	ASt_{6-2}	$\omega_f+6\omega_{R_2}$		768	
		0.7555	ASt_{5-2}	$\omega_f+5\omega_{R_2}$		768	
		0.8020	ASt_{4-2}	$\omega_f+4\omega_{R_2}$		768	
		0.8546	ASt_{3-2}	$\omega_f+3\omega_{R_2}$		768	
		0.9147	ASt_{2-2}	$\omega_f+2\omega_{R_2}$		768	
		0.9838	ASt_{1-2}	$\omega_f+\omega_{R_2}$		768	
		1.1589	St_{1-2}	$\omega_f-\omega_{R_2}$		768	
1.06415 (Fig. 10a)	$\approx\Vert a, \approx\Vert b$	0.8020	ASt_{4-2}	$\omega_f+4\omega_{R_2}$		768	
		0.8546	ASt_{3-2}	$\omega_f+3\omega_{R_2}$		768	
		0.9147	ASt_{2-2}	$\omega_f+2\omega_{R_2}$		768	
		0.9218	$ASt_{1-3}ASt_{2-2}$	$\omega_f+2\omega_{R_2}-\omega_{R_3}$		768	84
		0.9838	ASt_{1-2}	$\omega_f+\omega_{R_2}$		768	
		0.9920	$St_{1-3}ASt_{1-2}$	$\omega_f+\omega_{R_2}-\omega_{R_3}$		768	84
		1.0003	$St_{2-3}ASt_{1-2}$	$\omega_f+\omega_{R_2}-2\omega_{R_3}$		768	84
		1.0547	ASt_{1-3}	$\omega_f+\omega_{R_3}$			84
		1.0738	St_{1-3}	$\omega_f-\omega_{R_3}$			84
		1.0835	St_{2-3}	$\omega_f-2\omega_{R_3}$			84
		1.0935	St_{3-3}	$\omega_f-3\omega_{R_3}$			84
		1.1589	St_{1-2}	$\omega_f-\omega_{R_2}$		768	
1.06415 (Fig. 10b)	$\approx\Vert c, \approx\Vert b$	0.8546	ASt_{3-2}	$\omega_f+3\omega_{R_2}$		768	
		0.8929	ASt_{2-1}	$\omega_f+2\omega_{R_1}$	901		
		0.9147	ASt_{2-2}	$\omega_f+2\omega_{R_2}$		768	
		0.9711	ASt_{1-1}	$\omega_f+\omega_{R_1}$	901		
		0.9838	ASt_{1-2}	$\omega_f+\omega_{R_2}$		768	
		1.0738	St_{1-3}	$\omega_f-\omega_{R_3}$			84
		1.1589	St_{1-2}	$\omega_f-\omega_{R_2}$		768	
		1.1770	St_{1-1}	$\omega_f-\omega_{R_1}$	901		
1.06415 (Fig. 10c)	$\approx\Vert c, \Vert b$	0.7193	ASt_{5-1}	$\omega_f+5\omega_{R_1}$	901		
		0.7555	ASt_{5-2}	$\omega_f+5\omega_{R_2}$		768	
		0.7692	ASt_{4-1}	$\omega_f+4\omega_{R_1}$	901		
		0.8020	ASt_{4-2}	$\omega_f+4\omega_{R_2}$		768	
		0.8264	ASt_{3-1}	$\omega_f+3\omega_{R_1}$	901		
		0.8546	ASt_{3-2}	$\omega_f+3\omega_{R_2}$		768	
		0.8929	ASt_{2-1}	$\omega_f+2\omega_{R_1}$	901		
		0.9037	$ASt_{1-2}ASt_{1-1}$	$\omega_f+\omega_{R_1}+\omega_{R_2}$	901	768	
		0.9147	ASt_{2-2}	$\omega_f+2\omega_{R_2}$		768	
		0.9711	ASt_{1-1}	$\omega_f+\omega_{R_1}$	901		
		0.9838	ASt_{1-2}	$\omega_f+\omega_{R_2}$		768	
		1.1589	St_{1-2}	$\omega_f-\omega_{R_2}$		768	
		1.1770	St_{1-1}	$\omega_f-\omega_{R_1}$	901		
0.53207 (Fig. 11)	$\approx\Vert c, \Vert b$	0.5112	ASt_{1-2}	$\omega_f+\omega_{R_2}$		768	
		0.5507	$ASt_{1-1}St_{2-2}$	$\omega_f-2\omega_{R_2}+\omega_{R_1}$	901		

0.5547	St_{1-2}	$\omega_f-\omega_{R_2}$		768
0.5589	St_{1-1}	$\omega_f-\omega_{R_1}$	901	
0.5750	$ASt_{1-1}St_{3-2}$	$\omega_f-3\omega_{R_2}+\omega_{R_1}$	901	768
0.5794	St_{2-2}	$\omega_f-2\omega_{R_2}$		768
0.5839	$St_{1-1}St_{1-2}$	$\omega_f-\omega_{R_2}-\omega_{R_1}$	901	768
0.5885	St_{2-1}	$\omega_f-2\omega_{R_1}$	901	
0.6064	St_{3-2}	$\omega_f-3\omega_{R_2}$		768
0.6113	$St_{1-1}St_{2-2}$	$\omega_f-2\omega_{R_2}-\omega_{R_1}$	901	768
0.6164	$St_{2-1}St_{1-2}$	$\omega_f-\omega_{R_2}-2\omega_{R_1}$	901	768
0.6360	St_{4-2}	$\omega_f-4\omega_{R_2}$		768
0.6415	$St_{1-1}St_{3-2}$	$\omega_f-3\omega_{R_2}-\omega_{R_1}$	901	768
0.6470	$St_{2-1}St_{2-2}$	$\omega_f-2\omega_{R_2}-2\omega_{R_1}$	901	768
0.6687	St_{5-2}	$\omega_f-5\omega_{R_2}$		768
0.6747	$St_{1-1}St_{4-2}$	$\omega_f-4\omega_{R_2}-\omega_{R_1}$	901	768

* The first mark denotes excitation direction, the second electric vector direction for the excitation.

** Measurement accuracy is ± 0.0003 μm.

*** For example, the notation $ASt_{1-3}St_{1-2}$ is defined as the first anti-Stokes component with the third vibration mode ω_{R_3} of the second Stokes component with first vibration mode ω_{R_1}.

IIB. Srs In A-Al2o3 Single Crystal

With the same experimental setup (Fig. 2) we have observed enticing multiple Stokes (up to 7th order) and anti-Stokes generation in trigonal leucosapphare α-Al_2O_3 crystal under picosecond excitation ($\lambda_f = 0.53207$ μm) at 300 K (Fig. 17). On the basis of analysis of the SRS generation data we are able to attribute Stokes and anti-Stokes emission in this very important laser host-crystal to the only optical vibration A_{1g} of its atoms with a frequency of $\omega_{R1} = 419$ cm^{-1} [14,19]. Results of measurements steady-state SRS-gain coefficients and conversion efficiency for the first Stokes generation are given in Fig. 18 and Table 2. As will be seen from the last Figure for the α-Al_2O_3 single crystal along the a-axis with the polarization vector for fundamental and SRS emission along the c-axis, the conversion efficiency was measured to be about 1%.

IIC. New Optical Effects In Acentric Laser Host-Crystals Having High $X^{(2)}$ And $X^{(3)}$ Nonlinear Susceptibilities

Recently, in [16] was reported the first experimental observation of new manifestations of a nonlinear optical interaction under picosecond excitation in acentric trigonal $LaBGeO_5$ and orthorhombic β'-$Gd_2(MoO_4)_3$ laser host-crystals, which offer simultaneously relatively high $\chi^{(2)}$ and $\chi^{(3)}$ nonlinear effects. Among them are multiple anti-Stokes and Stokes SRS arising from intensive intrinsic SHG and multiple anti-Stokes generation in the $LaBGeO_5$ ferroelectric which converts one-micron fundamental emission to the visible. We registered the emission at $\lambda_{ASt10} = 0.5783$ μm wavelength from $\lambda_f = 1.06415$ μm and the SRS-active vibration mode of the crystal $\omega_{R2} = 790$ cm^{-1} (Fig. 19). Of course, such SHG-SRS effects can be observed in the $\chi^{(2)} + \chi^{(3)}$ -active crystals only in the direction of the phase-matching for their SHG. The same effects were observed quite recently with the novel nonlinear laser host-crystal $Ca_4Gd(BO_3)_3O$ [20].

In conclusion of this chapter, in Table 3 are listed the $\chi^{(3)}$ and $\chi^{(2)} + \chi^{(3)}$ -active laser host-crystals which are subjects of our SRS investigations.

Table 2

Steady-state SRS-gain coefficients of α-Al_2O_3 single crystals at 300 K [15]

Excitation conditions*		g**
Direction	Polarization	(cm/GW)
‖ a	‖ c	0.46
‖ a	‖ m	0.3
‖ m	‖ c	0.34
‖ m	‖ a	0.25
‖ c	‖ m	0.25
‖ c	‖ a	0.4

* Was measured at $\lambda_f = 0.53207$ μm wavelength with the use of picosecond ($\tau \approx 30$ ps) frequency-doubled Nd^{3+}:$Y_3Al_5O_{12}$ laser.

** Belongs to the 1-st Stokes component at $\lambda_{St_1} = 0.5442$ μm wavelength ($\omega_{R_1} \approx 419$ cm^{-1}, dephasing time $T_2 \approx 3.7$ ps).

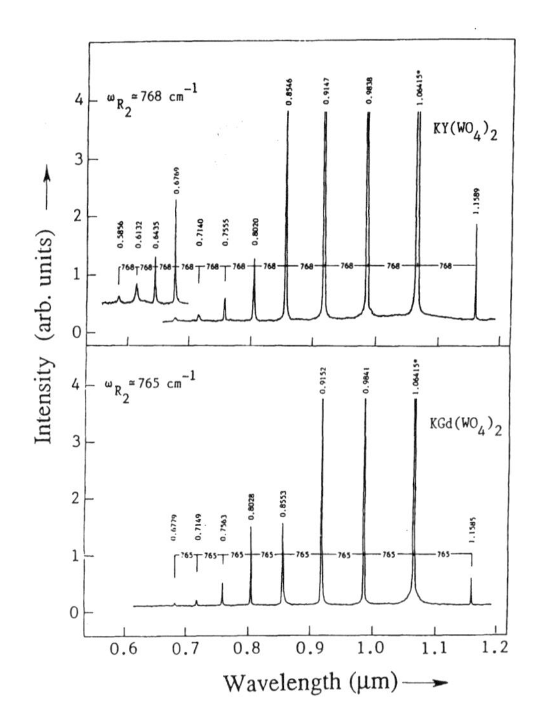

Figure 6. Orientational SRS spectra of the monoclinic $KY(WO_4)_2$ and $KGd(WO_4)_2$ single crystals recording in experimental condition at $\approx$‖ a, $\approx$‖ b; and 300 K. Notations and explanations as in Figure 5.

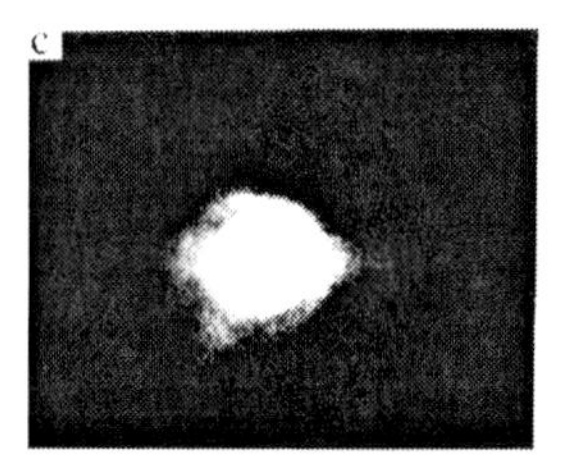

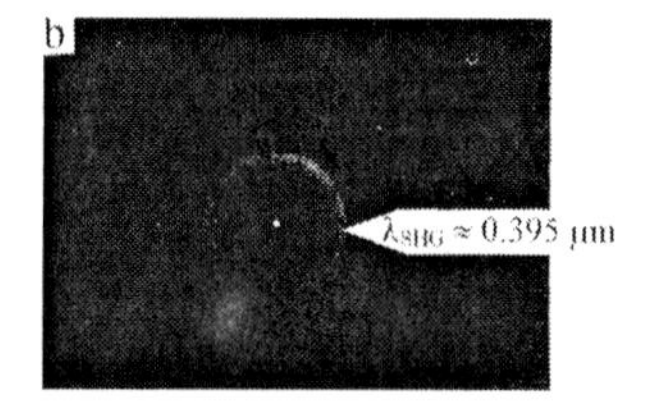

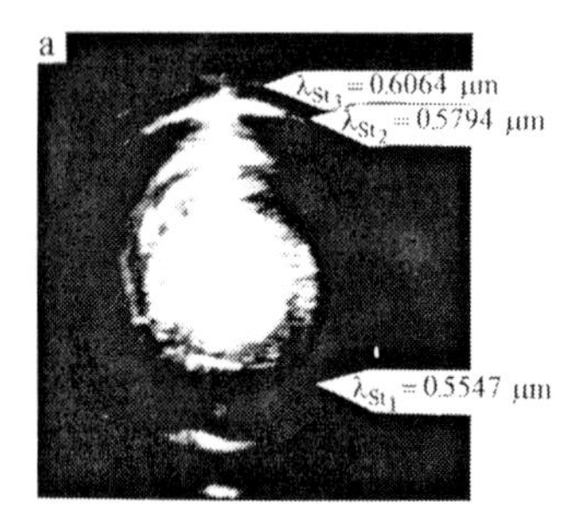

Figure 7. (a) Spatial beam pattern ("Raman rainbow") in the SRS generation of the $KGd(WO_4)_2$ single crystals pumped by the SHG of Nd^{3+} : $Y_3AL_5O_{12}$ ($\lambda_f = 0.53207$ μm) laser. (b) and (c) Spatial beam patterns ("Cerenkov rainbow") in the SHG of the $LaBGeO_5$ single crystal at monochromatic (Ti^{3+}:Al_2O_3 laser, (λ_f = 0.97 μm) and broadband (kr-hypercontinuum) pumping, correspondingly.

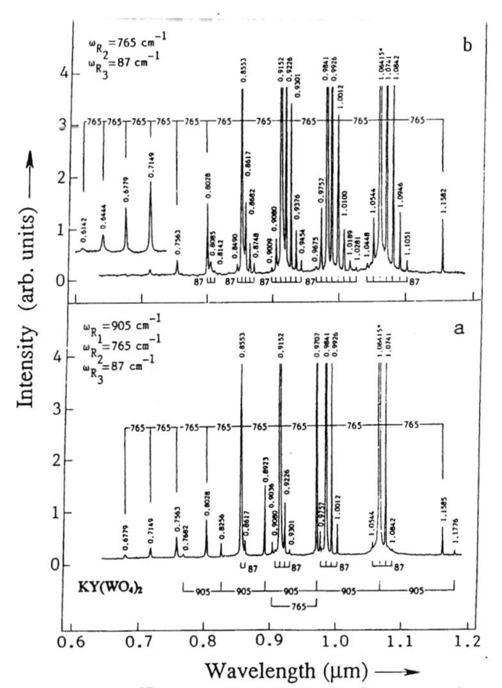

Figure 8. Orientational room-temperature SRS spectra of the monoclinic $KY(WO_4)_2$ single crystals recording in experimental condition at (a) ≈‖ c, ≈‖ b and (b) ≈‖ a, ≈‖ b. Notations and explanations as in Figure 5.

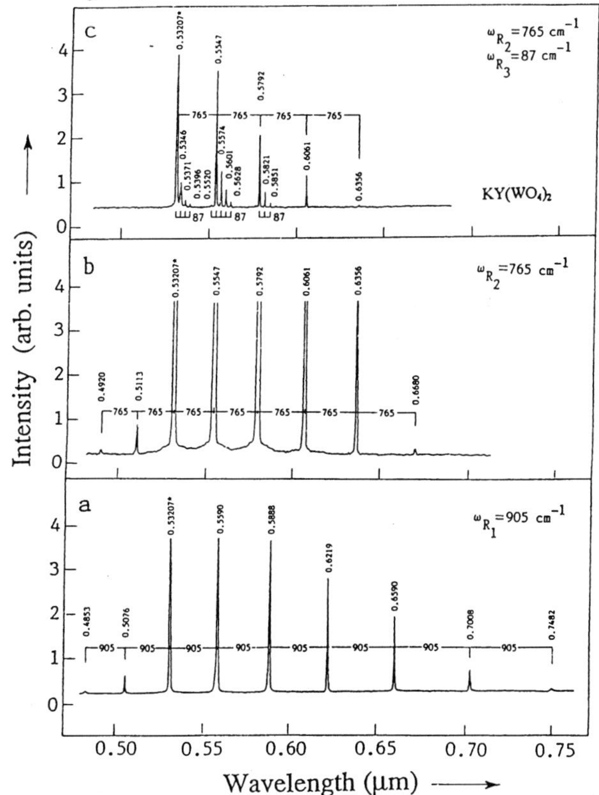

Figure 9. Orientational room-temperature SRS spectra of the monoclinic $KY(WO_4)_2$ single crystals recording in experimental condition at (a) ‖ b, ≈‖ a and (b) ≈‖ a, ‖ b, and (c) ≈‖ a, ≈‖ b. Notations and explanations as in Figure 5.

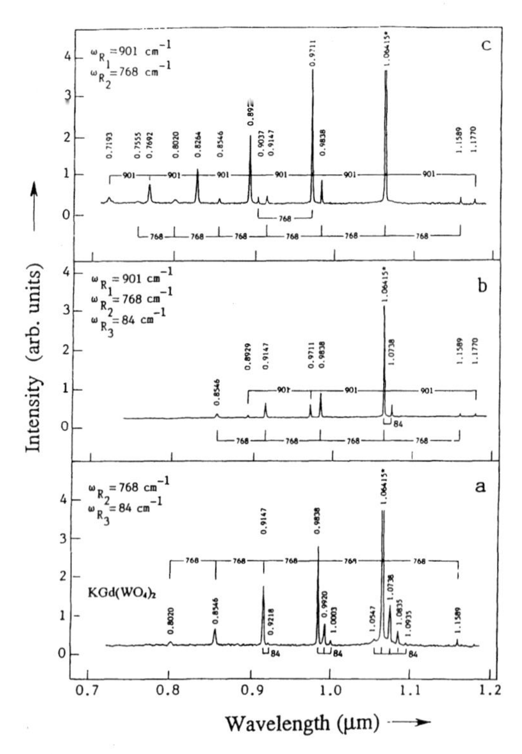

Figure 10. Orientational room-temperature SRS spectra of the monoclinic $KGd(WO_4)_2$ single crystals recording in experimental condition at (a) ≈‖a, ≈‖ b and (c) ≈‖ c, ≈ ‖ b, and (c) ≈‖ c, ‖ b. Notations and explanations as in Figure 5.

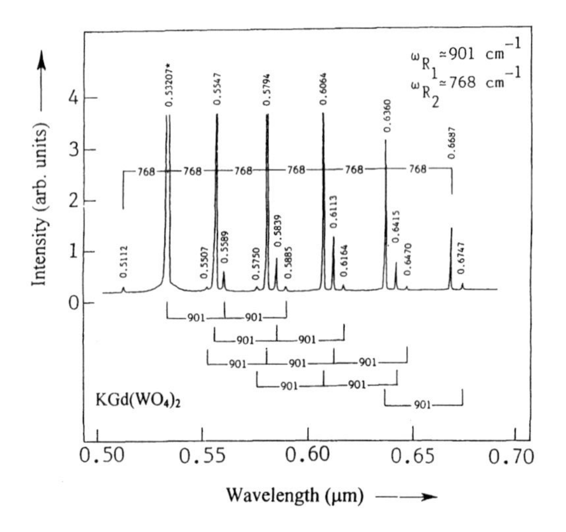

Figure 11. Orientational room-temperature SRS spectra of the monoclinic $KGd(WO_4)_2$ single crystals recording in experimental condition at ≈‖ c, ‖ b. Notations and explanations as in Figure 5.

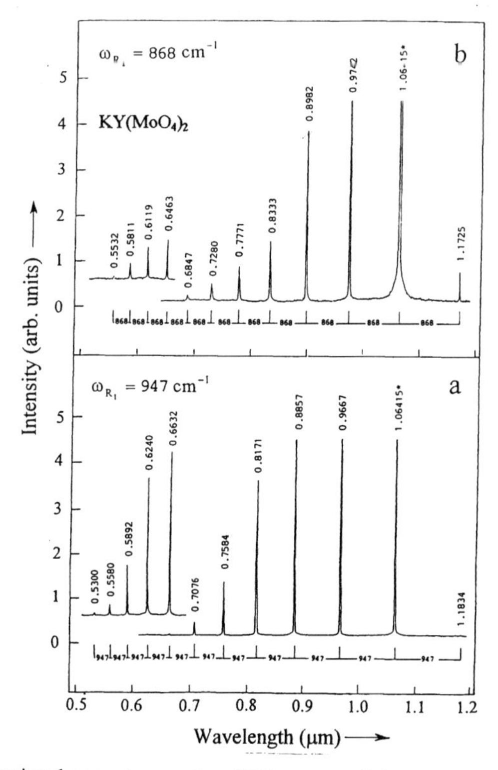

Figure 12. Orientational room-temperature SRS spectra of the monoclinic $KGd(WO_4)_2$ single crystals recording in experimental condition at (a) ≈‖b, ‖ a and (b) ≈‖ a, ≈ ‖ b. Notations and explanations as in Figure 5.

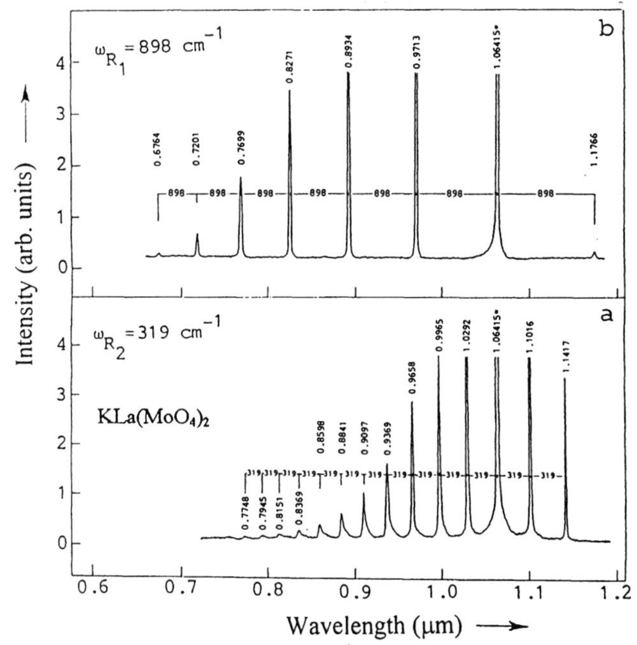

Figure 13. Orientational room-temperature SRS spectra of the monoclinic $KLa(MoO_4)_2$ single crystals recording in experimental condition at (a) ≈‖c, ≈‖ a and (b) ≈‖ a, ≈ ‖ b. Notations and explanations as in Figure 5.

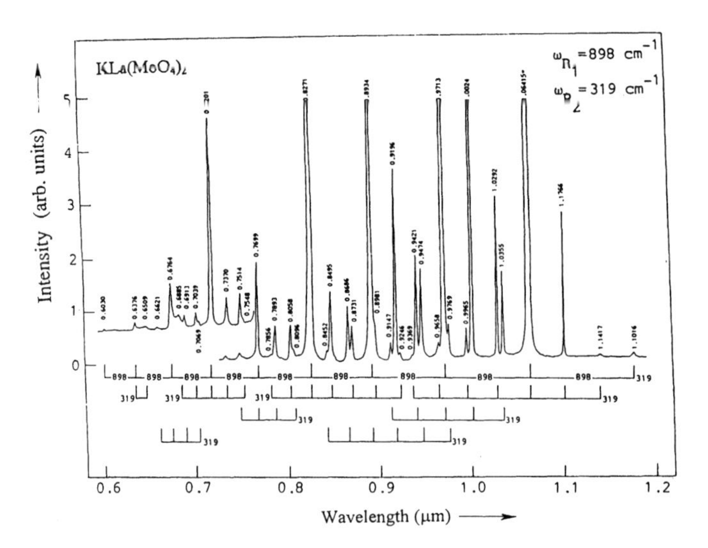

Figure 14. Orientational room-temperature SRS spectra of the monoclinic $KLa(MoO_4)_2$ single crystals recording in experimental condition at ‖ c, ≈ ‖ b. Notations and explanations as in Figure 5.

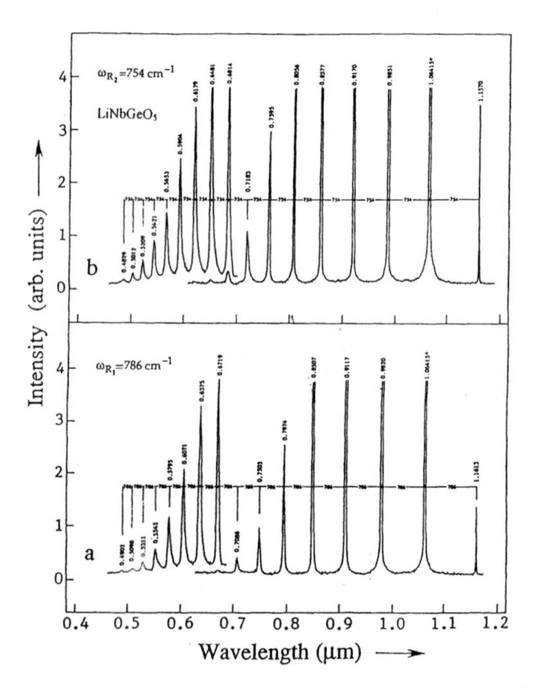

Figure 15. Orientational room-temperature SRS spectra of the monoclinic $KGd(WO_4)_2$ single crystals recording in experimental condition at (a) ‖ b, ‖ c and (b) ‖ c, ‖ b. Notations and explanations as in Figure 5.

Table 3

SRS-active laser host crystals

Crystal	Space group	Lasing ion*	Nonlinearity	SRS-active vibration mode (cm^{-1})**	SRS efficiency***
α-Al_2O_3	D_{3d}^{6}	TM	$\chi^{(3)}$	$\omega_{R_1} = 419$	l
β'-$Gd_2(MoO_4)_3$	C_{2v}^{8}	Ln	$\chi^{(2)} + \chi^{(3)}$	$\omega_{R_1} = 960$	h
				$\omega_{R_2} = 943$	
				$\omega_{R_3} = 857$	
				$\omega_{R4} = 100$	
$LiNbGeO_5$	D_{2h}^{16}	TM	$\chi^{(3)}$	$\omega_{R_1} = 786$	h
				$\omega_{R_2} = 754$	
				$\omega_{R_3} = 699$	
$NaBi(MoO_4)_2$	C_{4h}^{6}	Ln	$\chi^{(3)}$	$\omega_{R_1} = 877$	h
$NaBi(WO_4)_2$	C_{4h}^{6}	Ln	$\chi^{(3)}$	$\omega_{R_1} = 910$	h
$NaLa(MoO_4)_2$	C_{4h}^{6}	Ln	$\chi^{(3)}$	$\omega_{R_1} = 888$	h
				$\omega_{R_2} = 320$	
$KY(MoO_4)_2$	D_{2h}^{14}	Ln	$\chi^{(3)}$	$\omega_{R_1} = 947$	h
				$\omega_{R_2} = 868$	
$KY(WO_4)_2$	C_{2h}^{6}	Ln	$\chi^{(3)}$	$\omega_{R_1} = 905$	h
				$\omega_{R_2} = 765$	
				$\omega_{R_3} = 87$	
$KLa(MoO_4)_2$	C_{2h}^{4}	Ln	$\chi^{(3)}$	$\omega_{R_1} = 898$	h
				$\omega_{R_2} = 319$	
$KGd(WO_4)_2$	C_{2h}^{6}	Ln	$\chi^{(3)}$	$\omega_{R_1} = 901$	h
				$\omega_{R_2} = 768$	
				$\omega_{R_3} = 84$	
$Ca_4Gd(BO_3)_3O$	C_s^{3}	Ln	$\chi^{(2)} + \chi^{(3)}$	$\omega_{R_1} = 933$	h
$LaBGeO_5$	C_3^{2}	Ln, TM	$\chi^{(2)} + \chi^{(3)}$	$\omega_{R_1} = 803$****	h
				$\omega_{R_2} = 790$****	
				$\omega_{R_3} = 385$	

* Ln - trivalent lanthanide ion or TM - transition metal ion.

** Measurement accuracy is ± 3 cm^{-1}.

*** l - low and h - high.

**** Data should be refined.

III. CONTINUOUSLY BROADBAND TUNABLE SHG IN ORTHORHOMBIC β' -Gd_2 $(MoO_4)_3$:Nd^{3+} SINGLE CRYSTAL UNDER FEMTOSECOND HYPER-CONTINUUM COHERENT LIGHT PUMPING

In tunable SHG experiments with β' -Gd_2 $(MoO_4)_3$ and β' -Gd_2 $(MoO_4)_3$:Nd^{3+} crystals we used the hyper-continuum coherent radiation station, developed at the Institute for Laser Sciences of the University of Electro-Communications, Tokyo [18]. A 2 TW Ti^{3+}:Al_2O_3 laser was used as a short pulse generator as shown in Fig. 20. It produced $\approx$125fs pulses of about 250 mJ energy at $\lambda_f \approx 0.79$ μm wavelength, operating at 10 Hz. Multiple self-focusing spots appeared 3m behind the pulse-compressor (a grating pair) due to the nonlinear refractive index of air. The multi-spot beam was focused with a r = 10m gold mirror into a 8m traveling tube with atmospheric pressure Kr gas. The long focusing

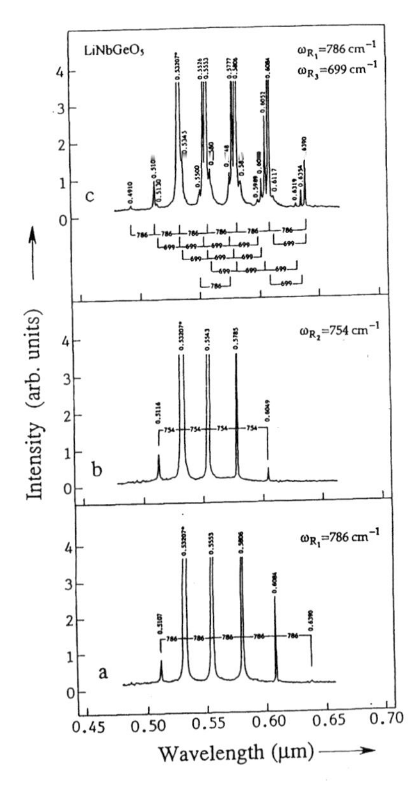

Figure 16. Orientational room-temperature SRS spectra of the monoclinic $LiNbGeO_5$ single crystals recording in experimental condition at (a) ‖b, ‖ c and (b) ‖ c, ‖ b, and (c) ‖ b, ‖ a. Notations and explanations as in Figure 5.

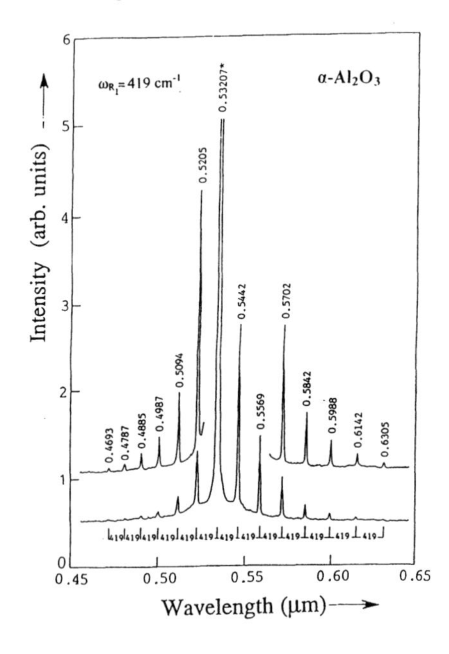

Figure 17. Orientational room-temperature SRS spectra of the trigonal α-Al_2O_3 single crystals recording in experimental condition at ‖ m, ‖ c [14]. Notations and explanations as in Figure 5.

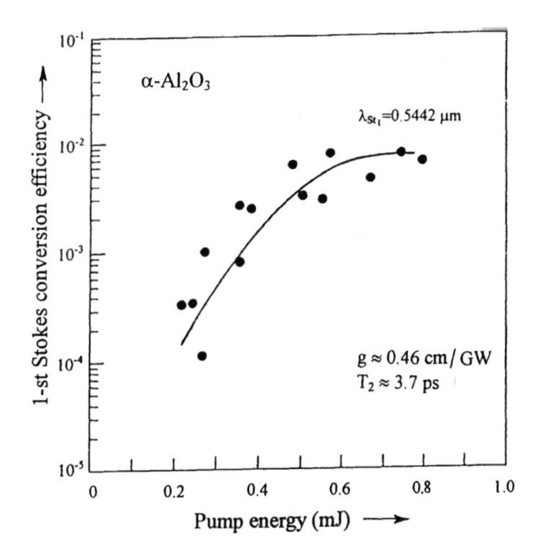

Figure 18. Conversion efficiency for first Stokes generation of pumping energy for Raman laser on the base of trigonal α-Al_2O_3 single crystals in excitation condition ‖a, ‖ c at $\lambda_f = 0.53207$ μm wavelength at 300 K [15]. See also Table 2.

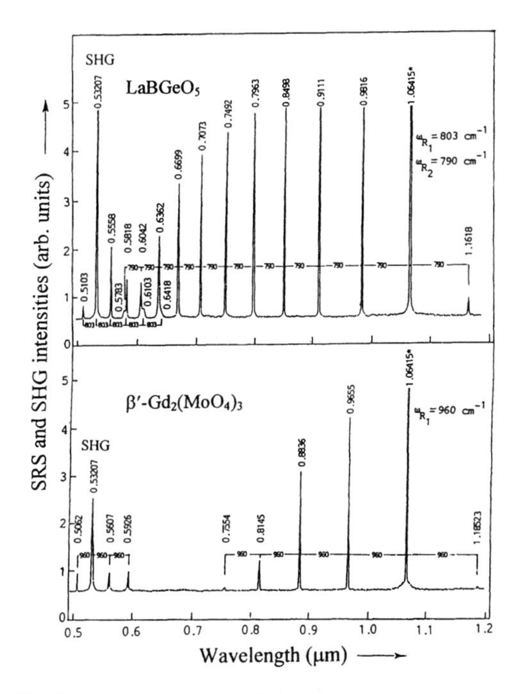

Figure 19. Orientational room-temperature SRS and SHG-SRS spectra of the Trigonal LaBGEO5 and orthorhombic β' -$Gd_2(MoO_4)_3$ single crystals under picosecond Nd^{3+}:$Y_3Al_5O_{12}$ laser excitation along their SHG phase-matching direction for the ee-o interaction [16]. Notations and explanations as in Fig. 5.

geometry produced multichannel self-trapped filaments behind the focal point. The self-phase modulation due to the optical Kerr effect in the Kr gas produced wide-band continuum radiation which extends from the vacuum-ultraviolet to the infrared region (Fig. 21). The hyper-continuum had a spectral intensity of about 1 GW/nm and a pulse duration of about 150 fs. Also, the hyper-continuum had an extremely short temporal coherence, but it had the same spatial and frequency coherency as an ultra-short pulses of Ti^{3+}:Al_2O_3 pumping laser. The beam divergency of the hyper-continuum light was measured to be less than 1 mrad. It was linearly polarized and it had the same polarization as the pump beam at $\lambda_f \approx 0.79$ μm. The phase-matching angle measurements for SHG in our β' -$Gd_2(MoO_4)_3$:Nd^{3+} crystals, a spectral component of the hyper-continuum was selected by a pair of BK-7 glass dispersion prisms and moving 2 mm optical slit. The prisms were placed to compensate the frequency chirp due the the self-phase modulation and the group velocity dispersion in the optics. When a beam from prisms was focused to the β' -$Gd_2(MoO_4)_3$ or β' -$Gd_2(MoO_4)_3$:Nd^{3+} crystals with an f = 500 mm lens, the waist diameter on the surface of the sample was about 200 μm. The spectral intensity and SHG wavelength were monitored by a fiber-coupled grating S-10 monochromator and multi-channel spectroanalyzer (Hamamatsu PMA-11) which was provided with a cooled Si-CCD array for the 0.3 - 0.9 μm spectral area.

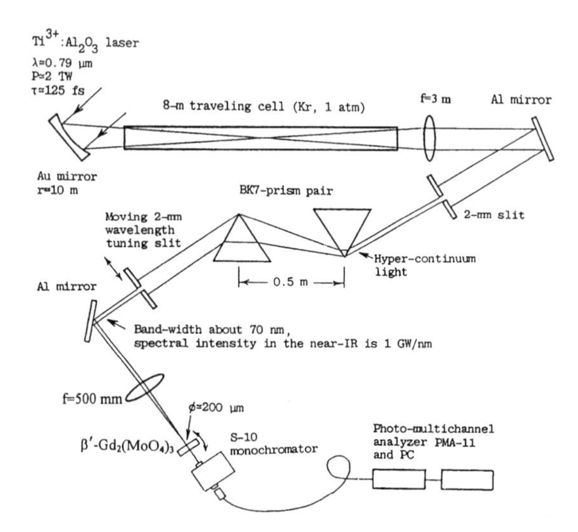

Figure 20. Schematic experimental set-up of the Kr hyper-continuum coherent-radiation station used for investigations of nonlinear interactions in the trigonal LaBGeO5 and orthorhombic β' - $Gd_2(MoO_4)_3$ single crystal [17]. For explanation see text.

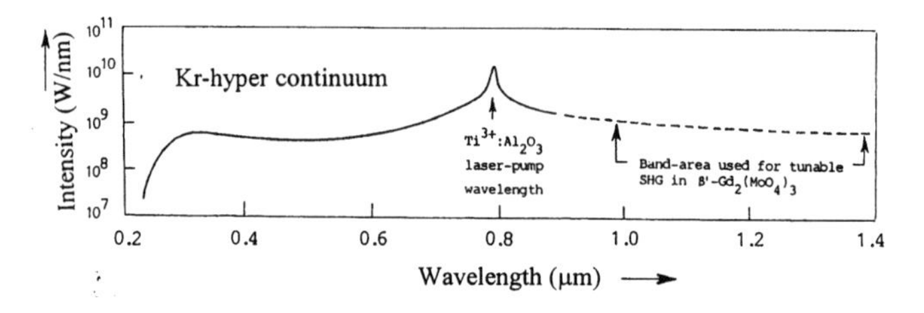

Figure 21. Spectral intensity of the Kr hyper-continuum coherent-radiation station [18]. Dashed line is indicated that the data should be refined.

Several SHG spectra and tuning curves for SHG wavelength versus incident angle for the 4 mm thickness β' -$Gd_2(MoO_4)_3$ crystalline element are shown in Fig. 22 and 23, respectively. The type I (ee-o) phase-matching for β' -$Gd_2(MoO_4)_3$ is possible in a spectral region above $\lambda_{SHG}^{min} \approx 0.487$ μm, i.e. $\lambda_f \approx 0.974$ μm for the fundamental pumping wavelength. In the area around 0.725 μm, the tuning curve is broken (see dashed part of the curve in Fig. 23) due to the decreased intensity of the pump radiation by OH-group absorption in the UV-SiO_2 windows of the Kr-traveling tube. The longest wavelength of about 0.79 μm was limited by the size of the particular β' -$Gd_2(MoO_4)_3$ sample used. In the same excitation condition we have also obtained tunable SHG in the blue-green spectral region in the 3 mm thickness β' -$Gd_2(MoO_4)_3:Nd^{3+}$ element. A detrimental effect on the

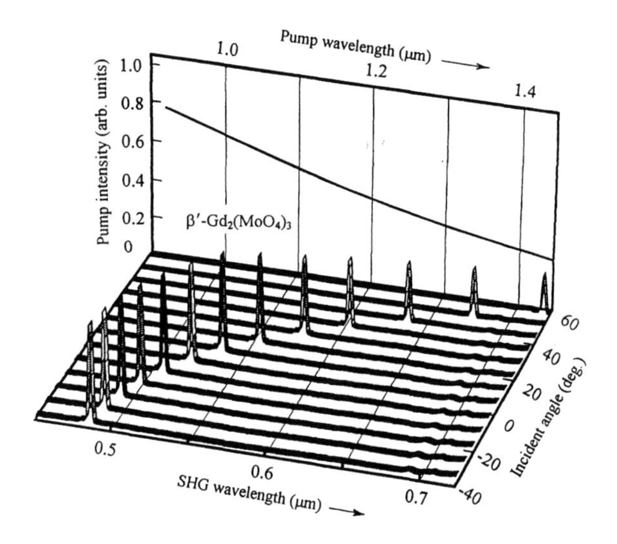

Figure 22. Femtosecond continuously tunable (several spectra) SHG over the entire visible range in the orthorhombic β' -$Gd_2(MoO_4)_3$ crystal at 300 K [17, 19]. For explanation see also text.

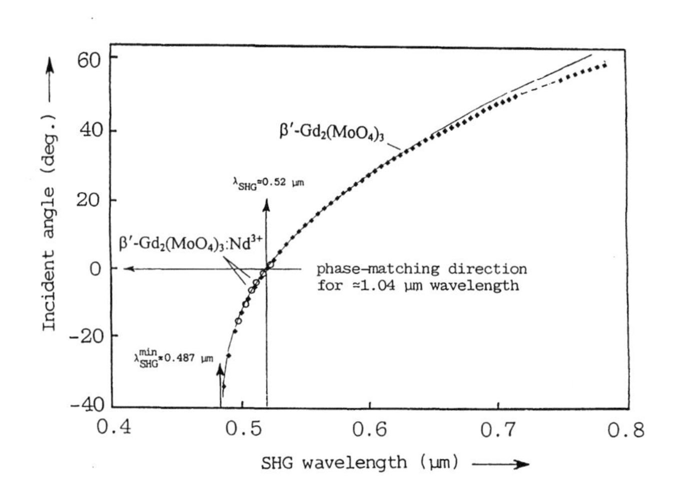

Figure 23. Phase-matching condition (tunable curve) for SHG over the entire-visible range in the orthorhombic β' -$Gd_2(MoO_4)_3$ and β' -$Gd_2(MoO_4)_3$: Nd^{3+} (open circles) crystals at 3000 K [17, 22].

SHG in the green for this crystal is an absorption in $^4I_{9/2}$(insert arrow)$^2K_{13/2}$, $^4G_{7/2,9/2}$ intermanifold transitions of Nd^{3+} ions requires special attention.

We have made the calculation of the values of λ_f^{min} in the β' -$Gd_2(MoO_4)_3$ crystal for the type I (ee-o) and type II (oe-o) SHG based on the conventional calculation as an uniaxial approximation (because for the β' -$Gd_2(MoO_4)_3$, $n_a = n_b$ [21]). The phase-matching angle θ_m for the type I SHG was calculated using the equation

$$\sin 2\theta_m = \frac{\left(n_o^{2\omega}\right)^{-2} - \left(n_o^{\omega}\right)^{-2}}{\left(n_e^{\omega}\right)^{-2} - \left(n_o^{\omega}\right)^{-2}} \tag{9}$$

Here $n_o = n_a = n_b$ and $n_e = n_c$. The type I phase matching-off wavelength is estimated to be 0.9747 μm. As may be seen, the data of our measurement and calculation (see the solid line in Fig. 23) for the ee-o interaction are in good agreement with each other. For the type II (oe-o) phase-matching wavelength-off is estimated as 1.372 μm.

Thus, we have demonstrated broadband, continuously tunable SHG in orthorhombic acentric β' -$Gd_2(MoO_4)_3$ and β' -$Gd_2(MoO_4)_3:Nd^{3+}$ single crystals using femtosecond hyper-continuum coherent light pumping. It is also shown that the hyper-continuum light source is useful for characterization of nonlinear properties of crystals.

IV. SHG WITH "CERENKOV"-TYPE PHASE MATCHING IN TRIGONAL ACENTRIC $LaBGeO_5$ CRYSTAL

Recently [23], we are able to discover new nonlinear potentiality of laser host-crystals: we observed efficient SHG with "Cerenkov"-type phase matching in the bulk trigonal $LaBGeO_5$ nonlinear compound with stillwellite structure, which we developed in 1990 [24-26]. This very attractive coaxial ring SHG was excited practically throughout the visible range (effect of laser "Cerenkov rainbow", Fig. 7c) under femtosecond broadband IR emission of the Kr hyper-continuum station (Fig. 20), as well as at fixed wavelengths in the UV and visible spectral areas under picosecond and nanosecond radiation field of a $Ti^{3+}:Al_2O_3$ (λ_f ~ 0.79 μm) and a Q-switched $Nd^{3+}:Y_3Al_5O_{12}$ (λ_f ~ 1.06415 μm) laser, respectively. The coaxial ring "Cerenkov"-type SHG in $LaBGeO_5$ crystals was photographed with a Nikon F90 camera from a semitransparent screen (Fig. 24).

Before, in [26] we have obtained efficient pulsed SHG in the $LaBGeO_5$ ferroelectric by ordinary type I (ee-o) phase-matching using the "birefringence" phenomenon. Conditions for SHG in this optically positive crystal are

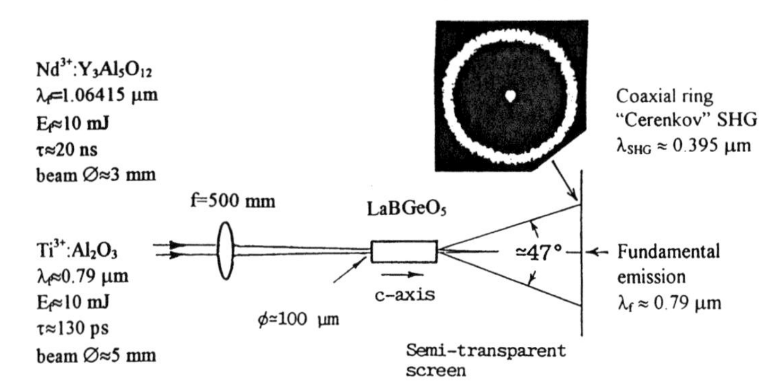

Figure 24. Schematic experimental set-up used in "Cerenkov"-type SHG in nonlinear $LaBGeO_5$ and nanosecond $Nd^{3+}:Y_3Al_5O_{12}$ lasers [23].

$$v^{2\omega} = v^{\omega} \text{ and } n_o^{2\omega} = n_e^{\omega} \tag{10}$$

where v^{ω}, $v^{2\omega}$ and n_e^{ω}, $n_o^{2\omega}$ are the phase velocities and refractive indices for fundamental and second harmonic emission. Fig. 25 explains several examples for the achievement of SHG in $LaBGeO_5$ crystals at 300 K. In particular, for the fundamental emission at λ_f = 1.06415 μm wavelength, the phase matching angle is θ_m ˜ 54°. It is evident from Fig. 25 also that for ee-o interaction in $LaBGeO_5$ crystals the phase matching-off wavelength is about 0.849 μm.

For coaxial ring "Cerenkov"-type SHG the phase-matching condition is very simple:

$$\cos \theta_c = \frac{v^{2\omega}}{v^{\omega}} = \frac{n^{\omega}}{n^{2\omega}} \tag{11}$$

(see also Fig. 25). It is almost always satisfied by nonlinear crystals with normal dispersion, and this is true in a wider spectral range than in the case of the above mentioned collinear "birefringence" SHG in the same material. In particular, as is seen from Fig. 24, we obtained with the $LaBGeO_5$ crystal oriented along the c-axis the "Cerenkov"-type SHG at λ_{SHG} ˜ 0.395 μm wavelength (λ_f ˜ 0.79 μm of a Ti^{3+}:Al_2O_3 laser).

The specific nonlinear properties and the ability of excitation with radiation from a different part of the spectrum of the Kr hyper-continuum station (Fig. 20) enabled us simultaneously to obtain in the single-pass experimental geometry its continuous tunability in the range 0.3 - 0.5 μm the ee-o type (incident angle dependent) and broadband (incident angle independent) "Cerenkov"-type SHG (Fig. 26). In these measurements the femtosecond coherent light from the Kr hyper-continuum station was resolved by a system of prisms into a continuous spectrum from which a moving slit (as in the case with the orthorhombic β' -$Gd_2(MoO_4)_3$ crystal) selected the required interval of about 70 nm

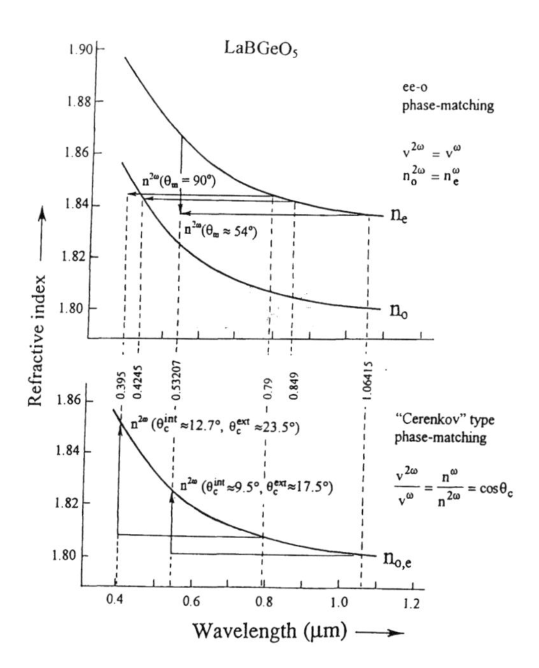

Figure 25. Physical conditions for SHG in the trigonal nonlinear $LaBGeO_5$ single crystal by "birefringence" (ee-o type) and "Cerenkov" phase-matching at 300K. For explanation see text.

halfwidth. The spectral composition of this series was governed by the orientation of the $LaBGeO_5$ crystal relative to the direction of the exciting beam (Fig. 26).

Thus, for the first time in the bulk nonlinear crystal the "Cerenkov"-type SHG was observed. To the best of our knowledge, various manifestations of SHG with "Cerenkov" phase matching have been investigated before only in thin-film $LiNbO_3$ crystalline waveguide structures (see, for example [27]).

V. SELF-FREQUENCY DOUBLING AND SELF-SRS IN LASER HOST-CRYSTALS

Self-frequency conversion phenomena in insulating crystals are old and, at the same time, new problems in laser crystal physics [1]. This chapter includes our new resultsj, which we obtained during the last years in this research direction.

V.A. Cw Self-Frequency Doubled Laser On The Base Of Orthorhombic B' -$Gd_2(Moo_4)_3$:Nd^{3+} Crystal

The first experiments on CW SHG in β' -$Gd_2(MoO_4)_2$ and β' -$Gd_2(MoO_4)_3$:Nd^{3+} crystals were performed in [28]. Due to the relatively high effective tensor element d_{eff} these crystals needed orientation (of thicknesses 3 and 4 mm, respectively) began to generate stable SHG when the fundamental emission power (CW Nd^{3+}:$LiYF_4$ laser, λ_f = 1.0530 μm) was about of 25 mW (Fig. 27). This investigation was the first step in our program to create CW self-frequency doubled laser with laser-diode pumping on the base of the β' - $Gd_2(MoO_4)_3$:Nd^{3+} crystals.

Quite recently, in [29] we have achieved CW self-frequency doubling with β' - $Gd_2(MoO_4)_3$:Nd^{3+} (C_{Nd} ˜ 3 at .% and 4x4x6 mm^3 size) crystal. A simplified operating scheme of this novel crystalline laser is given in Fig. 28. The experiemental set-up and pumping condition by the emission from GaAlAs laser-diode at 0.807 μm wavelength are

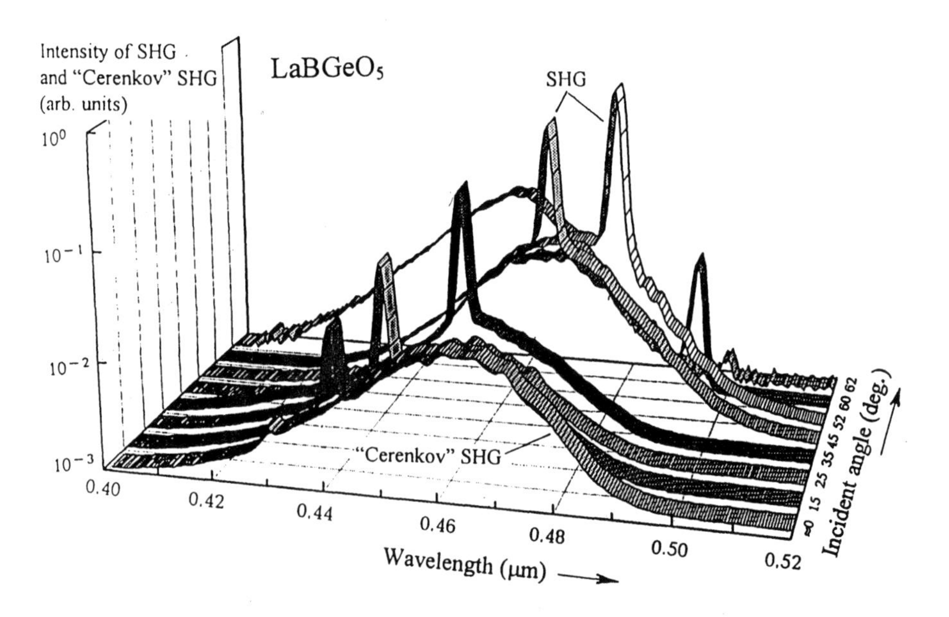

Figure 26. Spectrograms of the "birefringence" (narrow peaks, shifted with variation of the incident angle) and "Cerenkov"-type (broadband coutour) SHG in the trigonal $LaBGeO_5$ single crystal [23].

shown in Figs. 29 and 30, respectively. The CW π-polarized SE ($^4F_{3/2} \rightarrow {}^4I_{11/2}$ generating channel) of this nonlinear laser crystal was converted into SHG with an efficiency of about 0.25% (Fig. 31).

The modern status of this research is shown in Table 4, where are demonstrated some very important data of all known nonlinear-laser crystals doped with Nd^{3+} ions for self-frequency doubled lasers.

Table 4

Phase-matching angles θ_m and effective tensor elements d_{eff} for self-frequency doubling in nonlinear-laser crystals doped with Nd^{3+} ions

Crystal	Fundamental emission		Phase-matching condition for SHG at 300 K		$\|d_{eff}\|$ (pm/V)	Ref.
	λ_{SE} (μm)	Polarization	Type	Angles		
$LiNbO_3$*	1.0933	σ	oo-e	$\theta_m \approx 90°$ **	≈6	[2]
$YAl_3(BO_3)_4$	1.064	σ	oo-e	$\theta_m \approx 30°$; $\varphi_r = 0°$	1.4-2	[3]
	1.318	σ	oo-e	$\theta_m \approx 27°$; $\varphi_r = 0°$	1.9	[30]
$LaBGeO_5$	1.0482	π	ee-o	$\theta_m \approx 54°$; $\varphi_r \approx 39°$	0.3	[4]
	1.3141	π	ee-o	$\theta_m \approx 40°$; $\varphi_r \approx 39°$	0.5	[4]
β'-$Gd_2(MoO_4)_3$	1.0606	π	ee-o	$\theta_m \approx 60°$; $\varphi_r \approx 45°$	1.9	[29]
$Ca_4Gd(BO_3)_3O$	1.0600	σ	oo-e	$\theta_m \approx 46°$; $\varphi_r \approx 90°$	≈1	[31]

* Self-frequency doubling effect was achieved also with $LiNbO_3:Nd^{3+}$ crystals codoped with Mg^{2+} and Sc^{3+} ions. [32,33].

** At elevated temperature.

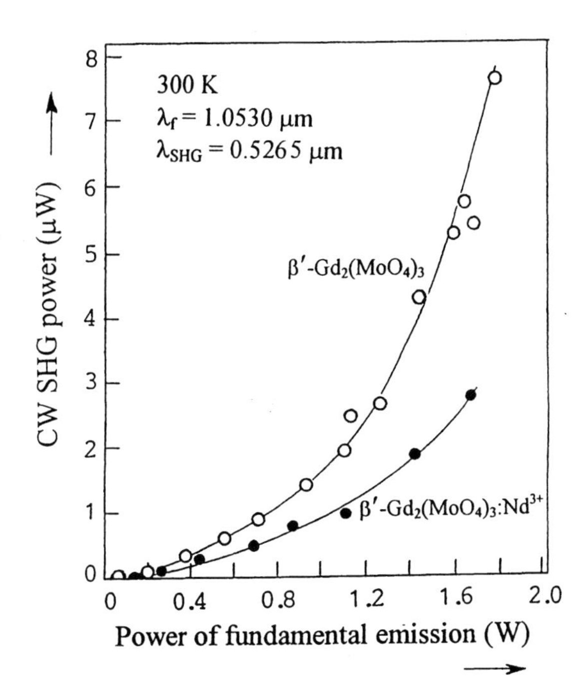

Figure 27. Dependence of the power of CW SHG in the β' -$Gd_2(MoO_4)_3$ and β' -$Gd_2(MoO_4)_3$: Nd^{3+} single crystals on the power of the fundamental Nd^3:LiYF4 laser emission in the single-pass excitation configuration [28].

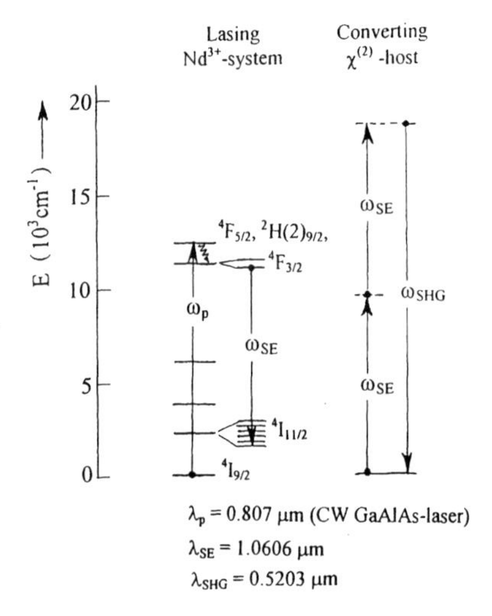

Figure 28. Simplified operating scheme of self-frequency doubled laser on the base of the orthorhombic and β' -$Gd_2(MoO_4)_3$: Nd^{3+} crystal with the $^4F_{3/2} \rightarrow {}^4I_{11/2}$ generating channel and CW GaAlAs laser pumping [29].

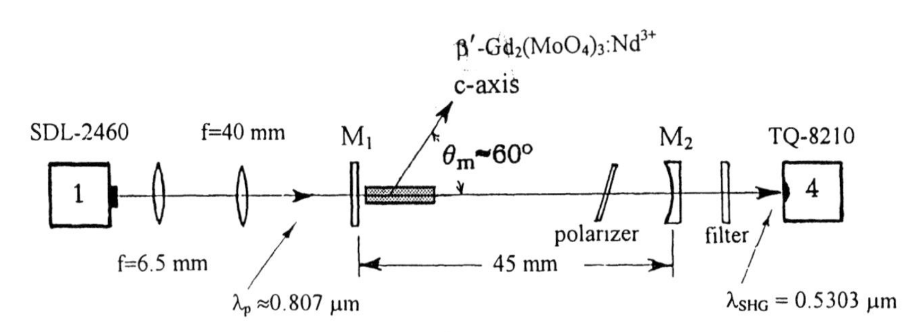

Figure 29. Schematic experimental set-up of the CW self-frequency doubled laser on the base of the β' -$Gd_2(MoO_4)_3$: Nd^3+ crystal with the laser-diode pumping at 300 K. For explanation see text.

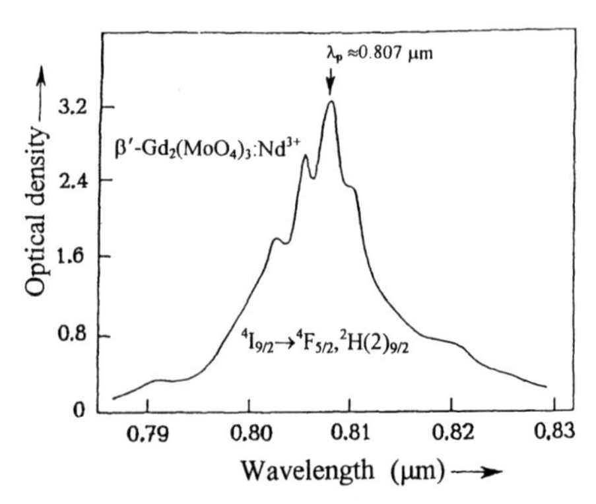

Figure 30. Fragment of the absorption spectrum ($^4I_{9/2} \rightarrow {}^4F_{5/2}$, $^2H(2)_{9/2}$) of the ($C_{nd} \approx 3$ at .% thickness ≈ 6 mm). The crystal is excited at $\lambda_p = 0.807$ μm wavelength (indicated by an arrow by emission of the GaAlAs laser diode.

V.B. Pulsed Self-Pump Frequency Doubled Laser On The Base Of A Trigonal Labgeo5:Nd3+ Crystal

The nonlinear-optical phenomenon of SHG also underline new principles for excitation SE in activated insulating crystals which we briefly consider here. However, this effect is used in completely different aspects. In the experiments which are mentioned in the previous section V.A., the SHG conversions resulted from intrinsic SE of lasing activated nonlinear crystals, and they led to an enrichment of the output spectra of generating crystals. In this new operating laser scheme of crystalline lasers developed in [34] SHG is called on to play a different role: to convert the emission ω_p of the external pumping laser in the doped crystal with high $\chi^{(2)}$ nonlinear susceptibility, so as to match the active absorption lines (bands) of its lasing ions, i.e. to create the conditions required for exciting the SE of these activators. One of such type self-pump frequency doubled laser on the base trigonal nonlinear-laser crystal $LaBGeO_5:Nd^{3+}$ is illustrated by Fig. 32.

V.C. Pulsed Self-Srs Laser On The Base Of Monoclinic Kgd(Wo4)2:Pr3+

For activated insulating laser crystals, SRS can take place in all their classes with and without a center of symmetry (see Chapter II and [1,35,36]). It is important that they have a high enough tensor and that the excitation of sufficiently powerful laser action at the fundamental wavelength of their generating activator ions is achieved. Studies of the last few years have been a success; several pico- and nanosecond crystalline lasers with self-SRS conversion of the $\omega_f = \omega_{SE}$ frequency have been demonstrated, both "down" conversion (Stokes emission or Stokes shift) and "up" conversion (anti-Stokes emission). Among them were the first based on monoclinic $KY(WO_4)_2$ and $KGd(WO_4)_2$ crystals doped with Nd^{3+} ions. Some information about these self-SRS lasers is presented in Table 5.

Quite recently, in [37] we reported priliminary results on the self-SRS conversion phenomenon in new pulsed one-micron laser on the base of monoclinic $KGd(WO_4)_2:Pr^{3+}$ ($^1D_2 \rightarrow {}^3F_4$ SE channel) [38-40]. The principle of this novel laser is explained in Fig. 33 and its set-up shown in Fig. 34. As illustrated in Fig. 34, the oriented (along the b-axis) rod

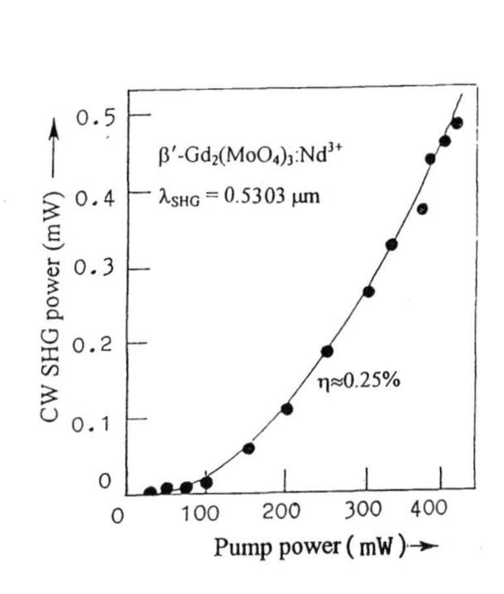

Figure 31. Input-output dependence of single mode CW self-frequency doubled laser on the base of the orthorhombic β'-$Gd_2(MoO_4)_3$: Nd^{3+} laser with laser-diode pumping.

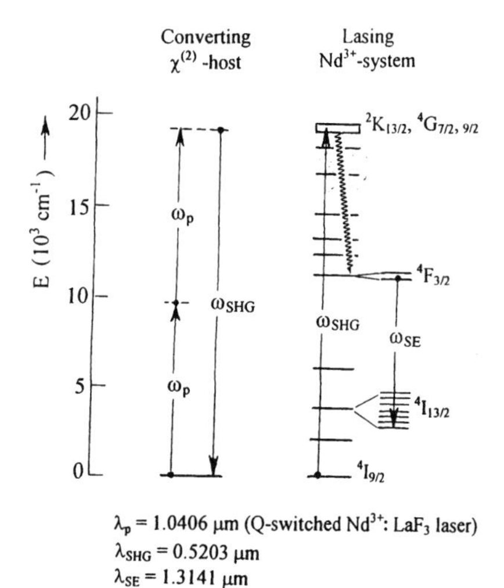

Figure 32. Simplified operating scheme of pulsed self-frequency doubled laser on the base of the trigonal $LaBGeO_5:Nd^{3+}$ crystal with the $^4F_{3/2} \rightarrow {}^4I_{13/2}$ generating channel and the Q-switched $Nd^{3+}:LaF_3$ one-micron laser [34].

Table 5

Pulsed self-SRS crystalline lasers

Crystal (orientation)	Lasing ion	SE generation: Channel	SE generation: λ_{SE} (μm)	SE generation: Mode	Stokes and anti-Stokes emission: Line	Stokes and anti-Stokes emission: Wavelength* (μm)	Ref.
$KY(WO_4)_2$ (‖ b-axis)	Nd^{3+}	${}^4F_{3/2}\rightarrow{}^4I_{11/2}$	1.0688	ps**	ASt_1	0.9746	[7]
					St_1	1.1833	
					St_2	1.3252	
$KGd(WO_4)_2$ (‖ b-axis)	Nd^{3+}	${}^4F_{3/2}\rightarrow{}^4I_{11/2}$	1.0672	ps**	ASt_1	0.9736	[6,7]
					St_1	1.1808	
					St_2	1.3213	
	Pr^{3+}	${}^1D_2\rightarrow{}^3F_4$	1.0657	μs***	ASt_1	0.9724	[37]
					St_1	1.1790	
					St_2	1.3191	

* Recalculated values taking into account of refined data on SRS-active optical vibration modes of monoclinic $KY(WO_4)_2$ and $KGd(WO_4)_2$ crystals of used orientation (see Table 1).

** Train of picosecond pulses.

*** Train of nanosecond pulses.

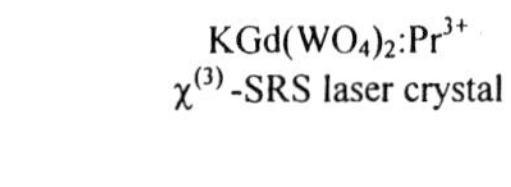

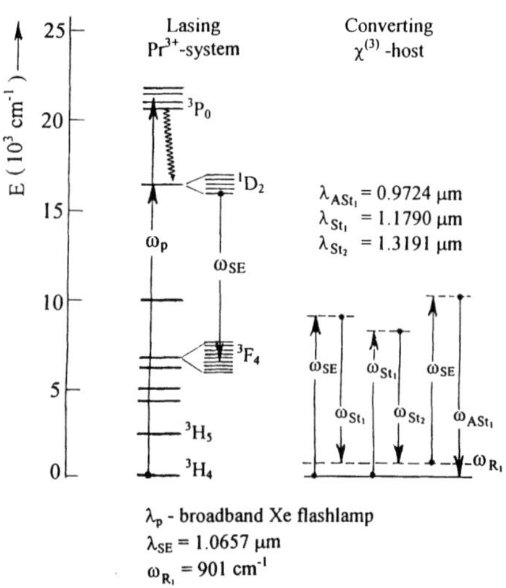

Figure 33. Simplified operating scheme of pulsed self-SRS laser on the base of the monoclinic $KGd(WO_4)_2:Pr^{3+}$ crystal with the ${}^1D_2\rightarrow{}^3F_4$ generating channel and the Xe-flash-lamp pumping at 300 K.

IZ-25 head of
Q-switched LTIPCh-7 laser

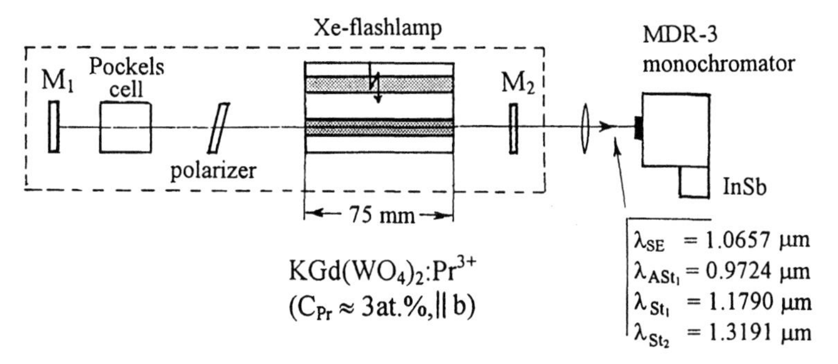

Figure 34. Schematic experimental set-up of self-SRS laser on the base of the monoclinic $KGd(WO_4)_2$:Pr^{3+} crystal with Xe-flash-lamp pumping. For explanation see text.

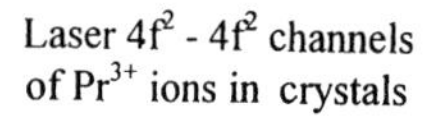

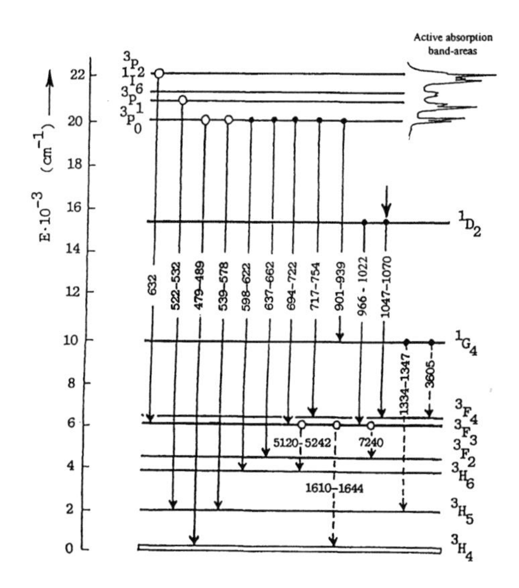

Figure 35. Laser intermanifold 4f2-4f2 channels of Pr3+ ions in crystals. Solid and dashed arrows are room-temperature and cryogenic laser transitions, light and dark circles indicate some of them with selective laser and broadband Xe-flash-lamp pumping, respectively. Generation spectral ranges of the SE channels are given in nm.

Table 6

Crystal-field splitting of $^{2S+1}L_J$ manifold of Pr^{3+} ions in monoclinic $KGd(WO_4)_2$ crystal at 77 K

$^{2S+1}L_J$	Stark-level energy (cm^{-1})	Number of levels		ΔE (cm^{-1})
		Theory	Exp.	
$^3H_4$	0, 39, 157, 181*, 190, 363, 616, 640	9	8	640*
$^3H_5$	2190, 2196, 2222, 2237, 2371*, 2392, 2415, 2553, 2670, 2702, 2725*	11	11	535*
$^3H_6$	4286, 4298, 4317, 4343, 4532, 4614, 4780	13	7	494*
$^3F_2$	5038, 5103, 5142, 5182, 5196	5	5	158
$^3F_3$	6455, 6466, 6474, 6487, 6522, 6533, 6565	7	7	110
$^3F_4$	6829, 6876, 6899, 6972, 6993*, 7045, 7105, 7120, 7135	9	9	306
$^1G_4$	9683, 9697, 9760, 9831, 99 5, 10045, 10105	9	7	422*
$^1D_2$	16346, 16655, 16745, 16865, 16880*, 17095	5	6	749*
$^3P_0$	20382	1	1	-
$^3P_1+^1I_6$	20840, 20897, 20939, 21014, 21024, 21210, 21310, 21425, 21710, 21758	16	10	(918*)

* Stark-level energy and complete splitting ΔE of the manifolds should be refined.

of the $KGd(WO_4)_2:Pr^{3+}$ (C_{Pr} ~ 3 at .%) placed into the IZ-25 head (instead of $Y_3Al_5O_{12}:Nd^{3+}$ laser crystal with the same size) of commercial Q-switched LTIPCh-7 type laser with Xe-flashlamp pumping generated two Stokes and one anti-Stokes components (ω_{R1} = 901 cm^{-1}). The maximum intercavity peak-power density at the fundamental emission (λ_f = 1.06576 μm) was estimated at about 30 GW/cm^2. But, the first Stokes emission arose rapidly with a SE intensity of about 2 GW/cm^2.

The discovered new one-micron SE channel ($^1D_2 \rightarrow {}^3F_4$) of Pr^{3+} ions in monoclinic tungstates [38] is very attractive for many applications, because its inter-Stark peak cross-section is higher than that of the main laser $^4F_{3/2} \rightarrow {}^4I_{11/2}$ channel of Nd^{3+} ions ub widely used $Y_3Al_5O_{12}$ crystal. The terminal $^3F_4$ state of this new one-micron SE channel is located at ~ 7000 cm^{-1} (Fig. 35), which is very important spectroscopic property of lasing Pr^{3+} ions. The new two Fig. 36 and 37 show the oriented luminescence spectra ($^1D_2 \rightarrow {}^3F_4$) of Pr^{3+} ions in monoclinic $KGd(WO_4)_2$ crystal (refined data) are listed in Table 6.

V.D. Pulsed Self-Pump Srs Three-Micron Laser On The Base Of Monoclinic Ky(Wo4)2 Crystal

The monoclinic SRS-active $KGd(WO_4)_2:Er^{3+}$ crystal had C_{Er} = 30 at .% and the orientation along its b-axis. In this case the pumping source also was a Q-switched laser, but on the base of $Al_2O_3:Ti^{3+}$ sapphire (Fig. 38). Other details of this experiment can be found in [34,41].

VI. ACKNOWLEDGMENTS

This work was carried out with partial financial support of the Russian Foundation for Basic Research, Russian State Scientific-Technical Programs "Fundamental Metrology" and "Fundamental Spectroscopy", as well as Scientific Cooperation Program DFG/RFFI

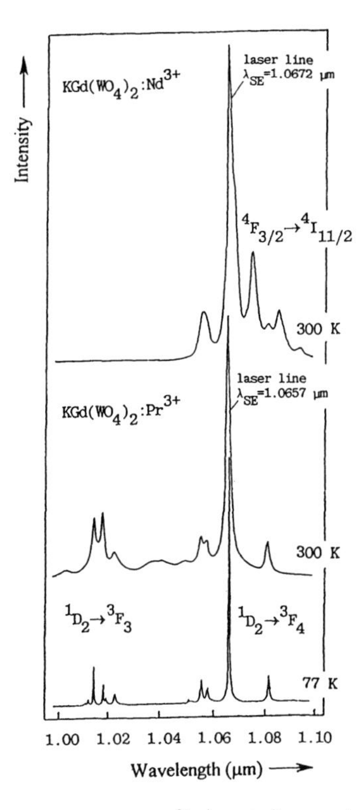

Figure 36. Oriented luminescence spectra of Pr^{3+} ($^1D_2 \rightarrow {}^3F_4$ and $^1D_2 \rightarrow {}^3F_3$ channel at 77 and 300 K) and Nd^{3+} ions ($^4F_{3/2} \rightarrow {}^4I_{11/2}$, 300K) in monoclinic $KGd(WO_4)_2$ crystals (along the b-axis) [38].

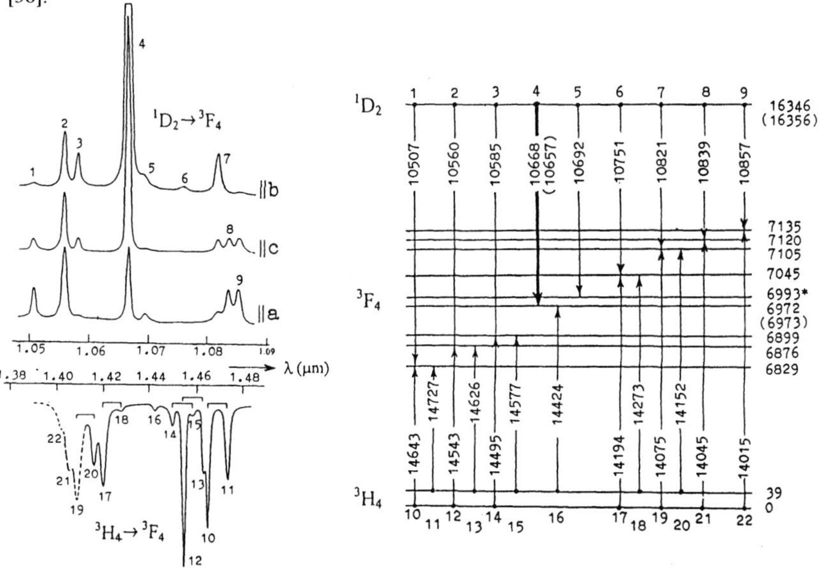

Figure 37. Luminescence ($^1D_2 \rightarrow {}^3F_4$ channel for three directions of a registration) and absorption ($^3H_4 \rightarrow {}^3F_4$) spectra, as well as the crystal-field splitting scheme for $^1D_2$,$^3F_4$ and $^3H_4$ manifolds of Pr^{3+} ions in $KGd(WO_4)_2$ crystals. The part of the absorption spectrum detected in a condition under atmospheric water absorption is shown by a dashed line. Stark-level energies in the scheme are given in cm^{-1}, and wavelengths of transitions between them in Å. The splitting of the ground $^3H_4$ state (39 cm-1) is shown by square brackets in the absorption spectrum. Thick arrow in the energy scheme indicated the laser transition, where the corresponding data for 300 K are also presented in parentheses. The asterisk indicates data which should be refined [40].

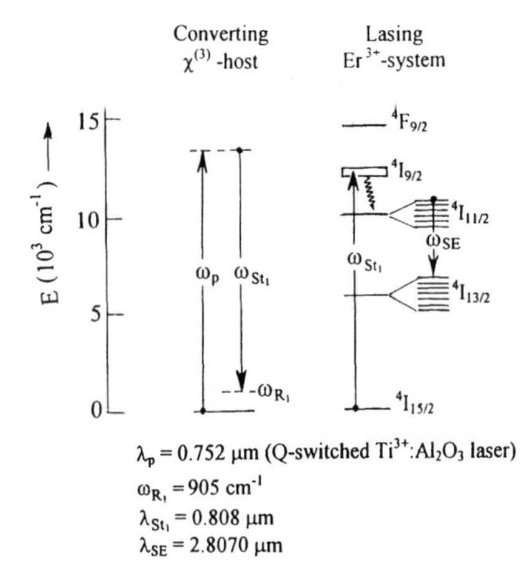

Figure 38. Simplified operation scheme of pulsed self-pump SRS laser on the base of the monoclinic KY(WO4)2Er3+ crystal with generating 4I11/2 ➔ 4I13/2 channel and with Q-switching Ti3+:Al2O3 laser pumping [10].

between Germany and Russia (grant No. 436 RUS 113/115 O(R)). The author also notes that the investigations were considerably progressed due to the cooperation with groups of Professors H. J. Eichler, S. N. Bagayev, K. Ueda, and A. Z. Grasyuk within Joint Open Laboratory for Laser Crystals and Precise Laser Systems.

REFERENCES

[1] A. A. Kaminskii, Crystalline Laser: Physical Processes and Operating Schemes, CRC Press, Boca Raton (1996).

[2] V.G. Dmitriev, R. E. Raevskii, N. M. Rubina, L. N. Rashkovich, O. O. Silichev, and A. A. Fomichev, Zh. Tekh. Fiz. Pis'ma 5, 1400 (1979).

[3] L. M. Dorozhkin, I. I. Kuratev, N. I. Leonyuk, T. I. Timoshenko, and A. V. Shestakov, Zh. Tekh. Fiz. Pis'ma 7, 1297 (1981).

[4] A. A. Kaminskii, S. N. Bagayev, A. V. Butashin, and B. V. Mill, Neorg. Mater. (Russia) 29, 545 (1993).

[5] M. J. Ki, M. deMicheli, Q. He, and D. B. Ostrowsky, IEEE J. Quantum Electronics 26, 1384 (1990).

[6] A. M. Ivanyuk, P. A. Shachverdov, V. D. Belyaev, M. A. Ter-Pogosyan, and V. L. Ermolaev, Opt. Specktroskopiya (Russia) 58, 967 (1985).

[7] K. Andryunas, Yu. Vishakas, V. Kobelka, I. V. Mochalov, A. A. Pavlyuk, G. T. Petrov, and B. Syrus, Zh. exper. teor. Fis. Pis'ma 42, 333 (1985).

[8] A. A. Kaminskii, Kvantovaya Electronika (Russia) 20, 532 (1993).

[9] A. A. Kaminskii, H. J. Eichler, D. Grebe, R. Macdonald, S. N. Bagayev, A. A. Puvlyuk, and F. A. Kuznetsov, Phys. Status Solidi (a)153, 281 (1996).

[10] A. A. Kaminskii, in Spectroscopy and Dynamics of Collective Excitations in Solids, B. Di Bartolo ed., Plenum Press, New York and London 1997, p. 493

[11] A. Penzkofer, A. Laubereau, and W. Kaiser, Prog. Quantum Electronics 6, 55 (1979).

[12] H. J. Eichler and B. Liu, Optical Materials 1, 21 (1992).

[13] A. A. Kaminskii, H. Nishioka, Y. Kubota, K. Ueda, H. Takuma, S. N. Bagayev, and A. A. Pavlyuk, Phys. Status Solidi (a) 148, 619 (1995).

[14] A. A. Kaminskii, A. A. Kaminskii, H. J. Eichler, D. Grebe, R. Macdonald, and A. V. Butashin, Phys. Status Solidi (b), 199, R3 (1997).

[15] A. Z. Grasyuk, S. B. Kubasov, L. L. Losev, A. P. Luzenko, A. A. Kaminskii, V. B. Semenov, Kvantovaya Elektronika (Russia) (in press).

[16] A. A. Kaminskii, H. J. Eichler, D. Grebe, R. Macdonald, A. V. Butashin, S. N. Bagayev, and A. A. Pavlyuk, Phys. Status Solidi (b), 198, K9 (1996).

[17] A. A. Kaminskii, A. V. Butashin, H. J. Eichler, D. Grebe, R. Macdonald, K. Ueda, H. Nishioka, W. Odajima, M. Tateno, J. Song, M. Musha, S. N. Bagayev, and A. A. Pavlyuk, Optical Materials 7, 59 (1997).

[18] H. Nishioka, W. Odajima, K. Ueda, and H. Takuma, Opt. Letters 20, 2505 (1995).

[19] A. A. Kaminskii, A. V. Butashin, S. N. Bagayev, H. J. Eichler, D. Grebe, and R. Macdonald, Kvantovaya Elektronika (Russia), 24, 750 (1997).

[20] A. A. Kaminskii, G. Aka, J. Findaisen, D. Vivien, H. J. Eichler, R. Macdonald, and D. Pelenc (in press).

[21] Handbook of Laser Science and Technology, Vol. 1, Part 1, M. J. Weber ed., CRC Press, Boca Raton 1986, p. 142

[22] H. Nishioka, W. Odajima, M. Tateno, K. Ueda, A. A. Kaminskii, A. V. Butashin, S. N. Bagayev, and A. A. Pavlyuk, Appl. Phys. Letters 70, 1366 (1997).

[23] A. A. Kaminskii, H. Nishioka, K. Ueda, W. Odajima, M. Tateno, K. Sasaki, and A. V. Butashin, Kvanovaya Elektronika (Russia), 23, 391 (1996).

[24] A. A. Kaminskii, B. V. Mill, and A. V. Butashin, Phys. Status Solidi (a), 118, K59 (1990).

[25] A. A. Kaminskii, B. V. Mill, and A. V. Butashin, Phys. Status Solidi (a), 118, K59 (1990).

[26] A. A. Kaminskii, A. V. Butashin, I. A. Maslyanizin, B. V. Mill, V. S. Mironov, S. P. Rozov, S. E. Sarkisov, and V. D. Shigorin, Phys. Status Solidi (a), 125, 671 (1991).

[27] M. P. DeMicheli, in Guided Wave Nonlinear Optics: Proceedings of the NATO Advanced Study Institute, D. B. Ostrowsky and R. Reinisch Eds., Vol. 214, Kluwer Academic Press, Dordrecht 1992, p. 147.

[28] A. A. Kaminskii, K. Ueda, S. N. Bagayev, A. A. Pavlyuk, D. Song, H. Nishioka, N. Uehara, and M. Musha, Kvantovaya Elektronika, 23, 389 (1996).

[29] A. A. Kaminskii, S. N. Bagayev, K. Ueda, A. A. Pavlyuk, and M. Musha, Kvantovaya Elektronika (Russia), (in press).

[30] L Baoshing, W. Jun, P. Hengfu, J. Minhua, L. Enquan, and H. Xueyuan, Chinesse Phys. Letters, 3, 413 (1986).

[31] G. Aka, A. Kahn-Harari, D. Vivien, J. M. Benitez, F. Salin, and J. Godard, Eur. J. Solid State Inorg. Chem. 33, 727 (1996)

[32] M. Gong, G. Xu, K. Hah, and G. Zhai, Electron. Letters, 26, 2062 (1990).

[33] J. K. Yamamoto, A. Sugimoto, and K. Yamagishi, Optics Letters, 19, 1311 (1994).

[34] A. A. Kaminskii, Kvantovaya Elektronika (Russia), 20, 532 (1993).

[35] Y. R. Shen, The Principles of Nonlinear Optics, Wiley, New York 1984.

[36] N. F. Nye, The Physical Properties of Crystals, Clarendon Press, Oxford 1960.

[37] A. A. Kaminskii, "XXI Vavilov Reading" (Lebedev Physica Institute, March 1997, Moscow) and "Cristaux Laser" (CPH, Mai 1997, Les Houches).

[38] A. A. Kaminskii, S. N. Bagayev, and A. A. Pavlyuk, Phys. Status Solidi (a), 151, K53 (1995).

[39] A. A. Kaminskii, S. N. Bagayev, L. Li., F. A. Kuznetsov, and A. A. Pavlyuk, Kvantovaya Elektronika (Russia), 23, 3 (1996).

[40] A. A. Kaminskii, L. Li., A. V. Butashin, V. S. Mironov, A. A. Pavlyuk, S. N. Bagayev, and K. Ueda, Optical Review, 4, 309 (1997)

[41] A. A. Kaminskii, J. de Physique IV, 4, 297 (1994).

UNIFICATION IN PARTICLE PHYSICS: SYMMETRIES AND HIDDEN DIMENSIONS

Giovanni Costa

Dipartimento di Fisica "Galileo Galilei", Università di Padova
Istituto Nazionale di Fisica Nucleare, Sezione di Padova
Via Marzolo, 8 – 35131 Padova (Italy)

ABSTRACT

The properties of matter at the atomic level are expressed in terms of a few fundamental constants, while, going down to the level subnuclear constituents, the present theoretical description requires the introduction of a large number of independent parameters. The main goal of elementary particle physics is to arrive at a complete theory which describes all subnuclear phenomena in terms of a unified interaction with the least number of fundamental constants. We discuss the lines of research in this field, which are mainly based on the implementation of higher symmetries. Grand Unified Theories unify the couplings of the fundamental interactions and, with the implement of family symmetry, some of the parameters of the basic fermions become related. Supersymmetry is an essential ingredient of the present theories and the introduction of hidden space–time dimensions appears to be a common feature. The gravitational interaction plays an important rôle in the unifying picture, linking low–energy parameters with the physics at the Plank scale.

1. INTRODUCTION

The description of the physical reality requires the introduction of different ingredients, which are adopted according to the energy scale (or length scale) of the phenomena. The investigation of the structure of matter shows a hierarchy of levels each of which is, to some extent, independent of the others.

At the first level, ordinary matter is described in terms of atoms composed by nuclei and electrons, and this is the level of atomic physics. At the second level, atomic nuclei reveal their structure, which is described in terms of nucleons (protons and neutrons).

At the third level, even nucleons appear to be composed of more elementary constituents: the quarks. At this level, ordinary matter seems to require the presence of only three ingredients: electrons and two kinds of quarks (*up* and *down*). As a

Ultrafast Dynamics of Quantum Systems: Physical Processes and Spectroscopic Techniques, Edited by Di Bartolo and Gambarota, Plenum Press, New York, 1998

matter of fact, one has to add a fourth particle, which is very elusive but is copiously produced around us: the neutrino. After the World War II many new particles were discovered in high energy reactions in cosmic rays and later at the accelerators. They belong to two different groups: *hadrons* and *leptons*. The former are bound states of quarks; the latter have similar properties to those of electrons and neutrinos; they show no structure, up to the present energies.

The analysis of the hadronic spectrum and of the high energy reactions revealed the existence of six different kinds of quarks, each of which is distinguished by a specific quantum number, called "flavour". On the side of leptons, three charged states and three types of neutrinos were discovered. Now we believe that there are three *families* or *generations* of elementary particles, which share similar properties.

In atomic physics all phenomena are described very well without the need of a detailed description of the atomic nucleus. Besides the two universal constants, $c =$ speed of light in vacuum and $\hbar =$ Planck constant, one has to introduce two other fundamental constants: the electric charge e and the mass m_e of the electron. These constants provide the natural units for atomic physics. The dimensionless constant $\alpha = e^2/4\pi\hbar c \simeq 1/137$ sets the strength of the electromagnetic interactions; the length scale is given by the Bohr radius $a_o = \hbar/mc\alpha \simeq 0.5 \times 10^{-8} cm$, while the typical energy scale is of the order of a few eV.

In terms of these constants, combining relativity with quantum mechanics, Dirac was able to derive from his equation (which includes spin and particle–antiparticle doubling) the electron magnetic moment $\mu = e\hbar/2m_e c$. Later on, Quantum Electrodynamics was built as a respectable quantum field theory, and it was possible to calculate higher order corrections for the electron magnetic moment μ as well as for the Lamb shift, providing precise tests in agreement with experiments.

Going from atomic physics to elementary particles, one has to introduce many more basic ingredients: three families of quarks and leptons which share two other kinds of couplings: the *strong* and *weak* interactions, besides the electromagnetic ones. All these ingredients are included in the so–called Standard Model (SM), which is a quantum field theory describing extremely well all the present experimental data up to energy scales of a few $TeV = 10^{12} eV$.

However, notwithstanding its remarkable phenomenological success, the SM cannot be considered the fundamental theory of elementary particles. In fact, there are some reasons for which it cannot be considered satisfactory from the theoretical point of view. In particular, it contains a large number of independent parameters which, in the minimal version, amounts to 19! Many theoretical investigations have been performed to build a more complete theory, either by implementing higher symmetries, or by expanding the space–time dimensions, or by going beyond field theory.

In the following sections, after a coincise review of the essential ingredients of the SM, we shall consider two main roads that may lead to a unified picture with a small number of fundamental constants: the first is the way to Grand Unified Theories (GUT's) and family symmetries, along which the micro–world appears more and more symmetric; the second is the Kaluza–Klein way, along which the micro–world reveals hidden extra space dimensions. Unfortunately, neither of these roads leads to the final goal, since the nice features are accompanied by serious difficulties that cannot be solved. Other questions remain unanswered: which are the fundamental constants in particle physics? What is the rôle of the different families?

We should point out that the phenomena occurring in the portion of the Universe that we can explore around us are explained, except for small perturbations, in terms of the first light family alone. The understanding of the rôle of the different flavours and of the seemingly "useless" heavy families represent a big challenge, and may provide important clues for solving fundamental questions.

Much hope is addressed to superstring theory that shares the nice features of both GUT's and Kaluza–Klein theories, but can solve their difficulties and might answer to the above questions.

2. THE BASIC CONSTITUENTS OF MATTER AND THEIR INTERACTIONS

There are two different cathegories of elementary particles: *hadrons* and *leptons*. In fact, hadrons are not elementary, but they are composed of more basic constituents: the *quarks*. They are separated into two distinct classes:*mesons* (with integer spin $S = 0, 1, 2...$), which are quark–antiquark bound states, and *baryons* (with half–integer spin $S = 1/2, 3/2...$) which are bound states of three quarks. Quarks interact strongly, and never appear as free particles; they are permanently bound inside hadrons. Leptons are spin 1/2 particles; they interact weakly and have no structure down to length scales of $10^{-16} \sim 10^{-17}$ cm. There are charged leptons, like the electron, and neutral ones: the neutrinos.

The basic constituents of matter are then *quarks* and *leptons*, all of which are spin 1/2 fermions. They are grouped in three similar "families" or "generations" as shown in Table I.

Table 1. Quarks and Leptons

	Q=2/3	Q= −1/3	Q=−1	Q=0
1st family	u (up) $m_u \sim 5$	d (down) $m_d \sim 10$	e $m_e = 0.511$	ν_e $m_\nu < 5 \times 10^{-6}$
2nd family	c (charm) $m_c \simeq 1500$	s (strange) $m_s \sim 200$	μ $m_\mu = 105.66$	ν_μ $m_\nu < 0.17$
3rd family	t (top) $m_t \simeq 175 \times 10^3$	b (bottom) $m_b \simeq 4.4 \times 10^3$	τ $m_\tau = 1.777 \times 10^3$	ν_τ $m_\nu < 24$

The values of the electric charge Q are given in units of the elementary charge e; the mass values [1] are given in $MeV = 10^6 eV$. Each family has exactly the same structure and the corresponding members have similar properties. We consider in more detail the first family:

$$\text{quarks} \begin{pmatrix} u_1 u_2 u_3 \\ d_1 d_2 d_3 \end{pmatrix} \qquad \text{leptons} \begin{pmatrix} e \\ \nu_e \end{pmatrix}$$

The quarks appear in two different states of electric charge $Q = 2/3, -1/3$. All together there are six kinds of quarks characterized by six different "flavours"; moreover each quark appears in a triplet of states which are distinguished only by an internal symmetry attribute called "colour". The other members of the family are two (singlet state) leptons: the electron ($Q = -1$) and its neutral partner, i.e. the neutrino. Ordinary matter contains only the members of the first family: atomic nuclei are made of quarks; atoms and molecules contain nuclei and electrons. Neutrinos are produced in radioactive decays of nuclei and in nuclear reactions in the interior of the stars.

The members of the second and third families are much heavier and are unstable. The μ–lepton and "strange" hadrons containing the s–quark have been discovered in the interactions of cosmic rays, while the τ–lepton and hadrons with "charm", "beauty", as well as the heaviest of all, the top–quark, have been discovered and detected at high energy accelerators.

There are three kinds of interactions which are relevant for elementary particles: electromagnetic, weak and strong. The fourth kind, the gravitational interaction, is negligible at the level of elementary particle phenomenology, but it is of great importance from the theoretical point of view. The main distinctive features of the different kinds of interactions are summarized in the following (For more details see ref.[2]).

Electromagnetic interactions

Except for neutrinos, quarks and leptons are sensitive to them. They are mediated by massless spin–1 photons. As already mentioned, the strength of the interaction is given by the dimensionless fine structure constant $\alpha \simeq 1/137$; the relevant length scale is the Bohr radius $a_0 = \hbar/m_e\alpha \simeq 0.53 \times 10^{-8}cm$, and the typical energy scale can be related to the ground state of the hydrogen atom: $E_0 = -\frac{1}{2}e^2/a_0 \simeq -13.6eV$.

Weak interactions

The effective coupling is weaker than the electromagnetic one. It is given by the so–called Fermi coupling G_F; to get a dimensionless constant, it is usually multiplied by the squared proton mass: $G_F m_p^2 \simeq 10^{-5}$.

All particles are sensitive to weak interactions; the "carriers" are the massive vector bosons $W^{\pm}$ and Z^0. The charged current reactions (like $d \rightarrow u + e^- + \nu_e$, which takes place in the neutron decay) are mediated by $W^{\pm}$, and the relevant coupling is given by $\alpha_W \simeq 1/31$; the neutral current reactions (like $\nu_\mu e^- \rightarrow \nu_\mu e^-$) are mediated by Z^0, and the coupling is $\alpha_Z = \alpha_W(M_Z/M_W)^2 \simeq 1/24$. The Fermi constant is related to α_W by $G_F \simeq \alpha_W/M_W^2$. The typical energy scale is of the order of 100 GeV ($1GeV = 10^9eV$), which is the order of magnitude of the $W^{\pm}$ and Z^0 masses: $M_Wc^2 \simeq 80GeV, M_Zc^2 \simeq 92GeV$; the corresponding length scale is given by $\hbar/M_Wc \simeq 2.4 \times 10^{-16}cm$.

A peculiar and distinctive feature of weak interactions is that they violate parity to the maximum degree. Specifically, the vector boson $W^{\pm}$ is coupled to quarks and leptons in a combination $V - A$ of vector and axial currents, which is said to correspond to left-handed "chirality", while the other combination $V + A$ correspond to right-handed chirality. To get a more physical insight into the situation, it is convenient to consider massless spin-1/2: in this case, chirality coincides with helicity: when spin is quantized along the direction of motion of the particle, one can have either spin component $+1/2$ or $-1/2$ and, correspondingly, one says that the particle is in a state of right-handed or left-handed helicity. The remarkable fact is that only left-handed neutrinos (ν_L) exist, while antineutrinos ($\bar{\nu}_R$) are only right-handed. This fact is a clear evidence of maximal parity violation, since symmetry under space reflection would transform a ν_L state into a non-existing ν_R state. Therefore, weak charged currents contain only left-handed chirality i.e., when the mass can be neglected as in the case of very relativistic particles, only left-handed helicity. The case of the neutral boson Z^0 is more complicated, since it is coupled to the neutral current which is a superposition of a $V - A$ term and a pure vector term.

Strong interactions

The coupling is strong and it varies with the energy scale. In fact, as it will be discussed later, all couplings vary with energy, but the dependence is, in general, very weak. Writing, in analogy with the electromagnetic coupling: $\alpha_s = g_s^2/4\pi\hbar c$, one has $\alpha_s(1GeV) \simeq 0.5$ and $\alpha_s(100GeV) \simeq 0.12$.

Strong interactions are responsible for binding quarks inside nucleons, i.e. proton (uud) and neutron (udd) and, in general, inside hadrons. The typical energy scale can be taken as the mass of the lowest $q\bar{q}$ bound state, which is the π–meson: $m_\pi c \simeq 140MeV$; the corresponding length scale is given by $\hbar/m_\pi c^2 \simeq 1.4 \times 10^{-13}cm$, which is the size of the nucleon.

Strong interactions are mediated by massless vector bosons, called *gluons*: they are coupled to colour, and then only to quarks. This kind of interactions is parity-conserving, since gluons are coupled only to pure vector currents.

Gravitational interactions

The coupling is given by the Newton constant $G_N/\hbar c \simeq 6.7 \times 10^{-39}(GeV/c^2)^{-2}$. Its value shows that this kind of interactions is extremely weak at the normal elementary particle scales. It becomes important at very high energies, of the order of the Planck mass $M_P = (\hbar c/G_N)^{1/2} \simeq 1.22 \times 10^{19}GeV/c^2$ and, correspondingly, at length scales of the order of the Planck length $\lambda_P = \hbar/M_P c \simeq 1.6 \times 10^{-33}cm$.

Concluding this schematic presentation, we stress again that all basic constituents of matter (quarks and leptons) are spin–1/2 fermions, and their fundamental interactions are mediated by spin–1 bosons, except for the case of gravitation which is mediated by spin–2 gravitons.

Moreover, we point out that there is a *doubling* of all particles states: for each particle (quark, lepton) there is a corresponding *antiparticle* (antiquark, antilepton) with the same "mechanical" quantities (mass, spin) and opposite internal quantum numbers (electric charge, colour). So for each quark ($u, d, ...$) there is an antiquark ($\bar{u}, \bar{d}, ...$), for each lepton ($e^-, \nu_e, ...$) an antilepton ($e^+, \bar{\nu}_e, ...$), while photons, gluons and Z^0 coincide with their antiparticles.

Only the members of the first family appears in the composition of the portion of the Universe around us. What is the rôle of the two other families? At present, they seem to be "useless", but maybe they played an essential rôle in the first istants after the Big–Bang.

Moreover, who tells us that there are three families and no more? Experimental information comes indirectly from the number of neutrino species. Assuming that the neutrino masses are less than $M_Z/2$, in agreement with the fact that known neutrinos are very light (the upper mass limits are given in Table I), one can infer the number $N_\nu = 3$ of neutrino species from the shape (height and width) of the Z^0–resonance in the reaction $e^+e^- \to Z^0 \to$ hadrons. However, in principle, other families with very massive neutrinos are not excluded.

All the interaction listed above, with the exception of the gravitational ones, are combined in a unified picture by a "gauge" field theory, named the *Standard Model* (SM). It is based on the local symmetry group

$$G_S = SU(3)_c \otimes SU(2)_I \otimes U(1)_Y \tag{1}$$

Such a theory is said to be "chiral", since it reproduces the asymmetry under parity required by weak interactions by assigning different transformation properties to left-handed and right-handed fermions.

In particle physics, extensive use has been made of local symmetries and gauge field theories. The requirement of a local symmetry on a field theory, i.e. of the invariance under a set of transformations performed independently at different space–time points, is very strong. In fact, it implies the introduction of a set of massless vector (gauge) bosons, in one–to–one correspondence with the generators of the symmetry group. (See the example I of the Appendix as an illustration).

In the SM, the first factor $SU(3)_c$ on the r.h.s. of eq. (1) represents the "colour" symmetry group: the corresponding gauge bosons are the 8 *gluons* that mediate the strong interactions. The gauge field theory based on $SU(3)_c$ is called Quantum Chromodynamics (QCD); it exhibits very peculiar features: at very high energies, the coupling becomes weaker and weaker and tends to the "asymptotic freedom"; at low energies, it becomes stronger and stronger, which should explain the "quark confinement", namely that quarks are permanently bound in hadrons, which are "colourless", i.e. singlets under $SU(3)_c$, and cannot be taken apart. The other two factors $SU(2)_I$ and $U(1)_Y$ conglobe the electroweak interactions, which are mediated by four gauge bosons: $W^{\pm}$, Z^0 and γ.

At high energy scales (above $1TeV = 10^3 GeV$), the G_S symmetry is assumed to be exact, and all particles (quarks, leptons and vector bosons) are massless. Below 1 TeV, the symmetry $SU(3)_c \otimes SU(2)_I \otimes U(1)_Y$ is broken "spontaneously", namely the system reaches a stable configuration which possesses a lower symmetry which is given by $SU(3)_c \otimes U(1)_Q$, where $U(1)_Q$ is the gauge symmetry of QED. The vector bosons $W^{\pm}$ and Z^0 acquire mass according to the Higgs mechanism, while the photon remains massless; in this process, also the masses of quarks and charged leptons are generated.

A necessary ingredient for the occurrence of this mechanism is the presence of a set of scalar fields in the theory, which trigger the spontaneous symmetry breaking. The phenomenon is analogous to a phase transition: the more symmetric phase is broken down, at low energies, to a less symmetric phase. A remnant of the higher symmetry is a physical scalar boson, called Higgs boson. (For details see e.g. ref. [3]).

The SM describes with great accuracy the present phenomenology, so it is considered a very successful theoretical model. However, from a more theoretical side, it cannot be considered completely satisfactory. In particular, it contains too many independent parameters. In the minimal version, with three fermion families, there are at least 19 of them: 3 coupling constants (related to the 3 factors in the G_S group); 9 fermions masses (considering massless neutrinos); 3 mixing angles and a phase in the scalar (Higgs) sector; finally, the last parameter controls the dangerous appearence of CP violation in the strong interactions.

Before going to the next section, we would like to comment the mixing among quarks. What happens is that weak interactions, in general, do not conserve flavour: the quark mass eigenstates are different from the states which are coupled to $W^{\pm}$. The two bases are related by a unitary matrix which, for a number n of families is characterized by $n(n-1)/2$ angles and $(n-1)(n-2)/2$ phases. The mixing is compensated in the neutral–current interactions, mediated by Z^0, which are flavour conserving. In particular, for $n = 2$ there would be only one angle, the so-called Cabibbo angle θ_c, while, for $n = 3$, there are 3 angles and one phase. What is important to notice is that a non–vanishing phase implies CP violation, which is one of the ingredients required to build a model explaining the matter–antimatter asymmetry of our universe. This is an indication for the rationale of having three families of fermions: three is the minimum number, in the SM, for having a non-vanishing phase in the mixing matrix.

3. GRAND UNIFICATION AND HIGHER SYMMETRIES

The Standard Model provides only a partial unification, since the group G_S is the factor of three simple groups and, correspondingly, there are three independent gauge couplings: α_1, α_2 (related to α_W and α_Z) and $\alpha_3 = \alpha_s$. The idea of Grand Unified Theories (GUT) [4] is to look for a gauge field theory based on a simple group (with a single coupling constant) which contains G_S as a subgroup. Since the rank (number of commuting generators) of G_S is 4, the rank of the GUT group has to be $r_G \geq 4$. The minimal solution is based on $SU(5)$ ($r_G = 4$); of course, other solutions exist for $r_G > 4$. In the literature, special rôle was played by the groups $SO(10)$ with $r_G = 5$ and E_6 with $r_G = 6$.

By going to higher ranks, also the order of the group increases, so that one has to deal with a larger number of gauge vector fields. On the other hand, the fermions of each family now belong to a single irreducible representation (or at most two in the case of $SU(5)$). In Table II we compare the situation of the above mentioned groups with the case of G_S, in which the representation content with respect to the subgroups $SU(3)$ and $SU(2)$ is specified.

Table II

Group	$G_S = SU(3) \otimes SU(2) \otimes U(1)$	SU(5)	SO(10)	E_6
vector bosons	(8,1)+(1,3)+(1,1)	24	45	78
fermions of each family	(3,2) + (1,2) + ($\bar{3}$,1) + ($\bar{3}$,1)+(1,1)	$\bar{5}$+10	16	27

The idea of grand unification is that the gauge couplings converge towards the same value at a very high energy scale M_{GUT}; beyond this scale, all members of the gauge vector multiplet can be interchanged, as well as all the fermions of each family. This higher symmetry produces new kinds of phenomena, which we are not going to discuss here. The general feature is that, at the scale M_{GUT} (of the order of $10^{14} \sim 10^{15} GeV$), the symmetry is broken spontaneously (directly or in more than one step) into G_S, by a generalization of the mechanism introduced in the SM.

The natural question is if we have some hint for grand unification. By assuming that no other particles, except those present in the SM, appear up to M_{GUT}, it is possible to predict the behaviour of the gauge couplings. They show a logarithmic energy dependence, so that they vary very slowly. Specifically, α_2 and α_3 increase with energy, while α_1 decreases with increasing energy: they approach similar values around 10^{14} GeV, but they do not meet at a point.

A much better result, i.e. the meeting of the three couplings, can be obtained in an extension of the SM, which is its (minimal) supersymmetric version: in this case the convergence occurs at the energy scale $\sim 10^{16}$ GeV (See ref.[3]). It implies the introduction of another ingredient in the game, that is "supersymmetry". This peculiar kind of *global* symmetry requires invariance under boson–fermion exchange, which implies a doubling of the spectrum of particles, as shown in Table III (The specific spin values are indicated in parentheses: S=0,1/2,1).

Table III	
particles in the SM	supersymmetric partners
quarks (1/2)	s–quarks (0)
leptons (1/2)	s–leptons (0)
vector bosons (1)	gauginos (1/2)
(photon, gluons, etc.)	(photino, gluinos, etc.)

The introduction of supersymmetry is welcome for theoretical reasons [5]: it softens the divergences arising in the scalar sector of the SM, and solves the problem of hierarchies at different energy scales. On the other hand, it is just the increase of the number of particles which reduces the slope in the energy dependence of the couplings, and produces a convergence to the same value.

Keeping both grand unification and supersymmetry, GUT's become susy GUT's, and the advantages of grand unification are preserved. In particular, one obtaines, at the unification scale, mass relations among masses of different fermions, such as quarks and leptons. A typical relation, which is in agreement with the low–energy data (when it is scaled down from the unification scale) is the following:

$$m_b = m_\tau. \tag{2}$$

Along this way, one is going in the direction of reducing the number of independent parameters, but not enough!

Recently, many theoretical investigations have been devoted to the analysis of the mass spectrum of quarks and charged leptons. The pattern of this mass spectrum is characterized by a strong hierarchy which can be approximately described in terms of $\lambda = sin\theta_c \simeq 0.22$; specifically, in the case of the down-type quarks, one gets:

$$m_d : m_s : m_b \approx \lambda^4 : \lambda^2 : 1 \tag{3}$$

and in the case of the up-type quarks:

$$m_u : m_c : m_t \approx \lambda^8 : \lambda^4 : 1 \tag{4}$$

The case of charged leptons is roughly similar to that of down-type quarks.

Such hierarchies lead to the ansatz of "zero textures", namely to the assumption that the mass matrices contain a number of null entries [6]. In turn, this feature was interpreted as emerging from an underlying "family" symmetry which implies invariance under the exchange of the corresponding members of different families, while GUT's require invariance under the exchange of members within the same family. Several schemes have been proposed, based on discrete or continuous groups. (For a review see ref. [7]). The Abelian group $U(1)$, either global or local, has been one of the most favoured: it can be interpreted as the remnant of a larger "family" or "flavour" symmetry group $U(3) \subset U(45)$, which acts among the three fermion families, or among all the different flavours.

An approximate flavour $U(2)$ symmetry, which applies only to the two lighter families, has recently been proposed [8] as an interesting framework for understanding the rôle of flavour breaking in supersymmetric theories. It is more promising than the complete $U(3)$ symmetry, which is badlly broken by the large top-quark mass, and it is more predictive than the Abelian $U(1)$ symmetry. While solving the supersymmetric flavour-changing problem, it leads to interesting relations among the masses and mixing angles, some of which appear to be well satisfied. It should be noted that these results are obtained within supersymmetric grand-unified models. More recently, the analysis based on $U(2)$ flavour symmetry has been extended to the sector of the neutrinos [9], which present very peculiar features with respect to the other fermions, and results of some interest for the neutrino phenomenology have been obtained.

In general, the introduction of a family symmetry gives rise to relations among the parameters of the SM, specifically among the quark masses and the mixing angles, i.e. elements the of the unitary matrix which mixes the quark flavours in the charged weak currents. In particular, all types of family symmetry lead to the well known relation

$$sin\theta_c = \sqrt{m_d/m_s} \tag{5}$$

Before leaving this section, we would like to point out an important advantage of supersymmetry: namely the fact that it allows the introduction of the gravitational interaction, which has not taken into account till now, but which can no longer be neglected above the GUT scale. This requires, however, to promote supersymmetry from a *global* to a *local* symmetry.

In the Appendix, we indicate the main steps which have to be taken when going from a global susy–Lagrangian to a local susy–one. In fact, it is necessary to introduce both a symmetric second rank tensor field $g_{\mu\nu}$ (describing spin 2 bosons) and its supersymmetric partner, which is a Rarita–Schwinger spin 3/2 field, called *gravitino*. It can be shown that the $g_{\mu\nu}$ tensor satisfies the Einsteins's equation [10]:

$$R_{\mu\nu} - \frac{1}{2} g_{\mu\nu} R = -8\pi G_N T_{\mu\nu}, \tag{6}$$

where $T_{\mu\nu}$ is the total energy–momentum tensor; $R_{\mu\nu}$ is the Ricci tensor related to the curvature Riemann tensor ($R_{\mu\nu} = g^{\rho\sigma} R_{\mu\nu\rho\sigma}$) and $R = g^{\mu\nu} R_{\mu\nu}$.
We note, in fact that a simple susy–model with a "supermultiplet"

$$\begin{pmatrix} \psi & (S = 1/2) \\ \phi & (S = 0) \end{pmatrix} \tag{7}$$

requires, when local supersymmetry is imposed, a gravity supermultiplet

$$\begin{pmatrix} \psi_\mu & (S = 3/2) \\ g_{\mu\nu} & (S = 2) \end{pmatrix} \tag{8}$$

This is the content of the theory of supergravity [11]. Starting from susy-GUT's and promoting global to local supersymmetry, one can extend the *pure* supergravity to different supergravity models. The picture is, in principle, very attractive; unfortunately, such supergravity theories have the disease that they are not renormalizable, namely that (higher order) quantum corrections cannot be evaluated.
This problem induces us to describe an alternative path, i.e. the one extending the space–time dimensions, as will be outlined in the following section.

4. UNIFICATION AND HIDDEN SPACE DIMENSIONS

The idea of obtaining unification of different interactions through the extension of hidden space–time dimensions is originally due to T.F.E. Kaluza who already in 1919, after a successful test of general relativity (on the occasion of a solar eclipse), tried to unify electromagnetism with gravity (which were the only types of interactions known at that time). The idea was improved a few years later by O. Klein, and now the theory goes under the name of Kaluza–Klein theory (See ref. [12]).

One starts from the Einsteins's equation for gravity, in which the metric tensor $g_{\mu\nu}$ is determined locally by the matter density, and one increases the space dimension by one unit:

$$x_\mu \longrightarrow z_M \equiv (x_\mu, x_5 \equiv y). \tag{9}$$

From $g_{\mu\nu}$ one goes to a metric tensor g_{MN} (M,N =1, ...5) which satisfies a generalized $d = 5$ Einstein's equation. Kaluza assumed that the fifth dimension is curled up into a tiny ring, so small that it could not be experimentally observed by any instruments. This should explain why we do not see the fifth dimension. Since the scale of gravity is of the order of the Planck length $\lambda_P \simeq 10^{-33} cm$, the fifth dimension would have curled up on a sphere of radius r approximately equal to λ_P.

The set of the four components $g_{\mu 5}$ of the d=5 metric tensor behaves as a four–vector which can be identified with the electromagnetic potential $A_\mu(x)$: it is remarkable that they satisfy the Maxwell equation. The component g_{55} behaves as a scalar field $\phi(x)$, which satisfies the Klein–Gordon equation. The situation can be summarized as follows:

$$g_{MN} \longrightarrow \begin{pmatrix} g_{\mu\nu} + \kappa^2 A_\mu A_\nu & \kappa A_\mu \\ \kappa A_\mu & \phi \end{pmatrix} \tag{10}$$

where $\kappa = \sqrt{G_N/\hbar c} = M_P^{-1}$.

Since the fifth dimension is periodic ($0 \leq y \leq 2\pi r$), one has $\phi(y) = \phi(y + 2\pi r)$, and one can expand ϕ in series:

$$g_{MN}(x, y) = \sum_{n=-\infty}^{+\infty} g^n_{MN}(x) e^{inmy} \tag{11}$$

where m stands here for the Planck mass. The $n = 0$ term corresponds to the massless modes, while the other terms with $n \neq 0$ correspond to massive particles with $m_n = nm$. The massless modes should acquire a small mass through some symmetry breaking mechanism: they would correspond to the observed physical particles. On the other hand, the $n \neq 0$ modes are very massive, i.e. of the order of the Planck mass. It is interesting to note that the electric charge is related to the mass by: $q_n = \kappa nm$, so that the lowest non–vanishing mass is given by $m \simeq e/\kappa \simeq 3 \times 10^{18} GeV/c^2$, where $e = \sqrt{4\pi\alpha}$.

More recently the Kaluza–Klein idea has been extended [13, 14] to include also the strong and weak interactions. The main idea is that the extra gauge fields, which mediate all the other interactions besides gravitation, would be generated by the components of the matrix tensor in the extra $k = d - 4$ dimensions. Due to the advantage of keeping supersymmetry in the framework, the starting point is supergravity rather than simple gravity: in this case, the introduction of fermions is not arbitrary, since they are related to the tensorial components.

The first question is: how much should one increase the space–dimensions? The maximum dimension is fixed by the fact that we do not want to deal with states corresponding to spin higher than 2, since no reasonable theory exists with fundamental fields with spin $S > 2$. One can make use of the following argument: in the supersymmetric theories formulated in $d = 4$ dimensions, one can increase the number of supersymmetric charges, going from $N = 1$ to a maximum value $N = n_{max}$. Since each charge generator changes helicity by $\Delta h = \pm 1/2$, starting from the lowest value

$h = -2$, one can reach the highest value $h = +2$ in 8 steps: so that $n_{max} = 8$.

One can show that a supersymmetric $N = 8$ theory corresponds to $N = 1$ theories with $d = 10$ or $d = 11$. It is then instructive to compare $N = 1$ supergravity theories in $d = 11$ and $d = 10$ with the case $d = 4, N = 8$. This is shown in the Table IV, where the numbers of independent components are indicated in parantheses.

Table IV

d=11 ,N=1	d=10 ,N=1	d=4 ,N=8
g_{MN} (44)	$g_{\mu\nu}$ (35)	$g_{\mu\nu}$ (2)
	A_μ (8)	$g_{\mu a}$ (14)
	φ (1)	g_{ab} (28)
A_{MNP} (84)	$A_{\mu\nu\rho}$ (56)	$A_{\mu\nu a}$ (7)
	$B_{\mu\nu}$ (28)	$A_{\mu ab}$ (42)
		A_{abc} (35)
Ψ_M (128)	Ψ_μ, Ψ'_μ (2x56)	Ψ^i_μ (16)
	χ, χ' (2x8)	Ψ^i_a (112)

In the case of $d = 11$ dimensions, besides the matrix tensor g_{MN} (spin 2), we need an antisymmetric third–rank tensor A_{MNP} (spin 0,1, 2) and a fermion field ψ_M (spin 1/2, 3/2): they amount to 128 bosonic and 128 fermionic degrees of freedom. We should point out that all fields are massless, and this reduces the number of independent components.

The transition from a $d = 11$ (or $d = 10$) world to our four–dimensional world is assumed to occur via a compactification of the extra 7 dimensions. This presumably occurs below the Planck scale. However, different kinds of compactification are possible, according to the topology of the manifold. The simplest one corresponds to the transition

$$SO(1,10) \to SO(1,3) \otimes SO(8), \tag{12}$$

i.e. the extra dimensions are compactified on a 7–dimensional sphere:

$$d = 11 \to M^4 \times S^7 \tag{13}$$

where M^4 is the Minkowsky space, invariant under the Lorentz group $SO(1,3)$ and S^7 is the $d = 7$ sphere, invariant under the rotation group $SO(7)$, of rank $r_G = 4$ and order $n = 28$.

After compactification, all the particles (all together 256 degrees of freedom) are grouped into irreducible representations (IR) of $SO(8)$, as shown in Table V.

One can check that these fields correspond exactly to those presented in Table IV in the case of $d = 4$, $N = 8$.

The present scheme possesses very nice general features: in particular, the internal symmetry emerges from supergravity. However, the specific features are not good:

IR	particles (spin)
1	graviton (2)
8	gravitino (3/2)
28	gauge bosons (1)
56	spinors (1/2)
35	scalars (0)
35	pseudoscalars (0)

Table V

– the symmetry group $SO(8)$ is not large enough, since it does not contain the standard group $G_S = SU(3) \otimes SU(2) \otimes U(1)$ (it contains only the subgroup $SU(3) \otimes U(1) \otimes U(1)$), and a-fortiori it is to small for GUT;
– there is no room for chiral fermions; in fact, the group $SO(8)$ has only real representations, while compex ones would be needed;
– the theory is not renormalizable: this is a common disease of supergravity theories.

Of course, other compactifications are possible, but none of them lead to acceptable results. What can one learn from this lesson? A possible way out will be shortly mentioned in the following section.

5. OUTLOOK

Starting with the problem of unifying the fundamental interactions and reducing the number of independent parameters of the Standard Model, we have explored two alternatives paths which have been investigated up to now: either by enlarging the internal symmetries to Grand Unification and beyond, or by extending the hidden space-time dimensions. In the former case, spontaneous breaking is necessary to go down to the low–energy regime; in the latter case, compactification is required to recover the experimentally detectable world.

Many nice features are obtained: both approaches show that gravitation can be joined to the other three fundamental interactions, provided local supersymmetry is included. However, theories which include gravity are not renormalizable, and this is a common bad feature.

This situation lead to the present development of *superstring* theory [15], which is hoped to provide a breakthrough to the solution of the main problems. In fact, superstring theory appears, up to now, the only one which:
– contains gravity;
– includes SM and GUT symmetries;
– contains chiral fermions;
– is renormalizable.

The theory is formulated in $d = 10$ dimensions, so that, also in this case, compactification is required. In a way, superstring theory shares some features with GUT's and with Kaluza–Klein theory. As a matter of fact, the internal symmetries are already present in the 10–dimensional formulation, but, in the building of the theory, also the Kaluza–Klein mechanism has been used. At d=10, the symmetry corresponds to the group $E_8 \otimes E_8$ but, in the compactification process to $d = 4$, it reduces to $E_6 \otimes E_8$, where E_6 would be the symmetry of the observable world, while

E_8 would be related to an invisible sector. In turns, one can have the breaking chain

$$E_6 \longrightarrow SO(10) \otimes U(1) \longrightarrow SU(5) \otimes U(1) \otimes U(1)$$

Alternative formulations of string theories are possible, with different compactification patterns and different symmetry breaking chains. A discussion of this matter is outside the scope of the present analysis; it is sufficient to point out that, through duality simmetries, the equivalence of the different string theories has been established (For a review on recent developments see ref. [16]). It was conjectured that all these theories should emerge from a not better specified M–theory (a quantum theory of membranes), whose low–energy effective action is $d = 11$ supergravity. Is this the final unified theory we are looking for? The future reserves for us fine promises but probably also big surprises.

APPENDIX

In this Appendix we would like to give some more explicit motivation of the rôle of gauge symmetry. We said that, in promoting a set of global transformations under a symmetry group G to *local* ones, a new situation occurs: the requirement of local invariance demands that one introduces a set of massless vector (gauge) fields, whose number equals the order of the group i.e. the number of generators of the group.

I. We give here a very simple example, related to QED. Let us start with the Lagrangian density of a free (massless) Dirac field $\psi(x)$:

$$\mathcal{L} = \frac{1}{2} i\bar{\psi}\gamma^\mu \partial_\mu \psi \tag{A1}$$

For the sake of simplicity, we use natural units ($\hbar = c = 1$). We assume that $\mathcal{L}$ is invariant under the global transformations

$$\begin{aligned} \psi(x) &\to e^{-i\epsilon}\psi(x) \\ \bar{\psi}(x) &\to e^{+i\epsilon}\bar{\psi}(x) \end{aligned} \tag{A2}$$

or, equivalently, under the infinitesimal transformations

$$\begin{aligned} \delta\psi(x) &= -\, i\epsilon\psi(x) \\ \delta\bar{\psi}(x) &= i\epsilon\bar{\psi}(x) \end{aligned} \tag{A3}$$

Due to this invariance, one can deduce that the four–current

$$J^\mu(x) = \frac{1}{2}\bar{\psi}(x)\gamma^\mu\psi(x) \tag{A4}$$

satisfies the equation

$$\partial_\mu J^\mu = 0 \tag{A5}$$

so that the total charge is conserved.

Let us suppose that the quantity ϵ is a function of the four–dimensional point x: now the infinitesimal variation of $\mathcal{L}$ is given by

$$\delta\mathcal{L} = \epsilon\partial_\mu J^\mu + J^\mu \partial_\mu \epsilon, \tag{A6}$$

and while the first term of the r.h.s. vanishes according to (A5), the second one does not. As well known, the invariance is recovered by adding the following term to (A1):

$$\mathcal{L}_A = -\frac{1}{2}e\bar{\psi}\gamma^\mu\psi A_\mu(x), \tag{A7}$$

where $A_\mu(x)$ is a four-vector field; its transformation properties related to (A3) are given by

$$A_\mu(x) \to A_\mu(x) + \frac{1}{e}\partial_\mu\epsilon. \tag{A8}$$

In fact, under (A3) and (A8):

$$\delta\mathcal{L}_A = -J^\mu\partial_\mu\epsilon \tag{A9}$$

so that $\delta\mathcal{L} = 0$. To recover (massless) QED (i.e. electron field interacting with EM field) one has simply to add the kinetic EM field term to the Lagrangian, so that $\mathcal{L}$ is replaced by

$$\mathcal{L}' = \frac{1}{2}i\bar{\psi}\gamma^\mu\partial_\mu\psi - \frac{1}{2}e\bar{\psi}\gamma^\mu\partial_\mu\psi A_\mu - \frac{1}{4}F_{\mu\nu}F^{\mu\nu} \tag{A10}$$

where $F^{\mu\nu} = \partial^\mu A^\nu - \partial^\nu A^\mu$.
The first two terms of (A10) correspond to the minimal coupling

$$\partial_\mu\psi \to (\partial_\mu + ieA_\mu)\psi. \tag{A11}$$

II. Next we consider a simple example of supersymmetry [17]. The supersymmetric (global) extension of (A1) is:

$$\mathcal{L} = \frac{1}{2}i\bar{\psi}\gamma^\mu\partial_\mu\psi + \partial^\mu\varphi^+\partial_\mu\varphi \tag{A12}$$

where the complex scalar field

$$\varphi(x) = \sqrt{\frac{1}{2}}(a(x) + ib(x)) \tag{A13}$$

is the supersymmetric partner of the Dirac field ψ.
We leave as an excercise the proof that the Lagrangian is invariant under the supersymmetric transformations

$$\begin{aligned} \delta a =& \bar{\xi}\psi \\ \delta b =& -i\bar{\xi}\gamma_5\psi \end{aligned} \tag{A14}$$

and

$$\begin{aligned} \delta\psi =& -i\gamma^\mu[\partial_\mu(a - i\gamma_5 b)]\xi \\ \delta\bar{\psi} =& i\bar{\xi}\gamma^\mu\partial_\mu(a - i\gamma_5 b), \end{aligned} \tag{A15}$$

where ξ is a constant spinor, and $\bar{\xi} = \xi^+\gamma^0$ its conjugate.
In fact, one can easily show that the variation of $\mathcal{L}$ is given by:

$$\delta\mathcal{L} = \bar{\xi}\partial_\mu\mathcal{J}^\mu \tag{A16}$$

where $\mathcal{J}^\mu$ is the four vector:

$$\mathcal{J}^\mu(x) = \frac{1}{2}\gamma^\mu\gamma^\nu[\partial_\nu(a - i\gamma_5 b)]\psi \tag{A17}$$

Its divergences vanishes due to the equation of motion of the fields, so that the current is conserved.

However, in the case in which $\xi(x)$ is space–time dependent, i.e. one imposes *local* supersymmetry, instead of (A16), one obtains

$$\delta\mathcal{L} = \bar{\xi}\partial_\mu\mathcal{J}^\mu + (\partial_\mu\bar{\xi})\mathcal{J}^\mu \tag{A18}$$

The second term on the r.h.s. of (A18) is now different from zero, so that the Lagrangian (A12) is not invariant. To re-establish invariance, one has to add two more terms to $\mathcal{L}$, i.e.:

$$\mathcal{L}' = \mathcal{L} + \mathcal{L}_N - g_{\mu\nu}T^{\mu\nu}, \tag{A19}$$

where:

$$\mathcal{L}_N = -k\bar{\psi}_\mu\gamma^\mu\gamma^\nu[\partial_\nu(a - i\gamma_5 b)]\psi \tag{A20}$$

and $T^{\mu\nu}$ is the energy–momentum tensor of the fields φ and ψ. Moreover the variations of $\bar{\psi}_\mu$ and $g_{\mu\nu}$ are given by:

$$\delta\psi_\mu = \frac{1}{k}\partial_\mu\xi \tag{A21}$$

$$\delta g_{\mu\nu} = k\bar{\psi}_\mu\gamma_\nu\xi. \tag{A22}$$

We notice that ψ_μ is a spin–3/2 field, and $g_{\mu\nu}$ is the symmetric rank–2 tensor field, corresponding to spin 2. The latter is the *graviton* field, while the former is the *gravitino*, i.e. the supersymmetric partner of the graviton. Both ψ_μ and $g_{\mu\nu}$ are necessary for obtaining invariance under local supersymmetric transformations; in fact, making use of (A21) and (A22) one obtains finally $\delta\mathcal{L}' = 0$.

REFERENCES

1. Particle Data Group, Phys.Rev. D54, 1 (1996).
2. G.Costa, in *Disordered Solids*, edited by B.Di Bartolo, Plenum Press, New York (1989).
3. G.Costa, in *Spectroscopy and Dynamics of Collective Excitations in Solids*, edited by B.Di Bartolo, Plenum Press, New York (1995).
4. G.G.Ross, *Grand Unified Theories*, Benjamin inc., Reading (1985).
5. S.Ferrara and P.Fayet, *Phys.Reports* 32C, 250 (1977);
   J.Wess and J.Bagger, *Supersymmetry and Supergravity*, Princeton University Press, Princeton (1982).
6. H.Fritzsch, *Phys.Lett.* B70, 436 (1977);
   P.Ramond, R.Roberts and G.G.Ross, *Nucl.Phys.* B404, 19 (1993).
7. S.Rabi, preprint OHSTPY-HEP-T- 95–024, published in *Trieste HEP Cosmology* (1995);
   L.J.Hall, preprint LBL-38110/UCB-PTH-96/10, published in *SLAC Summer Institute* (1995).

8. R.Barbieri, G.Dvali and L.J.Hall, *Phys.Lett.* B377, 76 (1996);
R.Barbieri and L.J.Hall, *Nuovo Cimento* A110, 1 (1997).
9. C.D.Carone and L.J.Hall, *Phys. Rev.* D56, 4198 (1997);
G.Costa and E.Lunghi, *Nuovo Cimento* 110A, 549 (1997).
0. S.Weinberg, *Gravitation and Cosmology. Principles and Applications of the General Theory of Relativity*, John Wiley and Sons, New York (1972).
1. P.van Nieuwenhuizen, *Phys.Reports* 68, 191 (1981).
2. M.J.Duff, in *Supersymmetry and Supergravity '84*, edited by B.de Wit, P.Fayet and P.van Nieuwenhuizen, World Scientific Publishing Co., Singapore (1984).
3. E.Cremmer, B.Julia and J.Scherk, *Phys.Lett.* 76B, 409 (1978).
4. E.Witten, *Nucl.Phys.* B186, 412 (1981).
5. M.B.Green, J.H.Schwarz and E.Witten, *Superstring Theory*, Cambridge University Press, Cambridge (1987).
6. S.Ferrara, *Proceed. of the 28th Int.Conf. on High Ehergy Physics,ICHEP'96*, Vol.I, ed. by Z.Ajduk and A.K.Wroblewsky, World Scientific Publ.Co.,Singapore (1997).
7. H.P.Nilles, *Phys.Reports* 110, 1 (1984).

MEASUREMENTS WITH AN OPTICALLY GATED SCANNING TUNNELING MICROSCOPE

J.R. Jensen

Mikroelektronik Centret
Technical University of Denmark
Lyngby
DENMARK

The optically gated scanning tunneling microscope is an instrument that attempts to combine the high spatial resolution of a STM (atomic resolution) with the high temporal resolution obtainable with optical pump-probe techniques (picosecond resolution). With the pump beam an electrical pulse is generated, that propagates to the STM tip and modulates the current in the tip. Using an optical switch integrated on the STM tip, the modulated current is sampled with the probe beam. Electrical signals varying in time and space can now be investigated in detail, by varying the delay between the pump and the probe (time) and by scanning the STM tip on the sample (space).

Examples of the first measurements of picosecond electrical pulses on a transmission line with simultaneous spatial *and* temporal resolution (spatio-temporal imaging) are shown. These "snapshots" of electrical pulses are very useful for determining the mode structure, and for studying reflection and transmission properties at discontinuities on the transmission line. It is found that the dominant interaction between the tip and the sample is due to capacitive coupling. Hence, the spatial resolution is determined by the spread of the electric field lines from the tip to the sample - and *not* by the extend of the tunneling region. For spatio-temporal imaging the spatial resolution is better than 5 microns and the temporal resolution is ~2 ps.

Finally, it is outlined how the optically gated STM can be used for studying other physical processes like surface plasmons on metallic films, optically excited semiconductor surfaces or electrical responses of devices integrated in transmission lines.

This work was done in collaboration with U.D. Keil and J.M. Hvam.

TIME RESOLVED SPECTROSCOPY OF DEEP EMISSIONS FROM LARGE BAND GAP SEMICONDUCTORS

R. Seitz

Departamento de Física
Universidade de Aveiro
PORTUGAL

II-VI semiconductors based on ZnSe and III-V semiconductors based on GaN are promising materials for optoelectronic devices working in the blue and UV wavelength range.

The samples used in this work are hetero-epitaxially grown mainly by metal organic vapour phase epitaxy or by molecular beam epitaxy. The layer thickness various from 500 nm up to several μm.

The photoluminescence spectra of both materials are characterised by emissions due to free and bound exciton and pair recombinations. But we can find even in good quality layers an emission with an energy much lower than the corresponding band gap. In the case of ZnSe which has a band gap of 2.8eV this emission is centered at 2.0eV. For GaN which has a band gap of 3.5eV this emission is centered at 2.2eV. It is important to understand the nature of these emissions because they are partly responsible for the quenching of the near band gap emissions.

Time resolved spectroscopy shows that the mid band gap emissions have life times in the ms range. They are usually superpositions of various bands which in some cases can be separated by varying the time delay after the exciting light pulse. From the temperature behaviour of the emission intensity and the life times we set up models which describe the dynamic processes of the involved centers.

This work was done in collaboration with C. Gaspar, T. Monteiro and E. Pereira.

CHEMICAL DYNAMICS OF UNI- AND BIMOLECULAR REACTIONS STUDIED WITH NONLINEAR OPTICAL TECHNIQUES IN THE FEMTOSECOND TIME DOMAIN

M. Motzkus

Max-Planck-Institut für Quantenoptik
D-85748 Garching
GERMANY

Collisions between electronically excited sodium atoms and molecular hydrogen are one of the most fundamental examples of excited atom-molecule systems. The interest of studying such systems is due to the fact that nonadiabatic energy transfer including chemical reactions takes place and detailed results have been already obtained on a ns time scale. Ultra short laser pulses in the fs time domain provide a novel tool for studying such collision processes and the relevant transition-complexes in real time by using pump-probe sequences. Quite recently, fs-DFWM (Degenerate Four-Wave-Mixing) was introduced as a new probe technique in order to improve the detection sensitivity and to investigate the dynamics of uni- and bimolecular collisions in the gas phase [1]. The well-known unimolecular dissociation process of NaI was chosen to test this technique and the experimentally observed wave-packet motion on the excited potential surface was in excellent agreement with the results obtained earlier by laser-induced fluorescence. At the MPI we want to apply this probe method to detect the transition complex of the bimolecular collision system, $Na+H_2$, and first measurements were already initiated. The experimental set-up contains a femtosecond Ti:sapphire laser system with 1 kHz repetition rate and two OPAs which provide the different wavelength necessary for the pump-probe scheme.

[1] M. Motzkus, S. Pedersen, and A.H. Zewail, *J. Phys. Chem.* **100**, 5620 (1996).

OPTICAL SPECTROSCOPY ON SINGLE MOLECULES IN SOLIDS

D. Walser

Physical Chemistry Laboratory
Swiss Federal Institute of Technology
Zurich
SWITZERLAND

Experiments with single quantum systems are at the frontier of modern physics. Probing single molecules allows for very precise control of light-matter interaction and for quantum optical studies. Moreover, a single guest molecule in a solid host matrix is a very sensitive probe of local dynamics in its "nano-environment". Fluorescence excitation spectroscopy on single molecules is a powerful tool for spectroscopic characterizations at highest resolution, as information can be revealed that is not accessible in experiments on ensembles of molecules. The detection of the fluorescence signal of a single absorber is achieved by exploiting the high spectral selectivity given by the ratio (10^{-4} - 10^{-5}) of homogeneous to inhomogeneous linewidths of optical transitions at cryogenic temperatures (zero-phonon transitions).

Two-photon fluorescence excitation, a fundamental nonlinear optical process, was observed on a single quantum system [1]. Single diphenyl-octatetraene molecules trapped in an n-tetradecane matrix were excited from the 1^1Ag ground (S_0) to the 2^1Ag lowest excited singlet (S_1) state by simultaneous absorption of two photons using a single-mode cw laser. Excited molecules were observed through weak S_1 -> S_0 one-photon emission resulting from a symmetry breaking induced by the crystal field. The fundamental principles and the experimental technique of single molecule spectroscopy will be introduced briefly and some particular single molecule experiments presented.

This work was done in collaboration with T. Plakhotnik, A. Renn and U.P. Wild

[1] T. Plakhotnik, D. Walser, M. Pirotta, A. Renn, U.P. Wild, *Science* **271**, 1703 (1996)

NANOSCOPIC INTERACTIONS PROBED BY SINGLE MOLECULE SPECTROSCOPY

J.M. Segura

Physical Chemistry Laboratory
Swiss Federal Institute of Technology
Zurich
SWITZERLAND

We are aiming at the study of interactions at a nanoscopic scale. Using a mesoscopic structure for disturbing, it is possible to influence the properties of a single molecule working as a nanoscopic detector. For probing we take advantage of the extreme sensitivity of single molecule spectroscopy. Single chromophores embedded in solid matrices at cryogenic temperature exhibit very narrow lines, whose position and width strongly depend on the local environment. In order to correlate these spectral properties to the location of the disturbance, we need the three-dimensional spatial imaging offered by confocal scanning optical microscopy (CSOM). The main part of the setup consists of an infinity-corrected microscope objective, that both focus the excitation CW laser beam and collect the single molecule fluorescence emission light. The objective can be adjusted by a stepper motor along the z direction, and motion of the sample is offered by an x-y bimorph scanner. The whole assembly is immersed in a He-liquid bath cryostat. The detection light is first filtered to remove the laser scattered beam, and finally focused on an avalanche photodiode.

An AFM tip has a defined structure and therefore represents an ideal disturbance, which can be very precisely controlled. The influence on the single molecule can be correlated to the three-dimensional position and motion of the tip. Moreover we get topographic imaging with high resolution.

The concept and the setup will be presented together with the first images of test samples using the CSOM at room temperature.

This work was done in collaboration with B. Sick, B. Hecht, A. Renn, and U. P. Wild.

FLUORESCENCE INTENSITY AND ANISOTROPY DECAYS MEASUREMENTS IN THE FREQUENCY DOMAIN

F. De Matteis

Department of Physics
University of Roma-Tor Vergata 00133
Rome
ITALY

Taking advantage of the harmonic contents of a pulsed laser beam, it is possible to measure relaxation times ranging from picoseconds to nanoseconds in the frequency domain. The response of the sample to the train of laser pulses brings all together the response to the different harmonics of the repetition frequency in the laser beam. Then the response to each harmonic can be extracted by means of a cross-correlation technique. The advantages of such a technique are that the contribution from a single harmonic is easily singled out from the overall response with no need to deconvolve the exciting pulse, and that one does not need extremely fast electronics in order to detect the signal. Indeed the only requirement is that the detection equipment can follow the modulation frequency of the light. On the other hand the interpretation of the data is not straightforward as in time domain measurements and the experimental errors can no longer be derived assuming a Poissonian distribution, but they rather have to be measured explicitly.

The lifetime of the fluorescence can be obtained from the curves of the phase and the amplitude modulation as functions of the modulation frequency. Moreover information on rotational correlation times can be obtained measuring the response to linearly polarized light. Namely one is able to measure the decay of the fluorescence anisotropy induced by the polarized exciting beam.

As an example , the fluorescence decay of fluorophores in GST W28H is shown with and without glutation, gathering valuable information about the protein configuration dynamics.

This work was done in collaboration with L. Stella.

NON-DESTRUCTIVE OPTICAL CHARACTERIZATION OF $IN_xGA_{1-x}AS$ EPITAXIAL LAYER COMPOSITION, THICKNESS AND OPTICAL QUALITY

X. Chen

Department of Physics and Astronomy
Wheaton College, Norton
MA 02766, USA

Structures incorporating $In_xGa_{1-x}As$ epitaxial layers have found wide applications in high-speed electronic and optical devices. The composition x, thickness, optical quality and their uniformity across the wafer are crucial in obtaining the desired device performance and high device yield. Here, we will report a rapid and non-destructive technique to determine the composition, thickness and optical quality across a wafer of a multilayer $In_xGa_{1-x}As$/InP stack by combining photoluminescence (PL) and spectroreflectance (SRL) measurements. Our results suggest that a system using a spectrograph and an infrared photodiode array would allow for quick, routing use of PL and SRL as a quality control tool. The advantages of the system are that: the measurements are much faster than using a scanning monochromator with a single detector, it can do both photoluminescence and optical reflectance; the whole process is automable and non-destructive. To demonstrate the usefulness of the proposed system, we will present the mappings of composition, thickness and PL intensity on a 50 mm wafer of a multilayer $In_xGa_{1-x}As$/InP stack.

This work was done in collaboration with David Weyburne, and Qing S. Paduano.
(Work is partially supported by Air Force Office of Scientific Research)

OPTICAL HOMOGENEOUS LINEWIDTH IN AMORPHOUS SOLIDS IN THE FRAMEWORK OF THE SOFT POTENTIAL MODEL

A.J. Garcia
Dpto. de Fisica Aplicada I, E. T. S.
Universidad del Pais Vasco
48013 Bilbao
SPAIN

It is well known that amorphous solids exhibit anomalous behaviours when compared with their crystalline counterparts [1]. These anomalies include low-temperature specific heat, thermal conductivity, ultrasound propagation, dielectric losses, and optical dephasing among others. Their most surprising characteristic is universality, i.e., the order of magnitude depends very weakly on the chemical composition of the solid.

In this work, the temperature dependence of the optical homogeneous linewidth (HLW) of an impurity embedded in a glass matrix is investigated in the framework of the Soft Potential model [2], which has allowed to explain some of these anomalies above 1 K in terms of soft anharmonic potentials, the so called soft modes. The HLW in glasses follows a T^{α} power law with temperature with α ranging from 1 to 2.6, depending on the material [3].

A model Hamiltonian which couples the impurity to the soft modes characteristic of glasses in the phonon field of the matrix is introduced. We find an excellent agreement between our model predictions and experimental data [4]. We are also able to reproduce supralinear behaviours in the low temperature regime without any adjustable parameter for the density of states. Finally, the quadratic behaviour extending to temperatures much below the Debye temperature of the matrix is explained in terms of the Bose peak equivalent temperature.

This work was done in collaboration with R. Balda and J. Fernandez.

[1] *Amorphous Solids: Low Temperature Properties*. Edited by W. A. Phillips (Springer, 1981)

[2] V.G. Karpov, M.I. Klinger, and F.N. Ignat'ev, *Sov. Phys*. JETP **57**, 439 (1983).

[3] *Optical Linewidths in glasses*. Guest Editor: Marvin J. Weber, *J. Lumin.* **36**, 4&5, p. 179-329 (1987)

[4] A.J. Garcia and J. Fernandez, *Phys. Rev. B* **56**, 1 (1997).

RAMAN SHIFTED LASER CRYSTALS WITH $\chi^{(3)}$ NON LINEARITY: TUNGSTATE AND MOLYBDATE CRYSTALS FAMILIES

F. Bourgeois

Laboratoire de Physico-Chemie Materiaux Luminescents
UMR 5620 CNRS
Villeurbanne
FRANCE

The stimulated Raman scattering (SRS) phenomenon is theoretically well known since the 30s but its application had to wait till 1960 for a powerful light source: the laser. Up to now, a very few crystals, which present high third order dielectric susceptibilities, have been studied and actually, only $Ba(NO_3)_2$ is commercialized. Others, like double tungstate and molybdate crystals have been noticed for their high Raman active modes (about 900 cm^{-1} and 960 cm^{-1} respectively) and their fair stability, but not entirely developed yet.

Combining Nd^{3+}:YAG laser emission and SRS, we expect to achieve all solid laser cavities in the eye safe emission region. For that purpose, it is necessary to examine all the possibilities offered by the tungstate and molybdate crystal families. Therefore, a full characterization has just been started. Using different sources and techniques, the growth and structural characterization should complete physical measurements such as transmission window, optical axes, $\chi^{(3)}$ susceptibility coefficients, vibrational active modes, Raman gain coefficient, phonon dynamics and of course, generation and study of Raman shifted laser beam.

Further studies will determine whether the Nd^{3+} doped crystal is suitable as a self Raman converter laser material. Among such studies, the spectroscopy of those crystals is vital. During this seminar, a few results of this beginning research were presented.

LUMINESCENCE QUANTUM EFFICIENCY IN CDSSE MIXED CRYSTALS

R. Westphaeling

Institut fuer Angewandte Physik
Universitaet Karlsruhe
D-76128 Karlsruhe
GERMANY

There are basically two different ways to determine absolute photoluminescence (PL) quantum efficiencies. With the calorimetric absorption spectroscopy (CAS) the heating of a sample under illumination due to nonradiative processes of excited carriers is measured. In combination with calorimetric transmission spectroscopy (CTS) it is then possible to calculate PL quantum efficiencies. In this indirect method it is useful to go to the mK regime to reach high sensitivity. We use the other, more direct method with an integrating or Ulbricht sphere fixed into a cryostat. With this setup PL quantum efficiencies can be determined from 5 K up to room temperature or even above.

At low temperatures we found in the binary systems CdS and CdSe efficiencies less than 25 % depending on the crystal quality within the main luminescence bands (arising from donor and acceptor bound excitons and donator acceptor pair recombination and their phonon replica). In contrast, CdSSe mixed crystals show an efficiency up to 70 % within the luminescence out of localized states originating from composition fluctuations of the alloy. Electron-hole pairs created above the mobility edge relax into localized states from where they recombine radiatively with low nonradiative losses.

With increasing temperature the efficiency is firstly unaffected, but decreases drastically when the temperature exceeds the equivalent of a certain thermal activation energy. Taking into account three parallel recombination channels, namely radiative, nonradiative and thermal activated nonradiative recombination, we can fit the efficiency as a function of temperature and deduce an activation energy which coincides with the localization depth of the excitons below the mobility edge.

This work was done in collaboration with C. Klingshirn.

LUMINESCENCE AND AFTERGLOW IN CSI:TL

H. Wieczorek

Philips GmbH Forschungslaboratorien
Weisshausstrasse 2
D - 52066 Aachen
GERMANY

Thallium doped Cesium Iodide (CsI:Tl) is a scintillator well suited for application in solid state X-ray detectors. Its main advantages are a high light output, an emission spectrum well adapted to photosensors like amorphous or crystalline silicon diodes, and the possibility of large area deposition at moderate temperature, resulting in a light guiding structure and excellent spatial resolution. A major drawback, however, is the strong afterglow which may last for minutes. There are hardly any literature data on that item, and afterglow is often considered an intrinsic property of CsI:Tl.

Using a new deposition method, we optimised the luminescence light output and the spatial resolution of CsI:Tl layers. Afterglow has been strongly reduced. We were able to show that it consists of at least three different effects: long term afterglow is caused mostly by an insufficiently low dopant concentration. An unnecessarily high Tl concentration, on the other hand, results in an enhanced afterglow at intermediate and long times. The short term signal decay, in the millisecond time range, seems to be an intrinsic property of CsI:Tl. It shows a build-up with the applied dose just like a trapping effect. The revealing of possible mechanisms that could account for the afterglow properties of CsI:Tl is an ongoing challenge for luminescence research.

POSITION SENSITIVE TIME OF FLIGHT MASS SPECTROMETRY OF MOLECULES

J. Larsen

Institute of Physics and Astronomy
University of Aarhus
8000 Aarhus C
DENMARK

We are currently designing and building a position sensitive time of flight mass spectrometer. The apparatus consists of a pulsed molecular beam which is perpendicular crossed with a femtosecond laser pulse. The idea is to employ a pump probe pulse arrangement where the probe pulse should be so intense that the molecules are multiple ionized in order to let them undergo a Coulomb explosion. The resulting kinetic energies of the fragments is a direct measure of the bonding distances due to Coulombs law.

The charged fragments is accelerated by an electric field towards a microsphere plate equipped with a phosphor screen. The fragments appear as small blobs of light on the screen. A CCD camera is used to measure the positions of the fragments whereas two photomultipliers is used to measure the time at which the fragments arrive at the detector.

We hereby get (x,y,t) information of the fragments which enables us to calculate the (x,y,z) coordinates of the fragments at the time of the Coulomb explosion. This technique will allow us to follow the dynamics of the investigated molecule as the delay between the pump and probe pulse is scanned.

This work was done in collaboration with N. Bjerre.

PROBING EXCITON DISTRIBUTIONS IN SEMICONDUCTOR QUANTUM LAYERS

J. Hoffmann

Institute of Applied Physics
University of Karlsruhe
D-76128 Karlsruhe
GERMANY

Analyzing time resolved luminescence from excitons to get information about the dynamic of the exciton distribution is a complicated job. To satisfy energy and wave vector conservation the excitons can recombine and emit a photon only where exciton dispersion crosses the light dispersion. But more polarly bound semiconductors, like the II-VI, offer a much better way to visualize the exciton dynamics. This is due to the strong exciton LO-phonon coupling: there is the possibility for the excitons to recombine from nearly anywhere on the 1s-dispersion under emission of an photon and a LO-phonon. The LO-phonon just consumes all the excess wave vector. Of course this 3 particle process has a lower probability e.g. in our samples about 1% than the direct recombination. But this is OK, as you just want to probe the distribution and not influence it strongly by this 3 particle recombination.

This means time resolved measurements of the 1st LO-phonon replica gives detailed information about the time evolution of any exciton distribution created in the quantum layer.

This work was done in collaboration with M. Umlauff, H. Kalt , W. Langbein and J. Hvam.

LIGHT INDUCED MOLECULAR STRUCTURE IN H_2^+

K. Saendig

Max-Planck-Institut fuer Quantenoptik
D-85748 Garching
GERMANY

Quantum Behavior of diatomic molecules as H_2^+ in strong laser fields are studied by observation of the energy distribution of fragments of photodissociation. An ion source is used, in which by electric discharging several kinds of Hydrogen-molecules are produced. The ions then are accelerated by several kV. The so generated ion beam is then mass-selected in order to extract only H_2^+. Crossing the collimated H_2^+ beam with the laser beam one can observe the dynamics of the dressed states by projecting the fragments on a phosphorscreen leading to luminescence. The spatial intensity distribution of the luminescence is proportional to the fragment distribution projected onto the screen and can be read out using a CCD-Camera. Possible effects, which could be observed are:

1) Tunneling through the barrier at the avoided crossing of the lower state.
2) Bond softening and bond tightening
3) Above Threshold Ionization (ATI)

The effect can be identified by the energy distribution of the photofragments which can be obtained by an inverse Abel-Transform of the intensity distribution of the signal read out by the CCD-camera.

POSTERS

TIME RESOLVED SPIN RELAXATION IN III-V SEMICONDUCTOR QUANTUM WELLS

R.S. Britton

University of Southampton
Highfield, Southampton, SO17 1BJ
UNITED KINGDOM

Quantum confinement of carriers in semiconductors quantum wells greatly affects the optical (linear and non-linear) properties of the material. In particular, the reduced dimensionality lifts the degeneracy of the k=0 light hole-heavy hole valence band states, modifying the optical selection rules so that it is possible to create a fully spin polarised population of photoexcited carriers using circularly polarised photons. Another important consequence of quantisation is the observation of excitonic resonances at room temperature.

The use of polarised light enables investigation of the spin dynamics of carriers. An understanding of carrier spin-relaxation is necessary to explain the results of polarisation sensitive measurements. Application of magnetic fields allows observation of quantum beating and accurate determination of Zeeman splitting and Lande g-factors (important for band structure theory and the Quantum Hall Effect). The dephasing of a coherently prepared electronic state allows insight into scattering by phonons, defects and carriers, processes which limit the capability of ultrafast optoelectronic devices. Technological applications include possible coherent control of optical properties. Ultrafast optical switching based on spin relaxation has already been demonstrated.

This work was done in collaboration with R. T. Harley.

ANALYSIS OF THE $4F^N$ ENERGY LEVEL STRUCTURE OF THE RE^{3+} IONS IN RE OXYHALIDES

R.J. Lamminmäki

University of Turku
Department of Chemistry
FIN-20014 Turku
FINLAND

The energy level structure of the trivalent rare earth ions (RE^{3+}) can be studied with a phenomenological model by determining the free ion and crystal field (c.f.) parameters from the spectroscopic information. Even though the luminescence of a RE^{3+} ion in solid state is relatively insensitive to the matrix, significant differences in the optical properties may be observed when the host has been changed. These effects can be investigated in the RE oxyhalide series where both the matrix anion and cation can be changed. The main factor responsible for the different spectral properties of the RE^{3+} activator is the crystal field effect.

For this purpose, the energy level schemes of the rare earth oxychlorides (REOCl or REOCl:RE^{3+} ; RE^{3+} = Pr^{3+}, Nd^{3+}, Sm^{3+}, Eu^{3+}, Tb^{3+}, Ho^{3+}, Er^{3+}, and Tm^{3+}), were studied and simulated with a phenomenological model accounting simultaneously for both the free ion interactions and the c.f. effect. The energy level schemes for the $4f^N$ (N = 2, 3, 5, 6, 8, and 10 - 12) electron configurations were deduced from the absorption and luminescence spectra and analyzed according to the C_{4v} point symmetry of the RE^{3+} site in RE oxychloride. Good correlation was obtained between the experimental and calculated energy level schemes. The evolution of the free ion and c.f. effects on the energy level scheme of the RE^{3+} ions is rather smooth and the systematic trends in parameter values were evaluated.

This work was done in collaboration with Jorma Hölsä and Pierre Porcher.

TEMPERATURE DEPENDENCE OF NEAR INFRARED TO VISIBLE UPCONVERSION LUMINESCENCE OF ER^{3+} IN YSGG

X. Chen

Department of Physics and Astronomy
Wheaton College, Norton
MA 02766, USA

There has been great interest in laser materials for efficient conversion of infrared light into visible light (called upconversion). One of the important applications of upconversion is to make lasers that emit short-wavelength light, pumped by long-wavelength light sources such as infrared diode lasers. We have investigated the near infrared (790 nm) to visible (400-700 nm) upconversion luminescence from Er^{3+}(30%) doped YSGG at different temperatures (77-300 K). The luminescence in the region of 400-700 nm, pumped by an Ar ion laser (488 nm), was also investigated and compared with the upconversion luminescence, pumped by a diode laser (790 nm). This work provides additional information on the upconversion processes in YSGG:Er^{3+}(30%) and describes how temperature affects these processes.

This work was done in collaboration with John Collins and B. Di Bartolo.

QUANTUM CONFINEMENT IN POROUS SILICON IN THE EFFECTIVE-MASS LIMIT

I.H. Libon

LMU Munich
Sektion Physik
D-80799 München
GERMANY

We report the controlled PL peak shifting of p^+ porous silicon (PoSi) by the method of atomic layer etching (ALEP). We hereby investigate the dependence of the crystallite size on the PL peak position of this material. By this method of repeated oxidation by H_2O_2 and stripping of the oxidized surface layer, we were able to reduce the size of the clusters layer by layer [1]. In all previous reports the PoSi PL displayed a natural lower energy limit of ≈ 1.4 eV. We are the first to report a continuous PoSi PL peak shift between 1.01 and 1.20 eV. From our experiments we draw several conclusions for the luminescence mechanism: we demonstrate the responsibility of geometrical quantum confinement in silicon crystallites for the efficient room-temperature PL in PoSi near the indirect bandgap of c-Si. For the relatively small shifts from the bulk Si bandgap we employ the effective-mass theory to calculate the cluster-diameter dependence of the bandgap. Along with observations of size-independent PL peaks around 1.6 eV in thermally oxidized samples our measurements indicate that the PoSi PL cannot be described by one origin alone. Both the existence of molecular centers and the geometrical quantum confinement are responsible for the PoSi PL in their specific range of etching and post-anodic treatment parameters.

This work was done in collaboration with C. Voelkmann, V. Petrova-Koch, and F. Koch.

[1] I.H. Libon, C. Voelkmann, V. Petrova-Koch, and F. Koch, in: *MRS Proceedings*, MRS Fall Meeting, Boston, 1996 (in press).

ULTRASHORT LASER PULSES MEASUREMENTS BY MEANS OF SELF DIFFRACTION IN POLYDISPERSED LIQUID CRYSTALS

A. Mazzulla
Physics Department
University of Calabria
87036 Rende
ITALY

Polymer dispersed liquid crystals can be efficiently used to measure light pulse [1,2] duration in the sub-picosecond range by inducing in these materials the formation of a transient thermal grating with consequent self diffraction of the incoming laser beams. This technique [3] involves the interaction of two laser beams in a nonlinear medium leading to self diffraction of the beams themselves. In order to measure the light pulse duration, the laser beam is spliced (amplitude sharing) to get two coherent waves able to interfere in the non linear medium where they cross each other. The time coherence between them is controlled by a variable delay that can be introduced by appropriate reflectors. In the presence of a non linear optical response of the medium they give rise to a transient phase grating, able to produce diffraction of the two beams and the efficiency of the diffraction is measured as the ratio between the diffracted intensity vs. the impinging one: $h=I_d/I_0$.

Usually, measurements are taken on the first order diffraction beam, which is the most intense one. By changing the optical delay between the two interacting beams one can vary the diffraction efficiency that will be maximum for zero delay (corresponding to complete overlap of the two laser pulses and it will be zero for a delay bigger than the pulse duration since in this case no overlap of the two pulses is obtained. It has been proved that diffraction efficiency is related to the short pulses fourth-order coherence function and its measurements vs. the pulses delay can provide both the pulse duration t_p and the coherence time t_c. This method has been successfully applied using semiconductor films and dyes as non linear materials; anyway, in these cases the data needed to be corrected since the fast relaxation time of used materials affected the experimental observations. On the other hand this technique was shown to be very reliable when using media with nonlinearities much slower than the pulse duration [4] since in this case the material relaxation doesn't affect the measurements. Moreover this technique has some advantages over the conventional second order autocorrelation that exploits second harmonic generation (SHG) in non linear crystals with a Michelson type interferometric geometry: it is broad band, background free and easy to align.

[1] G. Cipparrone, D. Duca, A. Mazzulla, C. Umeton and F. Simoni, *Optics Communications* **97**, 54 (1993).
[2] G. Cipparrone, D. Duca, A. Mazzulla, C. Umeton and F. Simoni, *Mol. Cryst. Liq. Cryst.* **263**, 521 (1995).
[3] H. J. Eichler, U. Klein and D. Langhans, *Applied Physics* **21**, 215 (1980).
[4] R. Trebino, C. C. Hayden, A. M. Johnson, W. M. Simpson and A. M. Levine, *Optics Letters* **15**, 1097 (1990).

OPTICAL NEAR-FIELD MEASUREMENTS ON A SINGLE GOLD NANOPARTICLE

T. Klar

Ludwig-Maximilians-Universität
München
GERMANY

We probe surface-plasmons on spherical nanoparticles embedded in dielectric matrices with high refractive index. The surface-plasmon resonance in far-field transmission spectra is inhomogeneously broadened due to variations in particle diameter or refractive index of the matrix. To determine the homogeneous linewidth of the surface-plasmon resonances we use a Scanning Near-Field Optical Microscope (SNOM), which enables us to probe a single nanoparticle. The near-field transmission spectrum of a single colloid shows an enhanced transmission near the plasmon resonance, in contrast to the resonant extinction observed in far-field experiments. We conclude that evanescent light modes near the SNOM tip stimulate the dipole mode of the surface plasmon on the nanoparticle. These dipoles emit propagating modes that can be detected in the far field. The emission resonance observed for a single colloid is much narrower (170 meV FWHM) than the resonance which we obtain in far-field transmission (350 meV FWHM) for a colloid ensemble. The peak positions of the resonances from different particles differ by 100-150 meV. Assuming that the observed linewidth can be directly converted into a dephasing time T_2, we obtain a value of $T_2 = 8$ fs, a reasonable value for surface plasmons.

This work was done in collaboration with M. Perner, S. Grosse, G. von Plessen, W. Spirkl, U. Lemmer and J. Feldmann.

SINGLE-ATOM CONDITIONAL MANIPULATION OF CONFINED QUANTIZED ELECTROMAGNETIC FIELD

A. Napoli

INFM
Istituto di Fisica dell'Universita'
Palermo
ITALY

It has been recently shown [1] that, under suitable conditions, the interaction between a three-level atom and the quantized field of a bimodal microcavity may be effectively reduced to the two-photon coupling between a single two-level atom and the electromagnetic field in the resonator. The corresponding Hamiltonian model is characterized by intensity-dependent Stark shifts due to the presence of an intermediate off-resonance atomic level. Assume the cavity field prepared at t=0 in an arbitrary state and let an initially excited atom cross the cavity with an appropriately selected velocity. It is possible to show that, exploiting a single conditional measurement of the atomic state immediately after the atom leaves the resonator, the bimodal field is projected into a new state characterized by the same photon number distribution present at t=0. More in detail, the experimental scheme we propose makes it possible manipulating only the phases of the initial probability amplitudes of finding a well defined population in the cavity. We prove that the state into which the cavity collapses after a successful atomic measurement is unitarily related to the initial field state and the explicit form of the accomplishing operator is constructed. We wish to remark that, beside its inherent theoretical interest, this simple experimental scheme might also be of relevance in the so-called field of quantum communication [2], where implementing unitary modifications of a quantum state signal represents an issue of topical and central importance. More in general, our proposal provides a systematic procedure for controlling selected properties of a quantum state. We in fact demonstrate that our experimental scheme may be reliably exploited to generate orthogonal states to a prefixed one or to control the squeezing direction.

[1] Shih-Chuan Gou, *Phys. Lett. A* **147**, 218 (1990); A. Napoli and A. Messina, *J. Mod. Opt.* **43**, 649 (1996).

[2] C.M. Caves and P. Drummond, *Reviews of Mod. Phys.* **66**, 481 (1994); A. Ekert and R. Josa, *Rev. Mod. Phys.* **68**, 733 (1996).

ITERATIVE NONIDEAL CONDITIONAL GENERATION OF NONCLASSICAL CAVITY FIELD

A. Napoli

INFM, Istituto di Fisica dell'Università
Palermo
ITALY

The generation of nonclassical quantum states of the electromagnetic field plays a central role in understanding the conceptual foundations of quantum mechanics. Over the last few years many experimental schemes for realizing Fock states of the field have been reported. Three reasons are at the origin of the attention toward these schemes: a) a number state, as prototype of a nonclassical state, provide a peculiar condition to test basic aspects of the quantum theory: b) the invention of the micromaser as well as the availability of sophisticated techniques for producing Rydberg atoms, has opened up the possibility of manipulating the microcavity field statistics exploiting appropriate single few-level-atom-radiation field coupling mechanisms; c) the applicative potentialities of these states for encoding and processing quantum information. We focus our attention on those number-state preparation projects based on the conditional measurement procedure, in the context of the micromaser theory. Within this approach the field is iteratively guided into the desired state with the help of an appropriate sequence of similar steps. Each one involves the flight of a Rydberg atom through the cavity and the detection of its internal state immediately after it leaves the resonator. By definition a conditional scheme requires an a priori well-defined sequence of N atomic outcomes and, consequently, of N field-collapse-processes inside the cavity. This means that the registration of even one different atomic outcome in the sequence has to be interpreted as failure of the preparation of the desired cavity state.

Here we propose an approach for predicting the probability of success of creating cavity Fock states with a controllable intensity, emphasizing the role played by a nonideal performance of the experimental setup. In general many ingredients concur to the success of experiments of this kind. One is related to the possibility of controlling sharply the interaction times between the cavity field and the atoms. Another one is the existence of an highly efficient selective state atomic detection apparatus. Moreover effects stemming from cavity damping have to be taken carefully into consideration. An attractive aspect of our way of analyzing this problem is that it turns out to be applicable to different experimental schemes based on conditional measurement. Considering two recently proposed projects for building up Fock states of a high-Q microcavity, we demonstrate that the presence of an imperfection in the apparatus may be a source of ruining effects both in the statistics of the final state of the cavity and in the probability of success of the experiment.

This work was done in collaboration with A. Messina.

MULTIPLE BRANCHING OF VECTORIAL SPATIAL SOLITARY WAVES IN NONLINEAR QUADRATIC MEDIA

G. Leo

University "Roma Tre"
Dept. of Electronic Engineering
Rome
ITALY

Propagation of spatial solitary waves associated to three-wave mixing in a non-centrosymmetric medium is numerically investigated, in the frequency-degenerate, polarization non-degenerate case (Type II second-harmonic generation). A pure up-conversion scheme is adopted, in which only the two orthogonally polarized gaussian beams at the fundamental frequency are input, and the process is phase-independent. Multiple branching of the solitary waves is shown to occur and to be associated to a sharp polarization switch, favoured by balanced input launch conditions. The branching, which takes place even in the absence of asymmetries and/or higher-order input beams, is explained in terms of a transverse modulation of the parametric gain across the three interacting fields. Numerical results are reported for the case of a KTP planar waveguide.

This work was done in collaboration with G. Assanto.

RELAXATION DYNAMICS OF LOCALIZED EXCITONS IN ZnSe-ZnSSe STRAINED-LAYER SUPERLATTICES

M.C. Netti

Physics Department
University of Bari
70126 Bari
ITALY

The exciton localization effects due to the unavoidable well width fluctuation or compositional disorder are of crucial importance in quantum well (QW) systems . The role of localized and free excitons on the radiative recombination dynamics has been recently investigated both theoretically and experimentally. On the contrary little is known about exciton dynamics at low carrier densities in II-VI QWs in spite of their relevance in the lasing processes.

We report a detailed study of the relaxation dynamics of localized excitons in a set of symmetric ZnSe-ZnSSe strained-layer superlattices having different well width ranging between 2 and 10 nm. Time resolved photoluminescence (TR-PL) measurements were performed at 10 K using a frequency doubled Nd:YAG-pumped dye laser (5 ps) at low photoexcited carrier densities (10^6-10^7 cm^{-2}) with an overall time resolution of 20 ps. The TR-PL were recorded at different energies throughout the inhomogeneously broadened cw PL line in order to monitor the different contribution of localized and free excitons.

Two clear distinguishable trends can be figured out in these samples. For well width (L_W) narrower than or comparable with the Bohr radius ($a_B \approx 4$ nm) the rise time of the TR-PL curves is higher at low emission energy reflecting the acoustic-phonon assisted hopping of the exciton population towards the localized states. At the opposite limit, for $L_W > a_B$ there is a very fast population of the low energy localization states. This different dynamics of free and localized excitons also reflects on the decay of the luminescence, being monoexponential or biexponential depending on emission energies and well width. In the case of biexponential decays (higher emission energies) the fast decay time of TR-PL is approximately equal to the rise time of the TR-PL at lower energy providing an estimate of the exciton localization time.

This work was done in collaboration with R. Tommasi, M. Lepore, M. Dabbicco, A. Vinattieri and M. Colocci

IDENTIFICATION OF SEVERAL ENERGY TRANSFER PATHWAYS IN THE LIGHT-HARVESTING COMPLEX II OF GREEN PLANTS USING CURRENT STRUCTURAL AND SPECTROSCOPIC INFORMATION

S. Ozdemir

Dept. of Physics
The Middle East Technical University
06531, Ankara
TURKEY

A set of procedures to run numerical experiments to investigate the excitation dynamics (energy transfer processes) in a structurally resolved network of molecules has been developed. The procedures have been used to simulate the energy transfer processes in the Light- Harvesting Complex II (LHC) of green plants. Energy transfer between a pair of molecules (chlorophylls) of the complex has been modeled by the Forster mechanism and time evolution of the excitations has been evaluated in the Pauli master equation formalism. The aim of the simulations has been to correlate some of the processes observed in the sub-picosecond time-resolved measurements with the current LHCII structural model. Several possibilities for the energy transfer pathways among the chlorophylls of the LHCII complex have been identified.

This work was done in collaboration with D. Gulen

TRANSIENT AND PERSISTENT SPECTRAL HOLE-BURNING IN CuBr SEMICONDUCTOR QUANTUM DOTS

J. Valenta

Department of Chemical Physics and Optics
Charles University
Prague
CZECH REPUBLIC

Pump and probe optical techniques are applied to study size-dependent excitonic dynamics, exciton-phonon interactions and the mechanism of persistent optical changes in CuBr nanocrystals (NCs) embedded in a borosilicate glass. In the samples containing big NCs (mean radius, *a*, of about 5 to 12 nm) we observe overall changes of excitonic absorption bands under resonant excitation, but in small NCs ($a < 5$ nm, i.e. about four times the excitonic Bohr radius) the same excitation produces selective changes of an absorption band around the position of excitation - so called spectral hole-burning (SHB). All spectral holes contain a central part and pronounced side bands due to strong exciton-LO phonon coupling. Beside the transient holes, which exist some hundreds of picoseconds, also the persistent holes, lasting for hours at low temperatures, are observed. Measurements of the dynamics of transient absorption changes and of the width of central holes give us information about the excitonic lifetime T_1 and the dephasing time T_2, respectively. We show that values of T_2 derived from SHB are comparable to those we measured by time-resolved degenerate four-wave-mixing, providing that the burning laser is spectrally narrow and the extrapolation to the zero intensity and zero temperature is considered.

This work was done in collaboration with J. Dian, J. Moniatte and P. Gilliot.

MULTIPHOTON EXCITATION OF GaN USING PICOSECOND PULSES

C. Voelkmann

Max-Planck-Institut für Quantenoptik
D-85740 Garching
GERMANY

We observed the UV bandgap photoluminescence (PL) from bulk wurtzite GaN by means of multiphoton excitation. The two-photon PL excitation spectrum near the bandgap agrees with a calculation of the theoretical two-photon absorption spectrum using a parabolic two-band model [1]. The pump intensity dependence of the PL becomes more nonlinear as the excitation photon energy is decreased as expected. The PL excitation spectrum in the infrared and the dependence of PL intensity on excitation intensity show the existence of midgap defect states 1 eV above the valence band maximum. We confirmed the energetic position of this deep level, which is responsible for the omnipresent yellow PL in GaN, with a two-color, two-photon excitation experiment [2].

This work was done in collaboration with I.H. Libon, D. Kim, V. Petrova-Koch and Y.-R. Shen.

[1] D. Kim, I.H. Libon, C. Voelkmann, Y.-R. Shen, and V. Petrova-Koch, *Phys. Rev. B* **55**, R4907 (1997).

[2] I.H. Libon, C. Voelkmann, D. Kim, V. Petrova-Koch, and Y.-R. Shen, in: *MRS Proceedings*, MRS Fall Meeting, Boston, 1996 (in press).

0.94 μm SPECTROSCOPY AND LASER DEVELOPMENT

B.M. Walsh

NASA Langley Research Center
Hampton
VA 23681 USA

Tunable lasers have played an important role in providing the technology necessary for active remote sensing of the atmosphere. Researchers at NASA Langley Research Center are developing various laser systems for remote sensing of the atmosphere and monitoring of environmental conditions affecting planet Earth. These include lidar (light detection and ranging). differential absorption lidar (DIAL), velocity sensing and ranging [1]. Tunable narrowband lasers operating in pulsed mode with high energy are ideal for use in lidar systems. In particular, DIAL techniques can be used to determine molecular constituent concentrations present in the atmosphere. Some molecular constituents which are of interest due to their impact on the environment are ozone (O_3), and greenhouse gases such as carbon dioxide (CO_2), methane (CH_4), carbon monoxide (CO), and water-vapor (H_2O).

In the DIAL technique, two lasers of slightly different wavelength are used. One laser is tuned to a wavelength corresponding to an absorption feature of the molecular species of interest (the on-line) and the second laser is tuned to a region where the molecular species does not absorb (the off-line). Both beams are transmitted, scattered and received, but the on-line beam is attenuated by absorption of the molecular species. The differential between the on-line and off-line signals is used to determine the molecular distributions as a function of distance. This provides a profile of molecular concentrations with altitude.

One molecular species, which is of interest at NASA as a greenhouse gas, is water-vapor. Injection seeded $Ti:Al_2O_3$ lasers are available to measure H_2O concentrations around 0.82 μm. However, for measurements in regions with low H_2O density such as the polar regions, upper troposphere and lower stratosphere, H_2O lines with stronger absorption features are required in order to get substantial absorption of the on-lone beam in the DIAL technique. H_2O lines in the 0.94 μm region are up to 30 times stronger than those in the 0.73 - 0.83 μm range. In addition, transition metal lasers like $Ti:Al_2O_3$ while widely tunable, they suffer from low gain or a short upper laser level lifetime. To obtain high gain and long lifetime, lanthanide series ions can be used although they are often not tunable.

The lanthanide ion Nd^{3+} exhibits lasing at 0.946 μm when doped into YAG, but is not very tunable. Therefore, other host materials can be investigated for doping with Nd^{3+} to produce lasing transitions which correspond to water-vapor lines. One such material is $Nd:Y_3Ga_xAl_{(5-x)}O_{12}$ (Nd:YGAG). YGAG is a mixed crystal with a garnet structure that does not suffer from stoichiometric constraints on the proportion of Ga to Al, since Ga^{3+} and Al^{3+} have approximately the same ionic radii. Changing the material composition results in a variation of the lattice parameters, and thus a variation in the crystal field strength at the site of the Nd^{3+} dopant ions, allowing different emission wavelengths depending on the ratio of Ga to Al.. This method of compositional tuning has the advantage of providing the required wavelength flexibility needed to tune to the water-vapor lines.

measurements include branching ratios, cross sections, peak wavelengths and lifetimes. The laser performance measurements include threshold and slope efficiency.

This work was done in collaboration N.P. Barnes and J.C. Barnes.

[1] N.P. Barnes, "Lidar systems shed light on environmental studies," *Laser Focus World*, pp87-94, April 1994.

SPECTRAL LINESHAPES OF AMMONIA TRANSITIONS OF THE ν_2 BAND VERSUS PRESSURE AND TEMPERATURE

F. Pelagalli

ENEA, INN-FIS
00044 Frascati (Roma)
ITALY

Lineshape study of molecular transitions have always been of great importance in the spectroscopic community since the lineshapes contain useful information on the nature of molecular interactions. Ammonia, among the molecular species, has been extensively measured not only because it is a component of the planetary atmosphere and, as a trace gas of the Earth atmosphere, but also owing to its large molecular dipole and its singular inversion energy spectrum. The latter properties are important for developing theoretical models which should then be validated against the experimental results. However this much needed procedure has not been performed in a satisfactory way up to a few years ago, because pressure broadenings and shifts in ammonia were known with experimental errors usually around 10% and more 30% respectively, and practically no measurements as a function of temperature were available. This unpleasant situation changed recently, when for a few transitions in the ν_2 band around 900 cm^{-1} the errors have been drastically reduced to less than 3% and 10% respectively [1,2] thanks to an apparatus based on tunable diode lasers and variable temperature cell in the range 180-380 K, and a semiautomatic analysis of the data, realized just for this task. Now these measurements have been extended to several other ammonia transitions, obtaining results which have been discussed in the frame of the slightly modified ATC theory [3,4]. In particular the aQ(9,9) line at 921.2550 cm^{-1} has been studied versus temperature also with foreign gases, N_2, O_2, H_2, He, Ar, as perturbers. The calculations concerning foreign gases are suitable when the long range interaction is so effective that the broadening cross-section is far larger than the kinetic one. This is the case of self broadening, because of the large dipole-dipole interaction, and, to a lesser extent, of N_2 broadening, because of the large quadrupole moment of nitrogen molecule.

The main results of this much closer comparison with respect to previous attempts, is that the theory describes very well the self broadening and the N_2 foreign broadening, has some limitations for self shift while improvements are needed for N_2 foreign shift, and does not agree well with the empirical power laws describing the temperature behavior, especially when wide temperature ranges are involved. Even more these empirical laws are completely at odds in those cases when the shift changes sign versus temperature as we have recently measured for the first time [5]. As far as the effects of the foreign gas collisions are concerned, we have been able to produce very fine measurements with errors as low as 2-3% for broadening parameters and often less than 10% for shift parameters. These results, which are very important also for an improvement of the theoretical model, have been obtained thanks also to surprisingly big foreign shifts compared to the foreign broadening parameters. Indeed, for foreign gas collisions, we found a shift/width ratio by far larger than for self-collisions.

This work was done in collaboration with G. Baldacchini, G. Buffa, F. D'Amato, M. De Rosa and O. Tarrini.

[1] Baldacchini, A. Ciucci, F. D'Amato, G. Buffa and O. Tarrini, *J. Quant. Spectrosc. Radiat. Transfer* **53**, 671 (1995).

[2] Baldacchini, F. D'Amato, M. De Rosa, G. Buffa and O. Tarrini, *J. Quant. Spectrosc. Radiat. Transfer* **55**, 745 (1996).

[3] Anderson, *Phys. Rev.* **76**, 647 (1949).

[4] Tsao and I. Curnutte, *J. Quant. Spectrosc. Radiat. Transfer* **2**, 41 (1962).

[5] Baldacchini, F. D'Amato, F. Pelagalli, G. Buffa and O. Tarrini, *J. Quant. Spectrosc. Radiat. Transfer*, **55**, 741 (1996).

MULTIPHOTON EXCITED LUMINESCENCE OF POROUS SILICON

C. Voelkmann

Max-Planck-Institut für Quantenoptik
D-85740 Garching
GERMANY

We employed PL spectroscopy to study multiphoton excitation of porous silicon (PoSi). When using picosecond mid-infrared pulses, up to nine photons were necessary to generate one visible PL photon which corresponds to a nonlinearity of the third order [1]. Because of the existence of electronic surface states in the PoSi bandgap it is a multi-resonant and multi-step process. The involved electronic surface states acting as intermediate resonant levels are surface tail states and unoccupied Si dangling bonds. After each step the system reaches a certain energy level; depending on the excitation intensity, it can be pumped further up or dropped to lower levels via luminescence or non-radiative decay.

This work was done in collaboration with I.H. Libon, D. Kim, V. Petrova-Koch and Y.-R. Shen.

[1] D. Kim, I.H. Libon, C. Voelkmann, V. Petrova-Koch, and Y.-R. Shen, in: 6th NEC Symposium on Fundamental Approaches to New Material Phases, Karuizawa, Japan, 1996, in: *Materials Sciences and Engineering B* (Solid State Materials for Advanced Technology), April 1997 (in press).

FIRST ROUNDTABLE DISCUSSION

The first roundtable discussion was held on June 19th, the fourth day of the lectures. Prof. Di Bartolo opened the discussion floor by inviting everybody to express their opinion and give positive and negative feedback about all aspects of the course including the lectures, the facilities, and the organization and to suggest ways of improvement.

The first comments were around the topic of student participation during the lectures. The lecturers stated that the students were not asking questions during the lecture and the students felt that they were not invited to ask questions.

The issue of student background level allignment with the level of the lecture was discussed. It was noted that the participants were asked about their background and that the information was communicated to the lecturers.

Suggestions were made as to how to stimulate student participation such as pointing out the ambiguities of the subject matter by the lecturer, more elaboration and in depth discussion of both the fundamentals and the advanced material rather than focusing on covering a lot of material, and avoidance of a confrontational style when making corrections or asking questions.

SECOND ROUNDTABLE DISCUSSION

The second roundtable discussion began at 6:00 p.m. on the evening of June 29th, the last day of the conference. Professor Di Bartolo opened the discussion with comments on some the organizational aspects of the meeting. He explained the effort on the part of the organizers to distribute the NATO grant funds in a way that was need-based, and he hoped that everyone was treated fairly. Also, he explained some of the considerations to be taken into account when responding to the question of the topic for the meeting in 1999. The main considerations were that the topic should be broad enough so as to be attractive to a large number of people, and also that the proceedings become a compilation of material that presents the subject in a unique way. Finally, he mentioned that books can be donated to the Erice Library, Erice (TP), Italy.

A question was asked about the long publication time of the proceedings. Professor Di Bartolo believes that there is little to be done to substantially speed up the publication process. There may be some benefit to submitting the manuscripts as they arrive, rather than in one set. It was also suggested that lecturers bring manuscripts to the school, but although this would be beneficial to the students, it is not always possible.

Another point was raised regarding the desire on the students part to be informed ahead of time of the possibility to present a poster or a short seminar at the meeting. The counterargument was that this is a school, and one should be careful not to give the impression of being a conference, and aslo that the bureaucratic demands on reviewing abstracts are to be avoided. A consensus seemed to emerge that students should be encouraged to prepare just a few transparencies telling about their institution and some generalities about their work. Also students should be told of the possibility of presenting a poster if they wish to present their work in greater detail. All participants were then asked to fill out an evaluation form for the school. The meeting adjourned at 7:30 p.m.

SUMMARY

(K.A. Nelson)

The NATO Advanced Study Institute dealt with optics and spectroscopy both generally and as related specifically to ultrafast time-resolved pulse generation, linear and nonlinear optical effects, and spectroscopic measurements of atomic, molecular, and solid state systems.

A number of speakers presented discussions at a fundamental level of ultrafast linear and nonlinear optics and spectroscopy. A treatment of ultrashort pulse propagation in linear and nonlinear optical media was presented by Dr. Di Bartolo. Dr. Hvam discussed coherent interactions between light and matter and how these interactions are exploited spectroscopically. Dr. Glezer presented a discussion of ultrafast optical measurement methods with examples of their use in various spectroscopic areas. Many aspects of ultrafast pulse generation and laser materials used for it were discussed. Dr. Ippen provided a review of the underlying mechanisms and practical methods of ultrashort pulse generation. Dr. Ferguson reviewed available ultrafast laser sources, and Dr. Kaminskii discussed nonlinear optics in laser crystals. Dr. Boulon discussed very-broadband nonlinear optics in laser crystals. Dr. Terzi, Dr. Collins and Dr. Baldacchini treated different aspects of ultrafast processes in laser materials and Dr. Auzel discussed relaxation processes and novel laser materials. Dr. Ronda treated Jahn-Teller effects and their consequences in laser crystals.

Applications of nonlinear optical effects were treated by several speakers. Dr. Mollenauer treated the formation and behavior of solitons in optical fibers, and applications of solitons in high bit transmission. Mr. Schaffer discussed ultrafast x-ray generation through up-conversion.

Several speakers discussed the fundamental mechanisms of ultrafast spectroscopy in gas phase or condensed matter systems. Drs. Demtröder, Klingshirn, and Benedek presented systematic treatments of the nature and spectroscopic elucidation of elementary excitations and dynamical processes characteristic of atomic and molecular systems, solids, and surfaces respectively. Dr. von Baltz treated interactions between light and semiconductors.

Several speakers focused on ultrafast spectroscopy of particular classes of materials or specialized methods in ultrafast spectroscopy. Dr. Cardona presented a description of spectroscopy of semiconductors and high-Tc superconductors in which electron-phonon interactions are of primary interest Dr. Mazur and his research group members discussed ultrafast spectroscopy of metal surfaces, including those in which chemical reactions or melting occur upon irradiation. Also presented were results of laser "writing" into dielectric media with submicron spatial resolution at high light intensities. Dr. Nelson discussed femtosecond pulse shaping and its spectroscopic applications. There were two discussions of topics beyond the main focus which provided some change of pace and outside interest. Dr. Barnes reviewed the NASA program on LIDAR development and Dr. Costa presented a review of fundamental constants in particle physics.

In all, a substantial range of ultrafast optics and spectroscopy was discussed in considerable depth. All those in attendance were treated to a comprehensive presentation of the fundamentals of ultrafast spectroscopy as well as a timely review of many of the most exciting areas in the field. The rigor and depth of the treatments and the extensive opportunities for informal discussions between the seminars provided a refreshing and productive environment.

Finally, a summary of the conference would be incomplete without mention of the beauty of Erice and its environs, which offered inspiration to all the participants.

List of Participants

Angel Garcia Adeva
Departamento de Fisica Aplicada I
Universidad del Pais Vasco
Alda - Urquijo S/N 48013 Bilbao
SPAIN
e-mail: wubgaada@bi.ehu.es

Low temperature anomalous properties of amorphous solids.

Daniela Alba[1]
Via Aloisio Juvara, 138
90142 Palermo
ITALY

English Language.

Francois Auzel
CNE
196, Ave. H. Ravera
92220 Bagneux
FRANCE
e-mail: auzel@bagneux.cnet.fr

Rare-earth doped glasses for optical amplifiers.

Rolindes Balda De la Cruz
Departamento de Fisica Aplicada I
Universidad del Pais Vasco
Alda - Urquijo S/N
48013 Bilbao
SPAIN
e-mail: wupbacrr@bi.ehu.es

Spectroscopy of solid-state laser materials.

Giuseppe Baldacchini
ENEA, Centro Ricerche Energia Frascati
Via E. Fermi, 27
00044 Frascati (Roma)
ITALY
e-mail: baldacchini@frascati.emea.it

Optical spectroscopy of solid-state laser materials, color centers and lasers. High resolution molecular spectroscopy and trace gas detection.

James Barnes
NASA
Langley Research Center, MS 474
Hampton, VA 23681-0001
USA
e-mail: j.c.barnes@larc.nasa.gov

Solid state laser materials, non linear materials, fiber optics and optics and fiber lasers, laser diode array pump technology, UV-NIR detectors spectroscopy of materials, an atmospheric constituents.

Paolo Bartolini
LENS Università di Firenze
Largo Fermi 2
50125 Firenze
ITALY
e-mail: bart@generale.lens.unifi.it

Linear and non-linear spectroscopy in complex systems, liquids and glasses. Ultrafast non-linear spectroscopy in liquids.

Giorgio Benedek
Dipartimento di Fisica
Universitá degli Studi
Via Celoria, 16
Milano 20133,
ITALY
e-mail: benedek@axpmi.infn.it

Surface dynamics, gas-surface interactions; carbon structures.

Georges Boulon
Université Claude Bernard-Lyon I
Lab Phys/Chim Matériaux Luminescents
UMR 5620 CNRS
Bat. 205, 43 boul, 11 Nov 1918
69622 Villeurbanne Cédex
FRANCE
e-mail: boulon@pcml.univ-lyon1.fr

Solid-state laser materials and nonlinear laser materials: Growth, structural characterization, spectroscopy of activator ions, optical properties.

Frederic Bourgeois
Université Claude Bernard Lyon
Lab Phys/Chim Matériaux Luminescents
UMR 5620 CNRS
Bat. 205, 43 boul, 11 Nov 1918
69622 Villeurbanne
FRANCE
e-mail: bourgeoi@pcml.univ-lyon1.fr

Non-linear laser crystals, Raman shifts.

Robert Britton
Physics Dept
University of Southampton
Highfield Southampton SO17 1BJ
UNITED KINGDOM
e-mail: rb@orc.soton.ac.uk

Time-resolved spectroscopy of exciton spin dynamics in low dimensional semiconductor heterostructures. Development of OPO's for the above.

Manuel Cardona
Max Planck Institut FKF
Heisenberg Strasse, 1
D-7000 Stuttgart 80,
GERMANY

Semiconductors, semiconductor nanostructures. High-Tc superconductors.

Xuesheng Chen
Wheaton College
Norton, MA 02766
USA
e-mail: xchen@wheatonma.edu

Optical and laser spectroscopy of solids.

John Collins
Physics Dept.
Wheaton College
Norton, MA 02766
USA
e-mail: jcollins@wheatonma.edu

Spectroscopy of solid state laser materials.

Aliki Collins[2]
Allied Signal
Eastmans Road 6
Parsippany, NJ
USA

Magnetic properties and applications of amorphous metallic alloys. Amorphous Metals

Giovanni Costa
Università degli Studi
Istituto di Fisica "Galileo Galilei"
Via F. Marzolo, 8
35100 Padova
ITALY

Elementary particles. High energy physics.

Timothy Crimmins
Chemistry Dept Rm. 2-052
MIT
Cambridge MA 02141
USA
e-mail: crimmins@mit.edu

Pulse shaping and ISRS measurement of ferroelectrics.

Fabio De Matteis
Dipto di Fisica
Università di Roma-Tor Vergata
Via della Ricerca Scientifica 1
00133 Roma
ITALY

Picosecond pump-probe spectroscopy on insulating crystals.
Optical properties of large molecules in sol-gel glass matrix.

Wolfgang Demtroder
Fachbereich Physik
Universitat Kaiserslautern
E. Schroedinger Strasse
67663 Kaiserslautern
GERMANY
e-mail: demtroder@physik.umi-kl.de

High resolution laser spectroscopy of molecules and clusters; time resolved molecular dynamics;applications of laser spectroscopy to medicine and technology.

Baldassare Di Bartolo
Department of Physics
Boston College
Chestnut Hill, MA 02167
USA

Luminescence spectroscopy; photoacoustic,spectroscopy; optical interactions in solids.

Daniel Di Bartolo[3]
Department of Physics
Boston College
Chestnut Hill, MA 02167
USA

Psychology.

Matteo Dominoni
Universita degli Studi di Milano
Dipto. di Scienza dei Materiali
Via Emanueli 15
20126 Milano
ITALY
e-mail:matteo.dominoni@mater.unimi.it

Numerical simulation of the first transient in the excited state of localized systems: non-adiabatic dynamics and coherent wavepacket of phonon propagation.

Allister I. Ferguson
Department of Physics
University of Straithclyde
John Anderson Building, 107 Rottenrow
Glasgow, Scotland, G4 ONG
UNITED KINGDOM

Laser sources for ultrafast spectroscopy.

Joaquin Fernandez
Departamento de Fisica Aplicada I
Universidad del Pais Vasco
Alda - Urquijo S/N 48013 Bilbao
SPAIN
e-mail: wupferoj@bi.ehu.es

Spectroscopy of solid state laser materials.

Richard Finlay
Gordon McKay Lab
9 Oxford St
Cambridge, MA 02138
USA
e-mail: rich_finlay@lucifer.harvard.edu

Energy exchange processes of adsorbate in metal systems following ultra-fast excitation; surface chemistry.

Jurgen Gallus
Lab for Physical Chemistry
Universitatsst. 22
ETH-Zentrum
CH-8092, Zurich
SWITZERLAND
e-mail: gallus@phys.chem.ethz.ch

Fs - two pulse photon echoes of dye molecules in a polymer matrix.

Giulio Gambarota[4]
Department of Physics
Boston College
Chestnut Hill, MA 02167
USA

Nuclear magnetic resonance spectroscopy

Juan Lopez Garriga
Chemistry Dept
University of Puerto Rico
Mayaguez Campus,
Mayaguez, P.R. 00680
PUERTO RICO
e-mail: sonw@caribe.net

Spectroscopic studies and dynamical events in biomolecules and sensitization of semiconductors.

Nuria Garro
Optoelectronics Group
Cavendish Lab, Madingley Rd
Cambridge CB3 0HE
UNITED KINGDOM
e-mail: ng206@cus.cam.ac.uk

Dynamics of excitations in GaAs quantum wells. Localization due to flucuations in the well width.

Eli Glezer
Division of Applied Sciences
Harvard University
9 Oxford Street
Cambridge, MA 02138
USA
e-mail: eli_glezer@lucifer.harvard.edu

Ultrafast electron and lattice dynamics in solids; ultrafast optics; laser driven micro-explosions in solids and liquids.

Christian Hess
Institut für Physikalische Chemie
Universitat Gottingen
Tammannstr 6
D 37077 Gottingen
GERMANY

Resonant SHG at interfaces examined by fs-pulses.

Jürgen Hoffmann
Institut für Angewandte Physik
Kaiserstr. 12, Postfach 6980
D-76128 Karlsruhe
GERMANY
juergen.hoffmann@phys.uni-karlsruhe.de

Spectroscopy of semiconductor nanostructures; II-VI materials.

Jorn Hvam
Mikroelektronik Centret
DTU, Bldg. 345 east
DK-2800 Lyngby
DENMARK
e-mail: hvam@mic.dtu.dk

Fabrication and optical characterization of semiconductor nanostructures; ultrafast nonlinear optical spectroscopy; scanning near-field optical spectroscopy.

Erich Ippen
Optics Group
Research Laboratory of Electronics, 36-319
Massachusetts Institute of Technology
Cambridge, MA 02139
USA
e-mail: ippen@mit.edu

Ultrafast optics: femtosecond spectroscopy of materials; ultrafast dynamics in devices; ultrashort-pulse fiber lasers and devices.

Jacob Riis Jensen
Mikroelektronik Centret
Technical University of Denmark
Buil. 345 East
DK-2800 Lyngby
DENMARK
e-mail: jacob@mail.mic.otu.dk

Optically gated scanning tunneling microscopy; fabrication and optical properties of quantum wires.

Alexander Kaminskii
Institute of Crystallography
Russian Academy of Sciences
Leninsky pr.59
Moscow 117333
RUSSIA

Physics and spectroscopy of lasers and non-linear laser crystals.

Thomas Klar
Universität München
Sektion Physik
LS Feldmann
Amalienstr. 54
80799 München
GERMANY

Near-field measurements (SNOM) on metal nanoparticles.

Claus Klingshirn
Institut für Angewandte Physik
der Universität Karlsruhe
Kaiserstr.12
D-76128 Karlsruhe
GERMANY

Semiconductor optics (linear, non-linear, time-resolved). Semiconductor epitaxy. II-VI compounds, alloys, quantum wells, wires and dots.

Richard Koehl
Chemistry Dept Rm. 2-052
MIT
Cambridge MA 02141
USA
e-mail:rkoehl@mit.edu

Spatiotemporal femtosecond pulse shaping. Terahertz radiation in systems of reduced dimensionality.

Ralf-Johan Lamminmaki
Department of Chemistry
University of Turku
FIN - 20014 Turku
FINLAND
e-mail: ralflamm@utu.fi

Simulations of the 4fn energy level structure of the RE3+ ions in RE oxyhalides (RE3+ =Pr3+, Nd3+, Sm3+, Eu3+, Tb3+, Ho3+, Er3+, and Tm3+).

Jakob Juul Larsen
Institute of Physics and Astronomy
University of Aarhus
Ny Munkegade
8000 Aarhus C
DENMARK
e-mail: juul@ufi.aau.dk

Time resolved molecular reaction dynamics studied using a position sensitive time of flight mass spectrometer.

Giuseppe Leo
Dip. Ing. Elettronica
Universitá Roma Tre
Via della della Vasca Navale 84
00146 Roma
ITALY
e-mail: leo@ele.uniroma3.it

NLO group: 2nd order nonlinear optical effects in bulk and waveguide structures. Spatial solitary waves, gap solitons, and applications to all-optical processing devices.

Imke Libon
Sektion Physics
Universitat Munchen
Amalienstr 54
80799 Munchen
GERMANY
e-mail: imke.libon@physik.uni-muenchen.de

Terahertz spectroscopy of semiconductors, liquids and biological systems.

Maria Cristina Martinoni
Corecom
Via Ampere 30
20131 Milano
ITALY
e-mail: martinoni@corecom.polimi.it

Optical elaboration and switching
Photon echo with femtosecond nanosecond lasers on rare earth doped crystals.

Eric Mazur
Division of Applied Sciences
Harvard University
Pierce Hall 225
Cambridge, MA 02138
USA
http://mazur-www.harvard.edu

Ultrafast processes in semiconductors; reactions at metal surfaces; Langmuir monolayers; 3D-optical data storage.

Alfredo Mazzulla
Physics Department
University of Calabria
87036 Rende (CS)
ITALY
e-mail: mazulla@fis.unical.it

Electro-optic properties of liquid crystals; linear and nonlinear optics on materials.

Antonino Messina
Istituto di Fisica
Universita degli Studi
Via Archirafi, 36 Palermo
I-90123
ITALY
e-mail: messina2@ist.fisica.unipa.it

Quantum Optics; nonclassical states (properties and generation)
spin-boson interaction.

Linn Mollenauer
AT&T Bell Lab, Rm 4c-306
101 Crawford Rd. P.O. Box 3030
Holmdel, New Jersey 07733
USA

Nonlinear pulse propagation in optical fibers and the engineering of high-bit-rate, ultra-long distance data transmission based on that basic physics.

Michel Mortier
Groupe Optique Terres Rares
CNET / CNRS
196, Ave. H. Ravera, BP 107
92225 Bagneux
FRANCE
e-mail: mortier@bagneux.cnet.fr

Rare-earth doped glasses for optical amplifiers.

Marcus Motzkus
Max-Planck-Institut für Quantenoptik
Hans Kopfermann St 1
D-85748 Garching
GERMANY
e-mail: mcm@mpq.mpg.de

Chemical dynamics of bimolecular reactions;nonlinear optical techniques (fs,ns);femtosecond chemistry.

Anna Napoli
Istituto di Fisica
Universita degli Studi
Via Archirafi, 36 Palermo
I-90123 ITALY
e-mail: messina@ist.fisica.unipa.it

Nonclassical states generation and detection; micromasers.

Keith A. Nelson
Department of Chemistry, Room 6-231
Massachusetts Institute of Technology
Cambridge, MA 02139
USA
e-mail: knelson@mit.edu

Time-resolved spectroscopy of condensed matter; collective structural rearrangements;structural phase transitions; solid state chemistry; liquid-glass transitions; thin films;fs pulse shaping; multiple pulse fs spectroscopy; optical control.

Maria Caterina Netti
Dipartimento di Fisica
Università degli Studi di Bari
Via E Orabona 4
70126 Bari
ITALY
e-mail: netti@axpba1.ba.infn.it

Linear and nonlinear optical properties of ZnSe-based heterostructures in frequency and time domains. Linear and multiphonon absorption spectroscopy. Pico- and femtosecond time-resolved spectroscopy.

Sevgi Ozdemir
Physics Dept.
Middle East Technical University
06531 Ankara
TURKEY
ozsev@newton.physics.metu.edu.tr

Identification of excitation energy transfer pathways in photosynthetic systems, especially light harvesting complex (LHCII) of green plants by using spectroscopic and structural information.

Francesco Pelagalli
Via Rio Fresco (GESCAL FS)
04023 Formia LT
ITALY

High resolution molecular spectroscopy by tunable diode lasers on gases of environmental and medical interest.

Paul C.M. Planken
FOM Institute Rijnhuzen
PO Box 1207
3430 BE Nieuwegein
THE NETHERLANDS

Time resolved spectroscopy on solid-state materials; nonlinear optics; electron dynamics; phonon dynamics (far infrared).

Loretta Pugh
Optoelectronics Group
Cavendish Laboratory, Madingley Rd.
Cambridge CB3 0HE
UNITED KINGDOM
e-mail: lp202@cus.cam.ac.uk

Study of excitons in GaAs/AlGaAs quantum wells using linear and nonlinear spectroscopic techniques.

Marilena Ricci
LENS Università di Firenze
Largo Fermi 2
50125 Firenze
ITALY
e-mail: ricci@generale.lens.unifi.it

Nonlinear spectroscopy of liquids; linear and nonlinear spectroscopy of complex systems like liquid crystals and glasses before transition phase.

Elisa Riedo
Corecom / Università di Fisica
Via Ampere 30
20131 Milano
ITALY
e-mail: reido@corecom.polimi.it

Ultrafast spectroscopy of quantum structures (quantum wells and doped s superlattice (III/V materials)) in particular; pump and probe experiments and time-resolved luminescence.

Cees Ronda
Philips GmbH
Forschungslaboratorien
Weisshausstrasse 2
Postfach 500185 D-52021 Aachen
GERMANY
e-mail: ronda@pfa.research.philips.com

Preparation and characterization of luminescent materials for application in lamps and displays new classes of luminescent materials; new luminesence mechanisms.

Karsten Sändig
Max Planck Inst. für Quantenoptik
Hans Kopfermann St 1
D-85748 Garching
GERMANY
e-mail: kss@mpq.mpg.de

Light induced molecular structure; dressed states in simple diatomic molecules as H2+using fs and ns lasers.

Chris B. Schaffer
Physics Department
Harvard University
9 Oxford St.
Cambridge, MA 02138
USA

Studies of highly nonlinear absorption, plasma formation, and resulting dynamics in transparant materials.

João-Manuel Segura
Lab fur Physikalische Chemie
Universitatstr 22, ETH Zentrum
CH 8092 Zurich
SWITZERLAND
segura@phys.chem.ethz.ch

Probing of external perturbations using single molecule spectroscopy.

Roland Seitz
Departamento de Física
Universidade de Aveiro
3800 Aveiro
PORTUGAL
e-mail: seitz@ideiafix.fis.ua.pt

Optical characterization (time-resolved - photo luminescence, photoreflectance, Raman spectroscopy) of ZnSe and GaN based semiconductor structures.

Sebastian Spörlein
Institut für Medizinische Optik
Ludwig-Maximilians-Universität München
Barbarastrasse 16
D-80797 München
GERMANY
e-mail: sebastian.spoerlein@physik.uni-muenchen.de

Protein folding; femtosecond spectroscopy of cylizied peptides using azobenzene as an optical switch

Nice Terzi
Dipt. Scienza dei Materiali
Universitá degli Studi di Milano
Via Emanueli 15, 20126 Milano
ITALY

Relaxation processes of centers in crystals; coherent phonons; electron-phonon interaction in fast dynamics and numerical simulation.

Jan Valenta
Dept. of Chemical Physics and Optics
Charles University
Ke Karlovu 3
CZ-12116 Prague 2
CZECH REPUBLIC
e-mail: valenta@karlov.mff.cuni.cz

Linear and nonlinear optical properties of low dimensional semiconductors (semiconductor doped glasses, Si-based nanostructures);pump and probe techniques; four-wave mixing; spectral hole burning.

Carsten Voelkmann
Max-Planck Inst. für Quantenoptik
D-85740 Garching
GERMANY
e-mail: voelkmann@mpq.mpg.de

Nonlinear optical spectroscopy of semiconductor surfaces; energy relaxation and dephasing of electronic surface states.

Ralph von Baltz
Institut für Theorie der Kondensierte Materie
Fakultät für Physik der Universität
Postfach 6980
D-76128 Karlsruhe
GERMANY
e-mail: baltz@tkm.physik.uni-karlsruhe.de

Optical properties of semiconductors, metals, semiconductor microstructures. plasmons; quasicrystals (electrical properties)and photogalvanic effect.

Daniel Walser
Lab fur Physikalische Chemie
Universitätstr 22, ETH Zentrum
CH 8092 Zürich
SWITZERLAND
e-mail: walser@phys.chem.ethz.ch

One- and multi-phonon spectroscopy on single molecules in solids; spectroscopic characterization of polyenes on the single molecule level; quantum optical studies.

Brian Walsh
NASA Langley Research Center, MS 474
Hampton VA 23681
USA
e-mail: b.m.walsh@larc.nasa.gov

Solid-state laser spectroscopy of RE ions in crystals and glasses. Laser engineering.

Ralf Westphäling
Institut für Angewandte Physik
Kaiserstr. 12 Postfach 6980
D-76128 Karlsruhe
GERMANY
e-mail: ralf.westphaeling@phys.uni-karlsruhe.de

Photoluminescence quantum efficiencies in II-VI and III-V semiconductors (bulk, quantum wells, quantum dots).

Herfried Wieczorek
Philips GmbH
Forschungslaboratorien
Weisshausst 2, D-52066 Aachen
GERMANY
e-mail: wieczorek@pfa.research.philips.com

Material development; basic characterization of luminescence and afterglow properties of CsI:Tl and other scintillators for x-ray detector systems.

Hong Zhang
Laboratorium voor Fysiche Chemie
Nieuwe Achtergracht 127
1018 WS Amsterdam
THE NETHERLANDS
e-mail: hong@fys.chem.uva.nl

Chemical reactions of molecules in liquids and ultrafast laser spectroscopy.

---

(1) Administrative Assistant
(2) Assistant to the Director
(3) Administrative Secretary of the Course
(4) Scientific Secretary of the Course

INDEX